SOLID-STATE POWER
CONVERSION HANDBOOK

SOLID-STATE POWER CONVERSION HANDBOOK

RALPH E. TARTER, P.E.

A Wiley-Interscience Publication

JOHN WILEY & SONS, INC.

New York • Chichester • Brisbane • Toronto • Singapore

Library of Congress Cataloging in Publication Data:

Tarter, Ralph E., 1934–
    Solid-state power conversion handbook / Ralph E. Tarter.
        p.   cm.
    "A Wiley-Interscience publication."
    ISBN 0-471-57243-8
    1. Power electronics—Handbooks, manuals, etc.   2. Electric
current converters—Handbooks, manuals, etc.   3. Electromagnetic
compatibility—Handbooks, manuals, etc.   I. Title.
TK7881.15.T362   1993
621.31′7—dc20   92-18179

Printed in the United States of America

10 9 8 7 6 5 4 3 2 1

*To
Marjorie,
Valerie, Kimberly, Jeffrey*

CONTENTS

Preface

1. Transient Analysis, Circuit Analysis, and Waveforms   **1**

  1.0 Introduction, 1
  1.1 Waveform Relations, 3
  1.2 Magnetic Fields, 13
  1.3 Dielectric Fields, 13
  1.4 The *RL* Circuit, 14
  1.5 The *RC* Circuit, 17
  1.6 The *RLC* Circuit, 20
  1.7 The *RLCR* Circuit with a dc Input, 25
  1.8 ac Circuit Analysis, 30
  1.9 Component Scaling, 40

2. Semiconductors, Resistors, and Capacitors   **42**

  2.0 Introduction, 43
  2.1 Rectifiers, 46
  2.2 Bipolar Power Transistors, 56
  2.3 MOSFET Power Transistors, 74
  2.4 Insulated-Gate Bipolar Transistors, 86
  2.5 Thyristors, 91
  2.6 Other Power Devices, 103
  2.7 Resistors, 105
  2.8 Capacitors, 115

3. Transformers, Inductors, and Conductors   **129**

  3.0 Introduction, 130
  3.1 Magnetic Circuit Review, 130

3.2   Magnetic Relations for Transformers, 133
3.3   Magnetic Relations for Inductors, 138
3.4   Hysteresis Loops, 139
3.5   Magnetic Materials, 140
3.6   Transformer Losses, Regulation, and Efficiency, 143
3.7   Copper and Other Conductors, 152
3.8   Windings, 164
3.9   Temperature Rise, 173
3.10 Audible Noise, 174
3.11 Laminations, 175
3.12 C and E Cores, 183
3.13 Toroid Tape Core Transformers, 188
3.14 Permalloy Powder Cores, 190
3.15 Iron Powder Core Inductors, 193
3.16 Ferrites, 194
3.17 Air Core Inductors, 200
3.18 Pulse Transformers, 202
      References, 208

4. Rectifiers and Filters   210

4.0   Introduction, 210
4.1   Rectifier Circuits, 212
4.2   Capacitor Input Filters, 227
4.3   Inductor Input Filters, 253
4.4   Three-Phase Voltage Doubler, 258
      References, 259

5. Phase Control Circuits   260

5.0   Introduction, 260
5.1   Single-Phase Power Circuits, 261
5.2   Three-Phase Power Circuits, 264
5.3   Six-Phase Dual Bridge, 270
5.4   Critical Inductance, 273
5.5   Power and Control Stages, 273
      References, 276

6. Transistor Inverters   277

6.0   Introduction, 277
6.1   Square-Wave Inverters, 278
6.2   Quasi-Square-Wave Inverters, 290
6.3   Inverters with Harmonic Cancellation, 298
6.4   High-Frequency, Pulse-Demodulated Inverters, 300
6.5   Inverters with Ferroresonant Transformers, 311
6.6   Three-Phase Inverters, 311
      References, 321

7. Thyristor Inverters   322

7.0   Introduction, 322
7.1   Classification of Inverter Circuits, 324
7.2   Parallel Inverter, 326
7.3   Half-Bridge Inverter, 330
7.4   Full-Bridge Inverter, 331
7.5   Impulse-Commutated Full-Bridge Inverter, Quasi–Square
      Wave, 333
7.6   Impulse-Commutated Full-Bridge Inverter, Pulse Width
      Modulated, 340
7.7   Three-Phase Inverters, 341
      References, 348

8. Switching Regulators   349

8.0   Introduction, 349
8.1   Topologies, 350
8.2   Pulse Width–Modulated Control Circuits, 353
8.3   Buck Regulator, Fixed Frequency, 364
8.4   Boost Regulator, 373
8.5   Buck–Boost Regulator, 380
8.6   Protection for Switching Regulators, 386
8.7   Input Filter for Switching Regulators, 388
      References, 392

9. dc–dc Converters   393

9.0   Introduction, 393
9.1   Topologies, 395
9.2   Blocking Oscillator, 403
9.3   Square-Wave Converters, 405
9.4   Flyback Converter, 407
9.5   Forward Converters, 413
9.6   Push–Pull Center-Tapped Converter, 416
9.7   Half-Bridge Converter, 420
9.8   Full-Bridge Transistor Converter, 422
9.9   SCR Converters, 425
9.10 Current-Fed Converters, 429
9.11 Multiple-Output Converters, 434
9.12 Input Filters for dc–dc Converters, 437
      References, 438

10. Resonant Mode Converters   439

10.0 Introduction, 439
10.1 Classification, 441
10.2 Resonant Mode Control Circuits, 442
10.3 Zero-Current Switching, 445
10.4 Series Resonance, 448

10.5 Parallel Resonance, 457
10.6 Series–Parallel Resonance, 460
10.7 Zero-Voltage Switching, 461
    References, 467

11. Modulator and Control Analysis   **468**

11.0 Introduction, 468
11.1 Amplifiers and Compensation, 469
11.2 Stability Analysis, 475
11.3 PWM Voltage Mode Control, 484
11.4 PWM Feedforward Control, 490
11.5 PWM Current Mode Control, 491
11.6 Resonant FM Mode, 497
    References, 498

12. Pulse-Forming Networks and Modulators   **499**

12.0 Introduction, 499
12.1 Pulse-Forming Networks, 500
12.2 Line-Type Modulators, 505
12.3 Magnetic Modulators, 509
    Reference, 513

13. Protection and Safety   **514**

13.0 Introduction, 514
13.1 Fuses, 515
13.2 Circuit Breakers, 520
13.3 Transient Protection, 522
13.4 Snubbers, 528
13.5 Safety Requirements, 538

14. Electromagnetic Compatibility and Grounding   **540**

14.0 Introduction, 540
14.1 Electromagnetic Compatibility, 541
14.2 Grounding Techniques, 563
    References, 573

15. Semiconductor and Equipment Cooling   **574**

15.0 Introduction, 574
15.1 Thermal Conduction and Resistance, 577
15.2 Natural Convection and Radiation, 588
15.3 Forced Air Cooling, 593
15.4 Forced Liquid Cooling, 600
15.5 Heat Pipes, 603
15.6 Thermoelectric Coolers, 604
    References, 604

16. Reliability and Quality   **605**

    16.0 Introduction, 605
    16.1 Terminology, 606
    16.2 Specifications and Documents, 608
    16.3 Reliability Equations, 608
    16.4 Mean Time between Failure, 609
    16.5 Confidence Levels, 614
    16.6 Reliability Modeling of Redundant Units, 616
    16.7 MTBF Example, 617
    References, 618

17. Regulated Power Supplies   **619**

    17.0 Introduction, 619
    17.1 Power Supply Topologies, 621
    17.2 Series-Regulated Power Supplies, 624
    17.3 Shunt-Regulated Power Supplies, 647
    17.4 Switch Mode Power Supplies, 651
    17.5 High-Power, Low-Voltage Power Supply, 657
    17.6 High-Power, High-Voltage Power Supply, 660
    17.7 Current-Regulated Power Supplies, 662
    References, 668

18. Uninterruptible Power Systems   **669**

    18.0 Introduction, 669
    18.1 UPS Equipment and Components, 670
    18.2 On-Line System, 670
    18.3 Off-Line System, 673
    18.4 Batteries, 674
    18.5 Battery Chargers, 681
    18.6 Inverters, 684
    18.7 Inverter Preregulator, 684
    18.8 Transfer Switches, 688
    18.9 Complete System Analysis, 693

19. Power Factor Correction   **695**

    19.0 Introduction, 695
    19.1 Power Factor Effects, 696
    19.2 Passive Correction, 698
    19.3 Active Correction, 702
    References, 705

Glossary   **706**

Index   **711**

PREFACE

This new handbook is the first to be devoted to the field of solid-state power conversion. The material in this book is to be used in engineering practice and is oriented toward application rather than theory. The purpose of the book is to assemble in a single volume all the pertinent and comprehensive information necessary to meet the growing demands placed upon solid-state power conversion equipment. These demands include increased efficiency, improved reliability, higher packaging density, improved performance, and meeting safety and electromagnetic compatibility (EMC) requirements. The material presented includes a thorough analysis of fundamental electrical and magnetic aspects of power conversion plus thermal, protection, and reliability considerations. Attention is focused on semiconductor and magnetic components and on analysis of various topologies.

In approaches to solid-state power conversion, the engineer or designer is faced with selecting a topology, conceptually correct and optimum, to meet the specifications of a particular custom application or the requirements of a planned standard product line. Block diagrams, circuit design, component selection, analysis, fabrication, producibility, and testing then follow. The analysis tasks include regulation, stability, thermal design, safety, EMC, reliability, and the effects of abnormal conditions. This latter parameter is often overlooked and is the primary reason some power conversion devices cause more problems than the equipment that they power. The units must properly interface with the intended input source and output load, normally dynamic in nature.

The handbook is organized into four sections. (1) Chapters 1–3 present the relations of various waveforms, transient analysis, and components with emphasis on power semiconductors and magnetic components. (2) Chapters 4–12 deal with single-level conversion of rectifier circuits, filters, inverters and converters, feedback and stability analysis, and modulators and pulse-forming networks. (3) Chapters 13–16 discuss ancillary topics related to safety, EMC, thermal management, and reliability. (4) Chapters 17–19 cover design and operation of power supplies and systems from a detailed building block standpoint.

The material discussed in this handbook represents considerable experience in the subject field by the author. Quantitative design information is presented to enhance and simplify procedures. Certain aspects of power conversion may be considered an art as well as a science. Some material is heuristic in nature but is intended to serve a practical engineering need.

RALPH E. TARTER

Saratoga, CA
January 1993

SOLID-STATE POWER
CONVERSION HANDBOOK

1

TRANSIENT ANALYSIS, CIRCUIT ANALYSIS, AND WAVEFORMS

1.0 Introduction
1.1 Waveform Relations
1.2 Magnetic Fields
1.3 Dielectric Fields
1.4 The *RL* Circuit
    1.4.1   dc Input
    1.4.2   ac Input
1.5 The *RC* Circuit
    1.5.1   dc Input
    1.5.2   Two Precharged Capacitors
    1.5.3   ac Input
1.6 The *RLC* Circuit
    1.6.1   dc Input
    1.6.2   ac Input
1.7 The *RLCR* Circuit with a dc Input
    1.7.1   Equations for Voltage
    1.7.2   Equations for Current
1.8 ac Circuit Analysis
    1.8.1   Passive Lag Networks
    1.8.2   Passive Lag–Lead Networks
    1.8.3   Passive Lead–Lag Networks
    1.8.4   Circuit Impedance and Graphs
    1.8.5   Typical Output Filter
1.9 Component Scaling
    1.9.1   Impedance Scaling
    1.9.2   Frequency Scaling

1.0 INTRODUCTION

This chapter is divided into three parts. The first portion (Sections 1.1–1.3) presents an analysis of waveforms typically encountered in solid-state power conversion equipment and the relation of voltage and current in various combinations of elements. The types of waveforms commonly generated are sine, square, exponential, triangular, sawtooth, and trapezoidal. Each waveform may have less than full-cycle or less than half-cycle conduction through an element, depending on the duty cycle D. The element may be an active device such as a rectifier, transistor, or thyristor or the element may be passive, such as a resistor, inductor, transformer, or capacitor. Most frequently, a particular circuit will have combinations of elements which, in turn, generate desired (and frequently undesired) waveforms.

The second portion of this chapter (Sections 1.4–1.7) deals with combinations of *RLC* circuits and their behavior under transient conditions. Effects of both direct current (dc) inputs and alternating current (ac) inputs are analyzed for transient and steady-state conditions.

TABLE 1.1. Notation

Notation	Interpretation
a, d	Time constant for *RLC* circuits
A	Gain (numeric or dB)
b	Angular function for *RLC* circuits
β	$jb, j = \sqrt{-1}$
C	Capacitance, farads
D	Duty cycle, t_{on}/T
e	Instantaneous voltage, volts
E_{pk}	Peak voltage, volts
E_{avg}	Average voltage, volts
E_{rms}	Root-mean-square voltage, volts
$\mathcal{E}$	Energy, joules or watt-seconds
f	Frequency, hertz
i	Instantaneous current, amperes
I_{pk}	Peak current, amperes
I_{avg}	Average current, amperes
I_{rms}	Root-mean-square current, amperes
L	Inductance, henrys
q	Instantaneous charge, coulombs
Q	Capacitor charge, coulombs
R	Resistance, ohms
t	Time, seconds
t_p	Pulse width, seconds
t_{pk}	Time to peak condition, seconds
T	Period
Z	Impedance, ohms
α	Angular function, radians or degrees
δ	Damping ratio
θ	Angular function, radians or degrees
λ	Angular function, radians or degrees
τ	Circuit time constant, seconds
ϕ	Angular function, radians or degrees
ψ	Angular function, radians or degrees
ω	Angular function, radians or degrees

The third portion of this chapter (Sections 1.8 and 1.9) deals with ac circuit analysis. Frequency response of various passive networks and amplifier circuits and compensation are presented. Detailed stability analysis for closed-loop systems is presented in the individual inverter and power supply chapters. Component value scalings for various conditions are presented, which may aid in the design of high-power systems.

Notations used in this chapter are listed in Table 1.1.

1.1 WAVEFORM RELATIONS

The most common waveforms are sine waves (as in power supplies or converters operating from the utility mains and in commutation circuits producing resonant characteristics) and square waves or rectangular waves (as in inverter switching via transistors and thyristors). Exponential waveforms are produced by combinations of resistance, inductance, and capacitance. Sawtooth waveforms are frequently produced for timing and sweep circuits and by constant current charging of an inductor or capacitor. Triangle waveforms, produced by integrating a square wave, are beneficial in providing symmetry around a fixed point in time.

Each waveform has its own interrelation to the peak, rms (root-mean-square), and average value of voltage or current. Current flowing in resistors and conductors produces heat, and the power dissipation is based on the rms value of current flowing in the resistor or the rms value of voltage impressed across the resistor. The rms value of current is also applicable to the current-carrying capacity or rating of conductors used in transformers and inductors, based on the dc resistance of the winding and, in some cases, the equivalent dc resistance which includes skin effects. In addition, the rms value of current is applicable to the current-carrying capacity or rating of capacitors based on the dissipation factor or ESR (equivalent series resistance) of the capacitor.

From an operating voltage standpoint, the resistor is rated for rms conditions, the transformer or inductor is rated for average conditions (applied to the magnetic circuit as opposed to the conductor), and the capacitor is rated for peak conditions (of voltage as opposed to the rms current rating). Various types of metering movements indicate different readings when nonsinusoidal waveforms are being measured. True rms meters operate via electrodynamometer movement or thermocouple heating means, the former being restricted to frequencies below 2500 Hz. Average indicating meters are just that. Rectifier meters respond to average voltage and are usually calibrated to indicate rms values but only for a sinusoidal waveform. Reference Section 6.1.4 and Fig. 6.6 for Fourier analysis of sine, square, and triangle waves and the resulting meter movement response. Good judgment must be exercised in measuring voltage and current parameters in the various elements. Common waveform relations of peak, rms, and average values of current or voltage are shown in Table 1.2. The relations are beneficial in determining the requirements of passive or active elements. The bipolar transistor is limited by its ability to conduct a peak value of current while maintaining a low saturation voltage, otherwise junction failure will result due to forward-biased second breakdown. The rectifier and thyristor have the ability to conduct currents in excess of their average or rms rating for short periods of time, and are therefore much more forgiving to transient or surge currents than are transistors.

Derivations of parameter relations of Table 1.2 are given below, where $T = 2\pi$ for Table 1.2, part A; $T = \pi$ for Table 1.2, part B; and $D = t_{on}/T$ = duty cycle for time-clipped waveforms. Symmetrical ac waveforms have an average value of zero over one complete cycle. A dc meter, if connected to measure current or voltage, would indicate zero. The magnitude of these waveforms is found by integrating the wave over a half-cycle period. For rectified waveforms, it is understood that the term *average* means "average of rectified" voltage or current.

1. *Sine Wave*

$$I_{avg} = \frac{2}{T}\int_0^{T/2} I_{pk} \sin \omega t\, dt = \frac{2}{\pi} I_{pk}$$
$$= 0.6366 I_{pk} \tag{1.1a}$$

$$I_{rms}^2 = \frac{1}{T}\int_0^T I_{pk}^2 \sin^2 \omega t\, dt$$
$$= \frac{I_{pk}^2}{T}\left[\frac{t}{2} - \frac{1}{4\omega}\sin 2\left(\frac{2\pi}{T}\right)t\right]_0^T = \frac{I_{pk}^2}{2}$$
$$I_{rms} = 0.7071 I_{pk} \tag{1.1b}$$

2. *Full-Wave Rectified Sine Wave*

$$I_{avg} = 0.6366 I_{pk} \tag{1.2a}$$
$$I_{rms} = 0.7071 I_{pk} \tag{1.2b}$$

3. *Half-Wave Rectified Sine Wave*

$$I_{avg} = \frac{1}{T}\int_0^{T/2} I_{pk} \sin \omega\, dt = \frac{1}{\pi} I_{pk}$$
$$= 0.3183 I_{pk} \tag{1.3a}$$

$$I_{rms}^2 = \frac{1}{T}\int_0^{T/2} I_{pk}^2 \sin^2 \omega t\, dt$$
$$= \frac{I_{pk}^2}{T}\left[\frac{t}{2} - \frac{1}{4\omega}\sin 2\left(\frac{2\pi}{T}\right)t\right]_0^{T/2}$$
$$= \tfrac{1}{4} I_{pk}^2$$
$$I_{rms} = 0.500 I_{pk} \tag{1.3b}$$

4. *Square Wave*

$$I_{avg} = \frac{1}{T}\int_0^T I_{pk}\, dt = \frac{1}{T}\left[I_{pk} t\right]_0^T$$
$$= I_{pk} \tag{1.4a}$$

$$I_{rms}^2 = \frac{1}{T}\int_0^T I_{pk}^2\, dt = I_{pk}^2$$
$$I_{rms} = I_{pk} \tag{1.4b}$$

5. *Rectified Square Wave*

$$I_{avg} = \frac{1}{T} \int_0^{T/2} I_{pk}\, dt = \frac{1}{T} \Big[I_{pk} t \Big]_0^{T/2}$$

$$= 0.50 I_{pk} \tag{1.5a}$$

$$I_{rms}^2 = \frac{1}{T} \int_0^{T/2} I_{pk}^2 = \frac{1}{T} \Big[I_{pk}^2 t \Big]_0^{T/2}$$

$$= \tfrac{1}{2} I_{pk}^2$$

$$I_{rms} = 0.7071 I_{pk} \tag{1.5b}$$

6. *Rectangular Wave*

$$I_{avg} = \frac{1}{T} \int_0^t I_{pk}\, dt = \frac{1}{T} I_{pk} t$$

$$= I_{pk} \frac{t}{T} \tag{1.6a}$$

$$I_{rms}^2 = \frac{1}{T} \int_0^t I_{pk}^2\, dt = \frac{1}{T} I_{pk}^2 t$$

$$I_{rms} = I_{pk} \sqrt{t/T} \tag{1.6b}$$

7. *Clipped Sinusoid, Full Wave*

$$I_{avg} = \frac{2}{T} \int_0^{2t} I_{pk} \sin \omega t\, dt$$

$$= \frac{2}{T} I_{pk} \frac{2t}{\pi}$$

$$= 1.273 I_{pk} \frac{t}{T} \tag{1.7a}$$

$$I_{rms}^2 = \frac{1}{T} \int_0^{2t} I_{pk}^2 \sin^2 \omega t\, dt = I_{pk}^2 \frac{2t}{2T}$$

$$I_{rms} = I_{pk} \sqrt{t/T} \tag{1.7b}$$

8. *Clipped Sinusoid, Half Wave*

$$I_{avg} = \frac{2}{T} \int_0^t I_{pk} \sin \omega t\, dt$$

$$= \frac{2 I_{pk}}{t} \frac{t}{\pi} = 0.6366 I_{pk} \frac{t}{T} \tag{1.8a}$$

$$I_{rms}^2 = \frac{1}{T} \int_0^t I_{pk}^2 \sin^2 \omega t\, dt$$

$$= I_{pk}^2 \frac{t}{4T}$$

$$I_{rms} = I_{pk} \sqrt{t/2T} \tag{1.8b}$$

TABLE 1.2. Common Waveform Relations[a]

Waveform[a]	Inverse	Peak	rms	Average
	A. T = 2π			
1. Sine	o–pk	1.000	1.414	1.571
	rms	0.7071	1.000	1.111
	avg	0.6366	0.900	1.000
2. Rectified sine (full wave)	o–pk	1.000	1.414	1.571
	rms	0.7071	1.000	1.111
	avg	0.6366	0.900	1.000
3. Rectified sine (half-wave)	o–pk	1.000	2.000	3.1416
	rms	0.500	1.000	1.571
	avg	0.3183	0.6366	1.000
4. Square	o–pk	1.000	1.000	1.000
	rms	1.000	1.000	1.000
	avg	1.000	1.000	1.000
5. Rectified square	o–pk	1.000	1.414	2.000
	rms	0.7071	1.000	1.414
	avg	0.500	0.7071	1.000

6. Rectangular pulse	o–pk	1.000	$\sqrt{\dfrac{T}{t}}$	$\dfrac{T}{t}$
	rms	$\sqrt{\dfrac{t}{T}}$	1.000	$\sqrt{\dfrac{T}{t}}$
	avg	$\dfrac{t}{T}$	$\sqrt{\dfrac{t}{T}}$	1.000
7. Clipped sinusoid (full wave)	o–pk	1.000	$\sqrt{\dfrac{T}{t}}$	$0.785\,\dfrac{T}{t}$
	rms	$\sqrt{\dfrac{t}{T}}$	1.000	$0.785\sqrt{\dfrac{T}{t}}$
	avg	$1.273\,\dfrac{t}{T}$	$1.273\sqrt{\dfrac{t}{T}}$	1.000
8. Clipped sinusoid (half-wave)	o–pk	1.000	$\sqrt{\dfrac{2T}{t}}$	$1.57\,\dfrac{T}{t}$
	rms	$\sqrt{\dfrac{t}{2T}}$	1.000	$1.11\sqrt{\dfrac{T}{t}}$
	avg	$0.6366\,\dfrac{t}{T}$	$0.9\sqrt{\dfrac{t}{T}}$	1.000

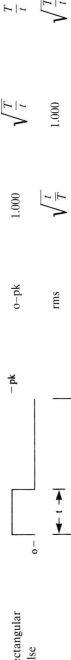
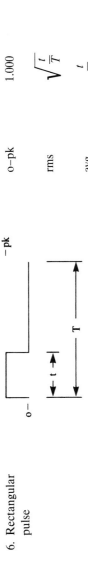
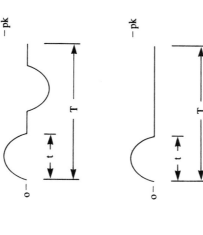
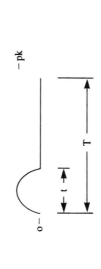

B. $T = \pi$

9. Sawtooth	o–pk	1.000	1.732	2.000
	rms	0.577	1.000	1.155
	avg	0.500	0.866	1.000

7

TABLE 1.2. (*Continued*)

Waveform[a]	Inverse	Peak	rms	Average
10. Clipped sawtooth	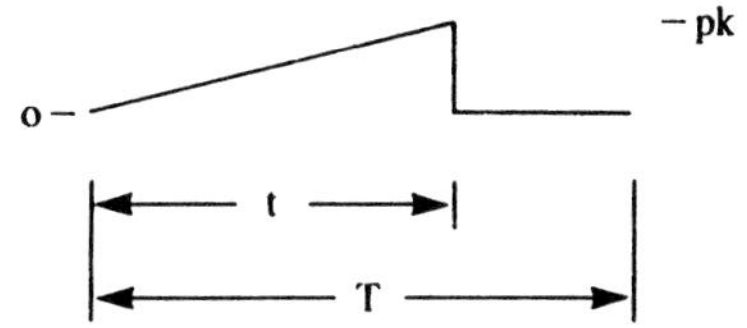 o–pk	1.000	$\sqrt{\dfrac{3T}{t}}$	$\dfrac{2T}{t}$
	rms	$\sqrt{\dfrac{t}{3T}}$	1.000	$\sqrt{\dfrac{T}{0.75t}}$
	avg	$\dfrac{0.5t}{T}$	$0.866\sqrt{\dfrac{t}{T}}$	1.000
11. Trapezoid	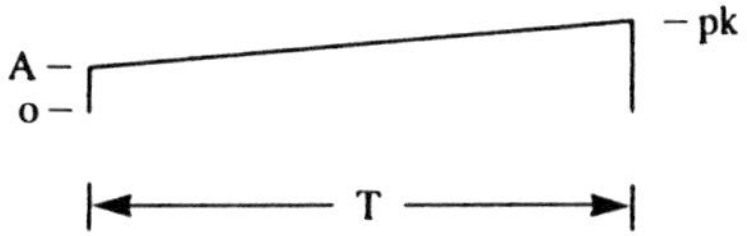 o–pk	1.000	$\sqrt{\dfrac{3}{A^2 + A\mathrm{pk} + \mathrm{pk}^2}}$	$\dfrac{2}{A + \mathrm{pk}}$
	rms	$\sqrt{\dfrac{A^2 + A\mathrm{pk} + \mathrm{pk}^2}{3}}$	1.000	b
	avg	$A + \dfrac{\mathrm{pk} - A}{2}$	c	1.000
12. Clipped trapezoid	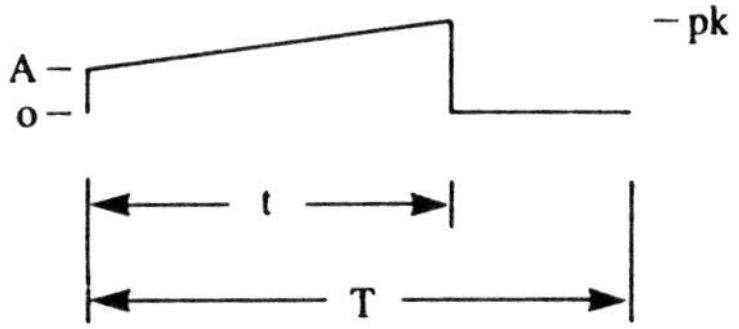 o–pk	1.000	$\sqrt{\dfrac{3T}{t(A^2 + A\mathrm{pk} + \mathrm{pk}^2)}}$	$\dfrac{T}{t\left[A + \frac{1}{2}(\mathrm{pk} - A)\right]}$
	rms	$\sqrt{\dfrac{t(A^2 + A\mathrm{pk} + \mathrm{pk}^2)}{3T}}$	1.000	b
	avg	$\dfrac{t}{T}\left(A + \dfrac{\mathrm{pk} - A}{2}\right)$	c	1.000
13. Triangle	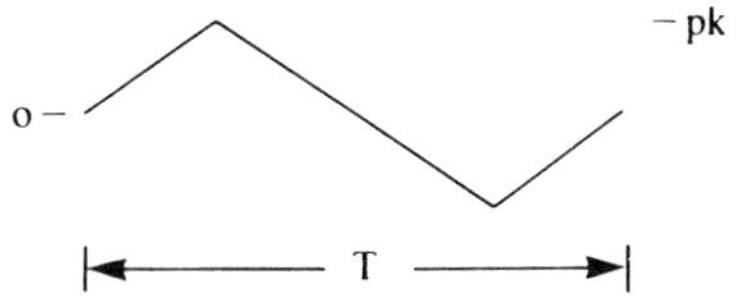 o–pk	1.000	1.732	2.000
	rms	0.577	1.000	1.155
	avg	0.500	0.866	1.000

8

14. Clipped triangle (half wave)	o–pk	1.000	$\sqrt{\dfrac{3T}{2t}}$	$\dfrac{2T}{t}$
	rms	$\sqrt{\dfrac{2t}{3T}}$	1.000	$\sqrt{\dfrac{T}{0.75t}}$
	avg	$\dfrac{0.5t}{T}$	$0.866\sqrt{\dfrac{t}{T}}$	1.000
15. Exponential decay $(T' = 5\tau)$	o–pk	1.000	$3.16\sqrt{\dfrac{T}{T'}}$	$5.025\dfrac{T}{T'}$
	rms	$0.316\sqrt{\dfrac{T'}{T}}$	1.000	$1.588\sqrt{\dfrac{T}{T'}}$
	avg	$0.199\dfrac{T'}{T}$	$0.63\sqrt{\dfrac{T'}{T}}$	1.000
16. Critically damped exponential $(T' > 4t_{pk})$	o–pk	1.000	$\dfrac{2}{\varepsilon}\sqrt{\dfrac{T'}{t_{pk}}}$	$\dfrac{T'}{\varepsilon t_{pk}}$
	rms	$\dfrac{\varepsilon}{2}\sqrt{\dfrac{t_{pk}}{T'}}$	1.000	$0.5\sqrt{\dfrac{T'}{t_{pk}}}$
	avg	$\varepsilon\dfrac{t_{pk}}{T'}$	$2\sqrt{\dfrac{t_{pk}}{T'}}$	1.000

[a] Waveform numbers correspond to equation numbers in text.

[b] $\dfrac{avg}{peak}$, $\dfrac{rms}{peak}$.

[c] $\dfrac{rms}{peak}$, $\dfrac{avg}{peak}$.

9

Example

The clipped sinusoid full wave is typical of a thyristor commutating capacitor current, while the clipped sinusoid half wave is typical of the thyristor current, the opposite half cycle of full-wave capacitor current being conducted by the opposite thyristor (Reference Section 9.9). For an operating frequency of 2 kHz, $T = 0.5$ ms, the resonant LC circuit produces a current pulse period of $t = 50$ μs, and the peak current is 500 A. Then the capacitor current rating at a 50-μs pulse width must be $I_{rms} = 500\sqrt{50 \times 10^{-6}/1 \times 10^{-3}} = 112$ A. The thyristor current rating for this circuit must be $I_{rms} = 0.5 \times 500\sqrt{50 \times 10^{-6}/0.5 \times 10^{-3}} = 79$ A.

9. *Sawtooth Wave*

$$I_{avg} = \frac{1}{T}\int_0^T \frac{I_{pk}}{T}t\,dt = \frac{1}{T}\left[\frac{I_{pk}}{T}\frac{t^2}{2}\right]_0^T$$

$$= 0.5I_{pk} \tag{1.9a}$$

$$I_{rms}^2 = \frac{1}{T}\int_0^T \frac{I_{pk}^2}{T^2}t^2\,dt = \frac{1}{T}\left[\frac{I_{pk}^2}{T^2}\frac{t^3}{3}\right]_0^T$$

$$= \tfrac{1}{3}I_{pk}^2$$

$$I_{rms} = 0.577I_{pk} \tag{1.9b}$$

10. *Clipped Sawtooth Wave*

$$I_{avg} = \frac{1}{T}\int_0^t \frac{I_{pk}}{T}t\,dt = \frac{1}{T}\left[\frac{I_{pk}}{2}\frac{t^2}{2}\right]_0^t$$

$$= 0.5I_{pk}\frac{t}{T} \tag{1.10a}$$

$$I_{rms}^2 = \frac{1}{T}\int_0^t \frac{I_{pk}^2}{T^2}t^2\,dt = \frac{1}{T}\left[\frac{I_{pk}^2}{T^2}\frac{t^3}{3}\right]_0^t$$

$$= \frac{I_{pk}^2}{3}\frac{t}{T}$$

$$I_{rms} = I_{pk}\sqrt{t/3T} \tag{1.10b}$$

11. *Trapezoid Wave*

$$I_{avg} = I_a + \tfrac{1}{2}(I_{pk} - I_a) \tag{1.11a}$$

$$I_{rms}^2 = \left[I_{pk}^2 - I_{pk}(I_{pk} - I_a) + \frac{(I_{pk} - I_a)^2}{3}\right]\frac{T}{T}$$

$$= \tfrac{1}{3}\left(3I_{pk}I_a + I_{pk}^2 - 2I_{pk}I_a + I_a^2\right)$$

$$I_{rms} = \sqrt{\tfrac{1}{3}\left(I_{pk}^2 + I_{pk}I_a + I_a^2\right)} \tag{1.11b}$$

12. *Clipped Trapezoid Wave*

$$I_{avg} = \left[I_a + (I_{pk} - I_a)/2 \right](t/T) \tag{1.12a}$$

$$I_{rms}^2 = \left[(I_{pk}^2 + I_{pk}I_a + I_a^2)/3 \right](t/T)$$

$$I_{rms} = \sqrt{(I_{pk}^2 + I_{pk}I_a + I_a^2)t/(3T)} \tag{1.12b}$$

Example

For the current waveform in 12 of Table 1.2, part B, let the initial step current be 8 A, the peak current be 16 A, and the duty cycle (t/T) be 45%. Then

$$I_{avg} = [8 + (16 - 8)/2] \times 0.45 = 5.4 \text{ A}$$

and $I_{rms} = \sqrt{(16^2 + 16 \times 8 + 8^2) \times 0.45/3} = 8.2 \text{ A} = 1.5$ times the average current.

13. *Triangle Wave (same function as 9)*

$$I_{avg} = 0.50 I_{pk} \tag{1.13a}$$

$$I_{rms} = 0.577 I_{pk} \tag{1.13b}$$

14. *Clipped Triangle Wave (same function as 10)*

$$I_{avg} = 0.50 I_{pk} t/T \tag{1.14a}$$

$$I_{rms} = I_{pk}\sqrt{t/(3T)} \tag{1.14b}$$

15. *Exponential Decay Wave*

$$I_{avg} = \frac{1}{T} \int I_{pk} \, dt$$

$$= \frac{1}{T} \int_0^T I_{pk} \varepsilon^{-t/\tau} \, dt$$

$$= \left[\frac{I_{pk}}{T} \frac{\varepsilon^{-t/T}}{-\frac{1}{\tau}} \right]_0^T$$

$$= \frac{I_{pk}\tau}{T}(1 - \varepsilon^{-t/\tau})$$

For $T^1 = 5\tau$, $T > T^1$:

$$I_{\text{avg}} = (I_{\text{pk}}/5)(1 - \varepsilon^{-5})(T^1/T)$$

$$= 0.199 I_{\text{pk}}(T^1/T) \tag{1.15a}$$

$$I_{\text{rms}}^2 = \frac{1}{T}\int_0^T I_{\text{pk}}^2 t\, dt$$

$$= \frac{1}{T}\int_0^T \left(I_{\text{pk}}\varepsilon^{-t/\tau}\right)^2 dt$$

$$= \frac{I_{\text{pk}}^2}{T}\left(\frac{-\tau}{2}\right)(\varepsilon^{-2t/\tau} - \varepsilon^0)$$

For $T^1 = 5\tau$, $T > T^1$:

$$I_{\text{rms}}^2 = \frac{I_{\text{pk}}^2}{5\tau}\left(\frac{-\tau}{2}\right)(\varepsilon^{-10} - 1)\frac{T^1}{T}$$

Since $\varepsilon^{-10} \cong 0$:

$$I_{\text{rms}} = I_{\text{pk}}\sqrt{T^1/T}\,/\sqrt{10}$$

$$= 0.316 I_{\text{pk}}\sqrt{T^1/T} \tag{1.15b}$$

Example

A series RC snubber network with a time constant of $\tau = 0.3\ \mu$s is connected across a device switching at 40 kHz ($T = 25\ \mu$s). The snubber must conduct a peak current of 10 A when the device turns off. Then the current is $I_{\text{rms}} = 0.316 \times 10\sqrt{5 \times 0.3 \times 10^{-6}/25 \times 10^{-6}} = 0.77$ A.

The power rating of the resistor is then $P = I_{\text{rms}}^2 R$, and this value should be multiplied by a factor of at least 2 for derating. Reference Section 13.4 for further discussion on snubbers.

16. *Critically Damped Exponential Wave (also reference Table 1.3)*

$$t_{\text{pk}} = 2L/R = \sqrt{LC} \qquad R = 2\sqrt{L/C}$$

$$I_{\text{pk}} = 2E/\varepsilon R$$

$$I_{\text{avg}} = (2E/2\varepsilon)\sqrt{C/L}$$

$$= (I_{\text{pk}}/\varepsilon)(t_{\text{pk}}/T^1) \tag{1.16a}$$

$$I_{\text{rms}} = E/R = \tfrac{1}{2}\left[I_{\text{pk}}\varepsilon\sqrt{t_{\text{pk}}/T^1}\right] \tag{1.16b}$$

From the above values and the values of Table 1.2, note the high values of rms current (relative to the peak current) associated with the sawtooth, trapezoid, and triangle waveforms as compared to the square or sine waves. These waveforms are frequently encountered in flyback and forward converters (see Chapter 8). The exponential wave also produces a high value of rms current, depending on the duty cycle. The constants given for the exponential wave provide an accuracy of 99% using five time constants.

1.2   MAGNETIC FIELDS

The current flowing in a conductor in a magnetic field supplied by a voltage source has the relations

$$e = -L\frac{di}{dt} = -N\frac{d\phi}{dt} \qquad i = \frac{1}{L}\int e\,dt \qquad \phi = \frac{1}{N}\int e\,dt \qquad (1.17)$$

where L is the inductance of the circuit and is equal to $N\,d\phi/di$ and $d\phi/di$ is the rate of change of flux with respect to the current.

The instantaneous power input to the magnetic field is $p = ei$, and the energy input during a time interval t (following the closure of the circuit) is

$$\varepsilon = \int p\,dt = \int L\,di/dt\,i\,dt = \tfrac{1}{2}Li^2 \qquad \text{J} \qquad (1.18)$$

The above relation is useful in the design of inductor input filters for rectified alternating current, either sine or square wave. The energy storage is also useful in the design of flyback converters. Reference Section 9.4 where a transistor, inserted in series with the inductor and a dc source, is turned on to commence current flow which increases linearly (as in the sawtooth waveform) until a time when the transistor is turned off. The energy stored in the magnetic field is then transferred to the load. In the above analysis, the dc resistance of the circuit is assumed negligible.

1.3   DIELECTRIC FIELDS

The current flowing in a dielectric field supplied by a voltage source has the following relations:

$$i = C\frac{dv}{dt} = \frac{dq}{dt} \qquad e = \frac{1}{C}\int i\,dt = \frac{1}{C}\int dq \qquad (1.19)$$

where C is the capacitance and is equal to q/e, q is the charge in coulombs, and $i = dq/dt$.

The instantaneous power input to the dielectric field is $p = ei$, and the energy input during time interval t (following the closure of a switch) is

$$\varepsilon = \int p\,dt = \int C\,de/dt\,e\,dt = \tfrac{1}{2}Ce^2 \qquad \text{J} \qquad (1.20)$$

The above relation is useful in timing circuits where a capacitor is charged at a constant current to produce a linear sweep of voltage, as in the sawtooth waveform. If the capacitor is also discharged with a constant current, a triangle voltage waveform is produced. The energy stored in the dielectric field is then transferred to the load, as in rectifier filters for power supplies. In the above analysis the dc resistance is assumed negligible.

1.4 THE *RL* CIRCUIT

When resistance is inserted in series with an inductor and the circuit is energized from a dc or an ac source, transient conditions of exponential nature occur until a steady-state condition is reached. In the case of a dc input, the steady-state current is limited only by the dc resistance of the circuit. In the case of an ac input, the steady-state current is a function of the dc resistance and the impedance of the inductor at the exciting frequency.

The general solution for a circuit with initial (E_i) and final (E_f) values of voltage is given by

$$e_0 = E_f + (E_i - E_f)\varepsilon^{-Rt/L} \qquad (1.21)$$

and will be used in the following sections. The term τ is the time constant L/R.

1.4.1 dc Input

Referring to the circuit of Fig. 1.1A, when switch S is closed to a, the equation of voltage is

$$E = Ri + L\,di/dt + RI_i \qquad (1.22)$$

and by integration, the equation of current is

$$i = (E/R)(1 - \varepsilon^{-Rt/L}) + E_i\varepsilon^{-Rt/L} \qquad (1.23)$$

where I_i is the initial current flowing and E_i is the initial inductor voltage.

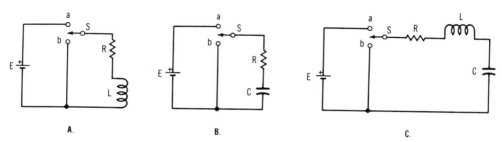

A.   B.   C.

Fig. 1.1. *RL*, *RC*, and *RLC* circuits.

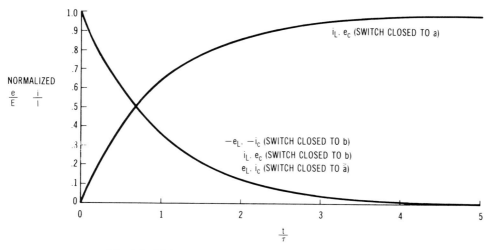

Fig. 1.2. Voltage and current in *RL* and *RC* circuits.

Since the current cannot change instantaneously in the inductor, the inductor voltage is

$$e_L = (E - I_i R)\varepsilon^{-Rt/L} \qquad (1.24)$$

The waveforms of current flow and inductor voltage for $I_i = 0$ are shown in Fig. 1.2 as a function of time. When $t = \tau$, the current has reached 63.2% of the final value, and the inductor voltage is 36.8% of the initial value. At 3τ, the current has reached 95% of the final value. Calculations for the exponential waveform of Table 1.1 used $\tau = 5$, at which time 99% of the final value is reached. When switch S of Fig. 1.1A is instantaneously transferred from point a to point b, the induced voltage in the inductor causes a discharge current to flow, and

$$i = I_i \varepsilon^{-Rt/L} \qquad (1.25)$$

where I_i is the current flowing before the switch was transferred and is equal to E/R if a steady-state condition existed before transfer.

Current and voltage relations for the circuit in Fig. 1.1 are graphed in Fig. 1.2.

1.4.2   ac Input

When the battery source of Fig. 1.1A is replaced by an alternating voltage source and switch S is closed to a, the resultant waveforms consist of a transient term as well as a steady-state term, where the transient term diminishes as time increases. If the switch is closed at time $t = 0$, the instantaneous input may be at any angle of the sine wave, and the resulting transient current waveform is a function of the angle at $t = 0$. The general equation is

$$E_{pk} \sin(\omega t + \lambda) = iR + L\, di/dt \qquad (1.26)$$

where E_{pk} is the peak of the sinusoid and λ represents the radians (or degrees) between the voltage crossover ($e = 0$) and the point at which S is closed ($t = 0$).

The solution for current in the above equation, after lengthy mathematical relations, is

$$i = (E_{pk}/Z)\sin(\omega t + \lambda - \theta) - (E_{pk}/Z)\sin(\lambda - \theta)\varepsilon^{-Rt/L} \qquad (1.27)$$

where

$$Z = \sqrt{R^2 + (\omega L)^2} \qquad \theta = \tan^{-1}(\omega L/R)$$

The left term is the steady-state condition and the right term is the transient condition. In this case, $-\theta$ signifies the current lags the voltage. The power factor is equal to $\cos\theta$. It is noted that the transient term equals zero when $\lambda - \theta = 0, \pi, 2\pi, \ldots$ and is a maximum when $\lambda - \theta = \pi/2, 3\pi/2, 5\pi/2, \ldots$. Physically, this means that in highly inductive circuits, where θ approaches $90°$, the transient effect is zero when the circuit is energized near the peak of the input voltage. Conversely, the transient term is maximum when the circuit is energized near the crossover of the input voltage. Both conditions are superimposed in Fig. 1.3. In energizing large power transformers, where, of course, the dc resistance is extremely low, the peak surge current is a minimum when the transformer is energized at $\lambda = 90°$. This surge current is also a function of the magnetizing level when voltage was previously removed from the transformer.

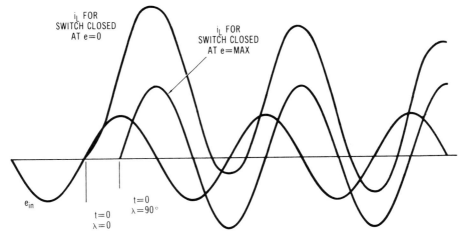

Fig. 1.3. An ac voltage instantaneously applied to an RL circuit.

1.5   THE *RC* CIRCUIT

When resistance is inserted in series with a capacitor and the circuit is energized from a dc or an ac source, transient conditions of exponential nature occur until a steady-state condition is reached. In the case of a dc input, the initial current is limited only by the dc resistance of the circuit. In the case of an ac input, the steady-state current is a function of the dc resistance and the impedance of the capacitor at the exciting frequency. Equation (1.21) still applies.

1.5.1   dc Input

Referring to the circuit of Fig. 1.1B, when switch S is closed to a, the equation of voltage is

$$E = Ri + \frac{q}{C} = Ri + \frac{1}{C} \int i\, dt + Q_0 \tag{1.28}$$

and by integration, the equation of current is

$$i = [(E - Q_0/C)/R]\varepsilon^{-t/RC} \tag{1.29}$$

where Q_0 is the initial charge on the capacitor.

If the initial charge is zero, the above equation becomes

$$i = (E/R)\varepsilon^{-t/RC} \tag{1.30}$$

Since the voltage cannot change instantaneously across the capacitor, the capacitor voltage with initial voltage E_0 is

$$e_C = E - [E - Q_0/C]\varepsilon^{-t/RC} = (E - E_0)(1 - \varepsilon^{-t/RC}) + E_0 \tag{1.31}$$

If the initial charge is zero, the preceding equation becomes

$$e_C = E(1 - \varepsilon^{-t/RC}) \tag{1.32}$$

The term τ is the time constant RC. The waveforms of current flow and capacitor voltage, for $I_i = 0$, are shown in Fig. 1.2 as a function of time. When $t = \tau$, the current has decayed to 36.8% of the initial value, and the capacitor voltage has reached 63.2% of the final value. At 3τ, the voltage has reached 95% of the final value. When switch S of Fig. 1.1B is instantaneously transferred from point a to point b, the stored capacitor charge causes a discharge current to flow, and

$$i = (E_C/R)\varepsilon^{-t/RC} \tag{1.33}$$

where E_C is the capacitor voltage before the switch was transferred and is equal to E if a steady-state condition existed before transfer.

The capacitor voltage, when the switch is closed to b, is

$$e_C = E_C \varepsilon^{-t/RC} \tag{1.34}$$

Current and voltage relations for the above circuit are graphed in Fig. 1.2. Calculations for the exponential waveform of Table 1.2 used $\tau = 5$, at which time 99% of the final value is reached. Thus for capacitor charging or discharging, as in the case of snubber circuits for transistors or thyristors, the power dissipation of the series resistor may be calculated from the rms value of current.

1.5.2 Two Precharged Capacitors

Referring to Fig. 1.4, C_1 and C_2 have been precharged with the polarities shown. When switch S is closed, each capacitor will discharge or charge through R and the final voltage on each capacitor will be the same value, since the final current will be zero. Initially, $e_1 = E_1$ and $e_2 = E_2$. At $t = \infty$, $i = 0$ and $e_1 = E_f = \pm e_2$ (observing correct polarity).

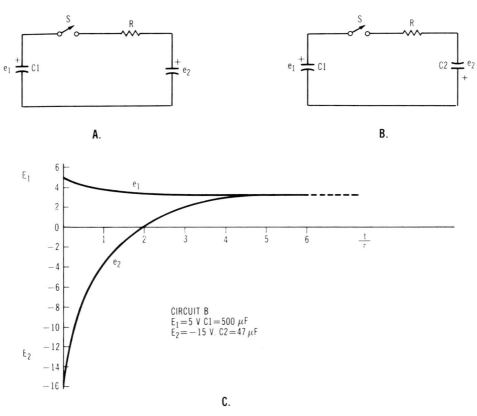

Fig. 1.4. Discharge characteristics of two precharged capacitors.

The final voltage on each capacitor will be the average of the initial voltage and the capacitance ratios, and

$$E_f = (E_1 C_1 + E_2 C_2)/(C_1 + C_2) \tag{1.35}$$

From Equation (1.30), the current flow when S is closed is

$$i = [(E_1 - E_2)/R]\varepsilon^{-t/RC'} \tag{1.36}$$

where

$$C' = C_1 C_2/(C_1 + C_2)$$

The dynamic equations for voltage are

$$e_1 = E_f + (E_1 - E_f)\varepsilon^{-t/RC'} \tag{1.37a}$$

$$e_2 = E_f + (E_2 - E_f)\varepsilon^{-t/RC'} \tag{1.37b}$$

Example

The circuit of Fig. 1.4B could be the output filter capacitors of a multioutput power supply with a common return. The switch could be an output-to-output short, where R is essentially zero. If either the preceding linear or switching stage has an automatic shutdown/restart feature when the short is applied, the dynamic and steady-state parameters may be calculated. For $E_1 = +5$ V, $C_1 = 500$ μF, $E_2 = -15$ V, $C_2 = 47$ μF, and $R = 0.1$ Ω; $E_f = (5 \times 500 - 15 \times 47)/(500 + 47) = 3.28$ V, and $C' = 500 \times 47/(500 + 47) = 43$ μF.

The dynamic voltages for the above values are plotted in Fig. 1.4C. Thus C_2 will be polarized in the opposite direction, which could cause a failure in the capacitor. For this reason, a reverse-polarity rectifier should be connected across C_2, and the final voltage across each capacitor would then be the forward drop of the rectifier. Conversely, if the product $E_2 C_2$ is greater than $E_1 C_1$, a rectifier should be connected across C_1.

1.5.3   ac Input

When the battery source of Fig. 1.1B is replaced by an alternating voltage source and switch S is closed to a, the resultant waveforms consist of a transient term as well as a steady-state term, where the transient term diminishes as time increases. If the switch is closed at time $t = 0$, the instantaneous input may be at any value of the sine wave and the resulting transient waveform is a function of the angle at $t = 0$. The general equation is

$$E_{\text{pk}} \sin(\omega t + \lambda) = iR + \frac{1}{C} \int i \, dt$$

where E_{pk} is the peak of the sinusoid and λ represents the radians or degrees between the voltage crossover ($e = 0$) and the point at which S is closed ($t = 0$).

In a manner similar to Section 1.4, the solution for current is

$$i = (E_{pk}/Z)\sin(\omega t + \lambda + \theta) - (E_{pk}/\omega CRZ)\cos(\lambda + \theta)\varepsilon^{-t/RC} \quad (1.38)$$

where

$$Z = \sqrt{R^2 + (1/\omega C)^2}$$

$$\theta = \tan^{-1}(1/\omega CR)$$

λ = radians (or degrees) between the voltage crossover ($e = 0$) and the point at which switch S is closed ($t = 0$)

The left term is the steady-state condition and the right term is the transient condition. Also, $+\theta$ signifies the current leads the voltage. The power factor is equal to cos θ. A study of the above equation reveals that there is no fixed relation between the transient and the steady-state component of current at time $t = 0$ (unlike the RL circuit of Section 1.4.2). Again, the peak current is limited only by the resistance and the voltage at the time S is closed. Reference Section 4.2.3 for conditions of in-rush currents to rectifier–filter circuits.

The equation for capacitor voltage is

$$e_C = -\frac{E_{pk}}{\omega CZ}\left[\cos(\omega t + \lambda + \theta) - \varepsilon^{-t/RC}\cos(\lambda + \theta)\right] \quad (1.39)$$

1.6 THE *RLC* CIRCUIT

When resistance, inductance, and capacitance are in series and the circuit is energized from a voltage source, transient conditions of exponential, and frequently sinusoidal, nature occur until a steady-state condition is reached. The damping ratio is defined as

$$\delta = (R/2)\sqrt{C/L} \quad (1.40)$$

Three current modes are possible: overdamped if $\delta < 1$, critically damped if $\delta = 1$, and oscillatory if $\delta > 1$. Each of these modes will be analyzed, but since it is desirable to minimize heating due to resistance in power conversion equipment, the oscillatory mode is most common as in commutation circuits and certain filters. However, critically damped or overdamped circuits are frequently used for semiconductor snubbers, for transient suppression, and for minimizing input surges to LC input filters at turn-on. An alternate to compromising both input surge and resonant charging of an input filter capacitor is the use of a swinging choke (discussed in Chapters 3 and 4) for the input inductor.

1.6.1 dc Input

Referring to the circuit of Fig. 1.1C, when switch S is closed to a, the equation of voltage is

$$E = L\frac{di}{dt} + Ri + \frac{1}{C}\int i\,dt \quad (1.41)$$

Differentiating,

$$0 = L\alpha^2 + R\alpha + 1/C \tag{1.42}$$

and the quadratic solution for α is

$$\begin{aligned} \alpha &= -(R/2L) \pm \sqrt{R/2L^2 - 1/LC} \\ &= -a \pm b \end{aligned} \tag{1.43}$$

1.6.1.1   Equations for Current   The complementary function of the above equation is

$$i = k_1 \varepsilon^{(-a+b)t} + k_2 \varepsilon^{(-a-b)t} + 0 \tag{1.44}$$

By various substitutions, the equation becomes

$$i = [(CE - Q_0)/LCb][\varepsilon^{(-a+b)t} - \varepsilon^{(-a-b)t}] \tag{1.45}$$

where Q_0 is the initial charge in coulombs.

But b may be real (overdamped), zero (critically damped), or imaginary (oscillatory).

Case 1. If b is real, Equation (1.45) may be simplified to

$$i = [(CE - Q_0)/LCb]\varepsilon^{-at} \sinh bt \tag{1.46}$$

Case 2. If b is zero, the equation for current is

$$i = [(CE - Q_0)/LC]t\varepsilon^{-at} \tag{1.47}$$

Case 3. If $1/LC > (R/2L)^2$, b is imaginary and it is replaced by β, where $\beta = \sqrt{1/LC - (R/2L)^2}$. The equation for current is

$$i = [(CE - Q_0)/\beta LC]\varepsilon^{-at} \sin \beta t \tag{1.48}$$

If the initial charge Q_0 is zero, Equation (1.48) is

$$i = (E/\beta L)\varepsilon^{-at} \sin \beta t \tag{1.49}$$

Case 4. If $1/LC \gg (R/2L)^2$ or R is negligible (as in the ideal case of a commutation circuit), $\beta = \sqrt{1/LC}$) which is the resonant frequency (in radians) of the circuit, and the equation for current is

$$i = (E/Z)\sin \beta t \tag{1.50}$$

where $Z = \sqrt{L/C}$ is the circuit impedance.

TABLE 1.3. Analysis Formulas for *RLC* Circuit

	Oscillatory (Underdamp) $\delta < 1$	Critically Damped, $\delta = 1$	Overdamped, $\delta > 1$	Undamped, $\delta \ll 1\,(R = 0)$
i_1	$\dfrac{E}{L\beta}\varepsilon^{-at}\sin\beta t$	$\dfrac{E}{L}t\varepsilon^{-at}$	$\dfrac{E}{Lb}\varepsilon^{-at}\sinh bt$	$\dfrac{E}{\omega L}\sin\omega t$
I_{pk}	$E\sqrt{\dfrac{C}{L}}\varepsilon^{-at_{pk}}$	$\dfrac{2E}{\varepsilon R}$	$E\sqrt{\dfrac{C}{L}}\,\varepsilon^{-at_{pk}}$	$\dfrac{E}{\omega L}=\dfrac{E}{Z}$
t_{pk}	$\dfrac{1}{\beta}\tan^{-1}\left|\dfrac{\beta}{a}\right|$	$\dfrac{1}{a}=\sqrt{LC}$	$\dfrac{1}{b}\tanh^{-1}\left|\dfrac{b}{a}\right|$	$\dfrac{\pi}{2}\sqrt{LC}=\dfrac{t_p}{2}$
$a = \dfrac{R}{2L}$		$b = \sqrt{\dfrac{R^2}{4L^2}-\dfrac{1}{LC}}$	$\omega = \dfrac{1}{\sqrt{LC}}=2\pi f$	
$\theta = \tan^{-1}\dfrac{\beta}{a}$				
$\phi = \tan^{-1}\dfrac{\beta}{-a}$		$\beta = \sqrt{\dfrac{1}{LC}-\dfrac{R^2}{4L^2}}$	$Z = \sqrt{\dfrac{L}{C}}$	

The peak current occurs at $\beta t = 90°$ and $i = E/Z$.

Parameters of current, peak current, and time to peak current are summarized for the above conditions in Table 1.3, and a typical waveform for an underdamped circuit is shown in Fig. 1.5. For the undamped case, waveforms of current and voltage are shown in Fig. 1.6 for 180° conduction, such as in a rectifier or thyristor. Note the capacitor voltage charges to twice the dc input voltage. The decay of current in an *RLC* circuit (final current and voltage are zero) may be calculated from Equations (1.45)–(1.48) by eliminating CE and, if desired, by substituting C_e for Q_0.

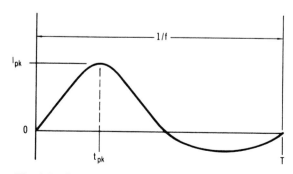

Fig. 1.5. Current waveform, underdamped circuit.

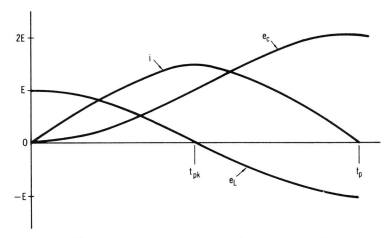

Fig. 1.6. Voltage and current waveforms in an undamped circuit.

1.6.1.2 Equations for Voltage   For simplicity and since the oscillatory case is most frequently encountered in power conversion designs, equations for voltage will be limited to the oscillatory case. The initial charge (or voltage) on the capacitor is assumed to be zero. Since $e = -L\,di/dt$, differentiating Equation (1.49) and multiplying by L give the inductor voltage as

$$e_L = \frac{E}{\beta\sqrt{L/C}}\,\varepsilon^{-at}\sin(\beta t + \phi) \tag{1.51}$$

Since $i = dq/dt$ and $q = CE$, integrating Equation (1.49) and dividing by C give the capacitor voltage as

$$e_C = E\left(1 - \frac{\varepsilon^{-at}\sin(\beta t + \theta)}{\beta\sqrt{LC}}\right) \tag{1.52}$$

where a, β, θ, and ϕ are given by equations in Table 1.3.

In the undamped case, for $R = 0$, Equation (1.51) becomes a cosine function and since $\theta = 90°$ and $\beta t = 180°$, $\sin 270° = -1$, which makes the voltage of Equation (1.52) equal to $2E$, as shown by Fig. 1.6.

1.6.2   ac Input

1.6.2.1 Equations for Current   If the dc source in Fig. 1.1C is replaced by an alternating voltage and the switch is closed, the basic voltage equation is

$$E_{pk}\sin(\omega t + \lambda) = L\frac{di}{dt} + Ri + \frac{1}{C}\int i\,dt \tag{1.53}$$

Differentiating and dividing by L give

$$\frac{\omega E_{pk}}{Z} \sin(\omega t + \lambda - \theta) = \frac{d^2 i}{dt^2} + \frac{R}{L}\frac{di}{dt} + \frac{i}{LC} \qquad (1.54)$$

Using the analysis in the above section, the expression for current is

$$i = \frac{E_{pk}}{Z} \sin(\omega t + \lambda - \theta) + c_1 \varepsilon^{(-a+b)t} + c_2 \varepsilon^{(-a-b)t} \qquad (1.55)$$

Further analysis will be limited to the oscillatory case since this condition is most prevalent in power conversion circuits. Evaluating the above equation for c_1 and c_2, the equation for current becomes

$$i = \frac{E_{pk}}{Z} \sin(\omega t + \lambda - \theta) + \frac{E_d}{\beta L}\varepsilon^{-at} \sin \beta t - \frac{E_{pk}}{Z} \sin(\lambda - \theta)\varepsilon^{-at} \cos \beta t \quad (1.56)$$

where

$$a = R/2L$$

$$\beta = \sqrt{1/LC - (R/2L)^2}$$

$$Z = \sqrt{R^2 + (\omega L - 1/\omega C)^2}$$

$$\theta = \tan^{-1} \frac{\omega L - 1/\omega C}{R}$$

$$E_d = E_{pk} \sin \lambda - \frac{Q_0}{C} - \frac{E_{pk}\omega L}{Z}\cos(\lambda - \theta) - \frac{E_{pk}R}{2Z}\sin(\lambda - \theta)$$

$$\lambda = \text{angle (radians) at which ac input is applied}$$

The left-side term is the steady-state term while both right-side terms are exponentially damped sine and cosine terms of like frequency. Thus the above equation consists of two sinusoidal terms. The frequency of the steady-state term, ωt, is the frequency of the input voltage. The frequency of the transient terms, βt and θ, may be less than, equal to, or greater than the input frequency, depending on R, L, and C. The transient terms disappear when the steady-state condition is reached and the current then becomes a sinusoid, lagging or leading the voltage depending on the values of L and C.

Also, the angle λ at which time switch S is closed has an effect on the initial transient current value and waveform. For instance, the peak current (and the capacitor voltage) will increase in value (each time the switch is closed) as λ increases. This inrush current, say to a rectifier circuit with an LC filter (where L is the filter or input line inductance and C is the filter capacitance), determines the inrush rating of the rectifier. Another application where this analysis may be used is described by referring to the circuit shown in Fig. 1.7. Input voltage at the utility main frequency is applied to a transformer having a leakage inductance or series inductance L. The secondary voltage is applied to a voltage doubler rectifier and filter, C_1 and C_2, which is connected to a load. The capacitors may be discharged on opposite and alternate half cycles into the load for short, high-energy bursts.

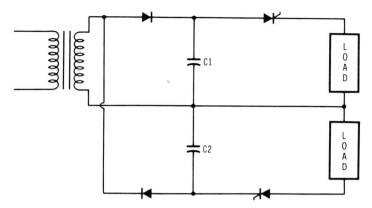

Fig. 1.7. Resonant charging circuit.

1.6.2.2 **Equations for Voltage**   From Equation (1.19), integrating Equation (1.56) will give the expression for capacitor voltage as

$$e_C = -\frac{E_{pk}}{\omega CZ}\cos(\omega t + \lambda - \theta) + \frac{E_d}{\beta}\varepsilon^{-at}(-a\sin\beta t - \beta\cos\beta t)$$

$$-\frac{E_{pk}L}{Z}\varepsilon^{-at}\sin(\lambda - \theta)(-a\cos\beta t + \beta\sin\beta t) \qquad (1.57)$$

Computer analysis of the above equation, as well as the equation for current, shows the capacitor voltage during the first half cycle of input voltage will increase as λ increases each time switch S is closed. To maintain a constant voltage on the capacitor, λ must decrease as the input voltage increases. This is opposite the steady-state conditions of normal phase control where a thyristor would be used to maintain a regulated voltage on a filter capacitor feeding a load. For this case, λ must increase as the input voltage increases, as discussed in Chapter 5.

1.7   THE *RLCR* CIRCUIT WITH A dc INPUT

This circuit is essentially a series *RLC* circuit but with an additional resistor shunting the capacitor, as in an inductive input filter with a load resistor, shown in Fig. 1.8. Whereas the preceding analysis utilized differential equations, this circuit is best analyzed by using Laplace transforms and admittance to establish equations. For the "left-hand" loop, the equation is

$$E_s = V_1(sLG_1 + 1) - V_2 sLG_1$$

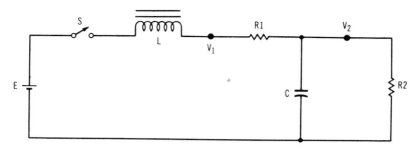

Fig. 1.8. Basic *RLCR* circuit.

For the "right-hand" loop, the equation is

$$0 = V_2(sC + G_1 + G_2) - V_1 G_1$$

where

$$G_1 = 1/R_1 \qquad G_2 = 1/R_2$$
$$G_1 + G_2 = (R_1 + R_2)/R_1 R_2$$

1.7.1   Equations for Voltage

Solving for V_1,

$$V_1 = [V_2(sC + G_1 + G_2)]/G_1 \tag{1.58}$$

Substituting,

$$E_s = \frac{V_2(sC + G_1 + G_2)(sLG_1 + 1) - V_2 sLG_1^2}{G_1}$$

Then

$$V_2 = \frac{E_s G_1}{LCG_1\left[s^2 + s(1/LG_1 + G_2/C) + (G_1 + G_2)/LCG_1\right]}$$

$$= \frac{E_s}{LC}\left(s^2 + s\left(\frac{R_1}{L} + \frac{1}{R_2 C}\right) + \frac{(R_1 + R_2)R_1}{R_1 R_2 LC}\right)^{-1} \tag{1.59}$$

which is of the form

$$1/s\left[(s + a)^2 + \beta^2\right]$$

Solving as the inverse transform and rearranging terms, the capacitor or load voltage is

$$V_2 = e_C = \frac{E}{LC(a^2 + \beta^2)}\left[1 - \frac{\sqrt{a^2 + \beta^2}}{\beta}\varepsilon^{-at}\sin(\beta t + \theta)\right]$$

$$= \frac{ER_2}{R_1 + R_2}\left[1 - \frac{\sqrt{(R_1 + R_2)/R_2}}{\beta\sqrt{LC}}\varepsilon^{-at}\sin(\beta t + \theta)\right] \quad (1.60)$$

where

$$a = \frac{1}{2}\left(\frac{R_1}{L} + \frac{1}{R_2 C}\right) \qquad \beta = \sqrt{\frac{R_1 + R_2}{R_2 LC} - a^2} \qquad \theta = \tan^{-1}\frac{\beta}{a}$$

Substituting V_2 into Equations (1.58),

$$V_1 = \left[\frac{E_s}{LC\left[s^2 + s(1/LG_1 + G_2/C)\right] + (G_1 + G_2)/LCG_1}\right]\left[\frac{sC + G_1 + G_2}{G_1}\right]$$

$$= \frac{E_s R_1}{L}\left[\frac{s + (1/R_1 C + 1/R_2 C)}{s^2 + s(R_1/L + 1/R_2 C) + (R_1 + R_2)/R_2 LC}\right] \quad (1.61)$$

which is of the form

$$(s + d)/s\left[(s + a)^2 + \beta^2\right]$$

Taking the inverse transform,

$$V_1 = \frac{ER_1}{L}\left[\frac{d}{a^2 + \beta^2} - \frac{1}{\beta}\sqrt{\frac{(d - a)^2 + \beta^2}{a^2 + \beta^2}}\varepsilon^{-at}\sin(\beta t + \phi)\right]$$

$$= E\left[1 - \frac{\varepsilon^{-at}\sin(\beta t + \phi)}{\beta\sqrt{LC}}\right] \quad (1.62)$$

where

$$d = \frac{R_1 + R_2}{R_1 R_2 C}$$

$$\frac{d}{a^2 + \beta^2} = \frac{L}{R_1}$$

$$\phi = \tan^{-1}\frac{\beta}{a} + \tan^{-1}\frac{\beta}{d - a}$$

In the above analysis, d was assumed to be greater than a (see the following section). Solving for the inductor voltage,

$$e_L = E - V_1 \quad (1.63)$$

1.7.2   Equations for Current

Using conventional terms, the equations for Fig. 1.8 are

$$E = L\frac{di_1}{dt} + R_1 i_1 + \frac{1}{C}\int i_1\, dt - \frac{1}{C}\int i_2\, dt$$

$$0 = R_2 i_2 + \frac{1}{C}\int i_2\, dt - \frac{1}{C}\int i_1\, dt$$

$$= (sL + R_1)i_1 + \frac{i_1}{sC} - \frac{i_2}{sC}$$

$$= \frac{i_2}{sC} + R_2 i_2 - \frac{i_1}{sC}$$

$$i_2 = \frac{i_1}{sC\left(R_2 + \dfrac{1}{sC}\right)}$$

$$0 = \left(sL + R_1 + \frac{1}{sC}\right)i_1 - \frac{i_1}{sC(R_2 + 1/sC)sC}$$

$$= \frac{s^3(R_2 + LC^2) + s^2(LC + R_1 R_2 C^2) + s(R_1 C + R_2 C)}{s(sR_2 C^2 + C)} \tag{1.64}$$

Then

$$I_s = \frac{Es(sR_2 C^2 + C)}{s\left[s^2 + (R_2 LC^2) + s(LC + R_1 R_2 C^2) + R_1 C + R_2 C\right]}$$

$$= \frac{E(s + 1/R_2 C)}{Ls\left[s^2 + s(1/R_2 C + R_1/L) + (R_1 + R_2)/R_2 LC\right]} \tag{1.65}$$

The bracketed denominator in the above equation is the same as the bracketed denominator in Equation (1.61), and the equation is of the form

$$(s + b)/s\left[(s + a)^2 + \beta^2\right]$$

Taking the inverse transform,

$$i_1 = \frac{E}{L}\left[\frac{b}{a^2 + \beta^2} + \frac{1}{\beta}\sqrt{\frac{(b - a)^2 + \beta^2}{a^2 + \beta^2}}\, \varepsilon^{-at}\sin(\beta t + \psi)\right] \tag{1.66}$$

where

$$a = \left(\frac{R_1}{L} + \frac{1}{R_2 C}\right) \quad \text{for } a > b$$

$$b = \frac{1}{R_2 C}$$

$$a^2 + \beta^2 = \frac{R_1 + R_2}{R_2 LC}$$

$$\beta = \sqrt{\frac{R_1 + R_2}{R_2 LC} - a^2}$$

$$\psi = \tan^{-1}\frac{\beta}{a} + \tan^{-1}\frac{\beta}{b - a}$$

It is apparent that a can be greater than or less than b. For $a > b$, Equation (1.66) is correct. For $b > a$, the equation for current is

$$i_1 = \frac{E}{L}\left[\frac{b}{a^2 + \beta^2} - \frac{1}{\beta}\sqrt{\frac{(b - a)^2 + \beta^2}{a^2 + \beta^2}}\; \varepsilon^{-at}\sin(\beta t + \psi)\right] \qquad (1.67)$$

Equations presented in the above sections are readily solved with a programmable calculator or computer and present an in-depth analysis of a circuit such as that of Fig. 1.8. The equations may be developed for a circuit with an ac input in a manner similar to that of Section 1.6.2.

Example

From the presentation in the above sections, a typical example of the circuit in Fig. 1.8 is shown in Fig. 1.9 with associated waveforms. For the circuit of Fig. 1.9A, a dc source of 28 V is applied when switch S is closed. Circuit values are $R_1 = 0.05\ \Omega$, $L = 2$ mH, and $C = 200\ \mu$F, and two values of load resistance are shown. At light load, $R_2 = 280\ \Omega$, which would apply to a circuit with a minimum load or to a soft-start mode in a switching converter or power supply. At full load, $R_2 = 7\ \Omega$, which is present when S is closed.

Figure 1.9B is a plot of the resulting loop currents, while Fig. 1.9C is a plot of the resulting component voltages. For $R_2 = 280\ \Omega$, note the highly oscillatory nature of the circuit. The reverse (negative) current flow could be eliminated with a rectifier in series with switch S. Capacitor C would then discharge into R_2 until a steady-state condition was reached.

For $R_2 = 7\ \Omega$, the damping produced by the load resistance results in a steady-state condition in approximately 10 ms. The steady-state load voltage is

$$E_0 = E \times R_2/(R_1 + R_2) = 27.8\text{ V} \qquad (1.68a)$$

and the steady-state load current is

$$I_1 = I_2 = E/(R_1 + R_2) = 3.97\text{ A} \qquad (1.68b)$$

1.8   ac CIRCUIT ANALYSIS

The ac circuit analysis presented is limited to dc-coupled networks since these circuits are most prevalent in power conversion equipment and specifically power supplies. Even the high-frequency pulse-demodulated inverter with sine wave output sensing (reference Section 6.4) must have a high dc gain loop to prevent a dc component in the output.

Networks commonly referred to as "lag" introduce a pole at some frequency and voltage attenuation as well as a lagging phase result. Networks referred to as "lead" introduce a zero at some frequency and voltage deattenuation as well as a leading phase result. Combinations of elements in conjunction with operational amplifiers form the heart of many gain and feedback loops and are used for compensation to ensure stable operation in closed-loop systems.

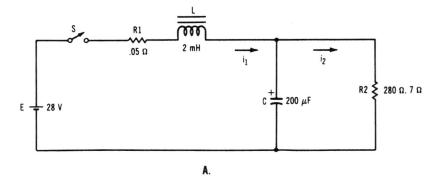

A.

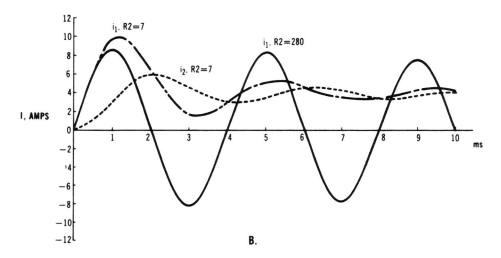

B.

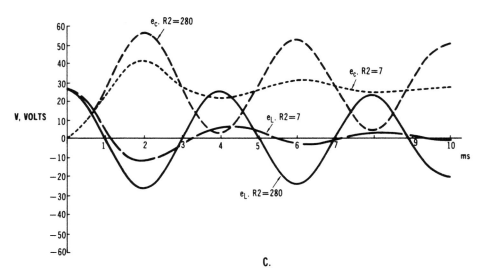

C.

Fig. 1.9. *RLCR* circuit values and waveforms.

1.8.1   Passive Lag Networks

Figure 1.10A is a simple *RC* lag network, and the transfer function is

$$\frac{e_0}{e_{\text{in}}} = \frac{X_c}{R + X_c} = \frac{1/sc}{R + 1/sc} = \frac{1}{1 + sRC} \tag{1.69a}$$

The Laplace *s* operator in the denominator indicates a pole. Substituting $j\omega$ for *s* ($j = \sqrt{-1}$) and $2\pi f$ for ω,

$$\frac{e_0}{e_{\text{in}}} = \frac{1}{1 + j(f/f_1)} \tag{1.69b}$$

where f is the input frequency in hertz and f_1 is the break frequency ($1/f_1 = 2\pi RC$).

This equation in cartesian or rectangular form may be converted to polar form as

$$\frac{e_0}{e_{\text{in}}} = \frac{1}{A\angle\theta} \tag{1.69c}$$

where $A = \sqrt{1 + (f/f_1)^2}$, and $\theta = -\tan^{-1}(f/f_1)$.

The segmented lines of Fig. 1.10A are a close approximation, while the exact values are (1) at $f = f_1$, $e_0/e_{\text{in}} = 1/\sqrt{2} = 0.707 = 20\log 0.707 = -3$ dB, and $\theta = -\tan^{-1}(1/1) = -45°$; (2) at $f = 0.1f_1$, $\theta = -\tan^{-1}(0.1) = -6°$; (3) at $f = 10f_1$, $\theta = -\tan^{-1}(10) = -84°$. The -1 slope indicates a roll-off of 20 dB per decade (6 dB per octave) above f_1 and a phase shift of 90° at higher frequencies, because the reactive impedance of the capacitor decreases linearly as frequency increases.

Figure 1.10B is the same circuit but with a resistive load which attenuates the output at direct current by the ratio $R_2/(R_1 + R_2)$. Figure 1.10C is a four-element double-pole circuit which introduces a pole of f_1 and another pole at f_2 and results in a 180° phase shift at higher frequencies. The circuit of Fig. 1.10D is common in SMPSs (switch mode power supplies) and sine wave inverter output filters. A double pole is introduced at the resonant frequency, f_0, resulting in a 40-dB-per-decade (12-dB-per-octave) roll-off, as indicated by the -2 slope, with a resulting 180° phase shift. A typical output filter circuit is analyzed in detail in Section 1.8.5.

1.8.2   Passive Lag–Lead Networks

The transfer function for the circuit of Fig. 1.11A is

$$\frac{e_0}{e_{\text{in}}} = \frac{R_2 + X_c}{R_1 + R_2 + X_c} = \frac{R_2 + 1/sc}{R_1 + R_2 + 1/sc} = \frac{1 + sR_2C}{1 + s(R_1 + R_2)C} \tag{1.70a}$$

The Laplace *s* operator in the denominator indicates a pole, while the *s* operator in the numerator indicates a zero. Again, substituting $j\omega$ for *s* and $2\pi f$

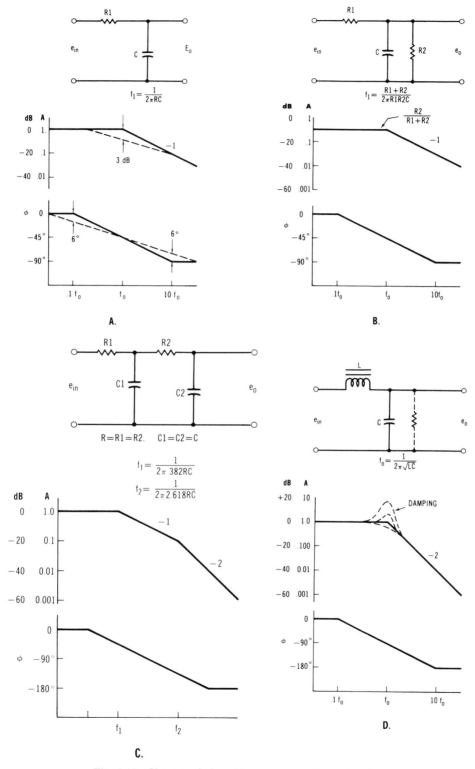

Fig. 1.10. Characteristics of lag-compensating networks.

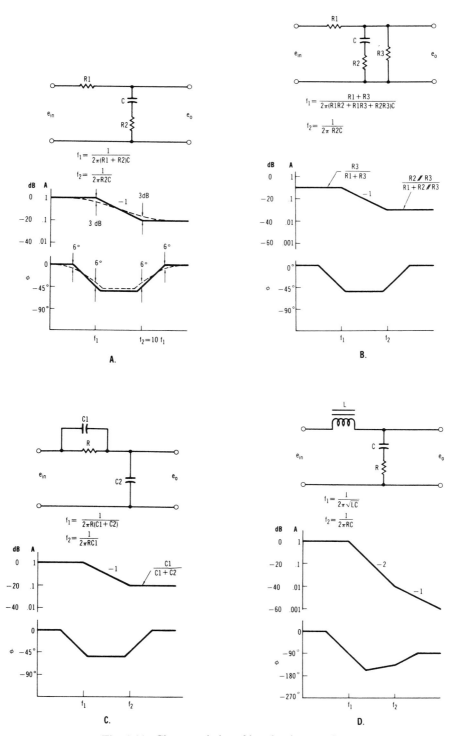

Fig. 1.11. Characteristics of lag–lead networks.

for ω,

$$\frac{e_0}{e_{in}} = \frac{1 + j(f/f_2)}{1 + j(f/f_1)} \tag{1.70b}$$

where f is the input frequency, $1/f_1 = 2\pi(R_1 + R_2)C$, and $1/f_2 = 2\pi R_2 C$.

This equation in cartesian or rectangular form may be converted to polar form as

$$\frac{e_0}{e_{in}} = \frac{b - jc}{d - je} = \frac{X \angle \theta_x}{Y \angle \theta_y} = A \angle \theta A \tag{1.70c}$$

where

$$X = \sqrt{b^2 + c^2} \qquad \theta_x = \tan^{-1}(c/b)$$
$$Y = \sqrt{d^2 + e^2} \qquad \theta_y = \tan^{-1}(e/d)$$
$$A = X/Y \qquad \theta_a = \theta_x - \theta_y$$

As an example, if $R_1 = 9R_2$, then $f_2 = 10f_1$, and the result of Equation (1.70) is shown in Fig. 1.11A by the dashed line.

Figure 1.11B shows the circuit with a resistive load added, with corresponding response. The "//" character in the figure indicates a parallel function. Figure 1.11C shows an additional series capacitor in shunt with R. The circuit has a transfer function similar to that of Fig. 1.11B except the output at high frequency is determined by the capacitive voltage divider, $C_1/(C_1 + C_2)$, as opposed to the resistive divider of Fig. 1.11B. The circuit of Fig. 1.11D is similar to Fig. 1.10D except the ESR of the capacitor is included, which introduces a zero at f_2 to bring the slope back to -1 or 20 dB per decade from the -2 slope after f_1. A typical output filter is analyzed in detail in Section 1.8.5. It should also be recognized that the ESR value usually has a wide tolerance and is quite temperature dependent. This is an important consideration in closed-loop operation, as discussed in Chapter 11.

1.8.3 Passive Lead–Lag Networks

Figure 1.12A produces a zero at f_1 and when the reactive impedance C equals that of the parallel resistor combination, a pole at f_2 results. The phase amplitude is again a function of the frequency break points. A similar response is shown in Figs. 1.12B and 1.12C. These circuits, or variations thereof, are useful in compensating double-pole output filters to maintain stable operation.

1.8.4 Circuit Impedance and Graphs

Circuits can be analyzed from the standpoint of impedance as opposed to the transfer functions of the above sections. Figure 1.13A is an interesting circuit with alternating "pole" and "zero" break frequencies every decade from 32 Hz to 320 kHz for the component values shown. The dc impedance is 10 kΩ and the

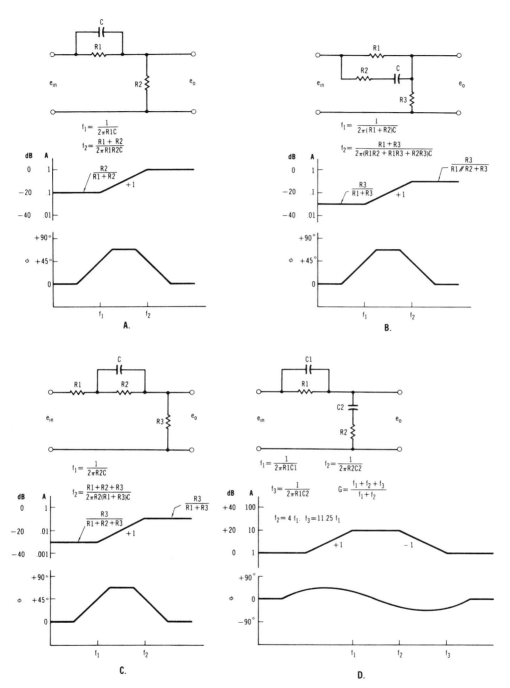

Fig. 1.12. Characteristics of lead–lag networks.

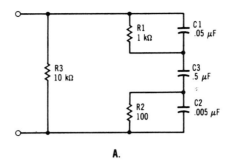

A.

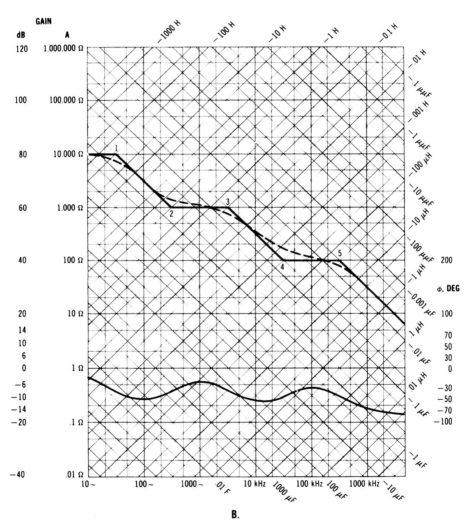

B.

Fig. 1.13. Circuit impedance graph.

impedance varies with frequency as

$$Z_t = \frac{R_2[R_1X_{C1}/(R_1 + X_{C1}) + X_{C2} + R_3X_{C3}/(R_3 + X_{C3})]}{R_2 + [R_1X_{C1}/(R_1 + X_{C1}) + X_{C2} + R_3X_{C3}/(R_3 + X_{C3})]} \quad (1.71)$$

Rather than solving the above equation or calculating the break frequencies, impedance charts may be used to analyze the circuit. In Fig. 1.13B, starting at $R_3 = 10 \text{ k}\Omega$ at dc, the $C_3 = 0.5 \text{ }\mu\text{F}$ line is intersected at point 1, or 32 Hz. At this point, R_3 and C_3 are in parallel. Capacitor C_3 dominates as frequency increases and the impedance decreases linearly to point 2, where $R_1 = 1 \text{ k}\Omega$ takes over until $C_1 = 0.05 \text{ }\mu\text{F}$ at point 3 is reached. The impedance decreases linearly to point 4 where $R_2 = 100 \text{ }\Omega$ takes over until $C_2 = 0.005 \text{ }\mu\text{F}$ at point 5 and the impedance decreases linearly with C_2 determining the circuit impedance at higher frequencies. The effect of C_1 and C_3 in series with C_2, which would decrease the effective capacitance value, was neglected at frequencies higher than 320 kHz.

The impedance chart (as well as the plots of Figs. 1.10–1.12) shows sharp break points which are smoothly tapered in actuality. The dashed line in Fig. 1.13B is obtained by solving Equation (1.71). However, use of the impedance chart shows how quickly a circuit's poles, zeros, and impedance can be determined with negligible error, as compared to the actual values given by the dashed line. Use of the impedance chart is most helpful in analyzing more complex circuits as in the following sections.

1.8.5 Typical Output Filter

An output filter for a power supply or sine wave inverter is shown in Fig. 1.14A. Resistor R_1 is the source, or dc, resistance of L; L is the series filter inductor; C is the shunt filter capacitor; R_2 is the ESR of the capacitor; and R_3 is the load resistance. At a no-load condition, R_3 becomes the series voltage divider resistance since some form of output sensing is normally used for voltage feedback in closed-loop systems.

The transfer function is

$$\frac{e_0}{e_{in}} = \frac{R_3(X_C + R_2)/(R_3 + X_C + R_2)}{R_1 + X_L + R_3(X_C + R_2)/(R_3 + X_C + R_2)} \quad (1.72a)$$

Using the Laplace transform, the transfer function is

$$\frac{e_0}{e_{in}} = \frac{R_3(1/sC + R_2)R_3 + 1/sC + R_2}{R_1 + sL + R_3(1/sC + R_2)R_3 + 1/sC + R_2}$$

$$= \frac{R_3(R_2Cs + 1)/[(R_2 + R_3)Cs + 1]}{(R_1 + sL)(R_2Cs + R_3Cs + 1 + R_2R_3Cs + R_3)} \quad (1.72b)$$

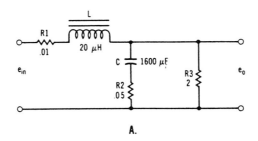

A.

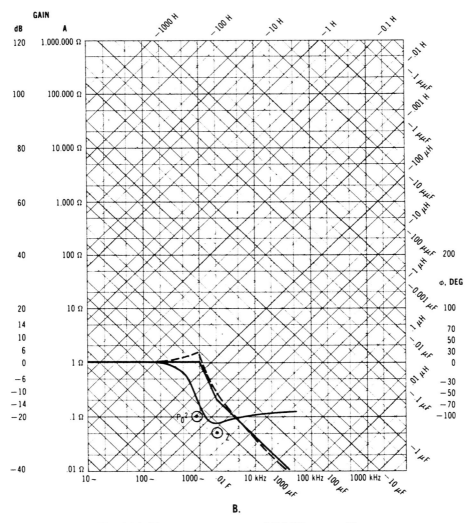

B.

Fig. 1.14. Frequency response of *RLCR* output filter.

For a highly efficient filter, $R_1/R_3 \ll 1$ and $R_2/R_3 \ll 1$ (even for low-voltage, high-current outputs) and the equation may be reduced to

$$\frac{e_0}{e_{\text{in}}} = \frac{1 + sR_2C}{s^2LC + s(R_1C + R_2C + L/R_3) + 1} \qquad (1.72c)$$

Substituting $j\omega$ for s and $-\omega^2$ for s^2, the transfer function is

$$\frac{e_0}{e_{\text{in}}} = \frac{1 + j(\omega/\omega_2)}{1 - (\omega/\omega_0)^2 + j(\omega/\omega_1 + \omega/\omega_2 + \omega/\omega_3)} \qquad (1.73a)$$

where

$$\omega = \text{input frequency, rad/s}$$
$$\omega_0 = 1/\sqrt{LC} = 2\pi f_0$$
$$\omega_1 = 1/R_1C = 2\pi f_1$$
$$\omega_2 = 1/R_2C = 2\pi f_2$$
$$\omega_3 = R_3/L = 2\pi f_3$$

Then

$$\frac{e_0}{e_{\text{in}}} = \frac{1 + f/f_2}{1 - (f/f_0)^2 + j(f/f_1 + f/f_2 + f/f_3)} \qquad (1.73b)$$

The equation can be expressed in cartesian or rectangular form as

$$e_0/e_{\text{in}} = (\pm b \pm c)/(\pm d \pm e) \qquad (1.73c)$$

and converted to polar form as

$$e_0/e_{\text{in}} = X \angle \theta_x / Y \angle \theta_y = A \angle \theta_a \qquad (1.73d)$$

where

$$X = \sqrt{b^2 + c^2} \qquad \theta_x = \tan^{-1}(\pm c/\pm b)$$
$$Y = \sqrt{d^2 + e^2} \qquad \theta_y = \tan^{-1}(\pm d/\pm e)$$
$$A = X/Y \qquad \theta_a = (\pm \theta_x) - (\pm \theta_y)$$

Thus, solving Equation (1.73b) as a function of input frequency and converting to Equation (1.73d) provide an attenuation (or gain) factor plus the phase shift in degrees for the filter. Since the constants in Equation (1.73b) are calculated or known, Equation (1.73d) can be readily solved with a programmable calculator or computer. Using the values $R_1 = 0.01\ \Omega$, $L = 20\ \mu\text{H}$, $C = 1600\ \mu\text{F}$, $R_2 = 0.05\ \Omega$, and $R_3 = 2\ \Omega$ for Fig. 1.14A, $f_0 = 890$ Hz, $f_1 = 9.9$ kHz, $f_2 = 2$ kHz, and $f_3 = 16$ kHz. Equation (1.73d) is plotted in Fig. 1.14B. Inspection of Equation (1.73b) shows a zero at f_2 and a double pole at f_0. Using the impedance chart of

Fig. 1.14B, the intersection of 20 μH and 1600 μF occurs at 890 Hz, the point circled P_0^2. The intersection of 0.05 Ω and 1600 μF occurs at 2 kHz, the point circled Z_2. Then, starting at unity gain (R_1 was assumed negligible), a horizontal line is drawn to 890 Hz, then at a -2 slope (-40 dB/decade) to 2 kHz, and then at a -1 slope to 40 kHz. The segmented line is a close approximation to the calculated curve, and the departure around 1 kHz is primarily due to damping.

1.9   COMPONENT SCALING

In high-power systems, the values of voltage and/or current may be quite large; the voltage may be several kilovolts and/or the current may be several kiloamperes. The passive component values involved in these systems may be analyzed by analytical or computer methods. Where high-power, high-voltage transformers are involved, the leakage reactance may also be included in the analysis. However, analysis by computer may be difficult if active devices are involved. Device models in computer software libraries are normally applicable to much lower voltage and current than the design to be analyzed.

Component scaling by *impedance* ratios offers a means of accomplishing the analysis at a lower voltage and/or current. For very high power systems, component values may be scaled to, say, one thousand times the actual impedance. The component scaling analysis offers an additional benefit for breadboard testing purposes. In this case, both active and passive components are scaled to values easily obtainable at minimum cost. The circuit may then be assembled with these components and tested to verify desired performance. Component changes can be readily made to observe effects.

1.9.1   Impedance Scaling

For component impedances, $R = E/I$, $X_L = 2\pi fL$, and $X_C = 1/(2\pi fC)$, and where R is negligible in LC circuits, $Z = \sqrt{L/C}$. For scaling purposes (at the same frequency)

$$E' = E/k_1 \tag{1.74a}$$

$$I' = I/k_2 \tag{1.74b}$$

$$R' = Ek_2/Ik_1 = Rk_2/k_1 = Rk_3 \tag{1.74c}$$

where the "prime" notation is the scaled value. Then

$$L' = LR'/R = Lk_3 \quad \text{and} \quad C' = CR/R' = C/k_3 \tag{1.74d}$$

Example

A power supply output of 150 V at 3000 A is desired to operate from a 60-Hz input, with a 12-pulse transformer/rectifier (reference Section 4.1.2.1). The output filter inductor is chosen as 10 μH and the output filter capacitor is chosen as 50,000 μF. The transformer is to be designed with 10% reactance. For scaling purpose, $E' = 15$ V and $I' = 3$ A. The actual output load resistance is $R =$

$150/3000 = 0.05$ Ω. Then $k_1 = 150/15 = 10$, $k_2 = 3000/3 = 1000$, and $k_3 = 150 \times 1000/(3000 \times 10) = 5$ from which $R' = 5/0.05 = 100$ Ω. Then $L' = 100 \times 10 \times 10^{-6} = 1$ mH, and $C' = 50{,}000 \times 10^{-6}/100 = 500$ μF. The transformer may be built with the scaled voltage and currents and leakage reactance. The circuit may then be assembled and tested for the desired output voltage and current, because the transformer reactance will have an effect on the output voltage.

1.9.2 Frequency Scaling

Operating frequency may also be scaled to a more readily available frequency for breadboard testing purposes. In this case

$$f' = fk_4 \tag{1.75}$$

and the inductance and capacitance values are affected accordingly while the resistance values remain unaffected by frequency.

Example

A full-scale sine wave inverter has the following desired output characteristics: $E = 120$ V, $I = 120$ A, and $f = 400$ Hz. The output filter inductance $L = 20$ μH and the output capacitance $C = 450$ μF. The output load resistance $R = 120/120 = 1$ Ω. It is desired to analyze an actual model of the output stage using a 60-Hz source (f'), and with $E' = 12$ V and $I' = 1.2$ A. Then, $R' = 12/1.2 = 10$, and $k_3 = 10/1 = 10$. From Equations (1.76d) and (1.77), $L' = (R'/R)(f/f')L = 10(400/60)20 \times 10^{-6} = 1.33$ mH, and $C' = (R/R')(f/f')C = 0.1(400/60)450 \times 10^{-6} = 300$ μF. Also, notice that the actual resonant frequency $f_0 = 1/(2\pi\sqrt{LC}) = 1678$ Hz, while the new resonant frequency $f_0'' = 1/(2\pi\sqrt{L'C'}) = 252$ Hz $= 1678 \times 60/400$. The actual time constants $\tau_1 = L/R$ and $\tau_2 = RC$ are related to the scaled time constants $\tau_1'' = L'/R'$ and $\tau_2'' = R'C'$ by the factor $f'/f = 60/400$. In this example, a 14.4-kVA, 400-Hz inverter was scaled to a 14.4-W, 60-Hz output.

2

SEMICONDUCTORS, RESISTORS, AND CAPACITORS

2.0   Introduction
2.1   Rectifiers
    2.1.1   Current and Voltage
    2.1.2   Voltage Breakdown
    2.1.3   Temperature Coefficient
    2.1.4   Rectifier Ratings
    2.1.5   Junction Capacitance
    2.1.6   Reverse Recovery
    2.1.7   Forward Recovery Time
    2.1.8   Rectifier Losses at High Frequency
    2.1.9   Surge Current Ratings
2.2   Bipolar Power Transistors
    2.2.1   Safe Operating Area
    2.2.2   Switching Waveforms
    2.2.3   Losses and Power Dissipation
    2.2.4   Base Drive Circuits
    2.2.5   Paralleling Transistors
2.3   MOSFET Power Transistors
    2.3.1   Device Parameters
    2.3.2   Capacitance Effects
    2.3.3   Safe Operating Area
    2.3.4   Switching Waveforms
    2.3.5   Losses and Power Dissipation
    2.3.6   Gate Drive Circuits
    2.3.7   Reverse Rectifier Characteristics
    2.3.8   Parallel MOSFETs
2.4   Insulated-Gate Bipolar Transistors
    2.4.1   Device Parameters
    2.4.2   Safe Operating Area
    2.4.3   Switching Waveforms
    2.4.4   Gate Drive Circuits
    2.4.5   Parallel IGBTs
2.5   Thyristors
    2.5.1   Gate Triggering
    2.5.2   Phase Control
    2.5.3   Inverter-Rated SCR
    2.5.4   Crowbar SCR

2.6 Other Power Devices
    2.6.1 Triac
    2.6.2 Gate Turn-off Thyristor
    2.6.3 Static Induction Transistor
    2.6.4 MOS-Controlled Thyristor
2.7 Resistors
    2.7.1 Carbon Composition Resistors
    2.7.2 Metal-Film Resistors
    2.7.3 Wire-Wound Resistors
    2.7.4 Resistance Wire Materials
2.8 Capacitors
    2.8.1 Aluminum Electrolytics
    2.8.2 Tantalum Capacitors
    2.8.3 Paper-and-Film Capacitors
    2.8.4 Oil-Filled Capacitors
    2.8.5 Ceramic Capacitors
    2.8.6 Mica Capacitors

2.0 INTRODUCTION

The fact that volumes have been written on each of the title subjects notwithstanding, these devices are discussed in this chapter from the standpoint of power conversion applications. Notations used in this chapter are listed in Table 2.1. The subscript notations apply primarily to semiconductors.

Power semiconductors normally have the highest actual failure rate of any component within a system or circuit. The failures are primarily due to junction overheating, neglecting the infant mortality failures which are normally "weeded out" during burn-in. The overheating may be due to prolonged high-ambient-temperature conditions but is frequently caused by transient power stress conditions. Power resistors also have a high failure rate, frequently due to a transient stress condition overlooked in the initial circuit design, and they contribute to heat generation within equipment. Capacitors can store high amounts of energy, especially for power filtering purposes, and can operate relatively failure free if the ac current (or ripple voltage) is maintained at a level low enough to prevent internal overheating. Transformers and inductors, discussed in Chapter 3, are rugged components and high reliability is achieved if the materials utilized are rated for the proper operating temperature, flux density, and current density. For high reliability and high operating confidence, Table 2.2 lists the recommended derating values for the subject components, plus others applicable to power conversion. The derating follows "good design practices" plus data adapted from NAVMAT P4855-1A, Department of the Navy, 1989.

To overcome semiconductor failures from transient stress conditions, the requirements of MIL-STD-1281, "Internal Transient Control for Solid State Power Supplies," is sometimes imposed on military equipment. The test requirements ensure adequate designs relative to safe operation of semiconductors under the

TABLE 2.1. Notation

			Subscripts
C	Capacitance, farad		
CR	Diode designation	A	Anode
$\mathcal{E}$	Energy, joules or watt-second	a	Ambient
F	Force, lb-ft	avg	Average
f	Frequency, hertz	B	Base
g	Transconductance	C	Cathode
h	Gain	C	Collector
I	Current, ampere	c	Case
i	Instantaneous current	D	Drain
J	Joules	D	Off-state, nontrigger
L	Inductance, henry	E	Emitter
P	Power, watt	F	Forward
Q	Charge, coulombs	FE	Direct current (gain)
R	Resistance, ohm	fe	Small signal (gain)
T	Period, second	fs	Small signal
T	Temperature, °C		(related to transconductance)
t	Time, second	G	Gate
V	Voltage, volt	H	Holding current
υ	Instantaneous voltage	j	Junction
X	Reactance, ohm	M	Maximum
Z	Impedance, ohm	O	Open circuit, third terminal
Q	Transistor, thyristor (SCR)	op	Operating
	designation	pk	Peak
α	Firing angle, degrees	R	Reverse, repetitive
θ	Thermal impedance, °C/W	REC	Recovery
Ω	Ohm	rms	Root-mean-squared
di/dt	Rate of rise of on state current	S	Source
dv/dt	Rate of rise of on state voltage	S	Surge, nonrepetitive
cmil	Circular mils	sat	Saturation
		stg	Storage
	Time Related	sus	Sustaining
t_c	Commutation	s/b	Second breakdown
t_d	Delay (current)	T	On-state, triggered
t_f	Fall (current)	TO	Turn-on
t_r	Rise (current)	W	Working
t_s	Storage (current)	X	Reverse bias, third terminal
t_{rr}	Reverse recovery	Z	Zener
t_{fr}	Forward recovery	cont	Continuous
t_{gt}	Gate-controlled turn-on		
t_p	Pulse width		*Capacitance Related*
t_x	Related to $I^2\sqrt{t}\sqrt{t_x}$	C_{iss}	Input capacitance
t_{pd}	Propagation delay	C_{ob}	Output capacitance
f_t	Gain bandwidth product	C_{oss}	Output capacitance
		C_{rss}	Reverse transfer capacitance

TABLE 2.2. Component Derating for High Reliability

Component	Voltage (%)	Current (%)	Power (%)	Temperature	Notes[a]
Rectifiers	65	60	50	$T_j = 110°C$ max	1
Transistors	75	60	50	$T_j = 110°C$ max	1, 2
Thyristors (silicon-controlled rectifiers)	65	70	50	$T_j = 110°C$ max	3
Linear integrated circuits	75	—	50		
Resistors	80	—	50		
Capacitors				10°C max rise	
Ceramic, film	50	—	—	above ambient,	
Aluminum electrolytic	70	70	—	all units	4
Tantalum, solid and foil	50	70	—		4
Tantalum, wet	60	70	—		4
Magnetic devices					
MIL Class V, 155°C	90% surge, 90% surge			75% hot spot	
IEEE Class F, 155°C	25% insulation breakdown				
Power relays (contacts)	—	50			5
Switches	50	30			
Connectors	25	50			
Component isolation to mounting	65% peak surge				6

[a]1. 125°C max for 200°C rated devices.
2. Applies to silicon, bipolar and metal–oxide–semiconductor field-effect transistor (MOSFET). Safe operating area (SOA) derated to 80%.
3. Do not derate di/dt, dv/dt, t_q for reduced junction temperature.
4. 50% ripple voltage rating.
5. 25% inductive, 80% surge, use arc suppression.
6. Does not include UL, IEC, and VDE safety standards. Reference Section 13.5.

stress of internal transients. The measurements determine the existence of transient voltage, current, or energy pulses generated by surges, spikes, or level changes resulting from combinations of the following: turn-on; turn-off; step changes in line; step changes in load; output short circuit and open circuit; plus stress tests at 20% greater than the specified high line input. If this specification is not imposed, performance of the defined guidelines is recommended during the design phase for commercial as well as military power conversion equipment.

Advances in semiconductor technology have contributed significantly to the success of modern power conversion equipment. Schottky rectifiers have improved the operating efficiency of high-current, low-voltage power supplies. Fast-switching, soft-recovery rectifiers have also improved efficiency and reduced electromagnetic interference (EMI) in all types of equipment. High-voltage capability, high-current capability, and high-speed switching have allowed bipolar transistors to intrude into areas formerly reserved for thyristors. The very fast switching and the high input impedance of metal–oxide–semiconductor field-effect transistors (MOSFETs) have allowed size and weight reduction of magnetic components through increased switching frequencies. Insulated-gate bipolar transistors (IGBTs) also have high input impedance and their conduction losses are lower than MOSFETs. Fast turn-off thyristors [silicon-controlled rectifiers (SCRs)] have minimized commutating components and improved efficiency in high-powered equip-

ment operating at moderate frequencies. Exotic devices such as gate turn-off
(GTO) thyristors, static induction thyristors (SITs), and MOS-controlled thyristors
(MCTs) are discussed in Section 2.6.

2.1   RECTIFIERS

The term *rectifier* applied to an individual component as opposed to a rectifier
circuit, is synonymous with the term *diode* and will be used because power devices
are discussed. An ideal rectifier would completely block a voltage of one polarity
and would allow current flow at no power loss when a voltage of opposite polarity
is applied. In reality, losses occur during the blocking, switching, and conducting
states. These factors have been well documented but are discussed briefly for
review.

2.1.1   Current and Voltage

The current relationship is given by

$$I = I_R(\varepsilon^{qV/nk\Gamma} - 1) \tag{2.1}$$

where

I_R = reverse junction current

kT/q = 26 mV at 300 K (27°C)

k = Boltzmann's constant, = 1.38×10^{-23} J/K

n = 1, 2 and increases as the current density increases (typically 1000 A/cm^2)

T = absolute temperature, K

q = electron charge, = $1.6 \times 10^{-19}Q$

V = voltage across junction

In germanium and Shottky rectifiers, I_R is primarily due to diffusion current I_D
produced by the majority carriers. In "standard" silicon rectifiers, I_R is primarily
due to the charge generation current produced by impurities in the depletion layer.
Also, a surface leakage current is present and appears as a resistance across the
rectifier.

The forward voltage drop is given by

$$V_F = (nkT/q)\ln[1 + (I_F/I_D)] + R_F I_F \tag{2.2}$$

where I_D is diffusion current and R_F is the resistive component.

Typical voltage drops for germanium, silicon, and Schottky rectifiers, each with
an average current rating of 30 A, are shown in Figs. 2.1A, B, and C, respectively.
The germanium and Schottky rectifiers are in the DO-4 stud mount package. The
standard silicon rectifier is in the DO-5 stud mount package. The larger DO-5
package is required due to the higher forward voltage drop resulting in a higher

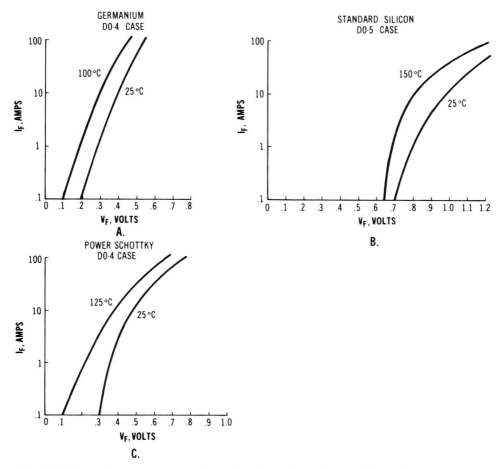

Fig. 2.1. Forward current versus forward voltage drop for rectifiers with 30-A average rating.

power dissipation in the standard silicon rectifier. Depending on the environment, germanium rectifiers may be used if the junction temperature does not exceed a derated value of 85°C and if the operating frequency is low. The penalty incurred, as compared to the silicon rectifier, is a larger heat sink for cooling, due to the reduced temperature required.

2.1.2   Voltage Breakdown

High reverse voltage ratings are achieved by an increase in material (semiconductor) resistivity, and the voltage gradient is typically 4×10^5 V/cm. Punch-through is not a problem in power rectifiers since avalanche breakdown occurs well below the punch-through voltage. The breakdown mechanism in low-voltage zener diodes is a tunneling effect rather than avalanche.

2.1.3 Temperature Coefficient

In forward conduction, the negative temperature coefficient decreases V_F as temperature increases, as shown in Fig. 2.1, and

$$dV_F/dT = -2.2 \text{ mV}/°\text{C} \quad \text{typical for silicon} \tag{2.3}$$

The reverse leakage current approximately doubles for every 10°C increase in temperature, or

$$+dI_R = 2^{+dT/10} \tag{2.4}$$

2.1.4 Rectifier Ratings

The term for reverse (frequently called inverse) voltage rating of a rectifier is V_{RRM} for "reverse repetitive maximum" (or V_{PIV} for "peak inverse voltage"). The forward current rating, I_F or I_{FM}, is normally stated in terms of average current for a rectified sine wave. Manufacturers' data may also show average current for a rectangular wave. Also, the dc current rating, which is the rms current rating, may be listed. The current rating is also a function of temperature. Data for a 100-A rectifier of the 1N3288-96 family is shown in Fig. 2.2. In Fig. 2.2A, the average current is plotted versus maximum stud temperature. Average current versus power dissipation is shown in Fig. 2.2B. Again, the dc current is the rms rating, the 1ϕ line is the average rectified single-phase sine wave current, the 3ϕ is a 6-pulse rectified current, and the 6ϕ is a 12-pulse rectified current (reference Chapter 4, Table 4.2). For instance, the ratio of average to rms for a sine wave is $2/\pi = 0.636$. Since the rms current rating for the rectifier of Fig. 2.2 is 157 A, the average current rating is $157 \times 0.636 = 100$ A.

Example

It is desired to select a rectifier for a three-phase, six-rectifier bridge delivering a dc output of 125 V at 200 A from an ac source. The average current per rectifier is the dc current divided by 3, or $200/3 = 66.7$ A. Referring to Fig. 2.2A, the rectifier will conduct 66.7 A at a case temperature of 155°C or less. From Fig. 2.2B, 66.7 A produces a power dissipation of 75 W. Derating the temperature to 80%, a heat sink must be selected to limit the case temperature to 124°C at a power dissipation of 75 W. Next, the voltage rating is determined. The minimum reverse voltage rating for a three-phase bridge is 1.05 times the dc voltage (from Table 4.2). For a derating to 65%, the required $V_{\text{RRM}} = 125 \times 1.05/0.65 = 200$ V, which is the rating of a 1N3289. Therefore, the bridge consists of six 1N3289 rectifiers, and the total power dissipation is $75 \times 6 = 450$ W. This is a considerable amount of loss, and would probably require forced convection cooling to the heat sinks, but since the output power is $125 \times 200 = 25$ kW, the rectifier losses are only $450/25000 = 1.8\%$ of the total output power. Since the forward drop is almost independent of V_{RRM}, the efficiency increases as the dc voltage increases (and vice versa).

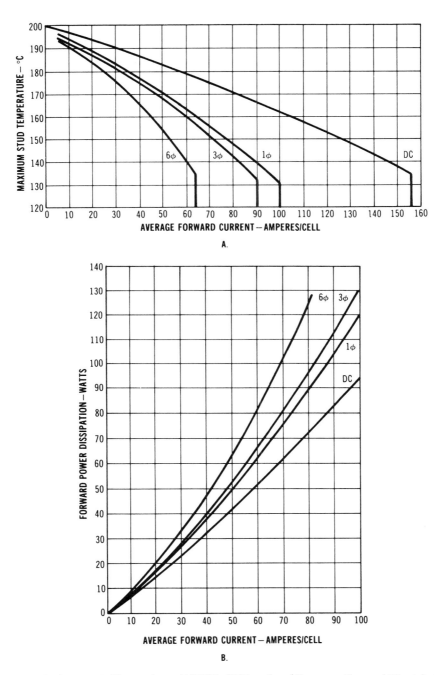

Fig. 2.2. Maximum rectifier ratings, 1N3289–3296 series. (Courtesy General Electric Co.)

2.1.5   Junction Capacitance

The capacitance is normally insignificant in standard or fast recovery silicon power rectifiers at $V_R > 30$ V but is important in Schottky power rectifiers due to the recovery transient (as distinguished from the reverse recovery). When a Schottky rectifier switches from a conduction mode to a reverse-blocking mode, current is required to charge the depletion capacitance, which has the same effect as the reverse recovery current in the fast recovery junction rectifier. At low values of V_R, the capacitance of the Schottky rectifier may be five times the capacitance of the standard rectifier. This higher capacitance in the Schottky rectifier may combine with circuit or other leakage inductances to produce resonances, causing excessive voltage overshoot, as discussed in the following section. The capacitance in standard rectifiers is practically independent of temperature but decreases as the reverse voltage V_R increases and is given by

$$C_T = k / \left(V_R^{0.33} \right) \qquad (2.5)$$

2.1.6   Reverse Recovery

The reverse recovery time and the peak reverse current amplitude of a rectifier are a function of charge storage in the junction and have become extremely important in high-frequency power supplies and converters. Since some form of inductance is present in the external circuitry, the rate at which the reverse current decays (abrupt or soft) also determines voltage transient amplitudes, because $e = -L\, di/dt$. These transients may cause noise and EMI and affect operation of other units.

Typical reverse recovery parameters are shown in Fig. 2.3A. A rectifier is initially conducting current I_{FM} at voltage drop V_F. When the rectifier turns off, the forward current decreases at rate di/dt, and reverse current flows until the junction recovers at time t_{rr} as the reverse voltage increases to V_{RM} and decays to V_R. A charge, Q_{rr}, is developed across the junction. Various manufacturers have different methods to measure recovery time. The waveform shape of the reverse current decay at rate $di_{R(REC)}/dt$ differs in many rectifiers. An accepted method for establishing t_{rr} is the time from the forward current crossing zero to the time a line from I_{RM} through $\frac{1}{4} I_{RM}$ crosses zero, as shown in Fig. 2.3A. Note that in this figure the rectifier becomes reverse biased after t_a and the rectifier dissipates power during t_b. The reverse leakage increases exponentially with temperature rise, as given by Equation (2.4), and in most rectifiers, the reverse recovery time also increases exponentially with temperature rise. In high-speed switching circuits, this action can lead to increased power dissipation and thermal runaway, resulting in rectifier failure. The thermal runaway conditions are most prevalent in very high-voltage, high-speed rectifiers. For medium-voltage applications, the 1N6620-6631 series (40 ns recovery) has a linear reverse recovery time with temperature increase and chances of thermal runaway are minimal.

Typical reverse recovery times for various rectifiers are shown in Fig. 2.3B. The standard rectifier is normally sufficient for utility or mains rectification, where the frequency is low. The fast recovery rectifier may be used for intermediate frequencies. Although the gold-doped process improves the reverse recovery characteristic

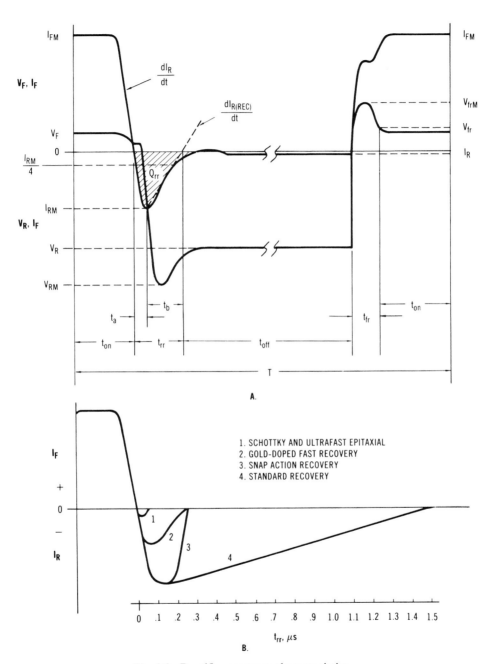

Fig. 2.3. Rectifier recovery characteristics.

of a rectifier, the gold doping tends to increase the forward voltage drop and reverse leakage. Ultrafast recovery rectifiers employ a planar, epitaxial structure to achieve reverse recovery time equivalent to Schottky devices and have maximum reverse-voltage ratings several times that of the Schottky rectifier. The reverse recovery time t_{rr} also increases with temperature. The equations governing recovery are

$$t_{rr} = \sqrt{2Q_R/(di/dt)} \qquad (2.6)$$

$$I_{RM(REC)} = \sqrt{2Q_R(di/dt)} \qquad (2.7)$$

Also shown in Fig. 2.3B is a fast recovery rectifier, with a snap action turn-off mechanism (as opposed to soft recovery) which may generate high transient voltages well above 1 MHz since the $di_{R(REC)}/dt$ factor is high. Those devices classified as fast, soft recovery have a low $di_{R(REC)}/dt$ and tend to minimize EMI emissions. However, the snap recovery devices will dissipate less power during recovery and are not as susceptible to thermal runaway as the soft recovery devices.

High-frequency converters providing low-voltage outputs employ a transformer for isolation, and the stepped-down secondary voltage is rectified and filtered. When Schottky rectifiers are used, the leakage inductance (reflected to the secondary) may resonate with the device capacitance, and the reverse recovery time is then

$$t_{rr} = \pi\sqrt{LC} \qquad (2.8)$$

where L is leakage inductance and C is junction capacitance. Since $Z = \sqrt{L/C}$, the peak reverse current is

$$I_{RM(REC)} = E_s\sqrt{C/L} \qquad (2.9)$$

where E_s is transformer secondary voltage.

Since the Schottky is a relatively low-voltage device, the resonant voltages produced could cause damage to the rectifier unless a "snubber" circuit, consisting of a series resistor and capacitor, is added across the rectifier. The snubber network also reduces EMI. Reference Section 13.4 for snubber circuit design.

2.1.7 Forward Recovery Time

When a step voltage is applied to a standard rectifier, current conduction is not instantaneous due to charge gradient transit time of the junction, and the forward voltage does not decrease to the steady-state forward drop until the forward recovery time t_{fr} occurs, as shown in Fig. 2.3A. The forward recovery time is a function of resistivity and of forward current. Since resistivity is higher in high-voltage rectifiers, t_{fr} is typically 0.1–0.4 μs for low-voltage devices and 0.3–1 μs for high-voltage devices. However, *lead inductance* may indicate a "higher than steady-state" forward drop at turn-on, because the voltage across the inductance cannot change instantaneously; this does not contribute to power dissipation.

Schottky rectifiers do not exhibit forward recovery time due to low-resistivity materials and to the absence of minority carriers. However, the junction capacitance must still be charged. For-high frequency converters, the current surge through the rectifier is reflected to the primary and causes an increase in the primary side transistor current as the device turns on. Additional transistor current at turn-on is also caused by charging the transformer interwinding capacitance. Both occurrences increase the turn-on switching losses as the transistor is driven to saturation.

2.1.8   Rectifier Losses at High Frequency

For high-frequency operation, total rectifier losses are

$$P_{CR} = P_{conduction} + P_{off} + P_{switching} \tag{2.10}$$

where

$$
\begin{aligned}
P_{conduction} &= V_F I_F t_{on}/T \\
P_{off} &= V_R I_R t_{off}/T \\
P_{switching} &= \text{turn-off plus turn-on losses} \\
t_{on} &= \text{time of rectifier conduction} \\
t_{off} &= \text{time of rectifier reverse bias} \\
T &= \text{total period time}
\end{aligned}
$$

The conducting and off-state losses are readily calculated from circuit parameters and manufacturers' data. The switching losses are more difficult to determine. The turn-off, or reverse recovery, losses may be calculated as follows. Referring to Fig. 2.3A, the reverse current may be assumed triangular in shape and the reverse voltage is applied at t_b. Typically, $t_a = t_b = \frac{1}{2}t_{rr}$. The peak power is $V_R I_R$, and the average power over one cycle is

$$P_{rr} = (V_{RM})(0.5 I_{RM})(t_{rr}/2)/T = 0.25 V_{RM} I_{RM} t_{rr}/T \tag{2.11a}$$

Since $Q = CV = It$ and $\varepsilon = \frac{1}{2}CV^2$, $\varepsilon = \frac{1}{2}QV = Pt$, and

$$P_{rr} = 0.5 Q_{rr} V_{RM}/T \tag{2.11b}$$

Referring to Fig. 2.3A, the forward recovery voltage is nearly sinusoidal in shape for time t_{fr}, and the average power over one cycle is

$$P_{fr} = (2/\pi)(V_{frM})(I_{FM}/2)(t_{fr})/T = 0.318 V_{frM} I_{FM} t_{fr}/T \tag{2.12}$$

Then Equation (2.10) becomes

$$P_{CR} = \frac{V_F I_{FM} t_{on}}{T} + \frac{V_R I_R t_{off}}{T} + \frac{0.25 V_{RM} I_{RM} t_{rr}}{T} + \frac{0.318 V_{frM} I_{FM} t_{fr}}{T} \tag{2.13a}$$

or

$$P_{CR} = \frac{V_F I_{FM} t_{on}}{T} + \frac{V_R I_R t_{off}}{T} + \frac{0.5 Q_{rr} V_R}{T} + \frac{0.318 V_{frM} I_{FM} t_{fr}}{T} \quad (2.13b)$$

Examples

Case 1. A typical 30-A DO-5 fast recovery rectifier has the following parameters for a desired circuit: the operating frequency is 10 kHz; $T = 100$ μs; $I_{FM} = 50$ A; $V_F = 1.2$ V; $V_{RRM} = 520$ V for an actual $V_{RM} = 400$ V; $I_R = 0.5$ mA; $t_{rr} = 0.2$ μs; $di/dt = 50$ A/μs; $Q_{rr} = 1$ μQ; $V_{frM} = 10$ V; $t_{fr} = 0.5$ μs; the duty cycle $D = 50\%$. From Equation (2.7), $I_{RM} = \sqrt{2 \times 1 \times 10^{-6} \times 50 \times 10^6} = 10$ A. Also, $T = t_{on} + t_{rr} + t_{off} + t_{fr}$. For $D = 50\%$, $t_{on} = t_{off} = \frac{1}{2}[T - (t_{rr} + t_{fr})] = (100 - 1.2)/2 = 49.4$ μs. Then $P_{cond} = 1.2 \times 50 \times 49.4/100 = 29.64$ W; $P_{off} = 400 \times 0.0005 \times 49.4/100 = 0.1$ W; $P_{rr} = 0.25 \times 400 \times 10 \times 0.1/100 = 1$ W; $P_{fr} = 0.318 \times 10 \times 50 \times 0.5/100 = 0.8$ W. Then the total power dissipated is $P_{CR} = 29.64 + 0.1 + 1 + 0.8 = 31.54$ W. From manufacturers' data, the maximum case temperature is $T_c = 100°C$. Also, using Equation (2.11b), $P_{rr} = 0.25 \times 1 \times 10^{-6} \times 400/(100 \times 10^{-6}) = 1$ W, which agrees with the result using Equation (2.11a).

Case 2. Use the rectifier from Case 1 except $f = 100$ kHz, $T = 10$ μs (example only, not recommended at this frequency). Find the total power dissipation. Now $t_{on} = t_{off} = (10 - 1.2)/2 = 4.4$ μs; $P_{cond} = 1.2 \times 50 \times 4.4/10 = 26.4$ W; $P_{off} = 400 \times 0.0005 \times 4.4/10 = 0.09$ W; $P_{rr} = 0.25 \times 400 \times 10 \times 0.1/10 = 10$ W; $P_{fr} = 0.318 \times 10 \times 50 \times 0.5/10 = 8$ W. Then $P_{CR} = 26.4 + 0.09 + 10 + 8 = 44.49$ W. The total dissipation is 1.4 times that of Case 1. From manufacturers' data sheets, a 100-A DO-8 rectifier is required, and the maximum case temperature is 125°C. For derating, the voltage rating is $520/0.65 = 800$ V. The example of Section 2.1.4 required a 100-A, 200-V rectifier (1N3289), and since 800 V is required for this "example," a 1N3294 would be "selected" if this were considered a viable design.

Case 3. It is desired to conduct the same 50 A in a low-voltage circuit using a Schottky rectifier. The parameters are $I_{FM} = 50$ A; $V_F = 0.7$ V; $V_R = 30$ V; $I_R = 50$ mA; $t_{rr} = 0.05$ μs; $I_{RM} = 2.5$ A. The forward recovery losses are zero and since t_{rr} is small, $t_{on} = t_{off} = 5$ μs; $P_{cond} = 0.7 \times 50 \times 5/10 = 17.5$ W; $P_{off} = 0.30 \times 0.05 \times 5/10 = 0.75$ W; $P_{rr} = 0.25 \times 30 \times 2.5 \times 0.025/10 = 0.05$ W. Then $P_{CR} = 17.5 + 0.75 + 0.05 = 18.3$ W. From manufacturers' data, a 30-A DO-4 Schottky rectifier could be used for a maximum case temperature of 100°C.

Comparing Case 1 (with a low-voltage rectifier) and Case 3, the total power dissipation of Case 1 is almost twice that of Case 3, even though the frequency is one-tenth that of Case 3.

2.1.9 Surge Current Ratings

The rectifier has the ability to sustain overload currents for a brief period of time without raising the junction temperature to a level which would cause failure. Two parameters are normally specified in the manufacturer data sheets. Forward surge maximum current I_{FSM} is the maximum peak half-cycle nonrepetitive current rating. This rating may be specified with V_{RRM} applied, or with V_{RRM} equal to zero after the surge, which would indicate a fuse or protective device had opened. The ampere-squared second (I^2t) rating is normally associated with protective fuse coordination, and the rating is usually given for 8.33 ms (60 Hz) with V_{RRM} applied, or with V_{RRM} equal to zero. For repetitive cycles, the number of cycles at a peak sinusoidal current is also given in the data sheets. Also, for times other than 8.3 ms, the term $I^2\sqrt{t}\sqrt{t_x}$ may be given, where t_x is the peak current duration and is a further aid for fuse selection. These parameters are best related by referring to the 1N3288 rectifier family (discussed in previous examples, although the A version of this family has higher surge ratings) with the following ratings:

1. I_{FSM} = 1600 A at one-half cycle at 60 Hz, V_{RRM} at rating.
2. I_{FSM} = 1900 A at one-half cycle at 60 Hz, V_{RRM} at zero.
3. I^2t = 10,500 A²s for t = 8.33 ms, V_{RRM} at rating.
4. I^2t = 15,000 A²s for t = 8.33 ms, V_{RRM} at zero.
5. $I^2\sqrt{t}\sqrt{t_x}$ = 165,000 A²s, 0.1–10 ms, V_{RRM} at zero.

Since $I_{rms} = 0.707I_{pk}$, using 1,

$$I^2t = (0.707I_{FSM})^2t = (0.707 \times 1600)^2 \times 0.00833 = 10,666$$

which closely agrees with the 10,500 rating of 3. Also, the rating for $I^2\sqrt{t}\sqrt{t_x}$ = 165,000 for V_{RRM} = zero. Again, for t_x = 8.33 ms, 165,000 × $\sqrt{0.00833}$ = 15,062, which closely agrees with the 15,000 rating of 4. If the $I^2\sqrt{t}\sqrt{t_x}$ rating is not given, I_{FSM} for times other than 8.33 ms may be calculated by

$$I_{FSM} = \sqrt{2I^2t_1/\left(\sqrt{t_1}\sqrt{t_2}\right)} \tag{2.14}$$

where I^2t_1 is the A²s rating for time t_1 and t_2 is the desired time duration.

Examples

Case 1. From the above discussion, it is desired to find I_{FSM} for the 1N3288 at 50 Hz (one-half cycle = 10 ms) from the 60-Hz (one-half cycle = 8.33 ms) I^2t rating of 10,500 A²s and for V_{RRM} applied. From Equation (2.14),

$$I_{FSM} = \sqrt{2 \times 10,500/(\sqrt{0.00833} \times \sqrt{0.01}\,)} = 152 \text{ A}$$

Case 2. Also, it is desired to find I_{FSM} for the 1N3288 where a fast interrupter or fuse will open at t_2 = 1 ms and V_{RRM} = 0. For this condition, I^2t = 15,000. Then, from Equation (2.14),

$$I_{FSM} = \sqrt{2 \times 15,000/(\sqrt{0.00833} \times \sqrt{0.001}\,)} = 3200 \text{ A}$$

2.2   BIPOLAR POWER TRANSISTORS

The physics of hole and electron flow plus doping and diffusion processes have been well covered in other texts. This section could be entitled "the care and feeding of the power transistor" and is oriented toward transistor selection and applications in power conversion equipment. Power transistor reliability has improved tremendously since the days of "purple plague" associated with early germanium devices and through the understanding of second-breakdown mechanisms. Although silicon transistors have become the mainstay and are discussed exclusively, the germanium transistor may still find applications in inverters and converters operating from 12-V sources (due to very low saturation voltage), switching at low frequencies (due to slow switching speed), and in limited environments (due to the low temperature rating of the junction).

Advances in silicon transistor technology and in processing techniques have produced a variety of device families suited for specific applications. The most notable achievements have been, and probably will continue to be, in power switching transistors. This is evidenced by the fact that several transistor manufacturers have assigned the prefix "SWITCH" to define various families. Knowledge of device characteristics aids in the optimum selection to meet these requirements. The four major families are as follows:

1. *Alloy.* Alloy process with molybdenum "hard solder" wafer. Highest energy capability, both $I_{s/b}$ and $\varepsilon_{s/b}$. Slow switching ($f_T = 1$ MHz typical). Limited voltage rating.

2. *Epitaxial.* Epitaxial base–collector. High-energy capability. Fairly high saturation voltage. Medium speed ($f_T = 8$ MHz typical).

3. *Triple Diffused.* Multiepitaxial collector. High-clamped-energy capability. Highest sustaining voltage. Fast switching ($f_T = 20$ MHz typical).

4. *Planar.* Triple-diffused planar. Low $I_{s/b}$. Low $\varepsilon_{s/b}$. Very fast switching ($f_T = 75$ MHz typical). Low $V_{CE,\text{sat}}$. High gain. The V_{BE} for reverse bias normally limited to -1.5 V.

The two limitations in power-handling ability of any transistor are junction temperature (average and instantaneous) and second breakdown. The junction temperature can be maintained to a safe or derated value by proper thermal design and heat sinking, because the steady-state and transient thermal impedance from junction to case is specified by the manufacturer. The second breakdown, although triggered by thermal runaway, can be prevented by observing manufacturer-designated safe operating area (SOA) for forward-biased and reverse-biased conditions.

2.2.1   Safe Operating Area

The SOA is one of the most important parameter ratings for reliable switching of power transistors when operating into inductive or transformer loads or when operating in a nonsaturated condition. In the first case, the energy content in leakage inductances can produce excessive transient voltages which would exceed the transistor rating and cause reverse-biased second breakdown if the voltage is not clamped or limited when the transistor turns off. The energy level is normally specified as $\varepsilon_{s/b}$ and the designation RBSOA (reverse-bias SOA) applies. In the latter case, the transistor power dissipation becomes excessive when the device is in the active region or at turn-on, and forward-biased second breakdown occurs if the current exceeds the specified $I_{s/b}$ rating at the particular transistor voltage. The designation FBSOA (forward-bias SOA) applies.

An FBSOA curve is shown in Fig. 2.4 for three transistors of similar voltage and current ratings but which have vastly different performance as well as process characteristics. The 2N2778 is a gold alloy–processed device with inherent ruggedness, has a high reverse-base voltage rating ($V_{BEO} = -15$ V), but is very slow switching ($t_s + t_f$ is typically 10 μs). The 2N6687 is a multidiffused epitaxial structure with a rugged emitter and is fast switching ($t_s + t_f$ is 1.2 μs maximum, and f_T is 100 MHz). The 2N6766 is a power MOSFET (reference Section 2.3.3), included for comparison, and is very fast switching (t_s does not occur in MOSFET, t_f is 0.2 μs maximum). Since few transistors are operated at $T_c = 25°C$, the curves are derated to $T_c = 75°C$ at maximum junction temperature. The dc operation as well as 1-ms nonrepetitive pulses are shown. The 2N2778 is power dissipation limited until I_C reaches $I_{s/b} = 3$ A and $V_{CE} = 67$ V. The 2N6687 is power dissipation limited until I_C reaches $I_{s/b} = 8$ A at $V_{CE} = 18$ V. Since the 2N6766 is a MOSFET and not subject to forward-biased secondary breakdown, the device is power dissipation limited by I_D and V_{DS}.

For the forward-biased conditions of V_{CE} between 30 and 70 V, the 2N2778 can conduct five times the collector current of the 2N6687. For the forward-biased conditions of V_{CE} and V_{DS} between 7 and 70 V, the 2N2778 can conduct twice the drain current of the 2N6766. However, for high-frequency operation, the switching losses of the 2N2778 would be enormous as compared to the other two devices. The 2N2778 is ideally suited for linear regulators and power supplies. More specifically, paralleled 2N2778s could be used in high-power current-regulated magnet power supplies where voltage compliance would range in increments from 10 to 100 V. In this case, the low-frequency characteristic is not a deterrent, since the time constant of the load is relatively long and the regulated current cannot change instantaneously in the highly inductive dc load. The MOSFETs could also be easily paralleled for high output current, perhaps more reliably than the bipolars, because the positive temperature coefficient of the on resistance inherently forces current sharing. Conversely, the 2N6687 could be used in push–pull converters operating from a 48-V dc source and switching at frequencies up to 50 kHz. In fact, two transistors could provide required outputs in the 500–600-W range. Also, two 2N6766 transistors could be used for the same converter with a switching frequency over 100 kHz and providing 300–400 W output.

An RBSOA curve for a typical fast switching transistor is shown in Fig. 2.5. The RBSOA extends from $V_{CEX} = 500$ V to $I_C = 16$ A. The narrow dashed line represents a transistor with a break at $V_{CEX} = 500$ V and $I_C = 10$ A, to $V_{CEO} = 400$ V and then extending to $I_C = 16$ A. Also shown is the FBSOA for dc operation and for a 10-μs pulse. Note the linear graph distorts the FBSOA and resolution is minimal except for the pulse condition which could occur at turn-on. A typical load line for a half-bridge converter (reference Section 9.9) is also shown at turn-on, saturated conduction, and turn-off.

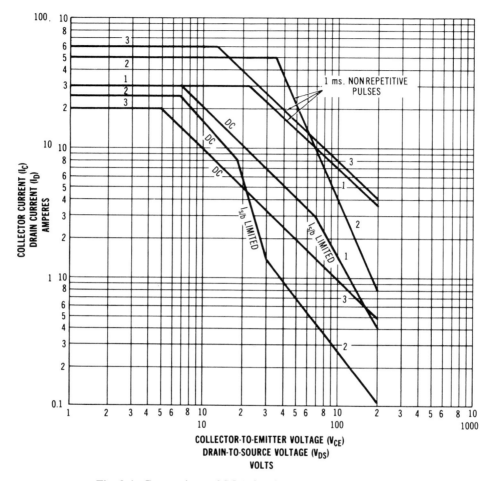

1. 2N2778. $T_C=75°C$ (ALLOY BIPOLAR)
2. 2N6687. $T_C=75°C$ (DIFFUSED BIPOLAR)
3. 2N6766. $T_C=75°C$ (POWER MOSFET)

Fig. 2.4. Comparison of SOA for three types of transistors.

Reverse-biased second breakdown is a complex situation, but $\mathcal{E}_{s/b}$ ratings are intended to define the amount of energy the device can absorb in the unclamped (reverse-biased avalanche) or clamped mode. As the transistor turns off, collector current is concentrated in a small region of the emitter. If the current is restricted to a very small central area by a large reverse-bias voltage on the base, the current density becomes very high, impurity gradients and device doping reduce the voltage rating of the junction, the transistor enters the sustaining region, hot spots develop which causes further regeneration, and second breakdown results. From a thermal standpoint, the high current density in the transistor at turn-off is somewhat analogous to the high current density in a thyristor at turn-on (di/dt limits). Generally, the higher the $\mathcal{E}_{s/b}$ rating, the slower the transistor switches.

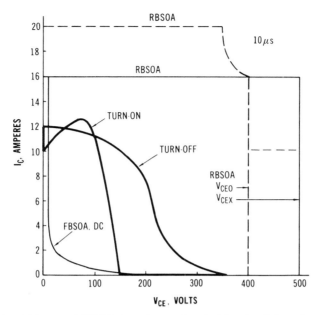

Fig. 2.5. Typical FBSOA and RBSOA rating and switching load line for a high-voltage transistor.

For instance, the 2N2778 of Fig. 2.4 has an enormous $\varepsilon_{s/b}$ rating of 1 J ($I_C = 5$ A, $L = 80$ mH), while the 2N6687 has a clamped $\varepsilon_{s/b}$ rating of less than 8 mJ. The clamped $\varepsilon_{s/b}$ rating is of primary importance in high-frequency switching, and additional circuitry such as snubbers and load line shaping must frequently be used to absorb the reactive energy and reduce the dv/dt across the transistor. Factors affecting $\varepsilon_{s/b}$ are as follows:

1. $-V_{BE}$. Increasing the reverse-bias base voltage of the transistor for turn-off decreases the $\varepsilon_{s/b}$ capability even though storage and fall times are reduced. For Darlington transistors, a reverse bias should be applied to each base. Normally, the $-V_{BE}$ value is specified by the manufacturer.
2. R_{BE}. Increasing the base-to-emitter resistance (a resistor is normally recommended) increases the $\varepsilon_{s/b}$ capability. The resistance value is normally specified by the manufacturer and is typically 50 Ω.

2.2.2 Switching Waveforms

Figure 2.6 shows some typical switching waveforms for a bipolar power transistor in the common-emitter configuration. Base current, base-to-emitter voltage, collector current, and collector-to-emitter voltage are shown in Figs. 2.6A, B, C, and D, respectively. The transistor conducts for period t_{on} and is nonconducting for period t_{off}.

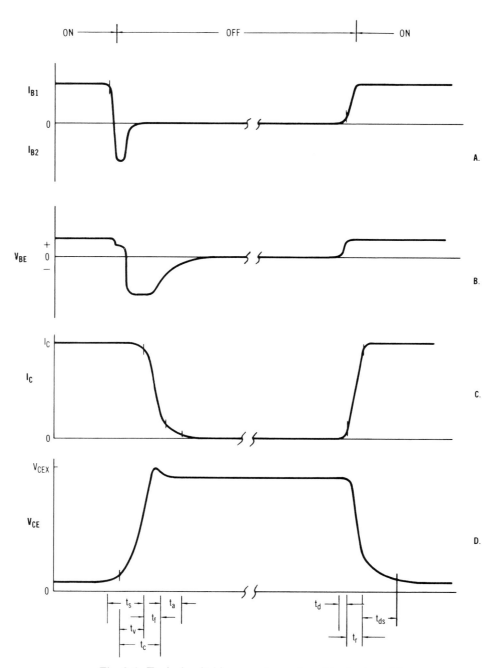

Fig. 2.6. Typical switching waveforms for a bipolar transistor.

The turn-off switching times are defined as

t_s, storage time from $0.9I_{B1,pk}$ to $0.9I_{C,pk}$;
t_v, voltage rise time from $0.1V_{CE,pk}$ to $0.9\,V_{CE,pk}$;
t_f, current fall time from $0.9I_{C,pk}$ to $0.1I_{C,pk}$;
t_c, commutation time from $0.1V_{CE,max}$ to $0.1I_{C,pk}$; and
t_a, current tail time from $0.1I_{C,pk}$ to $0.02I_{C,pk}$.

The turn-on switching times are defined as

t_d, delay time from $0.1I_{B1,pk}$ to $0.1I_{C,pk}$;
t_r, rise time from $0.1I_{C,pk}$ to $0.9I_{C,pk}$; and
t_{ds}, dynamic saturation time from $0.1V_{CE,max}$ to $1.1V_{CE,sat}$.

2.2.3 Losses and Power Dissipation

Referring to Fig. 2.6, t_v is a part of the storage time at turn-off. Even though dissipation may be considerable during $t_s - t_v$, negligible energy is absorbed by the transistor since this transition is very rapid. The majority of the power dissipation and energy occurs during $t_v - t_f$. This period is equal to t_c and is frequently specified by the manufacturer for a certain collector current with the collector voltage clamped to rated V_{CEX} and for a given inductance in the load. If the base current transition time (from $+I_{B1}$ to $-I_{B2}$) is very rapid, charge trapping in the collector-to-base capacitance occurs and the result is the often observed current tail during t_a. Since the collector voltage is near its maximum at this time, considerable power may be dissipated and the device may fail due to second breakdown. For this reason, a finite transition time in I_B is recommended which, in turn, decreases t_a. In fact, open-loop oscillations can occur due to storage time modulation. The storage time, which can be appreciable in Darlington transistors, may be reduced by antisaturation techniques such as the Baker clamp (CR_{1-3}) shown in Fig. 2.10B.

During turn-on, transition through the active region (t_r) is very rapid. However, in high-voltage transistors, the collector has a high resistivity and the collector region accumulates a charge as V_{CE} and f_T decreases. This results in a quasi-saturation state for period t_{ds}, as defined earlier and shown in Fig. 2.6. Even with high-speed transistors, this time may typically be 5 μs and considerable energy must be absorbed. For instance, if $V_{CE,max} = 160$ V, $I_C = 10$ A, $t_r = 0.3$ μs, $t_{ds} = 5$ μs, the energy during the turn-on time is

$$\varepsilon = \tfrac{1}{2}V_{CE}I_C t_r = 0.5 \times 160 \times 10 \times 0.2 \times 10^{-6} = 160\ \mu J$$

Since the voltage waveform is nearly exponential during t_{ds}, the average voltage is $0.2 \times 0.1V_{CE}$ (the 0.2 factor is from Chapter 1, Table 1.2, part B, waveform 15), and the energy during this time is

$$\varepsilon = 0.2 \times 16 \times 10 \times 5 \times 10^{-6} = 160\ \mu J$$

Thus, the energy during dynamic saturation can be as high, or higher, than the energy during the rise time. The dynamic saturation time may be decreased by overdriving the base (current peaking) at turn-on. However, analysis of the complete circuit must be investigated. For instance, if a rectifier in the circuit is turning off while the transistor is turning on (as is the case in most switching stages), the reverse current through the rectifier will cause an increase in the transistor current and in the saturation voltage, which may falsely appear as a dynamic saturation problem. When a transformer is used for isolation and voltage scaling (as in a converter stage), the charging of the interwinding capacitance of the transformer at turn-on will cause a transient increase in collector current and delay the effective transistor saturation.

Power dissipation in the switching transistor consists of (1) conduction losses during saturation; (2) leakage losses during nonconduction; (3) turn-on switching losses; (4) turn-off switching losses.

For resistive loads, the following formulas govern these power dissipations, respectively:

$$P_{on} = (V_{CE,sat}I_C + V_{BE,sat}I_B)t_{on}/T \tag{2.15a}$$

$$P_{off} = V_{CC}I_{co}t_{off}/T \tag{2.15b}$$

$$P_{sw\text{-}on} = V_{CC}I_C t_r/6T \tag{2.15c}$$

$$P_{sw\text{-}off} = V_{CC}I_C t_f/6T \tag{2.15d}$$

$$P_T = P_{on} + P_{off} + P_{sw\text{-}on} + P_{sw\text{-}off} \tag{2.16}$$

where P_T is total transistor loss, V_{CC} is supply voltage (or twice the supply voltage for push–pull topology), and I_{co} is collector leakage current (not necessarily negligible at high voltages and high temperatures). For inductive loads, P_{on} and P_{off} are as above, and

$$P_{sw\text{-}on} = 0.5V_{CC}I_C t_r/2T \tag{2.17a}$$

$$P_{sw\text{-}off} = 0.5V_{CC}I_C t_c/2T \tag{2.17b}$$

The total power dissipation is again Equation (2.16) with the switching losses of Equation (2.17).

In the preceding formulas, the numerator is an energy term (watt-seconds) while the denominator is the period time, or the reciprocal of the switching frequency. The switching times are normally given in manufacturer data. However, external snubbers or wave-shaping circuits usually lower the energy absorbed by the transistor. Referring to Fig. 2.7, straight-line approximations of the voltage and current during typical turn-off of a high-voltage transistor plus the corresponding instantaneous power are shown for two conditions. In Fig. 2.7A, the rate of collector voltage rise is very rapid and results in a high peak as well as average pulse power, as shown in Fig. 2.7B. However, the fast switching results in a nominal energy level. Turn-off switching losses at 25 and 40 kHz are 30 and 48 W, respectively, as shown in Fig. 2.7C. In Fig. 2.7D, the rate of collector voltage rise is halved, and the pulse power and energy are shown in Fig. 2.7E. Turn-off switching losses at the same 25 and 40 kHz are 21.5 and 34.4 W, respectively, as shown in

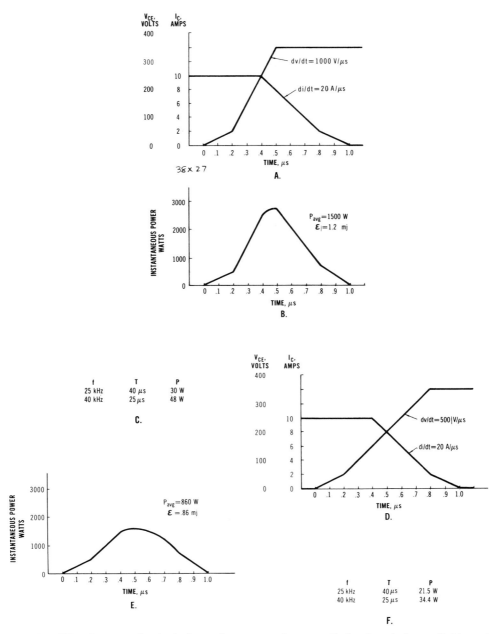

Fig. 2.7. Waveforms and calculations of energy and power dissipation during switching at different rates of dv/dt.

Fig. 2.7F. Thus, by decreasing the collector voltage dv/dt by a factor of 2, the turn-off losses are decreased approximately 40%.

For a particular circuit design (which should include some form of current sensing or limiting), an oscilloscope X–Y display of I_C versus V_{CE} can be readily implemented for SOA observation under the conditions of circuit start-up, normal operation, overload, short circuit, shutdown, transient line, and load changes, each at the specified system operating temperature conditions. By integrating an oscilloscope display of I_C and V_{CE} in the time domain, power dissipation and energy levels during switching and conduction can be readily calculated for each of the above conditions, and the thermal design and required heat sinking can be implemented accordingly. This system testing is also imperative for observing second-order effects, such as leakage inductance and rectifier recovery time. In this manner, high reliability should be achieved by operating the bipolar power transistor within the manufacturer's specified ratings for the worst-case operating conditions.

As a side observation, designers might wishfully hope switching losses in bipolar transistors were as easily obtained from manufacturers' data as the pulse conditions of inverter-rated SCRs, discussed in Section 2.4.2 and shown in Fig. 2.24. However, conditions such as reverse base current, reverse base voltage, rate of decay of forward-biased conditions, inductive loads, saturation voltage, wide operating frequencies, and snubber networks peculiar to various topologies make this desire a formidable task. But just as the SOA has been a boost to understanding secondary breakdown limits, data and graphs of pulse energy ratings for various operating conditions may be available in the future.

2.2.4 Base Drive Circuits

The forward-biased (conducting) state of the bipolar transistor is controlled by the amount of base current relative to the collector current. The dc gain is expressed as $h_{FE} = I_C/I_B$ at a given V_{CE}. But $V_{CE,\,sat}$, at a given I_C and I_B, is more informative for minimizing or calculating conduction losses. A power Darlington transistor is shown in Fig. 2.8A in the common-emitter configuration. For a forced gain of 50 total, typical values of current and voltage are as shown. The minimum saturation voltage is then

$$V_{CE,\,sat,\,Q2} = V_{CE,\,sat,\,Q1} + V_{BE,\,sat,\,Q2} = 0.8 + 1.5 = 2.3 \text{ V}$$

The power dissipation in the two transistors is

$$P = I_{load}V_{CE,\,sat,\,Q2} = 10 \times 2.3 = 23 \text{ W}$$

The input current drive can be reduced by the complementary circuit of Fig. 2.8B with a forced gain of 500 total and the same load current. The minimum saturation voltage is again

$$V_{CE,\,sat,\,Q3} = V_{EC,\,sat,\,Q2} + V_{BE,\,sat,\,Q3} = 0.8 + 1.5 = 2.3 \text{ V}$$

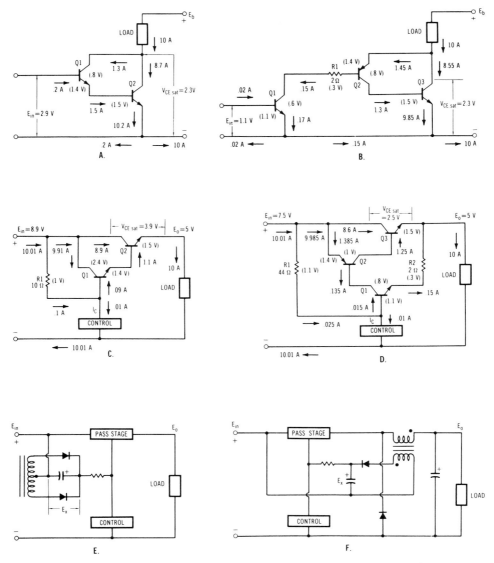

Fig. 2.8. Base drive circuits for driving bipolar transistors into saturation.

Resistor R_1 is added to limit the current through Q_1 since

$$V_{CE,\,\text{sat},\,Q3} = V_{EB,\,\text{sat},\,Q2} + V_{R1} + V_{CE,\,\text{sat},\,Q1}$$

Then

$$V_{R1} = 2.3 - 1.4 - 0.6 = 0.3 \text{ V}$$

For $I_{C,\,Q1} = 0.15$ A, $R_1 = 0.3/0.15 = 2 \ \Omega$.

In this circuit, as compared to Fig. 2.8A, the input current is reduced by a factor of 10 and the input voltage is less than half. Again, the total load current is equal to the sum of the collector currents.

Saturation voltages increase significantly in Darlington stage emitter-follower circuits operating from a single input. The circuit of Fig. 2.8C could be the pass transistor stage of a linear or switching regulator. In either case, minimum saturation voltage is desired to decrease power dissipation. However, the minimum saturation voltage is

$$V_{CE,\,\mathrm{sat},\,Q2} = V_{R1} + V_{BE,\,\mathrm{sat},\,Q1} + V_{BE,\,\mathrm{sat},\,Q2}$$

Resistor R_1 should be small in value to minimize voltage drop across the pass stage. But as R_1 is decreased, I_c increases and if the pass transistors are to be turned off, the power dissipated in R_1 is $P_{R1} = (E_{\mathrm{in}})^2/R_1$. If $R_1 = 10$ Ω and $E_0 = 5$ V,

$$V_{CE,\,\mathrm{sat},\,Q2} = (0.1 \times 10) + 1.4 + 1.5 = 3.9\text{ V}$$

$$E_{\mathrm{in},\,\mathrm{min}} = 5 + 3.9 = 8.9\text{ V}$$

$$P_{R1} = 8.9^2/10 = 7.9\text{ W}$$

Since $V_{CE,\,\mathrm{sat},\,Q2}$ and $V_{CE,\,\mathrm{sat},\,Q1}$ are governed by the voltage drop across R_1, the base currents decrease as the saturation voltage increases. Thus $I_{B,\,Q1}$ is less than that of Fig. 2.8A.

The power dissipation and efficiency are

$$P_{\mathrm{loss}} = (E_{\mathrm{in}} - E_0)I_0 = 3.9 \times 10 = 39\text{ W}$$

$$\text{Efficiency} = (5 \times 10)/(8.9 \times 10.01) = 56\%\text{ maximum}$$

As E_{in} increases, P_{R1} increases significantly. For instance, if E_{in} increases 33%,

$$P_{R1} = (1.33 \times 8.9)^2 \times 10 = 14\text{ W}$$

and

$$\text{Efficiency} = (5 \times 10)/(1.33 \times 8.9 \times 10.01) = 42\%$$

This problem can be overcome by the complementary circuit of Fig. 2.8D, which is the emitter follower of Fig. 2.8B. The minimum saturation voltage is

$$V_{CE,\,\mathrm{sat},\,Q3} = V_{R1} + V_{BE,\,\mathrm{sat},\,Q1} + V_{R2}$$

If $R_1 = 44$ Ω, $R_2 = 2$ Ω, and $E_0 = 5$ V

$$V_{CE,\,\mathrm{sat},\,Q2} = (44 \times 0.025) + 1.1 + (2 \times 0.15) = 2.5\text{ V}$$

$$E_{\mathrm{in},\,\mathrm{min}} = 5 + 2.5 = 7.5\text{ V}$$

$$P_{R1} = 7.5^2/44 = 1.3\text{ W}$$

$$P_{\mathrm{loss}} = (E_{\mathrm{in}} - E_0)I_0 = 2.5 \times 10 = 25\text{ W}$$

$$\text{Efficiency} = (5 \times 10)/(7.5 \times 10.01) = 67\%\text{ maximum}$$

If E_{in} increases 33%,

$$P_{R1} = (1.33 \times 7.5)^2/44 = 2.3 \text{ W}$$
$$\text{Efficiency} = (5 \times 10)/(1.33 \times 7.5 \times 10.01) = 50\%$$

Operating efficiency may be increased by a boost voltage to drive the pass stage. For a linear power supply (Fig. 2.8E), E_x is provided by rectifying and filtering an auxiliary secondary winding voltage. For a switching regulator (Fig. 2.8F), E_x is developed by flyback action of an auxiliary winding on the output filter inductor, and E_x will remain fairly constant since E_0 is constant. Reference Chapters 8 and 17, respectively.

For switching circuits, turn-on and turn-off considerations for the power transistor depend on the family type (described in Section 2.2.1) and on the particular topology of the circuit. Fixed base current, peaking base current at turn-on, the ratio of forward base current (I_{B1}) to reverse base current ($-I_{B2}$), and base current proportional to collector current are common approaches. The switching of Darlington transistor stages may require different base drive techniques than switching of individual transistors, as discussed in the above paragraph. Also, circuits employing antisaturation techniques may be used to reduce effects of storage time.

Typical direct-coupled base-drive circuits and accompanying waveforms are shown in Fig. 2.9. The load is shown as a "block" but is understood to be any conventional form, including the primary or one-half the primary of an output power transformer.

In Fig. 2.9A, base current into Q_1 turns on Q_1, turning off Q_2, and when Q_1 turns off, Q_2 turns on. For a required collector current in Q_2 and a forced gain of 12,

$$I_{\text{Bss},Q2} = I_{C,Q2}/12 \quad \text{(for steady state)}$$
$$= (E_a - V_{BE,\text{sat},Q2})/(R_1 + R_2 + R_3) - V_{BE,\text{sat},Q2}/R_4$$

When Q_2 turns on, current flows through C_1 for current peaking and

$$I_{\text{Bpk},Q2} = (E_a - V_{BE,\text{sat},Q2})/(R_1 + R_3) - V_{BE,\text{sat},Q2}/R_4$$

Then C_1 charges to $V_{C1} = I_{\text{Bss},Q2}R_2$. When Q_1 turns on, C_1 discharges through Q_1, R_4, Q_2 B–E, and R_3 to reverse bias Q_2. Also, C_1 discharges through R_2 and the value of C_1 may be large to provide the desired time constant. This circuit has a major disadvantage in that Q_1 must conduct the reverse base current of Q_2 as well as the current through R_1.

Example

Let $I_{C,Q2} = 12$ A, $V_{BE,\text{sat},Q2} = 1.8$ V, $V_{CE,\text{sat},Q1} = 0.8$ V, $E_a = 9$ V, and $R_4 = 47$ Ω. Then $I_{\text{Bss},Q2} = (9 - 1.8)/R_T + 1.8/47 = 1$ A. Then $R_T = 6.9$ Ω. For $I_{\text{Bpk},Q2} = 1.5$ A, $R_1 + R_3 = 4.5$ Ω. Then $R_2 = 2.4$ Ω and $V_{C1} = 2.4$ V. If $R_3 = 1$ Ω, $R_1 = 3.5$ Ω. When Q_1 turns on and if $I_{B1} = I_{B2}$ in Q_2, $I_{C,Q1} = (9 - 0.8)/3.5 + 1 = 3.34$ A, which is 28% of the collector current of Q_1. Thus the circuit is not gain efficient. In a steady-state condition of Q_1 on, or at low duty cycles of Q_2,

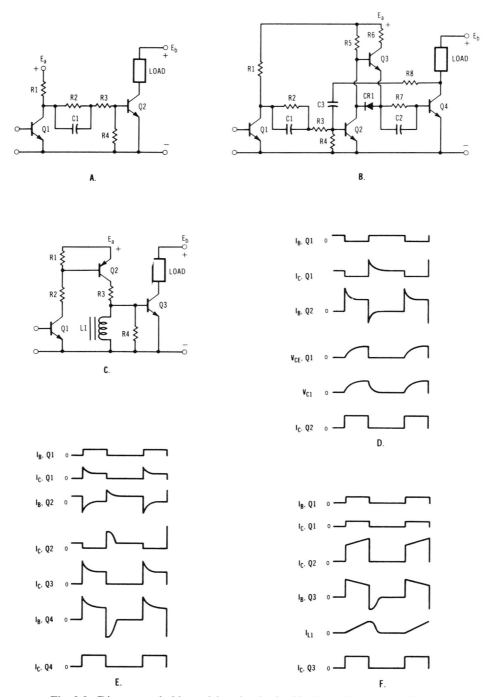

Fig. 2.9. Direct-coupled base drive circuits for bipolar switching transistors.

the power dissipation in R_1 is $P_{R1} = (9 - 0.8)^2/3.5 = 19$ W, and a power rating of 40 W should be used. However, if E_b is 200 V, the peak output power is $200 \times 12 = 2400$ W; the actual peak power delivered by E_a is $9^2/3.5 = 23$ W. Neglecting switching losses, the circuit efficiency is $2400/2423 = 99\%$.

The circuit of Fig. 2.9B overcomes the high-power dissipation in the drive resistors and is a modified totem pole. With Q_1 off, Q_2 is on, Q_3 is off, and Q_4 is off. The collector current in Q_2 is E_a/R_5. When Q_1 turns on, Q_2 turns off, and Q_3 turns on Q_4. Current peaking is provided by C_2. When Q_1 turns off, Q_2 turns on, turning off Q_3 and Q_4, and C_2 discharges through CR_1 (which reverse biases Q_3), Q_2, and Q_4 B–E. Resistors R_6 and R_7 are selected to supply Q_4 base current, with R_7 determining the voltage on C_2. Since R_5 supplies base current to Q_3, the value of R_5 may be high, and the collector current of Q_2 may be low. Resistor R_8 and C_3 provide a positive feedback to decrease the switching time of Q_2.

The circuit of Fig. 2.9C may be suitable for high-frequency, low-duty cycle switching. Without base current applied to Q_1, all transistors are off. When Q_1 turns on, Q_2 turns on Q_3. Inductor L_1 across Q_3 B–E initially supports the base-emitter voltage, but then current in L_1 increases, decreasing the base current. Then L_1 stores energy, and when Q_1 turns off, the voltage reversal across L_1 reverse biases Q_3 for a fast turn-off.

For voltage isolation in converters and power supplies, transformer-coupled drive stages, with waveforms, as shown in Figs. 2.10 and 2.11 represent a few choices. Various features of each circuit may be combined, such as Baker clamp antisaturation rectifiers and transformer-coupled proportional base drive. These circuits again show a collector load, which is normally a power transformer primary.

In Fig. 2.10A, when Q_1 turns on, a voltage is induced in the secondary of transformer T_1 to turn on Q_3. Current flows through Q_3 B–E and R_1, and also charges C_1 through R_3 and CR_2. Here Q_2 is reverse biased. When Q_1 turns off, the voltage reversal produces a current through R_1 and CR_1. Also, Q_2 is forward biased by C_1 and Q_2 conducts $-I_{B2}$ to rapidly turn off Q_3. This circuit may be limited to low duty cycles due to the reset time of T_1. In fact, the base of Q_2 may be capacitively coupled to conserve the charge on C_1 for low duty cycles. The current through R_1 should decay to zero before Q_1 is turned on again.

Figure 2.10B shows a push–pull primary. Normally Q_1 is on and R_1 limits the current through Q_1 since T_1 is in negative saturation for a steady-state condition. When Q_1 turns off and Q_2 turns on, the voltage induced in the secondary turns on Q_3. Current flows through CR_2, CR_3, Q_3 B–E, and R_2. Capacitor C_3 charges to approximately 2 V. When Q_3 is on, a portion of the secondary current flows through CR_1 and Q_3 C–E to maintain the base-to-collector junction in reverse bias. This Baker clamp effect decreases the storage time of Q_3 when Q_3 turns off, especially if Q_3 is a Darlington transistor. Also during this time, C_1 discharges through R_1. When Q_2 turns off and Q_1 turns on again, a current surge flows through Q_1 as C_1 charges. This secondary voltage reversal, plus the discharge of C_3, reverse biases Q_3 for fast turn-off. Capacitor C_2 is included for transient spike suppression on source E_a. The duty cycle is limited to 50% maximum due to the required reset of T_1 when Q_1 is on. This circuit is also applicable to SCR gate triggering where a gate signal is desired for the SCR conduction time to conduct reactive load currents (reference Section 2.5.1).

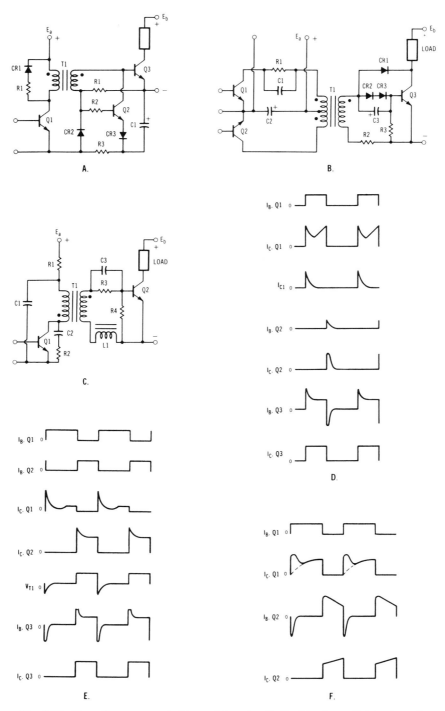

Fig. 2.10. Transformer-coupled base drive circuits for bipolar switching transistors.

The circuit of Fig. 2.10C provides a flyback drive and is frequently used to drive the power transistor in forward converters. Normally Q_1 is on and the current is limited by R_1. When Q_1 turns off, the energy stored in T_1 primary is transferred to the secondary to turn on Q_2. Current flows through R_3, charging C_3, Q_2 B–E, and L_1. The inductance of L_1 is normally set equal to the leakage inductance of T_1. Also, when Q_1 turns off, C_1 charges through R_1. When Q_1 turns on again, C_1 discharges across T_1 and the secondary voltage reverses to reverse bias Q_2 for turn-off. To increase the time for $-I_{B2}$ current flow, R_3 may be replaced by two series rectifiers or a low-voltage zener diode to prevent C_3 from discharging through R_3.

The circuit of Fig. 2.11A provides drive for a Darlington transistor in the output stage. When Q_1 is on, the constant-current source stores energy in the T_1 primary. When Q_1 turns off, the voltage reversal forward biases Q_3 and Q_4, and base current flows through CR_1 and C_1. Then Q_2 is reverse biased by CR_1. When Q_1 turns on again, the voltage of T_1 reverses and C_1 discharges through Q_3, Q_4, L_1, and Q_2. Inductor L_1 is a low value, which initially slows the turn-off of Q_3 and Q_4 to prevent charge trapping in the collector junction at low duty cycles. Also, at low duty cycles, partial charge on C_1 may be provided by an auxiliary dc source.

The transformer-coupled circuits discussed so far use an individual transformer to drive an individual output transistor. In Fig. 2.11B, a single-drive transformer with two secondaries drives two output transistors shown connected to a push–pull stage. The separate secondaries could therefore drive the output stage of half-bridge and full-bridge converters. The drive signals to Q_1 and Q_2 are supplied by a typical pulse width–modulated integrated circuit, such that Q_1 and Q_2 are conducting during the off time of the power stage. Thus a constant current rather than a constant voltage is used as the source supply. With Q_1 and Q_2 on, T_1 is effectively shorted. When Q_1 turns off, the current increases in Q_2 and current flows through C_1/R_1 and Q_3 turns on. When Q_1 again turns on the primary and secondary of T_1 are shorted and C_1 discharges to provide $-I_{B2}$, and Q_3 turns off. Next, when Q_2 turns off, the collector current in Q_1 increases and base drive is applied to Q_4, which turns on. The cycle repeats as shown. Again, R_1 and R_2 may each be replaced by two series rectifiers and a rectifier connected from C_1 to Q_3 collector plus a rectifier connected from C_2 to Q_4 collector for antisaturation to decrease storage time. This circuit is also ideal for proportional base drive wherein the collector or emitter lead passes through the core of T_1, observing the correct winding polarity.

Proportional base drive is shown in the circuit of Fig. 2.11C, but with a slightly different configuration than would be applied to Fig. 2.11B. In this case, Q_1 is on for a period greater than 50% of the output duty cycle. When Q_1 turns on, Q_2 is reverse biased as the current increases exponentially through T_1–N_1, and Q_1 and R_1 limit the peak current. When Q_1 turns off, the energy in N_1 is transferred to N_4, which forward biases Q_2. But this current would start to decay (as in Fig. 2.11A) at the rate

$$I_{B,Q2} = I_{\text{pk}}(N_1/N_4) - V_{BE,\text{sat},Q2}(N_1^2 t)/(L_p N_4^2)$$

where I_{pk} is current flowing in Q_1 at turn-off, L_p is primary inductance, and t is time.

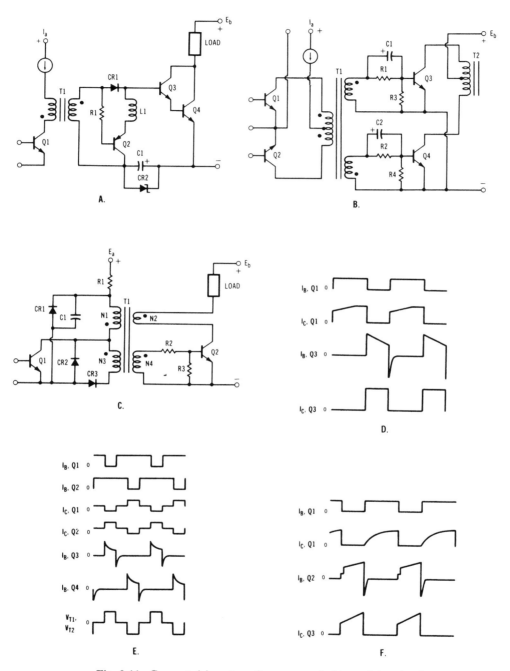

Fig. 2.11. Current-driven transformer-coupled base drive circuits.

But this initial base current causes collector current to flow through N_2. The base current then increases by $I_{B,Q2} = I_{C,Q2}(N_2/N_4)$.

During the off time of Q_1, C_1 charges through R_1. When Q_1 turns on again, N_3 is initially shorted. Then C_1 discharges through N_1 and all windings reverse in polarity. Then Q_2 is reverse biased and turns off with $-I_{B2} = I_{C,Q1}(N_1/N_4)$.

2.2.5   Paralleling Transistors

Paralleling bipolar transistors requires careful attention, both in the static conduction mode and in the dynamic switching mode. Current unbalances in the static linear mode may cause transistor failure from forward-biased second breakdown. In switching circuits, current imbalances at turn-off may cause failure from reverse-biased second breakdown. But single high-voltage transistors are available with current ratings of 200 A. High-current modules, where several chips or die are contained in a single package, are available from various manufacturers. These devices, and Darlington versions thereof, have somewhat precluded the need to parallel a group of lower current rated devices to conduct the desired high current. However, economics may play a role in the decision. Five transistors conducting 20 A each may be more economical than one 100-A transistor, even with external forced current sharing components for the five transistors. The total power dissipation in the transistors is essentially the same in either case. Spreading the heat produced in a group of transistors over a large surface area heat sink, as compared to a heat sink for the single transistor, may actually be more volumetrically efficient. Thermal bonding of the transistor cases directly to the heat sink (electrically isolated heat sink) helps equalize the case temperatures.

Figure 2.12 shows three methods to equalize currents in paralleled transistors. In each case $I_C = I_{c1} + I_{c2} + I_{c3}$, $I_B = I_{b1} + I_{b2} + I_{b3}$, $I_E = I_{e1} + I_{e2} + I_{e3}$, $I_c = I_b h_{FE}$, $R_{b1} = R_{b2} = R_{b3}$, and $R_{e1} = R_{e2} = R_{e3}$.

For Fig. 2.12A, $E_{in} = V_{BE,max} + I_{b,max}R_b$ and $V_{D,max} = V_{CE,max}$.

Resistor R_b should be as large as practical to prevent I_b imbalance and to minimize I_c imbalance. However, increasing R_b requires an increase in E_{in} and results in an increase in resistor power dissipation. Here $P_{Rb} = I_b^2 R_b$.

For Fig. 2.12B, $E_{in} = V_{BE,max} + I_{e,max}R_e$ and $V_{D,max} = V_{CE,max} + I_{e,max}R_e$. Resistor R_e provides a voltage drop due to I_e. An increase in I_{c3}, for instance, produces a voltage increase across R_{e3} which results in a decrease in I_{b3}, and I_{c3} then decreases. Here $P_{Re} = I_e^2 R_e$.

For Fig. 2.12C, $E_{in} = V_{BE,max} + I_{b,max}R_b + I_{e,max}R_e$ and $V_{D,max} = V_{CE,max} + I_{e,max}R_e$.

Both R_b and R_e are used to minimize collector current imbalance. The advantages of the previous two circuits are combined. Current sharing is a function of base voltage and base current. These parameters are not applicable to MOSFETs, and MOSFETs may not require current-balancing resistors, as discussed in Section 2.3.7.

Example

In the circuit of Fig. 2.12B, let $I_C = 60$ A; $h_{FE} = 15$ at $I_c = 20$ A at $V_{ce} = 2$ V; then, $I_b = 1.33$ A; $I_B = 4$ A. Now let Q_3 have a gain of 1.5 times the other

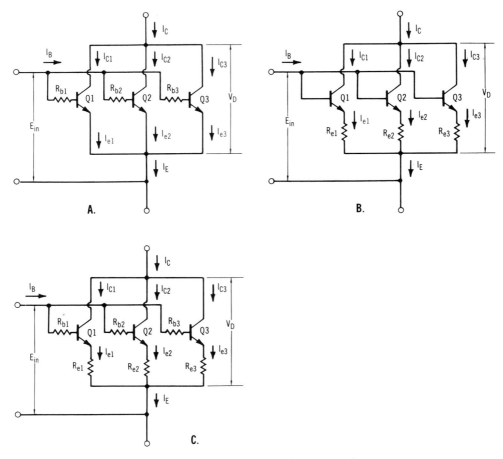

Fig. 2.12. Means of paralleling bipolar transistors.

transistors. Without balancing resistors, $I_{c1} = I_{c2} = 60/(1 + 1 + 1.5) = 17$ A; $I_{c3} = 60 - (17 + 17) = 26$ A. The emitter resistors are to be added to equalize the currents. Let $V_{Re} = 0.65$ V. Then, $R_e = 0.65/(20 + 1.33) = 0.03$ Ω. Now *if* $I_{c3} = 26$ A, $V_{Re3} = 0.78$ V, $V_{ce3} \cong 2.22$ V, and $V_D = 3$ V. But $V_{Re1} = V_{Re2} = 0.03 \times 17 = 0.51$ V, and $V_{ce1} = V_{ce2} = 3 - 0.51 = 2.49$ V. Also, Q_1 and Q_2 have a rated saturation voltage of 2V at 20 A; $V_D = 2 + (21.33 \times 0.03) = 2.64$ V. For $I_{c3} = 20$ A, I_{b3} must decrease by 1.5. Then $I_{b3} = 1.33/1.5 = 0.89$ A, and if $I_{c3} = 20$ A, then $V_{Re3} = (20 + 0.89) \times 0.03 = 0.627$ V and $V_{ce3} = 2.64 - 0.627 = 2.013$ V. Also, $I_B = 1.33 + 1.33 + 0.89 = 3.55$ A.

2.3 MOSFET POWER TRANSISTORS

The MOSFET power transistor may be considered the closest thing to a perfect switch. Performance characteristics and advantages, as compared to the bipolar transistor, are as follows:

1. Voltage control (instead of current control) with very high input impedance. Control and drive circuitry power is therefore reduced.
2. Majority carrier (instead of minority carrier) "junctions" provide very fast switching without storage time delay.
3. The positive temperature coefficient of resistance eliminates localized hot spots and provides ease of paralleling.
4. Switching time is independent of temperature.
5. Freedom from secondary breakdown failure since the SOA extends over the full current and voltage rating but is of course limited by the device power dissipation and pulse ratings.
6. The on-state forward voltage drop is higher than the bipolar transistor of equal die size but the voltage drop continues to decrease as technology improves.
7. However, as in the bipolar transistor, thermal runaway can occur. As the device temperature increases both the on resistance and the leakage current increase. This produces an increase in power dissipation which further increases the device temperature, which further increases leakage current and on resistance, which can result in the "snowball" effect and in failure due to excessive junction temperature.

N-channel MOSFETs are discussed herein due to popularity for the same reasons as the NPN bipolar. P-channel devices are used in complementary circuits and in applications where the bipolar PNP would be used for the same reasons as the MOSFET. True complementary characteristics may not be identical due to the higher resistivity of the P-type silicon, as compared to the N type. Technology and varying manufacturers' processes have produced designations such as DMOS, SMOS, VMOS, MOSPOWER, SIPMOS, TMOS, and HEXFET (which has a planar hexagonal cell structure).*

2.3.1 Device Parameters

The MOSFET (shown in Fig. 2.13A) terminals, as compared to the bipolar transistor, are drain (collector), gate (base), and source (emitter). The load to be controlled or switched may be located in the drain (common-source configuration) or in the source (source–follower configuration). With no voltage on the gate (with respect to the source), the device is off, as shown in Fig. 2.13B, and the only current flowing is the negligible leakage current. The MOSFET is turned on by a voltage applied to the gate with respect to the source. When the gate voltage exceeds the threshold voltage, $V_{GS(th)}$, drain current flows, as shown in Figs. 2.13B and C. Voltage $V_{GS(th)}$ ranges from 1 to 4 V, increasing as device power increases, and is typically 3 V. Thus the MOSFET provides better noise immunity than the bipolar transistor or the thyristor. However, the temperature coefficient of thresh-

---

*SIPMOS is a trademark of Siemens. TMOS is a trademark of Motorola. HEXFET is a trademark of International Rectifier.

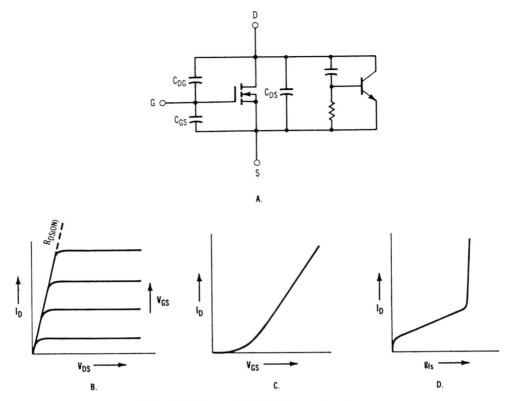

Fig. 2.13. Power MOSFET characteristics.

old voltage is approximately -5 mV/$^\circ$C and must be considered for wide operating temperature applications. Also, threshold voltage decreases when the MOSFET is exposed to the radiation effect of total dose.

Three operating regions occur when the MOSFET conducts, as shown in Fig. 2.13B. One is the constant-resistance mode where the drain current I_D increase is proportional to the drain–source voltage, $V_{DS(\text{on})}$, for a given V_{GS}. The second region is the transition between constant resistance and constant current. The third region is the constant-current or active mode, where I_D remains almost constant as V_{DS} increases and is the "nonsaturation" region. Being a voltage-controlled device, transconductance g_{fs} is a more ideal "figure of merit" parameter than the current gain h_{fe} parameter of the bipolar. Transconductance is the ratio of change of drain current to change in gate voltage and is plotted versus I_D in Fig. 2.13D. For switching circuits, $V_{DS(\text{on})}$ should be low to minimize power dissipation at a given I_D, and V_{GS} should be high enough to accomplish this. Then $R_{DS(\text{on})}$ becomes the conducting resistance in the "saturated" mode. The positive temperature coefficient of $R_{DS(\text{on})}$ is approximately $+0.6\%/^\circ$C.

The MOSFET also has an inherent parasitic *NPN* bipolar transistor from drain to source as shown in Fig. 2.13A. This transistor is a reverse rectifier with a forward current rating typically equal to the drain current rating of the MOSFET and a voltage rating equal to BV_{DS}. Originally perceived as a "free component" which could conduct reactive currents when the MOSFET was off, the rectifier has become a liability in certain topologies. The reverse-recovery time of the rectifier (a majority carrier) is much longer than the switching time of the MOSFET (a minority carrier). An overlap current flow through the rectifier of this device and through the drain to the source of a series-connected MOSFET can cause device failure, as discussed in Section 2.3.7.

2.3.2   Capacitance Effects

MOSFET switching times are determined primarily by device capacitances, stray capacitances, and the impedance of the gating circuit. Even though the device has a very high static input impedance, the gate-to-source capacitance must be charged at turn-on and discharged at turn-off. Thus the gating circuit must have high momentary peak current sourcing and sinking capability for switching the MOSFET. In Figs. 2.13A and 2.14A, assume the device is off. Capacitors C_{DG} and C_{DS} are charged to the supply voltage. At turn-on, the gating circuit must supply a current to charge C_{GS} plus the discharge current from C_{DG} which also flows through the gating circuit. A high value of R_{in} produces a long time constant and a slow rise in V_{GS}. When the device turns off, the gating circuit must sink the discharge current from C_{GS} plus the charging current through C_{DG}. For specified switching times, manufacturers' data sheets normally specify the gating circuit impedance. For these conditions, the switching times with a resistive load are specified as turn-on delay $t_{d(on)}$, which is the time from application of the gate voltage to $V_{GS(th)}$; rise time t_r, which is the time from $V_{GS(th)}$ to specified I_D; turn-off delay $t_{d(off)}$, which is the time for the gate voltage to fall to $V_{GS(th)}$; and fall time t_f, which is the time from $V_{GS(th)}$ to zero I_D.

Instead of terminal capacitances, which are dynamically proportional to V_{DS} and are related to charge on the capacitors, manufacturers' data normally specifies C_{iss} (input capacitance), C_{oss} (output capacitance), C_{rss} (reverse-transfer capacitance), where $C_{iss} = C_{GS} + C_{DG}$, $C_{rss} = C_{DG}$, and $C_{oss} = C_{DS} + C_{DG}$. These capacitances increase significantly as the device current rating (and to a lesser extent, as the voltage rating) increases. For high-power devices, careful attention must be paid to the design of the gating circuit, as discussed in Section 2.3.6.

An additional important reason for using a low-impedance gating circuit with high-voltage devices stems from the capacitor divider effect of C_{DG} and C_{GS} causing transient voltage spikes on the gate. Referring to Fig. 2.13, assume the gating circuit impedance is very high, the supply voltage is several hundred volts, and the device is turning off. Then $V_{GS} = V_{DS}[C_{DG}/(C_{DG} + C_{GS})]$. Since the absolute maximum V_{GS} rating is typically ± 20 V (otherwise permanent damage may result), precautions must be exercised. Due to the high switching speed, inductance in the external gating circuit (such as the leakage inductance of a drive transformer) could support a transient voltage exceeding 20 V and the inductance would isolate the MOSFET gate from the low static impedance of the drive circuit. There are two solutions to this problem. One solution, which is normally sufficient, is to connect an external resistor from gate to source, which forms a voltage divider to limit V_{GS}. The other solution is to connect a zener diode (approximately 15 V rating) from gate to source to clamp the gate voltage to a safe level.

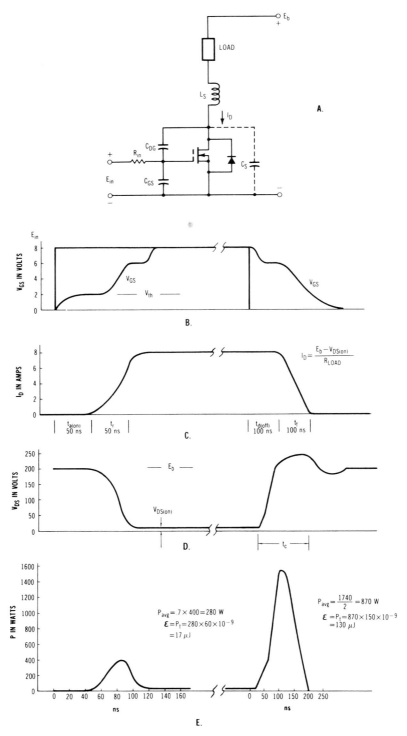

Fig. 2.14. Power MOSFET switching waveforms.

2.3.3 Safe Operating Area

An outstanding feature of the MOSFET is freedom from second breakdown and wide SOA, both forward biased (FB) and reverse biased (RB). The SOA comparison of MOSFET to bipolar devices is discussed in Section 2.2.1 and is shown in Fig. 2.4. The wide FBSOA allows the device to operate in the dc or linear mode at constant rated power dissipation for all conditions of I_D and V_{DS} and to operate at even higher values of I_D (up to I_{DM}) for pulse conditions. Obviously, rated BV_{DS} must not be exceeded. The rated power dissipation for a given device as a function of case temperature is stated in manufacturers' data sheets. The pulse ratings provide a more informative energy capability. For instance, Fig. 2.4 shows the 2N6766 can dissipate 800 W for 1 ms pulse duration and at a case temperature of 75°C. This is equivalent to 0.8 J, neglecting the energy absorbed by switching. For inductive loads, RBSOA extends to rated I_D and clamped V_{DS}.

2.3.4 Switching Waveforms

A typical switching circuit is shown in Fig. 2.14A with a load connected from drain to E_b, stray inductance in the drain, an input resistor, and the MOSFET with associated capacitance. Waveforms are shown in Figs. 2.14B, C, D, and E. Parameter scales are somewhat arbitrary. When an input signal E_{in} is applied through R_{in}, C_{GS} charges as V_{GS} increases. At $V_{GS(th)}$, I_D begins to flow and V_{DS} decreases. The change in V_{DS} produces a "Miller" effect due to C_{DG} and a discharge current flows through C_{DG} and also through R_{in}. Thus, the lower the drive impedance, the higher the rate of rise of V_{GS}, and the faster the turn-on time. Voltage V_{DS} decreases rapidly but a voltage "tail" results before $V_{DS(on)}$ is reached, as shown in Fig. 2.14D. This condition is analogous to the dynamic saturation of the bipolar transistor but is caused by the sudden increase in C_{DG} at low values of V_{DS}. The rise of V_{GS} is again delayed as current flows through the increasing C_{DG}. A steady-state or "saturated" condition then exists. Power dissipation during turn-on is 280 W peak, as shown in Fig. 2.14E. For the scale times shown, the turn-on energy is 17 μJ.

When E_{in} is "shorted," the device begins to turn off. But due to the inductance in L_s, I_D does not decrease until V_{DS} approaches E_b. As V_{DS} increases, C_{DG} charges, delaying the rate of fall of V_{GS}. Then the rapid decrease in I_D produces an induced voltage across L_s and V_{DS}, and the energy stored in L_s is absorbed by C_{DS} and the stray capacitance C_s. The power dissipation during turn-off is 870 W. For the scale times shown, the turn-off energy is 130 μJ.

2.3.5 Losses and Power Dissipation

Power dissipation in the switching MOSFET consists of (1) conduction losses; (2) leakage losses; (3) turn-on switching losses; (4) turn-off switching losses; and (5) gate transition losses. The latter three losses are proportional to frequency. Except for the gate losses (which may be negligible at frequencies below 500 kHz), the factors governing dissipation are also analogous to the bipolar transistor.

For resistive loads, the following formulas govern the above power dissipations respectively:

$$P_{\text{on}} = (I_{D,\text{rms}})^2 R_{DS(\text{on, rms})} t_{\text{on}} / T \tag{2.18a}$$

$$P_{\text{off}} = V_{DS,\text{max}} I_{\text{DSS}} t_{\text{off}} / T \tag{2.18b}$$

$$P_{\text{sw-on}} = V_{DS,\text{max}} I_D t_r / 6T \tag{2.18c}$$

$$P_{\text{sw-off}} = V_{DS,\text{max}} I_D t_f / 6T \tag{2.18d}$$

$$P_{t1} = \left[C_{\text{iss}} (V_{GS})^2 + C_{\text{rss}} (V_{DG})^2 \right] / 2T \tag{2.18e}$$

$$P_T = P_{\text{on}} + P_{\text{off}} + P_{\text{sw-on}} + P_{\text{sw-off}} + P_{t1} \tag{2.19}$$

where

P_T = total transistor loss
t_r = rise time of I_D
t_f = fall time of I_D
T = total period of time

Notice the conducting power dissipation is an I^2R function with rms values of current and resistance for true power. The rms values become important for triangular and trapezoidal waveforms, as given in Chapter 1, Table 1.2. Also, $R_{DS(\text{on})}$ has a positive temperature coefficient range of $+0.6$ to $+0.85\%/°C$, which *must* be considered as device temperature rises.

For inductive loads, P_{on}, P_{off}, and P_{t1} are as above and

$$P_{\text{sw-on}} = V_{DS,\text{max}} I_D t_r / 2T \tag{2.20a}$$

$$P_{\text{sw-off}} = V_{DS,\text{max}} I_D t_c / 2T \tag{2.20b}$$

where t_c is the crossover time from $10\%\ V_{DS,\text{max}}$ to $10\%\ I_{D,\text{max}}$.

In the above formulas, the numerator is an energy term (watt-seconds) while the denominator is the period time (seconds) or the reciprocal of the switching frequency. The switching times are normally given in manufacturer data. External snubbers or wave-shaping circuits are recommended to lower the energy absorbed by the transistor at turn-off, to reduce voltage spikes resulting from $L\,di/dt$, and to reduce EMI. Overall circuit power dissipation is not reduced, but a portion of the power is transferred from the MOSFET to the snubber. The snubber circuit must have low inductance in the leads as well as in the components. For instance, let the snubber inductance be 200 nH, and the MOSFET switches 20 A in 200 ns. Then a voltage spike of $-e = L\,dt/dt = 200 \times 20/200 = 20$ V results. The external lead lengths of the MOSFET, such as a TO-202 and TO-220, may produce voltage spikes in low-current devices which can switch in 10 ns. Even the internal lead bonds of the transistor may produce a voltage spike at the chip or die which cannot be observed.

Storage time and base drive parameters peculiar to the bipolar transistor do not exist in the MOSFET. Switching losses, energy per pulse and di/dt ratings will hopefully be obtainable from manufacturers' data in the future, just as the present inverter-rated SCRs discussed in Section 2.5.2.

Examples

In Fig. 2.14, the various parameters necessary to calculate power dissipation are given, except let $I_{DSS} = 1$ mA, $R_{DS(on)} = 0.2$ Ω, and $t_{on} = t_{off} = \frac{1}{2}T$.

Case 1. $T = 20$ μs ($f_s = 50$ kHz); $P_{on} = 8^2 \times 0.2 \times 10/20 = 6.4$ W; $P_{off} = 200 \times 0.001 \times 10/20 = 0.1$ W; $P_{sw\text{-}on} = 200 \times 8 \times 0.07/(2 \times 20) = 2.8$ W; $P_{sw\text{-}off} = 200 \times 8 \times 0.2/(2 \times 20) = 8$ W; $P_T = 6.4 + 0.1 + 2.8 + 8 = 17.3$ W.

Case 2. $T = 5$ μs ($f_s = 200$ kHz). Also, let $C_{iss} = 3000$ pF and $C_{rss} = 200$ pF; $P_{on} = 6.4$ W, $P_{off} = 0.1$ W; $P_{sw\text{-}on} = 200 \times 8 \times 0.07/(2 \times 5) = 11.2$ W; $P_{sw\text{-}off} = 200 \times 8 \times 0.2/(2 \times 5) = 32$ W. For the transition loss, $P_{t1} = 10^{-6}(3000 \times 8^2 + 200 \times 192^2)/(2 \times 5) = 0.8$ W. Then $P_T = 6.4 + 0.1 + 11.2 + 32 + 0.8 = 50.5$ W.

The total power dissipation for Case 2 is almost three times the total power dissipation of Case 1. For a TO-3 case MOSFET with a $\theta_{j\text{-}c} = 1$°C/W and with the junction temperature derated to 150°C $\times$ 0.8 = 120°C, the allowable case temperature is $T_c = T_j - (P_T\theta_{j\text{-}c}) = 120$°C $- (50.5 \times 1) = 69.5$°C. (Reference Chapter 15, Fig. 15.1, for thermal impedance.) Of course, these calculations assumed $R_{DS(on)} = 0.2$ Ω at $T_c = 70$°C. Otherwise, $R_{DS(on)}$, P_{on}, P_T, and T_c must be recalculated and the allowable T_c will be lower.

2.3.6 Gate Drive Circuits

Gating the MOSFET appears a simple task at first glance, due to the high input impedance. Actually, gating is similar to driving a capacitive reactance network, as discussed in Section 2.3.2. A low-impedance drive circuit is required to charge and discharge the device capacitances. Figure 2.15 shows two methods of direct-coupled drive and the pertinent waveforms. In Fig. 2.15A, a totem-pole circuit is used. When Q_1 turns on, current flows through Q_1, R_1 and C_{GS}. When $V_{GS(th)}$ is reached, Q_3 turns on. When Q_2 turns on, Q_3 turns off. Resistor R_1 is typically 5–50 Ω and R_2 may be several kilohms. Resistor R_1 could be a much higher value with a small capacitor in parallel. The capacitor would charge to approximately E_a when Q_1 is on. When Q_2 turns on, V_{GS} is driven negative, to speed turn-off. In Fig. 2.15B, Q_1 could be the output transistor of a PWM (pulse width modulator) integrated circuit (IC). When Q_1 turns on, current flows through Q_1, CR$_1$, and C_{GS}. To ensure that Q_2 remains off, Q_2, which was not previously conducting, is reverse biased by CR$_1$. When Q_1 turns off, V_{GS} turns on Q_2. Capacitor C_{GS} discharges rapidly through Q_2, and Q_3 turns off. In each of these circuits, the *NPN* transistor is not conducting current when the *PNP* turns on, and vice versa. Thus the turn-off delay due to storage time of the bipolar driver transistors is eliminated.

Transformer-coupled gate drive circuits which provide isolation between the control circuit and the power circuit are shown in Fig. 2.16. Figure 2.16A is a forward converter drive, and Fig. 2.16B is a flyback converter drive. Both of these circuits have transformer volt-second and duty cycle limitations common to these topologies (reference Chapter 9). In Fig. 2.16C, the totem-pole circuit is capacitively coupled to T_1 by C_1. Again, the volt-second capacity of the transformer limits the duty cycle to 50% maximum, and gate voltage is a function of duty cycle. The circuit is also prone to oscillations. Transformer T_1 primary and C_1 form a series resonant circuit while T_1 secondary and C_{GS} form a parallel resonant circuit. Since C_{GS} dynamically changes with changes in V_{DS}, switching instability may result unless the circuit is critically damped with R_2 and R_3.

The circuit of Fig. 2.16D overcomes the above problems. Here $Q_1–Q_4$ may be a dual totem-pole driver IC which applies voltage to the primary of T_1. Two secondaries are shown which make this circuit ideal for driving push–pull and bridge–converter stages. From the waveforms of Fig. 2.16E, Q_5 is on when Q_1 and Q_4 are on (t_1); Q_6 is on when Q_3 and Q_2 are on (t_3). During t_2 and t_4, Q_2 and Q_4 are on, which shorts the primary of T_1 and provides a high degree of noise immunity to maintain Q_5 and Q_6 in the off state.

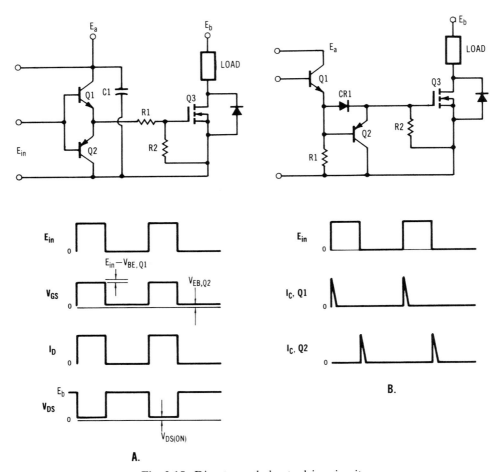

Fig. 2.15. Direct-coupled gate drive circuits.

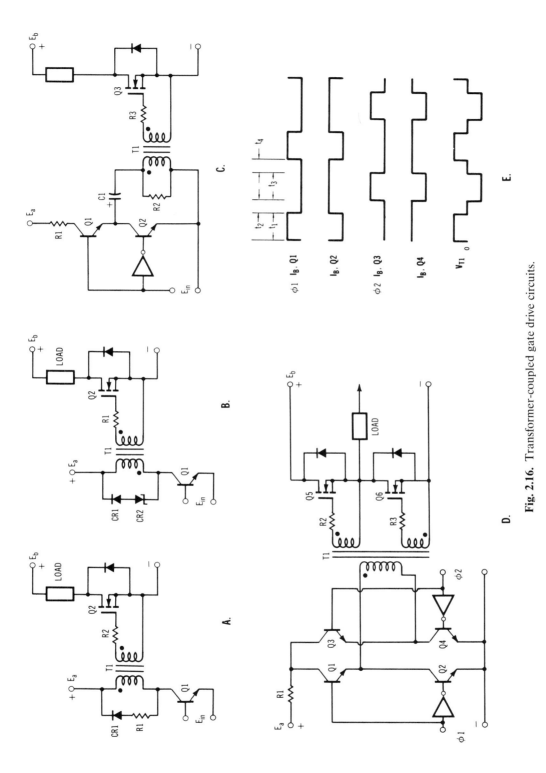

Fig. 2.16. Transformer-coupled gate drive circuits.

83

2.3.7 Reverse Rectifier Characteristics

The parasitic bipolar transistor of Fig. 2.13A is shown in the following figures as a reverse rectifier. This "body diode" has certain limitations, namely reverse recovery and dv/dt. In the single-ended circuits of Figs. 2.14, 2.15, and 2.16, the rectifier does not conduct during MOSFET switching. These figures would represent a drain load such as a flyback or forward converter stage or a source load such as a switching regulator. In push–pull or bridge stages, the rectifier frequently conducts reactive currents, especially where the load is a motor. Referring to Fig. 2.16D, assume a reactive current flows through the rectifier of Q_5. If this current is still flowing when Q_6 is turned on, the current then flows through Q_6. However, a reverse current flows through the rectifier of Q_5 and through Q_6 as the rectifier recovers. The high dv/dt switching of Q_6, the slow recovery time of the rectifier, and the high reverse current through the rectifier effectively shorts the dc bus, resulting in excessive energy in Q_6 which may cause device failure from reverse second breakdown. This phenomenon is frequently a degradation process and failure may occur after several hours or days of operation.

Methods to prevent reverse current through the internal rectifier require additional components. A Schottky rectifier, with inherent fast recovery, can be connected in series with each MOSFET and an ultrafast recovery rectifier connected in parallel with these devices. The Schottky rectifier prevents reverse current through the MOSFET's rectifier. The parallel rectifier must have the same blocking voltage rating as the MOSFET, but the inherent low forward-voltage drop allows use of a low blocking-voltage rating for the Schottky rectifier. Another approach would be connecting an external rectifier, with a very fast recovery and a lower forward-voltage drop than the internal rectifier, across the MOSFET. However, the required current rating of the rectifier would probably be higher than the rating of the MOSFET rectifier to achieve a lower forward drop. An alternate means to minimize overlap conduction is to install a center-tapped inductor between Q_5 and Q_6, with the center tap connected to the load. When Q_6 turns on, the voltage across the lower half of the inductor induces a voltage in the upper half which opposes current flow through the rectifier of Q_5. Also, a clamped inductor in series with the MOSFETs would prevent instantaneous high currents through the transistors.

Methods must also be employed to limit the dv/dt applied from drain to source when the transistor turns off. A typical dv/dt limit, assuming the reverse rectifier has been conducting, is 1 V/ns. With no current flow in the reverse rectifier, a typical dv/dt limit is 20 V/ns. However, specific manufacturers' data and information should be consulted for this parameter. Future semiconductor technology should overcome the present limitations and effects of the reverse rectifier.

2.3.8 Parallel MOSFETs

The inherent positive temperature coefficient of the on resistance is advantageous in paralleling MOSFETs. Current-sharing ballast resistors are not normally required. If two transistors are in parallel, the device with the lowest $R_{DS(on)}$ and with the highest g_{fs} will conduct the highest current. Since $P = I^2 R$, the device conducting the highest current will dissipate the most power, raising the junction

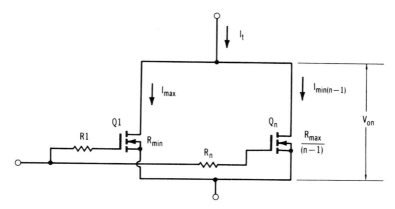

Fig. 2.17. Characteristics of parallel MOSFETs.

temperature, raising the on resistance, and forcing a better current balance. The MOSFET is not gain dependent as is the bipolar transistor, which has base current requirements affecting paralleling. Thus the analysis of parallel MOSFETs is straightforward. These observations apply to the static or conducting mode in switching circuits. For dynamic conditions, differences in switching time and differences in external inductances may produce gross imbalances in transient currents and transient voltages which affect device operation. The dynamic states are complex and are a function of many variables. As with high-current bipolars, multichip packages with current capabilities over 100 A may preclude the need for paralleling discrete devices, except for economic reasons.

Referring to Fig. 2.17 for the static mode, n number of transistors are connected in parallel and all transistors are driven on. The total current is I_T. It is assumed, for worst-case conditions, that one transistor has a minimum $R_{DS(on)}$, while the remaining transistors have a maximum $R_{DS(on)}$. The following relations apply:

$$V_{on} = I_{max} R_{min} \qquad (2.21a)$$

$$= I_{min}(n - 1) R_{max}/(n - 1) = I_{min} R_{max} \qquad (2.21b)$$

$$I_T = I_{max} + I_{min}(n - 1) \qquad (2.22a)$$

Then

$$I_{max} = I_T R_{max}/[R_{min}(n - 1) + R_{max}] \qquad (2.22b)$$

$$I_{min} = I_{max} R_{min}/R_{max} \qquad (2.22c)$$

Examples

Let $I_T = 60$ A. For approximately 15 A per device, choose $n = 4$ transistors. For Q_1, let $R_{min} = 0.1$ Ω; let $R_{max} = 0.15$ Ω. Then $I_{max} = 60 \times 0.15/[0.1(4 - 1) + 0.15] = 20$ A, and $I_{min} = 20 \times 0.1/0.15 = 13.33$ A. Voltage $V_{on} = 20 \times 0.1 = 2$ V $= 13.33 \times 0.15$. The power dissipation in Q_1 is $P_{Q1} = 20^2 \times 0.1 = 40$ W. The power dissipation in the remaining transistors is $P_{Q2,3,n} = 13.33^2 \times 0.15 = 26.67$

W each, and $P_T = 2 \times 60 = 120$ W. If five transistors were used, $I_{max} = 16.36$ A, $I_{min} = 10.91$ A, and $V_{on} = 1.636$ V. Then $P_{Q1} = 26.67$ W, and $P_{Q2,3,n} = 17.85$ W, for $P_T = 26.67 + (17.85 \times 4) = 98$ W. Thus, the total power dissipation is $120 - 98 = 22$ W less than with four transistors.

2.4   INSULATED-GATE BIPOLAR TRANSISTORS

The IGBT combines the advantages of MOSFET drive simplicity with current-handling capability of bipolar devices. The IGBT is constructed in a manner similar to the MOSFET except the device is fabricated by growing an N-type epitaxial layer on a heavily doped P-type wafer and then forming a gate using typical MOS techniques. The extra $P–N$ junction that is formed injects carriers (holes) into the N-epitaxial region. This process is called conductivity modulation and can significantly reduce the saturation voltage of the device. This in turn allows the device to conduct a much higher current than a comparable MOSFET for the same power dissipation. This characteristic comparison even improves for higher voltage ratings. Die size of the IGBT scales linearly with increasing voltage while die size of the MOSFET scales as the 2.5 power of voltage ratio. As another comparison, the saturation voltage ratio of the IGBT versus the MOSFET (or a bipolar) becomes smaller as device voltage ratings increase, assuming the same current density in each device. Although this minority carrier injection enhances forward conduction, it also slows the turn-off time, as discussed in Section 2.4.3. However, IGBTs have a much lower temperature coefficient of on-state resistance (lower saturation voltage) than MOSFETs.

2.4.1   Device Parameters

The industry standard symbol and equivalent circuit of the IGBT is shown in Figs. 2.18A and B, respectively. The symbol shown in Fig. 2.18E is frequently used for expendiency. The PNP bipolar and the N-channel MOSFET form a Darlington connection. A key parameter of the IGBT is that the high-resistivity PNP region, which is required for high-voltage blocking, exhibits low resistivity during forward conduction. The IGBT is turned on by applying a positive voltage to the gate, which in turn supplies base current to the PNP. However, the PNP is not meant to have gain, otherwise the regenerative action would cause the device to latch on when the collector current exceeded a certain level, as is true with thyristors. Similarities between Figs. 2.18B and 2.20 are evident. Elimination of any potential latch-up is paramount to successful operation and is a function of the manufacturing process. The IGBT includes a parasitic NPN, but unlike the MOSFET, this NPN is connected from the base of the PNP to the emitter terminal. Thus the IGBT does not conduct in the reverse direction and has a small reverse blocking capability of 5–10 V. Devices using single die packages require an external rectifier to conduct reverse currents normally encountered in power converters. High-power modules normally have a separate rectifier die built into the package to conduct these currents, and the rectifier switching characteristics are matched to the switching performance of the IGBT die.

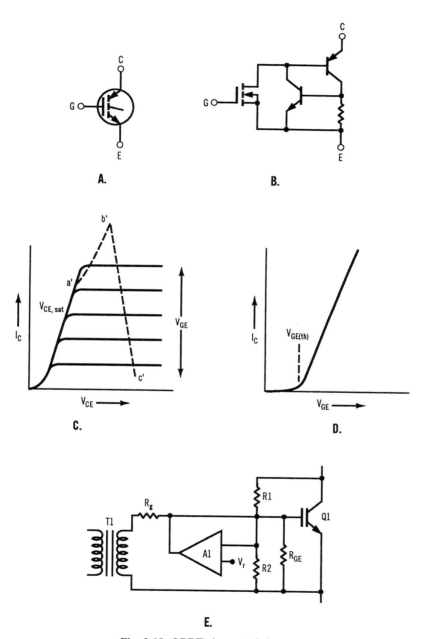

Fig. 2.18. IGBT characteristics.

The output characteristics are shown in Fig. 2.18C and show two distinct regions of operation similar to the bipolar transistor. In the saturation region, $V_{CE,\,\text{sat}}$ is a function of collector current, gate drive voltage, and temperature. As the device is voltage controlled, transconductance is more useful than the term *current gain*. Unlike the MOSFET, there is no initial current flow until $V_{CE,\,\text{sat}}$ exceeds the V_{BE} of the *PNP*. Typical gate control is shown in Fig. 2.18D with $V_{GE(\text{th})}$.

At low values of collector current, the temperature coefficient of saturation voltage is slightly negative, similar to a bipolar device. At rated collector current, this coefficient is typically $+0.1\%/°C$, or one-sixth that of the MOSFET. Another temperature-dependent parameter is the gate threshold voltage, typically $-0.2\%/°C$. Also, the collector current fall time is tempreature dependent, typically $+0.2\%/°C$.

2.4.2 Safe Operating Area

The FBSOA in the linear operating region is thermally limited. Like the MOSFET, the IGBT does not exhibit forward-biased second breakdown as is common to bipolar devices. Reference Fig. 2.4, curve 3, for a typical MOSFET; this in turn applies to a comparable IGBT. However, the die size is much smaller than that required for an equivalent MOSFET. By employing very large scale integration (VLSI) techniques for manufacturing large wafers, it is possible to obtain economic devices with dice of 1 cm^2. This size die could conduct 100 A. At a 1000-V rating, enormous power could be switched from a source to a load. Of course, the power dissipated as heat must be conducted from the die for thermal protection.

The IGBT is subject to a different RBSOA during turn-off as compared to the MOSFET. This is especially true under inductive load conditions. As shown in Fig. 2.18B, the device has a parasitic bipolar *NPN* structure. Under very high reapplied dV/dt conditions, typically greater than 2 V/ns, a displacement current caused by a high rate of current flow in the Miller effect capacitance of the *NPN* can force the *NPN* to conduct, resulting in loss of control and potential device failure. As is true with most active devices, snubbers may be added to reduce dV/dt. An alternate method of limiting dV/dt is to increase the series gate drive resistor R_g or increase any resistor between gate and emitter, R_{GE}. (Reference Fig. 2.18E). This tends to increase the turn-off time, which inherently reduces dV/dt. However, high values of R_g will significantly increase the turn-on losses. Also, low values of R_{GE} reduce the peak collector current capability of RBSOA at turn-off and may cause device latch-up. The dV/dt limit on the IGBT also applies to any reverse rectifier placed across the device to conduct reactive currents. Of course, limitations on the reverse recovery and stored charge of the rectifier apply when used with any active switching devices.

The short-circuit safe operating area (SCSOA) applies to intermittent operation at high stress levels. Normally the IGBT can be protected from an intermittent short-circuit condition by controlling the gate voltage, as discussed in Section 2.4.4.

2.4.3 Switching Waveforms

Typical switching waveforms are shown in Fig. 2.19. Parameter scales are arbitrary. When a gate drive circuit applies an input signal, C_{ies} charges as V_{GE} increases.

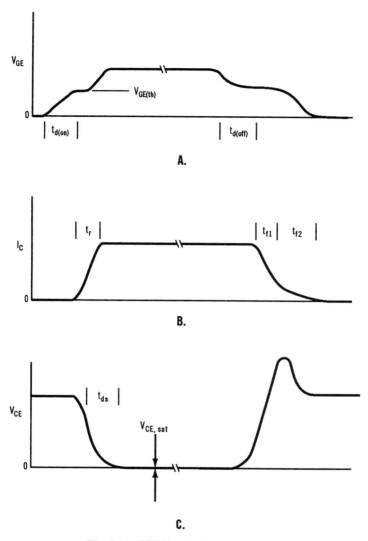

Fig. 2.19. IGBT switching waveforms.

After the turn-on delay $t_{d(\text{on})}$, the threshold voltage $V_{GE(\text{th})}$ is reached and I_C begins to flow and V_{CE} decreases. At this point, the collector current is governed by the product of gate voltage and transconductance. Both parameters decrease with increasing temperature, and thus the rate of rise in collector current, dI_C/dt, decreases with temperature. The collector current increases during t_r as the gate voltage increases to the steady-state input level. Like the bipolar device, the IGBT exhibits dynamic saturation t_{ds} due to the time required for the concentration of minority carriers to build. After this time, the device reaches $V_{CE(\text{sat})}$.

Three distinct intervals comprise the total turn-off of the IGBT. First is the turn-off delay $t_{d(\text{off})}$, which is dominated by the time required for the gate drive circuit to discharge the input capacitance to $V_{GE(\text{th})}$. The delay time is thus a function of the gate drive impedance. At this point, the collector voltage begins to rise. Second is the initial collector current fall time t_{f1}, which is the time required for the gate drive circuit to remove the charge injected into the gate by the gate–collector capacitance as V_{CE} increases during turn-off. During this interval, V_{GE} tends to remain at a plateau and di_C/dt is highest. Third, the collector current decays exponentially during t_{f2} and is controlled by minority carrier recombination in the bipolar *PNP*. The base region of this structure is not externally available to pull minority carriers out through reverse bias techniques as it is common in many bipolar circuits. During this time, V_{CE} has reached, or exceeded, the bus voltage.

The power dissipation caused by this current "tail" during t_{f2} is the major limitation on the highest switching frequency that IGBTs can be used in PWM topologies, where the collector current must fall from a peak value. The current tail can be reduced by optimizing the doping process in the collector layers and by cell geometry. Fortunately, these paramters also enhance latch-up protection.

The IGBT is considered an optimum switching device in quasi-resonant zero-current converters. In this topology (reference Section 10.3), the inherent half-sinusoid current reaches zero as the device turns off and the power dissipation is minimal. Turn-on and turn-off losses can be negligible. The low switching losses combined with the inherent low conduction losses allow IGBTs with high-voltage ratings to switch at frequencies well above the capabilities of bipolar transistors and comparable to swtiching frequencies of MOSFETs. The major switching frequency limitation then becomes the recovery time of an antiparallel rectifier placed across the device to conduct the flyback or ring-over current.

2.4.4   Gate Drive Circuits

Since the gate structure of the IGBT is similar to the MOSFET, the gate drive circuits discussed in Section 2.3.6 also apply to the IGBT. Normally, the gate drive resistance can be larger with IGBT circuits than with MOSFET circuits, due to the lower value of input gate capacitance of the IGBT. For short-circuit protection (SCSOA), active circuitry can be used to detect desaturation. This circuitry does not shut the IGBT off but reduces the gate voltage, which in turn reduces collector current. A simplified operation is illustrated in Figs. 2.18C and E. While the transistor is driven by T_1 and operating at point a', a load or ground fault occurs, causing the collector voltage to rise to point b'. This produces a very high instantaneous power dissipation in the device. A portion of this voltage is sensed by divider R_1 and R_2 and applied to A_1. The other input to A_1 is reference voltage V_r. When the divider voltage exceeds the reference voltage, A_1 sinks current from T_1 via R_g, which reduces the gate voltage. This in turn reduces the collector current and the device operates at point c', where the power dissipation is significantly reduced. Various manufacturers offer specific drivers (employing amplifiers and logic gates) for this desaturation and protection purpose.

2.4.5   Parallel IGBTs

The paralleling of devices requires the same careful attention as that of paralleling bipolar devices, especially in dynamic switching modes. The slightly positive temperature coefficient of $V_{CE(sat)}$ ($+0.1\%/°C$) at rated current is advantageous in minimizing current variations between devices in static conduction. Addition of emitter current-sharing resistors, as shown in Fig. 2.12 for bipolar devices, usually provide minimal sharing improvement. This is because the voltage drop across such resistors is a small percentage of V_{GE}, as compared to the percentage of V_{BE} in bipolar devices. Current sharing during turn-off is critical, due to the longer turn-off delay and current tail inherrent in rectnagular current pulses. Since external means to control this time are nonexistent (due to inaccessibility of the base of the *PNP*) and this time is a function of the manufacturing process, the manufacturers' recommendation should be followed.

2.5 THYRISTORS

Thyristors are four-layer (*PNPN*) semiconductor devices which can be switched from an off state to an on state and will continue to conduct as long as the external load current is greater than the holding current of the device. The devices include the SCR, triac, diac, silicon-controlled switch, reverse-blocking diode, thyristor, and programmable unijunction transistor. Discussion in this section is limited to the SCR as a gate-controlled power device. The triac is essentially a gate-controlled bidirectional ac switch and finds many applications in light dimmers, solid-state relays, and low-power phase-controlled power supplies. One limitation of the triac is the low commutating dv/dt parameter at device turn-off, which limits operating frequency to a few hundred hertz. High-power circuits normally employ two SCRs connected back to back to achieve the same function as the triac, since the SCR normally has a much higher commutating dv/dt rating than the triac.

The equivalent circuit of an SCR is the two-transistor analogy shown in Figure 2.20. With the gate open, or shorted to the cathode, the device is off and no current flows from anode to cathode, except for negligible leakage current. When an external positive pulse is applied to the gate, which is the base of the *NPN* transistor, this transistor turns on and the resulting collector current is the base current from the *PNP* transistor. Then the collector current of the *PNP* transistor supplies base current to the *NPN* transistor. This regenerative action then maintains the device in the conducting state and the gate signal may be removed. The device continues to conduct until the anode voltage is less positive than the cathode voltage.

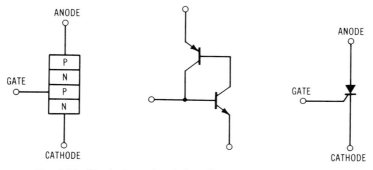

Fig. 2.20. Equivalent circuit for silicon-controlled rectifier.

2.5.1 Gate Triggering

Even though the SCR may be forced into conduction by a breakover voltage applied from anode to cathode, this normally uncontrolled feature is seldom used in power conversion equipment. Instead, the SCR is triggered on by a positive pulse applied to the gate. Figure 2.21 shows the gate characteristics of a General Electric C385 series SCR (further discussed in Section 2.5.3). The required V_{GT} and I_{GT} for turn-on are somewhat a function of anode voltage and load impedance. Also, higher gate voltage and current are required at low temperatures. At high temperature, the gate voltage required for triggering may be as low as 0.3 V. This makes the SCR susceptible to noise triggering, and a low-impedance gate-firing circuit should be used. With inductive loads, the gate current must be maintained for a time sufficient to allow the anode current to rise well above the holding current. With capacitive loads, the rise time of the gate current must be very rapid, while ensuring the di/dt rating of anode current is not exceeded.

A typical gate-triggering circuit and waveforms are shown in Fig. 2.22. The circuit is essentially a capacitor discharge technique and is suitable for triggering in phase control circuits and in inverter commutation circuits. Referring to Fig. 2.22A and the waveforms of Fig. 2.22B, initially C_1 is charged to B^+. When Q_1 is biased on, E_{C1} is applied across the pulse transformer T_1 and current flows through R_2 and the gate of Q_2, thus turning the SCR on. Then Q_1 is held in conduction for period t_p (typically 20 μs) until C_1 discharges. When Q_1 turns off, the time constant allows C_1 to again charge to B^+ through R_1 during t_{off}. One advantage of this circuit is the low peak current draw from B^+ since the energy stored in C_1 supplies power to the gate. The fast turn-on of Q_1 also provides a fast rise in gate current to Q_2. The circuit in Fig. 2.22C is similar except false triggering may result, as shown by the waveforms of Fig. 2.22D. The short on time of Q_1 does not allow

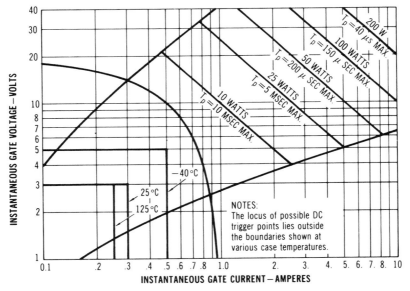

Fig. 2.21. SCR gate characteristics, type C385. (Courtesy General Electric Co.)

C_1 to discharge. When Q_1 turns off, CR_1 provides a current path for the energy trapped in T_1. The abrupt current interruption in CR_1 may cause a voltage reversal across T_1, which then induces a positive voltage in the secondary and false triggers SCR Q_2, as shown at time t_x. Regardless of the triggering method used, ringing in the primary or secondary side of T_1 must be avoided to eliminate the possibility of false triggering. The interaction of a capacitor connected across the gate-cathode terminals (for noise suppression) and the inductance of the pulse transformer must not cause a ringing effect.

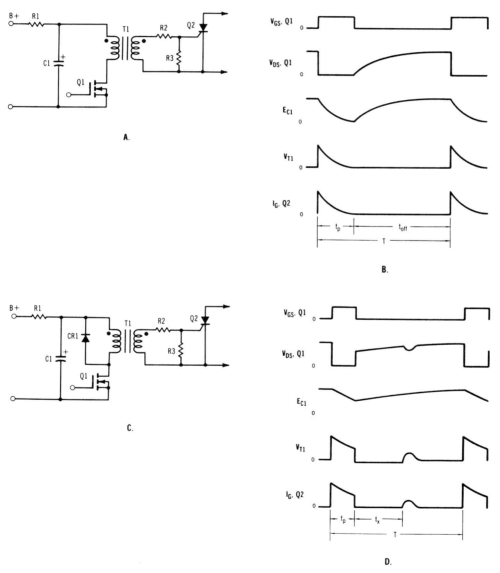

Fig. 2.22. SCR gate drive circuits.

Referring again to Fig. 2.22A, the primary and secondary of T_1 have a voltage E_p and E_s and a dc resistance of R_p and R_s, respectively. The following relations apply:

$$E_p = E_{C1} - I_p R_p - V_{DS,\,\text{sat}} = nE_s = n[I_s(R_s + R_2) + V_{GC}]$$

$$I_g = I_s - (V_{GC}/R_3) = [(E_s - V_{GC})/(R_s + R_2) - (V_{GC}/R_3)]$$

Solving for the gate current,

$$I_g = \frac{nR_3(E_{C1} - V_{DS,\,\text{sat}}) - n^2 V_{GC}(R_s + R_2 + R_3) - V_{GC} R_p}{R_3[n^2(R_s + R_2) + R_p]} \qquad (2.23)$$

Example

It is required to trigger an SCR having the gate characteristics of Fig. 2.21 at $-40°C$. The desired peak gate current is 1 A. To simplify Equation (2.23), assume the following conditions for the circuit of Fig. 2.22A: $B^+ = 20$ V, $R_p = R_s = 5\Omega$, $n = 1$, $R_2 = 0$, $R_3 = 39$ Ω, $V_{GC} = 6.6$ V, $V_{DS(\text{on})} = 1$ V. From Equation (2.23), $I_g = (741 - 290.4 - 33)/390 = 1.07$ A. Also, $I_p = I_s = 1.24$ A. Here Q_1 is selected for a drain current of $1.24(19 - 3.3)/(19 - 6.6) = 1.6$ A when V_{GC} decreases at high temperature. For $t_p = 20$ μs, let the discharge time constant of C_1 be $\tau = \frac{1}{3}t_p = 7$ μs. The resistance reflected in the primary is 10 Ω and then $C_1 = \tau/R_1 = 7 \times 10^{-6}/10 = 0.7$ μF; use a standard value of 0.68 μF. For a switching frequency of 2500 Hz, $T = 400$ μs and the recharge time of C_1 is $400 - 20 = 380$ μs, or 3 time constants. Then $\tau' = 380/3 = 127$ μs, $R_1 = \tau'/C_1 = 127/.68 = 186$ Ω; use a standard value of 180 Ω. Then the peak input current to charge C_1 is $20/180 = 0.11$ A, or approximately 10% of the actual gate current.

When triggering power stage SCRs conducting load currents in inverter circuits, it may be important to maintain a gate signal on the SCR during the desired conduction time. If the load is reactive, the anode current may reverse and flow through a reactive rectifier connected across the SCR during a portion of the cycle, and the SCR will turn off. When the voltage again reverses and the anode is positive later in the same half cycle, the SCR will not conduct if gate voltage is not present. In this case, a transformer capable of maintaining the gate signal for the full half cycle is required, just as in the drive circuitry for power transistors. However, the required gate current is much lower than the base current of bipolar power transistors used in a power stage providing the same total output power.

2.5.2   Phase Control

The term *phase control* normally applies to ac line commutation, or class F operation (reference Chapter 5), where the application of the gating signal varies within the positive half cycle of the ac source. Once the thyristor is turned on and is conducting current, turn-off is achieved by the ac input reversal during the negative half cycle, which reverses the voltage across the SCR, thereby preventing further conduction. The turn-off time rating for phase control devices is not critical (for normal utility or military ac input frequencies) since reverse voltage is applied for a full half cycle of the input frequency. By varying the point in the ac cycle where the SCR is gated on, the phase angle or firing angle may be controlled via a closed loop to regulate the output voltage independent of input voltage variations. The output voltage may also be adjusted by varying the firing angle. As the firing angle increases, the conduction period of the SCR decreases and the average current rating of the device decreases, with respet to the rms value of the current. The disadvantages of phase control are (1) the input current from the ac source is chopped and therefore contains harmonics and (2) input filters to reduce noise fed back to the source and to meet EMI requirements become large in size. These disadvantages are overcome somewhat by multiphase approaches discussed in Chapter 5.

Figure 2.23 shows the relation of maximum case temperature and maximum power dissipation as a function of device average current and conduction angle for the 2N1910 (TO-94 case) and the 2N1792 (TO-83 case) SCR families. These devices are rated at 70 A average and 110 A rms (the ratio of average to rms for single-phase circuits is $2/\pi$). For a fixed average current, the allowable case temperature decreases as the conduction angle decreases. This is due to the decreasing ratio factor of average to rms as the conduction angle decreases. For single-phase circuits with a resistive load and neglecting losses, the output voltage, from Chapter 5, Equation (5.4), is

$$E_{dc} = 0.45 E_{in}(1 + \cos \alpha) \tag{2.24a}$$

$$= 0.45 E_{in}[1 + \cos(180° - \theta)] \tag{2.24b}$$

Rearranging terms,

$$\theta = 180° - \cos^{-1}(2.22 E_0/E_{in} - 1) \tag{2.25a}$$

It is observed that α is a minimum (0°) and θ is a maximum (180°) when E_0 is a maximum and E_{in} is a minimum. Then for design purposes, the minimum input voltage is calculated to provide the maximum output voltage when $\theta = 180°$. Then the ratios of minimum output and the maximum input can be substituted to find the minimum θ. At $\theta = 180°$, $E_{0,max} = 0.9E_{in,min}$; then

$$\theta_{min} = 180° - \cos^{-1}\left[\frac{2 E_{0,min} E_{in,min}}{E_{0,max} E_{in,max}} - 1\right] \tag{2.25b}$$

Knowing the desired output current, where the average current (for single phase) of the SCR is one-half the output current, the maximum case temperature and power dissipation at that current can be obtained from manufacturers' data similar to that shown in Fig. 2.23.

Example

It is desired to provide a constant output current of 60 A to a load using two SCRs to switch the secondary voltage of a transformer in a single-phase, center-tapped circuit. The output voltage is to be adjustable over a 2-to-1 range. The input

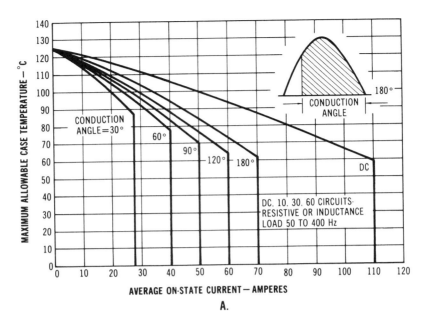

A.

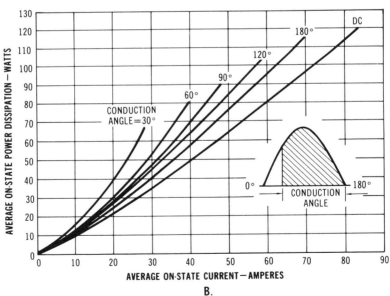

B.

Fig. 2.23. Maximum allowable case tempertaure and average power dissipation versus average current for 2N1910 and 2N1792 SCR series. (Courtesy General Electric Co.)

voltage varies by 1.25-to-1. Neglecting losses, let $\alpha = 0°$ and $\theta = 180°$ at minimum input voltage and maximum output voltage. For this condition, the average current per device is 30 A and from Fig. 2.21, the maximum case temperature is 103°C and the power dissipation is 40 W. Minimum θ is calculated from Equation (2.25b) by using the voltage ratios of $E_{0,\max} = 2E_{in,\min}$, $E_{in,\max} = 1.25E_{in,\min}$, and $E_{0,\max} = 0.9E_{in,\min}$ for $\theta = 180°$. Then, $\theta_{\min} = 180° - \cos^{-1}[2.22 \times 0.9/(2 \times 1.25) - 1] = 78°$.

From Fig. 2.23, the maximum case temperature is now 94°C and the power dissipation is 50 W. Even though the output current is the same, the output voltage is decreased and the input voltage is increased; the SCRs dissipate more power and the case temperature must be lowered as compared to $\theta = 180°$.

For simplicity, the calculation of $\theta_{\min}$ assumed no losses. The actual input voltage must increase to then overcome these losses. Also, the temperatures calculated are maximum allowable. To reduce the junction temperature to 110°C, the case temperature should not exceed $94 - (125 - 110) = 79°$C. To complete the design for required SCR heat sinking in a given ambient temperature, reference Section 15.3.3 and Fig. 15.10.

2.5.3 Inverter-Rated SCR

The term *inverter rated* applies to fast turn-off devices, normally employed in coverters operating at frequencies from 60 Hz to over 10 kHz. The source is normally direct current and the SCR switches the dc line across some load. Turn-off is achieved by some form of commutation, discussed in Chapter 7. As the turn-off time rating of a device decreases, the size and value of the required commutating components decreases and the operating efficiency normally increases.

Operation and definitions are best discussed by referring to Fig. 2.24. Initially the SCR is conducting current I_{TM} at a forward voltage drop V_{TM}. At some point, the SCR is to be turned off by some form of commutation. Then I_{TM} decreases at a di/dt rate and a reverse current (similar to the reverse current in a rectifier) flows in the SCR; the reversal of voltage to V_{RM} turns the device off. However, sufficient time must be allowed for the junction to recover, and if forward voltage is applied before full recovery, the device may again conduct without the application of a gate signal. Thus the "turn-off time" t_q is the minimum time from the forward current falling to zero to the time when forward voltage is again applied, as shown in Fig. 2.24. The rate of forward voltage rise, dv/dt, also determines the recovery characteristc. The turn-off time specified by thyristor manufacturers includes a maximum reapplied dv/dt rating. The dv/dt may be specified as a linear slope, to V_{DRM} at time "t" or, in the case of exponential voltage rise, as a linear slope to $0.63V_{DRM}$ at time "τ." Frequently, a rectifier is connected across the SCR to conduct reactive currents occurring in a circuit. In this case, V_R is limited to the forward voltage drop of the rectifier, as shown in Fig. 2.24C. When the rectifier ceases conduction and since some form of inductance is normally present, a voltage rise of $e = -L\,di/dt$ results in a very high dv/dt. For this condition, the required t_q will increase and is normally specified separately.

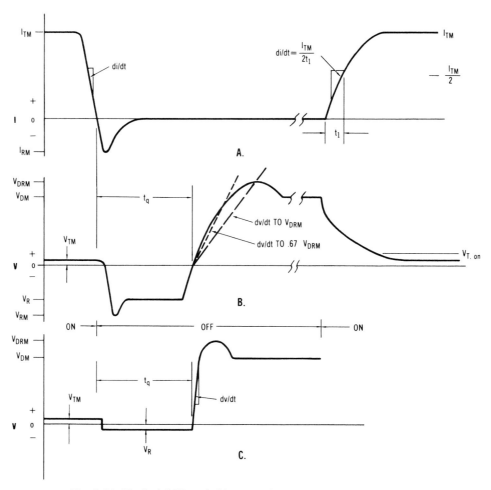

Fig. 2.24. Typical SCR switching waveforms at turn-off and turn-on.

When the device is again triggered on, as shown in Fig. 2.24A and B, the rate of rise of forward current, di/dt, must be limited. Initially, the current is concentrated in a small area of the junction and localized heating may produce hot spots which exceed the junction rating. The device then fails and, in effect, a short circuit results. The di/dt rating may apply to a linear current rise or, in the case of sinusoidal or exponential waveforms, the $di/dt = \frac{1}{2}I_{TM}/t_1$, where t_1 is the time to $\frac{1}{2}I_{TM}$. The rate of rise of the gate current should be high to ensure dynamic current spreading in the junction and should typically be 100 ns. The finite turn-on time t_{on} is the time required for the forward voltage to drop to $V_{T\text{-on}}$ and is somewhat analogous to the dynamic saturation of the bipolar transistor. For applications where the calculated rate of current rise exceeds the device rating, a saturable inductor may be used in series with the anode. When the SCR is gated on, a low current flows for a short period of time while the inductor supports the voltage. Before the inductor saturates, sufficient time (perhaps 1 or 2 μs) has elapsed to allow the current to spread across the junction. When the inductor saturates, the device forward drop is low enough to allow high current levels without instantanous excessive power dissipation. The inductor may be reset by an auxiliary source connected to a separate winding. This technique is effectively employed in high-power lasers, pulse modulators, and PFNs (pulse-forming networks) where the discharge of a capacitor across a coupling transformer is required to develop a fast voltage rise time.

Due to proprietary designs and processes of high-power inverter-rated thyristors, manufacturers' individual part numbers are frequently used instead of a JEDEC designation. Three similar devices, each with approximately 400 A rms current rating are Westinghouse T627-25; General Electric C384/385; and International Rectifier 240PAL/PAM. Forward- and reverse-voltage ratings are in excess of 800 V, and turn-off times from 10 to over 30 μs are available. The devices are packaged in the 0.5 × 1.6-in. "press-pak" or "hockey-puk" which allows double-sided cooling when mounted between two head sink surfaces. The devices' current capability is further divided into sine wave and rectangular (or trapezoidal) wave shapes. The sine wave data is beneficial for commutation circuits where a half-sinusoid current is conducted. The trapezoid wave (so called to show effects of di/dt at turn-on) data then apply to devices employed in the power stage which, in effect, conduct the load (or reflected load) current. For instance, each rating for the above-mentioned Westinghouse device is shown in Fig. 2.25. The case temperature is 90°C, $V_R \geq 50$ V, and a series RC snubber (with values listed in the graph) is connected across the device to limit dv/dt. Figure 2.25A shows that the peak sine wave current at high frequency is even less than the effective rms rating of the device! This is primarily due to the losses incurred during turn-on and turn-off. For the square (actually a rectangle or trapezoid) wave, the peak current is a function of duty cycle, which is shown in Fig. 2.25B. Thus for reliable operation, device data similar to that in Fig. 2.25 must be considered for a particular application and usually override device selection based solely on the rms current rating.

An additional parameter from which the device power dissipation may be obtained, thereby greatly simplifying thermal design, is shown in Fig. 2.26. Peak current is again plotted versus pulse width, but in this figure, energy per pulse values for the above-mentioned International Rectifier device are given. Since the energy is in joules or watt-seconds, power is equal to energy divided by time, or power equals energy times frequency. Since the switching frequency is known, the power dissipation is readily calculated from the maximum energy per pulse allowable. Data for sine wave and trapezoidal currents are shown in Figs. 2.26A and B, respectively.

Examples

Case 1. An inverter stage applies voltage to a power transformer. The switching frequency is 2500 Hz and the maximum duty cycle is 40%. Thus the pulse width $t_p = 0.40/2500 = 160$ μs. From Fig. 2.25B, a peak current of 245 A maximum is allowed if the case does not exceed 90°C. From Fig. 2.26B (assuming, for this example, the devices are the same), at 245 A and 160 μs, the maximum energy per pulse is 0.24 J. The power dissipation in the SCR is then $P = \mathcal{E}f = 0.24 \times 2500 = 600$ W. The appropriate safety margin, either in peak current or in case tempera-

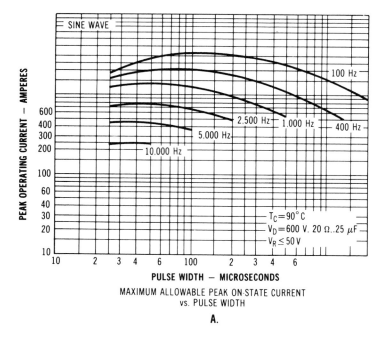

MAXIMUM ALLOWABLE PEAK ON-STATE CURRENT
vs. PULSE WIDTH

A.

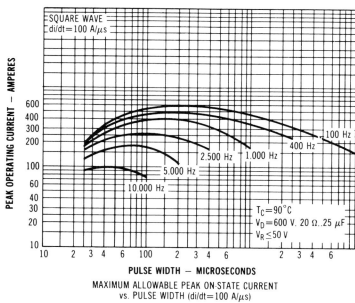

MAXIMUM ALLOWABLE PEAK ON-STATE CURRENT
vs. PULSE WIDTH (di/dt=100 A/μs)

B.

Fig. 2.25. Allowable peak current versus pulse width for T627 series of SCRs. (Courtesy Westinghouse Electric Corp.)

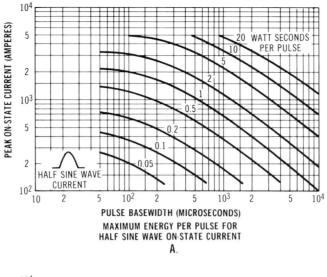

A.

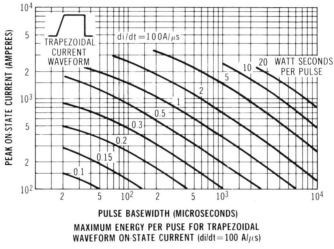

B.

Fig. 2.26. Maximum rated energy per pulse for 240PAL series of SCRs. (Courtesy International Rectifier.)

ture, may now be applied for derating. To show the effects of switching losses, using only the forward voltage drop of 2.3 V at 245 A, the average power dissipation for conduction is $2.3 \times 245 \times 0.40 = 225$ W.

Case 2. If the switching frequency in the above example is lowered to 400 Hz and the duty cycle is 45% for a pulse width of 1.12 ms, the peak current is 320 A (from Fig. 2.25B). The energy per pulse at 320 A and 1.12 ms pulse width is 0.9 J (from Fig. 2.26B). Then $P = \varepsilon f = 0.9 \times 400 = 360$ W dissipation.

For comparison of the two cases, assume the input voltage to be 215 V dc, which is the voltage near end-of-discharge for a 250-V battery. Four SCRs are used in a bridge inverter (reference Section 7.4 and Fig. 7.7). For Case 1, the average input current $I_{\text{in}} = 0.80 \times 245 = 196$ A, the average input power is $215 \times 196 = 42.14$ kW, and the total SCR power dissipation is $600 \times 4 = 2400$ W. Thus the SCR power dissipation (excluding commutation losses) is $2400/42140 = 5.7\%$ of the input power. For Case 2, the average input current is $0.90 \times 320 = 288$ A, the average input power is $215 \times 288 = 61.92$ kW, and the total SCR power dissipation is $360 \times 4 = 1440$ W. Thus, the SCR power dissipation is $1440/61920 = 2.3\%$ of the input power. Conversely, for the same type SCR in each case, lowering the switching frequency from 2500 to 400 Hz results in an increase in input power to the transformer of $(61920 - 1440) - (42140 - 2400) = 20.7$ kW (which essentially provides a 20-kW increase in transformer output power to a load), while at the same time reducing the power loss by $2400 - 1440 = 960$ W. Of course, the power transformer will be larger at 400 Hz as opposed to 2500 Hz, but if space permits, the reduced switching frequency will substantially reduce thyristor losses.

2.5.4 Crowbar SCR

Power supplies and converters frequently require built-in crowbar circuitry for overvoltage protection of critically powered loads from either internal or external causes. This is especially applicable to high-density transistor–transistor logic (TTL) and solid-state computer memory. In linear power supplies (reference Chapter 17) and buck regulators (reference Chapter 8), the load must be protected if the pass transistor or the switching transistor becomes shorted. In this case, an SCR connected across the dc output terminals is triggered when the voltage exceeds a preset value. The SCR then discharges the output filter capacitor and absorbs the circuit energy until an input fuse or circuit breaker opens to remove input power.

The overvoltage protection in switching power supplies (reference Chapter 17) and converters (reference Chapter 9) is somewhat inherent since the load is isolated by a transformer. Overvoltage protection may be accomplished by simply terminating the power transistor switching for a short interval. However, overvoltage protection may still be desired in multioutput units where an accidental short from a higher voltage output to a critical lower voltage output would cause catastrophic integrated circuit failures. In this case, an SCR connected across the low-voltage dc output is then triggered to crowbar that output to a safe level. At the same time, switching is terminated until the discharge current is well below the SCR holding current, and the unit may then automatically restart.

An SCR for crowbar circuits may be selected by I^2t rating although various manufacturers offer SCRs specifically designed for crowbar applications. For these devices, the peak capacitor discharge current rating is normally specified for time t_ω equal to five time constants of an exponentially decaying current pulse. The peak current rating is usually 25 times the I_{rms} rating of the SCR. However, the peak current may be limited only by the ESR (equivalent series resistance) of the capacitor and wiring impedance. A ferrite bead in the anode lead may be used to limit the di/dt when the SCR is triggered. Integrated circuits specifically designed for crowbar triggering, or other circuits, such as comparators, are recommended as opposed to a zener diode connected from the output to the gate. The ICs provide a fast-rise current pulse to the gate. The zener diode may only produce a "trickle" current to the gate which causes device failure due to the previously mentioned current crowding in the junction at turn-on.

2.6   OTHER POWER DEVICES

The triac and the GTO have been available for many years but have limited applications, as discussed below. The SIT and the MCT, discussed below, are modern devices for special applications.

2.6.1   Triac

The triac is essentially a *PNPN* device in parallel with a *NPNP* device. It also can be described as a pair of phase-controlled thyristors connected in inverse parallel. The symbol and characteristics are shown in Fig. 2.27A. The three-terminal device can be triggered on in both the positive and negative half cycles of supply voltage by applying positive and negative gate pulses, respectively. The turn-off time is inherently long due to storage charge effects and the maximum operating frequency is typically 400 Hz. For these reasons, the triac is limited to applications in lamp and heating controls, appliance-type motors, and solid-state relays.

2.6.2   Gate Turn-off Thyristor

The gate turn-off (GTO) thyristor is a regenerative device with substantial improvements in high-temperature turn-off capability compared to early devices. That device was also called a gate-controlled switch (GCS). The symbol and characteristics are shown in Fig. 2.27B. Since the GTO is fully controlled at the gate, commutation circuitry is not required, as compared to the SCR. However, snubbers to limit dv/dt are required and gate-assisted turn-off or partial commutation is sometimes used. The GTO is turned on by a positive pulse at the gate with respect to the cathode, in the manner similar to the SCR. Turn-off is accomplished by a negative pulse at the gate, which extracts a portion of the anode current while a snubber capacitor provides an additional path for diverting anode current. The GTO turn-off time (t_{gq}) is a different parameter than the SCR turn-off time (t_q) and the GTO cannot be retriggered on again until the end of a permissible off time, $t_{OFF(min)}$, following t_{gq}.

2.6.3   Static Induction Transistor

The static induction transistor (SIT) is essentially the solid-state version of a triode vacuum tube and is almost identical to the junction FET (JFET). The symbol and characteristics are shown in Fig. 2.27C. The SIT is normally on, that is, if V_{GS} is zero (or positive), current will flow from drain to source. The device is cut off when V_{GS} is negative. This characteristic can present problems when power is simultaneously applied to a power stage with SITs and to the control circuitry. The SIT has

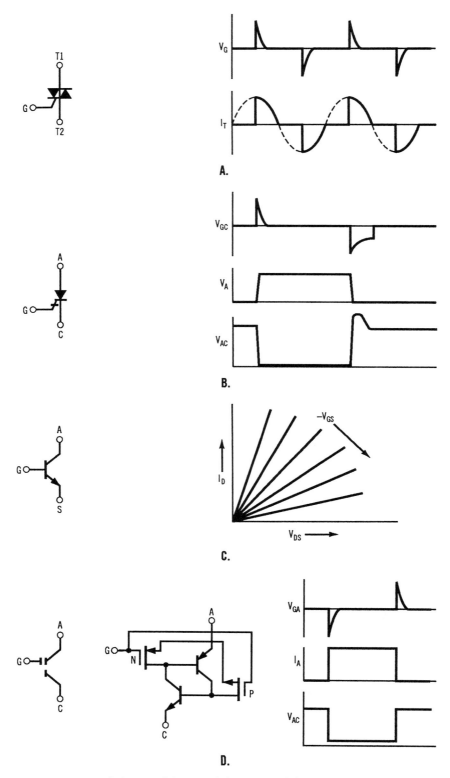

Fig. 2.27. (A) Triac, (B) GTO, (C) SIT, and (D) MCT characteristics.

very high speed capability and a high dv/dt rating and is suitable at RF frequencies for use in power amplifiers and induction heaters. The high-speed capability is achieved by the low device capacitance, as compared to the MOSFET. However, the device lacks the low saturation voltage necessary for high-efficiency conduction in power-switching circuits.

2.6.4   MOS-Controlled Thyristor

The MOS-controlled thyristor (MCT) has combined MOSFET and regenerative thyristor features. The symbol and characteristics are shown in Fig. 2.27D. The MCT has a *PNPN* structure with MOSFETs for gating. The device is turned on by a negative voltage pulse at the gate with respect to the anode. This voltage causes the *P*-FET to forward bias the *NPN* transistor and the MCT turns on by positive feedback, similar to the SCR. The conduction voltage drop is very low. The device is turned off by a positive voltage pulse at the gate with respect to the anode. This voltage causes the *N*-FET to "short circuit" the base–emitter of the *PNP* transistor and the MCT turns off. However, leakage current at very high temperature can be a problem. The positive temperature coefficient of on resistance in the *N*-FET can reduce the ability to maintain a low base–emitter voltage on the *PNP*. The MCT switches very fast due to the absence of Miller-type capacitances in the FETs.

Gate triggering with reference to the anode presents a new feature for gate-pulsed devices. The load circuit can be placed in the cathode and the gate circuitry return connected to the anode, or positive input source.

2.7   RESISTORS

The resistor has changed considerably since the days of Georg Simon Ohm, a German physicist of the early nineteenth centry. In 1881, the unit of electrical resistance was named the ohm and was defined to be the resistance of a column of mercury 106.3 centimeters long and having a mass of 14.45 grams at a temperature of 0°C. The modern definition is simply that resistance which will allow 1 ampere to flow in an item if 1 volt is impressed across its terminals.

The equivalent circuit of a resistor has, in addition to the dc resistance value, a series parasitic inductance and a parallel parasitic capacitance, as shown in Fig. 2.28. The inductance may be due to the unit's leads or to the manufacturing technique. The capacitance results from conducting particle plates and distributed capacitance. When the resistor value is measured by a dc potential, the true or equivalent value of R is found. When the resistor is measured by an ac potential,

Fig. 2.28. Equivalent circuit of a resistor.

the "impedance" of the resistor is

$$Z = j\omega L + \frac{R/(j\omega C)}{R + 1/(j\omega C)} \qquad (2.26a)$$

$$= \frac{[R + (j\omega L)(1 + j\omega RC)]}{1 + j\omega RC}$$

$$= \frac{R}{1 + \omega^2 R^2 C^2} + j\frac{\omega^2 L C^2 R^2 - \omega C R^2 + \omega L}{1 + \omega^2 R^2 C^2} \qquad (2.26b)$$

where the let-hand term is the real or ESR part; the right-hand term is the imaginary part, and frequency becomes a second-order factor.

The most common way to specify a resistor is by dc ohmic value, initial tolerance, power dissipation, and in the case of "precision" resistors, temperature coefficient and life stability. In addition, resistance variations due to changes in voltage, operating frequency (parasitic inductance and capacitance effects and skin effects), and exposure to temperature cycling, moisture, dielectric stress, transient power overload, radiation, shock, and vibration may be important considerations in choosing a resistor. Reliability and failure rate parameters, typically established by military specifications, are also important considerations in choosing a resistor. To reduce production costs, resistors to be installed on printed circuit boards may require a physical package which allows automatic insertion. Materials used in the resistor must withstand not only normal operating environmental conditions but also the application of cleaning solvents and the high temperature surges of wave soldering equipment.

There are two basic types of resistors, fixed and variable. Fixed resistors are discussed herein since the usage of variable resistors in most power conversion equipment is limited to potentiometers which set a desired operating parameter within the control circuitry. There are five basic types of fixed resistors: carbon composition, metal film, wire wound, carbon film, and metal oxide film. For purposes of this section, the first three types will be discussed. The carbon composition types normally use bulk, molded carbon as the resistive material and includes all resistors with an initial tolerance of 5% or greater, a long-term stability higher than 10% and a maximum power dissipation of 2 W. The resistive element in metal-film types (frequently referred to as precision) consists of a resistive film deposited on an insulating substrate, and these resistors have initial tolerances of 0.02–2% and are stable with time and temperature. Wire-wound (power) resistors utilize a resistance wire material which is wound on an insulating bobbin, and these resistors have power ratings greater than 1 W (up to 250 W) and have initial tolerances of 0.1–10%.

In addition to the considerations of the third paragraph of this section, resistors are rated for voltage as well as for power. This rating becomes important in high-voltage power supplies and converters where resistors may be employed in voltage dividers for sensing purposes and in snubber circuits for high-voltage transistors and thyristors. The critical resistance is

$$R_c = (V_r)^2/P_r \qquad (2.27)$$

where V_r is the maximum rated voltage and P_r is the maximum rated power.

Below R_c, the resistor is limited by power dissipation, and above R_c, the resistor is limited by the voltage applied to the terminals.

For practicality, the resistor types will be referred to by military designation wherever possible, since the various manufacturers typically have their own part numbering system for industrial types. For an application guide, MIL-STD-199C, "Resistors, Selection and Use of," is recommended.

2.7.1   Carbon Composition Resistors

The majority of resistor applications require 1 W or less of power dissipation and can function with 5% or greater tolerance. Therefore, the hot molded carbon composition resistor is normally chosen due to low cost, rugged construction, and high-surge-power capability. The major disadvantages are wide temeprature coefficient of resistance, poor stability, and an extreme sensitivity to humidity. For these reasons, as well as inspecting and testing (which are the secrets to high reliability) cost increases plus decreased costs of other types of resistors, most manufacturers have discontinued offering carbon composition resistors. For ease of installation and usage in somewhat controlled environments, the carbon composition resistor remains a viable component.

The characteristics of the mil-designated *RCR* series of resistors are given in Fig. 2.29: band coloring in Fig. 2.29A; color coding in Fig. 2.29B; part number sequence in Fig. 2.29C, style, power, resistance, range, and voltage rating in Fig. 2.29D; and decade resistance values in Fig. 2.29E. For example, a 7500-Ω, $\pm 5\%$, $\frac{1}{4}$W resistor would have color bands of violet, green, red, gold, and yellow. The yellow band signifies a 0.001% per 1000 h failure rate, which is normally the standard. From Fig. 2.29C, the part number is RCR07G752JS. A 1.2-Ω, $\pm 10\%$, $\frac{1}{2}$-W resistor would have color bands of brown, red, gold, silver, and yellow. The part number for this resistor is RCR20G1R2KS where "R" is the decimal.

The power ratings listed are for an ambient temperature up to 70°C at sea level. The power derates to zero at $T_{\text{amb}} = 150°C$. For increased reliability, the actual power dissipation should not exceed 50% of the rating and should be derated further (to 35% of rated power) for high-altitude applications. For instance, a $\frac{1}{2}$-W resistor operates at an ambient temperature of 85°C at 40,000 ft. The power dissipation should be limited to $\frac{1}{2} \times 0.5 \times 0.7 \times 0.8 = 0.14$ W. For the 2-W resistor, the actual power should not exceed one-third the rating if the resistor is mounted on a printed circuit board, and even then the 1-W and 2-W resistors should be mounted $\frac{1}{4}$ in. from the board to prevent excessive temperature rise in the printed circuit board.

Power overload occurrences can cause unsafe conditions in carbon composition resistors. The resistor may catch fire or explode. The pulse-withstanding capability of hot-molded carbon composition resistors for a single-pulse occurrence is shown in Fig. 2.30. This pulse-withstanding capability is primarily a function of energy in the pulse, not just the peak pulse power. For repetitive pulses, sufficient time must be allowed for the resistor to cool, and the minimum period between pulses is

$$T = t_p P_{\text{pk}}/P_r \qquad \text{s} \qquad (2.28)$$

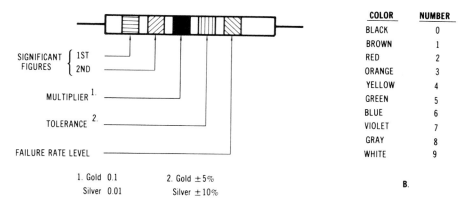

COLOR	NUMBER
BLACK	0
BROWN	1
RED	2
ORANGE	3
YELLOW	4
GREEN	5
BLUE	6
VIOLET	7
GRAY	8
WHITE	9

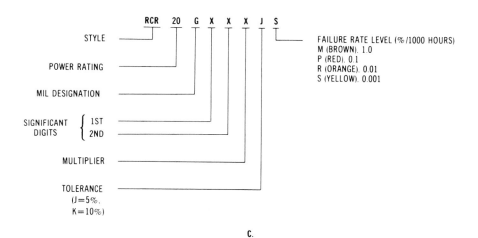

STYLE	POWER RATING @70°C	RESISTANCE RANGE	WORKING VOLTAGE DC OR RMS	Resistor values, each decade from minimum to maximum. Bold type are ±10%. All values available in ±5%				
RCR05	1/8 W	2.7 Ω-22 MΩ	150	**1.0**	1.6	**2.7**	4.3	**6.8**
RCR07	1/4 W	2.7 Ω-22 MΩ	250	1.1	**1.8**	3.0	**4.7**	7.5
RCR20	1/2 W	1 Ω-22 MΩ	350	1.2	2.0	**3.3**	5.1	**8.2**
RCR32	1 W	2.7 Ω-22 MΩ	500	1.3	**2.2**	3.6	**5.6**	9.1
RCR42	2 W	10 Ω-22 MΩ	500	**1.5**	2.4	**3.9**	6.2	**10.0**

D.   E.

Fig. 2.29. Characteristics of carbon composition resistors.

where

$$t_p = \text{pulse duration}$$
$$P_{\text{pk}} = \text{peak power dissipation}$$
$$P_t = \text{rated power of the resistor}$$
$$P_{\text{pk}}t_p = \text{energy absorbed}$$

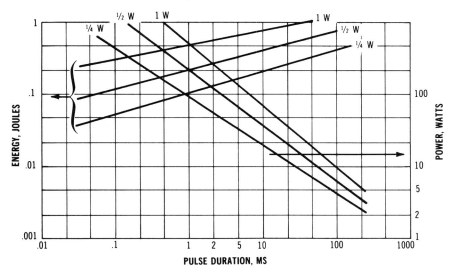

Fig. 2.30. Carbon composition resistor surge ratings.

Example

A 60-V pulse is applied to a 150-Ω, $\frac{1}{2}$-W resistor. The peak power is $P_{pk} = E^2 /$ $R = 60^2 / 150 = 24$ W. From Fig. 2.30, the maximum pulse duration is 18 ms and the energy is $24 \times 0.018 = 0.43$ J. From Equation (2.28), the minimum required period between pulses is $T = 0.43 / \frac{1}{2} = 0.86$ s and $T = 1.7$ s at a 50% derating.

High-voltage resistors with compounds containing carbon on extruded ceramic cylinders are ideal for many general-purpose applications such as current-limiting, capacitor-discharging (crowbar circuits), and high-frequency snubbers. Since the resistors are solid bodies, they are inherently noninductive. Average power dissipation and voltage ratings are a function of size and resistivity. High peak currents and high voltage gradients can be tolerated. However, these resistors are typically rated in terms of peak energy since the time duration is an important consideration. For short-duration pulse applications, the charging or discharging of large quantities of energy produces heat, causing high-temperature gradients, resulting in resistor failure. Strict adherence to manufacturers' data of energy ratings and pulse duration is imperative for reliable operation.

2.7.2 Metal-Film Resistors

In the manufacture of metal-film resistors, frequently referred to as thin-film, an alloy such as nickel–chrome (nichrome) or tantalum nitride is deposited on the surface of a glass or ceramic substrate by evaporation or sputtering. The film resistivities range from 3 to 25,000 Ω per square. After deposition, the unit is heat treated to oxidize the film surface and metal caps are pressed onto the ends of the unit for electrical contact. A helical groove is then cut in the film (frequently by laser beam) to provide a helixed resistive path from end to end. A serpentine pattern may be cut to minimize flux fields and to achieve minimum inductance. When the desired resistance value is achieved, the cut is stopped. An epoxy or

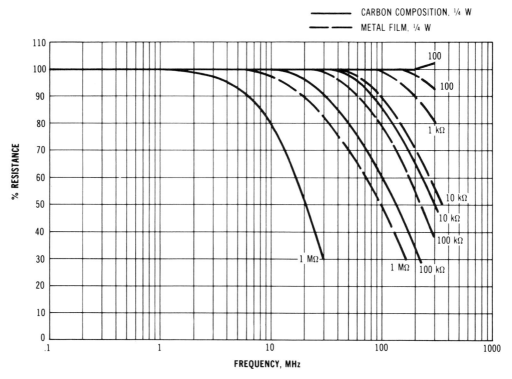

Fig. 2.31. High-frequency characteristics of resistors.

molding compound is then rolled onto the surface and the unit cured. Photo etching techniques are also used on bulk metal-film units. A silicone rubber coating is added for mechanical strength and the unit is encapsulated in an epoxy molding compound.

Metal-film resistors have excellent temperature stability, down to 5 ppm/°C (parts per million per degree Celsius); precise tolerances (down to 0.025%); good tracking (down to 2 ppm); excellent voltage coefficient (0.05 ppm/V, the maximum change in nominal resistance when a voltage change occurs); very low current noise (less than 0.02 μV/V or -34 dB at 1 MΩ); and low thermal or "white" noise. For long-term stability, a change of ± 35 ppm/yr is typical in ultraprecision units. Despite their helical construction, which might indicate metal-film resistors to be inductive, capacitive impedance predominates except for the very low values of resistance, as shown in Fig. 2.31. Typical high-frequency performance of the carbon composition resistor is shown for comparison.

2.7.2.1 Low-Voltage Resistors   The metal-film resistors are further divided into thin-film and thick-film types. The thick-film types, frequently referred to as Metal Glaze,* have the *RLR* mil designation with color codes the same as the *RCR* carbon composition with two exceptions. RED in the tolerance band indicates $\pm 2\%$. In the part number sequence RLRxxCxxxxGR, the first three digits are

---

*Metal Glaze is a trademark of TRW.

significant figures, and the fourth digit is the multiplier. G signifies $\pm 2\%$ tolerance. The $\pm 5\%$ parts (J) use a three-digit code. A 1200-Ω, $\pm 5\%$, 1-W resistor part number is then RLR32C122JR.

The characteristics of the thin-film types, such as the RNxxC and RNCxxH, are listed in Fig. 2.32. The RNCxxH series are established reliability types. These resistors find wide usage in voltage sensing, timing, and compensation circuits for the control networks of power conversion equipment. For instance, a

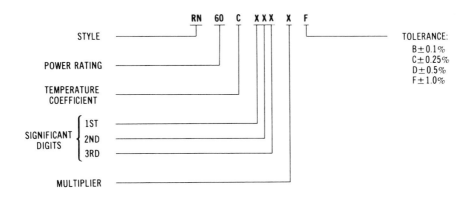

STYLE	POWER RATING		VOLTAGE RATING	RESISTANCE RANGE	TEMPERATURE COEFFICIENT
	@70°C	@125°C			
RN55	1/8 W	1/10 W	200 V	49.9 Ω-301 kΩ	B=0±500 PPM/°C
RN60	1/4 W	1/8 W	250 V	49.9 Ω-1 MΩ	C=0±50 PPM/°C
RN65	1/2 W	1/4 W	300 V	49.9 Ω-2 MΩ	D=0±200 PPM/°C
RN70	3/4 W	1/2 W	350 V		E=0±25 PPM/°C
RN75	W	1 W	500 V		

A.

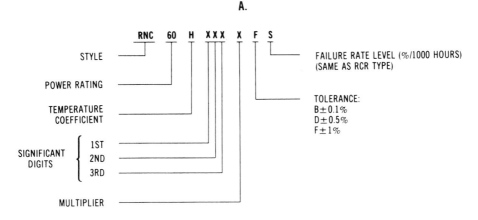

B.

Fig. 2.32. Characteristics of metal-film resistors.

RNC60J3322BS is a $\frac{1}{8}$-W, 33.2-kΩ resistor with a $\pm 0.1\%$ tolerance and a temperature coefficient of ± 25 ppm/°C.

2.7.2.2 High-Voltage Resistors

For precision high-voltage film resistors, a resistive film of proper pattern is fired directly onto a ceramic core or thermally bonded to the core, and the unit is fired at a high temperature. The film may be deposited in a serpentine pattern, as opposed to a spiral pattern, to minimize inductance. Insulation reistance becomes an important factor since it, in effect, parallels the resistive element it insulates. For solid cylindrical resistors with voltage ratings up to 25 kV, end caps with axial leads provide the terminations. A silicone encapsulant is added and the unit is cured. For voltage ratings higher than 5 kV and power ratings higher than 20 W, a hollow cylinder is normally used for the core and a conducting ferrule is added at each end of the cylinder for termination. An epoxy encapsulant is then added and cured. Configurations such as flat pack and chassis mount are also available.

Precision thick-film resistors are used in voltage dividers of high-voltage power supplies and converters providing regulated outputs. A portion of the output voltage is obtained from the resistive voltage divider and is applied to the control circuit. Depending on the high-frequency response requirements and the characteristics of the resistor, capacitors may also be added to provide frequency compensation. The high-voltage resistor may be tapped to form a voltage divider of desired resistance ratio, or multiple taps and terminations may be added. As compared to using discrete resistors, where the voltage across the resistor feeding the control circuit is invariably low, the high-voltage divider network has several advantages, most importantly, safety. The divider network is not susceptible to accidental opening of the connection between the two resistors and is less susceptible to the connection opening under operating conditions. With discrete resistors, personnel safety due to electrical shock is in jeopardy and component damage could result if the low-voltage resistor or the connection to the low-voltage resistor became an open circuit. There are two other advantages to the combined divider network. Due to equal temperature coefficients, the resistance values change proportionally for changes in surrounding temperature. Heat generated by power dissipated in the high-voltage portion is transmitted to the low-voltage end, producing a temperature and resistance ratio equilibrium.

In addition to the parameter considerations previously discussed, the suppression of corona in high-voltage equipment becomes important. Corona is caused by the electrostatic field established between a conductor and a nearby conductive element at voltage potentials of more than 750 V in air. The surrounding air becomes ionized and the dielectric strength decreases. Corona promotes arcing and generates ozone, which degrades the insulating properties of many materials. To minimize corona, sharp points in the resistor terminations and in the electrical connections to the resistor must be avoided. Reference Section 3.8.2 for further discussion on corona.

2.7.3 Wire-Wound Resistors

Perhaps to the chagrin of wire-wound power resistor manufacturers, the following statement is made. The requirement of power resistors in power conversion equipment notwithstanding, their resultant usage is the production of heat and should therefore be avoided wherever possible. Of course, this statement does not apply to resistors used for test loads where the heat can be removed by forced convection or other means. From a reliability view, wire-wound power resistors have a high failure rate. The majority of failures appear to be caused by a poor weld joint between the resistance element and the resistor lead or terminal. Heat produced by soldering large-size external leads to the resistor terminal plus the mechanical strain to the terminal caused by the external leads may cause failure of even a proper weld.

Power resistors are manufactured in various shapes. The most common are axial lead (1–10 W), hollow cylinder (5–225 W), and chassis mount (5–120 W). The resistor coating plays a key role in durability. Vitreous enamel is used on industrial, nonprecision resistors, while silicone compounds are used on mil-type resistors. The enamel must be cured at about 500°C, which may damage small-diameter resistive material, but for high-power resistors, the enamel can withstand higher temperatures than the silicones. Enamel chips and cracks more easily than the flexible silicones. The silicone can be cured at about 350°C, thereby reducing damage to the resistive element, and silicone places less strain on the resistive element. For very high power (kilowatt) applications, the resistance element is wound on edge, as in the rib-wound and POWR-RIB.*

Body temperature of chassis-mount resistors can reach 275°C, causing hot spots, and these resistors can also radiate heat to adjacent components. The internal hot spots are located at the center of the resistor element, which is the farthest point from the heat-sinked housing. Therefore the core material must withstand high temperature, have good dielectric strength, and be extrudable in various shapes. Steatite material is a common core material, although aluminum oxide (Al_2O_3) has five times the thermal conductivity of steatite; beryllium oxide (BeO), though fragile, has eight times the thermal conductivity of aluminum oxide. To reduce resistor size, some manufacturers use BeO cores on low-power resistors. The rib-wound types normally use heavy-resistance wire, edge wound on a ceramic core. The ceramic provides very high temperature capability and is inherently flameproof.

Characteristics of common mil-designated power resistors are given in Fig. 2.33. Resistance tolerances vary from 0.1 to 1.0%. Transient voltage-withstanding capability is 500 V for RW80 and RW81; 1000 V for RWR89, RWR74, RWR84, RWR78, RER60, RER65, and RER70; 2000 V for RE75; and 4500 V for RE77 and RE80.

Frequently, applications in power conversion equipment require "noninductive" resistors, especially where high di/dt (rate of current change with time) is encountered. Examples are resistors for snubber networks and for current sensing in high-frequency switching circuits. Minimum inductance is achieved by spiral-winding techniques, with adjacent wires conducting current in opposite directions. The RWRxxN series (as opposed to the RWRxxS series), for instance, specifies 0.25–1.2 μH maximum at 1 MHz at resistance values above 50 Ω and typically half this value below 50 Ω. Military and industrial resistor part numbers normally include the lattern "N" to indicate noninductive.

---

*POWR-RIB is a trademark of Ohmite Manufacturing Company.

STYLE	POWER RATING U	POWER RATING V	VOLTAGE RATING
RW81, RWR81	1 W		33 V
RW80, RWR80, RWR71	2 W		80 V
RW69		3 W	150 V
RW79, RWR89	3 W		150 V
RW74, RWR74	5 W		350 V
RW55, RWR84		7 W	
RW78, RWR78	10 W		850 V

STYLE	POWER RATING MIL	POWER RATING INDUSTRIAL	POWER RATING FREE AIR	VOLTAGE RATING
RE60, RER60	5 W	7.5 W	3 W	160 V
RE65, RER65	10 W	12.5 W	6 W	265 V
RE70, RER70	20 W	25 W	8 W	550 V
RE75, RER75	30 W	50 W	10 W	1250 V
RE77	75 W	100 W		1900 V
RE80	120 W	250 W		2300 V

Fig. 2.33. Characteristics of wire-wound resistors.

2.7.4 Resistance Wire Materials

Rather than being classified as a conductor, resistance wire is used in wire-wound power resistors and ribbon form wire is used in high-power rib-wound resistors. Resistance wire may also be used as the conductor in small pulse transformers where the increased resistance provides damping and eliminates the need for an external resistor. The resistance wire has a high resistivity and a low temperature coefficient of resistance change. Common materials and characteristics are as follows*

CUPRON: 55% Cu, 35% Ni; 300 Ω-cmil/ft resistivity (as compared to 10.6 for Cu); temperature coefficient of resistance ± 40 ppm/°C. For example B&S #18 (0.04-in.-diameter) wire has a resistance of 0.1837 Ω/ft and requires 4.32 A to raise the wire temperature 100°C.

MANGANIN: 84% Cu, 12% Mn, 4% Ni; 290 Ω-cmil/ft resistivity; ± 15 ppm/°C temperature coefficient of resistance; ideal for semiprecision shunts.

EVANOHM: 825 Ω-cmil/ft resistivity, ± 1 ppm/°C temperature coefficient of resistance; ideal for precision resistors and current sensing over a wide temperature range.

Ribbon resistance wire may be used to construct an economical high-current, high-frequency shunt. A length of ribbon is formed in a closed U shape, with a mica insulator placed between the ribbon. A BNC coaxial connector is welded across the open ends of the U for shielded sensing purposes. The U shape and the mica insulator form a capacitor which cancels the inductive reactance of the ribbon. The high-current connection is made to the extremities of the ribbon outside the coaxial connector. The coax shunt is calibrated by measuring the

*CUPRON and EVANOHM are trademarks of W. B. Driver Co., AMAX, Inc.

voltage drop across the BNC connector with the shunt conducting a known dc current, as measured by an external low-frequency precision shunt.

Precision high-current and high-frequency shunts are also manufactured from solid brass with bus bar power connections and BNC signal connections. These shunts are normally specified in peak current rating and in energy rating for pulse applications. Typical resistance for a 5000-A shunt is 50 $\mu\Omega$, and typical frequency response is 50 kHz.

2.8   CAPACITORS

A capacitor is a component that stores electrical charge or energy for release at some predetermined rate at some predetermined time. In its simplest form it consists of two conductive surfaces separated by an insulator or dielectric, as shown in Fig. 2.34A. The basic equation for capacitance is

$$C = KA_p/(4\pi d) \qquad \text{statfarads} \qquad (2.29a)$$

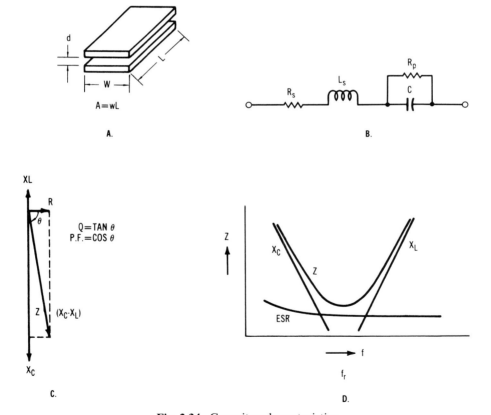

Fig. 2.34. Capacitor characteristics.

where

K = dielectric constant
A_p = surface area of plates, cm^2
d = distance between plates, cm
1 farad = 9×10^{11} statfarads = 10^{12} pF

Since the farad is a large quantity, microfarads and picofarads are generally used. The equation then becomes

$$C = 0.0884KA_p(n - 1)/d \qquad \text{pF} \tag{2.29b}$$

where n is the number of plates.
For dimensions in *inches*, the equation is

$$C = 0.2246KA_p(n - 1)/d \qquad \text{pF} \tag{2.29c}$$

The dielectric constant of the insulator material is a measure of its ability to store electrons. The high dielectric constant of ceramics produces a high-capacitance value in a small size, as discussed in Section 2.8.4.
Capacitors may be divided into the following types:

1. *Electrolytic.* Aluminum and tantalum.
2. *Paper.* Paper and metalized paper.
3. *Paper–film.* Paper–mylar and metalized paper–mylar.
4. *Films.* Polycarbonate, mylar, polystyrene, polystyrene–mylar, polypropylene, Teflon, metalized polycarbonate, and metalized Teflon.
5. *Ceramic.* Low-voltage ceramic, general-purpose ceramic, temperature-compensating ceramic, plus glass, and porcelain.
6. *Mica.* Mica, reconstituted mica.

In selecting a capacitor, one must evaluate such circuit parameters as capacitance value, peak voltage, rms voltage, peak current, rms current, pulse width, repetition rate, maximum and minimum temperature, and all other tpyical environmental conditions. One must also determine how a specific capacitor will react under circuit conditions. Manufacturers' data sheets must be consulted for parameters such as capacitance vs. temperature, insulation resistance vs. temperature, power factor vs. temperature, voltage vs. frequency, curent vs. frequency, temperature rise vs. power dissipation, dissipation vs. physical size, leakage current, insulation resistance, and where applicable, ESR. As with most components, capacitor life decreases with increased temperature. Increased reliability is achieved by voltage derating, current derating, and temperature derating. The power dissipated is equal to the rms current squared times the ESR. This power heats the capacitor and the temperature rise should normally be limited to 10 or 15°C maximum.

The equivalent circuit of a capacitor is shown in Fig. 2.34B. Resistor R_s is the ESR, L_s is the ESL, and R_p is the parallel resistance which causes a leakage current or self-discharge of the capacitor (R_{dc1}). The impedance of the capacitor is

$$Z = R_s + j\omega L + (R_p/j\omega C)\big/\big[R_p + 1/(j\omega C)\big] \tag{2.30a}$$

$$= R_s + j\omega L + R_p\big/(1 + j\omega R_p C)$$

$$= R_s + \frac{R_p}{1 + \omega^2 R_p^2 C^2} + j\frac{\omega L - \omega R_p^2 C + \omega R_p^2 L C^2}{1 + \omega^2 R_p^2 C^2} \tag{2.30b}$$

where the two left-hand terms are the real, or ESR, part; the right-hand term is the imaginary part; and frequency becomes a squared factor.

Eliminating the imaginary j ($j = \sqrt{-1}$), the phasor diagram of Fig. 2.34C relates the reistive and reactive components of the capacitor to

$$Z = \sqrt{(R_s)^2 + (X_C - X_L)^2} \tag{2.31a}$$

$$Q \text{ (quality factor)} = (X_C - X_L)/R_s = \tan^{-1}\theta \tag{2.31b}$$

$$\text{DF (dissipation factor)} = 1/Q \tag{2.31c}$$

$$\text{PF (power factor)} = R_s/Z = \cos^{-1}\theta \tag{2.31d}$$

At low values of L_s (ideal capacitor) and for $R_s = $ ESR

$$\text{DF} = 2\pi f C(\text{ESR}) \tag{2.31e}$$

The capacitor impedance varies with frequency, as shown in Fig. 2.33D. The impedance is minimum at the resonant frequency of

$$f_r = 1\big/\big(2\pi\sqrt{L_s C}\big) \tag{2.31f}$$

At frequencies higher than f_r, the capacitor acts as an inductor. For this reason, capacitors used for EMI suppression must have very low inductance. Reference Fig. 14.7 for typical resonant frequencies of various capacitors. For low values of capacitance, the crossover between X_C and X_L at f_r is sharply defined. The high capacitance values associated with aluminum electrolytics have a lower capacitive impedance at lower frequencies. Since the minimum impedance is limited by ESR, the minimum impedance base widens and does not change appreciably over a decade, or more, of frequency change.

Many manufacturers produce several types of commercial capacitors, each with their code, or part-numbering system. Space does not permit complete listing, although specific lines are mentioned later. Various manufacturers also produce capacitors to military specifications which then become a standard denominator for tabulation. Typical military capacitor types, descriptions, and charcteristics commonly used in power conversion equipment are given in Table 2.3. For an applications guide, MIL-STD-198E, "Capacitors, Selection and Use of," is recommended.

TABLE 2.3. Typical Capacitor Types

Capacitor Type	Military Specification	Prefix Type	Notes
Aluminum electrolytics	MIL-C-39018	CU	Operation to 125°C, stacked foil available
		CUR	Established reliability
Tantalum	MIL-C-39003	CSR	Solid, established reliability
	MIL-C-3965	CL	Plain, etched foil
	MIL-C-39006	CLR	Wet-slug, established reliability
Paper, film	MIL-C-18312	CH	Various films
	MIL-C-30922	CHR	Established reliability
	MIL-C-19978	CQ	Hermetic seal, high voltage available
		CQR	Established reliability
	MIL-C-83421	CRH	Metalized polycarbonate, established reliability
Ceramic	MIL-C-39014	CKR	Radial lead, established reliability
Mica	MIL-C-39001	CMR	Various temperature coefficients established reliability

2.8.1 Aluminum Electrolytics

Aluminum electrolytics are perhaps the most widely used capacitor in power conversion equipment, especially in power supplies, both linear and switching. Their primary use is in filtering, due to the high-energy storage per unit volume. Their popularity stems primarily from their low cost. The term *electrolytic capacitor* is applied to any capacitor in which the dielectric layer is formed by an electrolytic method. Aluminum foil ribbon is the anode, aluminum oxide (electrochemically formed on the surface of the ribbon) and fiber separators are the dielectric, and the electrolyte is the cathode. High capacitance per unit volume is achieved by the high dielectric constant (typically $K = 9$ for Al_2O_3 as compared to $K = 3$ for paper-and-film capacitors). New ac etching methods provide physical integrity and higher capacitance per unit volume. The electrolyte is an aqueous solution of ethylene glycol, conductive salts, and water. For low ESR, the electrolyte must have good conductivity but not so high as to cause sparking or scintillaiton, which will result in failure. At low temperature, the electrolyte tends to crystallize, resulting in higher ESR. Capacitors utilizing nonaqueous amide electrolytes have a very low ESL, have a lower ESR at low temperature, and have a longer shelf life but are generally limited to 200-V ratings. Electrolytic capacitors in chip form minimize printed circuit board real estate. So-called supercaps consist of individual cells of carbon particles in an organic electrolyte. These capacitors have the highest farad value per unit volume but also have a relatively high ESR.

Capacitance tolerances of $\pm 50\%$, -10% are easily tolerated in brute force filtering of rectified ac voltages. For switching regulators and high-frequency converters, the change in capacitance plus the change in ESR may change the frequency of the output filter double pole (L, C) and the zero (C, ESR) sufficiently to cause instability in the feedback loop. These factors should be considered in the modeling and design stage for worst-case conditions.

2.8.1.1 Cleaning Solvents Care must be exercised in using cleaning solvents on capacitors, especially commercial and computer grade types, as these reagents may attack the end seals and cause physical leakage of the electrolyte over a prolonged period of time, which then results in failure of the capacitor. Analyses by manufacturers have shown that on certain capacitor types, 90% of the units returned because of failure have resulted from improper cleaning solvents being applied. Unless epoxy or protective end seals are applied in the manufacturer's process, solvents such as halogenated hydrocarbons [freon, methyl ethyl ketone (MEK), carbon tetrachloride, etc.] should not be used for cleaning purposes. Permissible cleaners are ethyl and propyl alcohol and water.

2.8.1.2 Life, Temperature, and Ripple Current Reliability handbook MIL-STD-217D treats capacitors more favorably than the "rule of thumb" that life is halved for every 20°C increase in temperature (reference Chapters 15 and 16). In either case, temperature rise must be limited. Temperature rise may be due to a high ambient temperature or to excessive ripple current and high ESR causing $I_r^2 R$ losses. Thermal runaway can also occur, especially in higher voltage capacitors, and is primarily due to the increase in leakage current as the temperature increases. This causes additional losses of $E_C I_{dc1}$, which further increase the core temperature. The leakage current also generates internal gasing which may increase internal pressures causing blowout of the safety vent.

The permissible ripple current is determined by the allowable ripple voltage on the capacitor, the ripple frequency, and the power which can be safely dissipated. The ripple voltage parameter is normally specified for tantalum foil capacitors, while the ripple current is specified for aluminum electrolytics. The allowable power dissipation is then

$$P = I_r^2 \text{ESR} = kA_s(T_c - T_a) \tag{2.32}$$

where

$$I_r = \text{rms current in capacitor}$$

$$\text{ESR} = \text{equivalent series resistance}$$

$$k = \text{thermal conductivity}$$

$$= 0.006 \text{ W}/°\text{C} - \text{in.}^2 \text{ for Al electrolytics}$$

$$A_s = \text{surface area of case in inches, } d(l + \tfrac{1}{2}d)$$

$$d = \text{capacitor case diameter}$$

$$l = \text{capacitor case length}$$

$$T_c = \text{capacitor case (or core) temperature}$$

$$T_a = \text{ambient temperature}$$

Then

$$I_r = \sqrt{P/\text{ESR}} \tag{2.33}$$

The surface area (capacitor size) is proportional to capacitance value and voltage rating. For medium-voltage ratings, the following formulas provide a typical ripple current rating for the respective capacitor types for C in microfarads.

For common aluminum electrolytics (CU17 and CU71) types with 120-Hz ripple,

$$I_r = C^{0.66}/45 \tag{2.34a}$$

For stacked foil aluminum electrolytics (CU81) types with 1-kHz ripple,

$$I_r = C^{0.66}/30 \tag{2.34b}$$

As a comparison, for tantalum wet-slug capactiros (CLR79) types with 40-kHz ripple,

$$I_r = C^{0.3}/3 \tag{2.34c}$$

Manufacturers frequently specify the maximum temperature differential for a given capacitor type at a rated temperature. However, this does not imply that the temperature differential could increase if the ambient temperature is well below the rated temperature. For instance, if a capacitor rated for 85°C ambient and 10°C rise is operated at a 40°C maximum ambient, the allowable temperature would probably be less than $95 - 40 = 55$°C.

2.8.1.3 *Series and Parallel Operation*

Electrolytics operated in series, as in filter capacitors of half-bridge switch-mode power supplies and converters, should have a resistor connected across each capacitor. This provides a discharge path when power is off and enhances personnel safety by eliminating possible electrical shocks. The resistor value should allow three times the maximum capacitor leakage current to flow in the resistor. The resistor also balances the voltage across each of the series capacitors.

Example

Two 500-μF, 250-V capacitors having a leakage current of 1.5 mA at 65°C are to be connected in series. The maximum voltage on each capacitor is 175 V and the desired current through the parallel resistor is 4.5 mA. The resistance value is $175/0.0045 = 39$ kΩ and the power dissipation in each resistor is $175^2/39,000 = 0.8$ W; use a 2-W rating. Also, the discharge time constant is 20 s and the capacitors will discharge to a safe voltage in less than 1 min after the input is turned off.

Capacitors are frequently paralleled to decrease ripple voltage. Variations in ESR will produce a higher ripple current in the capacitor with the lowest ESR and thus a higher power dissipation and temperature rise. Normally capacitors of the same type or series can be paralleled without current-balancing networks. If a large number of large-size capacitors are paralleled in a "capacitor bank," it may be desirable to insert an appropriate fuse in series with each capacitor (reference Section 13.1 for fusing). In this manner, an individual capacitor failure will open the fuse and will not disrupt system operation. The fuse and the capacitor can be replaced at a more convenient time.

2.8.1.4   Capacitors with Threaded Insert Terminals   Care must be exercised when connecting multiple leads to cylindrical capacitors with threaded inserts. The proper torque (20 to 25 lb-in. for no. 10-32) must be applied to minimize I^2R losses at the terminals. The threaded aluminum inserts are also mechanically weak, as compared to the inserted screw material. Reaction due to dissimilar metals may cause problems over a prolonged period. Low source and load impedance is desirable to minimze transients, and current-carrying conductors "going to" and "coming from" the capacitor should be physically terminated at the capacitor terminals. For instance, in Fig. 2.35, the filter of a switching stage provides 20 A output at an input frequency of 100 kHz. AWG 14 wire is used from the rectifier to the output. In Fig. 2.35A, wire nos. 2 and 3 connect to the output terminals; then wire nos. 4 and 5 (also AWG 14) connect from the output terminals to the capacitor and are 4 in. long. The inductance in these two wires may be 0.5 μH. At 100 kHz, the wire impedance is $2\pi fL$, or 0.3 Ω. This may be an order of magnitude higher than the ESR of the capacitor and the high-frequency ripple at the output terminals will be high. In Fig. 2.35B, wire nos. 2 and 3 connect directly to the capacitor and wire nos. 4 and 5 connect from the capacitor to the output terminals. Now the low ESR of the capacitor is utilized to full advantage and the high-frequency portion of the output ripple may be a factor of 10 lower than that of Fig. 2.35A. It should be obvious that the recommended connections are not restricted to screw terminal electrolytics but apply to all capacitor configurations in high-frequency circuits and even to conductor-clad layout on printed circuit boards. This latter point is one reason radial lead, low-profile snap-in capacitors have replaced the axial lead capacitors for printed circuit board mounting. Most radial lead capacitors also have a higher energy density and a lower ESR than the axial lead types.

The stacked foil capacitors are ideally suited for filtering high-frequency ripple voltages. Typical ESR is 2 mΩ for large-capacitance, low-voltage units. Maximum ESL is tyically 2 nH and is independent of capacitance and voltage. Multiple tabs are ultrasonically welded to the aluminum foil for low-resistance connections and improved reliability. The flat terminal surfaces with threaded inserts are ideal for attaching current-conducting bus bars.

2.8.2   Tantalum Capacitors

Tantalum capacitors are more expensive than aluminum electrolytics but have better performance characteristics and are normally smaller in size, since tantalum oxide has a dielectric constant about twice that of aluminum oxide. The tantalum capacitors have considerably longer operating life and longer shelf life, since aluminum oxide tends to deteriorate with time. These factors have made tantalum capacitors a standard choice in military equipment. Due to the higher specific

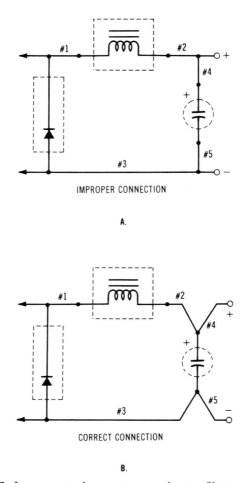

Fig. 2.35. Improper and proper connections to filter capacitors.

gravity, tantalum capacitors are heavier than aluminum capacitors of the same volt-microfarad rating. However, much of the discussion on aluminum electrolytics and applications also applies to tantalum electrolytics.

The sintered, porous-anode, liquid-electrolyte (sulfuric acid) capacitor, normally referred to as "wet-slug" type, has the highest volumetric efficiency of any "electrolytic" but is presently limited to less than 150 working volts. Excessive ripple currents in silver-cased wet slugs may cause the silver to plate onto the anode, causing failure. For this reason, units with tantalum cases (such as the mil-type CLR79) are required for certain applications. These capacitors may be hermetically sealed by laser welding or other techniques.

The porous-anode, solid-electroltye capacitor (such as the mil-type CSR13 and the low-ESR mil-type CSR21) is generally less expensive and has a lower ESR than the wet slug. Since there is no liquid in the solid type (manganese dioxide electrolyte), it is possible to use a conventional hermetic seal. The voltage ratings of solid types are generally less than the wet slugs.

Foil types are available in plain or etched foil, polarized and nonpolarized. The use of high-purity tantalum provides the lowest leakage current of any electrolytic and also allows higher operating voltage. The plain foil types normally have the lowest capacitance change with temperature and frequency. The etched-foil types have a higher capacitance per unit volume. Nonpolarized types are ideal in ac circuits where high capacitance is required, as in filtering sine wave harmonics in power filters.

2.8.3 Paper-and-Film Capacitors

Paper-and-film capacitors find usage in a wide range of applications, from low voltage to high voltage, both ac and dc. They may be plastic sealed, hermetic sealed, and/or oil filled. Metalized capacitors use an evaporated metal film as an electrode, instead of the conventional aluminum foil. This process provides higher working transient voltage stress, considerably lower volume, and a self-healing feature if a low-impedance circuit and sufficient volt-amperes are supplied. The paper-and-film types are symmetric (nonpolar) and are well suited for ac applications. They have low dissipation factors and high insulation resistance. Teflon is perhaps the best polymer dielectric available, having both high-temperature capability (to 250°C) and flat temperature coefficient of capacitance. Polystyrene is limited to 85°C but is especially useful in tuned circuits where the linear positive temperature coefficient directly offsets the linear negative temperature of Permalloy powder core inductors. Polycarbonates have a low dissipation factor and are small in size. Polypropylene has excellent temperature stability (to 105°C) and is less expensive than Teflon. Metalized polypropylene capacitors are specifically useful in switch mode power supplies due to their high current and frequency capability, excellent temperature stability, low ESR (very low dissipation factor), and considerably lower volume than other film types. These low-impedance characteristics are also useful in snubber networks and in input EMI filters for reducing conducted emissions. Typical dissipation factor, impedance, and resonant frequency are 0.1%, 0.01 Ω, and 0.5 MHz, respectively, for a 3-μF, 400-V dc polypropylene capacitor.

2.8.4 Oil-Filled Capacitors

Oil-filled capacitors have applications in high-voltage circuits since the oil impregnates the dielectric, increases the insulation resistance, and fills air gaps to reduce corona. Other than mineral oil or transformer oil, the designation "oil" is a convenient term as opposed to "biodegradable liquid." These capacitors do not contain PCB (polychlorinated biphenyl) materials. Typical impregnants are ECCOL, GECONOL, and DYKANOL.*

These capacitors find wide usage in power factor correction, energy storage, and high-power pulse systems. High-voltage types have ceramic bushing standoffs for

---

*ECCOL is a trademark of Sprague Electric. GECONOL is a trademark of General Electric. DYKANOL is a trademark of Cornell-Dubilier.

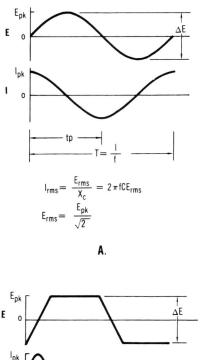

$$I_{rms} = \frac{E_{rms}}{X_c} = 2\pi fCE_{rms}$$

$$E_{rms} = \frac{E_{pk}}{\sqrt{2}}$$

A.

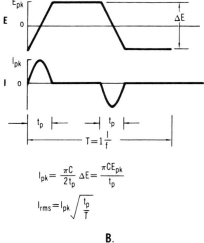

$$I_{pk} = \frac{\pi C}{2t_p} \Delta E = \frac{\pi CE_{pk}}{t_p}$$

$$I_{rms} = I_{pk}\sqrt{\frac{t_p}{T}}$$

B.

Fig. 2.36. Waveforms and ratings for ac capacitors.

insulation to the case, which is normally grounded. Energy densities of 8 J/in³ are available.

Capacitors with a low DF are ideal for high-power and pulse applications of thyristor (SCR) commutation circuits due to the high rms current rating and low inductance. For instance, the General Electric 97F series of polypropylene capacitors has a dissipation factor of 0.02%. A typical 2-μF, 800-V capacitor has an rms current rating of 60 A at 55°C ambient. However, this current rating is for a 50-μs pulse width, and the allowable rating for other conduction times must be calculated. The commutating capacitor types may also be rated in terms of volt-amperes (VA or $E_{rms}I_{rms}$). Fig. 2.36A shows waveforms for a sinusoidal voltage. The voltage and current relations are

$$E_{rms} = E_{pk}/\sqrt{2} \tag{2.35a}$$

$$I_{rms} = E_{rms}/X_C = 2\pi f C E_{rms} \tag{2.35b}$$

For the trapezoidal voltage waveform of Fig. 2.36B, the voltage (assuming $t_p \ll T$) and current relations are

$$I_{pk} = \pi C E_{pk}/t_p \tag{2.36a}$$

$$I_{rms} = I_{pk}/\sqrt{t_p/T} = I_{pk}\sqrt{t_p f} \tag{2.36b}$$

where

C = capacitance, F
t_p = pulse width, s
T = period time, s
f = operating frequency, Hz

The rating factor for pulse periods other than 50 μs is

$$k_t = \sqrt{50/t_p} \tag{2.37}$$

where t_p is the actual pulse period in microseconds.

For ambient temperatures other than 55°C, the allowable rms current decreases approximately 2% per °C increase and

$$k_T = 1 + 0.02(55°C - T_a) \tag{2.38}$$

Examples

Case 1. Ten-microfarad 400-V capacitors are to be used as an output filter for a 200-V, 3ϕ, 400-Hz inverter at an ambient temperature of 65°C. From Equation (2.35b), $I_{rms} = 2 \times \pi \times 400 \times 10^{-5} \times 200 = 5$ A. At 400 Hz, $t_p = 1/800 = 1250$ μs. Then $k_t = \sqrt{50/1250} = 0.2$. At 65°C, $k_T = 1 + 0.02(55 - 65) = 0.8$. The required rms current rating, based on pulse duration and temperature, is then $I_{rated} = 5/(0.2 \times 0.8) = 31.25$ A.

Case 2. A 5-μF, 800-V capacitor is used in a commutation circuit of a system operating at 5 kHz. The peak capacitor voltage is 500 V, the current pulse width is 30 μs, and the ambient temperature is 45°C. From Equation (2.36a), $I_{pk} = \pi \times 5 \times 10^{-6} \times 500/(30 \times 10^{-6}) = 262$ A, and $I_{rms} = 262\sqrt{5000 \times 30 \times 10^{-6}} = 102$ A; $k_t = \sqrt{50/30} = 1.29$ and $k_T = 1.2$. The required rms current rating, based on pulse duration and temperature, is then $I_{rated} = 102/(1.29 \times 1.2) = 66$ A. If the frequency were lowered to 2500 Hz, $I_{rated} = 66\sqrt{2500/5000} = 47$ A.

TABLE 2.4. Commercial Ceramic Capacitor Classifications

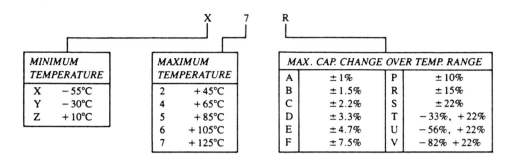

GENERAL APPLICATION AND HIGH-K CAPACITORS

MINIMUM TEMPERATURE		MAXIMUM TEMPERATURE		MAX. CAP. CHANGE OVER TEMP. RANGE			
X	−55°C	2	+45°C	A	±1%	P	±10%
Y	−30°C	4	+65°C	B	±1.5%	R	±15%
Z	+10°C	5	+85°C	C	±2.2%	S	±22%
		6	+105°C	D	±3.3%	T	−33%, +22%
		7	+125°C	E	±4.7%	U	−56%, +22%
				F	±7.5%	V	−82% +22%

TEMPERATURE STABLE AND TEMPERATURE COMPENSATING CAPACITORS

TEMP COEFF., ppm/°C		MULTIPLIER		TOLERANCE, ppm/°C	
		0	−1		
C	0.0	1	−10	G	±30
M	1.0	2	−100	H	±60
P	1.5	3	−1000	J	±120
R	2.2	5	+1	K	±250
S	3.3	6	+10	L	±500
T	4.7	7	+100	M	±1000
U	7.5	8	+1000	N	±2500

2.8.5 Ceramic Capacitors

Ceramic capacitors are used in coupling, decoupling, bypassing, and timing circuits. The common multilayered ceramics have a small internal inductance, and
the operating impedance actually decreases above 1 MHz. These capacitors are
available in various ranges of capacitance tolerance and temperature stability. EIA
designations for general-purpose and temperature-stable capacitors are listed in
Table 2.4. The X7R is specified for operation over a wide temperature range, and
maximum capacitance is typically 0.22 μF at 100 V. The Z5U is specified for
operation over a narrow temperature range, and maximum capacitance is typically
4.7 μF at 50 V. The COG series has an ultrastable temperature characteristic from
−55 to +125°C, and maximum capacitance is typically 0.01 μF at 100 V. Due to
the inherent low dissipation factor (typically 0.19%), the COG series is ideally
suited for EMI suppression in all types of equipment. The COG series was
formerly designated as NPO (negative–positive–zero). Part numbering sequence

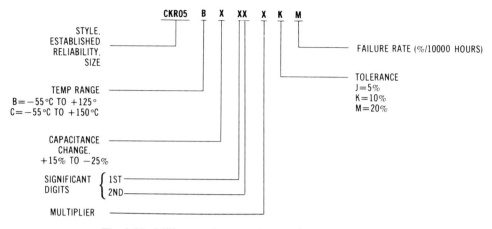

Fig. 2.37. Military-style ceramic capacitor part numbers.

for a popular military series is listed in Fig. 2.37, although this series is typically desingated as MIL-C-39014/1C/xxxx, where the xxxx signifies the capacitance and the failure rate. The BX characteristic is eqiuvalent to the commercial X7R characteristic.

2.8.6 Mica Capacitors

Mica capacitors have low loss, good capacitance stability, and a low-temperature coefficient. Low-voltage units are normally reserved for control circuitry timing and

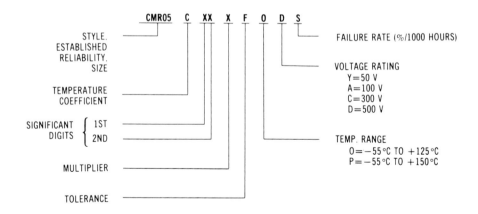

SYMBOL	TEMP. COEFFICIENT	SYMBOL	TOLERANCE
C	±200 ppm/°C	D	±0.5 pF
E	−20 TO +100 ppm/°C	F	±1%
F	0 TO 70 ppm/°C	G	±2%
		J	±5%

Fig. 2.38. Military-style mica capacitor part numbers.

compensation, where the low-capacitance tolerance and high stability are advantageous. Part numbering sequence for a military series per MIL-C-39001 is listed in Fig. 2.38.

Reconstituted mica (not recycled) is a natural capacitor grade mica of high purity and serves as a dielectric between the aluminum foil conductor. These high-voltage capacitors are ideally suited for PFNs in magnetic modulators, and for transmitter applications, as discussed in Chapter 12. Energy densities of 0.3 J/in.^{3} are typical. Dissipation factor is typically 0.5% at 125°C.

3

TRANSFORMERS, INDUCTORS, AND CONDUCTORS

3.0   Introduction

3.1   Magnetic Circuit Review

3.2   Magnetic Relations for Transformers

3.3   Magnetic Relations for Inductors

3.4   Hysteresis Loops

3.5   Magnetic Materials

3.6   Transformer Losses, Regulation, and Efficiency
      3.6.1   Magnetizing Current
      3.6.2   Core Loss
      3.6.3   Copper Loss, Reactance, and Regulation
      3.6.4   Transformer Efficiency and Power Factor
      3.6.5   Leakage Inductance
      3.6.6   Transformer Capacitance

3.7   Copper and Other Conductors
      3.7.1   Copper
      3.7.2   Litz Wire
      3.7.3   Aluminum
      3.7.4   Skin and Proximity Effects in Conductors
      3.7.5   Skin and Proximity Effects in Windings
      3.7.6   Conductor Loss

3.8   Windings
      3.8.1   Insulating Materials
      3.8.2   Voltage Isolation
      3.8.3   Three-Phase Windings and Connections

3.9   Temperature Rise

3.10   Audible Noise

3.11   Laminations
      3.11.1   Power Transformers
      3.11.2   Inductors
      3.11.3   Swinging Chokes

3.12   C and E Cores
      3.12.1   C Core Transformers
      3.12.2   C Core Inductors

3.13   Toroid Tape Core Transformers

3.14   Permalloy Powder Cores
      3.14.1   Inductors with ac Voltage
      3.14.2   Inductors with dc Current

3.15 Iron Powder Core Inductors
3.16 Ferrites
    3.16.1 Core Geometries
    3.16.2 Power Transformers
    3.16.3 Inductors
3.17 Air Core Inductors
3.18 Pulse Transformers
    3.18.1 Charcteristics and Measurements
    3.18.2 Design Parameters
    References

3.0 INTRODUCTION

This chapter initially presents a review of magnetic relations as applied to transformers and inductors. Magnetic materials and design parameters are discussed for these devices. The discussion of conductors applies to the current-conducting element within the transformer or inductor. Transformers are required for increasing or decreasing an input voltage to a desired output voltage, and to provide isolation between input and output. Low-frequency transformers operating from the utility line or mains input are required in various types of electronic equipment as well as in power conversion equipment. In switching circuits, operation at higher frequencies can reduce the size and weight of transformers as well as the inductors. The inductors are used as energy storage devices, for filtering, and for increasing the impedance of various networks. Operating life and reliability are increased by limiting the internal temperature to a safe value. The magnetic material and the electrical conductor can inherently withstand high temperatures, but temperature derating is required for the insulating materials.

Notations used in this chapter are listed in Table 3.1. The terminology of magnetic systems is given in Table 3.2, and Table 3.3 lists the conversion factors for the various systems. The centimeter–gram–second (CGS) system is used unless otherwise specified.

3.1 MAGNETIC CIRCUIT REVIEW

When an electrical conductor is placed around a magnetic material, lines of magnetic flux, ϕ, are produced when a current, I, flows in the conductor. The direction of current and flux is given by the "right-hand rule", and the amount of flux is directly proportional to the current. Normally, additional turns of the conductor are added and a magnetomotive force, F, is produced and is directly proportional to the ampere turns, or gilberts. The magnetic material offers an opposition to the establishment of the flux, and this opposition is termed reluc-

TABLE 3.1. Notation

A_c	Core cross-sectional area, in.^{2} or cm^2
A_e	Effective core cross-sectional area
A_L	Inductance index, mH/1000 turns
A_T	Surface area, in.^{2} or cm^2
A_w	Window or winding area, in.^{2} or cm^2
B	Induction or flux density, gauss
C	Capacitance, farads
E	Voltage, volts
f	Frequnecy, hertz
F	Magnetomotive force, gilberts
H	Magnetizing force, oersteds
I	Current, amperes
J	Current density, A/in.^{2}, A/cm^2, cmil
k	Constant (various)
l_c	Magnetic path length
l_g	Effective gap length
l_m	Mean turn length
L	Inductance, henrys
n	Turns ratio
N	Number of turns
p	Number of layers
P	Power, watts
PF	Power factor
P_c	Core loss, watts
P_{cu}	Copper loss, watts
Q	Quality factor
R	Resistance, ohms
R	Reluctance
S	Stack height
t	Time, seconds
T	Temperature, °C
T_w	Tongue width, laminations
μ_Δ	Incremental permeability
μ_e	Effective permeability
V	Volume, in.^{3} or cm^3
VA	Volt-amperes
W_t	Weight pounds or kilograms
X	Reactance, ohms
ω	Angular velocity, $2\pi f$
ϕ	Flux, maxwell or line
δ	Skin depth factor
ρ	Resistivity
ε	Dielectric constant
μ	Permeability

TABLE 3.2. Magnetic Systems

Unit	Definition	CGS System	MKS System	English System
ϕ	Flux	Maxwell or line	Weber	Maxwell or line
B	Flux density	Gauss or lines/cm^2	Webers/m^2 (tesla)	Lines/in.^{2}
F	Magnetomotive force	Gilbert	Ampere turns	Ampere turns (NI)
H	Magnetizing force	Oersted or gilbert/cm	NI/m	NI/in
A	Area	cm^2	m^2	in.^{2}
l	Length	cm	m	in.
V	Volume	cm^3	m^3	in.^{2}
μ	Permeability	1	$0.4\pi \times 10^6$	3.192

TABLE 3.3. Magnetic System Conversions

To convert	Into	Multiply by	Conversely, Multiply by
Ampere turns	Gilberts	0.4π	0.796
Ampere turns/in.	Oersteds	0.495	2.02
Ampere turns/m	Oersteds	0.004π	79.6
Lines/in.^{2}	Gauss	0.155	6.45
Webers	Lines	10^8	10^{-8}
Webers/m^2 (tesla)	Gauss	10^4	10^{-4}
in.^{2}	Circular mils	1.273×10^6	0.785×10^{-6}
cm^2	Circular mils	0.197×10^6	5.063×10^{-6}

tance as given by

$$R = l_c/(\mu_e A_e) \qquad \text{(dim. in cm)} \qquad (3.1a)$$

$$= 0.313 l_c/(\mu_e A_e) \qquad \text{(dim. in in.)} \qquad (3.1b)$$

where l_c is magnetic path length, μ_e is effective permeability, and A_e is cross-sectional area of the core.

The above equation is analogous to $R = \rho l/A$ for the electrical conductor. The permeability, μ, is a measure of the magnetic conductivity of the material and is the ratio of induction to magnetizing force. Thus the *Ohm's law* for the magnetic circuit is $F = \phi R$ gilberts. However, in most magnetic material the permeability varies with flux, and the performance curve for the specific magnetic material is normally consulted to obtain permeability figures. When the reluctance drop is distributed along the length of the magnetic path, the magnetizing force is given by Ampere's law as

$$H = 0.4\pi NI/l_c \qquad \text{Oe} \qquad (3.2)$$

where N is number of turns and I is current in amperes.

Rearranging the terms of the above equation, the term for flux is

$$\phi = F/R = 0.4\pi N I \mu_e A_e / l_c = BA_e \qquad (3.3)$$

where B is flux density in gauss.

The flux density or induction, is then the ratio of flux to the cross-sectional area of the magnetic material. [In keeping with the CGS system, the units of gauss are used instead of the more modern units of tesla used in the meter–kilogram–second (MKS) system.] When the winding is excited by a sine wave input, $\phi = \phi_m \sin(\omega t)$ and the voltage is given by Faraday's law as

$$e = -N \times 10^{-8} \, d\phi/dt \qquad (3.4)$$

3.2 MAGNETIC RELATIONS FOR TRANSFORMERS

From the above equations, the voltage which a transformer can support is given by

$$E_{rms} = (2\pi/\sqrt{2})\phi_m N f \times 10^{-8} = 4.44 B_m A_e N f \times 10^{-8} \qquad (3.5)$$

where ϕ_m is maximum flux, f is exciting frequency in hertz, and B_m is maximum flux density.

Rearraning the above equation and substituting B_{ac} for B_m, the number of turns required to support the volt-second product for a *sine* wave is

$$N = \frac{E_{rms} \times 10^8}{4.44 B_{ac} A_c f k_s} \qquad (A_c \text{ in cm}^2) \qquad (3.6a)$$

$$= \frac{3.49 E_{rms} \times 10^6}{B_{ac} A_c f k_s} \qquad (A_c \text{ in in.}^2) \qquad (3.6b)$$

When a *square*-wave voltage is applied to the winding, the form factor is eliminated and

$$N = \frac{E_{avg} \times 10^8}{4 B_{ac} A_c f k_s} \qquad (A_c \text{ in cm}^2) \qquad (3.7a)$$

$$= \frac{3.88 E_{avg} \times 10^6}{B_{ac} A_c f k_s} \qquad (A_c \text{ in in.}^2) \qquad (3.7b)$$

The stacking factor k_s is the ratio of the effective cross-sectional area to the total cross-sectional area A_e/A_c. Manufacturers' data lists A_e for certain material configurations and Table 3.4 lists values of k_s for typical "lamination" material configurations.

The number of turns required to support the volt-second product of various waveforms at given flux densities is shown in Table 3.5. Notice that the square wave and pulse waveforms use the peak voltage in the numerator. The ac voltage imposed on inductors used for output filtering is given as a function of the dc output. For instance, the full-wave-rectified single-phase sine wave has a form

TABLE 3.4. Magnetic Materials

Code	Trade Names or Trademark Designations
A	Hypersil, Microsil, Silectron, Orthosil, Magnesil
B	Deltamax, Orthonol, Orthonik, Hypernik V, Sqaure mu-49
C	Permalloy 45, 48 Alloy, Carpenter 49, 4750, Hypernik
D	Supermalloy, Square mu-79, Hymu 80, Square Permalloy 80, Mumetal, Hy Ra 80
E	Supermendur, Permendur, Hyperco
F	Molybdenum permalloy powder (MPP)
G	Carbonyl, Iron oxides, Crolite, Polyiron
H	Ferrite, Ferramic, Ceramag, Siferrit, Ferroxcube

Code	Shapes	Code	Thickness	Stacking Factor, k_s
1	Laminations	1*	0.025 (24 gage)	0.97
1a	EE, EI, FB,	2	0.0185 (26 gage)	0.96
	Strip Stampings	3	0.014 (29 gage)	0.95
1b	DU, L	4	0.012	0.95
2	C cores (cut),	5	0.006	0.90
	E cores (3 ∅)	6	0.004	0.90
3	Toroids	7	0.002	0.85
3a	Tape wound	8	0.001	0.78
3b	Bobbin wound	9	0.0005	0.50
3c	Stamped ring	10	0.00025	0.375
3d	Powder	11	0.000125	0.25
4	Toroid; cup core,	*	0.031 (22 gage)	0.97
	E, U, slug	*	0.028 (23 gage)	0.97
5	Toroid; pot core, Tube, E, H, U, X, cube core (ferrite)			

factor of 1.11 and a ripple factor of 48.3%, as given by Equation (4.6) and Table 4.2. The rectified ripple frequency is $2f_s$. Substituting these into Equation (3.6a), $N = 0.483E_{rms} \times 10^8/(4.44B_{ac} A_e 2f_s) = E_{dc} \times 10^8/(18.4B_{ac} A_e f_s)$.

The voltage, current, power, and impedance ratios for the single-phase transformer shown in Fig. 3.1 are related by

$$\frac{N_p}{N_s} = \frac{E_p}{E_s} = \frac{I_s}{I_p} = \sqrt{\frac{Z_p}{Z_s}} = n \tag{3.8a}$$

$$\text{VA} = E_p I_p = E_s I_s \tag{3.8b}$$

The equation assumes no losses in the transfomer and states: primary volts per turn equals secondary volts per turn, primary ampere turns equals secondary ampere turns, primary volt-amperes equals secondary volt-amperes, and secondary impedance reflected to the primary is the ratio of the turns squared.

High-power circuits normally employ three-phase transformers to reduce current values. Referring to Fig. 3.2, the following relations apply, assuming no losses:

$$\text{VA} = \sqrt{3E_L I_L} = \sqrt{3E_L I_p} = 3E_p I_p \tag{3.9}$$

TABLE 3.5. Turns Required for Various Waveforms

Sine wave (steady state)		$N = \dfrac{E_{\text{rms}} \times 10^8}{4.44 B_{\text{ac}} A_e f_s}$
Square wave (symmetrical)		$N = \dfrac{E_{\text{pk}} \times 10^8}{4 B_{\text{ac}} A_e f_s}$
Quasi-square wave		$N = \dfrac{E_{\text{pk}} t_p \times 10^8}{2 B_{\text{ac}} A_e}$
Half-sine wave pulse		$N = \dfrac{E_{\text{pk}} t_p \times 10^8}{1.57 B_{\text{ac}} A_e}$
Rectangular pulse		$N = \dfrac{E_{\text{pk}} t_p \times 10^8}{B_{\text{ac}} A_e}$
Full-wave-rectified single-phase sine wave		$N = \dfrac{E_{\text{dc}} \times 10^8}{18.4 B_{\text{ac}} A_e f_s}$
Half-wave-rectified three-phase sine wave		$N = \dfrac{E_{\text{dc}} \times 10^8}{76 B_{\text{ac}} A_e f_s}$
Full-wave-rectified three-phase sine wave		$N = \dfrac{E_{\text{dc}} \times 10^8}{664 B_{\text{ac}} A_e f_s}$

where

E_L = line-to-line voltage

I_L = line current

E_p = phase voltage

I_p = phase current

For three-phase transformers, Y(wye) and Δ (delta) winding connections may be made Y–Δ, Δ–Y, and Δ–Δ. The Y–Δ is very useful for stepping down the

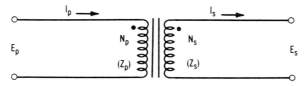

Fig. 3.1. Single-phase transformer schematic diagram.

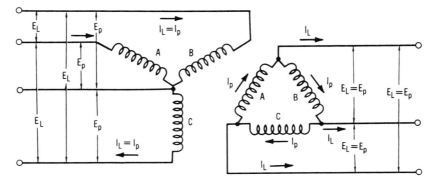

Fig. 3.2. Three-phase transformer schematic diagram.

voltage, as shown in Fig. 3.2. The primary coil need only support the line-to-neutral (phase) voltage. For high-voltage step-up, the Δ–Y is most often used. Again, the volts per turn and insulation required in the secondary is only concerned with phase voltage, whereas a secondary must support line voltage. The Y–Y connection is seldom used because large third-harmonic voltages may appear from phase to neutral. The Δ–Δ connection is sometimes used where the power source is a local generator.

Voltage isolation between a given input and a desired output may not always be required. In these cases, an autotransformer, as shown in Fig. 3.3, is used. The input voltage is stepped down (or up) by tapping the winding at the desired point. Depending on the input to output voltage ratio, a significant reduction in size and weight can be achieved, as commpared to an isolation transformer.

Example

Let $E_{in} = 120$ V and $E_0 = 100$ V at 600 VA. Neglecting losses, $P_{in} = E_{in}I_1 = P_0 = E_0I_2$. Then $I_1 = 600/120 = 5$ A, $I_2 = 600/100 = 6$ A, and $I_3 = I_2 - I_1 = 1$ A. The voltage across N_1 is 20 V and the voltage across N_2 is 100 V. The VA rating of $N_1 = N_2 = 100$ VA. Thus, the autotransformer need only be sized for 100 VA, whereas a 600-VA isolation transformer would be required, as in Fig. 3.1.

A magnetic device is capable of storing energy. The area contained within the hysteresis loop (reference Fig. 3.4 and Section 3.4) is the *BH* product. The

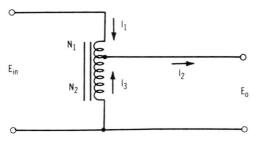

Fig. 3.3. Autotransformer schematic diagram.

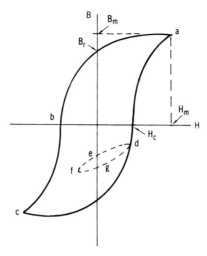

Fig. 3.4. Hysteresis loop for magnetic material.

gauss-oersted is also a unit of energy per unit volume with the following relation:

$$1 \text{ GOe} = 1 \text{ erg}/4\pi \text{ cm}^3 \tag{3.10}$$

where

$$1 \text{ erg} = 10^{-7} \text{ W-s (J)}$$

From the above equations, the magnetic induction at one turn can also be expressed by $1 \text{ V-s/cm}^2 = 10^{-8}$ G and the magnetizing force by $1 \text{ A/cm} = 0.4 \pi$ Oe. The energy density may then be expressed in watt-seconds per cubic centimeters. The power rating of the magnetic device is then a function of energy density, frequency, and core volume. The effective core volume is $V_e = A_e l_c$ (cm^3), and the power rating is

$$\text{VA} = \frac{\text{BHf}V_e}{2} \left(\frac{\text{V-s}}{\text{cm}^2} \times \frac{\text{A}}{\text{cm}} \times \text{Hz} \times \text{cm}^3 \right) \tag{3.11}$$

Introducing J for current density (amperes per square centimeters) and adding turns to the magnetic core with a window of area A_w (square centimeters), the power rating is

$$\text{VA} = JA_e A_w B_{\text{ac}} f \times 10^{-8} \tag{3.12}$$

Empirical formulas for VA ratings have been developed through studies, designs, and experience for various magnetic materials and configurations. These short-cut methods for transformers using laminations, C cores, tape-wound toroids, and ferrite materials are discussed in Sections 3.10.1, 3.11.1, 3.12, and 3.14.1, respectively.

3.3   MAGNETIC RELATIONS FOR INDUCTORS

Inductors are frequently required for filtering, reactance, and commutation purposes. The inductor may be required to carry a dc current, in which case an air gap may be introduced to prevent saturation. The air gap has the effect of increasing the reluctance and:

$$R = \frac{l_c}{\mu_e A_e} = \frac{l_g}{A_e} + \frac{l_c}{\mu A_e} \tag{3.13}$$

where μ is dc permeability, l_c is magnetic path length, and l_g is air gap length, and

$$\mu_e = \frac{l_c \mu_\Delta}{l_c + l_g \mu_\Delta} = \frac{\mu_\Delta}{1 + \mu_\Delta l_g / l_c} \tag{3.14}$$

Frequently, l_g is large compared to l_c/μ and $B = \mu H$ becomes $B \cong H$. Because of fringing of flux around the gap, an average of $0.85B$ crosses over the gap [2]. Now the dc flux density is

$$B_{dc} = \frac{0.4\pi NI_{dc}}{0.85 l_g} = \frac{1.48 NI_{dc}}{l_g} \qquad \text{(dim. in cm)} \tag{3.15a}$$

$$= \frac{0.58 NI_{dc}}{l_g} \qquad \text{(dim. in in.)} \tag{3.15b}$$

The inductor may also be required to support an ac voltage comonent and B_{ac} is found from previous equations. The total flux density, $B_{ac} + B_{dc}$, should be less than B_m, the saturation flux density of the core material.

The expression for inductance is derived from the following:

$$B = \frac{0.4\pi NI}{l_g + l_c/\mu_\Delta} \qquad E = N\frac{d\phi}{dt} = L\frac{di}{dt}$$

$$L = \frac{BA_e N \times 10^{-8}}{I}$$

$$L = \frac{0.4\pi N^2 A_e \times 10^{-8}}{l_g + l_c/\mu_\Delta} \qquad \text{(dim. in cm)} \tag{3.16a}$$

$$L = \frac{3.19 N^2 A_e \times 10^{-8}}{l_g + l_c/\mu_\Delta} \qquad \text{(dim. in in.)} \tag{3.16b}$$

With large air gaps, l_g is much greater than l_c/μ_d and inductance is nearly independent of permeability. With small air gaps, permeability controls the inductance. For core configurations of ferrite and permalloy powder, the above equation becomes

$$L = 0.4\pi\mu N^2 A_e \times 10^{-8}/l_c \tag{3.17}$$

where μ and l_c are given by manufacturers' data.

Further simplification may be obtained by using the *inductance index* A_L (frequently given in manufacturers' data) with units of millihenrys per thousand turns. The inductance equation then becomes

$$L = N^2 A_L \times 10^{-6} \text{ mH} \tag{3.18a}$$

For a desired inductance value, the required turns are

$$N = 1000\sqrt{L/A_L} \tag{3.18b}$$

where L is desired inductance in millihenrys and A_L is in millihenrys per 1000 turns.

Empirical formulas for inductor ratings are discussed in Sections 3.10.2, 3.11.2, 3.13, and 3.14.2 for laminations, C cores, permalloy powder cores, and ferrites, respectively. The A_L factor for ferrite core materials is normally given in terms of nanohenrys per turn squared (nH/N^2). This term is frequently more convenient due to the small number of turns involved.

3.4 HYSTERESIS LOOPS

The discussion in Section 3.1 assumed a certain value of current produced a corresponding value of flux, which is correct if the material has a constant permeability. However, when the material has a permeability which varies with B, a given H produces a flux whose value may vary widely. If the magnetic circuit is energized with a voltage and the inductance is measured as the magnetizing force is varied, the induction does not retrace the same values but produces a hysteresis loop as shown in Fig. 3.4. The significant points labeled are defined by American Society for Testing and Materials (ASTM) designation A-340 as follows:

B_m Maximum induction or flux density.
H_m Maximum magnetizing force.
B_r Retentivity or residual induction when the magnetizing force has been reduced from H_m to zero.
H_c Coercivity or coercive force necessary to reduce the residual induction to zero.

As the frequency of applied voltage is increased (assuming a constant flux density), the hysteresis loop becomes wider due to the increase in eddy currents flowing in the core material. A minor hysteresis loop, *defgd* of Fig. 3.4, results when the magnetizing force is cycled over an increment of H smaller than H_m.

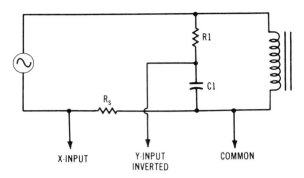

Fig. 3.5. Circuit for observing hysteresis loop on an oscilloscope.

The slope of the line from d to f is the incremental permeability, μ_Δ, and is the ratio of ΔB to ΔH.

It is possible to display the hysteresis loops of a magnetic device on an X–Y oscilloscope as shown in Fig. 3.5. When the circuit is excited with an alternating voltage, the voltage across R_s is proportional to the current. Since the number of turns and magnetic path length is probably known or measured, the magnetizing force in ampere turns per inch (or oersteds) may be calibrated on the X axis (abscissa). Since $E = N \, d\phi/dt$, the $R_1 C_1$ integrator produces a voltage across C_1 which is proportional to flux. Since the cross-sectional area is known or measured, the flux density in lines per square inches (or converted to kilogauss) may be calibrated on the Y axis (ordinate). To minimize harmonics in the integrated voltage of flux, the value of R_1 should be at least 10 times the reactance of C_1 at the excitation frequency.

3.5 MAGNETIC MATERIALS

Magnetic materials are manufactured under various trade names, available in numerous shapes, consist of various alloys to tailor performance, and thus have varying characteristic properties. Trade names, shapes, and material thickness are listed and coded in Table 3.4, and performance of various materials is listed in Table 3.6.

Figure 3.6 shows dc magnetization hysteresis loops for frequently used materials. As can be seen from the curves, Supermendur has the highest flux density, Deltamax has a very sharp saturation, and Supermalloy has the lowest core loss. Of course, operating frequency must also be considered. Silicon–iron materials are available in various grades, designated by American Iron and Steel Institute (AISI) type, and are listed in Table 3.7. Cost is, of course, an important consideration in choosing a core material. Deltamax, Supermalloy, and other nickel-content materials are relatively expensive. Supermendur, which contains nickel, also contains cobalt, and is therefore quite expensive. Ferrite core materials are relatively economical due to the sintering process during manufacture and to their usage at high frequencies which, in turn, results in a low weight. Silicon–iron materials are most economical, resulting from the high volume stamping for manufacture.

TABLE 3.6. Material Characteristics

Material	Trade Name	Percentage of Alloy	Characteristic Property	Typical Application	Initial Permeability	B_m, kG	B_r, kG	H_c, Oe	Shapes	Thickness	Typical Frequenc
Silicon–iron	AISI Grade	4% Si	Low cost	Transformers and reactors	400 typical	19	12	0.3–6	1a	1–3	To 100 Hz
Silicon–iron	A	3% Si	Low cost, grain oriented	Low-loss transformers and reactors	1500 typical	18	13.6	0.1–0.3	1a, b, 2, 3c	3, 4 5, 6 7, 8	To 100 Hz 100–1000] 2–5 kHz
Nickel–iron	B	45–50% Ni	Rectangular hysteresis loop	Saturating transformers and reactors	1000 typical	15–16	12–15	0.1–0.3	1a, b 3a 3b	3–6 6–8 8–11	100–1000] 1–8 kHz 10–50 kHz
Nickel–iron	C	45–50% Ni	Semioriented, soft	Combined permeability and flux density characteristics	7500 typical	12–15	6.2–8	0.03–0.1	1a, b 3a 3b		
Nickel–iron	D	70–80% Ni, 3–5% Mo	High permeability, low core loss, high stablity	High-frequency reactors	20,000–120,000	7–8	5–6	0.01–0.06	1a, b 3 3	3 6–8 9–11	To 500 Hz 2–20 kHz 40–100 kH
Cobalt–iron	E	3–50% Co (2% V)	High flux density, high cost	Small size and weight saturating	800 typical	22	21	0.1–0.3	1a, b	6–8	1–5 kHz
Powder Permalloy	F	—	High stability, selectable permeability	transformers High-Q filters	14–550	3–6	—	0.01–0.6	3a 3d	— —	10 kHz–1]
Powdered iron	G	—	Stable Q and permeability	Radio-frequency transformers, wave filters	10–100	9–10	—	0.01–2.2	4	—	10 kHz–microwa
Ferrites	H	Mn Zn Ni Zn	Wide variety of shapes and sizes, lower cost than nickel–iron, high resistivity	High-frequency pulse and power transformers and reactors	250–5000	2–4	1–2	0.1–0.3	5	—	10 kHz–microwa

Note: See Table 3.4 for definitions of codes under "Trade Name" and "Shapes."

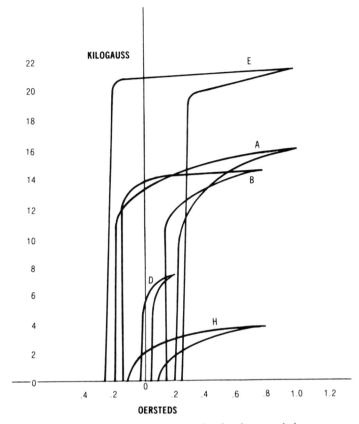

Fig. 3.6. Direct-current magnetization hysteresis loops.

TABLE 3.7. Applications of Standard Grades of Silicon–Iron Material

ASTM Type	Former AISI Type	Nominal Thickness (in.)	General Use
35G066	M-6X	0.014	Lowest core losses at high induction, for use in power and distribution transformers. High permeability at low induction for audio transformers. Grain oriented.
36F145	M-15	0.014	Low core loss for transformers and high-efficiency rotating machines.
47F174	M-19	0.0185	Most suitable for small transformers, inductors, and large rotating machines.
47F185	M-22	0.0185	Intermittent duty transformers, induction motor stators.
64F225	M-27	0.025	Intermittent duty motors and small transformers at moderate induction.

3.6 TRANSFORMER LOSSES, REGULATION, AND EFFICIENCY

Figure 3.7 shows a transformer having a primary winding which is connected to an ac source and a secondary which is connected to a load. The directions of voltage, current, and flux as well as leakage flux are shown for clarity in the loosely coupled windings. Figures 3.8A and B represent an equivalent circuit and phasor diagram of the transformer [3, 4].

3.6.1 Magnetizing Current

The magnetizing current I_m flows through the primary core reactance L_p and lags the induced voltage E_p by 90°, where L_p is a function of the primary turns squared, magnetic path length, cross-section area, and effective permeability. Since an increase in effective air gap (of C cores, devices using laminations, and ferrite cores) decreases the permeability and the winding inductance, the air gap is minimized during design and manufacture of the transformer. However, in the case of high-frequency ferrite-core transformers used in switch mode power supplies, an intentional air gap may be installed to prevent core saturation due to the possibility of unbalanced switching characteristics. Reference Section 9.6.

3.6.2 Core Loss

The core loss current I_c flows through the core resistance R_c and is the sum of eddy current and hysteresis current. Hysteresis loss, as discussed in Section 3.3, is the power required to magnetize the core first in one direction and then in the other on alternate half cycles. Eddy current loss is caused by the circulating current in the core material and is proportional to the square of the *lamination* thickness, frequency, and induction. Ferrite materials have a very high resistivity and therefore almost negligible eddy current loss. Figures 3.9 and 3.10 depict typical core losses as a function of flux density for various materials at various operating frequencies. These power dissipation losses produce heat in the core, and temperature rise, as discussed in Section 3.8, must be limited.

Manufacturers' data for ferrite core losses is usually presented in mW/cm^3 (milliwatts loss per unit volume). For comparison purposes, curves F and G of Fig. 3.10 are plotted in watts per pound by applying the appropriate volume and

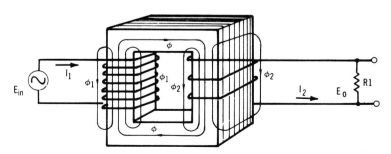

Fig. 3.7. Voltage, current, and flux in a transformer.

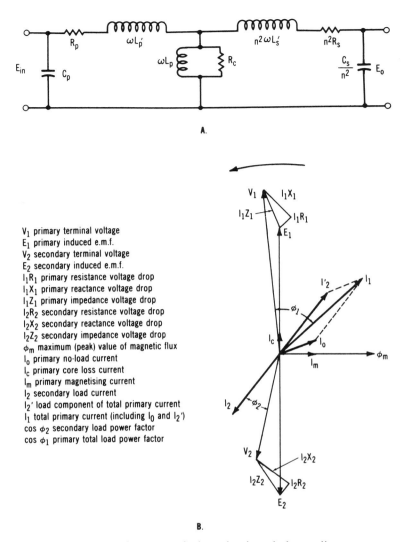

Fig. 3.8. Transformer-equivalent circuit and phasor diagram.

density values. Manufacturers' data is also available for apparent core loss (VA/lb), which is the power (VA) necessary to magnetize the core at no load.

Grain-oriented silicon–iron laminations have higher core losses at higher flux densities than C cores since the flux must traverse the back of the E lamination, which is perpendicular to the easy direction of magnetization. Interleaving E's and I's is standard practice to decrease reluctance, increase permeability, and minimize the magnetizing current by decreasing the effective air gap.

Hysteresis losses dominate powder core materials, which have the greatest slope, as shown in Fig. 3.10. Eddy current losses dominate tape-wound material. This is because eddy current losses increase as the square of the frequency while hysteresis losses increase by a lower exponent.

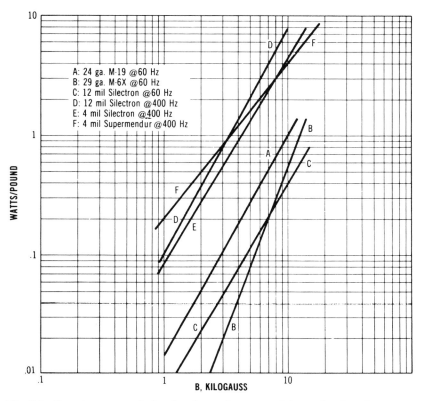

Fig. 3.9. Core loss versus induction for various materials used at low frequency.

3.6.3 Copper Loss, Reactance, and Regulation

Under load conditions the primary and secondary currents in the transformer produce an IR drop and an IX drop, as shown in Fig. 3.8B, which affect the voltage regulation. Also, I^2R losses (P_{cu}) produce heat in the windings, which affects temperature rise. At unity power factor, transformer regulation is defined as the difference in output voltage at no load and at full load by

$$\text{Percent Regulation} = (E_{in} - E_0) \times 100/E_0 \qquad (3.19)$$

where

$$E_0 = E_{in} - IR_{xfr} \text{ for low leakage inductance}$$

and by

$$\text{Percent regulation} = \frac{E_{in} - \left(\sqrt{E_{in}^2 - E_x^2} - IR_{xfr}\right)}{\sqrt{E_{in}^2 - E_x^2} - IR_{xfr}} \qquad (3.20)$$

if leakage inductance cannot be neglected (as in Fig. 3.12).

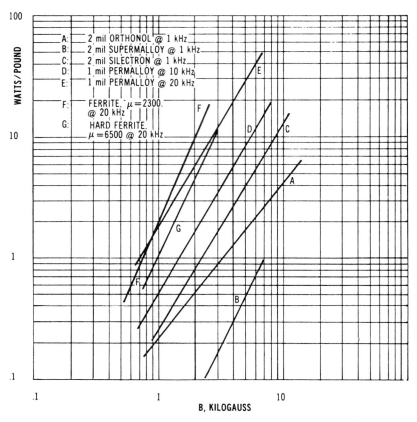

Fig. 3.10. Core loss versus induction for various materials used a high frequency.

The total input current to the primary is

$$I_1 \cong \sqrt{\left(i_p + I_c\right)^2 + I_m^2} \tag{3.21}$$

where I_p is load current referred to the primary.

For power supplies, copper loss and VA ratings increase as a function of rectifier configurations, as discussed in Section 4.1.

3.6.4 Transformer Efficiency and Power Factor

The transformer efficiency is defined as

$$\text{Percent efficiency} = \left(P_0/P_{\text{in}} \right) \times 100 \tag{3.22}$$

Power factor is

$$\text{P.F.} = P_{\text{in}}/\text{VA}_{\text{in}} \tag{3.23}$$

In general, highest transformer efficiency is obtained when copper losses are equal to core losses. In 50/60-Hz low-power transformers, copper losses invariably exceed core losses for straight resistive loads. However, when a transformer feeds a rectifier circuit the rated transformer VA must incresae (relative to the dc output power), as presented in Chapter 4, Table 4.2, and the secondary rms current will usually decrease (relative to the dc output current) resulting in lower copper loss.

For highest efficiency and energy conversion practices, the relation of copper loss (P_{cu}) to core loss (P_c) is now investigated. Referring to the core loss curves of Figs. 3.9 and 3.10 (preferably manufacturers' data), the equation of the linear portion (or a line tangent to an operating point) is

$$P_c = bB^a$$

where b is the P_c intercept at $B = 1$ kG and a is the slope of the line.

In order to determine the relation of VA, B, P_c, and P_{cu}, the differential of VA with respect to flux density is made, and the equation set to zero to obtain the optimum condition. From Equation (3.4), $E \cong B$ and $I \cong \sqrt{P_{cu}}$. Thus VA $\cong EI \cong B\sqrt{P_{cu}}$. Substituting $P_{cu} = P_{loss} - P_c$ and $P_c \cong B^a$,

$$VA \cong B\sqrt{P_{loss} - B^a}$$

$$(VA)^2 = P_{loss}B^2 - B^{a+2}$$

$$d(VA)^2/dB = 2P_{loss}B - (a+2)B^{a+1} = 0$$

$$2P_{loss} = 2P_c + 2P_{cu} = (a+2)B^a = (a+2)P_c$$

$$2P_c + 2P_{cu} = aP_c + 2P_c \qquad aP_c = 2P_{cu}$$

Then

$$P_c/P_{cu} = 2/a \qquad (3.24)$$

Thus, from this equation, optimum transformer efficiency is obtained when the ratio of core loss to copper (or conductor) loss is equal to $2/a$, where a is the slope of the core loss to flux density line at a point tangent to the operating flux density. For low-frequency transformer materials, the value of a typically ranges from 1.8 to 2.7.

3.6.5 Leakage Inductance

Due to the leakage flux, it is impossible to obtain complete flux linkage between windings of a transformer, as shown in Fig. 3.7. The result of this imperfect coupling is a leakage inductance between windings, which is undesirable in most cases. The transformer configurations of Fig. 3.11 have been analyzed for relevant leakage inductance calculations via the following formulas [2, 5, 6–7].

The geometry of Figs. 3.11A, B, and C is a stacked-layer winding and the leakage inductance, referred to the primary, is given by

$$L_1 = \frac{0.4\pi N_p^2 l_m}{w\,s^2}\left[\sum t + \sum \frac{h_p + h_s}{3}\right] \times 10^{-8} \qquad \text{H} \qquad (3.25a)$$

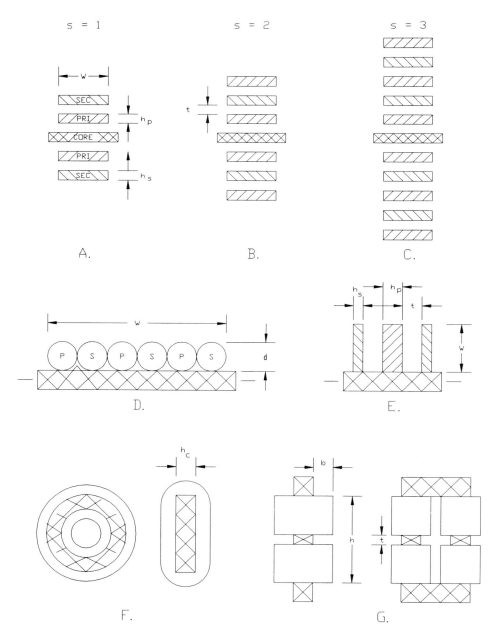

Fig. 3.11. Transformer configurations for leakage inductance analysis.

where l_m, h, t, and w are linear dimensions in centimeters (as are b, d, and l_c); N_p is the number of primary turns (leakage reflected to the primary); and s is a function of winding interleaving.

As noted by s^2 in the denominator, interleaving of the windings will substantially reduce the leakage inductance. This technique, as shown in Figs. 3.11B and C, would normally require additional insulation between windings and would reduce the window available for the conductors.

In push–pull inverter–transformer cirucits, bifilar winding of the primaries or secondaries is frequently mandatory to reduce leakage inductance and, therefore, voltage spikes produced by $L\,di/dt$. Figure 3.11D shows a transformer with bifilar windings and the leakage inductance is given by

$$L_1 = \frac{0.3\pi N_p^2 L_m d}{w} \times 10^{-8} \quad \text{H} \tag{3.25b}$$

Figure 3.11E is a transformer with pi windings, used for RF coils and high-frequency or high-voltage isolation, and the leakage inductance is given by

$$L_1 = \frac{0.1\pi N_p^2 L_m}{w}\left[\sum t + \sum \frac{h_p + h_s}{3}\right] \times 10^{-8} \quad \text{H} \tag{3.25c}$$

Figure 3.11F is a transformer wound on a toroid cure and the leakage inductance is given by

$$L_1 = \frac{2\pi N_p^2 d(2h + d/4)}{3l_c} \times 10^{-8} \quad \text{H} \tag{3.25d}$$

Figure 3.11G is a quad-winding transformer, which is universally popular for meeting voltage isolation and safety requirements. The windings normally consist of two primaries for series–parallel 115/230-V connection, and two secondaries for series–parallel, full-wave center-tap or bridge rectification. The leakage inductance is given by

$$L_1 = \frac{0.265 N_p^2 l_m(4\tau + h)}{b} \times 10^{-8} \quad \text{H} \tag{3.25e}$$

For transformers with high-voltage isolation, dimension t of Figs. 3.11A and E may be appreciable. In such cases the reactive impedance is expressed as a percentage of total impedance. Figure 3.12 represents a transformer with 30% reactance (X_r/Z) and $R = \cos\theta = \cos[\sin^{-1}(X_r/Z)] = 0.954$. The regulation due to the transposed IR drop is $100 - 95.4 = 4.6\%$ and must be added to the actual drop of the conductors. Leakage inductance and rectifier commutation also affect transformer regulation for single-phase and for three-phase rectified secondary voltages. This topic is discussed in Section 4.1.3.

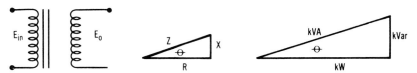

Fig. 3.12. Leakage reactance in high-voltage isolation transformer.

3.6.6 Transformer Capacitance

In high-frequency and/or high-voltage step-up transformers, capacitance effects can be detrimental to semiconductors switching the primary voltage. Each time the primary is abruptly switched, the capacitance which had been charged in one polarity must now be discharged and charged in the opposite polarity. The load appears instantaneously capacitive, and high peak currents can be produced in the primary switching devices due to $C\,dV/dt$.

Capacitance between surfaces can be expressed by [6]

$$C_0 = \varepsilon_0 \varepsilon A_s / t_i \quad \text{F} \tag{3.26a}$$

(with dimensions in meters) and

$$C_0 = 0.225 \varepsilon A_s / t_i \quad \text{pF} \tag{3.26b}$$

(with dimensions in inches), where

$\varepsilon_0 = 8.85 \times 10^{-12}$

ε = dielectric constant of insulating material

$A_s = w l_m$

w = winding width

l_m = mean length turn

t_i = insulation thickness

(For coils without layer insulation, t_i is equal to the insulation thickness on the conductor wire.)

Figure 3.13A illustrates capacitance in a step-up transformer of single-layer primary and secondary. For uniformly varying voltage:

$$C_{\text{eff}} = \frac{C_0 \left(E_1^2 + E_1 E_2 + E_2^2 \right)}{3E^2} \tag{3.27}$$

where E_1 is minimum voltage across C, E_2 is maximum voltage across C, and E is reference voltage of C.

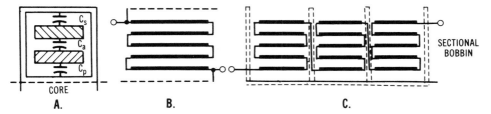

Fig. 3.13. Transformer and winding capacitance.

If the voltage at one end is zero (normally the case), $E_1 = 0$, $E_2 = E$, and $C_p = \frac{1}{3}C_0$. Then for a turns ratio $n = N_p/N_s$:

$$C_s = \tfrac{1}{3}C_0 + (n - 1)^2 C_p \qquad (3.28)$$

With a shield between primary and secondary, C_a may be omitted. For multi-layer windings, as depicted in Fig. 3.13B, a capacitance exists between layers (C_1) and from layer to core (C_0). Then

$$C_p = C_0/3p^2 + 4C_1(p - 1)/3p^2 \qquad (3.29)$$

where C_1 is measured layer capacitance and p is number of layers.

From the equation of C_0, capacitance is a function of winding configuration and may be reduced by using a tall winding and short width. However, this configuration increases leakage inductance, as discussed in the previous section. The capacitance may also be reduced by sectional windings, as shown in Fig. 3.13C and the right-hand portion of the above equation is divided by the number of sections. Minimum capacitance is obtained with an odd number of layers, and where coils are wound in the same rotational and traverse directions with starts at one end and finishes at the other side to obtain the same ac potential to the core.

The majority of transformers are vacuum impregnated, and high-frequency high-voltage transformers should also be vacuum encapsulated for isolation and environmental protection. Insulation materials should have a low dielectric constant to reduce stray capacitance. Epoxy resins have a dielectric constant of 3 to 4, but this value may be reduced with appropriate fillers. The dielectric constant of Teflon, polypropylene, and polyethylene is 2.1 to 2.3; this along with their high dielectric strength makes these materials most suitable for bobbins.

When a step voltage is initially impressed across the winding, current flows from the first few turns through the capacitance to the core. After a short interval, the current flows inductively through the remaining turns. Between the initial and final current distribution, oscillations due to leakage inductance and winding capacitance may occur, which extends the initial volts per turn from the first few turns into some of the remaining turns. This effect may be minimized by wrapping a thin copper sheet before the first layer and after the last layer. The first sheet is electrically connected to the finish, as shown in Fig. 3.13B. In this manner, the capacitance is more evenly distributed throughout the winding. Of course, the

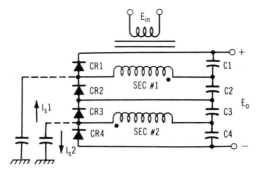

Fig. 3.14. Dual voltage-doubler with canceling capacitive secondary currents.

start–finish of each sheet must not make electrical contact, or else a shorted turn results.

For transformers whose secondary voltage is rectified to produce a dc output, interwinding capacitance may be reduced by the multiwinding techniques discussed above. Each winding output is indiviudally rectified, and the rectified outputs are connected in series. Dual voltage-doubler circuitry may be used to reduce the number of secondary turns, layers, and capacitance. Figure 3.14 shows a technique where the secondary capacitive current flow is also reversed [8].

3.7   COPPER AND OTHER CONDUCTORS

The dc resistance of a conductor is given by

$$R = \rho l / A \qquad \Omega \qquad (3.30)$$

where ρ is the resistivity of the material in Ω-cmil/ft (μ-Ω-cm), l is the length of the conductor in feet (cm), and A is the cross-sectional conductor area in cmil (cm^2).

Table 3.8 lists characteristics of common conductors. "Enameled" copper wire (sometimes called magnet wire) is used in the majority of magnetics due to its high conductivity, low cost, winding ease, and turn insulation. Aluminum may be used in large magnetics where high conductor area is available and low conductor weight is required. Silver-plated copper wire is sometimes used in high-frequency and high-current conductors to minmimize skin effect losses. Copper-clad steel wire may be used for high-strength requirements.

3.7.1   Copper

Characteristics of copper wire gages are tabulated in Tables 3.9A and B and 3.10 for bare, single Formvar (SF) and heavy armored polythermaleze (HAPTZ), two typical insulation coatings. Round wire is commonly used to conduct low to medium currents. Square wire, for medium to high currents, provides higher cross-sectional area and better winding utilization. Rectangular wire is normally

TABLE 3.8. Properties of Conductor Materials[a]

Material	Resistivity		Density		Thermal Conductivity	
	$\mu\Omega$-cm	Ω-cmil/ft	lb/in.^{3}	g/cm^3	Btu/h°Fft^2/ft	W/°C-cm
Aluminum	2.72	16.4	0.0967	2.7	126	2.18
Brass	3.9	23.4	0.31	8.65	63	1.08
Cadmium	7.5	45.1	0.31	8.65	53	0.92
Chromium	2.6	15.6	0.256	7.14	40	0.69
Copper			0.321	8.96	227	3.93
Annealed	1.724	10.37				
Hard drawn	1.77	10.65	0.321	8.96	227	3.93
Gold	2.42	14.6	0.692	19.32	171	2.96
Nickel	6.9	41.5	0.32	8.9	52.3	0.9
Silver	1.62	9.74	0.376	10.49	243	4.2
Steel	13–22	78–132	0.282	7.87	38	0.66
Tin	11.5	69.2	0.261	7.3	37	0.64
Zinc	6.1	36.7	0.255	7.13	64	1.1

[a]At 20°C.

used for high currents if skin effects and eddy current losses are negligible. Current ratings and resistance are graphed in Fig. 3.15. As a "rule-of-thumb", the 500-cmil/A rating (line A) means that the voltage drop is 2% per 100 ft of wire length.

Cross-sectional area for round wire is given in terms of circular mils (cmils), square millimeters, and square inches since current density ratings employ one of the terms. The following relations apply:

$$cmil = \pi/4 \text{ square mils}$$
$$= (\pi/4) \times 10^{-6} \text{ in}^2$$
$$= 5.036 \times 10^{-6} \text{ cm}^2 \tag{3.31}$$

The following relations presented to speed the selection of a wire size of required area (if tables are not convenient) are accurate to 0.5%, and apply to wire sizes AWG 0 and higher:

$$A = 10.55 \times 10^4 \times 0.793^{(\text{AWG})} \quad \text{cmil} \tag{3.32}$$

$$\text{AWG} = -10 \times \log(\text{cmil}/10.55 \times 10^4) \tag{3.33}$$

Copper magnet wire insulation characteristics and insulation class for various "hot spot" temperature ranges are listed in Table 3.11.

Thin copper sheets or strips, cut to the full window width minus margins, are frequently used to carry high current, especially in switch mode power supplies providing high currents at 5 V output. The copper strip or strips, normally 10 to 50 mils (0.2 to 1.2 mm) thick are wound on a mandrel, one turn per layer, with insulation material wound simultaneously between the turns. For large devices carrying very high currents, copper tubing of the appropriate cross-sectional area and dimaeter is used, and a liquid coolant is circulated through the tubing to maintain an acceptable temperature rise.

TABLE 3.9A. Wire Table, Standard Annealed Copper

Gage	Diameter (mils)	Cross Section cmils	Cross Section in.²	Ω/1000 ft	ft/Ω	lb/1000 ft	ft/lb	Ω/lb	lb/Ω
0000	400.0	211 600	0.1662	0.04901	20,400	640.5	1.561	0.00007652	13,070
000	400.6	167 800	0.1318	0.06182	16,180	507.8	1.969	0.0001217	8,215
00	364.8	133 100	0.1045	0.07793	12,830	402.8	2.482	0.0001935	5,169
0	324.9	105 600	0.08291	0.09825	10,180	319.5	3.130	0.0003075	3,252
1	289.3	83 690	0.06573	0.1239	8,070	253.3	3.947	0.0004891	2,044
2	257.6	66 360	0.05212	0.1563	6,398	200.9	4.978	0.0007781	1,285
3	229.4	52 620	0.04133	0.1971	5,074	159.3	6.278	0.001237	808.3
4	204.3	41 740	0.03278	0.2485	4,024	126.3	7.915	0.001967	508.5
5	181.9	33 090	0.02599	0.3134	3,190	100.2	9.984	0.003130	319.5
6	162.0	26 240	0.02061	0.3952	2,530	79.44	12.09	0.004975	201.0
7	144.3	20 820	0.01635	0.4981	2,008	63.03	15.87	0.007902	126.5
8	128.5	16 510	0.01297	0.6281	1,592	49.98	20.01	0.01257	79.58
9	114.4	13 090	0.01028	0.7925	1,262	39.62	25.24	0.02000	49.99
10	101.9	10 380	0.008155	0.9988	1,001	31.43	31.82	0.03178	31.47
11	90.7	8230	0.00646	1.26	793	24.9	40.2	0.0506	19.8
12	80.8	6530	0.00513	1.59	629	19.8	50.6	0.0804	12.4
13	72.0	5180	0.00407	2.00	500	15.7	63.7	0.127	7.84
14	64.1	4110	0.00323	2.52	396	12.4	80.4	0.203	4.93
15	57.1	3260	0.00256	3.18	314	9.87	101	0.322	3.10
16	50.8	2580	0.00203	4.02	249	7.81	128	0.514	1.94
17	45.3	2050	0.00161	5.05	198	6.21	161	0.814	1.23
18	40.3	1620	0.00128	6.39	157	4.92	203	1.30	0.770
19	35.9	1200	0.00101	8.05	124	3.90	256	2.06	0.485
20	32.0	1020	0.000804	10.1	98.7	3.10	323	3.27	0.306
21	28.5	812	0.000638	12.8	78.3	2.46	407	5.19	0.193
22	25.3	640	0.000503	16.2	61.7	1.94	516	8.36	0.120
23	22.6	511	0.000401	20.3	49.2	1.55	647	13.1	0.0761
24	20.1	404	0.000317	25.7	39.0	1.22	818	21.0	0.0476
25	17.9	320	0.000252	32.4	30.9	0.970	1,030	33.4	0.0300
26	15.9	253	0.000199	41.0	24.4	.765	1,310	53.6	0.0187
27	14.2	202	0.000158	51.4	19.4	.610	1,640	84.3	0.0119
28	12.6	159	0.000125	65.3	15.3	0.481	2,080	136	0.00736
29	11.3	128	0.000100	81.2	12.3	0.387	2,590	210	0.00476
30	10.0	100	0.0000785	104	9.64	0.303	3,300	343	0.00292

154

31	8.9	79.2	0.0000622	131	7.64	0.240	4,170	546	0.00183
32	8.0	64.0	0.0000503	162	6.17	0.194	5,160	836	0.00120
33	7.1	50.4	0.0000396	206	4.86	0.153	6,550	1,350	0.000742
34	6.3	39.7	0.0000312	261	3.83	0.120	8,320	2,170	0.000460
35	5.6	31.4	0.0000246	331	3.02	0.0949	10,500	3,480	0.000287
36	5.0	25.0	0.0000196	415	2.41	0.0757	13,200	5,480	0.000182
37	4.5	20.2	0.0000159	512	1.95	0.0613	16,300	8,360	0.000120
38	4.0	16.0	0.0000126	648	1.54	0.0484	20,600	13,400	0.0000747
39	3.5	12.2	0.00000962	847	1.18	0.0371	27,000	22,800	0.0000438
40	3.1	9.61	0.00000755	1,080	0.927	0.0291	34,400	37,100	0.0000270
41	2.8	7.84	0.00000616	1,320	0.756	0.0237	42,100	55,700	0.0000179
42	2.5	6.25	0.00000491	1,660	0.603	0.0189	52,900	87,700	0.0000114
43	2.2	4.84	0.00000380	2,140	0.467	0.0147	68,300	146,000	0.00000684
44	2.0	4.00	0.00000314	2,590	0.386	0.0121	82,600	214,000	0.00000467
45	1.76	3.10	0.00000243	3,350	0.299	0.00938	107,000	357,000	0.00000280
46	1.57	2.46	0.00000194	4,210	0.238	0.00746	134,000	564,000	0.00000177
47	1.40	1.96	0.00000154	5,290	0.189	0.00593	169,000	892,000	0.00000112
48	1.24	1.54	0.00000121	6,750	0.148	0.00465	215,000	1,450,000	0.000000690
49	1.11	1.23	0.000000968	8,420	0.119	0.00373	268,000	2,260,000	0.000000443
50	0.99	0.980	0.000000770	10,600	0.0945	0.00297	337,000	3,570,000	0.000000280
51	0.88	0.774	0.000000608	13,400	0.0747	0.00234	427,000	5,710,000	0.000000175
52	0.78	0.608	0.000000478	17,000	0.0587	0.00184	543,000	9,260,000	0.000000108
53	0.70	0.490	0.000000385	21,200	0.0472	0.00148	674,000	14,300,000	0.0000000701
54	0.62	0.384	0.000000302	27,000	0.0371	0.00116	859,000	23,200,000	0.0000000431
55	0.55	0.302	0.000000238	34,300	0.0292	0.000916	1,090,000	37,400,000	0.0000000267
56	0.49	0.240	0.000000189	43,200	0.0232	0.000727	1,380,000	59,400,000	0.0000000168

Source: National Bureau of Standards, *Handbook 100 Copper Wire Tables*.

Note: American Wire Gage; English units; values at 20°C. The fundamental resistivity used in calculating the tables is the International Annealed Copper Standard, viz., 0.153.28 $\Omega\text{-g}/\text{m}^2$ at 20°C. The temperature coefficient for this particular resistivity is $\Omega t_0 = 0.00393$ per °C, or $\alpha_0 = 0.00427$. However, the temperature coefficient is proportional to the conductivity, and hence the change of resistivity per °C is a constant, 0.000597 $\Omega\text{-g}/\text{m}^t$. The "constant-mass" temperature coefficient of any sample is

$$\alpha_t = \frac{0.000597 + 0.000005}{\text{resistivity in } \Omega\text{-g}/\text{m}^2 \text{ at } t°C}$$

The density is 8.89 g/cm^2 at 20°C.

The values given in the table are only for annealed copper of the standard resistivity. The use of the table must apply the proper correction for copper of any other resistivity. Hard-drawn copper may be taken as about 2.5% higher resistivity than annealed copper.

TABLE 3.9B. Characteristics of Round, Insulated, Annealed Copper Magnet Wire

AWG Gage	Diameter (mils)		Cross Section		$\Omega/$ 1000 ft., 20°C	Turns/in.^{2}	ft/lb
	SF	HAPTZ	cmil	in.^{2}			
8	131	134	16,510	0.01297	0.6282	57	20.0
9	116	119	13,090	0.01028	0.7921	72	25.2
10	104	107	10,380	0.00816	0.9989	90	31.6
11	93	96	8,234	0.00647	1.26	113	39.9
12	83	86	6,530	0.00513	1.588	141	50.3
13	74	77	5,178	0.00407	2.003	176	63.4
14	66	68	4,107	0.00322	2.525	220	79.9
15	59	61	3,257	0.00256	3.184	273	100.7
16	53	55	2,583	0.00203	4.016	350	126.8
17	47	49	2,048	0.00161	5.064	432	159.5
18	42	44	1,624	0.00128	6.385	540	201.7
19	37	40	1,288	0.00101	8.051	668	254
20	33	35	1,022	0.0008	10.15	850	320
21	30	32	810.1	0.00064	12.8	1,045	404
22	26	28	642.4	0.0005	16.14	1,300	508
23	24	26	509.5	0.0004	20.36	1,650	638
24	21	23	404.0	0.00032	25.67	2,030	806
25	19	21	320.4	0.00025	32.37	2,500	1,010
26	17	19	254.1	0.002	40.81	3,160	1,280
27	15	17	201.5	0.00016	51.47	3,880	1,610
28	13	15	159.8	0.00013	64.90	4,770	2,033
29	12	13	126.7	0.0001	81.83	5,920	2,555
30	11	12	100.5	0.00008	103.2	7,300	3,225
31	9.7	11	79.7		130.1	9,260	4,050
32	8.8	10	63.2		164.1	11,100	5,050
33	7.9	9	50.1		206.9	13,900	6,370
34	7.0	8	39.8		260.9	16,900	8,050
35	6.3	7	31.5		329.0	22,300	10,200
36	5.6		25.0		414.8	26,900	12,880
37	5.1		19.8		523.1	33,100	16,320
38	4.5		15.7		659.6	40,000	20,400
39	4.0		12.5		831.8	51,800	25,770
40	3.6		9.9		1049.0	66,200	32,640

3.7.2 Litz Wire

The use of Litzendraht (or Litz) wire can be an effective means of reducing skin effect conditions in high-frequency devices, as discussed in Sections 3.7.4 and 3.7.5. The skin effect is caused by current crowding at or near the surface of the conductor and increases the effective resistance of the conductor. Litz wire overcomes this problem through multistrand construction of fine film wires, each wire insulated from the other to prevent formation of a solid conductor. The construction may be spiraled or braided, which positions each wire near the outer surface at regular intervals.

However, Litz wire has approximately one-half the copper cross-sectional area of a solid magnet wire of equal overall diameter. Therefore, to benefit from the

TABLE 3.10. Characteristics of Square and Rectangular (Twin) HAPTZ Insulated, Annealed Copper Wire

Wire Size	Maximum Dimension in.	Allowance per turn	Nominal cmils	$\Omega/$ 1000 ft, 20°C	ft/lb
Square Wire					
5	0.188	0.195	39,730	0.267	8.01
6	0.168	0.175	32,450	0.327	10.2
7	0.150	0.155	25,440	0.416	12.9
8	0.134	0.140	19,960	0.531	16.4
9	0.120	0.125	16,400	0.645	20.0
10	0.107	0.110	12,950	0.817	25.3
11	0.096	0.100	10,210	1.04	32.1
12	0.086	0.090	8,050	1.32	40.8
13	0.078	0.082	6,440	1.66	50.5
14	0.070	0.074	5,070	2.10	61.9
Rectangular or Twin Wire					
3	0.236 × 0.458	0.245 0.463	133,000	0.084	2.55
4	0.211 × 0.418	0.219 0.422	108,000	0.104	3.22
5	0.188 × 0.372	0.195 0.375	81,800	0.134	4.00
6	0.168 × 0.332	0.175 0.335	66,700	0.169	5.10
7	0.150 × 0.296	0.155 0.300	52,300	0.208	6.45
8	0.134 × 0.264	0.140 0.270	41,000	0.266	8.20
9	0.120 × 0.235	0.125 0.240	33,600	0.323	10.0
10	0.107 × 0.210	0.110 0.215	26,600	0.409	12.6
11	0.096 × 0.187	0.100 0.190	20,900	0.520	16.1
12	0.086 × 0.167	0.090 0.170	16,600	0.646	20.0
13	0.078 × 0.156	0.082 0.159	13,000	0.831	25.2
14	0.068 × 0.132	0.072 0.135	10,500	1.05	30.9

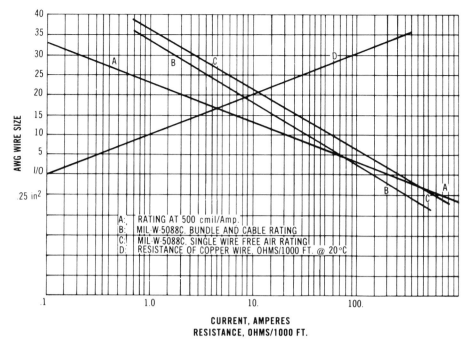

Fig. 3.15. AWG wire size versus current-carrying capacity and resistance of copper wire (round).

TABLE 3.11. Thermal Classification and Insulation Characteristics

"Hot Spot" Limiting Insulation Temperature (°C)	IEEE Class	NEMA Standard (MW 1000)	Military MIL-W-583 Class	Military MIL-T-27 Class	Characteristics
85	—	—	—	Q	Enamel; good moisture resistance
90	0	—	—	—	Enamel; good moisture resistance
105	A	MW 1, 2, 3, 15, 19, 29	105; Type E, T	R	Formvar, polyurethane, nylon; excellent flexibility and windability, good adhesion and abrasion resistance
130	B	MW 9, 28	130; Type B	S	Epoxy, polyurethane nylon; oil and moisture resistant
155	F	MW 5	155; Type L	V	Polyester; exceptional adhesion and flexibility, excellent moisture resistance in open construction
170	—	—	—	T	—
180	H	MW 24, 26, 30	180; Type H	U	Polyester nylon, polyester imide, Anatherm; excellent flexibility, scrape and abrasion resistance
200	—	MW 35	200, Type K	U	Polyester amide imide, anaclad, armored polythermaleze; resists thermoplastic flow, tough winding surface
220	C	MW 16	220, Type M		Polyimide[a], resistant to temporary overloads, resistant to radiation

[a]Dupont trademark.

use of Litz wire, the ratio of ac to dc resistance (R_{ac}/R_{dc}) must be greater than 2 to 1 to overcome the reduction in conductor cross-sectional area. As a guide, the use of Litz wire should be considered when the ratio of diameter to skin depth d/δ [reference Equation (3.35)] is greater than $2:1$. Conversely, at frequencies greater than 5 MHz, Litz wire becomes less practical due to capacitance effects between strands.

As the operating frequency of switch mode power supplies increases, skin effects become more prevalent. High-frequency transformers (in the range of 100 kHz) with ferrite cores are a typical example. In flyback converter transformers (unlike transformers used in forward converters), the primary and secondary

TABLE 3.12. Litz Wire Parameters

Litz Cable			Single Celanese, Single Polyurethane			Single Nylon, Single Polyurethane			Nominal Turns/in.^{2} (500–1000 cmil/A)
Strands		Nominal Wire Cable Area	Nominal Wire			Nominal Wire			
			Mean OD			Mean OD			
No.	AWG	(cmil)	(in.)	(lb/1000 ft)	(ft/lb)	(in.)	(lb/1000 ft)	(ft/lb)	
5	40	48.05	0.013	0.188	5,330	0.011	0.174	5,757	11,000
6	40	57.66	0.013	0.222	4,510	0.012	0.207	4,824	9,00
7	40	67.27	0.014	0.256	3,911	0.013	0.240	4,162	7,600
8	40	76.88	0.015	0.289	3,455	0.014	0.273	3,662	6,700
9	40	86.49	0.016	0.323	3,096	0.014	0.306	3,273	5,800
10	40	96.10	0.017	0.357	2,799	0.015	0.338	2,955	5,200
15	40	144.15	0.020	0.527	1,898	0.018	0.502	1,994	3,300
20	40	192.20	0.022	0.693	1,442	0.020	0.663	1,508	2,300
25	40	240.25	0.024	0.854	1,171	0.022	0.823	1,215	1,900
30	40	288.30	0.027	1.01	988	0.024	0.980	1,020	1,500
40	40	384.40	0.030	1.35	743	0.028	1.30	767	1,100
50	40	480.50	0.033	1.68	596	0.031	1.63	615	900
60	40	576.60	0.036	2.01	497	0.034	1.95	513	750
75	40	720.75	0.040	2.50	400	0.037	2.43	412	600
100	40	961.00	0.045	3.32	301	0.043	3.23	310	450
125	40	1201.25	0.050	4.12	243	0.047	4.00	250	350
150	40	1441.50	0.054	4.90	204	0.052	4.78	209	300
175	40	1681.75	0.058	5.68	176	0.056	5.56	180	250
5	44	20.00	0.009	0.0798	12,539	0.008	0.0730	13,691	28,000
6	44	24.00	0.010	0.0941	10,624	0.008	0.0862	11,598	25,000
7	44	28.00	0.010	0.108	9,222	0.009	0.0994	10,057	21,000
8	44	32.00	0.011	0.122	8,171	0.009	0.112	8,891	18,000
9	44	36.00	0.011	0.136	7,335	0.010	0.126	7,966	16,000
10	44	40.00	0.012	0.150	6,658	0.010	0.139	7,218	14,000
15	44	60.00	0.013	0.218	4,591	0.012	0.204	4,910	9,000
20	44	80.00	0.015	0.284	3,516	0.014	0.269	3,722	6,700
25	44	100.00	0.016	0.351	2,848	0.015	0.333	3,007	5,200
30	44	120.00	0.017	0.418	2,392	0.016	0.397	2,519	4,000
40	44	160.00	0.020	0.551	1,815	0.018	0.525	1,904	3,000
50	44	200.00	0.022	0.680	1,470	0.020	0.652	1,533	2,300
60	44	240.00	0.023	0.808	1,238	0.022	0.778	1,286	1,900
75	44	300.00	0.027	0.997	1,003	0.024	0.965	1,036	1,500
100	44	400.00	0.030	1.33	754	0.028	1.28	779	1,100
125	44	500.00	0.033	1.65	606	0.031	1.60	624	900
150	44	600.00	0.036	1.98	506	0.033	1.92	522	700
175	44	700.00	0.038	2.30	434	0.036	2.23	448	600

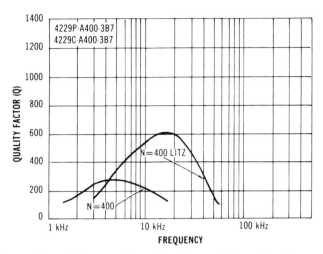

Fig. 3.16. Q factor of an inductor with solid wire and Litz wire. (Courtesy Ferroxcube, Linear Ferrite Materials and Components Catalog, Sixth Edition.)

currents flow during different parts of the cycle, preventing cancellations of perpendicular flux and thus increasing the skin effect. The inherent air gap in nontoroid core inductors also produces a strong magnetic field which increases the skin effect.

Typical Litz wire parameters are given in Table 3.12. Although base wire gages for the multistrand conductors range from AWG 30 to AWG 48, only AWG 40 and AWG 44 are shown due to space limitations. For comparison purposes with solid wire, 20/40 Litz wire has a cross-sectional area of 192 cmil and a diameter of 22 mils. From Table 3.9, the equivalent conductor of the same area is AWG 27 with a diameter of 15 mils. Conversely, if the requried turns of 20/40 Litz wire will fit in the winding bobbin then a solid wire of equivalent diameter, AWG 24, will also fit. The ratio R_{ac}/R_{dc} must then be analyzed to make a decision of which wire type to use for the application.

Figure 3.16 is an example of the effect on the Q factor ($Q = \omega L/R$) of an inductor wound with 400 turns of AWG 24, versus 400 turns of Litz 50/44 on a ferrite pot core by Ferroxcube. At frequencies up to 4 kHz, the Q of the solid conductor is higher than the Litz conductor. This is primarily due to the increased dc resistance of the Litz. Above 4 kHz, the Q with Litz wire increases while the Q with the solid wire decreases. The ratio of Q is thus greatly affected by the magnetic circuit as opposed to the ratio of Q for a straight conductor. For the given core and number of turns, the inductance is 64 mH for either wire type.

3.7.3 Aluminum

Aluminum strips or tubing may be used in the same manner as copper discussed above. The resistivity of aluminum is 1.6 times that of copper but requires more cross-sectional area than calculations using this figure to maintain the same winding resistance, since the increased area requires a larger core window which increases the mean turn length. The larger core window requires additional core material, thus increasing weight and cost. From a total cost consideration aluminum becomes attractive as the ratio of aluminum cost to copper cost decreases. Also, the density of aluminum is 0.3 times the density of copper, resulting in substantial weight savings.

3.7.4 Skin and Proximity Effects in Conductors

When an alternating current flows in a conductor, the current density varies from instant to instant. The changing current produces unequal electromotive forces (emf's) in the conductor cross section. These emf's oppose the flow of current and are highest at the center of the conductor. The current is thus forced toward the surface, producing a *skin effect*. The skin depth is the distance below the surface where the current density has diminished to $1/\varepsilon$ or 37% of its value at the surface. Skin effect causes an increase in the effective resistance of the conductor, as compared to the dc resistance, and $R_{ac} = FR_{dc}$ where F is a function of conductor diameter, frequency, and resistivity. The skin depth is

$$\delta = \sqrt{\lambda \rho / \pi \mu_r c} \quad \text{m} \tag{3.34}$$

where

c = velocity of light, $= 3 \times 10^8$ m/s $= 4\pi \times 10^{-7}\mu_r$

μ_r = permeability of conductor, $= 1$ for nonmagnetic conductors

ρ = resistivity, $\Omega - $m, $= 1.72 \times 10^{-8}$ for copper

λ = free space wavelength, $= 300 \times 10^6/f$ meters

Substituting and using copper as a conductor:

$$\delta = (1/2\pi)\sqrt{\rho \times 10^9/f}$$

$$= 6.61/\sqrt{f} \quad \text{(d cm)} \tag{3.35a}$$

$$= 2.60/\sqrt{f} \quad \text{(d in.)} \tag{3.35b}$$

The resistance per square for copper is

$$R_{sq} = \rho/\delta = 2.61 \times 10^{-7}/\sqrt{f} \quad \Omega \tag{3.36}$$

The resistance per unit length for copper is

$$R/l = \rho/A = 10.37 \times 10^{-6}/d^2 \quad \Omega/\text{ft} \tag{3.37a}$$

$$= 2.2 \times 10^{-6}/d^2 \quad \Omega/\text{cm} \tag{3.37b}$$

Since $d = k/\sqrt{R}$ and $d\sqrt{f} = k\sqrt{f/R}$, $F = R_{ac}/R_{dc}$ is plotted versus $d\sqrt{f}$ and $\sqrt{f/R_{dc}}$ for copper wire in Fig. 3.17A. Also shown is the proximity effect on

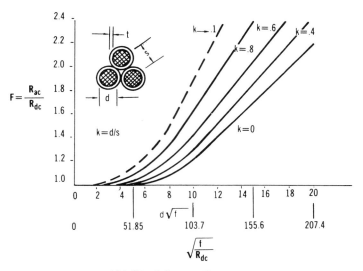

(A) Straight conductors.

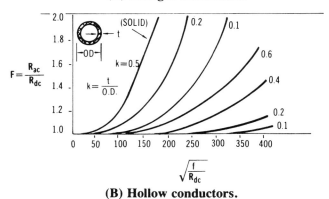

(B) Hollow conductors.

Fig. 3.17. Skin effect ratio.

resistance for a three-conductor cable with various ratios of $k = d/s$, where the insulation thickness of each wire is $t = \frac{1}{2}(s - d)$.

Increasing the wire gage (decreasing the diameter) and adding extra or paralleled conductors (bifilar, trifilar, quadfilar) to maintain the same overall cross-sectional area is obviously beneficial in reducing skin effects. For very high switching frequencies applied to conductors, especially in transformers and sometimes in inductors. Litz wire may be used, as discussed in Section 3.7.2, with parameters as given in Table 3.12. However, where space and cost permit, decreasing the wire gage to increase the cross-sectional area will *always* reduce the effective resistance. Since $R_{dc} = k_1/d^2 = k_2 R_{ac}/d$, doubling the wire diameter decreases the dc resistance by one-fourth, and the dc/ac resistance ratio by one-half. Thus the effective resistance R_{ac} always decreases as wire gage is decreased (wire diameter is increased).

For high-current requirements where an increase in the cross-sectional area is not a practical method to reduce skin effects, hollow conductors (tubes) may be used. A liquid is pumped through the tube for cooling purposes. The thickness of a tube is $t = \frac{1}{2}(\text{OD} - \text{ID})$ and the cross-sectional area of a tube is

$$A_{cu} = (\pi/4)(\text{OD}^2 - \text{ID}^2) = \pi t^2 = \rho l / R_{dc} \tag{3.38}$$

Skin effect on hollow conductors is plotted in Fig. 3.17B for various thickness ratios.

3.7.5   Skin and Proximity Effects in Windings

As discussed in the previous section, high-frequency currents produce a skin effect and a proximity effect which increases the dc resistance of a given conductor. In magnetic windings (discussed in the following section), these effects are more pronounced due to the high magnetic field environment, which further increases the eddy currents in the conductor. The following analysis includes a summary on the subject aptly treated by Snelling [6].

In a typical transformer, the primary and secondary volts per turn and ampere-turns are nearly balanced and the primary occupies nearly the same space as the secondary. Referring to Fig. 3.11, the following parameters for each winding must be evaluated: p, the number of layers; N_l, number of turns per layer; w, copper winding width; h, conductor height; b, conductor width. The latter two parameters are for square wire. For round wire, $h = 0.866d$. The copper layer factor is a dimensionless term:

$$F_l = N_l b / w = 0.866 N_l d / w \tag{3.39}$$

From the above equations, the following term may be calculated:

$$\frac{h\sqrt{F_l}}{\delta} = \frac{0.866 d\sqrt{F_l}}{\delta} \tag{3.40}$$

where δ is given by Equation (3.35).

In Fig. 3.18, $F = R_{ac}/R_{dc}$ is plotted versus $h\sqrt{F_l}/\delta$ for various numbers of layers p. In Fig. 3.11B, the mmf (magnetomotive force) is zero midway through the secondary, and each half of each winding could produce an equivalent half layer if the total number of layers is an odd integer. For this reason, half layers are shown in Fig. 3.18. Since the factor F is a function of diameter and height, Litz wire may be used to decrease F. Reference Table 3.12 for Litz wire characteristics.

3.7.6   Conductor Loss

Conductor loss, mentioned in Section 3.6.3 for copper, is the I^2R loss in the coil which produces heat and voltage drop in the winding. Conductor loss may be calculated by

$$P_{cu} = \frac{I_{rms}^2 \times F \times R' \times l_m \times N}{12000} \quad \text{W} \tag{3.41}$$

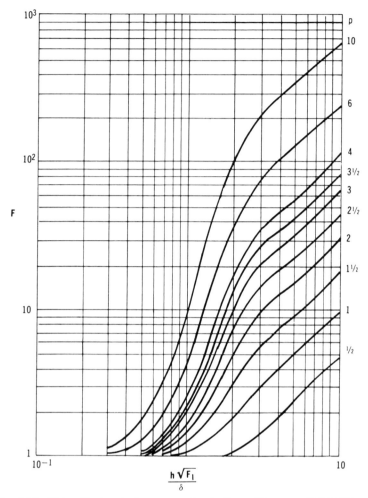

Fig. 3.18. Skin effect ratio for magnetic windings. (Courtesy ILIFFE.)

where F is the skin effect factor, l_m is the mean turn length, in., R' is $\Omega/1000$ ft, and N is number of turns.

In a transformer, l_m will be different for the primary and secondary(s). If the primary is wound first, it will have a lower power loss than the secondary, assuming an equal current density in the winding. In many cases, the higher copper loss in the secondary may be desirable since the "hotter" secondary can more readily conduct heat to the exterior surface of the winding.

3.8   WINDINGS

Coils wound by lathe-type equipment on bobbins or mandrel-tubes for EI and DU laminations, C cores, and ferrite E, U, and pot cores are considered most economical. Toroid shapes require loading shuttles (essentially requiring a "double-winding" operation), which may be limited in wire size handling capacity for some facilities.

Bobbins are available in numerous shapes and materials for small to medium coils. Some bobins have molded sections which can provide isolation between windings, or decrease interwinding capacitance. Bobbins with molded pins offer easy winding terminations, and are ideal for printed circuit board insertion.

Medium to large coils are usually wound on a paper tube or higher temperature material, with the internal dimensions slightly larger than the magnetic core dimensions over which the coil is to be placed. The tube is then placed over a mandrel. Since the tube has no shoulders to hold the winding, tape is used as shown in Fig. 3.19. The tape is placed on one end of the tube in several places, sticky side up, with the inner portion taped to the tube. The conductor is then started on that end, allowing an appropriate margin, and one or two turns are wound over the tape. The loose ends of the tape are then folded over the turns

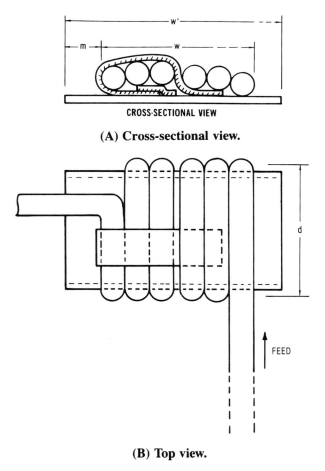

(A) Cross-sectional view.

(B) Top view.

Fig. 3.19. Methods of starting initial coil winding.

and taped to the tube. The next turn then passes over the tape, holding the first turns securely. The same procedure is followed at the other end of the tube. Insulation is then placed over the first layer and the taping process is repeated until the winding is complete. Coils with several layers and the same number of turns per layer can be wound in this manner. Multicoils can be wound simultaneously on machines so equipped.

The winding must fit into the window area of width w and height h. The tube width w' should be slighlty smaller than h. After choosing the wire size to carry the required current, the number of turns per layer (rounded to the lower integer) that will fit within the width is

$$N_l = [(w' - 2m)/d] - 1 \qquad (3.42)$$

where N_l is rounded to the lower integer, w' is tube width, m is the winding margin (Fig. 3.19), and d is the diameter of wire, including insulation.

The number of layers (rounded to the higher integer), which is a function of conductor diameter and the layer to layer insulation thickness, and for a large number of layers may be quite appreciable in high voltage windings, is

$$p = \frac{h - t_{p-s} - t_b - t_w}{d + t_l} \quad \text{maximum} \qquad (3.43)$$

where

$\quad\quad h$ = height of bobbin or window (build)

$\quad\quad t_b$ = thickness of bobbin or tube

$\quad t_{p-s}$ = primary to secondary insulation

$\quad\quad d$ = diameter of wire, including insulation

$\quad\quad t_l$ = thickness of layer insulation

$\quad\quad t_w$ = thickness of final wrap

An initial method for determining the window fill of a magnetic core, after the wire gage is determined, is from data shown in Fig. 3.20. Here, the number of turns N is plotted versus window area A_w for various AWG wire gages of single insulated magnet wire. For instance, if AWG 20 is used and the desired turns are 80, the required core window area is 0.1 in.^{2} For exact window fill calculations, observing the desired layer insulation and margin requirements, the above equations may then be solved to verify the winding will fit the core. This approach is valid for all core shapes except toroids where sufficient winding area must be allowed for the shuttle.

A novel technique for low-power toroid core magnetics involves placing the core on a printed circuit board which has the desired conductor-clad pattern and hole layout and forming the turns by "jumper wires" around the core to clad. For toroids where N_s is much greater than N_p, or vice versa, another technique is to wind the primary with a shuttle, split a bobbin and place it around the core and primary, reseal the bobin, and machine rotate the bobbin to wind the secondary.

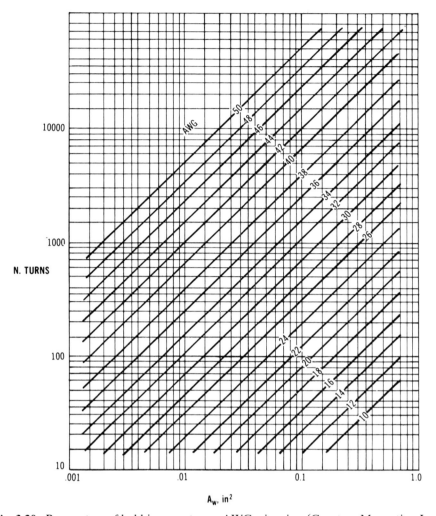

Fig. 3.20. Parameters of bobbin area, turns, AWG wire size. (Courtesy Magnetics, Inc.)

For high-voltage secondaries, the volts per turn are the same as the primary but the voltage gradient from start to finish may be quite high. If the start and finish of the secondary winding of a toroid core transformer are physically close after winding, voltage breakdown could occur between these points. A method to overcome this problem is shown in Fig. 3.21. After the primary is wound and the insulation from primary to secondary is wound, two secondary windings are then wound. Referring to Fig. 3.21A, winding 1 is wound from S_1 (start 1) in a CW rotation, finishing at F_1. Winding 2 is then wound from S_2 in a CCW rotation, finishing at F_2. In this manner, the voltage gradient between S_1 and S_2 and between F_1 and F_2 is zero, if both windings have the same number of turns. The two windings may be paralleled, where the smaller wire size of each winding (as compared to that of a single winding) also reduces skin effects. This technique may

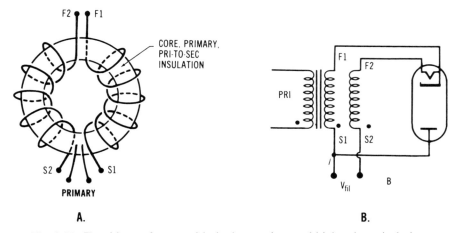

Fig. 3.21. Toroid transformer with dual secondary and high-voltage isolation.

be applied to transformers powering high-voltage vacuum tubes such as a pulsed magnetron. Referring to Fig. 3.21B, S_1 of winding F_1 is connected to the anode and to one side of an ac voltage, with F_1 connected to the cathode and S_2 of winding 2 is connected to the other side of the ac input, with F_2 connected to the filament. The ac source and the anode are then at a low potential with respect to *ground* while the cathode may be several kilovolts negative with respect to *ground*. In this manner, a high-voltage pulse transformer and filament transformer are combined in one device. The wire size for the secondary windings must be selected to conduct the continuous filament current as well as the pulsed cathode current.

3.8.1 Insulating Materials

The selection of insulating materials is based primarily on temperature rise in the transformer or inductor. Table 3.13 lists industrial and military thermal classifications of insulations. Materials are available in various grades, thicknesses, and trade names. Kraft paper is commonly used up to class B, Nomex may be used for class F and H, and Kapton for class C, as examples. Mica materials are especially corona resistant, operate at high temperatures, but have a relatively high dielectric constant. Kapton is quite expensive and has a dielectric constant of approximately 3.6. Nomex has a dielectric constant of approximately 2.3, but multilayers of insulation are difficult to wind and compress due to "cold flow" of the Nomex.

3.8.2 Voltage Isolation

The amount of insulation required to provide the desired voltage isolation between layers and between the primary and secondary is a function of internal voltage gradients, impregnation processes, and the specified isolation for a particular application. Typical "hipot" isolation is 1500 V, or two times the maximum internal voltage, whichever is greater. The isolation requirement of certain European safety standards is 3750 V (reference Chapter 13, Table 13.1). Three or more

TABLE 3.13. Thermal Classification of Insulations

MIL-T-27 Class	IEEE Class		Limiting Insulation Temperature "Hottest Spot" (°C)
Q (85°C)	O	Consists of cotton, silk, paper and similar organic materials when neither impregnated nor immersed in a liquid dielectric	90
R	A	Consists of (1) cotton, silk, paper, and similar organic materials when either impregnated or immersed in a liquid dielectric; (2) molded and laminated materials with cellulose filler, phenolic resins and other resins of similar properties; (3) films and sheets of cellulose acetate and other cellulose derivatives of similar properties; and (4) varnishes (enamel) as applied to conductors	105
S	B	Consists of materials or combinations of materials such as mica, glass fiber, asbestos, etc., with suitable bonding substances; other materials or combinations of materials, not necessarily inorganic, may be included in class B, if by experience or accepted tests they can be shown to be capable of operation of 130°C temperatures	130
V	F	Consists of materials or combinations of materials such as mica, glass fiber, asbestos, etc., with suitable bonding substances; other materials or combinations of materials, not necessarily inorganic; may be included in class F if by experience or accepted tests they can be shown to be capable of operation at 155°C	155
T (170°C)	H	Consists of materials or combinations of materials such as silicone elastomer, mica, glass fiber, asbestos, etc., with suitable bonding substances such as appropriate silicone resins; other materials or combinations of materials may be included in class H if by experience or accepted tests they can be shown to be capable of operation of 180°C temperatures	180
U (> 170°C)	C	Consists entirely of mica, porcelain, glass, quartz, and similar inorganic materials	Over 220

Note: It is important to recognize that other charcteristics in addition to thermal endurance such as mechanical strength, moisture resistance, corona endurance, etc., are required in varying degree in different applications for the successful use of insulating materials.

TABLE 3.14. Voltage Spacing of Dielectrics

Dielectric	dc Voltage Rating (kv)	Oversurface Spacing (in.)	Impregnation
Air	1	0.188	Vacuum varnish
	2	0.25	
	3	0.31	
	5	0.38	
	6	0.50	
	10	1.12	
Air	5	0.25	Vacuum varnish
	10	0.56	and Epoxy
	15	1.25	end fill
	20	1.75	
Oil	30	1.5	Vacuum varnish
	50	2.25	
	70	2.75	
	100	3.0	
	120	3.5	
	150	4.5	

layers of 0.1 mm insulation are normally required between the primary and secondary to meet these specifications, even with insulated magnet wire and with an impregnation process.

After being wound, the coil may be impregnated or it may be placed around the magnetic core and the entire assembly impregnated. Vacuum impregnation with a solvent varnish is most common but may leave voids or air pockets after curing. In high-voltage windings, these voids may cause high-voltage stress levels across the air pockets (dielectric constant of air equals one), inducing corona or failure after extended use. In these cases, solventless materials, such as epoxy resin, should be used where facilities permit.

The impregnation process offers several advantages. It protects the conductors and the core material (or laminations if used) from movement, thus reducing potential noise and mechanical damage. It provides environmental protection. It increases the dielectric strength of the coil insulating materials. It assists in thermal conduction and heat dissipation from the coil.

For high-voltage isolation in high-power magnetic devices, the complete magnetic assembly is normally placed in a container of transformer oil. The oil has approximately 2.5 times the dielectric strength of air. Table 3.14 lists required voltage spacing for air and oil insulation. If weight is an important factor, the assembly may be placed in a container of sulfur hexafluoride (SF_6). This gas is chemically inert below 250°C and has a low toxicity. The specific gravity is approximately five times that of air and the dielectric strength is twice that of air.

3.8.3 Three-Phase Windings and Connections

High-power equipment and transformers usually operate from three-phase power for improved efficiency and lower line-voltage drops. In power conversion equipment, the input line voltage or the voltage scaled by a transformer may be rectified to provide power to an inverter stage, such as in a frequency converter or uninterruptible power system. To reduce the filter requirements for rectifier circuits, the transformer frequently utilizes separate Δ and Y secondaries, as shown in Fig. 3.22A. The output from each winding is rectified to produce the individual 6-pulse phasor diagrams and waveforms shown in Figs. 3.22D and E. For high-voltage applications, the rectified voltage sources are connected in series (Fig. 3.22B) to produce a 12-pulse waveform, where $E_0 = E_1 + E_2$ and $I_1 = I_2$. For high-current applications, the rectified voltage sources are connected in parallel, but through an interphase transformer (Fig. 3.22C), and $E_0 = E_1 = E_2$ and $I_0 = I_1 + I_2$. The interphase transformer balances the difference in voltage by inducted voltage in the windings, and the exciting current is the difference of the currents to be combined. As discussed in Section 4.1.2, the primary advantage of the 12-pulse rectification is the reduction in ripple voltage. From Fig. 3.22E, the 6-pulse waveform has a peak-to-peak ripple of $(1 - \cos 30°) \times 100 = 13.4\%$, while the 12-pulse waveform has a peak-to-peak ripple of $(1 - \cos 15°) \times 100 = 3.41\%$.

The above discussion is introductory to the following analysis. In Fig. 3.22, all secondary line voltages and currents were assumed equal. However, the phase current of the Δ winding is $1/\sqrt{3}$ times the phase current in the Y winding, and the phase voltage of the Y winding is $1/\sqrt{3}$ times the phase voltage of the Δ winding. In the first case, the cross-sectional area of the conductor in the Y winding must be $\sqrt{3}$ times the area of the conductor in the Δ winding. More importantly, in the latter case, the turns in the Δ winding must be $\sqrt{3}$ times the turns in the Y winding. Since the number of turns is usually an integer (half-turn may be obtained from EI cores), designers are faced with the $\sqrt{3}$ factor, especially where the turns ratio, N_p/N_s, is high. Secondary turns ratio and resulting percent voltage imbalance are tabulated as

$4N:7N$	7.2%
$11N:19N$	5.2%
$15N:26N$	1.9%
$56N:97N$	0.5%

The voltage balance and current equalization in the secondary can be achieved with an *extended delta* approach. Phasor diagrams are shown in Figs. 3.23A and B. In winding each coil of the transformer, the primary is wound first. The two secondaries are wound simultaneously with a total of $(1 + \sqrt{3})N_s$ turns and tapped at $\sqrt{3}N_s$ turns. Figure 3.23C represents one coil for a particular phase. The other two coils are wound in the same manner. For ultimate balance and where facilities permit, all coils (three primaries and then six secondaries) may be simultaneously wound.

Example

The advantage of the extended delta appears in analyzing the phasor diagrams with phantom neutral. In Figs. 3.23A and B, each line and phase voltage is equal in magnitude and 120° apart. However, the winding connections produce a 30° phase shift from a to a'. When the outputs from each winding are rectified and "paralleled," the waveform of Fig. 3.22E results, producing a 15° voltage separation. The $\sqrt{3}$ turns factor is no longer a voltage or current imbalance problem and

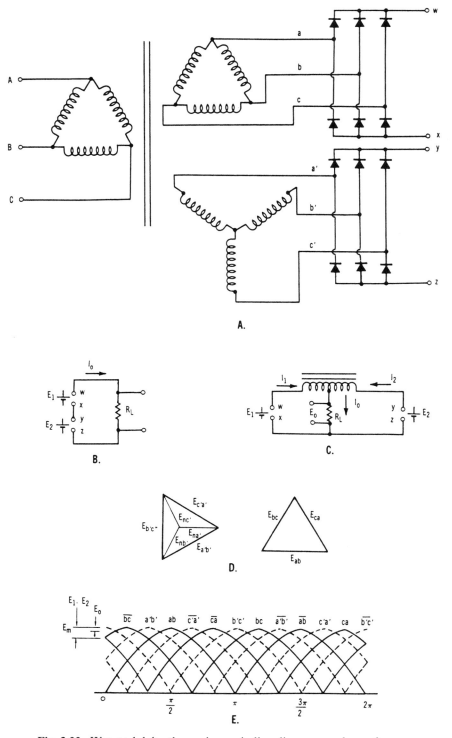

Fig. 3.22. Wye and delta three-phase winding diagrams and waveforms.

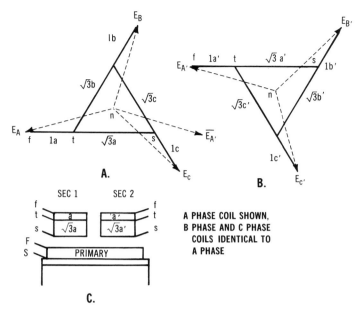

Fig. 3.23. Extended delta three-phase winding diagrams and winding sequence.

affects the phase balance slightly. Instead of $(\tan^{-1} 1/\sqrt{3}) = 30°$ phase shift, the $4/7$ ratio produces $(\tan^{-1} 4/7) = 29.74°$ shift from $a'b'$ to ab, and $30.26°$ from ab to $c'a'$. The resultant 3.47% peak-to-peak ripple voltage is insignificant as compared to the normal 3.41%. In fact, for very high current requirements such as magnet power supplies and in filament and arc power supplies for fusion sources, a turns ratio of $2:1$ in the extended delta secondary produces $26.6°$ and $33.4°$ phase shifts, and a peak-to-peak ripple voltage of only 4.22%.

3.9 TEMPERATURE RISE

Core and copper losses in a transformer or inductor produce heat, causing a temperature rise which must be limited to a compatibility with other materials, efficiency, and environment. Due to the physical complexity, temperature rise is difficult to predict with complete accuracy, but various methods provide a close approximation [2, 5–7]. The heat transfer equations of Chapter 15 may also be considered for analysis.

For open-construction transformers and inductors, the following analysis is reasonably accurate. Heat produced in the core is conducted to the edge surfaces and dissipated. Heat produced in the coils is conducted through the copper and insulation to the exposed surfaces of the coil and dissipated. Temperature rise is thus a function of the power dissipation and the surface area available for dissipation. Surface area may be calculated by the mehtods discussed in Sections 3.11 and 3.12. Dividing the calculated power loss by the surface area produces a surface dissipation, and temperature rise may be obtained from the curves of Fig. 3.24 [9].

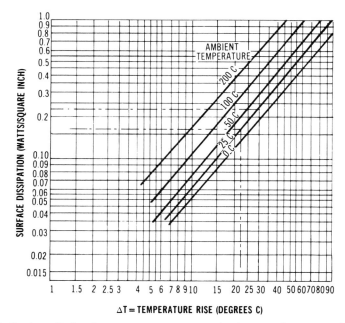

Fig. 3.24. Surface dissipation versus temperature rise for transformers and inductors.

The use of high-temperature materials usually results in a nominal cost penalty. However, the overall size of the magnetic device is decreased, resulting in iron and copper cost reduction. A cost-effective design should therefore operate at the highest temperature reasonable within the material and environmental confines.

3.10 AUDIBLE NOISE

Audible noise in magnetic devices can range from an objectionable 50-Hz hum to a piercing 10-kHz whistle. Since the peak sensitivity of the human ear is around 2 kHz, operating frequencies from 400 Hz to 8 kHz can be irritating. Intensity levels of equipment housing magnetic devices should not exceed 70 to 75 dB (decibels above 10^{-16} W/cm^2). Frequencies above 18 kHz are considered beyond the audible range.

To minimize noise in transformers and inductors, the following procedure or versions thereof may be used. In the case of coils and laminations, the assembly is vacuum impregnated using a low-viscosity varnish and then baked. The low-viscosity liquid penetrates the winding, securing the conductors and insulation as well as the space between the laminations. The process is then repeated using a higher viscosity varnish which fills all voids within the winding. The coil ends and the space between the coil and core are then filled with epoxy resin and cured. When a C core is used, the gap space (and gap material for inductors) is cemented and the core is banded to a recommended force, depending on the size of the core.

In the case of tape-wound toroids, the casing is normally sealed and the impregnation process prevents movement within the winding. The sintering process inherent in manufacturing ferrites prevents core material movement except under high vibration. Halves of pot cores, especially those with sensitive air gaps in the center post, should be thoroughly cleaned and cemented. Potting materials for ferrite cores should have the same thermal coefficient of expansion as the ferrite material to prevent core cracking.

For those interested in pursuing the subject of noise reduction in magnetic devices, the following article contains over 200 papers on the subject: "Bibliography on Transformer Noise," *IEEE Transactions on Power Apparatus and Systems*, Volume PAS-87, No. 2, February 1968

3.11 LAMINATIONS

Laminations, as presented in Tables 3.4, 3.6, and 3.15, are available in various shapes, sizes, and materials. Silicon–iron grades are the most common and economical due to the vast requirements for 50- and 60-Hz transformers. In over

TABLE 3.15. Dimensional Data of Popular Lamination Sizes

Lamination Designation	A (in.)	B (in.)	D (in.)	T (in.)	l_c (in.)	$A_c{}^a$ in.^{2}	A_w in.^{2}	Square Stack Weight (lb)
			Common Single-Phase Laminations					
EI-375[b]	$\frac{3}{16}$	$\frac{5}{16}$	$\frac{3}{4}$	$\frac{3}{8}$	2.875	0.141	0.234	0.11
EI-50	$\frac{1}{4}$	$\frac{1}{4}$	$\frac{3}{4}$	$\frac{1}{2}$	3.00	0.250	0.187	0.20
EI-21[b]	$\frac{1}{4}$	$\frac{5}{16}$	$\frac{13}{16}$	$\frac{1}{2}$	3.25	0.250	0.254	0.22
EI-625	$\frac{5}{16}$	$\frac{5}{16}$	$\frac{15}{16}$	$\frac{5}{8}$	3.75	0.39	0.293	0.40
EI-75	$\frac{3}{8}$	$\frac{3}{8}$	$1\frac{1}{8}$	$\frac{3}{4}$	4.5	0.562	0.422	0.68
EI-87	$\frac{7}{16}$	$\frac{7}{16}$	$1\frac{5}{16}$	$\frac{7}{8}$	5.25	0.766	0.574	1.09
EI-100	$\frac{1}{2}$	$\frac{1}{2}$	$1\frac{1}{2}$	1	6.0	1.0	0.75	1.62
EI-125	$\frac{5}{8}$	$\frac{5}{8}$	$1\frac{7}{8}$	$1\frac{1}{4}$	6.75	1.27	1.17	3.16
EI-138	$\frac{11}{16}$	$\frac{11}{16}$	$2\frac{1}{16}$	$1\frac{3}{8}$	8.25	1.88	1.42	4.22
EI-150	$\frac{3}{4}$	$\frac{3}{4}$	$2\frac{1}{4}$	$1\frac{1}{2}$	9.0	2.25	1.69	5.49
EI-175	$\frac{7}{8}$	$\frac{7}{8}$	$2\frac{5}{8}$	$1\frac{3}{4}$	10.5	3.05	2.30	8.7
EI-200	1	1	3	2	12.0	4.0	3.0	13.1
EI-212	$1\frac{1}{16}$	$1\frac{1}{16}$	$3\frac{3}{16}$	$2\frac{1}{8}$	12.75	4.52	3.38	15.4
EI-225	$1\frac{1}{8}$	$1\frac{1}{8}$	$3\frac{3}{8}$	$2\frac{1}{4}$	13.5	5.06	3.80	18.4
EI-19[b]	$\frac{7}{8}$	$1\frac{3}{4}$	3	$1\frac{3}{4}$	13.0	3.06	5.25	10.7
EI-250	$1\frac{1}{4}$	$1\frac{1}{4}$	$3\frac{3}{4}$	$2\frac{1}{2}$	15.0	6.25	4.69	25.4
EI-251[b]	$1\frac{1}{4}$	2	$5\frac{1}{2}$	$2\frac{1}{2}$	20.0	6.25	11.0	34.1
EI-3	$1\frac{1}{2}$	$1\frac{1}{2}$	$4\frac{1}{2}$	3	18.0	9.0	6.7	43.7
EI-4	2	2	6	4	24.0	16.0	12.0	105
EI-5	$2\frac{1}{2}$	$2\frac{1}{2}$	$7\frac{1}{2}$	5	30.0	25.0	18.7	206

TABLE 3.15.   (*Continued*)

Lamination Designation	A (in.)	B (in.)	D (in.)	T (in.)	$A_c{}^a$ (in.)^2	$A_w{}^c$ (in.)^2	Square Stack Weight^d (lb)
				Common Three-Phase Laminations			
EI-$\frac{1}{2}$	$\frac{1}{2}$	$\frac{5}{8}$	$1\frac{3}{8}$	$\frac{1}{2}$	0.25	0.86	0.75
EI-1	1	1.5	3	1	1.0	4.5	6.38
EI-1.5	1.5	1.5	$3\frac{3}{4}$	1.5	2.25	5.62	16.3
EI-1.8	1.8	1.8	4.5	1.8	3.24	8.1	27.9
EI-2.4	2.4	2.4	6	2.4	5.77	14.4	66.4
EI-3.6	3.6	3.6	9	3.6	12.96	32.4	224

Lamination Designation	A (in.)	B (in.)	C (in.)	E (in.)	F (in.)	G (in.)	l_c (in.)	A_c (in.)^2	A_w (in.)^2
					DU Lamination				
1 DU	$2\frac{1}{2}$	$\frac{7}{8}$	$\frac{1}{2}$	$\frac{1}{4}$	$\frac{3}{8}$	$1\frac{1}{2}$	$5\frac{1}{4}$	0.062	0.563
37 DU	3	$1\frac{1}{2}$	$\frac{3}{4}$	$\frac{3}{8}$	$\frac{3}{4}$	$1\frac{1}{2}$	$6\frac{3}{4}$	0.141	1.125
50 DU	4	2	1	$\frac{1}{2}$	1	2	9	0.25	2.0
75 DU	6	3	$1\frac{1}{2}$	$\frac{3}{4}$	$1\frac{1}{2}$	3	13.5	0.562	4.5
87 DU	7	$3\frac{1}{2}$	$1\frac{1}{4}$	$\frac{7}{8}$	$1\frac{3}{4}$	$3\frac{1}{2}$	15.75	0.766	6.12
125 DU	9	$4\frac{1}{2}$	$2\frac{1}{2}$	$1\frac{1}{4}$	2	4	19.5	1.562	8.0

^{a}Multiply by stacking factor k_s for A_e (square stack) and for actual weight.
^{b}Nonscrapless laminations.
^{c}Total window area. Divide by $\cong 2$ for each coil.
^{d}Multiply by k_s for actual weight.

95% of the applications the laminations are annealed after stamping for stress relief. This process may decrease core losses and increase permeability by 15% each as compared to nonannealed laminations. In fact, nickel–iron alloy laminations cannot be used without a final high temperature, dry hydrogen anneal [10]. This process increases permeability and decreases core losses by as much as 25 times.

The most popular laminations are of EI configuration and offer stacking ease around a single coil. The FB, FG, and EE laminations are available in small sizes. DU and L-shaped laminations are effective in obtaining a low profile transformer where height is limited, such as for printed circuit board applications. The DU lamination is also ideal for providing a high isolation between primary and secondary (the primary is wound on one leg and the secondary is wound on the opposite leg) to meet UL, CSA, FCC, and VDE safety and leakage requirements (reference Section 13.5). Custom applications for large magnetic devices and especially high-voltage types usually employ stampings or strips which are then stacked inside and around a winding to provide the magnetic path.

Dimensional data for common single-phase, three-phase, and ferroresonant laminations are listed in Table 3.15. In a scrapless lamination, the I pieces are

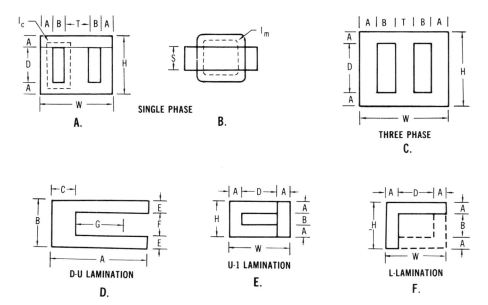

Fig. 3.25. Dimensional characteristics of single-phase and three-phase laminations.

obtained by the stamping operation from the window of the E. The nonscrapless laminations have a larger window area than the scrapless laminations and will hold more turns of the conductor. Since inductance is proportional to turns squared, these laminations are quite applicable for inductor designs.

For the single-phase scrapless lamination of Fig. 3.25, the following relations apply: $A = B$, $T = 2A$, $D = 3A$, $W = 6A$, and $H = 5A$.

The effective core area A_e equals the gross area A_c times the stacking factor k_s of Table 3.15. For a square stack of dimension S, $S = T = 2A$ and $A_w = 0.75A_c$. In those cases where a desired power rating requires additional core area, an over-square stack is considered a better choice than a smaller stack of larger laminations due to the smaller mean turn length.

The dashed line of Fig. 3.25A is the magnetic core path length l_c and is equal to $6T$ for scrapless laminations. The mean turn length l_m of Fig. 3.25B and for any lamination size is

$$I_m = 2[T + S + (2\pi B/4)] \tag{3.44a}$$

For a square stack where $T = S = 2B$,

$$l_m = 5.57T \tag{3.44b}$$

However, the tube for larger size devices must allow for lamination stacking ease and $l_m = 6T$ is probably a better figure. The above analysis is adequate for inductors but transformers have different l_m for the primary and secondary(s). Assuming the primary and secondary occupy an equal area, the primary is $1.1B/4$

from T, the secondary is $0.9 \times 3B/4$ from T and for a square stack,

$$l_m(\text{pri}) = 5.1T \qquad l_m(\text{sec}) = 6.7T$$

In open-construction transformers operating in air, the heat dissipated is convected from the surface. The total surface area consists of the exposed iron and coil surfaces of Fig. 3.25. However, most of the heat is conducted from the individual laminations to the edge surfaces and limited heat is conducted from lamination to lamination, and the front and rear EI surfaces are therefore neglected. This leaves the top, bottom, and two side surfaces. For scrapless laminations and from the previous relations, the iron surface area is found to be

$$A_{\text{iron}} = 2(6AS + 5AS) = 22AS = 11TS$$

The front and rear coil surfaces each consists of a plane surface and two quarter-cylinder surfaces. The top and bottom surfaces are neglected since limited heat is conducted from turn to turn. The coil surface area is

$$A_{\text{coil}} = 2(2A3A) + 4(2A3A/4) = 3T^2 + 3T^2/4 = 7.71T^2$$

The *total* surface area of the open construction transformer is

$$A_T = 11TS + 7.71T^2 \tag{3.45}$$

Magnetic path length, mean turn length, and surface area for three-phase and ferroresonant laminations may be calculated by methods similar to the above.

The DU, UI, and L laminations are shown in Figs. 3.25D, E, and F, respectively. The use of these laminations, with prewound coils, offers obvious advantages over toroid winding costs. The DU lamination provides a low-reluctance magnetic circuit and may be stacked for a desired core area and dimensional data as shown in Table 3.15. Except for the 1 DU and 125 DU, $A = 2G$, $B = G = 2F$, and $C = F = 2E$. Dimension G is the available winding width due to the stack interleave of dimension C. Additional choices of sizes are offered in the UI and L laminations, which are stacked and interleaved in the same manner as the EI. Surface area and I_m may be calculated in a manner similar to the EI.

3.11.1 Power Transformers

The VA rating for a transformer is given by Equation (3.12). From size relations for scrapless laminations, short-cut mehods may be employed to calcualte the required core size for a given power and frequency rating. The core size required to support a 60-Hz sine wave voltage is given by

$$A_e = k\sqrt{\text{VA}} \qquad \text{in.}^2 \tag{3.46}$$

where k is listed in Fig. 3.26 for various operating conditions.

For frequencies other than 60 Hz, the VA under the radical should be multiplied by $(60/f)^{0.76}$ as suggested by Remis [11].

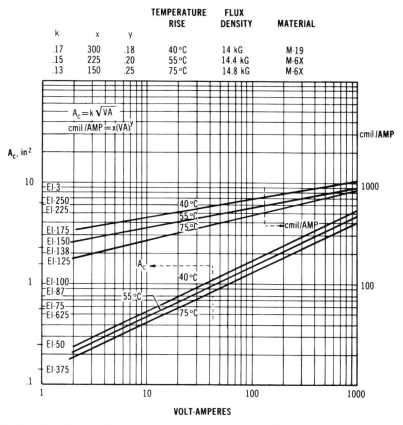

k	x	y	TEMPERATURE RISE	FLUX DENSITY	MATERIAL
.17	300	.18	40 °C	14 kG	M-19
.15	225	.20	55 °C	14.4 kG	M-6X
.13	150	.25	75 °C	14.8 kG	M-6X

Fig. 3.26. Required lamination core area and current density as a function of transformer power rating and temperature rise for 60-Hz operation.

Current density J in the winding is normally decreased as VA increases, since the larger A_c value produces a larger l_m and therefore increased resistance per turn. Using the inverse current density relation, cmil/ampere, where

$$500 \text{ cmil}/\text{A} = 1/(500 \times 0.785 \times 10^{-6})$$

$$= 2548 \text{ A}/\text{in.}^2 \tag{3.47a}$$

$$= 395 \text{ A}/\text{cm}^2 \tag{3.47b}$$

The approximate relation for cmil/ampere is

$$\text{cmil}/\text{A} = x(\text{VA})^y \tag{3.48}$$

where x and y are listed in Fig. 3.26 for various operating conditions.

Example

A rectifier transformer with EI laminations is required for the following parameters: $E_{dc} = 18$ V at $I_{dc} = 12$ A at nominal input; $E_{in} = 115/230$ V at 50/60 Hz.

Then $P_{dc} = 216$ W. Due to the low output voltage a center-tapped secondary will be used with a full-wave rectifier. For a rectifier voltage drop of 1.5 V, the power loss is 18 W and the transformer P_s is 234 W. From Chapter 4, Table 4.2, $VA_s = 1.57\,P_{dc}$ and $VA_p = 1.11\,P_{dc}$. The required transformer VA is then $(1.57 \times 234 + 1.11 \times 234)/2 = 314$ VA. The transformer design parameters will be: 55°C temperature rise, 1500 V isolation and class B (130°C) materials. From Fig. 3.26, use 29 gage M6-X at 14.4kG. Then for $P = 314$ VA, $A_c = 2.7$ in.^{2}, $J' = 610$ cmil/A. Using EI-150, the required stack height $= 2.7/1.5 = 1.8$ in., use 2 in. Again, from Table 4.2, I_s, rms $= 0.707 I_{dc} = 8.5$ A requiring $8.5 \times 610 = 5185$ cmil. From Table 3.9, use AWG 13, 0.074 in diameter. For the primary $I_p = 314/230 = 1.36$ A requiring $1.36 \times 610 = 830$ cmil. Use AWG 21, 0.03 in diameter. From Fig. 3.25A and Table 3.15, the winding width for EI-150 is 2.25 in. and the winding height is 0.75 in.

Next, the turns per layer and number of layers which will fit the window are calculated. The primary will be wound with two windings, each capable of supporting 115 V for parallel/series connection of the 115/230 V input. For the center-tapped secondary, two windings will also be wound. The primary turns per layer, allowing 0.25 in. margin (0.125 in. on each end), are $N_{p/l} = (2.25 - 0.25)/0.03 - 1 = 65 T/l$. The secondary turns per layer are $N_{s/l} = (2.25 - 0.25)/0.074 - 1 = 26 T/l$. From Equation (3.6b) and designing for 125/250 V maximum at 50 Hz and 14.4 kG, the primary and secondary turns are: $N_p = (3.49 \times 125 \times 10^6)/(14.4 \times 10^3 \times 3.0 \times 50 \times 0.95) = 212 T$; $N_s = E_s N_p/E_p = (18 + 1.5)212/115 = 36 T$.

The number of primary and secondary layers are: $N_{pl} = 212/65 = 3.3$, use 7 layers total; $N_{sl} = 36/26 = 1.4$, use 3 layers total.

Using a winding tube of 0.03 in. thickness, 0.007 in. layer insulation, 0.02 in. insulation between primary and secondary and 0.02 in. final wrap, the total build is height $= 0.03 + (0.03 + 0.01)7 + 0.02 + (0.074 + 0.01)3 + 0.02 = 0.60$ in. (which provides a 0.15-in. margin).

Notice the final layer of each first winding on both the primary and secondary is approximately one-half layer. Thus the first layer of each second winding on both the primary and secondary can be started in the middle of the winding. Since the secondary is low voltage the voltage gradient is low, but for the primary, the start of the second winding will be near the finish of the first winding. For this reason, it may be advisable to use eight layers for the primary. The 0.01-in. insulation between windings is primarily for structural support as opposed to voltage isolation and may be changed to 0.007 in. to allow more height margin.

3.11.2 Inductors

The design of inductors which carry a dc current is sometimes arbitrary and experimental. However, the size relations for scrapless laminations allow short-cut methods to be employed, which eliminate a "select another core and proceed" method. The relation between the parameters of an inductor and its size and weight is [5, 12]

$$S = K\left(\frac{LI_{dc}}{R^{0.6}}\right)^{1.2} \quad \text{(size or weight)} \qquad (3.49)$$

where K is a proportinality constant.

The energy storage is given by LI^2 (as opposed to $\frac{1}{2}LI^2$). Since A_c is a function of inductance, A_c is plotted versus LI^2 and $LI/\sqrt{R}$ (a close approximation to the above equation) in Fig. 3.27 for 29-gage M-19 material. Since the ac flux density and frequency are low (usually less than 5 kG and 500 Hz), the more expensive grain-oriented laminations may not be required unless size and weight are highly important factors.

Knowing the inductance, current, temperature rise, and desired resistance of the inductor, a lamination and core size may be selected from Fig. 3.27 and the wire size, in cmil, from Fig. 3.26, or AWG from Tables 3.9 and 3.10. For a selected flux density, l_g may be found from Equation (3.15), and N may be found from Equation (3.16) for a desired inductance. From Equation (3.5) B_{ac} is then summed

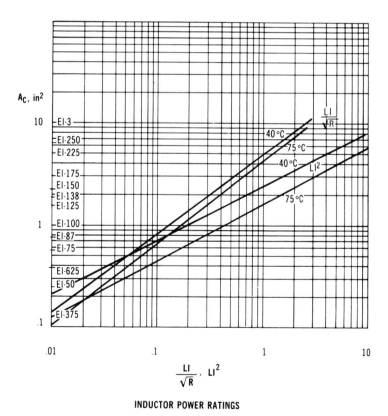

INDUCTOR POWER RATINGS

Fig. 3.27. Required lamination core area as a function of inductor energy storage and temperature rise.

with B_{dc} from Equation (3.15) and to ensure linearity, the total flux density must not exceed B_m.

Example

An inductor will now be designed (and a capacitor will be selected) to filter the output of the rectified voltage from the transformer example of the preceding section. The required output parameters are $E_{dc} = 18$ V at $I_{dc} = 12$ A at nominal input, and peak-to-peak output ripple = 20% maximum, or 3.6 V.

Since the output ripple waveform is a close approximation of a sine wave [reference Chapter 4, Equation (4.31)], the rms ripple is $3.6/2.8 = 1.3$ V or 7.2%. From Equation (4.30a), $LC = 0.829(60/50)^2/0.072 = 16.6$, where C is in microfarads. From Equation (4.33), $L/C = 0.1 \times 1.5^2 = 0.225$. Solving for the capacitance, $C = \sqrt{16.6 \times 10^{-6}/0.225} = 8600$ μF. Then $L = 0.225 \times 8.6 \times 10^{-3} = 1.9$ mH. The "energy" value is $LI^2 = 1.9 \times 10^{-3} \times 12^2 = 0.274$. From Fig. 3.27 and using a 55°C temperature rise, choose a square stack of EI-100. Also note that $LI/\sqrt{R} = 0.13$ from which the dc resistance and copper loss should be $R = (1.9 \times 10^{-3} \times 12/0.13)^2 = 0.031$ Ω; $P_{cu} = I^2R = 4.5$ W.

From Fig. 3.25A and Table 3.15, the winding width for EI-100 is 1.5 in. and the winding height is 0.5 in. Choose the wire size in the following manner: From Fig. 3.26, EI-100 intersects the 50°C point at 40 VA, going vertical to 50°C and right to the ordinate, the required "current density" is 530 cmil/A and the copper area is $530 \times 12 = 6360$ cmil. From Tables 3.9 and 3.10, either AWG 12 or AWG 13 square could be used. AWG 13 square will be used due to the smaller width which allows more turns. Allowing 0.15-in. margins on each end of the winding tube and using 0.082 in. allowance per turn and filling the window, the turns per layer are $N/l = (1.5 - 0.30)/0.082 - 2 = 12$.

The number of layers, with insulation, are $p = (0.5 - 0.125 - 0.03)/(0.082 + 0.007) = 3.75$; use 4.

Then the total build height is $0.089 \times 4 + 0.03 + 0.02 = 0.4$ in. for a 0.1-in. height margin. For total turns, $N = 48$. From Equation (3.5), the minimum air gap for a dc flux density of 10 kG (allowing for a 4 kG of ac) is $l_g = 0.58 \times 48 \times 12/10,000 = 0.033$ in.

From Equation (3.16b), the air gap for the required inductance of 1.9 mH is $l_g = 3.19 \times 48^2 \times 1 \times 10^{-8}/(1.9 \times 10^{-3}) = 0.039$ in.

This air gap results in a dc flux density of 8.6 kG. The average inductor voltage is $(18 - 1.3)/2 = 8.4$ V and from Equation (3.6b), the ac flux density is $B_{ac} = (3.49 \times 8.4 \times 10^6)/(1 \times 48 \times 100) = 6.1$ kG.

The total flux density is then 14.7 kG and 29 gage (0.014 in.) M-6X material could be used for the laminations. From Equation (3.44b), the mean turn length l_m is 5.57 in. From Table 3.10, the resistance of AWG 13 square is 1.66 Ω/1000 ft, and from Equation (3.41), the dc resistance and the copper loss (as compared to the values calculated from Fig. 3.27) are $R_{dc} = 1.66 \times 5.57 \times 48/12,000 = 0.037$ Ω; $P_{cu} = I^2R = 12^2 \times 0.037 = 5.3$ W.

The primary difference in the dc resistance, for the same number of turns, is the choice of AWG 13 square at 1.66 Ω/1000 ft versus the AWG 12 at 1.59 Ω/1000 ft.

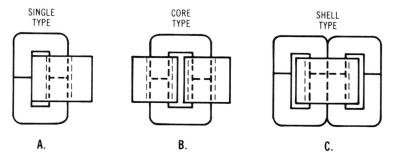

Fig. 3.28. C core and winding construction.

3.11.3 Swinging Chokes

For inductor input filters following a rectifier, a critical inductance may be required to prevent voltage peaking on the filter capacitor at light loads (reference Section 4.3). At full load, the filtering is accomplished by the capacitor, assuming the capacitor rms ripple current rating is not exceeded, and inductor saturation is usually immaterial. In these cases, a *swinging* choke may be desirable. The inductor is then designed at the mean inductance and maximum current and will be physically larger than an inductor calcualted for either initial condition.

3.12   C AND E CORES

For C cores, three types of construction are common, as shown in Fig. 3.28. The single type (A) consists of a single coil and core and is ideal for inductor design if the overall width can be tolerated. The core type (B) consists of a single core and double coil. Better space utilization is achieved and winding time is reduced by machine winding the coils simultaneously. The shell type (C) consists of a double core and single coil. Magnetic fields in the winding are suppressed and the core area is doubled. The shell type is not suitable for inductors due to the difficulty of air gap alignment and the stray flux produced in the exterior gaps.

In the manufacture of the cores, a silicon–iron strip (for instance) is treated on both sides with a finish to "electrically" insulate the surface but maintain good thermal characteristics. As previously noted, thin strips reduce eddy currents and the strips must not short together. The strip is then wound on a mandrel, annealed, and cut to produce the two core halves. The core faces are then processed and lapped to obtain a very small air gap l_g (0.001 in. for small cores and 0.002 in. for large cores), at the butt joint. A small air gap is important in the design of transformers operating from a sine wave, since a decrease in air gap increases the permeability and decreases the reluctance and excitation current. However, a small air gap may be desirable in medium frequency square-wave or pulse width–modulated transformers to prevent saturation and high peak currents resulting from unbalanced half cycle conduction periods.

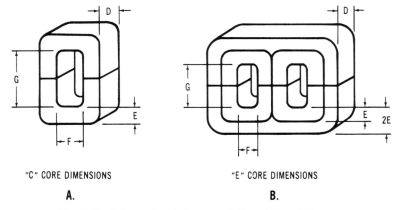

"C" CORE DIMENSIONS   "E" CORE DIMENSIONS

A.   **B.**

Fig. 3.29. Dimensional characteristics of C and E cores.

Manufacturers' data (similar to Figs. 3.10 and 3.11) provides core loss values in watts per pound and apparent core loss values in VA per pound for the chosen material as a function of flux density and frequency. Cores are normally dimensioned in the manner shown in Fig. 3.29. The nominal weight of a silicon–iron core may be calculated by

$$\text{Wt} = k_s DE(2F + 2G + 2.9E) \qquad \text{lb} \qquad (3.50)$$

where density (pounds per cubic inch) is typically 0.276 for silicon–iron material. The magnetic path length is

$$l_c = 2F + 2G + 2.9E \qquad \text{in.} \qquad (3.51)$$

For inductors using cores with large window area and small depth, the large value of l_c can have an appreciable effect on inductance if the air gap is small. The mean turn length for the single (A) type is

$$l_m = (2E + 2D + 4F) \times (\% \text{ build}) \qquad (3.52a)$$

The mean turn length for the core type (B) is

$$l_m = (2E + 2D + 2F) \times (\% \text{ build}) \qquad (3.52b)$$

The mean turn length for the shell type (C) is

$$l_m = (4E + 2D + 4F) \times (\% \text{ build}) \qquad (3.52c)$$

Surface area differs in the above configurations. For the single type, the coil area is $G(8F + 2E + D)$ and the iron area is $2[2(2E + F)E + GE]$ and

$$A_T = 8E^2 + 4EF + 8FG + DG \qquad \text{in.}^2 \qquad (3.53a)$$

For the core type, the coil area is $G(6F + 4E + 2D)$ and the iron area is $4(2E + F)E$ and

$$A_T = 8E^2 + 4EF + 4EG + 6FG + 2DG \qquad \text{in.}^2 \qquad (3.53b)$$

For the shell type, the coil area is $G(8F + 4E)$ and the iron area is $4[2(2E + F)E + GE]$ and

$$A_T = 16E^2 + 8EF + 8EG + 8FG \qquad \text{in.}^2 \qquad (3.53c)$$

In the above calculations, the surface area of facing coils is neglected and the iron surface covered by the coils is neglected. Also, square core edges and square coil surfaces are assumed.

As with laminations, the core loss may be obtained from manufacturers' data and the copper losses calculated knowing the current, wire size, and l_m. The sum of the core and coil losses divided by the surface area A_T produces a power dissipation density in watts per square inch. The temperature rise may then be obtained from Fig. 3.24.

The E cores are ideal for compact or large three-phase transformers. Construction consists of the C cores, enclosed by a wider C core of the same depth, as shown in Fig. 3.29. Equal cross-sectional area is obtained in each "leg" and the transformer is operated at the same flux density as a single-phase transformer of the same material thickness. However, due to third-harmonic flux, the core losses may be 20% greater than for a single phase core [13].

After the winding is assembled on the core, the two core halves are usually held together with one or more banding straps, depending on the core width. In the final assembly, the cores should be cemented together, banded at a force recommended by the manufacturer, and the complete assembly cured by vacuum impregnation.

3.12.1   C Core Transformers

Relative power handling capacity, in terms of $A_w A_c$, for C cores is frequently tabulated in manufacturers' data for 60-Hz, 15-kG operation. The power handling capacity, considering frequency, flux density, and current density, may be calculated by [14]

$$\text{VA} = 4.55 J A_w A_c B_{ac} f k_s \times 10^{-8} \qquad (3.54)$$

where

$$J = \text{current density, A/in.}^2$$
$$A_w = \text{GF, window area, in.}^2$$
$$A_c = \text{DE, gross core area, in.}^2$$
$$B_{ac} = \text{flux density, G}$$
$$f = \text{frequency, Hz}$$
$$k_s = \text{stacking factor}$$

Example

A 480- to 120-V step-down transformer is required for an output of 2500 VA at 60 Hz with a regulation of 2%. The input voltage variation is 10%. The required operating ambient temperature is 50°C. The flux density is chosen as 16.5 kG at 528 V. Using a 55°C temperature rise and interpolating Fig. 3.26, the current density chosen is 1000 cmil/A or 1275 A/in.^{2} From Equation (3.54), $A_w A_c = 2500 \times 10^8/(4.55 \times 1275 \times 16500 \times 0.95 \times 60) = 46$.

The construction will be shell type, per Fig. 3.28C. In this case, the shell type requires two cores, but since the copper loss will probably be higher than the core loss at 60 Hz, the extra core is justified. At a higher required operating frequency, the core type might be justified. Thus the $A_w A_c$ product for each core will be 23. From manufacturers' data sheets, an A-480 might be usable but for 2% regulation and 10% input voltage variation, an A-519 ($A_w A_c = 30.8$) is probably a better choice. The following parameters are listed for the A-519: weight = 14.1 lb, $D = 2.8125$ in.; $E = 1.094$ in.; $F = 2$ in.; $G = 5$ in.; $A_c = 3.08$ in.^{2}, $A_w = 10$ in.^{2}

For $E_o = 120$ V at $P_0 = 2500$ VA, $I_s = 2500/120 = 20.83$ A. For 1000 cmil/A, $A_{cu} = 20,830$ and from Table 3.10, AWG 11 rect. will be used for the secondary. Allowing 0.1 in. margin on each end of the coil, the turns per layer are $N_s/l = (5 - 0.2)/0.19 - 1 = 24$.

Using half the winding height for the secondary and allowing 0.2 in. margin for tube, insulation, and assembly, the number of layers l is $(1 - 0.2)/0.1 = 8$, and the total secondary turns $N = 8 \times 24 = 192$.

Using 95% efficiency, the primary current is $I_p = 2500/(480 \times 0.95) = 5.48$ A, and the required copper area is 5480 cmil. For the subject transformer, the secondary will be wound first since the primary will have a much higher mean turn length; AWG 13 square will be used. The turns per layers are $N_p/l = (5 - 0.2)/0.08 - 1 = 59$.

The number of layers is $p = (1 - 0.2)/0.08 = 10$, and the total turns are $20 \times 59 = 590$.

The required primary turns, from Equation (3.6b), are $N_p = (3.49 \times 532 \times 10^6)/(16500 \times 3.08 \times 60 \times 0.95) = 641$.

Therefore, more volume will be required for the primary than for the secondary. For 59 turns per layer, 641 turns require 11 layers, with 59 turns on the first 10 layers and 51 turns on the eleventh layer. The secondary turns are calcualted with a 2% regulation allowance, and $N_s = 641 \times 120 \times 1.02/480 = 164$.

Summarizing, the secondary is wound first, then insulation is added, and the primary is wound, and $N_s = 164$ turns AWG 11 rect., $24T/l$, 7 layers (20 turns on the seventh layer); $N_p = 641$ turns AWG 13 square, $59T/l$, 11 layers (51 turns on the eleventh layer).

Since the mean turn length for the primary will be higher than the secondary $l_{m,\,sec} = 4E + 2D + 3F = 16$ in., and $l_{m,\,pri} = 4E + 2D + 5F = 20$ in.

Then for 0.52 $\Omega/1000$ ft with AWG 11 rect. and 1.66 $\Omega/1000$ ft with AWG 13 square: $R_{sec} = 0.52 \times 16 \times 164/12000 = 0.114$ Ω; $R_{pri} = 1.66 \times 20 \times 641/12{,}000 = 1.77$ Ω; $P_{cu,\,sec} = 20.83^2 \times 0.114 = 49.5$ W; $P_{cu,\,pri} = 5.48^2 \times 1.77 = 53.2$ W; $P_{cu} = 49.5 + 53.2 = 102.7$ W total.

Using a flux density of $480 \times 16.5/528 = 15$ kG, the core loss at 60 Hz from manufacturers' data is 0.9 W/lb and the total core loss, for the two cores, is $P_c = 0.9 \times 14.1 \times 2 = 25.4$ W.

The total power loss is 102.7 + 25.4 = 128.1 W. The total surface area and power dissipation density is $A_T = 16E^2 + 8EF + 8EG + 8FG = 160$ in.^{2}; watts per square inch = 128/160 = 0.8.

From Fig. 3.24, the temperature rise at 50°C ambient is 70°C. Thus, the transformer temperature will be 120°C and allowing for hot spot conditions, wire insulation and winding insulating materials should be class F (155°C) or higher.

The output voltage at full load (neglecting the magnetizing current and core loss which are both quite low compared to the primary current) is $E_o = (E_{in} - IR_{pri})(N_s/N_p) - IR_{sec} = [(480 - (5.48 \times 1.77)](164/641) - (20.83 \times 0.114) = 118$ V (or 1.7%).

3.12.2 C Core Inductors

Due to the various configruations of core and window area available for C cores (as opposed to definite dimensional relations for laminations) plotting inductor characteristics is difficult. Table 3.16 therefore lists common 4-mil core numbers and a modified relation of Equation (3.49), $(LI/\sqrt{R})$ based on data from the cited references [15, 16]. Thinner core material may be effectively used for output filter inductors in switch mode converters. As compard to ferrites, the operating flux density can be a factor of 10 or greater. Since the ac voltage across the inductor is low, core losses are low and the flux density due to dc current predominates.

Example

The inductor design of Section 3.11.2 required 1.9 mH at 12 A. The $LI/\sqrt{R}$ factor was 0.15. The design used an EI-100 core with 48 turns AWG 13 square with a 0.039-in. air gap. Using the same $LI/\sqrt{R}$ factor, Table 3.16 shows an H-16 C core could be used. However, due to the low ripple frequency, a 12-mil core will be used. From manufacturers' data sheets, an A-10 core has approximately the same $A_w A_c$ product as the H-16 and the EI-100. However, the H-10 core has more than two times the window area, and since the turns will be higher than for the EI-100, a wire size with two times the cross-sectional area, or AWG 11 square, will be used. Also, for convenience the single core construction will be used. The turns per layer for the A-10 core are $N/l = (2.31 - 0.03)/0.1 - 2 = 20$. The number of layers are $l = (0.75 - 0.125 - 0.03)/(0.1 + 0.007) = 5.6$; use 5. The total build will be height = $(0.1 + 0.007)5 + 0.03 + 0.02 = 0.59$ in. The winding height margin is $0.75 - 0.59 = 0.16$ in. For $N = 100$ turns, the required air gap is $l_g =$

TABLE 3.16. C Core Selection for Inductors

Core No.	$\dfrac{LI}{\sqrt{R}}$	Core No.	$\dfrac{LI}{\sqrt{R}}$	Core No.	$\dfrac{LI}{\sqrt{R}}$
H-38	0.0135	H-11	0.046	H-16	0.136
H-39	0.0192	H-6	0.062	H-27	0.165
H-1	0.023	H-8	0.078	H-29	0.213
H-3	0.029	H-12	0.085	H-31	0.318
H-9	0.0325	H-13	0.091	H-34	0.42
H-5	0.041	H-15	0.119	H-155	0.77

$3.19N^2A_c \times 10^{-8}/L = 3.19 \times 100^2 \times 0.468 \times 10^{-8}/0.0019 = 0.078$ in. Then, $B_{dc} = 0.58NI/l_g = 0.58 \times 100 \times 12/0.078 = 8.9$ kG. Also, $B_{ac} = 3.49E \times 10^6/(ANf) = 3.49 \times 8.4 \times 10^6/(0.468 \times 100 \times 100) = 6.3$ kG. The total flux density is then 15.2 kG.

From Equation (3.52a), $l_m = 0.464$ in. Since the resistance of AWG 11 square is 1.04 Ω/1000 ft, the dc resistance and copper loss is then $R = 1.04 \times 4.64 \times 100/12,000 = 0.04$ Ω; $P_{cu} = I^2R = 12^2 \times 0.04 = 5.8$ W.

These values compare favorably with the example of Section 3.11.2.

3.13   TOROID TAPE CORE TRANSFORMERS

Tape cores are normally "wound" from nickel–iron material of a given width and thickness in an "overlapping" process to produce a toroid shape. The high permeability material is mechanically strain sensitive, and square or rectangular shapes are therefore avoided. After "winding," the cores are encased in a "box" filled with inert silicone for further damping and strain relief, and also to protect against pressure from the coil winding. The box material may be phenolic or aluminum, with insulation for the winding and to prevent a shorted turn.

The dimensional data as well as the overall dimensions, effective cross-sectional area, window area, and W_aA_c product are usually given by manufacturers' data. The shuttle winding process limits the window area which can be filled by the winding. However, at high frequencies, cores with an unused window produce increased core losses due to the unnecessary magnetic path length l_c of the core.

Saturating transformers utilize the square loop core material, such as Deltamax or Orthonol, in many types of inverter circuits. Typical examples are (1) high-frequency, flux-switching dc–dc converters; (2) high-frequency magnetic amplifiers for post regulation of multioutput switch mode power supplies; (3) inverter drive stages; (4) output transformers and switching reactors for pulse modulators.

The ideal transformer characteristics are small size, high efficiency, low cost, and wide environmental operation. Unfortunately most of these characteristics conflict and the designer must therefore make decisions based on the application and conditions. (Reference Tables 3.4 and 3.6.)

The following formulas may be used to select a core for a saturating inverter transformer [17]:

1. Small size, high flux, Supermendur material:

$$W_aA_e = 5.25P_0 \times 10^6/f \qquad (3.55a)$$

where B_m is 21 kG, density is 0.31 lb/in.^{3}, and the core is 50% cobalt, magnetic annealed.

2. Square hysteresis loop, Deltamax or Orthonol:

$$W_aA_e = 7.6P_0 \times 10^6/f \qquad (3.55b)$$

where B_m is 14.5 kG, density is 0.31 lb/in.^{2}, and the core is 50% nickel, grain oriented.

3. Low core loss, Supermalloy material:

$$W_a A_e = 15.7 P_0 \times 10^6 / f \tag{3.55c}$$

where B_m is 7 kG, density is 0.33 lb/in.^{3}, and the core is 80% nickel, high permeability.

In the above equations, W_a is the available core window area in circular mils (A_w is the winding area), A_e is the effective cross-sectional area in square centimeters, current density is 750 cmil/A, and 85% efficiency is assumed.

The winding space factor is

$$k = N A_w / W_a = \text{between 0.2 and 0.4}$$

Volume and weight are related by

$$V = 0.75(\text{OD}^2 - \text{ID}^2) \times \text{height} \qquad \text{in.}^2 \tag{3.56a}$$
$$\text{Weight} = k_s \times V \times \text{density} \qquad \text{lb} \tag{3.56b}$$

For saturating transformers used in pulse-forming networks and magnetic modulator applications, the saturated inductance becomes important in terms of pulse duration when the magnetic field collapses. When the core saturates, the inductance is effectively that of an air-core coil since the core can no longer support voltage. Manufacturers' data sheets list various parameters of cores such as dimensional data, magnetic path length, window area, effective cross-sectional area, and $W_a A_c$ (usually for 2 mils thickness). The multitude of available cores prevetns a complete listing of these parameters here. However, additional data pertinent to saturated conditions is given in Table 3.17 for a few core sizes. The volt-μs/turn values are based on a nominal B_m of 14 kG. From computer analysis of inductance in air core coils, the factor k_0 has the following relation to saturated inductance:

$$L_0 = N^2 k_0 \times 10^{-10} \qquad \text{H} \tag{3.57}$$

For nonsaturating transformers such as those used in voltage driven or clocked inverter stages, low core loss is important at high frequencies. The low operating flux density penalty of 80% nickel cores (such as Supermalloy) is more than offset by the inherent low core loss, as shown in Fig. 3.6. These cores are also excellent for precision current transformers. A single-turn primary conductor is passed through the toroid over which the secondary has previously been wound. The primary ampere-turns equals the secondary ampere-turns and the rated VA is a function of desired secondary voltage and load impedance (burden).

Example

A current transformer is desired to sense an alternating trapezoid current of 10 A initial step and 20 A peak with a duty cycle D of 75%, and develop a secondary voltage of 5 V initial step and 10 V peak. To conserve power, let the burden resistor dissipate 0.5 W so a 1-W rated resistor may be used. From Chapter 1,

TABLE 3.17. Tabulation of Square Loop Core Characteristics[a]

Core Part Number by Manufacturer[b]			Core Volume (in.³)	Core Weight (oz)	V-μs/ turn	K_0
1	2	3				
8043	50056	47	0.014	0.048	10.0	6.08
5340	50000	2	0.031	0.11	20.5	7.83
5515	50002	5	0.038	0.13	20.5	6.41
7699	50004	30	0.110	0.38	42.7	8.70
5502	50061	79	0.086	0.30	42.7	9.30
5958	50076	37	0.090	0.31	47.5	10.64
5504	50106	7	0.104	0.36	47.5	9.13
5651	50007	3	0.120	0.42	63	13.0
4635	50029	10	0.175	0.61	63	8.69
5800	50032	39	0.245	0.85	85	10.1
5387	50030	13	0.295	1.03	85	8.66
5233	50026	11	0.368	1.3	127	12.8
6847	50038	62	0.491	1.7	170	17.8
7441	50035	29	0.589	2.0	170	14.6
4178	50017	17	0.884	3.1	170	11.1
4180	50031	18	1.08	3.8	170	9.46
5772	50425	75	0.718	3.2	190	14.7
5320	50001	15	1.57	5.5	337	21.3
6110	50103	76	1.96	6.8	337	15.4
6100	50128	19	2.36	8.2	337	14.5
8027	50022	58	3.93	13.8	680	24.2
5468	50042	20	4.71	16.8	680	26.4
5690	50100	22	9.20	32.3	1270	41.5
5611	50112	25	18.2	64.2	1700	40.2
9260	50426	78	31.4	110	2700	55.1

[a]Parameters are for 1 mil, square loop cores.
[b]1. Arnold Engineering; 2. Magnetics, Inc.; 3. Magnetic Metals, Inc.

Equation (1.12a) and Table 1.2, the rms secondary voltage is $E_{rms} = [(10^2 + 10 \times 5 + 5^2)0.75/3]^{0.5} = 6.6$ V. Then, $R = E^2/P = 6.6^2/0.5 = 87$ Ω. Choosing $R = 100$ Ω, $P = 0.4375$ W. The peak secondary current $= 10/100 = 0.1$ A. For a single-turn primary, the secondary turns are $N_p I_p/I_s = 1 \times 20/0.1 = 200$ turns.

Tape-wound bobbin cores are frequently used for operating frequencies of 100 kHz and higher. Ultrathin tape (from 1 to 0.125 mil thickness) of a high permeability magnetic material is wound over a ceramic bobbin to provide support during the high-temperature annealing cycle. Manufacturers' data sheets give dimensional data as well as flux (maxwells) ratings for various sizes of cores.

3.14   PERMALLOY POWDER CORES

Toroid-shaped cores using basic raw materials of iron, nickel, and molybdenum are referred to as molybdenum Permalloy powder (MPP) cores. These cores are characterized by high resistivity, low hysteresis and eddy current losses, and exellent inductance stability under high dc bias conditions. The cores are ideal choices for high-Q inductors used in resonant circuits, for electromagnetic interference (EMI) filters, and for carrying dc current. These cores are characterized by medium flux density and high magnetizing force, and are typically 80% nickel composition. Thus permeability is lower than in previously discussed magnetic materials. Since the "gap" is not variable, as with some other matierals, various permeabilities are available which provide an effective air gap. Inductance is calculated by Equations (3.17) and (3.18). However, the A_L factor, given by manufacturers, greatly simplifes the inductor design. Permeability, and thus inductance, has a positive temperature coefficient which is a function of core stability. Cores are available in unstabilized (for general-purpose power inductors), stabilized, and linear types. Precision-tuned circiuts, capable of high-frequency stability over wide operating temperatures, may be obtained by matching the positive temperature coefficient of a linear core with a polystyrene capacitor which inherently has a linear negative temperature coefficient. As flux density increases, high-permeability cores have a higher rate of permeability increase than low-permeability cores. The B/NI data is therefore limited to flux densities which produce a 20% maximum permeability increase.

Manufacturers' data is usually consulted to obtain the maximum Q for a given operating frequency and inductance. Larger cores provide a higher Q since flux density and core losses are lower. In high-frequency circuits, maximum Q is obtained with lower permeability cores. Core loss is usually presented as

$$P_c = KB^2 R' \times 10^{-6} \qquad \text{W/lb} \tag{3.58}$$

where R' (in ohms per millihenry) is plotted versus frequency at various flux densities and K is inversely proportional to permeability.

3.14.1 Inductors with ac Voltage

When inductors are used in filter circuits with ac voltage applied, a specific value of inductance is required to provide a desired attenuation factor. Flux density and magnetizing force must be analyzed, as well as the change in permeability. The following relations apply:

$$B_{ac} = E_{rms} \times 10^8 / 4.44 A_e N f \tag{3.59}$$

$$H_{ac} = B_{ac}/\mu \tag{3.60}$$

$$NI = H l_c / (0.4\pi) \tag{3.61}$$

$$L = 0.4\pi N^2 A_e \times 10^{-8}/l_c$$

$$= N^2 A_L \times 10^{-6}$$

Substituting,

$$NI = 0.18 E_{rms} l_c \times 10^8 / A_e N f$$

$$E_{rms} = 4.44 L I f$$

$$B/NI = 0.4\pi\mu/l_c$$

$$N = 1000\sqrt{L/A_L} \tag{3.62}$$

where A_L is in millihenrys per 1000 turns (nanohenrys per turn).

To minimize design effort, manufacturers' data usually tabulates μ, A_L, and H/NI for specific cores and shapes. Reference the example at the end of the following section.

High-flux powder cores (made from 50% nickel–50%iron alloy) are available for applications in EMI filters and inductors carrying high dc currents. Here B_{max} is typically 15 kG as compared to 7 kG for the MPP cores.

3.14.2 Inductors with dc Current

Due to the inherent high magnetizing force (especially at low permeability), MPP cores may be effectively used in dc bias applications. However, a decrease in permeability results, which should be limited to 20% maximum reduction. For a 50% window fill and a current density of 500 cmil/A, the core selection procedure is simplified by the following method [18]:

1. Knowing the desired inductance, L in millihenrys, and the desired maximum dc current, compute LI^2.

2. From Fig. 3.30, follow the abscissa to the first line intersected to obtain the required permeability. On the ordinate, select the first core size that lies within the family of part numbers above the intersection. This is the smallest core that can be used.

3. From manufacturers' data of core listings select the actual core part number matching the permeability desired, and from the A_L value for the chosen core, calculate the required number of turns from Equation (3.18b).

4. Select the wire gage from Tables 3.10 and 3.11 for the desired current density. Multiply the conductor cross-sectional area by the total number of turns, and compare this area to the available core window area. The window fill should not exceed approximately 60% for ease of winding.

Example

A desired inductance of 50 μH is required to carry 14 A peak dc current. Then $LI^2 = 0.05 \times 14^2 = 9.8$. From Fig. 3.30, select core size 55548 but use $\mu = 60$. The correct core part number is 55071. Here $A_L = 61$ mH/1000 turns and $H/NI = 0.154$. From Equation (3.18b), the required turns are $N = 1000\sqrt{0.05/61} = 29$. For a current density of 500 cmil/A, use AWG 12. Calculating the total copper area versus the core window area, the window will be approximately 35% full. The magnetizing force is $H = 0.154 \times 29 \times 14 = 62.5$ Oe. The flux density is $B = \mu H = 60 \times 62.5 = 3.8$ kG. Considering the winding, the dc resistance is $R_{\text{dc}} = 0.008$ Ω, the inductor volume is V $= 1.85$ in.³ and the surface area is $A_s = 9$ in.² Thus the copper loss is $P_{\text{cu}} = 1.6$ W and the temperature rise (from Fig. 3.24) at 0.18 W/in.² at 25°C ambient is $\Delta T = 26$°C. Of course, the use of AWG 12 does not consider skin effects at high frequencies and the above power loss does not consider core loss. Knowing the average ac voltage across the inductor and the operating frequency, the total ac and dc flux density can be determined, and the core loss can be calculated from manufacturers' data and by Equation (3.58). The total dissipation and the temperature rise can then be

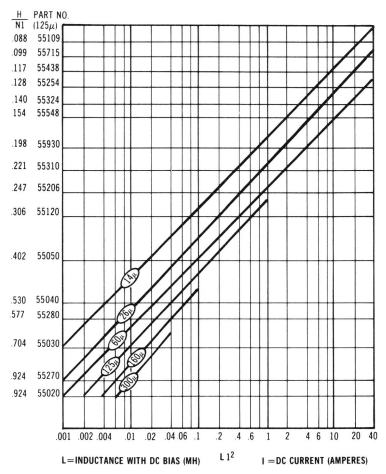

Fig. 3.30. A dc bias core selector chart; energy content versus size for various permeability values. (Courtesy Magnetics, Inc.)

determined. (Compare this inductor design example to the ferrite core inductor example of Section 3.16.3.)

3.15 IRON POWDER CORE INDUCTORS

Iron powder toroid cores have characteristics similar to those of the MPP cores discussed in the preceding section with one exception—cost is 20–25% that of MPP cores. Thus, these cores are ideal for high-volume production inductors where economy is a prime factor. Applications include dc filter inductors and EMI inductors. Selection is made from manufacturers' data sheets graphing energy storage versus percent saturation for various materials and A_L values [19]. The required turns are then calculated from Equation (3.18b) and a wire gage is chosen. Core loss, in units of milliwatts per cubic centimeter (also typical of units for ferrite core loss) is normally graphed versus flux density, from which the power dissipation may be calculated when the copper loss is summed with the core loss. Data by Micrometals, Inc. also tabulates inductance, number of turns, and wire size for a desired current for various iron powder core part numbers.

3.16   FERRITES

Ferrites are dense ceramic structures of ferromagnetic oxides (F_2O_3) mixed with one or several oxides such as NiO, MnO, and ZnO. They are commonly die pressed in various sizes and shaped and kiln fired or sintered for curing. The resultant core is quite brittle and very hard. This latter attribute provides minimum wear characteristics in magnetic recording heads. The oxygen ions in the ferrite insulate the metallic ions and provide a very high material resistivity. For this reason, thick shapes may be obtained with negligible eddy current loss. The NiO produces a material useful in the megahertz frequency range, while the MnO material is useful in the 2- to 200-kHz range, over which most switch mode converters and power supplies operate.

Due to the characteristics of ferrite, several parameters may be evaluated, which are not commonly considered in other magnetic materials. Complete analysis is beyond the scope of this book and only those characteristics relative to high-frequency power transformer and filter inductors are discussed. Contrary to some manufacturers' terminology, the nomenclature and symbology listed in Table 3.1 and the CGS system will be used throughout (except for the more common English terms for current density and AWG wire gage).

3.16.1   Core Geometries

Due to the molded or sintered process, various core geometries are available in ferrite. The more common geometries are shown in Fig. 3.31 and their description follows.

1. Pot cores comply with IEC dimension standards for interchangeability between manufacturers. The shape is cylindrical and consists of two identical halves. The major advantage of the pot core is the excellent shielding from EMI, since the winding is enclosed within the core. Also, bobbin winding of the coil and ease of assembly reduce cost. However, the core is more expensive than other geometries and the power rating is limited.

2. Square cores offer the advantages of pot cores plus minimized volume. The "footprint" is cubical in shape with a section cut off on either side. The winding is normally wound on a bobbin and connected to pins which protrude from the notches to allow ease of installation in printed circuit boards. Notched corners also allow better heat transfer from the winding than do the pot cores. A similar shape is the X core with four notched sides.

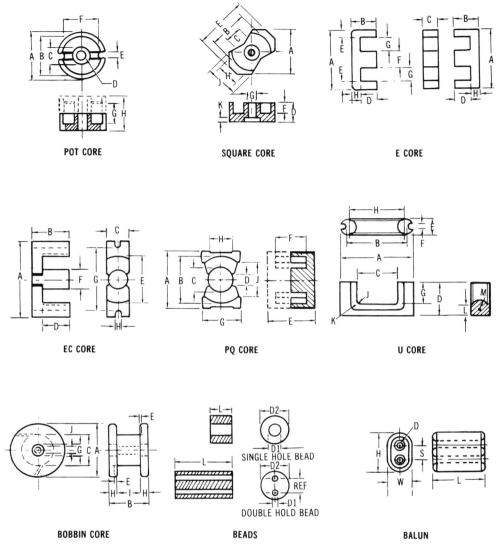

Fig. 3.31. Ferrite core geometries.

Square and X cores are ideal for pulse transformers firing thyristors, and for low-power transformers driving transistors.

3. E cores available in sizes corresponding to EI lamination dimensions (up to EI-75) but are not specficially limited to these shapes. E cores are less expensive than pot cores. Bobbins offer winding and assembly ease. However, E cores do not provide shielding of the winding for EMI suppression.

4. EC cores are a combination of pot core and E core. Like the pot core, the center post is round which reduces the mean turn length of the winding, as

compared to the square center post of the E core. The shorter mean turn length reduces conductor losses and enables the cores to handle higher output power. Most EC cores also have a grooved edge guide for ease of clamping the assembly to a mounting surface with a U-shaped bracket.

5. PQ cores are specifically designed for switch mode converters. These cores resemble EC cores, but the geometry provides an optimized ratio of volume, winding area, and surface area. As a result, PQ cores offer maximum power rating for a given weight and volume, as compared to other geometries.

6. U-shaped cores are available in large cross-section core area plus a wide window dimension. The square-shaped U cores allow dual windings similar to the strip-laminated iron C core geometry, and offer high-frequency performance at power ratings of several kilowatts. The round leg cores with wide window allow high-voltage isolation and are ideal for high-voltage switch mode converters and power supplies. In fact, the round leg U cores are used extensively in high-frequency converters providing power to cathode ray tubes (CRTs).

7. Gapped cores are available in pot core and EC core geometries. An air gap is located in the round center leg and the inductance factor is standardized to specific A_L values and effective permeabilities. This configuration greatly simplifies the design of inductors and transformers requiring a specific inductance value, such as the flyback transformer. Also, the magnetic field in the gap is contained within the assembly, as opposed to air gaps installed in the outside legs or periphery.

8. Toroid cores are economical to manufacture and are, therefore, the least costly, but the winding process normally requires a toroid winding machine with a shuttle to hold the conductor during winding. Toroids have a uniform cross-sectional area without an air gap, and are available in varous permeability and A_L values, depending on ferrite material type. The toroid is ideal for current transformers wherein a secondary winding is first wound and the primary consists of a single turn, frequently a large wire size, passed through the center. In fact, a center-tapped secondary and two single-turn primaries, passed through the core in opposite directions, provide an ideal current transformer for push–pull and bridge converter topologies.

9. Bobbin cores are ideal for inductive noise suppression and filtering in power circuits. Slots molded into the core allow ease of winding start and finish. For a desired inductance, the overall size will be less but the weight may be heavier than with an air core inductor.

10. Beads and baluns provide EMI suppression without attenuating dc and low-frequency signals. The single-hole bead is essentially a miniature toroid which can provide up to 10 dB insertion loss from 1 to 100 MHz. The double-hole bead is an effective balun for suppressing common mode high-frequency noise. In this latter case, the signal line passes through one hole and the signal return line passes through the other hole and in the same physical direction. Current flow then enters the bead at one end and exits the bead at the same end. The balun may also be used in series with each rectifier lead connected to a center-tapped high-frequency transformer

secondary winding for suppression of rectifier noise, which normally occurs in the 30-MHz region.

3.16.2 Power Transformers

The cylindrical center post EC and PQ cores have made a favorable impact on power transformers. Use of these cores produces (1) a more economical winding than toroids; (2) minimal external fields as compared to U shapes; (3) reduced mean turn length; and (4) termination ease when the winding is wound on a bobbin with molded pins.

The VA or power rating for a transformer is given by Equation (3.12) and is a function of several variables. For ferrite core transformers, operating frequency and core material become the predominant parameters. With current density and flux density fixed, either or both the core area or the window area can be decreased if the switching frequency is increased. Rather than proceed with what may be a laborious design task, various short-cut methods have been developed to aid in proper core selection. Also, empirical data and computer analysis from various sources provide information on transformer design. However, unlike the 60-Hz transformer graph shown in Fig. 3.26, ferrite cores are available in almost unlimited geometries and applications are quite varied. For switching circuits, the frequency may range from 10 to 500 kHz. The environment plays a significant role and varies from benign to hostile. A power supply transformer located in equipment in an air-conditioned room has less stringent requirements than a converter transformer located in equipment in an uninhabited area of a military aircraft. For these reasons, core selection will be based on empirical data and experience.

The effective volume (V_e) of a ferrite core is the product of magnetic path length (l_e) and cross-sectional area (A_e), and is related to power-handling capability. Rather than list the power rating for all geometries, Fig. 3.32 shows the effective volume required for a desired power rating as a function of typical switching frequencies. For simplicity, the data is considered independent of geometry and core material. The curves apply to a transformer used in a full-bridge inverter stage bridge (reference Chapter 9). For forward and flyback converter topologies, the power rating will be lower (reduced to approximately 70% of that shown in Fig. 3.32) due to the decrease in flux swing. The parameters having the greatest influence on these curves are flux density and temperature rise. These are tabulated as conditions for the respective switching frequencies. Also, high-voltage isolation from primary to secondary and from primary to core may decrease the power ratings shown. Once V_e is found as a starting point, the desired core geometry and material may be selected. After completing the transformer design for required number of turns and wire size, the regulation, losses, and temperature rise can then be calculated by methods discussed earlier in this chapter.

3.16.3 Inductors

Pregapped cores, both pot cores and EC cores, provide ideal "characteristics" for inductors. The gap is located in the center post of the core. This shields the air gap and minimizes external magnetic fields. The gap increases the reluctance and

CURVE	f, kHz	B, kG	ΔT, °C
A	20	1.6	40
B	40	1.4	45
C	100	1.2	50
D	200	1.0	55

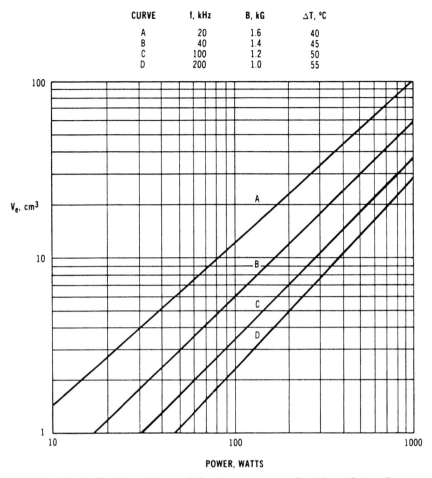

Fig. 3.32. Required effective volume of ferrite cores as a function of transformer power rating, operating frequency, and temperature rise.

prevents saturation of the core when a dc bias current flows in the winding. Manufacturers' data sheets frequently list values of A_L and μ_e for the gapped cores. Other data may be consulted, such as energy per unit volume versus magnetizing force, as developed by Hanna [20]. Just as the required power rating of a transformer is known, so is the energy content of an inductor. Empirical data is presented in Fig. 3.33 for core selection. Knowing the desired LI^2 (mH-A^2) value, V_e is chosen for a particular A_L. These curves apply to pot cores and EC cores and are intended to provide a starting point for core selection. The A_L value terms are in millihenrys per thousand turns, which is equivalent to nanohenrys per turn. The rapid rise in the slope of the curves above a certain effective volume indicates that ferrite cores are limited in energy content, and further volume increases in this region will not appreciably increase the stored energy capability.

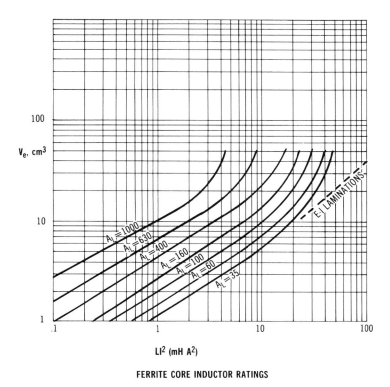

V_e, cm³

LI² (mH A²)

FERRITE CORE INDUCTOR RATINGS

Fig. 3.33. Required effective volume of ferrite cores as a function of energy storage and permeability factor.

As a reference point, the mean data of Fig. 3.27 for inductors with EI laminations is also shown in Fig. 3.33.

Example

A desired inductance of 50 μH is required to carry 14 A peak dc current. Choose a ferrite pot core and find the required turns and other characteristics. (This is the same requirement as the example of Section 3.14.2, which used an MPP core.) From Fig. 3.33, choose $V_e = 10$ cm³. Core data sheets show a 36×22 pot core has a V_e of 10.65 cm³ (the V_e of an EC41 is 10.8 which should also suffice). Also, $A_L = 100$ mH/1000 turns. Then the required number of turns is $N = 1000\sqrt{0.05/100} = 23$. For this small core, operate at 300 cmil/A. The required conductor area is $A_{\text{wire}} = 300 \times 14 = 4200$ cmil; use AWG 14 at 4107 cmil. The copper area is $A_{\text{cu}} = 23\pi \times 0.066^2/4 = 0.079$ in.² The bobbin area of the 36×22 core is $A_w = 0.116$ in.² for a 69% fill. If the bobbin were filled to 80%, the number of turns would increase to 27, which will be used since the optimum design is a full window. For the 36×22 core, $A_e = 2.01$ cm² $= 0.31$ in.² From Equation (3.16b), the air gap is $l_g = 3.19N^2A_e \times 10^{-8}/L = 3.19 \times 27^2 \times 0.31 \times 10^{-2}/50 = 0.144$ in. From Equation (3.15), $B_{\text{dc}} = 0.58NI/l_g = 0.58 \times 27 \times 14/0.144 = 1.5$ kG.

Further calculations reveal R_{dc} = 0.167 Ω, P_{cu} = 3.3 W, V = 1.5 in.^{3}, and A_s = 7.5 in.^{2} The expected temperature rise (from Fig. 3.24) at 0.44 W/in.^{2} at 25°C ambient is ΔT = 50°C. Comparing this design to the MPP core inductor of Section 3.14.2, the pot core inductor has 80% of the volume, twice the copper loss, and twice the temperature rise.

3.17 AIR CORE INDUCTORS

In some high-frequency pulse applications, air core inductors offer performance (as well as cost) advantages over iron core inductors. This is especially true for many commuting inductors which limit the rate of current rise (di/dt) in high-power thyristor (SCR) circuits and for output inductors in switch mode power supplies. Usually the required inductance is in the microhenry range. The low value of inductance in commutating circuits is due to high values of commutating capacitance, low impedance, and high frequency. The resonant frequency may range from 5 to 50 kHz, depending on circuit design and SCR turn-off time. See Chapters 7 and 9 for a more detailed discussion and circuit applications. In switch mode power supplies where volume is adequate, the air core inductors may be used as part of a second stage LC filter for noise suppression.

The air core inductor may be more suitable than iron core inductors by the following analysis. From Equation (3.16), the inductance of an iron core inductor is proportional to turns squared and to core area, and is inversely proportional to air gap. Thus for small values of inductance, the number of turns is low. However, the low number of turns may cause saturation of the iron [Equations (3.5) and (3.15)]. Adding turns to support the required voltage requires an increase in air gap to maintain the desired inductance. Pursuing this effect, very large air gaps may result and core material permeability becomes ineffective. Therefore, the solution is to eliminate the iron completely.

Various empirical formulas have been developed to calculate inductance values of an air core coil. The inductance of a single-layer coil (reference Fig. 3.19), accurate to 1% and for $w > 0.4d$, is given by [3]

$$L = d^2N^2/(18d + 40w) \qquad \mu H \qquad (3.63)$$

where d is mean coil diameter in inches and w is coil width in inches.

In the above equation, w is a function of N and wire size, assuming the conductor is wound with minimum turn spacing. This equation may be cumbersome when determining an inductor form factor. Inductance versus turns is plotted in Fig. 3.34 for various values of d and w for an appreciation of Equation (3.63).

If the coil diameter is first selected for the desired inductance value, an empirical formula which relates form factor regardless of dimension ratios and number of layers is given by

$$K = 100L/dN^2 \qquad (3.64)$$

where K is a numeric factor given by Fig. 3.35, L is inductance in microhenrys, and d is the mean coil diameter in inches.

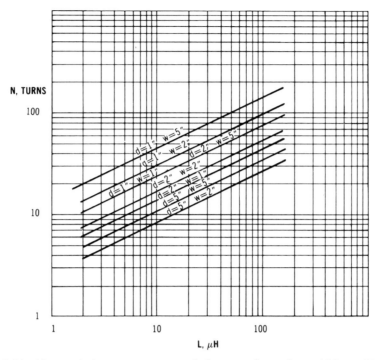

Fig. 3.34. Air core inductors: turn versus inductance for various widths and lengths.

The width-to-diameter ratio is obtained from Fig. 3.35, and the required width may be determined. Of course, the number of turns and wire size required to conduct the desired current must fit within the calculated width.

Examples

Case 1. An inductance value of 20 μH is to be used in an SCR commutating circuit. The rms current is 45 A. The inductor is to be wound on a 3-in. diameter coil. Since the coil is "open," use 400 cmil/A for a desired conductor area of 18,000 cmil. From Table 3.10, AWG 8 square is chosen. The allowance per turn is 0.14 in. and if the width is chosen at 3 in., the number of turns is 21. The width/diameter ratio is 1 and $K = 1.7$. Then $L = 1.7 \times 3 \times 21^2/100 = 22.5$ μH. Reducing N to 20, $L = 20.4$ μH. The mean turn length is $(3 + 0.14)\pi = 9.86$ in. and the resistance is 0.531 Ω/1000 ft. For 20 turns, $R = 9.86 \times 20 \times 0.531/12,000 = 0.0087$ Ω. The power dissipation is then $P = 45^2 \times 0.0087 = 17.7$ W. Also, the outside surface area is $A_s = 3(3 + 0.28)\pi = 30.9$ in.^{2} From Fig. 3.24, at $17.7/30.9 = 0.57$ W/in.^{2} and in a 50°C ambient, the temperature rise is $\Delta T = 54$°C. The inductor should be cooler than 104°C since some heat will be convected from the inside of the coil. For support, the coil may be wound on a 3-in. diameter mandrel. If a close coupled center-tapped air core coil is desired, a second layer may be wound over the first layer, starting at the finish of the first layer but with 18 or 19 turns on the second layer to achieve equal inductance.

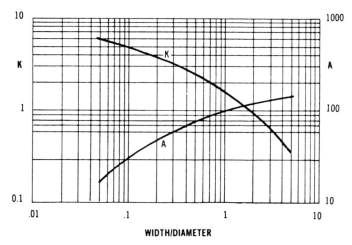

Fig. 3.35. Constants for air core inductors.

Case 2. A 6-μH core inductor is desired to conduct 10 A dc in a switching power supply noise filter. The diameter is restricted to 0.75 in. Using 300 cmil/A, AWG 15 is chosen and the turns per inch are 16. Using a 2-in. width, $N = 32$ and $K = 0.8$. From Equation (3.64), $L = 0.8 \times 0.75 \times 32^2/100 = 6.1$ μH. The dc resistance is $R_{dc} = (0.75\pi)32 \times 3.18/12{,}000 = 0.02$ Ω. The power dissipation is $P = 10^2 \times 0.02 = 2$ W. In this case, it may be advantageous to use a cylindrical powered iron core to reduce the number of turns and to reduce the overall length.

In half-sinusoid commutating coils, a high Q and low loss are desirable. Skin effect should therefore be considered for large conductors. In di/dt inductors, a high Q may cause unwanted ringing, and the coil is usually damped with a parallel power resistor. The Q of single layer air core coil is given by [3]

$$Q = Ad\sqrt{f}/31.6 \tag{3.65}$$

where A is the ordinate coefficient from Fig. 3.35, f is frequency in kilohertz, and d is the mean coil diameter in inches.

For instance, the 20-μH, 45-A inductor in the previous example had a width and a diameter of 3 in. each. From Fig. 3.35, $A = 100$ at $w/d = 1$. Also, the operating frequency is known to be 5 kHz. Then $Q = 100 \times 3\sqrt{5}/31.6 = 21$.

3.18 PULSE TRANSFORMERS

Pulse transformers have wide usage in power electronic circuits. Applications range from small, dual in-line packaged units used to trigger a power MOSFET at 1 MHz to very large units used to provide a rectangular output pulse of many kilowatts to a high-power vacuum tube. The latter application is discussed in Section 12.2. Small size units with repetitive, high-frequency operation usually employ ferrite materials for low core loss. Large units frequently utilize square-loop tape-wound core materials where the saturation characteristics are used to advantage. The pulse transformer is normally driven by a unipolar voltage and the circuit design does not provide a reversal of flux in the core. However, large or high-power units frequently utilize a tap on the secondary winding whereby a reset current is applied through a bias reactor.

3.18.1   Characteristics and Measurements

The equivalent circuit of a pulse transformer is shown in Fig. 3.36A and differs somewhat from that of Fig. 3.8A, which applies primarily to power transformers. In the pulse transformer, core loss is normally low due to operation at low flux density or low duty cycle, and the dc winding resistances may be considered negligible relative to source and load resistances. Again, the output voltage is inversely proportional to the transformer turns ratio $n:1$. The output voltage departs from the desired or ideal rectangular pulse and the shape is determined by the volt-second rating, the primary inductance, leakage inductance, and distributed capacitance. The latter two parameters have a major influence on the rise and fall time of the pulse. A typical output waveform is shown in Fig. 3.36B with performance parameters defined in order of occurrence.

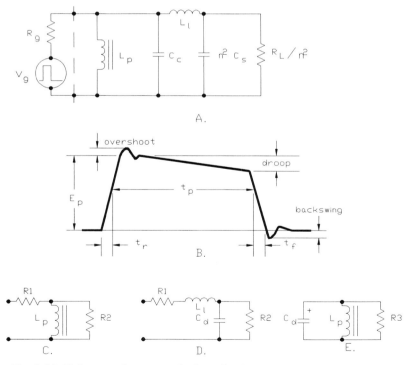

Fig. 3.36. Pulse transformer equivalent circuits and waveform parameters.

Rise Time. Time required for the pulse voltage to increase from 10 to 90% amplitude.

Overshoot. Percentage increase above the nominal or desired output amplitude due to parasitic effects.

Droop. Percentage tilt or decrease in amplitude during the top portion of the pulse.

Fall Time. Time required for the pulse voltage to decay from 90 to 10% amplitude.

Backswing. Percentage reversal in amplitude, which may be significant if the load impedance increases dynamically as the pulse voltage decreases.

The low-frequency equivalent circuit, shown in Fig. 3.36C, indicates that primary inductance must support the required voltage for a desired period. The output voltage droop, or tilt, during the pulse is proportional to the ratio of magnetizing current to peak pulse current. Again, the magnetizing current is inversely proportional to primary inductance by $i_m = E_p t_p / L_p$. The output voltage amplitude is a function of the ratio of the reflected output load resistance to the input, or source resistance. High-frequency performance parameters, such as rise time, fall time, overshoot, and backswing, are dominated by parasitics such as leakage inductance and winding capacitance, as shown in Fig. 3.36D. In fact, reflected load capacitance produces a decrease in the rise time of the output voltage, as given by Equation (3.70).

Test circuits for measuring transformer parameters are shown in Fig. 3.37, and these parameters are discussed prior to a presentation of actual design. Leakage inductance L_1 is measured by an impedance bridge connected across the primary with the secondary shorted, as shown in Fig. 3.37A. The leakage inductance may also be calculated by Equation (3.25). Winding capacitance C_w is measured by an impedance bridge connected from primary to secondary with each winding shorted, as shown in Fig. 3.37B. The distributed, or intrawinding, capacitance C_d may be determined indirectly from resonance measurements, as shown in Fig. 3.37C. A sweep generator in series with an input resistor is connected across the primary, along with a voltmeter or oscilloscope. First, the secondary is open circuited and the frequency range is scanned for maximum primary voltage. This occurs at parallel resonance with C_d and L_p. Thus the distributed capacitance can be calculated from $C = 1/[(2\pi f)^2 L]$. Second, the secondary is short circuited and the frequency range is scanned for minimum primary voltage. This occurs at series resonance with C_d and L_l. Again, the distributed capacitance can be calculated. In general, the winding capacitance is greater than the distributed capacitance.

When determining rise time, fall time, and turns ratios, the primary and secondary must be properly terminated. Such a circuit is shown in Fig. 3.37D. A pulse generator, in series with R_1, applies the desired voltage waveform to the primary. The secondary is terminated with $R_2 = R_1/n^2$. Oscilloscope probes are connected across the input and across the ouput to observe the waveforms.

Common-mode rejection (CMR) is a measure of the isolation between windings. This parameter is of major consideration in ultra-isolation transformers operating in noisy environments or requiring safety features. In pulse transformers, CMR is dependent on frequency and load resistance. For standardization, the primary is

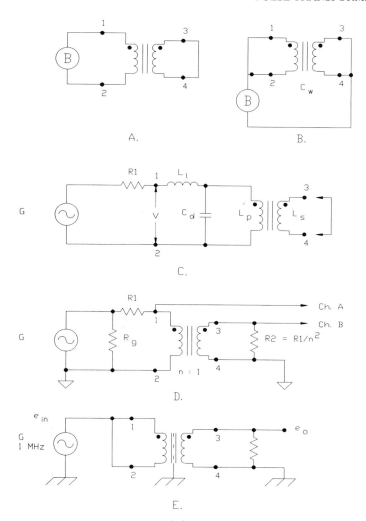

Fig. 3.37. Pulse transformer test circuits: (A) leakage inductance, (B) winding capacitance, (C) resonant measurements, (D) waveform observation, and (E) CMRR.

shorted and a 1-kΩ load resistor is connected across the secondary, as shown in Fig. 3.37E. The winding shield and one end of the secondary are connected to ground. A 1-MHz signal e_{in} is applied from the primary to ground. The output voltage e_0 across the load resistor is measured. The CMR (or CMRR) is equal to e_{in}/e_0.

3.18.2 Design Parameters

The desired parameters of pulse width, input voltage, output voltage, and output current are normally given for a particular requirement. The rise time, fall time, and flat top pulse characteristics must also meet predetermined requirements. The basic design of core selection and conductor size may be accomplished by means discussed earlier in this chapter. However, the numerator of Equation (3.7b) must be increased for a unipolar pulse (reference Table 3.5 for rectangular pulse) and the required number of turns is

$$N = 15.5 E_p t_p \times 10^6 / \Delta B A_e \qquad \left(A_e \text{ in in.}^2 \right) \qquad (3.66)$$

The primary inductance may be calculated from Equation (3.16). The flux density relates to the volt-second product where $Et = E_{pri} t_p$, expressed in terms of volt-microseconds. The inductance must be sufficient to support the input voltage and meet the voltage droop requirement.

In small pulse transformers, the retention of a rectangular pulse is more important than efficiency of operation. The core loss will inherently be minimized due to high primary inductance and low flux density. However, the winding resistances may be permitted to increase, perhaps to 10% of the generator or load impedances. Small wire size wound at close spacing will minimize leakage inductance. This method is preferred if the generator and load resistances are small in value. If the generator and load resistances are large in value, the windings may be spaced farther apart to minimize the effective capacitance.

The rectangular pulses are best analyzed by transient response methods, as opposed to Fourier analysis. This latter method involves resolving the pulse into the sum of a large number of odd harmonics of a sine wave, and is best applied to analysis of pulse forming networks, as discussed in Section 12.1.

3.18.2.1 Pulse Top and Droop   The pulse top and droop characteristics are governed by the low-frequency equivalent circuit of Fig. 3.36C. The required primary inductance is related to primary voltage, pulse width and droop by

$$L_p = 100 E_p t_p / D I_p \qquad \text{H} \qquad (3.67)$$

where

E_p = peak primary voltage, V

t_p = pulse width, s

I_p = peak primary current, A

D = percent droop

This value of inductance must then be equal to or less than the inherent inductance calculated by Equation (3.16). The droop, or tilt, in the output waveform is also proportional to the ratio of magnetizing current to primary current as

$$D = 100 i_m / I_p \quad \text{percent} \qquad (3.68)$$

where $i_m = E_p t_p / L_p$.

A high primary or source resistance cannot be neglected and in fact may have a significant impact on the required primary inductance in an exponential ratio. If the droop is less than 15%, the relation is simplified and the required inductance is

$$L_p = 100 R_{\text{eff}} t_p / D \quad \text{H} \tag{3.69}$$

where $R_{\text{eff}} = R_g n^2 R_L / (R_g + n^2 R_L)$, R_g is generator plus primary resistance, and R_L is load plus secondary resistance.

3.18.2.2 Rise Time The rise time is governed by the equivalent high-frequency, voltage step-up circuit of Fig. 3.36D. This damped circuit is in turn equivalent to Fig. 1.8 and the analysis discussed in Section 1.7 applies. Rise time is dependent on the damping factor k and can be reduced if slight overshoot or oscillations can be tolerated. For instance, if $k = 1$ for critically damping, the rise time, normalized to T is

$$t_r = 0.53T = 0.53 \times 2\pi \sqrt{\frac{L_l C_{\text{eff}} R_2}{R_1 + R_2}} \tag{3.70a}$$

where

$$C_{\text{eff}} = C_d + C_L / n^2 \qquad R_1 = R_g + R_p \qquad R_2 = R_s + n^2 R_L$$

Since the value of T is itself a function of k, a detailed calculation must be performed for precise analysis. However, if $R_2 \gg R_1$ and if k is reduced to 0.6 for a 10% overshoot, the rise time is approximated by

$$t_r = \tfrac{1}{2}\pi\sqrt{L_l C_{\text{eff}}} \tag{3.70b}$$

This equation shows the effect of secondary load capacitance, such as the gate input to a MOSFET or IGBT, on rise time. This load parameter is fixed by the device and the detailed design must focus on minimizing the leakage inductance to achieve a fast rise time.

3.18.2.3 Fall Time The fall time is best analyzed by the equivalent circuit of Fig. 3.36E. The voltage source is not shown since it has been turned off, but the capacitor is charged to E_g. The fall time is a function of damping factor, primary inductance, and R_3, which may be dynamic in nature. Lowering the droop during the pulse top improves the fall time. The damping factor k is typically less than 1 and the circuit is underdamped, producing the backswing voltage. If the overshoot at turn-on is 10% for the circuit of Fig. 3.36D, the backswing at turn-off will be approximately 5% and the fall time will be approximately twice the rise time.

The dynamic nature of R_3, as with a magnetron, may result in a significant backswing voltage after the fall time. In these cases, a saturating inductor (tailbiter) may be added to the secondary to clamp the voltage or a reverse polarity rectifier in series with a resistor may be added to the primary to rapidly dissipate the unused energy.

Example

A pulse transformer is required to meet the following parameters: $E_p = 500$ V, $E_s = 10$ kV, $I_s = 10$ A, $t_p = 5$ μs, pulse repetition frequency (PRF) = 500 Hz. The peak power is thus 100 kW, but the average power is $100,000 \times 5 \times 10^{-6} \times 500 = 250$ W. A 2-mil silicon steel core is chosen with $l_e = 8$ in., $A_e = 0.5$ in.^{2}, permeability = 5000, flux density = 15 kG. The core window has sufficient area for conductors and high-voltage isolation. From Equation (3.66), the required primary turns are $N_p = 15.5 \times 500 \times 5/15,000 \times 5 = 5$. From Equation (3.16), the primary inductance is $L_p = 3.19 \times 5^2 \times 1 \times 5 \times 10^{-8}/(8/5000) = 250$ μH. From Equation (3.8), $N_s = 5 \times 10,000/500 = 100$ and $n = \frac{1}{20}$, and $I_p = 20 \times 10 = 200$ A. From Equation (3.67), the droop is $D = 500 \times 5/200 \times 250 = 5\%$. As a check, $i_m = 500 \times 5/250 = 10$ A and $D = 10/200 = 5$ percent.

The following parameters are measured after the transformer is assembled: $L_1 = 1$ μH and $C_d = 0.05$ μF. The high reflected capacitance is due to the large number of layers in the secondary and the high turns ratio. Actually, $C_s = C_d/n^2 = 0.05$ μF$/20^2 = 125$ pF. Since the source resistance and winding resistance are considered negligible and the transformer is slightly underdamped, Equation (3.70b) gives the rise time as $t_r = \frac{1}{2}\pi\sqrt{10^{-6} \times 0.05 \times 10^{-6}} = 0.35$ μs. The fall time is estimated at $3t_r$ or 1 μs. This transformer example is also used in the example of a pulse forming network and magnetic modulator in Section 12.2.

REFERENCES

1. G. B. Finke, "Selecting Magnetic Materials for High Flux Applications," *Solid-State Power Conversion* (August 1975).

2. R. Lee, *Electronic Transformers and Circuits* (John Wiley & Sons, New York, 1955).

3. *Reference Data for Radio Engineers*, 6th ed. (Howard W. Sams & Co., Indianapolis, IN, 1975).

4. S. A. Stigant and A. C. Franklin, *The J & P Transformer Book* (John Wiley & Sons, Butterworth & Co., New York, 1973).

5. N. R. Grossner, *Transformers for Electronic Circuits* (McGraw-Hill, New York, 1967).

6. E. C. Snelling, *Soft Ferrites* (ILIFFE Books, London, 1969).

7. I. Richardson, "The Technique of Transformer Design," *Electro-Technology* (January 1961).

8. H. Roth, "Applying High Frequency Technology to Low Current, Regulated HIgh Voltage Supplies," Proceedings of Powercon 2, October 1975.

9. J. H. Davis, "Fast Optimization of Transformer Design," *Electrical Design News* (November 1962).

10. *Electrical Materials Handbook*, Chapter XI, Allegheny Ludlum Steel Corp., Pittsburg, PA, 1961.

11. I. Remis, "Simplified Design of High-Temperature Transformers," *Electrical Manufacturing* (November 1955).

12. R. G. Luebben, "How to Design Inductors for Electronic Equipment," *Electronic Equipment Engineering* (September 1959).

13. Bulletin SC-107B, Arnold Silectron Cores, Arnold Engineering Company (August 1981).

14. Bulletin CC1, Magnetic Metals (April 1973).

15. R. Harriman, "Specifying Inductors that Carry D-C," *Electronic Products* (September 1968).

16. Robert A. Booth, "Simplify Magnetic-Core Coil Design," *Electronic Design*, Vol. 10 (May 10, 1967).

17. How to Select the Proper Core for Saturating Transformers, Bulletin TWC-S210A-5, Magnetics Inc., Division of Spang Industries.

18. Molypermalloy Powder Cores, Catalog MPP-303T, Magnetics, Division of Spang Industries.

19. Iron Powder Cores, Catalog 4, Micrometals.

20. C. R. Hanna, "Design of Reactances and Transformers Which Carry Direct Current," *AIEE Journal*, Vol. 46 (February 1927).

4

RECTIFIERS AND FILTERS

4.0   Introduction
4.1   Rectifier Circuits
    4.1.1   Single-Phase Circuits
    4.1.2   Polyphase Circuits
    4.1.3   Transformer Reactance and Rectifier Commutation
4.2   Capacitor Input Filters
    4.2.1   Single-Phase Rectifier Filters
    4.2.2   Polyphase Rectifier Filters
    4.2.3   Turn-On Currents and Surge Limiting
    4.2.4   Operating Currents and Voltages
    4.2.5   Voltage-Doubler Circuits
    4.2.6   Voltage Multipliers
    4.2.7   Operational Holdup Time
4.3   Inductor Input Filters
    4.3.1   Critical Inductance
    4.3.2   Turn-On Currents and Surge Limiting
    4.3.3   Output Ripple
4.4   Three-Phase Voltage Doubler
    References

4.0   INTRODUCTION

This chapter initially presents characteristics of various rectifier circuits, operating in an off-line mode or preceded by a transformer. Capacitor and inductor input filters are presented which enable the designer to choose components to meet desired ripple reduction. Effect of transformer leakage inductance or input line inductance on rectifier performance is discussed. Presentation is aided by the notations used in this chapter in Table 4.1.

Electrical power generation and distribution is usually accomplished in the form of alternating current due to simplicity and economy. However, many types of electrical equipment operate from direct current "sources." The ac voltage must therefore be rectified and in most cases filtered to provide a desired dc output voltage at a required current or power. The ac input may be single phase for low-

TABLE 4.1. Notation

λ	Angular function, degrees
α	Angular function, degrees
μ	Commutation angle, degrees
C	Capacitance, farads
I	Inductance, henrys
R	Resistance, ohms
FF	Form factor-rms/avg
CF	Crest factor-peak/rms
E_{dc}	Average dc, volts
E_{do}	Open-circuit dc, volts
E_p	Primary voltage, volts
E_s	Secondary voltage, volts
E_{L-N}	Line-to-neutral voltage, volts
E_{L-L}	Line-to-line voltage, volts
E_{rms}	Root-mean-square voltage, volts
I_{avg}	Average current, amperes
I_{dc}	Average dc current, amperes
I_F	Rectifier dc current, amperes
I_{pk}	Peak current, amperes
I_{rms}	Root-mean-square current, amperes
I_p	Primary current, amperes
I_s	Secondary current, amperes
I_L	Line current, amperes
P	Power, watts
t	Time, seconds
V_F	Forward rectifier drop, volts
V_{PIV}	Peak inverse voltage, volts
VA	Volt-amperes
X	Reactance, ohms
k	Ratio factor
f_r	Rectified ripple frequency, hertz
f_s	Source frequency, hertz
n	Turns ratio; number of series rectifier devices
p	Number of rectified pulses per cycle
r	Ripple factor

to medium-power requirements or three phase for medium- to high-power requirements. In high-power systems, regulation is frequently accomplished by controlling the phase or conduction time of the ac wave by thyristors (SCRs). Phase control parameters and circuits are discussed in Chapter 5. In low-power systems, such as ac–dc power supplies, regulation is usually achieved by either a transistor post regulator after the isolation transformer/rectifier for linear types or by pulse width modulation for switching types operating from the rectified mains.

A transformer may be used ahead of the rectifier to scale the voltage and isolate the input, as in linear power supplies and battery chargers. Frequently, the input voltage is directly rectified (off-line method), and the dc bus is then switched at a high frequency across a load or transformer supplying the load. A typical example of the latter case is switching power supplies and step-up frequency converters. In either case, the magnetics operate at a higher frequency for size, weight, and cost reduction.

4.1   RECTIFIER CIRCUITS

Characteristics and design information for common rectifier circuits are shown in Table 4.2. Here E_{do} is given as the open circuit dc voltage (or the dc voltage neglecting losses). The actual output voltage under load is approximately

$$E_{dc} = \frac{E_{do}}{1 + X_t/k + R/100} - nV_F \tag{4.1}$$

where

X_t = % reactive voltage drop in the line and/or transformer
R = % resistive voltage drop in the line and/or transformer
k = ratio factor, depending on the circuit (see Table 4.2)
n = number of rectifier elements in the current path
V_F = forward voltage drop of each rectifier element

Here E_{do} is expressed in terms of $E_{dc, pk}$ and $E_{s, rms}$ where the rms values are line-to-line (LL) *in all configurations*. Two figures are shown in the columns for rectifier current I_F. The upper value is for a resistive load and the lower value is for an inductive load. Effects of capacitor loads are discussed in Section 4.2.

Average and rms current (or voltage) values for a sine wave, as given in Chapter 1, Equation (1.1) are

$$I_{avg} = \frac{2}{T} \int_0^{T/2} I_{pk} \sin \omega t \, dt = 0.6366 I_{pk} \tag{4.2}$$

$$I_{rms} = \sqrt{\frac{1}{T} \int_0^T I_{pk}^2 \sin^2 \omega t \, dt} = 0.7071 I_{pk} \tag{4.3}$$

An alternate method of calculating I_{avg} (or E_{avg}) for various configurations is given by [1]

$$I_{avg}/I_{pk} = (p/\pi)\sin(\pi_r/p) \qquad (p \neq 1) \tag{4.4}$$

where p is the number of pulses per cycle (f_r/f_s) and $\pi_r = 180°$ for $p = 1$, $\pi_r = 90°$.

Applying the above solution to Equation (4.3) provides an alternate method of calculating I_{rms} (or E_{rms}) by

$$\frac{I_{rms}}{I_{pk}} = \sqrt{\frac{p}{2\pi}\left(\frac{\pi}{p} + \frac{1}{2}\sin\frac{2\pi_r}{p}\right)} \tag{4.5}$$

TABLE 4.2. Properties of Rectifier Circuits[a]

Circuit Type	Figure No.	$\dfrac{E_{dc,pk}}{E_{do}}$	$\dfrac{E_{s,rms}}{E_{do}}$	$\dfrac{I_{F,avg}}{I_{dc}}$	$\dfrac{I_{F,rms}}{I_{dc}}$[b]	$\dfrac{I_{F,pk}}{I_{dc}}$	$\dfrac{I_{s,rms}}{I_{dc}}$	$\dfrac{V_{PIV}}{E_{s,rms}}$	$\dfrac{V_{PIV}}{E_{dc}}$	$\dfrac{V_{PIV}}{E_{s,rms}}$ (Rec)[c]	$\dfrac{V_{PIV}}{E_{dc}}$ (Rec)[c]	$\dfrac{VA_s}{P_{dc}}$[b]	$\dfrac{VA_p}{P_{dc}}$[b]	K Factor[d]	% pk-pk Ripple Voltage	% rms Ripple Voltage	$\dfrac{f_r}{f_s}$
Single phase Half wave	4.1	3.14	2.22	1.00	1.57	3.14	1.57	1.414	3.14	2.5	6	3.49	2.47	200	100	121	1
Full wave (center tap)	4.2	1.57	2.22	0.5	0.785 / 0.707	1.57 / 1.00	0.707	1.414	3.14	2.5	6	1.74 / 1.57	1.23 / 1.11	200	100	48.3	2
Full-wave bridge	4.3, 4.4	1.57	1.11	1.00	0.785 / 0.707	1.57 / 1.00	1.00	1.414	1.57	2.5	3	1.23 / 1.11	1.11 / 1.11	200	100	48.3	2
Three-phase Half wave	4.5, 4.6	1.21	1.48	0.333	0.587 / 0.577	1.21 / 1.00	0.577	1.414	2.09	2.5	4	— / 1.48	— / 1.21	191	50	18.3	3
Full-wave bridge	4.7, 4.8	1.05	0.74	0.333	0.579 / 0.577	1.05 / 1.00	0.816	1.414	1.05	2.5	2	— / 1.05	— / 1.05	200	13.4	4.2	6
Double wye with interphase transformer	4.9	1.05	1.71	0.167	0.293 / 0.289	.525 / .500	0.289	1.414	2.42	2.5	4	— / 1.48	— / 1.05	141	13.4	4.2	6
Star, Full wave (center tap)	4.10	1.05	1.48	.167	0.409 / 0.408	1.05 / 1.00	0.408	1.414	2.09	2.5	4	— / 1.81	— / 1.28	57.7	13.4	4.2	6
Parallel bridge without interphase transformer	4.11	1.01	0.715	0.167	0.409 / 0.408	1.01 / 1.00	0.577	1.36	1.01	2.5	2	— / 1.43	— / 1.05	200	3.41	1.19	12
Parallel bridge with interphase transformer	4.12	1.05	0.74	0.167	0.293 / 0.289	0.525 / 0.500	0.408	1.414	1.05	2.5	2	— / 1.05	— / 1.01	200	3.41	1.19	12
Series dual extended delta (except series bridge)	4.11	1.05	0.37	0.333	0.579 / 0.577	1.05 / 1.00	0.816	1.414	0.524	2.5	1	— / 1.05	— / 1.01	200	3.41	1.19	12

[a] $E_{s,rms}$ is line-to-line voltage.
[b] Top number applies to resistive load. Bottom number applies to inductive load.
[c] Recommended voltage rating. Rectifier voltage rating.
[d] Impedance factor. See Equation (4.1).

The form factor FF is the ratio of the rms component to the average component and may be significant when rectifiers must carry high peak currents for short intervals, as with capacitor filters, battery chargers and phase control circuits. The form factor also allows a calculation of the rms ripple by

$$\text{Percent rms ripple} = \sqrt{FF^2 - 1} \times 100 \qquad (4.6)$$

The crest (or peak) factor CF is the ratio of the peak value to the rms value and may be important in phase control circuits, commutating circuits, and dielectric testing.

The peak-to-peak ripple, relative to the peak voltage, is given by

$$\text{Percent pk–pk ripple} = \left[1 - \cos(\pi_r/p)\right] \times 100 \qquad (4.7)$$

The peak-to-peak ripple, relative to the average output, may be found by multiplying the above value by $E_{dc, pk}/E_{do}$.

TABLE 4.3. Common Mathematical Relations

Angle, degrees	Relation Factor	Decimal Equivalent	Decimal Inverse
$1 - \sin 75°$	—	0.034	29.35
$1 - \sin 60°$	—	0.134	7.46
	$1/2(\sqrt{3} + 1)$	0.183	5.46
$\sin 15°$	$(\sqrt{3} - 1)/2\sqrt{2}$	0.259	3.86
	$2\sqrt{2}/\pi^2$	0.287	3.49
	$1/\pi$	0.3183	3.1416
	$1/\sqrt{6}$	0.408	2.45
	$\pi/3\sqrt{6}$	0.4275	2.34
	$\sqrt{2}/\pi$	0.450	2.22
	$\sqrt{2}/3$	0.471	2.12
	$3/2\pi$	0.477	2.09
$\sin 30°$	$\frac{1}{2}$	0.500	2.00
	$\sqrt{3}/\pi$	0.551	1.81
$\tan 30°, 1/\tan 60°$	$1/\sqrt{3}$	0.577	1.732
	$2/\pi$	0.6366	1.57
	$3\sqrt{2}/2\pi$	0.675	1.48
$\sin 45°$	$1/\sqrt{2}$	0.7071	1.414
	$\pi/3\sqrt{2}$	0.74	1.35
	$\pi/4$	0.785	1.27
	$\sqrt{\frac{2}{3}}$	0.8165	1.225
	$3\sqrt{3}/2\pi$	0.827	1.21
$\sin 60°$	$\sqrt{3}/2$	0.866	1.155
	$2\sqrt{2}/\pi$	0.900	1.11
	$3/\pi$	0.955	1.05
$\sin 75°$	$(\sqrt{3} + 1)/2\sqrt{2}$	0.966	1.035
	$6\sqrt{2}/\pi(\sqrt{3} + 1)$	0.9886	1.0115
$\tan 45°$		1.00	1.00

Throughout this chapter, various quantities as related to waveforms and angles are used. These quantities and the numeric relations are tabulated in Table 4.3.

4.1.1 Single-Phase Circuits

The single-phase half-wave rectifier is shown in Fig. 4.1 with voltage and current relations and current waveforms for an *inductive load*. Since only one-half of the wave is utilized, regulation is poor, efficiency is low, and secondary current flow may cause transformer saturation. Capacitive filtering may be used to increase the output voltage and reduce ripple. Inductor input filters are impractical since continuous current flow cannot normally be maintained in the inductor. The half-wave circuit is usually limited to direct line rectification or a combination of low current and high voltage. From previous equations,

$$I_{\text{avg}} = 0.318 I_{\text{pk}} \qquad I_{\text{rms}} = 0.5 I_{\text{pk}} \qquad \text{FF} = 1.57$$

The above analysis is not applicable if the applied voltage is switched across the primary by an active device. In this case, the circuit resembles the familiar forward converter topology commonly used in switch mode power supplies.

The single-phase full-wave rectifier is universally common and two types of circuits are available. The center-tapped circuit (Fig. 4.2) is normally used for high current low voltage, while the bridge circuit (Fig. 4.3) is normally used for medium to high-voltage output. In the center-tapped circuit, the *extra winding* on the secondary increases the secondary VA requirement and the rectifier reverse voltage rating, but only one rectifier forward drop results. In the bridge circuit, the secondary VA and the rectifier reverse voltage are reduced, but two rectifier forward drops result.

The center-tapped circuit is extremely useful in supplying dual output voltages, as shown in Fig. 4.4. With center tap E_2 at "common," $E_{\text{do1}} = -E_{\text{do2}}$. Addition of filter capacitors and voltage regulators produces a power supply with output

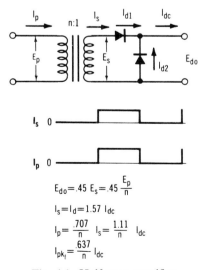

Fig. 4.1. Half-wave rectifier.

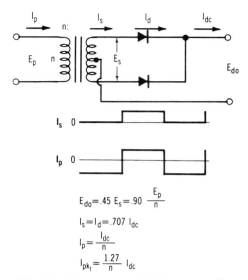

$$E_{do} = .45\ E_s = .90\ \frac{E_p}{n}$$

$$I_s = I_d = .707\ I_{dc}$$

$$I_p = \frac{I_{dc}}{n}$$

$$I_{pk_f} = \frac{1.27}{n}\ I_{dc}$$

Fig. 4.2. Center-tapped full-wave rectifier.

voltages of ± 5 V and higher. With the most negative side, E_3, at common, $E_1 = 2E_2$ to produce outputs of $+5$ V, $+10$ V, and higher. From previous equations,

$$I_{avg} = 0.637 I_{pk} \qquad I_{rms} = 0.707 I_{pk} \qquad FF = 1.11$$

Voltage, current, power, and ripple data are given in Table 4.2.

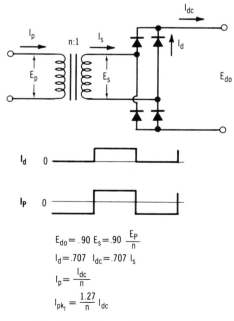

$$E_{do} = .90\ E_s = .90\ \frac{E_p}{n}$$

$$I_d = .707\ I_{dc} = .707\ I_s$$

$$I_p = \frac{I_{dc}}{n}$$

$$I_{pk_f} = \frac{1.27}{n}\ I_{dc}$$

Fig. 4.3. Full-wave bridge rectifier.

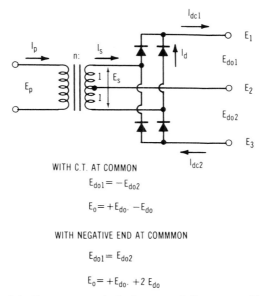

WITH C.T. AT COMMON

$$E_{do1} = -E_{do2}$$

$$E_o = +E_{do\cdot} - E_{do}$$

WITH NEGATIVE END AT COMMMON

$$E_{do1} = E_{do2}$$

$$E_o = +E_{do\cdot} + 2\,E_{do}$$

Fig. 4.4. Center-tapped, dual-output, full-wave rectifier.

4.1.2 Polyphase Circuits

Polyphase power (normally three-phase) is used in medium- to high-power applications and may be applied to an off-line rectifier, say to produce a nominal 280 V dc from a 120/208 V line (factor of 1/.74 from Table 4.2), or to a transformer whose secondary or secondaries are rectified in various manners. The advantages of using three-phase power as compared to single-phase power are (1) higher efficiency; (2) higher output voltages; (3) lower ripple, higher input power factor; and (4) reduced harmonic distortion of the input current. Voltage, current, and power

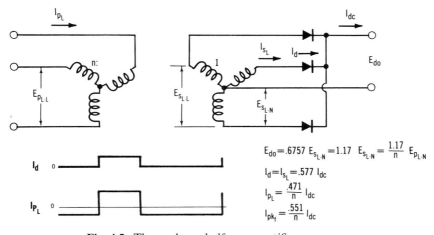

$$E_{do} = .6757\,E_{s_{L\cdot N}} = 1.17\,E_{s_{L\cdot N}} = \frac{1.17}{n}\,E_{p_{L\cdot N}}$$

$$I_d = I_{s_L} = .577\,I_{dc}$$

$$I_{p_L} = \frac{.471}{n}\,I_{dc}$$

$$I_{pk_f} = \frac{.551}{n}\,I_{dc}$$

Fig. 4.5. Three-phase, half-wave rectifier, wye–wye.

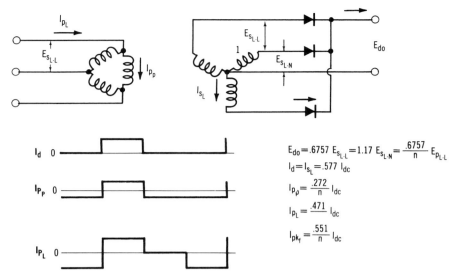

$E_{do} = .6757\ E_{s_{L\cdot L}} = 1.17\ E_{s_{L\cdot N}} = \dfrac{.6757}{n}\ E_{p_{L\cdot L}}$

$I_d = I_{s_L} = .577\ I_{dc}$

$I_{p_\rho} = \dfrac{.272}{n}\ I_{dc}$

$I_{p_L} = \dfrac{.471}{n}\ I_{dc}$

$I_{pk_f} = \dfrac{.551}{n}\ I_{dc}$

Fig. 4.6. Three-phase, half-wave rectifier, delta–wye.

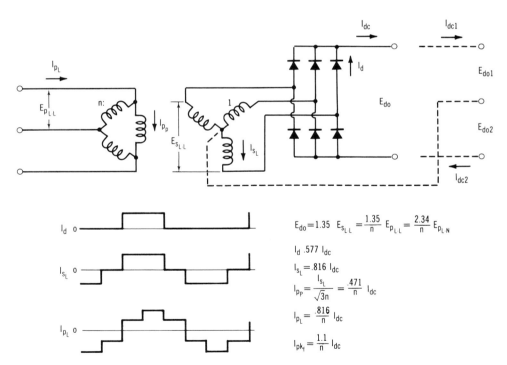

$E_{do} = 1.35\ E_{s_{L\cdot L}} = \dfrac{1.35}{n}\ E_{p_{L\cdot L}} = \dfrac{2.34}{n}\ E_{p_{L\cdot N}}$

$I_d\ .577\ I_{dc}$

$I_{s_L} = .816\ I_{dc}$

$I_{p_P} = \dfrac{I_{s_L}}{\sqrt{3}n} = \dfrac{.471}{n}\ I_{dc}$

$I_{p_L} = \dfrac{.816}{n}\ I_{dc}$

$I_{pk_f} = \dfrac{1.1}{n}\ I_{dc}$

Fig. 4.7. Three-phase, full-wave rectifier, delta–wye.

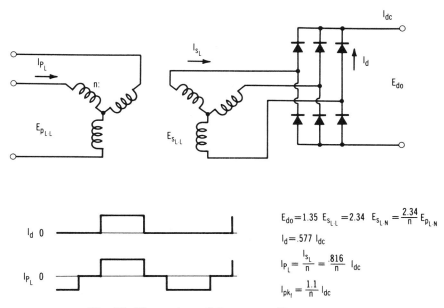

$$E_{do} = 1.35 \ E_{s_{L \cdot L}} = 2.34 \ E_{s_{L \, N}} = \frac{2.34}{n} \ E_{p_{L \, N}}$$

$$I_d = .577 \ I_{dc}$$

$$I_{P_L} = \frac{I_{s_L}}{n} = \frac{.816}{n} \ I_{dc}$$

$$I_{pk_f} = \frac{1.1}{n} \ I_{dc}$$

Fig. 4.8. Three-phase, full-wave rectifier, wye–wye.

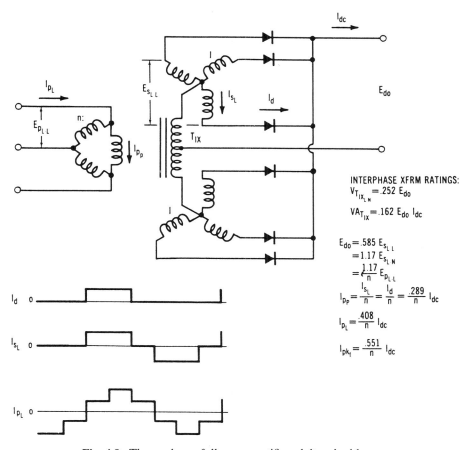

INTERPHASE XFRM RATINGS:

$$V_{T_{IX_{L \, N}}} = .252 \ E_{do}$$

$$VA_{T_{IX}} = .162 \ E_{do} \ I_{dc}$$

$$E_{do} = .585 \ E_{s_{L \, L}}$$
$$= 1.17 \ E_{s_{L \, N}}$$
$$= \frac{1.17}{n} \ E_{p_{L \cdot L}}$$

$$I_{pp} = \frac{I_{s_L}}{n} = \frac{I_d}{n} = \frac{.289}{n} \ I_{dc}$$

$$I_{p_L} = \frac{.408}{n} \ I_{dc}$$

$$I_{pk_f} = \frac{.551}{n} \ I_{dc}$$

Fig. 4.9. Three-phase, full-wave rectifier, delta–double wye.

relations are given in Chapter 3 by Equation (3.9), and transformer connections are discussed in Section 3.8.3.

For the three-phase half-wave rectifier, reasons similar to those for the single-phase half-wave circuit generally restrict this circuit to low currents. Figures 4.5 and 4.6 show relations and current waveforms for a Y–Y and Δ–Y circuit, assuming an inductive load. Note the offset line current in the Y–Y circuit. From previous equations,

$$I_{avg} = 0.827 I_{pk} \qquad I_{rms} = 0.841 I_{pk} \qquad FF = 1.017$$

As with the single-phase center-tapped transformer, three additional rectifiers may be added to produce a dual half-wave bridge with a negative and a positive output with respect to the neutral, as shown by the dashed lines of Fig. 4.7.

The three-phase full-wave bridge rectifier, shown in Figs. 4.7 and 4.8, offers a substantial improvement in rectification efficiency, lower transformer ratings, and lower output ripple than the previous circuits. As an off-line rectifier operating from a typical 120/208-V line, the bridge produces a nominal 280-V dc bus and for a typical 220/380-V line, the bridge produces a nominal 514-V dc bus. The waveform in Fig. 4.7 shows the advantage of a Δ–Y (or Y–Δ) connection over the Y–Y connection of Fig. 4.8. From previous equations,

$$I_{avg} = 0.955 I_{pk} \qquad I_{rms} = 0.9558 I_{pk} \qquad FF = 1.0009$$

To overcome two rectifier forward drops in the bridge circuit, the rectifiers of Figs. 4.9 and 4.10 may be used. The double wye produces a lower voltage than the

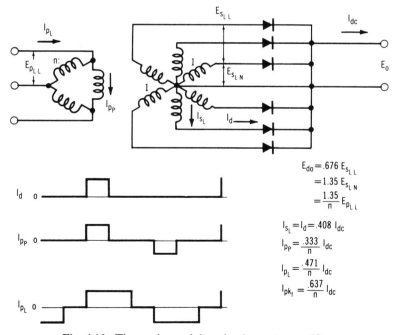

$$E_{do} = .676\, E_{S_{LL}}$$
$$= 1.35\, E_{S_{LN}}$$
$$= \frac{1.35}{n}\, E_{P_{LL}}$$

$$I_{S_L} = I_d = .408\, I_{dc}$$
$$I_{PP} = \frac{.333}{n}\, I_{dc}$$
$$I_{P_L} = \frac{.471}{n}\, I_{dc}$$
$$I_{pk_f} = \frac{.637}{n}\, I_{dc}$$

Fig. 4.10. Three-phase, delta, six-phase star rectifier.

bridge, and an interphase transformer is usually required to balance the currents from each half-wave rectifier. The six-phase star is essentially three center-tapped single phases and incurs a severe penalty in terms of tranformer VA rating, as given in Table 4.2. However, for very high currents, the decreased power dissipation in the rectifiers could warrant its use. It is also noted that a Y primary winding should not be used unless a tertiary winding is added to carry circulation currents.

4.1.2.1 *Common Six-Phase Full-Wave Circuits*

A further reduction in output ripple and transformer VA rating is possible with a 12-pulse rectifier or a 6-phase full-wave rectifier. Two circuits are shown in Figs. 4.11 and 4.12. For high-voltage

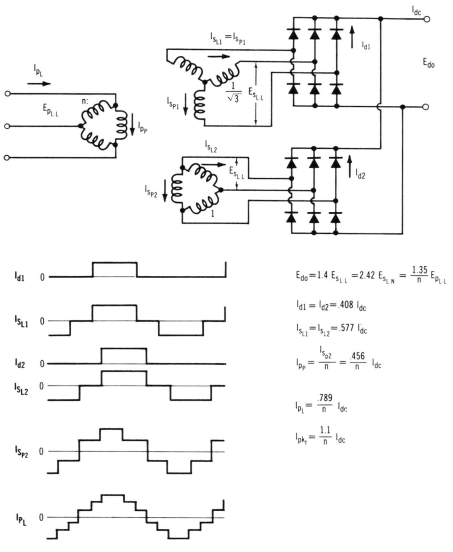

$$E_{do} = 1.4\, E_{s_{LL}} = 2.42\, E_{s_{LN}} = \frac{1.35}{n}\, E_{p_{LL}}$$

$$I_{d1} = I_{d2} = .408\, I_{dc}$$

$$I_{s_{L1}} = I_{s_{L2}} = .577\, I_{dc}$$

$$I_{p_P} = \frac{I_{s_{p2}}}{n} = \frac{.456}{n}\, I_{dc}$$

$$I_{p_L} = \frac{.789}{n}\, I_{dc}$$

$$I_{pk_f} = \frac{1.1}{n}\, I_{dc}$$

Fig. 4.11. Three-phase, delta–dual parallel bridge rectifier.

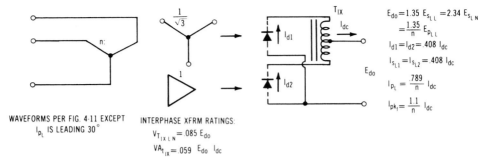

Fig. 4.12. Three-phase, wye–dual parallel bridge with interphase transformer.

output, the series bridges of Fig. 4.11 should be used. For high current output, the parallel bridges of Fig. 4.12 should be used, wherein each bridge supplies one-half the output current, via the interphase transformer if required. A Δ or Y primary may be used with either circuit, and the input line current waveforms are identical except for a 30° phase shift. From Equations (4.4) and (4.5)

$$I_{avg} = 0.988616 I_{pk} \qquad I_{rms} = 0.988668 I_{pk} \qquad FF = 1.00005$$

4.1.2.2   Other Six-Phase Full-Wave Circuits   Characteristics of the dual extended delta secondary (as discussed in Section 3.8.3 and by Fig. 3.23) are ideally the same as the Δ–Y secondaries. For high ratios of step-down voltage the extended delta offers advantages by eliminating the major dependence on the $n : \sqrt{3}$ factor, where n is the turns ratio.

An additional 12-pulse rectifier circuit is the dual zig-zag wye secondary, with each secondary applied to a bridge rectifier. Performance is ideally the same as above, except the secondary VA rating is $1.17 P_{dc}$ as opposed to $1.05 P_{dc}$ for the above circuits.

4.1.3   Transformer Reactance and Rectifier Commutation

Leakage inductance in a transformer, as discussed in Chapter 3, affects the voltage regulation of the transformer. When the transformer windings are connected to a rectifier circuit, the leakage reactance of the transformer and the reactance of the ac lines appear as inductance in series with the rectifier device. These reactances prevent the dc load current from instantaneously transferring from one conducting rectifier to the other. Figure 4.13 shows a single-phase center-tapped rectifier with transposed reactances L_1 and associated waveforms. As voltage e_{3-4} goes positive, current i_{CR1} cannot flow immediately due to L_{11}, and a voltage "appears" across L_{11}. During this latter period, CR_2 is forward biased, but cannot immediately conduct due to L_{12}.

The period of commutation μ is the time required for current transfer from one branch to the other, and during this period the output voltage is zero. The period is dependent on load current, input frequency, and the reactance of the transformer.

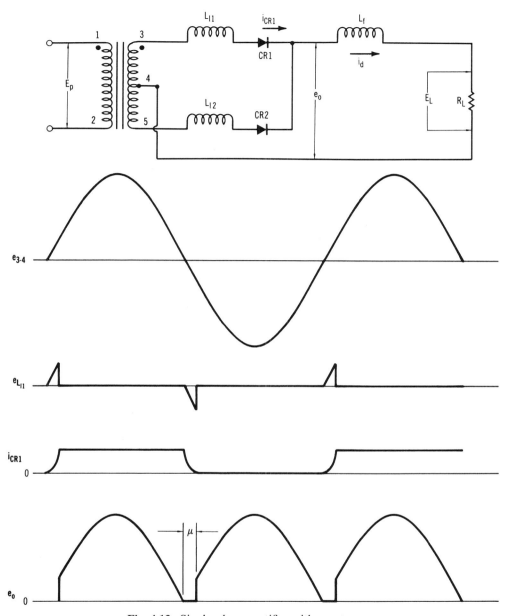

Fig. 4.13. Single-phase rectifier with reactance.

The expressions for output voltage and current are [2]

$$e_0 = \frac{1}{\pi} \int_{\mu}^{\pi} e_{3-4\text{pk}} \sin \omega t \, dt = \frac{E_{3-4\text{pk}}}{\pi} (1 - \cos \mu)$$

$$i_d = \frac{1}{L_{11}} \int_0^{\mu/\omega} e_{3-4\text{pk}} \sin \omega t \, dt = \frac{E_{3-4\text{pk}}}{L_{11}} (1 - \cos \mu)$$

Solving for $\cos \mu$ and substituting,

$$\cos \mu = 1 - \frac{I_d \omega L_{11}}{E_{3-4\mathrm{pk}}}$$

$$e_0 = \frac{E_{3-4\mathrm{pk}}}{\pi} \left(1 + 1 - \frac{I_d \omega L_{11}}{E_{3-4\mathrm{pk}}} \right)$$

$$= \frac{2 E_{3-4\mathrm{pk}}}{\pi} - \frac{I_d \omega L_{11}}{\pi}$$

The left-hand portion of the equality is the term for the average voltage, and the right-hand portion is the voltage reduction due to leakage reactance.

From Section 3.6.5, the reactance is usually expressed as a percent of VA rating. Normalizing the above equation as a per unit reactance and then transposing the reactance to the primary:

$$E_{0,\,\mathrm{pu}} = \frac{2\sqrt{2}}{\pi} - \frac{2 X_{lp}}{\pi} = \frac{2\sqrt{2}}{\pi} \left(1 - \frac{X_{lp}}{\sqrt{2}} \right) \tag{4.8}$$

Note the constant $2\sqrt{2}/\pi = 0.90$ is the value of $E_{\mathrm{do}}/E_{\mathrm{pk}}$ in Fig. 4.2 and the factor K in Table 4.2 is $141 = \sqrt{2} \times 100 \ (\%)$.

For the three-phase rectifier of Fig. 4.7, transformer reactance appears as a leakage inductance in series with each line to the bridge rectifier. The equation derived for E_0 is [2]

$$E_{0,\,\mathrm{pu}} = \frac{3\sqrt{3}\,\sqrt{2}}{\pi} \left(1 - \frac{X_{lp}}{2} \right) \tag{4.9}$$

Again, the constant $(3 \times \sqrt{3} \times \sqrt{2})/\pi = 2.34$ is the value of E_{do}/E_{pL-N} in Fig. 4.7, and the K factor in Table 4.2 is $200 = 2 \times 100 \ (\%)$. Transformer reactance is usually given in percentage of VA rating, from which X_{lp} in ohms may be calculated.

Further analysis of regulation in the three-phase full-wave or six-pulse rectifier is found from the work by Witzke [3]. Figure 4.14 shows the relation of output voltage to open circuit voltage as a function of $I_d X_l/E_s$. Note that at $I_d X_l/E_s = 1$, $E_{\mathrm{dc}}/E_{\mathrm{do}} = 0.5$, which compares favorably with the value in parentheses of Equation (4.9) for $X_{lp} = 100\%$. However, at higher values of $I_d X_l/E_s$, the slope increases and reaches a maximum at output short circuit. This relation may be utilized in very high power rectifier circuits where the output current at short circuit may be limited by purposely designing a transformer with a high reactance. The commutation angle is found by [3]

$$\mu = \cos^{-1} \left(1 - \sqrt{\frac{2}{3}} \frac{I_{\mathrm{dc}} X_l}{E_s} \right) \tag{4.10}$$

where I_{dc} is dc output current, X_l is leakage reactance, and E_s is the rms secondary voltage.

For high-power applications requiring a high dc output voltage, the 12-phase circuit, as previously discussed, is used. However, instead of the circuit of Fig. 4.11, an additional wye primary is wound which is coupled to one of two delta secondaries. Operation of this rectifier circiut is affected to a greater extent by common reactance than is the operation of the 6-pulse rectifier circuit [4]. The ratio of output voltage to open-circuit voltage as a function of reactance factors is shown in Fig. 4.15. The factor K is the ratio of reactance common to both 6-phase rectifiers to total commutating reactance. For $K = 0$, the 12-pulse operation is identical to the 6-pulse operation. The ratio K is determined by the commutation angle, the various conductive modes of operation, and the reactance factor or coupling between all windings. As the desired high-voltage output increases, the isolation between primary and secondary will increase, thus increasing the common reactance between windings and adversely affecting load regulation.

It is interesting to note the low value of K for the six-phase star, Fig. 4.10 and Table 4.2. This configuration is repeated in Fig. 4.16, showing leakage reactance and waveforms. The low value of K is primarily due to the 60° conduction of the rectifier as opposed to 120° for other six-pulse configurations. The heavy line represents the output voltage waveform, μ is the commutation angle, and

$$e_o = \frac{3}{\pi} \int_{\pi/3}^{2\pi/3} \left[e_{1-n} \sin \omega t - e_{6-n} \sin(\omega t - 120°) \right] d(\omega t)$$

$$- \frac{3}{2\pi} \int_{\pi/3}^{2\pi/3+\mu} \left[e_{1-n} \sin \omega t - e_{5-n} \sin(\omega t - 240°) \right] d(\omega t)$$

$$= \frac{3}{\pi} e_{1-n} \left[-\cos \omega t + \cos(\omega t - 120°) \right]_{60°}^{120°}$$

$$- \frac{3}{2\pi} e_{1-n} \left[-\cos \omega t + \cos(\omega t - 240°) \right]_{60°}^{120°+\mu}$$

$$= \frac{3 \times 1.5 e_{1-n}}{\pi} - \frac{3 \times 1.5 e_{1n}}{2\pi} - \frac{3\sqrt{3} \, e_{1-n} \sin \mu}{2\pi}$$

$$= \frac{3}{2\pi} e_{1-n} (1.5 - \sqrt{3} \, \sin \mu)$$

The expression for rectifier current is

$$i_F = \frac{1}{L_{11}} \int_{\pi\omega/3}^{(2\pi/3+\mu)\omega} \left[e_{1-n} \sin \omega t - e_{5-n} \sin(\omega t - 240°) \right] d(\omega t)$$

$$= \frac{e_{1-n}}{\omega L_{11}} (\sqrt{3} \sin \mu + 1.5)$$

Solving for μ and substituting,

$$e_0 = \frac{3 e_{1-n}}{2\pi} \left(1.5 - \frac{e_{1-n}}{\sqrt{3}} \right)$$

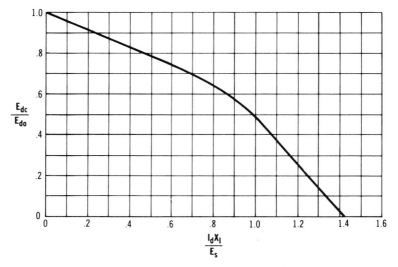

Fig. 4.14. Voltage regulation curve for six-phase rectifier. (Copyright © 1953 IEEE.)

Again, normalizing the unit reactance and transposing

$$E_{o,pu} = \frac{3\sqrt{3}}{2\pi} - \frac{3X_{lp}}{2\pi} = \frac{3\sqrt{3}}{2\pi}\left(1 - \sqrt{3}X_{lp}\right)$$

$$= \frac{3\sqrt{3}}{2\pi}\left(1 - \frac{X_{lp}}{0.577}\right) \tag{4.11}$$

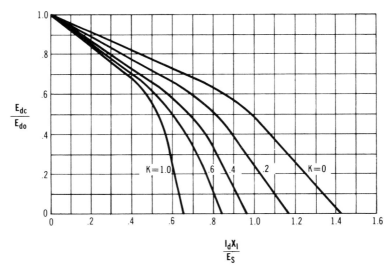

Fig. 4.15. Voltage regulation curve for 12-phase double-wye rectifier. (Copyright © 1953 IEEE.)

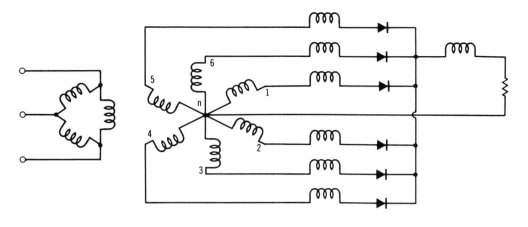

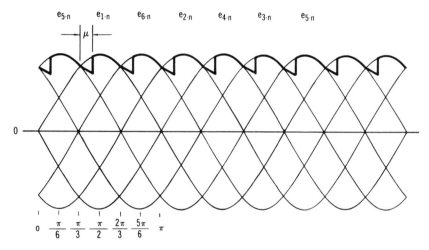

Fig. 4.16. Six-phase star rectifier with reactance.

Note the constant $(3 \times \sqrt{3})(2\pi) = 0.827$ is multiplied by $\sqrt{2/3}$ for rms and line-to-line factors, and equals 0.676, which is the value of E_{do}/E_{sL-L} in Fig. 4.10, and the factor K in Table 4.2 is $57.7 = 0.577 \times 100$ (%).

4.2   CAPACITOR INPUT FILTERS

Capacitor input filters remain the most volumetrically efficienct means of filtering rectified sine waves and storing energy. In the case of linear or brute force power supplies, an input transformer provides isolation and the desired secondary voltage, which is rectified and filtered. The size and weight factors of transformers for various power ratings are discussed in Chapter 3. For electrolytic capacitors, the volume required by the capacitor is approximated by

$$V_c = \frac{10}{nf_s DE_c} \quad \text{in.}^3/\text{W} \qquad (4.12)$$

where

n = power supply efficiency

f_s = power line frequency

$D = CV$ product per unit volume, F $\times$ V/in.^{3}

E_c = capacitor voltage

Since f_s is normally fixed and D is given by the manufacturer ($D = 0.035$ typically for aluminum electrolytics) an increase in efficiency and/or an increase in E_c will reduce the capacitor volume. Both of these latter characteristics are inherently provided by the off-line rectifier/filter utilized in high-frequency switch mode power supplies. The high-frequency transformer further aids in overall size and weight reduction.

The ripple voltage across the filter capacitor is a function of the amount of filter capacitance, the input frequency, and the load resistance or load current. Considerable detail is given to calculations of ripple amplitude because this parameter influences other design parameters for power conversion devices. In the case of linear power supplies the minimum capacitor voltage E_c at low line (minimum input voltage) must be equal to the output voltage plus the minimum voltage which the pass regulator can tolerate while maintaining a regulated output. At high line, the voltage across the pass regulator increases and the regulator must dissipate substantial power. For this reason, the typical operating efficiency of linear power supplies is approximately 50% (for a $\pm 15\%$ line variation) unless the filter capacitance is very large, which adds cost and volume.

In switch mode power supplies, higher ripple voltage may be tolerated since the pulse width modlator will average the variations without an increase in power dissipation. However, unless the gain of the feedback loop is very high at the second harmonic (for single-phase input), the predominant output ripple frequency will be harmonics of the input frequency, and the filter capacitance must be calculated accordingly. The filter capacitance may be reduced (for a desired output ripple) or the output ripple may be reduced (for a desired capacitance) by feed-forward techniques discussed in Section 8.6. In many cases, the power supply can then operate with a 25% peak-to-peak ripple voltage across the input filter capacitor, and a line variation of $\pm 15\%$.

4.2.1   Single-Phase Rectifier Filters

Typical single-phase rectifier circuits are shown in Figs. 4.1–4.4. In Fig. 4.17, a capacitor filter has been added, and the resultant waveforms are shown in

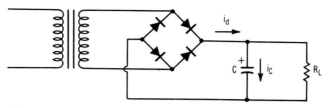

Fig. 4.17. Single-phase, full-wave rectifier, capacitive filter.

Fig. 4.18. The heavy line represents the output waveform, and

$$E_r = E_{\text{pk}} - E_2 \qquad e_r = E_r/E_{\text{pk}}$$
$$1 - e_r = (E_{\text{pk}} - E_r)/E_{\text{pk}} \qquad (4.13)$$

where

$$e_r \times 100 = \text{percentage of peak-to-peak ripple}$$
$$E_1 = E_{\text{pk}} \cos \theta$$
$$E_2 = E_1 \varepsilon^{-t/R_L C} = E_{\text{pk}} \cos \theta \varepsilon^{-t/R_L C}$$
$$= E_{\text{pk}} \sin \alpha = E_{\text{pk}} - E_r$$
$$\cos \theta \varepsilon^{-t/R_L C} = \sin \alpha = 1 - e_r$$

where

$$t = t_{B-D} \qquad \theta = \tan^{-1}(t/R_L C)$$
$$\alpha = \sin^{-1}(1 - e_r)$$
$$\varepsilon^{t/R_L C} = \cos \theta/(1 - e_r)$$

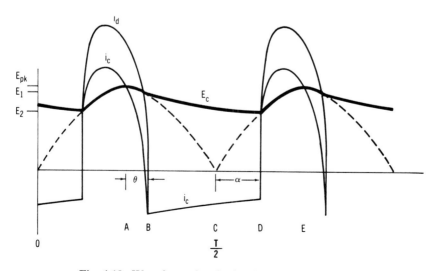

Fig. 4.18. Waveforms for single-phase capacitive filter.

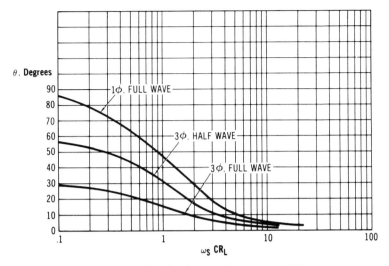

Fig. 4.19. Conduction angle versus $\omega_s CR_L$.

Solving for t_{B-D},

$$t_{B-D} = R_L C \ln\left(\frac{\cos\theta}{1-e_r}\right) = t_{B-C} + t_{C-D} = (90° - \theta) + \alpha$$

$$= \frac{T}{4 \times 90°}(90° - \theta + \alpha) = \frac{T}{4}\left(1 - \frac{\theta - \alpha}{90°}\right)$$

$$R_L C = \frac{T\{1 - [\tan^{-1}(t/R_L C) - \sin^{-1}(1-e_r)]/90°\}}{4\ln\{\cos[\tan^{-1}(t/R_L C)]/(1-e_r)\}}$$

Rearranging terms provides the relation of peak-to-peak ripple as a function of $\omega_s CR_L$ where

$$\omega_s CR_L = \frac{\pi\{1 - [\tan^{-1}(1/\omega_s CR_L)]/90° + \sin^{-1}(1-e_r)/90°\}}{2\ln\{\cos[\tan^{-1}(1/\omega_s CR_L)]/(1-e_r)\}} \quad (4.14)$$

for a single-phase, *full-wave* rectifier and capacitor filter. Unfortunately, the above equation cannot be solved directly, but is graphed in Fig. 4.19 for θ versus $\omega_s CR_L$ and in Fig. 4.20 for E_r versus $\omega_s CR_L$. Inspection of these figures indictaes θ is considered negligible in the commonly used ranges of allowable ripple.

For a given load resistance ($R_L = E_{in}/I_o$) and a desired or minimum ripple voltage, the required filter capacitance may be calculated for the minimum mains input frequency.

When the load resistance is large in value, a single-phase *half-wave* rectifier/filter may be used to conserve cost. Following the analysis above, the

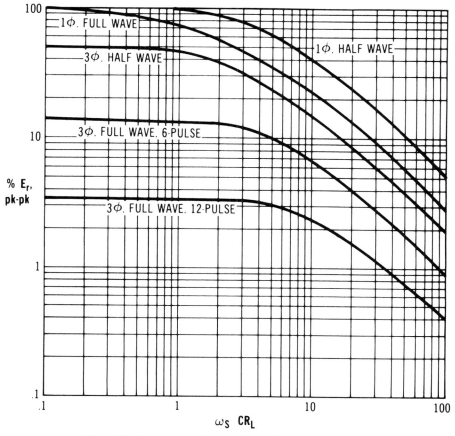

Fig. 4.20. Percent peak-to-peak ripple versus $\omega_s CR_L$.

peak-to-peak ripple as a function of $\omega_s CR_L$ is

$$\omega_s CR_L = \frac{\pi\{3 - [\tan^{-1}(1/\omega_s CR_L)]/90° + \sin^{-1}(1 - e_r)/90°\}}{2\ln\{\cos[\tan^{-1}(1/\omega_s CR_L)]/(1 - e_r)\}} \quad (4.15)$$

4.2.2 Polyphase Rectifier Filters

As previously mentioned, three-phase power is usually available where high-power equipment is involved. Power stages should always be preceded by an inductor input filter to minimize peak currents, as discussed in Sections 4.2.3 and 4.3. However, power for the control electronics may be obtained by the transformer–rectifier circuits of Figs. 4.6–4.8 followed by a capacitor input filter.

Waveforms for a three-phase, half-wave rectifier/filter and a three-phase, full-wave rectifier/filter are shown in Figs. 4.21 and 4.22, respectively. Following the analysis for the single-phase case, relations for three-phase *half-wave* rectifier

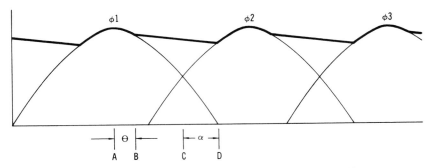

Fig. 4.21. Waveforms for a three-phase, half-wave rectifier.

are

$$E_2 = E_1 \varepsilon^{-t/R_L C} = E_{pk} \cos \theta \varepsilon^{-t/R_L C}$$

$$= E_{pk} \sin(\alpha + 30°)$$

Solving for t_{B-D} where $\alpha = \sin^{-1}[(1 - e_r) - 30°)]$,

$$t_{B-D} = t_{B-C} + t_{C-D} = [(60° - \theta) + \alpha]$$

Rearranging terms,

$$\omega_s CR_L = \frac{\pi[0.5 - \tan^{-1}(1/\omega_s CR_L)/60° + \sin^{-1}(1 - e_r)/60°]}{3 \ln\{\cos[\tan^{-1}(1/\omega_s CR_L)]/(1 - e_r)\}} \quad (4.16)$$

and is graphed in Fig. 4.20 for E_r versus $\omega_s CR_L$.

Relations for three-phase *full-wave* circuits are

$$E_2 = E_1 \varepsilon^{-t/R_L C} = E_{pk} \cos \theta \varepsilon^{-t/R_L C}$$

$$= E_{pk} \sin(\alpha + 60°)$$

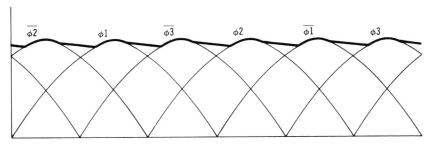

Fig. 4.22. Waveforms for a three-phase, full-wave rectifier.

Solving for t_{B-D}, where $\alpha = \sin^{-1}[(1 - e_r) - 60°]$,

$$t_{B-D} = \frac{T}{12 \times 30°}(30° - \theta + \alpha) = \frac{T}{12}\left(1 - \frac{\theta + \alpha - 60°}{30°}\right)$$

Rearranging terms,

$$\omega_s C R_L = \frac{\pi\left[\sin^{-1}(1 - e_r)/30° - \tan^{-1}(1/\omega_s C R_L)/30°\right]}{6\ln\left\{\cos\left[\tan^{-1}(1/\omega_s C R_L)\right]/1 - e_r\right\}} \qquad (4.17)$$

and is graphed in Fig. 4.20 for E_r versus $\omega_s C R_L$.

4.2.3 Turn-On Currents and Surge Limiting

The rectifier and transformer current relations given in Table 4.2 are for inductive and resistive loads. With a capacitor input filter, current flow becomes quite different due to the low capacitive impedance. If the ac power is turned on at or just before the peak of the input voltage waveform, the peak current is limited only by R_s (equivalent transformer and line resistance), R_{ESR} (equivalent series resistance of the capacitor), and X_L (the leakage reactance of the transformer and/or input line). The total equivalent resistance of the transformer looking into the secondary is

$$R_{sec} + R_{pri}\left(N_s^2/N_p^2\right)$$

and may be calculated from the regulation formula of Chapter 3, Equation (3.19), by

$$R_{xfr} = (E_{in} - E_0)I_0$$

The total source resistance is

$$R_s = \left(R_{line} + R_{pri}\right)\left(N_s^2/N_p^2\right) + R_{ESR}$$

The total source inductance L_s is a function of the inductance in the input line and the leakage reactance of the transformer referred to the secondary. Figure 4.17 is repeated in Fig. 4.23 showing an off-line bridge rectifier with the ac voltage applied at angle λ. The effect of L_s will be repeated by first assuming $L_s = 0$ in the following analysis, and then referring to the analysis of input inductance filters of Section 4.3.2. In many cases, the inductance of an EMI filter provides sufficient surge protection for the off-line rectifier circuit. The dynamic equation with $L_s = 0$ is

$$E_{pk}\sin(\omega t + \lambda) = iR_s + \frac{1}{C}\int_0^T i\,dt \qquad (4.18)$$

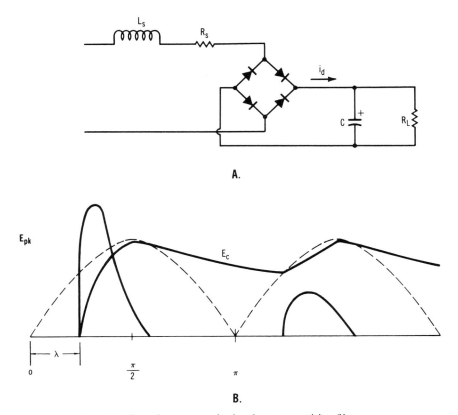

Fig. 4.23. Inrush-current, single-phase, capacitive filter.

Solving for current,

$$i = \frac{E_{pk}}{R_s\sqrt{1 + (1/\omega_s CR_S)^2}}\left[\sin(\omega t + \lambda + \theta) - \frac{\varepsilon^{-\omega t/\omega CR_S}}{\omega CR_S}\cos(\lambda - \theta)\right] \quad (4.19)$$

The left-hand term in the brackets is the steady-state portion, while the right-hand term in the brackets is the transient term. The equation can be reduced to

$$i_d = E_{pk}k/R_s \quad (4.20)$$

where k is plotted versus λ for various values of $\omega_s CR_s$ in Fig. 4.24.

Analysis of Equation (4.19) and Fig. 4.24 shows that at $t = 0$, the rectifier current is approximated by $i_d = E_p(\sin \lambda)/R_s$, and i_d decreases to zero when

$$\sin(\omega t + \lambda + \theta) = \left[\cos(\lambda + \theta)\varepsilon^{-t/R_s C}\right]$$

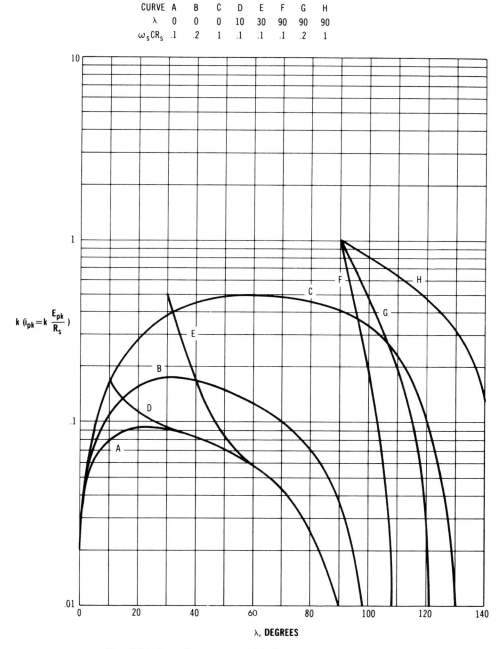

CURVE	A	B	C	D	E	F	G	H
λ	0	0	0	10	30	90	90	90
$\omega_s CR_s$	.1	.2	1	.1	.1	.1	.2	1

$k \left(i_{pk} = k \dfrac{E_{pk}}{R_s} \right)$

λ, **DEGREES**

Fig. 4.24. Inrush-current multiplier versus turn-on angle.

The value of k is a function of $\omega C R_s$ and also is dependent on the time in the cycle at which the power supply is turned on (reference Section 1.5.2).

For low values of R_s, the peak current may be reduced by a factor of 10 if the power supply is turned on at the zero crossing of the input wave, as compared to turning on the power supply at the peak of the input wave. In the design of the rectifier circuit, the desired output voltage and current are usually specified or calculated as discussed in Section 4.2.4. This allows a calculation of R_L and then C to meet the desired ripple. The value of R_s then determines the starting inrush current which the rectifier is required to handle. The peak current and duration can be calculated from Eqaution (4.19) and Fig. 4.24, and then compared to the I_{FSM} (peak half-cycle current) or $I^2 t$ (ampere^2-seconds) rating of the selected rectifier.

In many off-line rectifiers, the line voltage produces a substantial current surge which may be more economical to limit than choosing a higher current rectifier, or the mains may be incapable of delivering high currents without a substantial voltage drop. Figure 4.25 shows five methods of limiting the surge. In the step-start circuit (Fig. 4.25A) R_s is in series with R_1 through which current flows when S_1 is closed, thereby limiting the inrush current. The energizing time of K_1 is typically from one to three cycles of the input frequency which allows the filter capacitor to be charged, after which R_1 is shorted by K_1 contacts to eliminte power dissipation and voltage drop.

A highly negative temperature coefficient power thermistor is shown in Fig. 4.25B. At turn-on the "several ohms" resistance in TH_1 limits the inrush current which, in turn, produces power in the thermistor. The power raises the temperature of the device and the resistance drops to a low value for normal operation. However, the thermal time constant of the thermistor must be considered. If a power supply has been operating at near no load for some time, the resistance of TH_1 has increased. When full load is applied, the voltage drop across TH_1 may cause the output to drop substantially or in the case of a regulated supply, the output may drop out of regulation. Conversely, if the power supply is operating at full load where TH_1 is very low in resistance and a short power interruption occurs (long enough for the input filter capacitor to discharge), when the mains voltage returns a high inrush current will occur which negates the purpose of the thermistor.

The circuit in Fig. 4.25C utilizes a zero-crossing switch, which may be in the form of a triac or solid-state relay, to limit the inrush current for the condition $\lambda = 0$ in Fig. 4.24. This circuit has the advantage of being activated remotely by standard logic signals, which may be desirable from a systems operation.

The circuit in Fig. 4.25D again utilizes a triac (Q_1), as in Fig. 4.25C. However, the soft-start control circuit is phase controlled at turn-on to limit inrush current. Initially, the conduction peroid of each half cycle is very short but then increases with time until full conduction is reached. Three precautions should be observed with this circuit. First, the conduction period should increase linearly each half cycle at turn-on to prevent a dc component in the output, if the circuit is supplying power to a transformer or to an ac motor. Second, the control circuit should have mains interruption sensing, with a time constant sufficient to permit soft-start operation when the mains again returns; otherwise, a high inrush current will occur. Third, the control circuit should provide a constant signal (or a train of

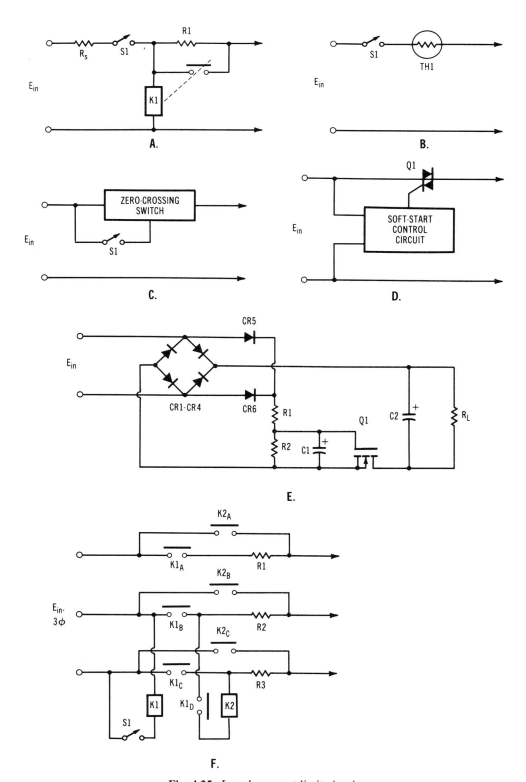

Fig. 4.25. Inrush-current limit circuits.

pulses derived from a one-shot multivibrator) during desired full conduction. Otherwise, Q_1 will turn off since the input current is normally discontinuous, as in the case of an input rectifier followed by a capacitive filter.

An input rectifier and capacitor filter are shown in Fig. 4.25E with an N-channel power MOSFET transistor inserted in the dc return line. Rectifiers CR_1 through CR_4 rectify the input to charge C_2 and supply power to R_L as Q_1 is turned on. The MOSFET has two main advantages over a bipolar transistor. First, the high input impedance of the gate allows high resistance values of biasing resistors, thus minimizing power dissipation. Second, the MOSFET is immune to forward-biased secondary breakdown, although the SOA (safe operating area) must still be observed. At turn-on, C_1 charges exponentially through R_1 to a final steady-state value determined by $R_2/(R_1 + R_2)$. Capacitor C_1 charges to a few volts before the gate threshold voltage is reached, at which time Q_1 begins to conduct and C_2 reaches full charge in a few cycles, determined by the time constant of $R_1 R_2 C_1/(R_1 + R_2)$. The addition of diodes CR_5 and CR_6 serves to decouple C_1 from C_2 during operation, and allows C_1 to discharge through R_2 when input power is removed. If CR_5 and CR_6 were omitted and R_1 were connected to C_2, the voltage across C_2 (as C_2 discharges) would maintain a charge on C_1 even though input power was momentarily removed. Thus, when input power was again applied, the charge on C_1 could be sufficient to bias Q_1 into full conduction. Complete discharge of C_1 during power interruptions could be achieved by paralleling C_1 with a *PNP* transistor with the base biased by a voltage divider from $CR_{5,6}$ to dc return. With power on, the transistor would be reverse biased but with power off, the transistor would conduct to rapidly discharge C_1.

Inserting a resistor between the source terminal of Q_1 and the negative terminal of C_1 will limit the peak inrush current at turn-on. As source current increases, V_{GS} decreases, thus increasing V_{DS}.

Example

In a typical circuit of Fig. 4.25E, let $E_{C1} = 0.1 E_{C2} = 0.1 \times 160 = 16$ V at steady state. Choosing $R_1 = 120$ kΩ and $R_2 = 13$ kΩ the resulting gate voltage is $160 \times 13/133 = 15.6$ V, and the power dissipation of R_1 is $(160^2 - 15.6^2)/120000 = 0.2$ W, allowing the use of a $\frac{1}{2}$-W resistor. Let the time required for E_{C1} to reach a 2-V threshold be 8.3 ms, or a half cycle. Then $2 = 15.6(1 - \varepsilon^{-t/\tau})$, from which $\tau = 0.06$ and $C_1 = 0.06(R_1 + R_2)R_1 R_2 = 5$ μF. When power is interrupted, $\tau = R_2 C_2 = 0.065$ and the time required for E_{C1} to fall to 2 V is 0.13 s. This time should be considerably less than the time for C_2 to discharge to 20 V if CR_5 and CR_6 were omitted, and R_1 were connected to C_2.

When a transformer is placed ahead of the rectifier, the transformer leakage inductance and equiavlent series resistance are usually sufficient to limit the inrush current to a safe value, regardless of the dc current or rectifier current rating. For low output currents, and thus low rectifier current ratings, the transformer impedance increases due to an increase in the number of turns (which increases leakage inductance) and an increase in wire gage. However, this condition may not be valid for transformers with a high step-up voltage ratio placed ahead of the rectifier. In higher power circuits, R_s may be moderate, but the ratio of E_{sec} to E_{pri} and thus i_d may be very high. For these applications, the resistor step-start circuit of Fig. 4.25A is ideal.

For high-voltage, moderate power outputs operating from a three-phase line, the circuit of Fig. 4.25F solves the inrush current problem associated with the transformer magnetizing current, as well as the inrush current caused by a capacitive filter which follows a bridge rectifier on the secondary of the transformer. This circuit is essentially a dual contactor–resistor step-start approach. Turn-on is accomplished by closing switch S_1 which energizes K_1. Initial inrush line current, determined by the resistance values of R_1, R_2, and R_3 and the input line-to-line voltage, then flows through R_1, R_2, and R_3 to the load, and power is applied to energize K_2. Since the pull-in time of K_2 is normally two cycles or more, sufficient delay is introduced such that when K_2 closes to short the resistors, the step change in current is no greater than the initial inrush current value. Thus, active time delay circuits are not required. In the case of a rectifier and capacitive filter on the secondary, the three-phase rectified ripple is inherently low, and the output reactor normally associated with inductor input filters may be eliminated.

4.2.4 Operating Currents and Voltages

The previous analysis of ripple and inrush characteristics of input capacitor filters leads to the output current and voltage obtained, or desired, for a given applica-

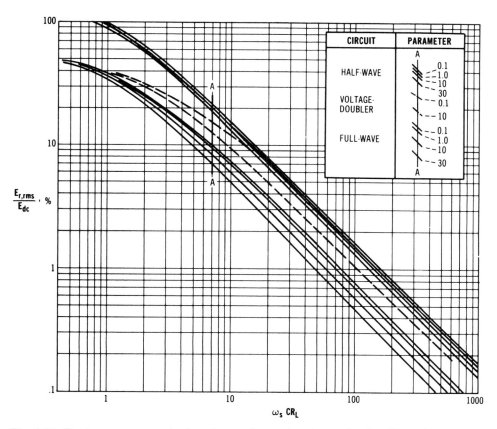

Fig. 4.26. Root-mean-square ripple voltage of capacitor input circuits. (Copyright © 1943 IEEE.)

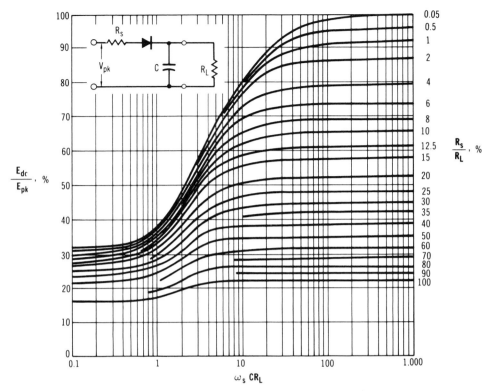

Fig. 4.27. Relation of applied alternating peak voltage to direct output voltage in half-wave, capacitor input circuits. (Copyright © 1943 IEEE.)

tion. These calculations are achieved with the aid of design graphs based on data presented by Schade [5]. In designing a rectifier capacitor–filter circuit, the allowable ripple must first be established. From Fig. 4.20 or 4.26, ωCR_L is selected from which C is calculated since $R_L = E_{dc}/I_{dc}$ and $\omega = 2\pi f_s$. The output voltage (or required input voltage for a desired output voltage) may be determined from Figs. 4.27 and 4.28 for half-wave and full-wave circuits, respectively, where R_s is determined by the discussion of Section 4.2.3.

The relation of rms and peak currents to average rectifier current is shown in Fig. 4.29. Many manufacturers' data sheets for rectifier ratings plot maximum case or lead temperature versus average current and continuous dc (rms) current. The average current is usually for a resistive load and not for a capacitive load unless specified. In this case, the rms current may be calculated from Fig. 4.29A and the rectifier chosen accordingly.

The PIV (peak inverse voltage) applied to the rectifier is also different with a capacitive filter than with a resistive or inductive load. Assuming the capacitor charges to E_{pk} during the "positive" half cycle, the voltage across the rectifier at the peak of the "negative" half cycle is $2E_{pk}$. When the effects of ωCR_L and R_s/R_L are considered, the PIV may be reduced, as shown in Fig. 4.30. However, all designs should consider line voltage variations and transients affecting PIV, and a safety factor for reverse-voltage rating should be chosen accordingly. Reference

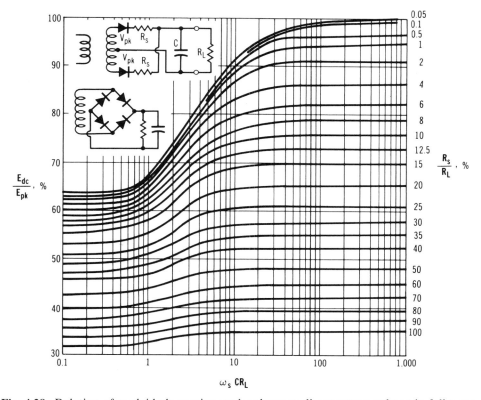

Fig. 4.28. Relation of appleid alternating peak voltage to direct output voltage in full-wave, capacitor input circuits. (Copyright © 1943 IEEE.)

Table 4.2 for recommended ratings. Transient suppression and snubber networks, as a means to reduce voltage spikes, are discussed in Chapter 13.

The capacitor ripple current (reference Fig. 4.18) flowing in the ESR (equivalent series resistance) of the capacitor generates heat and increases the internal temperature of the capacitor. Manufacturers' data sheets usually specify a maximum rms ripple current at one or more frequencies for various capacitors (reference Section 2.3 for an analysis of capacitor ripple current).

Example

A transformer, rectifier, and capacitive filter are to meet the following requirements:

Input voltage: 120 V (105–130 V range), 60 Hz.
Output: 24 Vdc at 4 A dc, 10% peak to peak ripple.

The full-wave center-tapped rectifier of Fig. 4.2 will be used. From Equation (4.1), neglecting reactance and assuming 1% resistive drop and using 1.1 V forward drop for the rectifier, $E_{do} = (E_{dc} + 1.1)1.01 = 25.4$ V.

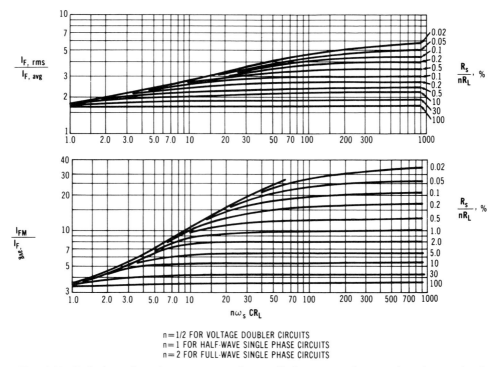

n = 1/2 FOR VOLTAGE DOUBLER CIRCUITS
n = 1 FOR HALF-WAVE SINGLE PHASE CIRCUITS
n = 2 FOR FULL-WAVE SINGLE PHASE CIRCUITS

Fig. 4.29. Relation of peak, average, and rms diode current in capacitor input circuits. (Copyright © 1943 IEEE.)

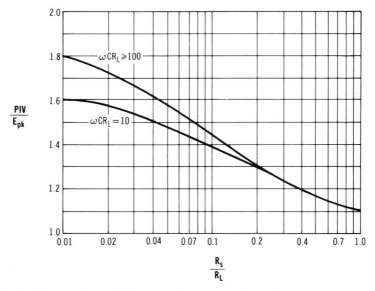

Fig. 4.30. Ratio of operating peak inverse voltage to peak applied ac for rectifiers used in capacitor input, single-phase filter circuits. (Courtesy Motorola, Inc.)

From Fig. 4.20, the value of $\omega_s C R_L$ for 16% peak-to-peak ripple is 16. Then from Fig. 4.28, the ratio of E_{dc}/E_{pk} is 90% for $R_s R_L = 1\%$. Then $E_{pk} = E_{dc}/0.9 = 25.4/0.9 = 28.2 V_{pk}$, or $E_{rms} = 28.2/\sqrt{2} = 20$ V.

From Chapter 3, Fig. 3.25, choose a square stack of EI-125 lamination and operate at 500 cmil/A "current density." This will produce a temperature rise of less than 50°C when using 29-gage (0.014-in.) M-6X material at 14.4 kG. The transformer must not saturate at 130 V input from which the primary turns, from Equation (3.6), are $N_p = (3.49 \times 130 \times 10^6)/(14400 \times 1.56 \times 60 \times 0.95) = 354$.

Thus the required secondary turns at nominal input are $354 \times 20/120 = 59$, or 118 turns center tapped.

The filter capacitance (where $R_L = E_{dc}/I_L$) is $C = 16/(\omega_s R_L) = 16 \times 4/(377 \times 24) = 7073\ \mu F$, and the maximum capacitor voltage (at no load and maximum input voltage) is $E_C = 130 \times \sqrt{2} \times 59/354 = 30.6$ V (use a 50-V rating capacitor).

The average current through the rectifier is $4/2 = 2$ A and from Fig. 4.29, the rms rating required for the rectifier (at $\omega_s C R_L = 16$ and $R_s/R_L = 1\%$) is

$$I_{rms} = 2.4 I_{avg} = 2.4 \times 2 = 4.8 \text{ A} \quad \text{(use 6 A rectifier rating)}$$

4.2.5 Voltage-Doubler Circuits

The dc output voltage may be "increased" with voltage-doubler or multiplier circuits. In the conventional voltage doubler (Fig. 4.31A), capacitors C_1 and C_2 are each charged (during alternate half cycles) to approximately the peak input ac. The capacitors then discharge in series into R_L. The output voltage value is a function of R_s/R_L and $\omega C R_L$, and may be calculated from Fig. 4.32.

In the cascade voltage doubler shown in Fig. 4.31B, C_1 is charged to E_{pk} through CR_2 during the "negative" half cycle, and discharges in series with E_{pk} through CR_1 during the "positive" half cycle to charge C_2 to approximately $2E_{pk}$. The cascade circuit has poorer regulation than the conventional doubler and the ripple frequency is at f_s, thus requiring a larger capacitor to achieve the same ripple as shown in the comparison waveforms. The cascade circuit does have the advantage of a common input and output terminal. This characteristic may be important in multiplier circuits where additional rectifiers and capacitors are added for voltage multiplication, shown by the dashed lines of Fig. 4.31B.

The bridge voltage doubler, shown in Fig. 4.31C, has achieved importance in off-line switch mode power supplies where operation from either 120 V or 240-V ac sources is desired. The operation is essentially that of a full-wave voltage doubler (CR_3, CR_4 and C_1, C_2) for 120 V and that of a full-wave bridge (CR_1 through CR_4 and C_1, C_2) for 240-V operation. However, as a full-wave bridge, the effective output filter capacitance is halved, since $C_{eff} = C_1/2 = C_2/2$.

Example

Choosing $\omega C R_L = 20$ and $R_s/R_L = 1\%$ and referring to Fig. 4.27, the output voltage for 240 V input is $E_{dc} = 0.922 \times \sqrt{2} \times 240 = 313$ V. Referring to Fig. 4.29, the output voltage for 120 V input is $E_{dc} = 1.66 \times \sqrt{2} \times 120 = 282$ V. For a $\pm 10\%$ input voltage variation on each input, the output voltage range is $282 \times 0.9 = 254$ V minimum to $313 \times 1.1 = 344$ V maximum (at no load, $E_{dc} = \sqrt{2} \times 240 \times 1.1 = 373$ V maximum), which is well within the operating ranges of switch mode or pulse width modulation (PWM) power supplies. The

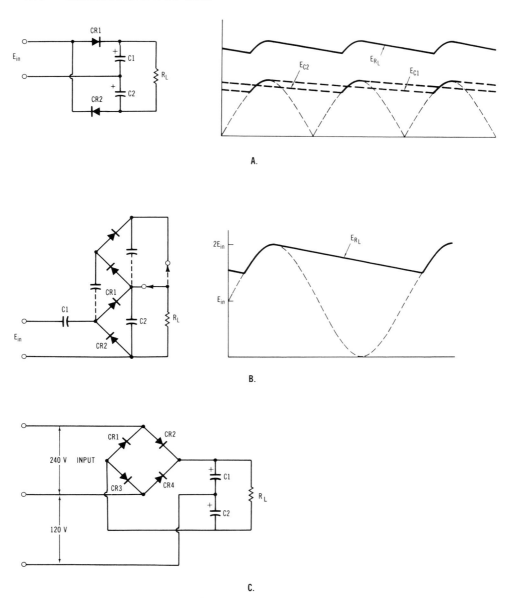

Fig. 4.31. Voltage-doubler circuits and waveforms.

circuit is ideal for 230-V, 50-Hz operation and if $R_s/R_L = 1\%$, $\omega C R_L = 20 \times 50/60 = 16.7$; then $E_{dc} = 0.915 \times \sqrt{2} \times 230 = 298$ V nominal.

4.2.6   Voltage Multipliers

Voltage multiplier circuits are used for high-voltage, low-current applications when a high-voltage transformer winding is inconvenient. Besides the full-wave doubler and the cascade doubler of Fig. 4.31, voltage triplers, quadruplers (half wave and

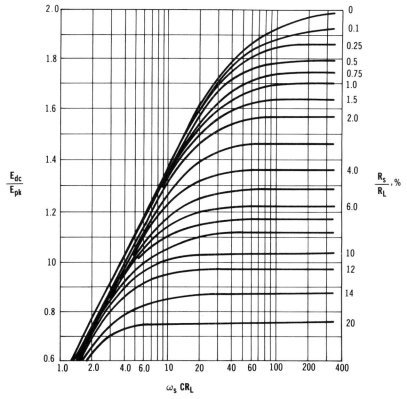

Fig. 4.32. Output voltage as a function of filter constants for full-wave voltage doubler. (Courtesy Motorola, Inc.)

full wave), quintuplers, and sextuplers are used in many designs. However, load regulation of these multipliers rapidly degrades as the multiplication ratio increases. In these cases, large values of capacitance, large ohmic values of load resistance, and/or very low source resistance values are required to achieve reasonable output voltage regulation. Multistage multipliers find applications in powering image tubes and low-power CRTs.

Historical analysis of voltage multipliers normally assumed a sine wave input, usually from a transformer secondary winding, and at a frequency supplied by the utility or mains. Modern switch mode converters employing PWM techniques produce an output voltage of variable pulse width with fast rise and fall times. This topology places stringent demands on the rectifiers (requiring very fast recovery devices) and on the capacitors (requiring high rms current and high-frequency capability). In voltage multipliers, these requirements are even more demanding since high-voltage rectifier stacks inherently have longer recovery time, t_{rr}, due to doping techniques, and reduced capability to withstand a high rate of rise of reverse voltage, dV_R/dt. Converters employing various forms of series or parallel quasi-resonance high-frequency switching techniques (reference Chapter 10) produce half cycles of fixed pulse width, sinusoidal current in the secondary winding, at a frequency modulated rate. The reduced di/dt inherent with these topologies

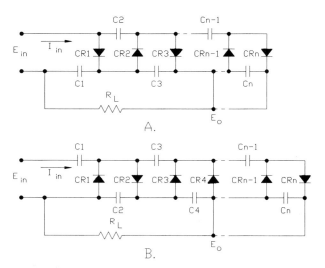

Fig. 4.33. Half-wave voltage multipliers: (A) odd number and (B) even number.

eases the stringent switching requirements placed on the rectifiers and reduces high-frequency conducted emissions associated with rectifier switching. For high-power converters and power supplies, the quasi-resonant topology significantly reduces the size and weight of the unit, as compared to operation at utility frequency. The capacitors are normally of mica, polypropylene, or ceramic material. When operating at high frequencies, the ceramic capacitors should be low dissipation COG types, as discussed in Section 2.8.5. Ideally, all capacitors in any multiplier are of equal value, which eases packaging constraints.

4.2.6.1 Half-Wave Analysis   Basic half-wave voltage multipliers are shown in Fig. 4.33A (odd-number multiplication) and Fig. 4.33B (even-number multiplication). Positive output voltage is assumed but negative output voltage is obtained if all rectifier polarities are reversed. Multiplication is designated by n, the number of capacitors or rectifiers. These circuits have uniform stress per stage on rectifiers and the applied reverse voltage is twice the peak input. The input capacitor C_1 is charged to the peak input, and all other capacitors are charged to twice the peak input. Also, each circuit has the advantage of a common input and output terminal. The voltage tripler (quintupler shown by the dashed lines) of Fig. 4.33A is ideal for asymmetrical waveforms, such as those produced by PWM topologies. The voltage quadrupler (sextupler shown by the dashed lines) of Fig. 4.33B is ideal for symmetrical waveforms. Note the location of the first capacitor in the odd versus even multipliers. This arrangement is necessary to assure the output filter capacitors are connected across the load. Otherwise, the dc output would have an extremely high ripple, produced by the unfiltered voltage of the last section, and would require an additional output filter capacitor.

In Fig. 4.33A, C_1 is charged to $E_{\text{in-pk}}$ through CR_1 during the positive half cycle, and discharges in series with $E_{\text{in-pk}}$ through CR_2 during the negative half cycle to charge C_2 to approximately $2E_{\text{in-pk}}$. The charge delivered to the last capacitor, C_3 or C_n, and to the load must first work its way up the diode ladder. In-rush current, when voltage is first applied, may be quite high as is typical of all capacitor input filters and several cycles may be required to reach equilibrium or full output voltage.

In Fig. 4.33B, C_1 is charged to $E_{\text{in-pk}}$ through CR_1 during the negative half cycle, and discharges in series with $E_{\text{in-pk}}$ through CR_2 during the positive half cycle to charge C_2 to approximately $2E_{\text{in-pk}}$. Again, several cycles may be required to achieve full output voltage on C_4, or C_n.

Assuming ideal components, the maximum output voltage is given by [6]

$$E_0 = nE_{\text{in}} - (n/6)(n^2/2 + 1)(I_0/Cf_s) \tag{4.21a}$$

for $n \equiv$ even and

$$E_0 = nE_{\text{in}} - (n/12)(n^2 - 1)(I_0/Cf_s) \tag{4.21b}$$

For $n \equiv$ odd, where

n = number of stages
C = capacitance, F (each capacitor)
I_0 = output load current, A
f_s = input source frequency, Hz

Rearranging in terms of R_L, for $R_L = E_0/I_0$,

$$E_0 = \frac{nE_{\text{in}}}{1 + \left[n(n^2/2 + 1)/6f_sCR_L\right]} \tag{4.22a}$$

for $n \equiv$ even and

$$E_0 = \frac{nE_{\text{in}}}{1 + \left[n(n^2 - 1)/12f_sCR_L\right]} \tag{4.22b}$$

for $n \equiv$ odd.

The maximum attainable output voltages using these equations are shown in Fig. 4.35 using the familiar ωCR_L term for comparison with other rectifier circuits. The poor regulation of several stages supplying a moderately high output current is dramatically evident since the equivalent source resistance increases as the cube of the multiplication factor. For instance, the output voltage is the same value for the tripler, the quadrupler, and the quintupler for identical input voltages and for ωCR_L approximately 600. Output ripple for the doubler and quadrupler is shown in Fig. 4.40. Full-wave performance is shown for comparison.

4.2.6.2 Full-Wave Analysis Basic full-wave voltage multipliers are shown in Fig. 4.34. The multiplier factor n is even. The doubler circuit of Fig. 4.34A is

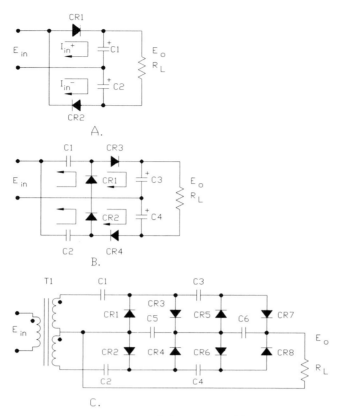

Fig. 4.34. Full-wave voltage multipliers: (A) doubler, (B) parallel quadrupler, and (C) center-tapped quadrupler.

discussed in Section 4.2.5, and performance parameters are shown in Fig. 4.32 as a function of R_s/R_L and in Figs. 4.35 and 4.36 for comparison to other multipliers. Also, two voltage doubler circuits may be connected in series and transformer secondaries connected in alternate polarity to minimize the effects of transformer winding capacitance, as shown in Fig. 3.14.

The quadupler of Fig. 4.34B is a common circuit that has good regulation and low output ripple. During the first positive half cycle, C_2 is charged through CR_2 to $E_{in\text{-pk}}$. Capacitor C_1 is charged through CR_1 during the negative half cycle. Also, during this half cycle, the voltage on C_2 is series aiding the input and current flows from C_2 through the input to charge C_4 via CR_4. During the next positive half cycle, C_2 is again recharged and the voltage across C_1 is series aiding the input and current flows from the input through C_1 and CR_3 to charge C_3. After a few cycles, each output capacitor is charged to twice the input voltage, and the sum of the output is four times the peak input voltage. Notice that current from the input flows through two of the capacitors each half cycle, as opposed to the half-wave multipliers that charge only one capacitor each half cycle.

Typical quadrupler regulation is shown in Fig. 4.35 and ripple characteristics are shown in Fig. 4.36. Data for these figures was developed by computer analysis

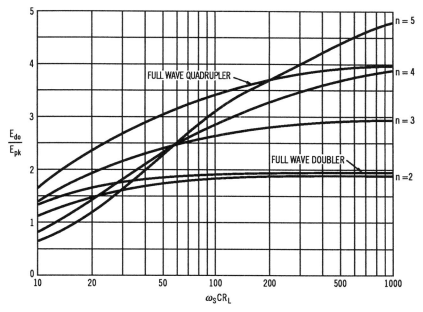

Fig. 4.35. Ratio of applied alternating voltage to direct output voltage in half-wave and full-wave voltage multipliers.

using $R_s/R_L = 0.1\%$ and typical high-voltage diode models. The average rectifier current is equal to the average output current. The peak current in CR_3 and CR_4 is smaller than the peak current in CR_1 and CR_2 because the conduction time of CR_3 and CR_4 is longer than that of CR_1 and CR_2. At low values of ωCR_L, operating modes change. For ωCR_L less than 9.5 (indicating a heavy output load), the voltage across C_1 and C_2 reverses during part of each cycle. Here, nonpolarized capacitors should be used. For $\omega CR_L < 3.7$, CR_1 and CR_3 conduct simultaneously and CR_1 must then carry not only the charging current of C_1 but also the load current.

An additional full-wave multiplier, known as the Cockcroft–Walton generator, is shown in Fig. 4.34C. The circuit operates from a center-tapped transformer secondary and consists of two half-wave quadruplers. Four additional rectifiers and two additional capacitors are required for this quadrupler. Stages may be added for further voltage multiplication. The primary advantage of this circuit is the common terminal for input and output return, which reduces secondary voltage isolation requirements in the transformer. Performance is essentially identical with that of other full-wave circuits.

For high-power, very high voltage applications, the transformer, rectifiers, and capacitors are normally placed in an oil-filled tank for voltage isolation. Several voltage doublers or quadruplers may be connected in series to obtain the desired output. A three-stage quadrupler is shown in Fig. 4.37. This topology has two advantages, as compared to a n-stage multiplier operating from a single secondary winding. First, the voltage regulation is significantly improved and the output ripple is reduced. Second, the effects of transformer winding capacitance is

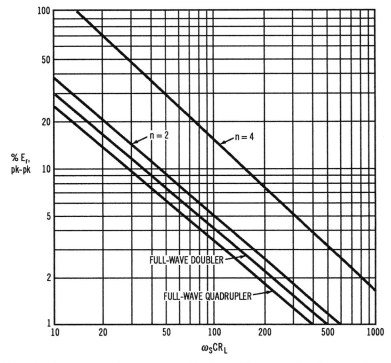

Fig. 4.36. Ratio of output ripple versus $\omega_s C R_L$ for half-wave and full-wave voltage multipliers.

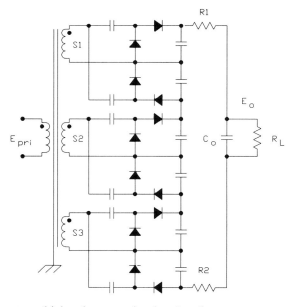

Fig. 4.37. Three-stage high-voltage quadrupler charging an energy storage capacitor.

minimized. As discussed in Section 3.6.6, capacitance is reduced by using a tall winding and short width. The secondaries may be wound in sections (reference Fig. 3.13). Also, universal pi-wound techniques (common in RF coils) may be used to reduce winding capacitance. This technique is even more beneficial as operating frequency increases, such as in high-frequency converters used to charge energy storage capacitors for lasers and to power microwave tubes, such as klystrons and traveling-wave tubes (TWTs).

4.2.7 Operational Holdup Time

In output voltage–regulated power supplies, the ac input is rectified and filtered, and the load seen by the output filter capacitor is some form of regulator stage which supplies an external load. Requirements are frequently encountered where the external load must be powered for a short period of operation if the ac input is momentarily interrupted. This period is normally referred to as holdup time and is defined as the time during which output voltage regulation will be maintained after interruption of input power. As will be discussed, the holdup time is a function of various parameters, one being the input voltage at the time of interruption. Switch mode power supplies inherently operate at high efficiencies and over a wide input voltage range as compared to linear mode power supplies. To achieve a corresponding holdup time in linear power supplies, the efficiency at nominal input voltage would be considerably less than 50%. For this reason, only switching power supplies will be discussed in this section. However, the analysis presented is also applicable to switching dc–dc converters which utilize a filter or storage capacitor between the dc input and the switching stage.

Referring to Fig. 4.38, the ac input is rectified and filtered to dc by capacitor C_1. At a fixed output load, the input power $P_{dc, in}$ to the converter is fixed. The dc voltage across C_1 is directly proportional to the input voltage. Thus the average dc current to the converter stage is inversely proportional to the input voltage. If input power is interrupted, the energy stored in C_1 supplies power to the converter for a period of time while the capacitor voltage decreases and the capacitor current increases. At some point in time, the capacitor voltage reaches a minimum value below which output voltage regulation cannot be maintained. If the external load includes a computer memory, additional circuitry may be utilized to perform an orderly shut-down or a transfer of data to a storage disk during the holdup time. If operating during power outage is required, an uninterruptible power system, as discussed in Chapter 18, is normally employed.

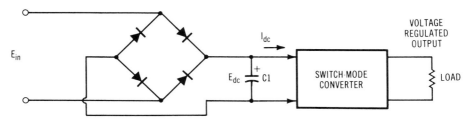

Fig. 4.38. Block diagram, switch mode power supply.

The capacitance value required to supply the current during the holdup time is given by

$$C = i\,dt/dv$$
$$= I_{avg}t_h/(E_{nom} - E_h) \qquad F \qquad (4.23a)$$

where

$I_{avg} = P_{dc,\,iN}/E_{avg}$, A

$E_{avg} = \frac{1}{2}(E_{nom} + E_h)$

E_{nom} = nominal capacitor voltage at nominal input

E_h = capacitor voltage at time t_h

t_h = holdup time, s

The following equation eliminates the calculation of I_{avg} by utilizing the required energy storage:

$$C = 2\varepsilon/E^2$$
$$= 2P_{in}t_h/(E_{nom}^2 - E_h^2) \qquad F \qquad (4.23b)$$

The assumption that the input power is constant if the output power is constant is true if the efficiency is constant. The dc voltage decrease and corresponding converter duty cycle increase as a function of t_h should introduce negligible change in efficiency.

From the discussion of Section 4.2.1, the capacitance value of C_1 is normally calculated to limit the ripple voltage to a desired value. The resulting holdup time is then

$$t_h = C(E_{nom}^2 - E_h^2)/2P_{in} \qquad s \qquad (4.24)$$

Rearranging the terms to find the minimum capacitor voltage, at the time of input power interruption, which will provide a desired holdup time for a given capacitance:

$$E_{min} = \sqrt{E_h^2 + [(2P_{in}t_n)/C]} \qquad V \qquad (4.25)$$

Example

A switching power supply delivers 390 W output at a converter efficiency of 78% and operates from a 120 V ac single-phase input. The dc power is $P_{in} = 390/0.78 = 500$ W, and the nominal dc voltage is $E_{dc} = 120 \times 0.88 \times \sqrt{2} = 149$ V_{nom}. The load resistance seen by the capacitor is then $R = E^2/P = 149^2/500 = 44.4$ Ω. For an operating ripple voltage of 10% peak-to-peak, the required value of $\omega_s C R_L$, from Fig. 4.20, is 27. For a 60 Hz input, $C = 27/(2\pi \times 60 \times 44\ 4) = 1613$ μF. Assuming the converter is designed to provide a regulated output at 90 V ac, the minimum dc voltage is then equivalent to the voltage at t_h, $E_h = 90 \times 0.88$ $\times \sqrt{2} = 112$ V. From Equation (4.24), the holdup time is $t_h = 1613 \times 10^{-6}(149^2 - 112^2)/(2 \times 500) = 15.5$ ms.

Conversely, if the desired holdup time is 30 ms, the filter capacitance value would be 1613 $(30/15.5) = 3122$ μF.

If the input voltage is 105 V ac at the time of power interruption and the capacitance is 3122 μF, $E_{nom} = 105 \times 0.88 \times \sqrt{2} = 131$ V, and the holdup time is

$$t_h = 3122 \times 10^{-6}(131^2 - 112^2)/1000 = 14.4 \text{ ms}$$

Not considered in the above analysis was the effect of the capacitor ripple voltage (which would lower the nominal voltage calculations at the time of power interruption) and the capacitance tolerance. Thus, to guarantee a required holdup time, the nominal capacitance value should be appropriately increased to offset these two factors.

4.3   INDUCTOR INPUT FILTERS

To provide lower output ripple, especially from high-power and polyphase rectifiers, an inductor is placed between the rectifier and the capacitor filter. Since current cannot change instantaneously in the inductor, inrush currents at turn-on are also reduced. A single-phase full-wave circuit and waveforms are shown in Fig. 4.39. During the peak portions of the voltage waveform, energy is stored in the inductor and during the valley portion of the voltage waveform, the energy is transferred to the capacitor and load. Assuming infinite inductance, the current waveforms are as shown in Fig. 4.2.

Referring to Fig. 4.39, the rectified waveform contains an average voltage, from Equation (4.2), of

$$E_{dc1} = \frac{E_{pk}}{\pi} \int_0^\pi \sin \omega t \, dt = \frac{2E_{pk}}{\pi}$$

and all even harmonics of n

$$E_{dcn} = \frac{2E_{pk}}{\pi} \int_0^\pi \sin \omega t \cos n\omega \, dt$$

from which the Fourier series is

$$E_{dco} = \frac{2E_{pk}}{\pi} \left(1 - \frac{2}{3} \cos 2\omega t - \frac{2}{15} \cos 4\omega t - \frac{2}{35} \cos 6\omega t \cdots \right) \quad (4.26)$$

Integrating for E_{dcn}, the ripple constant is

$$E_{dcn} = \frac{2E_{pk}}{\pi} \left(\frac{-2}{n^2 - 1} \right) \quad (4.27)$$

where n is the number for the even harmonics.

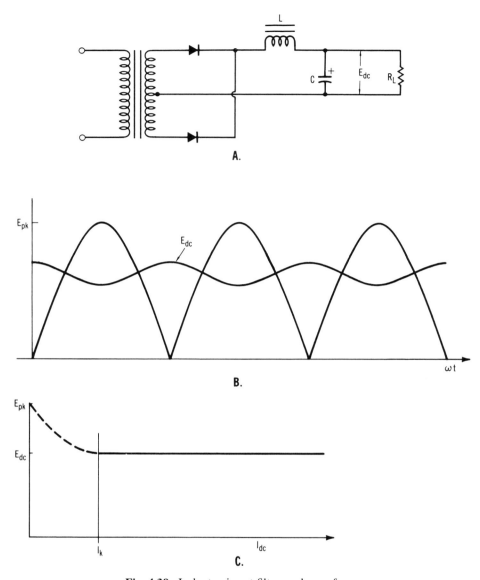

Fig. 4.39. Inductor input filter and waveforms.

4.3.1 Critical Inductance

To achieve the desired voltage regulation of the filter, it is necessary to maintain continuous current flow in the inductor. When the load resistance is very large in ohmic value (approaching open circuit), the output voltage will tend to rise toward the peak input voltage, as shown in Fig. 4.39C. Current I_k is the value at which the inductor current becomes discontinuous and each rectifier conducts for less than 180°. The inductor has a dc component and an ac component of current. Neglecting losses and voltage drops, the average or dc component is $E_{\text{in, avg}}/R_L$ and the ac component is $E_{pk}/n\omega L$. For the current to be continuous, the dc component, $2E_{pk}/R_L$, must not exceed the predominantly second harmonic component, $4E_{pk}/\pi(2^2 - 1)(2\omega_s L_k)$, and therefore $2E_{pk}/\pi R_L \geq 4E_{pk}/(3\pi 2\omega_s L_k)$, from which $L_k \geq R_L/3\omega_s$.

The general equation for critical inductance in single and polyphase rectifiers is then

$$L_k \geq \frac{2R_L}{p(p^2 - 1)\omega_s} \qquad \text{H} \tag{4.28}$$

where p is the number of pulses per cycle and ω_s is input source frequency in radians per second.

The calculated critical inductance, however, may be impractical to obtain considering the inductor design requirement at high load currents. In this case, a swinging choke, as discussed in Section 3.11.3 may be used. As an alternate compromise, a bleeder resistor could be added across the filter capacitor, where $L < L_k$. In this case, the power dissipation and rating of the resistor and reduction in overall efficiency must be considered. The value of the bleeder resistor for a single-phase full-wave rectifier is

$$R_B = E_{dc}/I_k = E_{dc} \times 3\pi \times \omega_s L/(2E_{pk}) = 4.7\omega_s L$$

The general equation for the bleeder resistor is then

$$R_B = p(p^2 - 1)\omega_s L/2 \qquad \Omega \tag{4.29}$$

4.3.2 Turn-On Currents and Surge Limiting

The discussion of Section 4.2.3 and Equation (4.20) assumed zero input inductance. Considering an input inductance, the dynamic equations presented in Section 1.6.2.2 and Equation (1.36) may be used to solve for the first half-cycle inrush current. An important consideration with inductor input filters (either from an ac source or a dc source) is the frequently underdamped nature of the circuit at turn-on. This is especially true of soft-start switch mode power supplies. In this case, the input filter capacitor will charge to a voltage higher than the peak input voltage and "downstream" switching transistors will experience a higher-than-normal voltage when switching commences.

4.3.3 Output Ripple

For the circuit of Fig. 4.39A, the relation of output voltage to input voltage is

$$\frac{E_0}{E_{in}} = \frac{X_C R_L/(X_c + R_L)}{X_L - X_c R_L/(X_c + R_L)}$$

The inductor has a high reactance and the capacitor has a low reactance at the even harmonics and the following discussion considers only the predominant second-harmonic effect, with negligible error in analysis. Also, the capacitive reactance, $1/2\omega_s C$, is normally much less than R_L so

$$\frac{E_0}{E_{\text{in}}} = \frac{X_C}{X_L - X_C} = \frac{1}{2\omega_s C(2\omega_s L - 1/2\omega_s C)}$$

$$= \frac{1}{4\omega_s^2 LC - 1} \tag{4.30}$$

The impedance to the filter is

$$Z = 4\omega_s^2 LC - 1/(2\omega_s C)$$

From the analysis in Section 4.3.1, the effective second harmonic is

$$I_{2\text{nd}} = \frac{E_{\text{dc}}}{\sqrt{2} Z_2} = \frac{4E_{\text{pk}} 2\omega_s C}{3\pi\sqrt{2}\left(4\omega_s^2 LC - 1\right)}$$

$$= I_R + I_C$$

Since $X_C \ll R_L$, $I_C = I_2$. Thus, $I_R/I_C = X_C/R_L = 1/2\omega_s C R_L$, where $I_R = E_{\text{dc}}/R_L = 2E_{\text{pk}}/\pi$. Substituting and solving for the ripple factor where $e_{r,\text{rms}} = E_{r,\text{rms}}/E_{\text{dc}} = I_{r,\text{rms}}/E_{\text{dc}} = I_{r,\text{rms}} X_C(I_R \times R_L)$:

$$e_{r,\text{rms}} = \frac{8E_{\text{pk}}\omega_s C}{3\pi\sqrt{2}\left(4\omega_s^2 LC - 1\right)} \times \frac{\pi}{E_{\text{pk}}} \times \frac{R_L}{2\omega_s C R_L}$$

$$= \frac{0.471}{4\omega_s^2 LC - 1} \tag{4.31a}$$

Figure 4.40 shows $e_{r,\text{rms}}$ versus $\omega_s^2 LC$. The rms ripple voltage is $E_{r,\text{rms}} = e_{r,\text{rms}} \times E_{\text{dc}}$ and percent rms ripple $= e_{r,\text{rms}} \times 100$. From Equations (4.30) and (4.31) it is obvious that the condition $\omega^2 LC = 0.5$ should be avoided to prevent oscillations and instability in the filter. In most cases, $\omega^2 LC \gg 1$, and the equation reduces to

$$e_{r,\text{rms}} = 0.471/(4\omega^2 LC) \tag{4.31b}$$

The preceding analysis was presented for the single-phase, full-wave rectifier shown in Fig. 4.39. Assuming $\omega^2 LC$ is much greater than 1, the general expression for output ripple in single-phase and polyphase inductor input filters is

$$\% e_{r,\text{rms}} = \frac{\sqrt{2} \times 100}{(p^2 - 1)(p^2\omega_s^2 LC)} \tag{4.32}$$

where

    p = number of pulses per cycle
    ω_s = source frequency, rad/s
    L = input inductance, H
    C = output capacitance, F

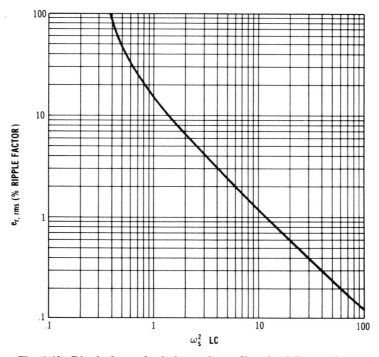

Fig. 4.40. Ripple factor for inductor input filter for full-wave input.

Rather than calculating the ripple from Fig. 4.40 or from Equation (4.32), the following "short form" equations may be used to calculate $e_{r,\text{rms}}$ where L is in henrys, C is in microfarads, and f_s is in hertz:

Single-phase, full-wave:

$$e_r = 0.829(60/f_s)^2/LC \qquad (4.33a)$$

Three-phase, half-wave:

$$e_r = 0.138(60/f_s)^2/LC \qquad (4.33b)$$

Three-phase, full-wave:

$$e_r = 0.0079(60/f_s)^2/LC \qquad (4.33c)$$

Six-phase, 12-pulse, full-wave:

$$e_r = 0.00048(60/f_s)^2/LC \qquad (4.33d)$$

Since the ripple voltage is a close approximation of a sine wave, the peak-to-peak ripple voltage (frequently of more interest) may be obtained by

$$E_{r,\text{pk-pk}} = 2\sqrt{2}\,e_{r,\text{rms}} = 2.82\,e_{r,\text{rms}} \qquad (4.34)$$

The preceding equations and the information of Fig. 4.40 provide the LC product, but individual component values must still be determined. The ratio of L and C may be calculated by letting the no-load filter impedance $(Z = \sqrt{L/C})$ be equal to 10% of the actual load resistance (R_L), or

$$L/C = 0.1 R_L^2 \tag{4.35}$$

Then, for a required value of LC (or $\omega_s^2 LC$) to meet the desired ripple, the values of L and C may be calculated, since R_L is known or calculated.

4.4   THREE-PHASE VOLTAGE DOUBLER

Transformer–rectifiers may also be employed to charge a high-voltage filter capacitor which stores energy. The capacitor may then be rapidly discharged into a load such as a flash tube or laser. When recharging occurs again, the output capacitor presents a very low output impedance, and high inrush currents may occur unless limited to a safe value. High-power systems also require some means of limiting surge voltage spikes, which is accomplished by snubbers.

One means of meeting the above requirements is through the three-phase voltage doubler shown in Fig. 4.41 and the discussion which follows [7]. Initially K_2 is energized and I_p (primary current) flows through R_1 and K_2 until K_1 is energized. This step-start limits the initial inrush current to magnetize the transformer. The voltage applied to the primary is induced into the secondary, and I_s flows through C_L, the rectifier bridge, C_0, and back to the secondary neutral. On negative half cycles, C_L is discharged and reverse charged to await the positive half cycle.

Since E_0 is initially zero and may be considered a short circuit, current is limited by the reactance of C_L. Primary capacitor C_p and secondary snubbers C_s and R_s limit voltage transients caused by the transformer reactance.

The output voltage is

$$E_0 = 2\sqrt{2}\,kE_s \qquad \text{V} \tag{4.36}$$

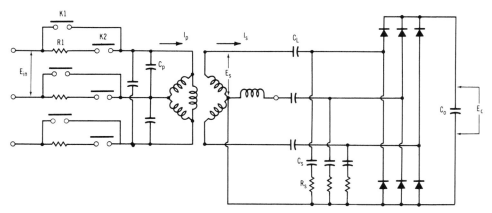

Fig. 4.41. High-power three-phase voltage doubler with short-circuit protection.

where k is a constant less than 1, otherwise the time required to charge the capacitor would be infinite. The secondary current is

$$I_s = \omega C_L E_s \qquad (4.37)$$

and

$$\text{Transformer VA rating} = 3E_s I_s \qquad \text{VA} \qquad (4.38)$$

from which $I_p = \text{VA}/(\sqrt{3}\,E_{\text{in}})$.

The time required to charge C_0 to a desired output voltage is given by

$$t = \frac{-\ln\left(1 - E_0/2\sqrt{2}\,E_s\right)}{3f\,\ln(C_1/C_0 + 1)} \qquad (4.39)$$

Example

Let $E_0 = 50$ kV and $C_0 = 2\ \mu$F for an energy storage of $\varepsilon = \frac{1}{2}CE^2 = 2500$ J. The input is 480 V at 60 Hz. The input power limit is 10 kVA. Then $I_p = 10{,}000/(\sqrt{3} \times 480) = 12$ A. For $k = 0.68$, $E_s = 50{,}000/(0.68 \times 2 \times \sqrt{2}) = 26$ kV. Current $I_s = 10{,}000/(3 \times 26{,}000) = 0.128$ A. Calculating the secondary series capacitor, $C_L = 0.128/(377 \times 26{,}000) = 0.013\ \mu$F. Thus the time required to charge the output capacitor is $t = -\ln(1 - 5000/2\sqrt{2} \times 26000)/[3 \times 60\ln(0.013/2 + 1)] = -\ln 32/180(\ln 1.0065) = 0.985$ s.

Since the output capacitor is capable of charging to $2 \times \sqrt{2} \times 26$ kV $= 73.5$ kV, a means of terminating the charge when the capacitor reaches 50 kV must be used. This may be accomplished by sensing the output voltage and deenergizing the input contactors when C_0 reaches the desired voltage. It is also interesting to note that for test purposes, the output capacitor may be indefinitely shorted and the system energized to monitor currents and voltages.

REFERENCES

1. J. Schaefer, *Rectifier Circuits, Theory and Design* (Wiley, New York, 1965).

2. B. D. Bedford and R. G. Hoft, *Principles of Inverter Circuits* (Wiley, New York, 1964).

3. R. L. Witzke et al., "Influence of AC Reactance on Voltage Regulation of 6-Phase Rectifiers," *AIEE Transactions on Communications and Electronics* (July 1953).

4. R. L. Witzke et al., "Voltage Regulation of 12-Phase Double-Way Rectifiers," *AIEE Transactions on Communications and Electronics* (November 1953).

5. O. H. Schade, "Analysis of Rectifier Operation," *Proceedings of the IRE*, Vol. 31, No. 7 (July 1943).

6. J. S. Brugler, "Theoretical Performance of Voltage Multiplier Circuits," *IEEE Journal of Solid-State Circuits* (June 1971).

7. B. T. Merritt, R. E. Tarter, and J. B. Button, "A 2.0 MVA Voltage Double Power Supply for Capacitor Bank Charging," 9th Symposium on Engineering Problems of Fusion Research, October 1981.

5

PHASE CONTROL CIRCUITS

5.0   Introduction

5.1   Single-Phase Power Circuits
      5.1.1   Resistive Load
      5.1.2   Inductive Load

5.2   Three-Phase Power Circuits
      5.2.1   Half-Wave Control
      5.2.2   Full Wave Half Control
      5.2.3   Full Wave Full Control

5.3   Six-Phase Dual Bridge

5.4   Critical Inductance

5.5   Power and Control Stages
      5.5.1   Single-Phase Input
      5.5.2   Three-Phase Input
      References

5.0   INTRODUCTION

In Chapter 4, we saw that the rectified dc output voltage was dependent on the amplitude of the ac input voltage and the condition of the load. The output voltage was, therefore, unregulated. As the input voltage increased, so did the output voltage. As the load approached an open circuit, the output voltage also increased. This chapter presents phase control topologies which may be employed to provide a regulated output. Line and load reegulation are achieved by using thyristors to control the conduction angle of the sine wave input in response to some desired control. Triacs are used to control ac load voltages, as in light dimmer assemblies. For high-power ac loads, parallel back-to-back SCRs can be used. Power supplies utilize rectifiers and triacs or SCRs as the control element to provide a desired dc output from single-phase or three-phase inputs. Notations for this cahpter are the same as those of Chapter 4, Table 4.1.

The general equation for dc output voltage to a *resistive* load when operating from a single-phase or from three-phase inputs (such as those of Figs. 5.1 and 5.3) is given by

$$E_{\mathrm{do}} = E_{\mathrm{rms}} \frac{\sqrt{2}\,p}{\pi} \frac{1 + \cos\alpha}{2} \sin\left(\frac{\pi}{p}\right) \tag{5.1}$$

where p is number of rectifier pulses per cycle and α is the trigger delay angle, subject to the limitations discussed in later sections.

The general equation for dc output voltage to an *inductive* load when operating from single-phase or three-phase inputs (such as those of Figs. 5.1 and 5.3), assuming continuous load current, is given by

$$E_{do} = E_{rms} \frac{\sqrt{2}\,p}{\pi} \cos(\alpha) \sin\left(\frac{\pi}{p}\right) \tag{5.2}$$

The above relations again assume that the current in the input line transfers instantaneously between successively conducting SCRs or rectifiers. Due to reactance, which is always present in the ac line or in a transformer supplying power to the phase control circuit, some rectifier commutation period will be involved. The discussion presented in Section 4.1.3 is also appropriate to circuits with SCRs. Generally, the average voltage lost due to commutation which must be *subtracted* from the calculated value of average output voltage is given by

$$V_x = E_{rms} \frac{p}{2\pi} (1 - \cos \mu) \tag{5.3}$$

where μ is the commutation angle.

This condition is discussed in Section 5.2 and the effect is shown in Figs. 5.5 and 5.6.

5.1 SINGLE-PHASE POWER CIRCUITS

Various single-phase power circuits are shown in Fig. 5.1. The full-wave center tap with resistive load is used primarily for low-voltage outputs, since a single forward SCR drop results. The resistive load of Fig. 5.1A is replaced by an input inductor filter in Fig. 5.1B. In this case, a clamping rectifier is inserted to provide a path for load current when the SCRs turn off. The bridge circuits of Figs. 5.1C, D, E, and F are used primarily for higher output voltage. Figures 5.1A–D have the advantage of common cathodes, which permits a single firing pulse to be applied to both SCRs simultaneously or from a pulse transformer with a single secondary if isolation is required. However, individual gate resistors to each SCR are recommended. The anode terminal is normally the case of the SCR in stud-mounted devices. The circuit of Fig. 5.1E allows the SCRs to be mounted to a common heat sink without isolation, while the rectifiers (cathode is normally the case of the device) are mounted to another heat sink without isolation. The two SCRs are connected in series in Fig. 5.1F which has the advantage of eliminating the clamping rectifier since the two bridge rectifiers are in series. However, isolated trigger pulses to the SCRs must be provided for the circuits in Figs. 5.1E and F.

5.1.1 Resistive Load

From Equation (5.1), the equation for dc voltage in single-phase circuits with a resistive load is

$$E_{do} = E_{rms} \frac{2\sqrt{2}}{\pi} (1 + \cos \alpha) \tag{5.4}$$

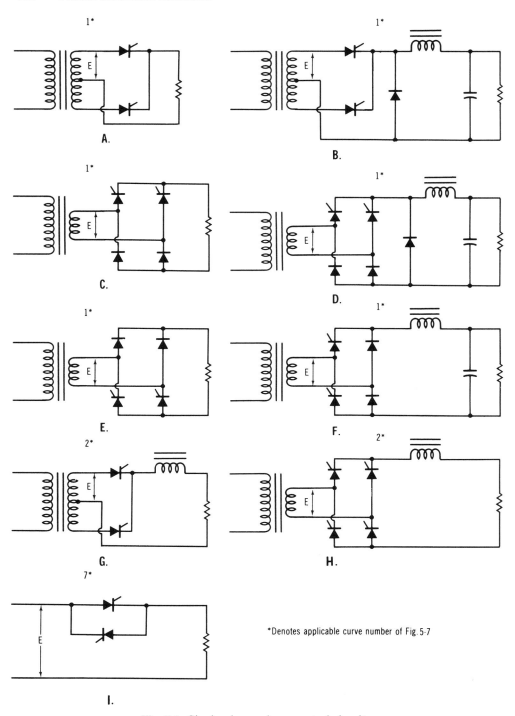

Fig. 5.1. Single-phase, phase control circuits.

FIGURE	α, DEGREES	VOLTAGE WAVEFORM	$\dfrac{E_{do}}{E_{S,pk}}$	$\dfrac{E_{do}}{E_{S,rms}}$	CURRENT WAVEFORM	$\dfrac{I_{F,rms}}{I_{dc}}$	$\dfrac{I_{F,pk}}{I_{dc}}$	%* RIPPLE VOLTAGE	% RMS RIPPLE CURRENT
5-1A. 5-1C. 5-1E	0		.637	.90	SAME AS VOLTAGE	.785	1.57	48	48.3
	30		.594	.84		.832	1.68	56	60.7
	60		.477	.675		.940	2.09	65	88.2
	90		.318	.45		1.11	3.14	62	121
	120		.159	.225		1.39	5.44	43	170
	150		.043	.06		1.99	11.72	18	264
5-1B. 5-1D. 5-1F	0		.637	.90		.707	1.00	48	
	30		.594	.84		.775	1.20	56	
	60		.477	.675		.866	1.50	65	
	90		.318	.45		1.0	2.0	62	
	120		.159	.225		1.225	3.0	43	
	150		.043	.06		1.73	6.0	18	
5-1G. 5-1H	0		.637	.90		.707	1.0		
	30		.551	.78					
	60		.318	.45					
	90		0	0					
	120		−.318	−.45					
	150		−.551	−.78					
	180		−.637	−.90					

*% Ripple voltage relative to peak voltage at $\alpha = 0°$

Fig. 5.2. Waveform and parameter relations for single-phase circuits.

Equation (5.4) is plotted in Fig. 5.7, curve 1.

An inductive load is shown in Figs. 5.1G and H. As stated previously, the analysis assumes a continuous current flow in the load. In Fig. 5.1G, the load current, which is reactive, must flow through the secondary of the transformer. This places additional restraints on the design of the transformer and on commutation of the SCRs. This handicap is overcome by the circuit in Fig. 5.1H, which utilizes four SCRs. Here, output polarity reversal and inductive load current are obtained.

5.1.2   Inductive Load

From Equation (5.2), the equation for output voltage with an inductive load is

$$E_{do} = E_{rms} \frac{2\sqrt{2}}{\pi} \cos \alpha \tag{5.5}$$

Equation (5.5) is plotted in Fig. 5.7, curve 2, Figure 5.2 shows a typical waveforms and parameter relations for various single-phase circuits and delay angles.

The circuit shown in Fig. 5.11 has a number of applications. In low-power requirements, the SCRs may be replaced by a triac to control the voltage to the load as in conventional light dimmers. For high-power requirements, the parallel back-to-back SCRs offer additional current capability and overcome the problem of low commutating dv/dt associated with triacs. If the resistive load is replaced by a transformer whose secondary voltage is rectified and filtered, a phase-controlled power supply results and regulation is achieved by controlling the firing angle via feedback circuitry from the dc output to the triac or SCR gates, as shown in Fig. 5.9 and discussed in Section 5.4. This technique is common for a product line of power supplies and a family of power levels where the components on the primary side (including the size of the transformer) are identical for a certain output power and only the secondary winding, rectifiers, output filter components, and output sensing voltage divider change for the various output voltages.

5.2   THREE-PHASE POWER CIRCUITS

Various three-phase power circuits are shown in Fig. 5.3. The half-wave circuits of Figs. 5.3A and B are seldom used due to poor utilization of transformer VA ratings but are presented for reference. The bridge circuits of Figs. 5.3C–G are more common. The circuit of Fig. 5.3H is discussed in Section 5.4.2.

5.2.1   Half-Wave Control

The equation for output voltage for the half-wave circuit is

$$E_{do} = E_{rms} \frac{3\sqrt{2}}{2\pi} \cos \alpha \quad \text{for } 0° < \alpha < 30° \tag{5.6a}$$

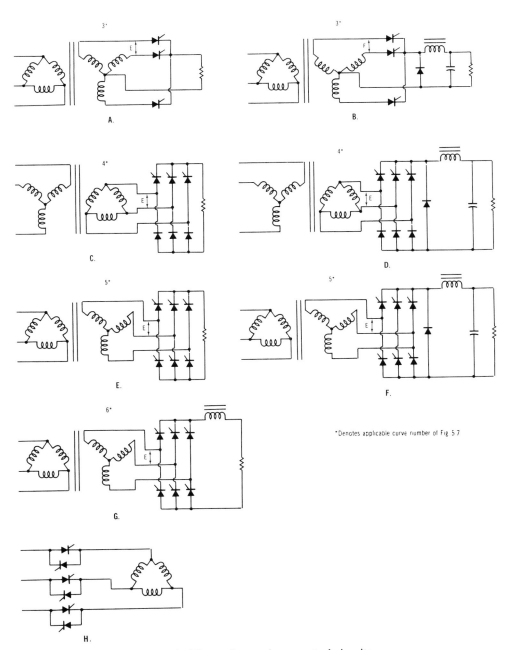

*Denotes applicable curve number of Fig. 5.7

Fig. 5.3. Three-phase, phase control circuits.

and

$$E_{do} = E_{rms}\frac{3\sqrt{2}}{\pi}\left[1 + \cos(\alpha + 30°)\right] \quad \text{for } 30° < \alpha < 150° \qquad (5.6b)$$

Equations (5.6a) and (5.6b) are plotted in Fig. 5.7, curve 3.

5.2.2 Full Wave, Half Control

The circuits in Figs. 5.3C and D are full-wave bridges with half control (only three SCRs are required) and are suitable for resistive loads where no regeneration is needed. The SCR cathodes are common which allows a single firing stage for trigger purposes. However, the output ripple frequency is reduced from six times the line frequency to three times the line frequency at delayed angles α (reference Fig. 5.4). The equation for output voltage for the full-wave half-control circuit is

$$E_{do} = E_{rms}\frac{3\sqrt{2}}{2\pi}(1 + \cos\alpha) \qquad (5.7)$$

Equation (5.7) is plotted in Fig. 5.7, curve 4.

FIGURE	α	VOLTAGE WAVEFORM	$\dfrac{E_{do}}{E_{S,pk}}$	$\dfrac{E_{do}}{E_{S,rms}}$	CURRENT WAVEFORM	$\dfrac{I_{F,rms}}{I_{dc}}$	$\dfrac{I_{F,pk}}{I_{dc}}$	% * RIPPLE VOLTAGE	% RMS RIPPLE CURRENT	LOAD
5·3C. 5·3D	30°	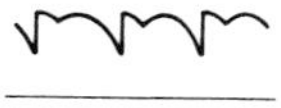	.891	1.26	SAME AS VOLTAGE	.584	1.12	11	17.3	R
										L+CR
	60°	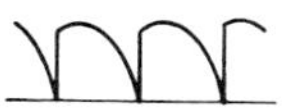	.716	1.01	SAME AS VOLTAGE	.615	1.377	29	35.2	R
					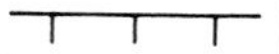					L+CR
	90°		.478	.675	SAME AS VOLTAGE	.729	2.07	40	75	R
										L+CR
	120°		.239	.337	SAME AS VOLTAGE	1.11	3.58	32	122	R
					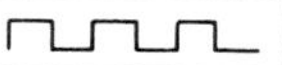					L+CR
	150°		.064	.090	SAME AS VOLTAGE	1.35	7.94	14	208	R
										L+CR
5·3E. 5·3F	30°	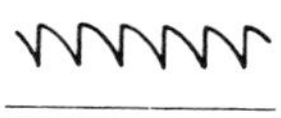			SAME AS VOLTAGE			13		R
										L+CR
	60°				SAME AS VOLTAGE			26		R
					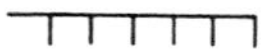					L+CR
	90°				SAME AS VOLTAGE			17		R
					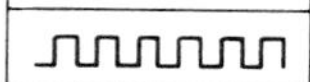					L+CR

*% Ripple voltage relative to peak voltage at $\alpha = 0°$

Fig. 5.4. Waveform and parameter relations for three-phase circuits.

5.2.3 Full Wave, Full Control

Full-wave bridges with full control (six SCRs) are shown in Figs. 5.3E and F. The firing sequence is more complicated than that of Figs. 5.3C and D since the SCRs require two gate signals each cycle with a gate pulse duration in excess of 60°, and the gate signals to the SCRs are 60° apart. The input, output, and gate firing sequence for $\alpha = 0°$ and for $\alpha = 45°$ are shown in Fig. 5.5. The equation for output voltage for the full-wave full-control *resistive load* is

$$E_{\text{do}} = E_{\text{rms}} \frac{3\sqrt{2}}{\pi} \cos \alpha \quad \text{for } 0° < \alpha < 60° \qquad (5.8a)$$

and

$$E_{\text{do}} = E_{\text{rms}} \frac{3\sqrt{2}}{\pi} [1 + \cos(\alpha + 60°)] \quad \text{for } 60° < \alpha < 120° \qquad (5.8b)$$

Equation (5.8) is plotted in Fig. 5.7, curve 5.

The output waveform of Figs. 5.5 and 5.6 also show the effects of line inductance or transformer leakage inductance L on device commutation. The average output voltage decreases due to μ (in this case, $\mu = 15°$) as given by Equation (5.3). Voltage and current parameter relations and circuit waveforms for other values of α are shown in Fig. 5.4 for the three bridge with half control and with full control. Again, note the lower output ripple with six SCRs for full control.

When the load is *inductive*, the load current will also be continuous beyond $\alpha = 60°$. The inductor then acts as a source during the time that the output voltage would ordinarily be zero; but in this case, the output voltage is negative for the period of α greater than 60°. This action decreases the output voltage, as seen by comparing curves 2 and 5 of Fig. 5.7. As α increases, the output voltage decreases, becoming zero at $\alpha = 90°$, and then increases in a negative direction as α increases further.

When the load is *capacitive* as in the case of an output filter for a power supply such as Fig. 5.3F, the capacitor charges to the peak value of the ac input (neglecting losses). The capacitor voltage remains at the peak until α increases to 30° and then the capacitor voltage decreases as α increases further. The equation for output voltage is

$$E_{\text{do}} = \sqrt{2}\, E_{\text{rms}} \quad \text{for } 0° < \alpha < 30° \qquad (5.9a)$$

and

$$E_{\text{do}} = \sqrt{2}\, E_{\text{rms}} [\cos(\alpha - 30°)] \quad \text{for } 30° < \alpha < 120° \qquad (5.9b)$$

Equation (5.9) is plotted in Fig. 5.7, curve 8.

To minimize high inrush currents due to the capacitive load, soft-start circuitry should be employed. At initial turn-on, the control circuitry (discussed in Section 5.4) sets α at a maximum, and then α gradually decreases until the required output voltage is achieved.

The circuit of Fig. 5.3H is frequently used for a family or product line of power supplies in the range of 5 kW output and higher. As described in Section 5.1, the

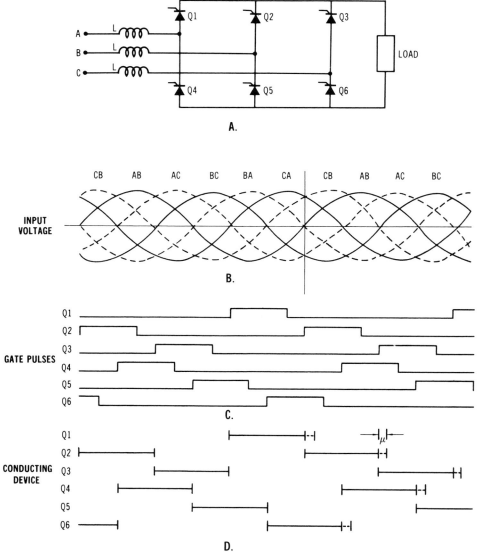

Fig. 5.5. Circuit and waveforms for a six-SCR bridge, with $\alpha = 0°$.

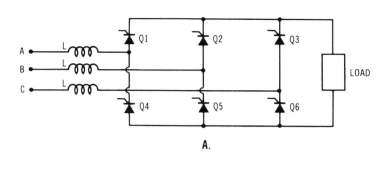

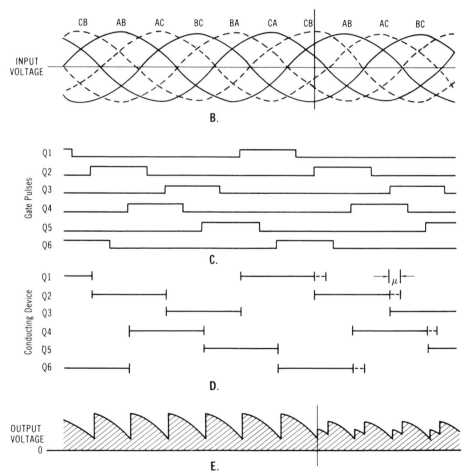

Fig. 5.6. Circuit and waveforms for a six-SCR bridge, with $\alpha = 45°$.

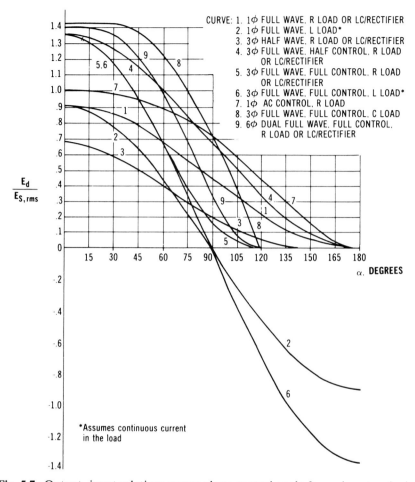

Fig. 5.7. Output–input relations versus phase control angle for various topologies.

components on the primary side of the transformer are identical for a certain output power level, and the secondary side components are designed for the desired output voltage and corresponding output current. The advantages of three-phase operation, as compared to single-phase operation, are higher operating power factor, lower mains input current distortion and reduced output filter component size.

5.3 SIX-PHASE DUAL BRIDGE

A further reduction in input harmonic current and in output ripple may be achieved with a 6-phase or 12-pulse dual bridge, as shown in Fig. 5.8A. This topology is common in very high power systems such as battery chargers for uninterruptible power systems. The three-phase input is isolated and voltage

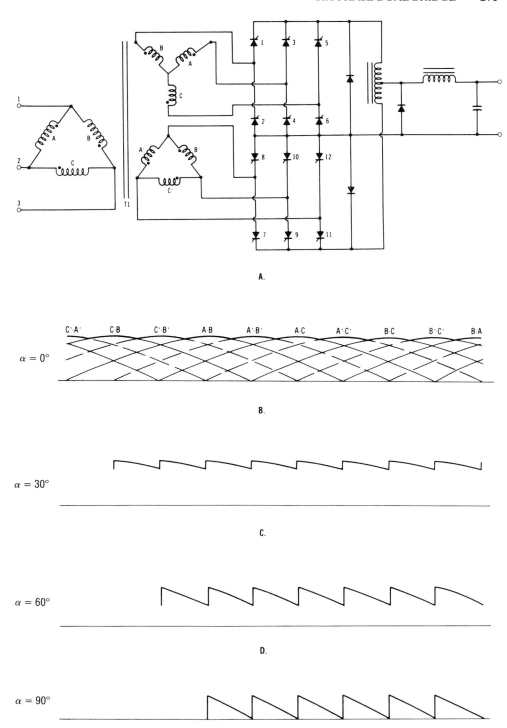

A.

B.

C.

D.

E.

Fig. 5.8. Circuit and waveforms for six-phase dual bridge with full control.

scaled by transformer T_1, which has a delta and a wye secondary, each of which is applied to an SCR bridge with full phase control. The output of each bridge is summed by the interphase transformer for balancing purposes. Reference Section 4.1.2.1 and Figs. 4.11 and 4.12 for current ratings and current waveforms as applying to straight rectifier circuits. For the circuit of Fig. 5.8A, the output voltage waveforms are shown in Figs. 5.8B, C, D, and E for $\alpha = 0°$, $\alpha = 30°$, $\alpha = 60°$, and $\alpha = 90°$, respectively. The equations for output voltage for a resistive load are

$$E_{do} = E_{rms}(12/\pi)\sin 15° \quad \text{for } 0° < \alpha < 15° \tag{5.10a}$$

$$E_{do} = E_{rms}(6/\pi)\left[\left(1 - \tfrac{1}{2}\sqrt{3}\right)\sin \alpha + \tfrac{1}{2}\cos \alpha\right] \quad \text{for } 15° < \alpha < 90° \tag{5.10b}$$

and

$$E_{do} = E_{rms}(6/\pi)\left[1 + \cos(\alpha + 60°)\right] \quad \text{for } 90° < \alpha < 120° \tag{5.10c}$$

where α is the delayed firing angle in degrees and E_{rms} is the line-to-line secondary voltage.

The preceding equation is plotted in Fig. 5.7, curve 9. It is interesting to compare curve 9 with curve 5 which is the plot for a three-phase bridge with full control. For the same output power and at $\alpha = 60°$, the six-phase circuit produces approximately 46% higher output voltage than the three-phase circuit. Since two bridges are used for the six-phase circuit, the current in each bridge is approximately 34% of the current in the three-phase bridge.

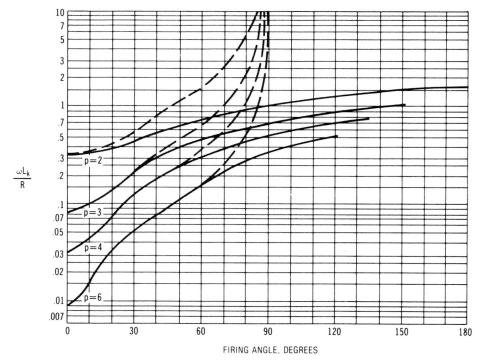

Fig. 5.9. Critical inductance as a function of firing angle. Solid lines represent circuit with a flyback diode; dashed lines represent circuit without flyback diode (Copyright 1965 IEEE.)

5.4 CRITICAL INDUCTANCE

To achieve the desired output voltage regulation of SCR power circuits with inductor input filters, resistive loads, and clamping rectifier of Figs. 5.1 and 5.3, it is necessary to maintain a current flow in the inductor. When the load resistance is very large in ohmic value, the output will tend to rise toward the peak input voltage, as with the capacitor filter in Section 5.2.3. Regulation may still be achieved by varying α via the feedback and control circuit but transient response may be poor, especially if the load becomes near open circuit.

This problem may be overcome by designing the inductor for critical inductance as in Section 4.3.1. The general equation for critical inductance, Chapter 4, Equation (4.23), is true for $\alpha = 0°$. The parameters governing critical inductance are shown in Fig. 5.9 for various integer values of pulses per cycles [1]. The effect of the clamping or flyback diode becomes apparent. Without the diode, the critical inductance L_k approaches infinity as α approaches 90°.

5.5 POWER AND CONTROL STAGES

To control the output voltage of a phase-controlled converter or power supply, it is necessary to control the firing angle to the thyristors. The firing pulses must be symmetrical each half cycle to eliminate a dc component in the load seen by the thyristor, such as a transformer in a power supply. A discussion of control circuits for single-phase and three-phase inputs follows.

5.5.1 Single-Phase Input

Many types of circuits have been developed for controlling the firing pulses to triacs or SCRs with single-phase inputs. These range from diac (voltage breakover diode) triggering, UJT (unijunction transistor) triggering, to blocking-oscillator triggering. The latter has the advantage of providing a continuous pulse string to the SCR during the desired conduction time. However, each approach usually depends on the charging of a capacitor to develop a triangular ramp voltage for timing purposes. The ramp voltage is reset at or near each crossover of the ac input waveform to form a "ramp and pedestal" technique.

The circuit of Fig. 5.10A shows a typical approach used in power supplies, and the pertinent waveforms are shown in Fig. 5.10B. The ac input is applied to Q_3 and to T_1. The secondary voltage of T_1 is full-wave rectified, and following R_1, this voltage is clamped by zener diode CR_1, normally in the 12- to 20-V range. Near the crossover points, Q_1 is conducting, C_2 is discharged, and point A is high. When Q_1 turns off, C_2 charges through R_2 and since the voltage across C_2 cannot change instantaneously, the voltage across R_2 decays exponentially as C_2 charges. The voltage across R_2 is thus an inverted ramp with a pedestal at the supply voltage level, as shown in Fig. 5.10B.

Output current is sensed by U_1 and an amplified error signal results at the output of U_1. Output voltage is sensed by U_2 and an amplified error signal results at the output of U_2. These two signals are OR'ed (for constant voltage/constant current operation) and applied to a soft-start circuit whereby the output (point B)

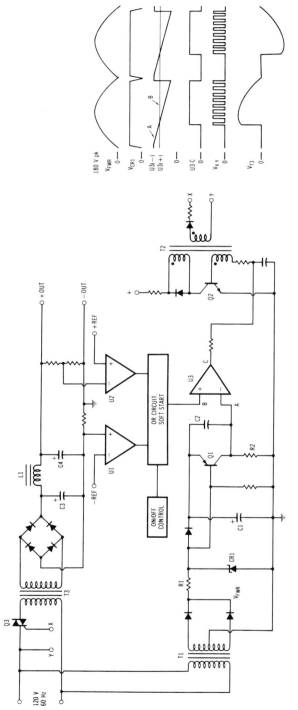

Fig. 5.10. Control circuit and waveforms for a regulated output from a single-phase input.

is controlled by an on/off signal and is initially low. As point B rises in voltage to intersect the voltage level of point A, the comparator output (point C) goes high for the amplifier and comparator polarities shown. When U_3 goes high, this signal triggers the blocking oscillator (Q_2, T_2); Q_2 conducts and is driven to saturation by the regenerative action of T_2. The collector current increases until the base drive can no longer maintain Q_2 in saturation. The voltage induced in the feedback winding then decreases, and Q_2 is rapidly driven off. (Reference Chapter 9 for blocking-oscillator designs.) Since the output of U_3 is still high, Q_2 again turns on and then off to provide the gate firing pulses to triac Q_3. When Q_3 turns on, the primary voltage across T_3 is induced into the secondary and is rectified and filtered to provide the dc output. If the output voltage (or output current) tends to increase, the output of U_2 (or U_1) decreases and the output of U_3 is delayed which retards the firing pulse to, in turn, decrease the output voltage (or current) to maintain a regulated output.

5.5.2 Three-Phase Input

Phase-controlled firing of three-phase circuits becomes more complicated than for single-phase circuits, and various methods have been analyzed by Pelly [2]. The most common means of pulse timing is by integral control, phase-locked oscillator, or cosine control. However, the integral control sometimes has the tendency to settle into an asymmetrical mode with timing pulses occurring at irregular intervals, and the phase-locked oscillator may produce a slow transient response. The cosine control requires a phase shift in the input-to-control waveform, and the control circuit is subject to transients on the input line plus the input frequency must be stable to minimize phase shift. However, the cosine control has the advantage of providing a linear relation between a reference voltage and the output voltage since the cosine of the firing angle is proportional to the reference voltage. For this reason, the following discussion pertains to the cosine control.

A typical three-phase controlled-power and drive stage for a high current power supply is shown in Fig. 5.11. The three-phase input is applied to the SCRs and to T_1. The secondary voltage of T_1 is clipped in the negative half cycle by R_1 and CR_1, while R_2, R_3, and C_3 provide a 60° phase shift in the ac waveform. The circuit is repeated in each of the alternate phases and the signals are applied to the driver and pulse transformer stage to trigger the SCRs, as shown in the diagram. The control signal level sets the dc level of each of the ac waveforms to maintain a balanced firing sequence. The control signal is developed by error amplifiers similar to that of Fig. 5.10.

The firing of Q_1–Q_6 applies a voltage to T_2 whose secondary is rectified and filtered by L_1, C_1, L_2, and C_2 to provide the dc output. The insertion of R_c in series with C_1 is frequently necessary to prevent the ripple current in C_1 from exceeding the maximum rating. An *electronic ripple filter* may be added to reduce the output ripple. Voltage source E_A is provided by an auxiliary transformer, rectifier, and filter. A portion of the output voltage is capacitively coupled to U_1 whose output is 180° out of phase with the output ripple. At the output ripple valley points, Q_7 is turned on to aid in boosting the output voltage and to reduce the output ripple.

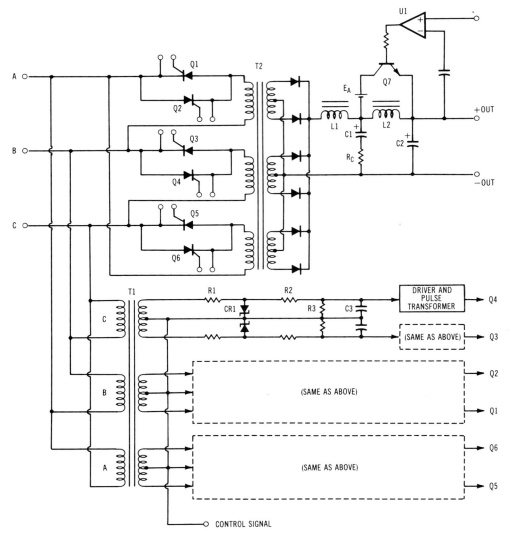

Fig. 5.11. Control circuit for a regulated output from a three-phase input.

REFERENCES

1. R. J. Distler and S. G. Munshi, "Critical Inductance and Controlled Rectifiers," *IEEE Transactions on Industrial Electronics and Control Instrumentation* (March 1965).

2. B. R. Pelly, *Thyristor Phase-controlled Converters and Cycloconverters* (Wiley-Interscience, New York, 1971).

6

TRANSISTOR INVERTERS

6.0   Introduction

6.1   Square-Wave Inverters
    6.1.1   Saturating Output Transformer
    6.1.2   Saturating Drive Transformer
    6.1.3   Clocked Inverters
    6.1.4   Waveform Analysis and Load Characteristics
    6.1.5   Filtering for Sine Wave Output

6.2   Quasi-Square-Wave Inverters
    6.2.1   Power Stages
    6.2.2   Output Waveform Analysis and Filter
    6.2.3   Combined Transformer and Harmonic Filter
    6.2.4   Current Limiting and Short-Circuit Protection

6.3   Inverters with Harmonic Cancellation

6.4   High-Frequency Pulse-Demodulated Inverters
    6.4.1   Push–Pull Single-Edge Modulation
    6.4.2   Double-Edge Modulation
    6.4.3   Control Functions and Features
    6.4.4   Modulation Analysis
    6.4.5   Sine Wave Output Filter
    6.4.6   HFPD Inverter Performance Characteristics

6.5   Inverters with Ferroresonant Transformers

6.6   Three-Phase Inverters
    6.6.1   Square-Wave and Quasi-Square-Wave Inverters
    6.6.2   Inverters with Harmonic Cancellation
    6.6.3   Step-Wave Inverters
    6.6.4   HFPD Inverters

References

6.0   INTRODUCTION

The function of an inverter is to change a dc input voltage to a symmetrical ac output voltage of desired magnitude and frequency and to supply a load with the desired power. The inverter may supply a load directly or form the inversion stage of a converter or power supply. For this reason, the inverters discussed in this

TABLE 6.1. Notation

A	Cross-sectional area, in.^{2} or cm^2
C	Capacitance, farads
E	Voltage, volts
f	Frequency, hertz
I	Current, amperes
L	Inductance, henrys
n	Turns ratio
N	Turns
P	Power, watts
Q	Quality factor
R	Resistance, ohms
t	Time, seconds
V	Component or element voltage
X	Reactance, ohms
Z	Impedance, ohms
η	Efficiency
ω	Angular function, radians
θ	Angular function, degrees
δ	Damping ratio

chapter provide ac outputs in the 50–500-Hz frequency range. Typical single-phase outputs are (1) 220 V at 50 Hz; (2) 120 V at 60 Hz; and (3) 115 V at 400 Hz. For high-power three-phase systems, typical outputs are (1) 220/380 V at 50 Hz; (2) 120/208 V at 60 Hz; (3) 115/200 V at 400 Hz; and (4) 120/208 V at 415 Hz.

The ac output waveform may be a square wave, suitable for some lower power applications, or a low-distortion sine wave, more commonly desired in medium-power applications and required in high-power applications. For a sine wave output, the odd harmonics inherent in the step function switching to convert dc to ac are filtered or minimized, thus reducing load heating effects due to harmonics. Square-wave inverters, simulated and synthesized sine wave inverters, filtering, and wave-shaping methods are presented.

The input may be a battery, fuel cell, solar cell, thermionic converter, or other dc source. The input voltage may range from 12 V, as from an automotive battery, to 250 V, as in an uninterruptible power system.

Notations for this chapter are given in Table 6.1. A comparison of the most useful types of inverters discussed in this chapter is presented in Table 6.2. Advantages, features, and disadvantages provide an initial selection or choice for a particular application.

6.1 SQUARE-WAVE INVERTERS

6.1.1 Saturating Output Transformer

The basic Royer [1] or flux-switching inverter, shown in Fig. 6.1, is the simplest and most economical form of a two-transistor push–pull inverter. Transformer core selection is a function of frequency. At 50 and 60 Hz, 0.014 in. silicon–iron is used,

TABLE 6.2. Inverter Characteristics

Inverter Type[a]	Advantages	Disadvantages
Square-Wave Output		
Saturating output transformer (6.1.1)	Most economical	No voltage regulation, no frequency regulation (unless preceded by a preregulator); output short-circuit protection difficult to achieve
Saturating base drive transformer (6.1.2)	Higher operating efficiency than above inverter	Same as above
Clocked inverter (6.1.3)	Provides inherent frequency regulation; low switching losses; high efficiency	No voltage regulation (unless preceded by a preregulator); output short-circuit protection difficult to achieve
Sine Wave Output, Single Phase and Three Phase		
Quasi–square wave PWM (6.2)	Excellent frequency regulation; good voltage regulation; high operating efficiency; short-circuit protection easy to implement	Limited to low and medium power due to output filter size
Harmonic cancellation (6.3)	Same as above, except output filter is smaller	Requires complex logic drive circuitry to implement sequenced half-cycle switching
High frequency pulse demodulated (6.4)	Excellent voltage and frequency regulation; low distortion output; fast response to line and load changes; small size and light weight	Complex logic and control circuitry; higher switching losses reduce efficiency; limited ability to clear load faults in high-power units
Ferroresonant (6.5)	Most economical; excellent frequency regulation; good voltage regulation; simple control and drive circuitry; inherent overload and short-circuit protection	Efficiency limited by transformer losses; not applicable to three-phase operation due to output phase shift with load change
Step wave (6.6.3)	Excellent frequency regulation; good voltage regulation; low distortion output; applicable for high-power outputs	Multiple power-switching stages typically restrict usage to high-power, three-phase applications

[a]Numbers in parentheses are section numbers.

and at 400 Hz, 0.004 in. silicon–iron is used (reference Chapter 3, Table 3.5). Ferrite materials are not recommended due to the roundness of the B-H loop, the low flux density, and the saturation flux density change with temperature.

When E_{in} is applied, current flows through R_1 and either through N_{d1}–R_2–$Q_{1,BE}$ or through N_{d2}–R_3–$Q_{2,BE}$. Assuming Q_2 has sufficient base current to start conduction, current flows through N_{p2} and $Q_{2,CE}$. The voltage induced in N_{p2} induces a voltage in N_{d2} to further forward bias Q_2 and a voltage in N_{d1} to reverse bias Q_1. When the base current from N_{d2} drives Q_2 into saturation, the voltage across Q_1 is $V_{CE} = E_{in} + V_{Np1} + L_1 di/dt$, where L_1 is the leakage inductance between N_{p1} and N_{p2}. Base current flows from N_{d2} through $Q_{2,BE}$ and CR_1. The conduction time of the transistor is determined by the number of turns and by E_{in} from $\phi = (10^8/N)\int e\, dt$. When the flux can no longer support the volt-seconds, the transformer saturates, developing a high current in Q_2, which pulls out of saturation. As Q_2 turns off, a voltage is induced in N_{d1} which forward biases Q_1, and current flows through N_{p1} and $Q_{1,CE}$. This induces a voltage in N_{d2} to reverse bias Q_2, ensuring turn-off. After this half cycle, the process repeats on a continuing basis, switching from $+B_m$ to $-B_m$ to develop an output voltage across N_s. From Chapter 3, Equation (3.7), the operating frequency is

$$f_0 = (E_{in} - V_{CE,sat})10^8/(4B_m A_c N_p k_s) \qquad \text{Hz} \qquad (6.1)$$

where B_m is a function of core material in gauss, A_c is the cross-sectional area in square centimeters, and k_s is the stacking factor. Then

$$E_0 = (N_s/N_p) \times (E_{in} - V_{CE,sat})$$

and

$$E_d = (N_d/N_p) \times (E_{in} - V_{BE,sat})$$

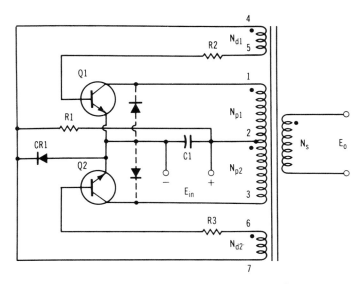

Fig. 6.1. Inverter with saturating output transformer.

For a desired output power

$$I_c = P_0 / \left[(E_{\text{in}} - V_{CE,\text{sat}}) \eta \right]$$

where η is the efficiency of the inverter.

The efficiency is a function of the transformer losses, base drive power, transistor conduction, and switching losses. From the transistor forced gain characteristics, I_b can be calculated. Then

$$R_2 = R_3 = (E_d - V_{BE} - V_F)/I_b \qquad (6.2a)$$

where E_d is high enough to provide sufficient base drive when E_{in} is a minimum and V_F is the forward voltage drop across CR_1.

Substituting CR_1 for a resistor, as compared to the original concept, allows lower power dissipation in R_1 and

$$R_1 = \left[(E_{\text{in}} - V_{BE}) h_{FE} / I_c \right] - R_2 \qquad \Omega \qquad (6.2b)$$

Typical waveforms are shown in Fig. 6.2.

The major advantages of this approach are simplicity and low cost. The disadvantages are no frequency regulation, no voltage regulation, high transistor peak currents, and no protection from overloads and short circuits on the output. However, the latter may be overcome by adding an extra winding on one of the outer legs (if an EI-type core is used). This winding is in series with N_s and serves as a leakage-type inductance between the secondary and the load to support the secondary voltage if the load is shorted. The inverter then operates at a much higher frequency but will not suffer damage, even in a prolonged output short.

To minimize primary leakage inductance, the primary should be wound bifilar (N_{p1} and N_{p2} wound simultaneously two-in-hand), and the drive windings would bifilar and distributed across one layer of the winding length. Because the trans-

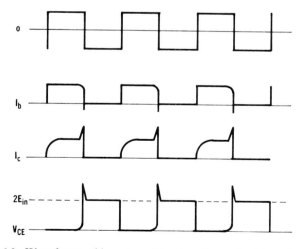

Fig. 6.2. Waveforms of inverter with saturating output transformer.

former saturates, core losses increase due to the increase in flux density, and the temperature rise may be 10% higher than that of a nonsaturating transformer.

Examples

An inverter has the following desired parameters:

Input: 12 V dc, 11–15 V range.
Output: 110-V, 60-Hz square wave, 200 VA at 12 V dc in.

A. Calculations and Transistor Selection. Let the design efficiency be 72%. Then P_{in} = 200/0.72 = 278 W and I_{in} = 278/12 = 23.2 A. Allowing for current peaking at transformer saturation, choose a transistor capable of switching a peak current of 23.2 × 1.25 = 29 A. For the push–pull circuit, the transistor collector voltage is $2E_{in}$, plus voltage spikes produced due to the leakage inductance of the transformer. Choose a transistor capable of 15 × 2 × 1.5 = 45 V. Due to the low-voltage input, expected low operating ambient temperature, and low switching frequency, germanium transistors such as the 2N1522 could be used.

B. Transformer Design. For the transformer, a power rating of 275 VA is required. Referring to Chapter 3, Fig. 3.26 and choosing a temperature rise of 55° C, a square stack of El-150, M6X laminations with a conductor rating of 650 cmil/A should work. From Table 3.15, the El-150 core area is 2.25 in.^{2} and the window area is 1.69 in^2. Also, the coil winding width D = 2.25 in. and the coil winding height B = 0.75 in, which will be of more benefit when calculating the conductor turns per layer of fit. Since 29-gage M6X material is used, the saturating flux density is chosen as 16 kG. The required primary turns, from Equation (3.7b), are N_p = 3.88(E_{in} − $V_{CE, sat}$)10^6/($B_m A_c f k_s$) = 3.88(15 − 0.5)10^6/(16 × 10^3 × 2.25 × 60 × 0.95) = 22. The required secondary turns, from Equation (3.8), are N_s = $E_{out}N_p$/(E_{in} − $V_{CE, sat}$) = 120 × 22/(12 − 0.5) = 230.

Choosing a value of E_d = 3 V, the required drive turns are n_d = $E_d N_p/E_p$ = 3 × 20/11.5 = 5.2. The turns must be an integer (although with EI cores, a half turn is possible) and choosing N_d = 5, E_d = 2.88 V.

From Table 1.2A, part A, waveform 5, the primary rms current is 23.2 × 0.7071 = 16.4 A, and the secondary rms current is 200/120 = 1.7 A. The total primary consists of two conductors sized for 16.4 A each, which will require additional window area as compared to a single primary conductor sized for 23.2 A. For 650 cmil/A, the required primary conductor is 650 × 16.4 = 10,600 cmil. From Tables 3.9 and 3.10, no. 10 round is less than adequate, no. 10 square is more than adequate, or no. 14 rectangular is adequate. For ease of winding, no. 14 rectangular, 0.072 × 0.135, will be used. Allowing for 0.125-in. margins on each end and winding the primary bifilar, the turns per layer are N_p/1 = (2.25 − 0.25)/(2 × 0.135) − (2 × 0.135) = 7.

Thus, 3 layers of 7 turns per layer means 21 turns will fit, whereas 22T are required to provide 60 Hz operation. However, 21T will be used and the stack height will be increased to (2.25 × 22)/(1.5 × 21) = 1.57 in. The secondary requires 650 × 1.7 = 1105 cmil, and from Table 3.9, no. 19 is chosen. For a wire diameter of 0.04 in., the secondary turns per layer are N_s/1 = (2.25 − 0.25)/0.04 − 1 = 49.

Thus, $230/59 = 4.7$, so 5 layers are required. Since no. 19 is used for the secondary, no. 19 will also be used for the drive winding to minimize winding time since enough room should exist for this one-layer winding. To verify height fit, allow 0.05 in. for coil winding tube, 0.01 in. insulation thickness between layers, 0.02 in. between feedback and secondary, and 0.02 in. final wrap. The total build height is then $0.05 + 3(0.072 + 0.01) + 0.04 + 0.02 + 5(0.04 + 0.01) + 0.02 = 0.626$ in.

For an available coil height of 0.75 in., 0.125 in. of height margin should allow stacking ease. Since the stack height was increased to 1.57 in., the inside dimensions of the coil tube should be 1.5×1.625 in.

C. Resistor and Rectifier Selection. For a gain of 20 in the power transistor, the required base current is $23.2/20 = 1.16$ A. From the text equations, $R_2 = R_3 = (E_d - V_{BE,\text{sat}} - V_F)/I_b = (2.88 - 0.4 - 0.7)/1.16 = 1.5$ Ω, and the resistor power is $I_{\text{rms}}^2 R = (0.707 \times 1.16)^2 \times 1.5 = 1$ W. Use a 3-W rating to allow for increased base current at 15 V input. For the rectifier, any low-voltage, 3-A device should work. For starting, h_{FE} may be greater than the previously forced gain value, say a gain of 50 and, $R_1 = 50(12 - 0.4)/23.2 = 25$ Ω, with a power rating of $E^2/R = (15 + 0.7)^2/25 = 9.9$ W. Use a 20-W rating.

A combined linear regulator-saturating transformer approach may be used to achieve voltage and frequency regulation and minimize the number of components, even though the total power dissipation is the same as a series regulator followed by a flux-switching inverter. In Fig. 6.3, R_1 and R_2 are starting resistors and the zener diode voltage of CR$_3$ is less than the minimum value of E_{in}. Assuming Q_2 turns on, current flows through Q_2 and N_{p2}. The voltage induced in N_{d2} drives Q_2 on. However, when $E_{p2} + E_{d2} + V_{R4}$ exceeds $V_{F(CR2)} + V_Z$, CR$_3$ conducts to draw current flowing through R_4 from the base of Q_2. Then Q_2 comes

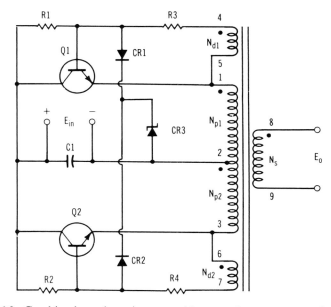

Fig. 6.3. Combined regulator-inverter with saturating output transformer.

out of saturation, and since V_{BE} is small, E_p and E_0 remain almost constant. When the collector current increases, Q_2 turns off and Q_1 turns on. Since Q_1 and Q_2 are nonsaturating, care must be exercised to ensure operation within the SOA (safe operating area) of the forward-biased rating of the transistors.

6.1.2   Saturating Drive Transformer

Improved efficiency, especially as frequency increases, is achieved with the two-transformer or Jensen [2] inverter shown in Fig. 6.4. Transformer T_1 operates in the linear region of the B-H loop, while T_2 saturates each half cycle. Thus the core material of T_1 may be silicon–iron or nickel–iron, while the core material of T_2 should be square loop. The gain of the transistors reduces the VA rating of T_2 to a fraction of the VA rating of T_1. Since T_1 does not saturate, the high transistor peak currents inherent in the inverter of the previous section are eliminated.

   Inverter starting is achieved by current flow through R_1 and the base of one of the transistors. Assuming Q_2 turns on, the voltage induced in N_{p1} is applied across R_4 and N_{p2}, inducing a voltage in N_{d2}, which saturates Q_2. The conduction time of the transistors is determined by the number of turns in N_{p2}, the value of R_4 and by E_{in}. As T_2 saturates, the rapidly increasing current in N_{p2} causes an increase in V_{R4}, which further reduces E_{Np2} and E_{Nd2} to turn off Q_2. The decrease in $I_{c,Q2}$ causes a reversal of voltage polarities across all windings, which turns on Q_1. But Q_1 does not conduct until T_2 has been reverse magnetized. Thus, neither transistor conducts during the switching time and common-mode, or overlap, currents are minimized.

   The inverter design should consider T_2 first and then T_1. For a required output power, $I_c = P_0/\eta E_{in}$ where η is the overall efficiency (as opposed to T_1 efficiency) since I_c includes the current in N_{p2}. Current I_b can be calculated from the transistor forced-gain characteristics. Then, $R_2 = R_3 = (E_d - V_{BE} - V_F)/I_b$. The reverse-bias voltage on the off transistor can approach $2E_d$ and should not exceed the secondary breakdown energy rating of the transistor under worst case conditions. The VA capacity of T_2 is calculated by $P_{T2} = E_d I_b \times 1.15$, where the 15%

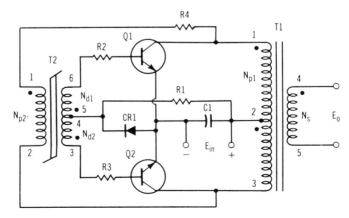

Fig. 6.4. Inverter with saturating drive transformer.

overage allows for reverse base current pulses and T_2 losses. A square-loop core material of sufficient size to handle the power is then selected. It is usually ideal to set E_{Np2} equal to E_{in} and since Q_1 and Q_2 voltages swing from zero to $2E_{in}$, R_4 drops E_{in} also [3]. The selected operating frequency determines T_2 primary turns from Equation (6.1). Because E_{Np2} drops sharply when T_2 saturates, the magnetizing current increases and Ampere's law must also be considered; $H = 0.4\pi N_p I_m / l_m$, where the magnetizing current $I_m = I_{Np2}$ and l_m is the magnetic path length of the core. The value of H, in oersteds, is typically 5–10 times the value of H_c. Once N_{p2} is calculated, $N_{d1} = N_{d2} = E_d N_{p2} / E_{in}$. The primary current is

$$I_{Np2} = P_{T2}/E_{in} = I_{R4} \quad \text{and} \quad R_4 = E_{in}/I_{Np} \tag{6.3}$$

At higher input voltages, N_{p2} may be a large number. The turns may be reduced (and conductor area increased) by adding a low-voltage winding to T_1, with the series combination of R_4 and N_{p2} connected across the added winding.

Summing the output power and expected T_1 losses, a core material of sufficient size can be selected for T_1, say from Fig. 3.26. From Equation (6.1), the primary turns are calculated except B_m must be decreased to B_{ac} (the desired flux density) and the secondary turns depend on the desired output voltage. Since T_1 is switched by Q_1 and Q_2 which are symmetrically driven by T_2, the volt-second/flux ratio should be the same each half cycle. However, this may not always be the case, due to different transistor saturation voltages and transistor switching times. With either of these unsymmetrical conditions, the flux change may "walk up" the B-H loop toward saturation, causing a high current in one transistor and inducing forward-biased second breakdown. This phenomenon may be eliminated by placing a small air gap in the core of T_1, but increased magnetizing current results. These parameters should be considered in the design and component selection. See the discussion in Section 17.4 on this subject. Saturating base drive transformers or magnetic amplifiers may also be used to control the output voltage. Reference Section 6.2 for quasi-square-wave operation.

6.1.3 Clocked Inverters

Where output frequency is critical, the inverter shown in Fig. 6.5 may be used. The astable multivibrator and flip-flop (Fig. 6.5A) drive the driver stage, which in turn drives the output stage (Fig. 6.5B or C). In the half-bridge inverter (Fig. 6.5B), only two transistors are required, and $E_{pri} = \frac{1}{2}E_{in} - V_{CE,sat}$. In the full-bridge inverter (Fig. 6.5C), which has the same primary drive as the circuit in Fig. 6.5B, $E_{pri} = E_{in} - 2V_{CE,sat}$. In each case, care must be exercised to minimize common-mode current by providing rapid reverse bias for turn-off and turn-on delay to overcome the storage time of the conducting transistor if bipolar types are used. Clocked inverters, as part of a converter, are discussed in detail in Chapter 9.

6.1.4 Waveform Analysis and Load Characteristics

Characteristics and Fourier analysis of symmetrical square, sine, and triangle waveforms are shown in Fig. 6.6. Quasi-square, step wave, and PWM waveforms

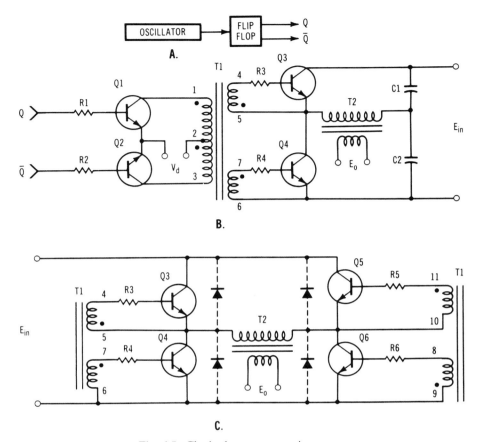

Fig. 6.5. Clocked square-wave inverters.

are discussed in following sections. The selection of an output voltage is a function of the type of load, the rated load voltage, and the type of voltmeter used for measurement. For example, with sine or triangle waves, resistive loads respond to rms values; inductive loads respond to average values; and capacitive loads respond to peak values. For a square wave, rms, average, and peak are identical.

6.1.5   Filtering for Sine Wave Output

The task of the output filter is to reduce the nominal 43% THD (total harmonic distortion) inherent in the square wave to a desirable level with minimum loss and complexity. A low-pass filter may consist of a series element (to attenuate harmonic voltages) and a shunt element (for harmonic current flow), as shown in Fig. 6.7. Attenuation is thus dependent on the ratio of the impedance of parallel and shunt elements, and the transfer function is

$$\frac{E_0}{E_{\text{in}}} = \frac{Z_L Z_2/(Z_L + Z_2)}{Z_1 + Z_L Z_2/(Z_L + Z_2)} \tag{6.4}$$

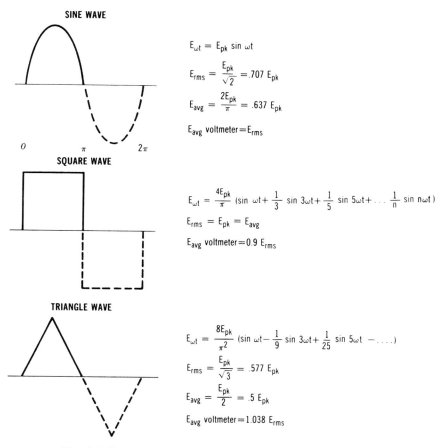

SINE WAVE

$$E_{\omega t} = E_{pk} \sin \omega t$$

$$E_{rms} = \frac{E_{pk}}{\sqrt{2}} = .707\, E_{pk}$$

$$E_{avg} = \frac{2E_{pk}}{\pi} = .637\, E_{pk}$$

$$E_{avg}\ \text{voltmeter} = E_{rms}$$

SQUARE WAVE

$$E_{\omega t} = \frac{4E_{pk}}{\pi}\ (\sin\ \omega t + \frac{1}{3}\ \sin\ 3\omega t + \frac{1}{5}\ \sin\ 5\omega t + \dots\ \frac{1}{n}\ \sin\ n\omega t\ )$$

$$E_{rms} = E_{pk} = E_{avg}$$

$$E_{avg}\ \text{voltmeter} = 0.9\, E_{rms}$$

TRIANGLE WAVE

$$E_{\omega t} = \frac{8E_{pk}}{\pi^2}\ (\sin\ \omega t - \frac{1}{9}\ \sin\ 3\omega t + \frac{1}{25}\ \sin\ 5\omega t\ - \dots)$$

$$E_{rms} = \frac{E_{pk}}{\sqrt{3}} = .577\, E_{pk}$$

$$E_{avg} = \frac{E_{pk}}{2} = .5\, E_{pk}$$

$$E_{avg}\ \text{voltmeter} = 1.038\, E_{rms}$$

Fig. 6.6. Fourier analysis and characteristics of ac waveforms.

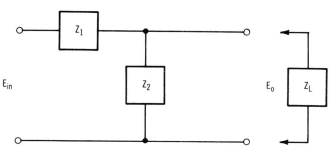

Fig. 6.7. Basic output filter.

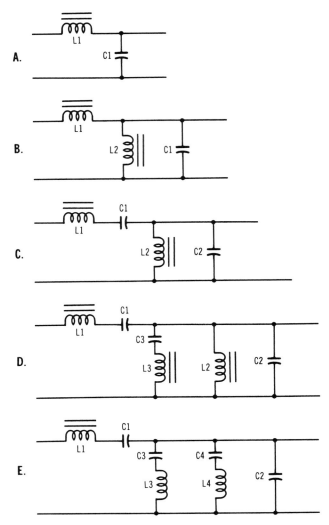

Fig. 6.8. Inverter filters for sine wave output.

The simplest form of an output filter is the single section LC filter shown in Fig. 6.8A. From Section 1.8.5, the transfer function in normalized form is

$$\frac{E_0}{E_{\text{in}}} = \frac{1}{1 - (\omega/\omega_0)^2 + j(2\delta\omega/\omega_0)} \tag{6.5}$$

and is plotted in Fig. 6.9. The damping ratio δ is greatly influenced by Z_L, and the filter may produce substantial gain or overshoot at light loads. The single-section LC filter provides very little attenuation at the predominant third harmonic. An additional inductor may be added in shunt with C_1 for further harmonic reduction and damping, as shown in Fig. 6.8B. This filter is most applicable to step wave filtering as discussed in Section 6.6.3.

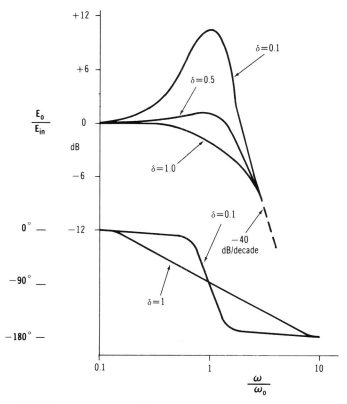

Fig. 6.9. Filter response and phase shift versus normalized frequency.

An improvement in regulation and harmonic attenuation in the square wave is achieved by the circuit of Fig. 6.8C. Inductor L_1 and C_1 are tuned to the fundamental, and only I^2R losses are present, while harmonics are attenuated at -40 db/decade. The series resonant impedance must be less than the minimum load impedance and the following relations apply:

$$\sqrt{LC} = 1/\omega_0 \quad \text{and} \quad Z_1 = \sqrt{L/C} \quad \text{at resonance} \tag{6.6}$$

Since the series path is inductive at the harmonics, $Z_{3rd} = 3Z_1$, $Z_{5th} = 5Z_1$, etc. Assuming $Z_1 = \frac{1}{2}Z_L$, the third-harmonic content is 12.7% and the fifth-harmonic content is 5.6% of the fundamental output voltage. Additional filtering such as the third-harmonic series trap of L_3–C_3 in Figs. 6.8D and E and a fifth-harmonic series trap of L_4–C_4 in Fig. 6.8E would further reduce the harmonics of the square-wave inverter.

Besides resonant frequencies designed into these output filters, other resonances are possible due to the interaction of the passive elements. Below the fundamental frequency, the series path is capacitive and the shunt path is inductive. Above the fundamental frequency, the series path is inductive and the shunt path is capacitive. Thus at certain frequencies, amplification (limited only by the Q

of the filter) instead of attenuation may result. Due to unsymmetries in the control circuit, different switching times in the power stage, and ripple voltage (from rectified ac sources) on the dc input, even harmonics may be produced in the inverter. Resonances at these frequencies must be avoided.

The losses in the output filter must be added to the output power when designing the output transformer. It may be desirable to "detune" the series path from the fundamental resonance to limit short-circuit current. The filter and simulated load characteristics may be scaled with respect to impedance and frequency and a low-power circuit built whereby the filter performance can be tested, as discussed in Section 1.9. It is also observed that a secondary winding could be added to L_2 of Figs. 6.8B and C to sample the output voltage for control purposes.

Ordinarily, the purpose for choosing a square-wave inverter is low cost and simplicity for supplying loads capable of square-wave operation. The filters just discussed provide an appreciation for the less stringent filter requirements for synthesized waveforms converted to a sine wave, as discussed below.

6.2 QUASI-SQUARE-WAVE INVERTERS

Variable pulse width control may be employed in the inverter to provide a regulated ac output voltage in a manner similar to the PWM (pulse width modulation) topology used in dc–dc converters and in switch mode power supplies. However, PSM (phase shift modulation) is normally required to provide a path for reactive current flow in the power stage. The block diagram of Fig. 6.10 shows two approaches. In Fig. 6.10A, the dc input is switched across a transformer, which provides isolation and voltage scaling, to produce a quasi-square-wave output. The output is sensed, say by a stepdown transformer, rectified, and applied to an input of an operational amplifier, the other amplifier input being the reference. The resulting error signal is applied to a one-shot multivibrator triggered by an oscillator. The oscillator and multivibrator trigger a dual flip-flop, which powers the drive stage. However, these functions could be accomplished by a single-PSM integrated circuit such as the UC1875 or the ML4818, which are discussed in Section 10.2.3.

The square-wave signals from the drive stages (reference Fig. 6.12) switch the power stage to produce the waveforms of Fig. 6.11A. The multivibrator controls the phase difference and subsequently the pulse width to maintain a constant output voltage. In the diagram of Fig. 6.10A, the master switching signal occurs at fixed points in time (A and D) and the other switching signal is delayed by 2θ (C and E). As can be observed in Fig. 6.11A, the zero crossover of the output sine wave (dotted waveform) shifts in phase as a function of 2θ.

In some cases, phase locking or synchronizing is desirable whereby the zero crossover (Fig. 6.11B) remains fixed in time. This is especially true in the design of three-phase inverters to maintain the 120° phase relation inherent in the logic and control functions. Under this condition, the approach of Fig. 6.10B is used. The power, drive, output, and sensing stages are the same as in Fig. 6.10A. The dc error signal and a triangular waveform are applied to a comparator stage. The comparator output and the oscillator output trigger a dual flip-flop to again

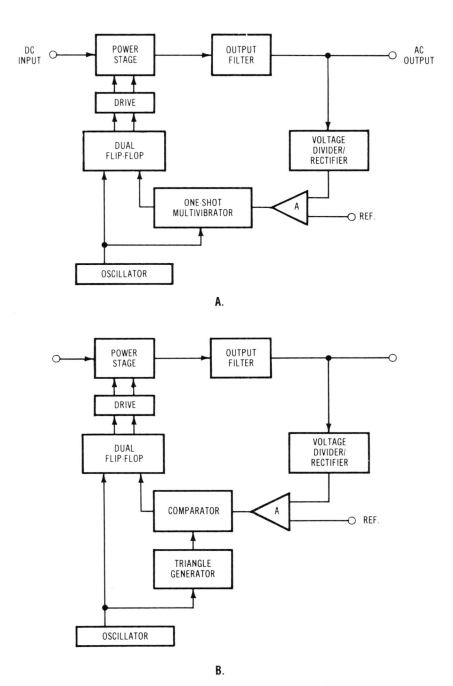

Fig. 6.10. Block diagram, quasi-square-wave inverter.

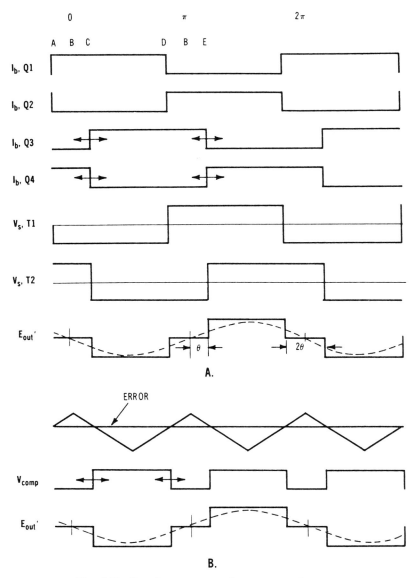

Fig. 6.11. Quasi-square-wave inverter waveforms.

provide the drive signals. The switching and delay occur symmetrically about the zero crossover at θ.

6.2.1 Power Stages

At low dc input voltages, push–pull stages, as shown in Fig. 6.12, may be used for low- to medium-power levels. Transistors Q_1 and Q_2 are driven from the fixed square waves, while Q_3 and Q_4 are driven from the delayed square waves shown in

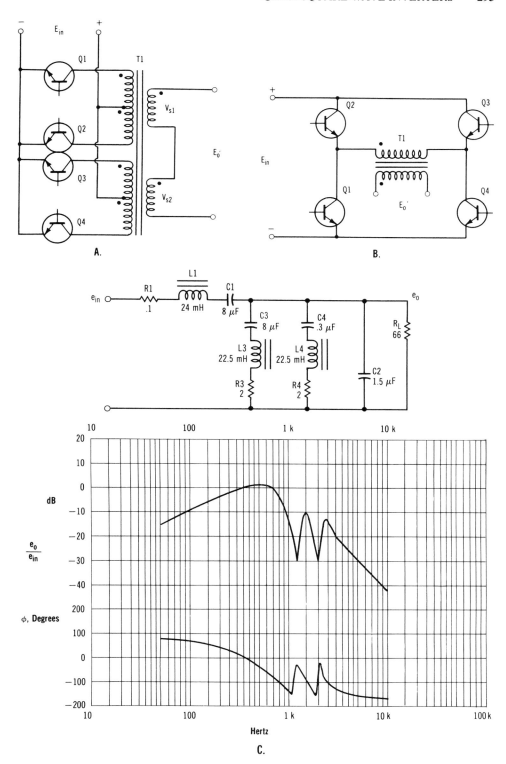

Fig. 6.12. Quasi-square-wave inverter power stages, 400 Hz output filter and filter response.

Fig. 6.11A. When T_1 is switched in phase with T_2, a square-wave output results. When the transformers are switched at 180°, the secondary voltages cancel. Output current limiting and short protection can be readily provided by increasing the phase delay or dwell time (2θ) to approach 180°. The disadvantage of the circuit of Fig. 6.12A is that two transformers are required, even though each transformer is rated at one-half the total output power and each transistor current is one-half the peak input current. The advantages of the circuit are one forward semiconductor drop results and low voltage transistors may be used.

Since the inverter load may be reactive, it is imperative that the secondary "see" a low impedance, and the square wave switching of each primary provides this. For example, if one push–pull power stage were used wherein both transistors were cut off during 2θ, reactive currents in the output filter and load could induce voltages in the open-circuit secondary which, when reflected to the primary, could exceed the voltage rating of the transistors.

At higher dc input voltages, the bridge transistor stage, as shown in Fig. 6.12B, may be used. The drive waveforms to the designated transistors are those of Fig. 6.11. The positive half cycle of E_0 is produced when Q_2 and Q_4 are on, and the negative half cycle is produced when Q_1 and Q_3 are on. When either Q_2 and Q_3 or Q_1 and Q_4 are on, a virtual short (reactive diodes across the transistors are assumed) exists across the primary, and load currents flow unimpeded through the secondary.

For load regulation, 2θ is a minimum at rated load or overload. For line regulation, 2θ is proportional to the input voltage. Thus the power transformer need only support an essentially constant volt-second product, independent of input voltage. This reduces the NA_e product of the transformer as compared to clocked square-wave inverter transformers, which must support the "high line" voltage for a full 180°.

6.2.2   Output Waveform Analysis and Filter

The desired output sine wave voltage E_0 is a function of the width of the quasi–square wave and can be expressed by the Fourier series

$$E\omega_t = (4E_{pk}/\pi)\big[\sin \omega t \cos \theta$$
$$+ \tfrac{1}{3} \sin 3\omega t \cos 3\theta + \tfrac{1}{5} \sin 5\omega t \cos 5\theta$$
$$+ \cdots (1/n)\sin n\omega t \cos n\theta\big] \qquad (6.7)$$

Inspection of this equation shows that θ controls the fundamental voltage as well as the odd harmonics. The equation is plotted in Fig. 6.13 with all values normalized to the peak of the wave. At $\theta = 0°$, a square wave exists with harmonics (third through seventh), as shown. For filtering purposes, the third harmonic is most difficult to eliminate. However, as seen in Fig. 6.13, the third (and ninth) harmonic is zero at $\theta = 30°$. This relation becomes very effective when the operating point at full load and nominal line is chosen near 30°. For example, if $\theta = 35°$, a $\pm 15\%$ input voltage change yields a worst case third harmonic of 15% and THD of 22%. The output filter may be designed from the guidelines of Section 6.1.5.

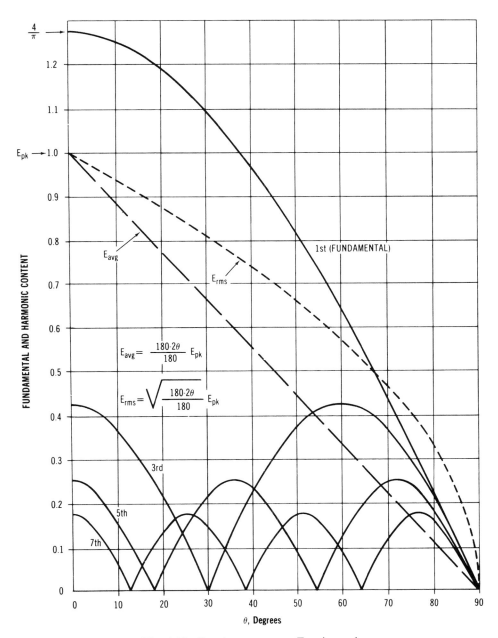

Fig. 6.13. Quasi-square-wave Fourier series.

Example

A 200-VA, 115-V, 400-Hz output is desired from the inverter of Fig. 6.12A. The load resistance is $R_L = E_0^2/P_0 = 66$ Ω. The filter will be detuned, say, to 360 Hz for overload protection, and the impedance will be $83\% \times 66$ $\Omega = 55$ Ω. Then $Z = \sqrt{L_1C_1} = 55$ Ω, and $1/f_0 = 2\pi\sqrt{L_1C_1}$. Then $C_1 = \sqrt{1.96 \times 10^{-7}/55^2} = 8$ μF, and $L_1 = 8 \times 10^{-6} \times 55^2 = 24$ mH. Since L_1 will impede third harmonics, use $Z_3 = 3Z = 165$ Ω. Then $L_3/C_3 = 165^2$, and $1/f_3 = 2\pi\sqrt{L_3C_3}$. Then $C_3 = \sqrt{1.76 \times 10^{-8}/165^2} = 0.8$ μF, and $L_3 = 22$ mH. For the fifth harmonic, use $Z_5 = 5Z = 275$ $\Omega = \sqrt{L_4C_4}$, and $1/f_5 = 2\pi\sqrt{L_4C_4}$. Then $C_4 = 0.3$ μF and $L_4 = 23$ mH. The output filter and filter response are shown in Fig. 6.12C. Both the third and fifth harmonic will be attenuated by more than 30 dB if the operating point of $\theta = 30°$ is chosen at nominal conditions of line and load. The design and complexity of this filter may be compared to the filter example in Chapter 7, Fig. 7.5, for the same load requirements of voltage and frequency.

6.2.3   Combined Transformer and Harmonic Filter

The inverter transformer may be constructed to provide harmonic filtering in addition to isolation and voltage scaling. The transformer and equivalent circuit are shown in Fig. 6.14. The quasi–square wave developed across N_p produces a sine wave at E_0. Since the ratio of capacitor volume to stored energy decreases as capacitor voltage increases, N_{s2} is added to N_{s1} to utilize a 660-V ac capacitor. The transformer is gapped to produce a total secondary inductance L_s, which resonates with C_1 at the fundamental operating frequency. In the equivalent

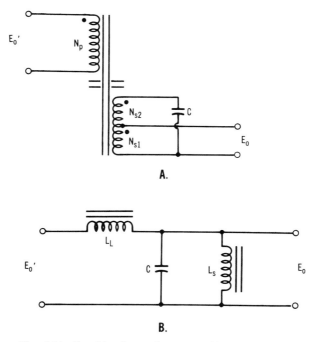

Fig. 6.14. Combined transformer and harmonic filter.

circuit of Fig. 6.14B, the leakage inductance L_1 is produced by the physical separation of the primary and secondary windings. Thus the combined transformer-filter acts as a voltage scaler with a low-pass filter stage and a resonant output stage. The capacitor current flowing in N_{s2} and C_1 is $I_C = E_c / X_c$. The current in N_{s1} is $I_{s1} = \sqrt{I_0^2 + I_c^2}$. If $E_c = 600$ V, a nominal value of C_1 is 0.01 μF per VA output at 60 Hz. This value of C_1 is considerably less than that required by inverters using ferroresonant transformers (see Section 6.5).

6.2.4   Current Limiting and Short-Circuit Protection

Any inverter design should provide for current limiting and short-circuit protection for reliable operation under abnormal conditions. In some cases, it is desirable to supply an output current in excess of rated output for tripping circuit breakers or fault clearing. In other cases, where load protection is not provided, foldback

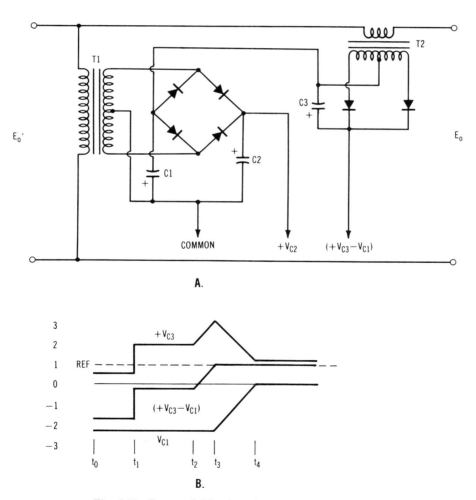

Fig. 6.15. Current foldback and waveforms for inverter.

current limiting, similar to that used in a typical linear or switch mode power supply, is desirable. Figure 6.15 shows a voltage- and current-sensing circuit with current foldback and typical waveforms. The voltage-sensing transformer T_1 may be a part of the output filter as previously discussed. The stepped-down secondary voltage is rectified and filtered to yield a positive and a negative voltage, $+E_{C2} = -E_{C1}$, which are proportional to E_0. Here T_2 is a current transformer and the voltage $+E_{C3}$ is proportional to output current. Voltage $+E_{C3}$ is summed with $-E_{C1}$ and fed back to a current control amplifier. Voltage $+E_{C2}$ is fed back to a voltage control amplifier.

Operating waveforms are shown in Fig. 6.15B. From t_0 to t_1, the inverter is delivering partial-load current, and then full-load current from t_1 to t_2. During this time $-E_{C1}$ is constant, therefore $+E_{C2}$ and E_0 are constant. At t_2, an overload, eventually developing a short circuit, appears at E_0. The output voltage is constant until t_3, at which time the increase in output current has produced an increase in $+E_{C3} - E_{C1}$, which is the same level as the current control amplifier reference. The inverter switches from voltage control to current control and the inverter output pulse width decreases, thus decreasing E_0. Since the sum of $E_{C3} - E_{C1}$ remains constant, $+E_{C3}$ decreases and the output current decreases. At t_4, $-E_{C1}$ is zero, the output is shorted, and the inverter is delivering less than rated load current. By scaling capacitor voltages or the reference voltages, the output current into a shorted load can be a small percentage of rated load current. It should be noted this technique is not limited to this inverter approach.

6.3 INVERTERS WITH HARMONIC CANCELLATION

Additional switching schemes may be employed to cancel (as opposed to attenuate as in the previous section) harmonics in the output and further minimize filter requirements. As previously stated, in a single-stage push–pull inverter both transistors should never be cut off at the same time. The power stage of Fig. 6.12A, with Q_1, Q_2, T_1, and $E_0 = E_{s1}$, could be switched with the waveforms of Fig. 6.16A, where $\theta = 20°$, to provide a waveform with no third harmonic [4]. However, the waveform of Fig. 6.11A contains no third harmonic for $\theta = 30°$ and does not contain negative voltages during the positive half cycle, and vice versa, that is evident in Fig. 6.16A.

The output waveform of Fig. 16.B contains four half-cycle switching angles and is of the form

$$E(n) = (4E_{pk}/\pi)[(1 - 2\cos n\theta_1 + 2\cos n\theta_2)/n] \qquad (6.8)$$

where n is the number of harmonics (or fundamental).

In order for this waveform to contain no third and fifth harmonic, the following equality applies: $1 - 2\cos 3\theta_1 + 2\cos 3\theta_2 = 1 - 2\cos 5\theta_1 + 2\cos 5\theta_2$, from which $\theta_1 = 23.6°$ and $\theta_2 = 33.3°$ [5]. However, the clock frequency may be chosen at 32 times the fundamental, which allows proper counters and logic to provide $\theta_1 = 22.5°$ and $\theta_2 = 33.75°$.

Six half-cycle switching angles offer the possibility of eliminating the third, fifth, and seventh harmonics [5]. For $\theta_1 = 12°$, $\theta_2 = 30°$, and $\theta_3 = 36°$, the third har-

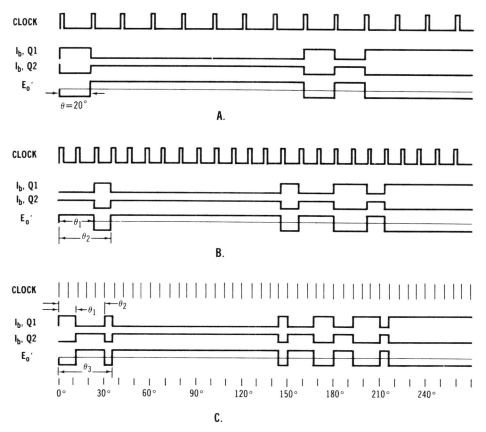

Fig. 6.16. Waveforms with harmonic cancellation.

monic is 0, the fifth harmonic is attenuated 24 dB, and the seventh harmonic is attenuated 25 dB. For a clock frequency of 60 times the fundamental, the exact angles are easily obtained with appropriate counters and logic. Waveforms are shown in Fig. 6.16C.

By itself, the push–pull single-stage inverter offers no regulation or voltage control. By using the two push–pull stages, where drive waveforms are delayed, as discussed in the previous section, the summed voltages of T_1 and T_2 would yield a constant output for line and load changes.

For the bridge inverter of Fig. 6.12B, switching sequences shown in Fig. 6.17 could be used for harmonic reduction [5]. This waveform contains no 3rd, 5th, 9th, or 15th harmonic. The 7th, 11th, and 13th harmonic content is 11.8, 22.2, and 22.7%, respectively. The clock frequency is 30 times the fundamental and $\theta = 12°$. Since drive transformers may be impractical for switching the power transistors, optical isolators may be used. This requires a separate bias supply around Q_{1-4}, Q_2, and Q_3. Power requirements for the bias and drive supplies are significantly reduced if MOSFET or IGBT devices are used instead of bipolar transistors. The absence of negative-output transformer voltages during the positive half cycle, and

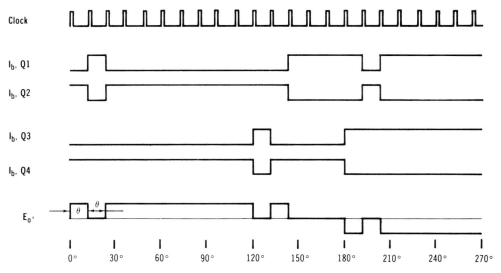

Fig. 6.17. Harmonic attenuation waveforms for bridge inverter.

vice versa, reduces the VA burden on the output transformer, as compared to the waveforms of Fig. 6.16.

The above techniques are readily applied to three-phase output inverters. Section 6.6.2 discusses a push–pull, three-phase, wye output inverter and a full-bridge, three-phase, delta output inverter with harmonic cancellation.

6.4 HIGH-FREQUENCY PULSE-DEMODULATED INVERTERS

High-frequency pulse-demodulated (HFPD) approaches to sine wave inverters represent significant advantages as compared to other designs. A high-frequency carrier is used for switching and the power magnetics operate at this frequency. The high-frequency carrier drastically reduces the size and weight of the output transformer and output filter. A sine wave reference, as opposed to a dc reference, is utilized for "real-time" comparison to achieve low distortion and fast dynamic response. Discussions of several techniques for HFPD inverters follow.

6.4.1 Push–Pull Single-Edge Modulation

The basic HFPD inverter power stage and waveforms are shown in Fig. 6.18. The base drives to Q_1 and Q_2 are square waves at the switching frequency. With the rectifier bridges shown, Q_3 and Q_4 form a bilateral switch which conducts current in either direction. The base drives to Q_3 and Q_4 are square waves modulated by a sine wave. This technique will be compared with double-edge modulation in the next section.

The major disadvantage of this circuit is the requirement for center-tapped primary and secondary windings resulting in high transistor voltages, especially for

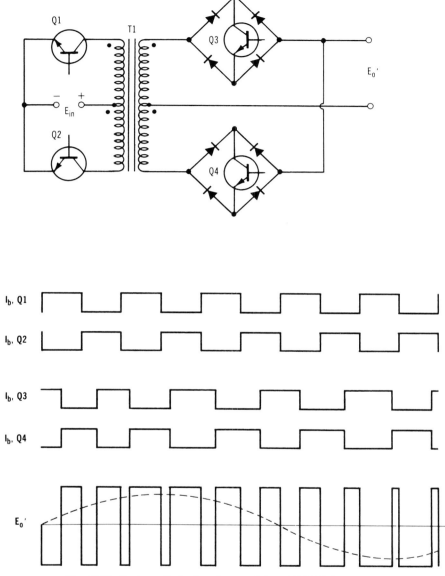

Fig. 6.18. Push–pull single-edge modulated HFPD inverter.

Q_3 and Q_4. Also, the negative output voltages during the desired positive half cycle add to the VA requirement of the transformer.

6.4.2 Double-Edge Modulation

An improvement in the inverter of Fig. 6.18 is achieved by double-edge modulation, as shown in Fig. 6.19. The primary side of the power stage (Fig. 6.19A) may

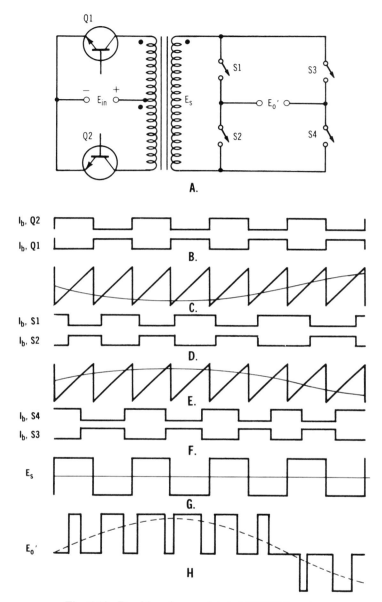

Fig. 6.19. Double-edge modulated HFPD inverter.

be push–pull, half bridge, or full bridge since the transformer is switched by the square waves (Fig. 6.19B) at the carrier frequency. Figure 6.19C represents a sawtooth waveform and sine wave reference (or error signal) as the inputs to a comparator, with outputs shown in Fig. 6.19D. In Fig. 6.19E, the inverted sine wave is used to provide comparator outputs (Fig. 6.19F). The transformer secondary voltage (Fig. 6.19G) is switched by S_1–S_4 to produce the waveforms of Fig.

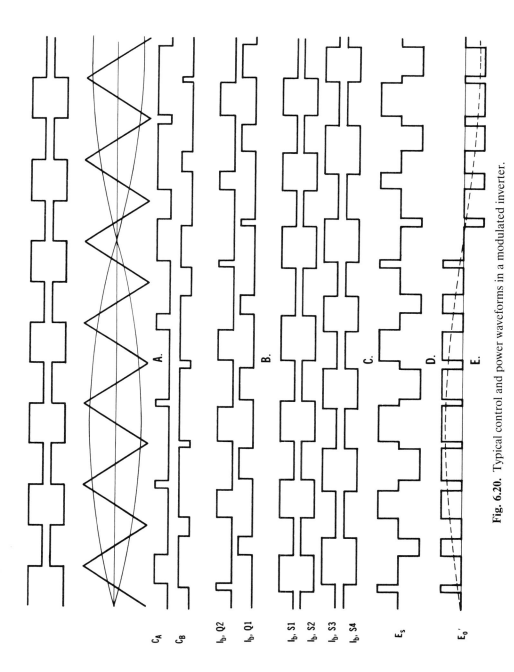

C_A

C_B

I_b, Q2

I_b, Q1

I_b, S1

I_b, S2

I_b, S3

I_b, S4

E_s

E_0'

A.

B.

C.

D.

E.

Fig. 6.20. Typical control and power waveforms in a modulated inverter.

6.19H, with superimposed sine wave resultant. Here S_1–S_4 represent bilateral switches, as shown in Fig. 6.18.

To minimize losses due to common-mode current spikes in the primary stage, the circuit of Fig. 6.19A may be operated with the drive waveforms of Fig. 6.20C to produce the secondary voltage (Fig. 6.20D) which is again switched by S_1–S_4 to obtain the desired output waveform (Fig. 6.20E). It should be noted that the output pulse train frequency of the above waveforms is twice the basic switching frequency, thus reducing output filter requirements. The triangle waveform used in Fig. 6.20 provides symmetrical modulation with respect to a fixed point in time. The comparator output waveforms (Fig. 6.20B) are obtained by comparing the differential error signals to the triangle waveform.

The block diagram for a typical inverter is shown in Fig. 6.21. The master oscillator establishes the switching frequency. The counter supplies the fundamental frequency, which is then locked to the carrier frequency. A low-distortion sine wave generator supplies an active reference for the differential amplifier. The output voltage is sampled and fed back to the differential amplifier. The resultant error signals and the triangle wave feed the comparators whose outputs are gated to produce the desired drive waveforms.

If sufficient dc input voltage is available (or a dc–dc converter may also be used to provide the desired dc input plus isolation), the circuit of Fig. 6.22 may be considered. The advantages of this circuit are elimination of the rectifier bridges

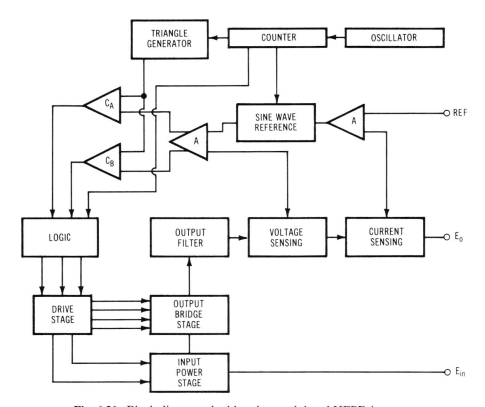

Fig. 6.21. Block diagram, double-edge modulated HFPD inverter.

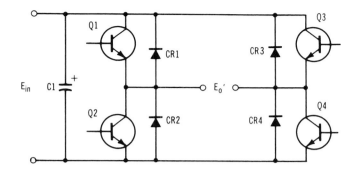

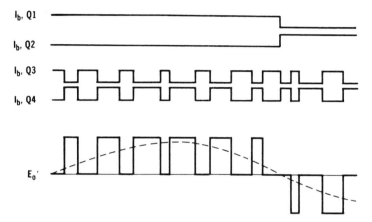

Fig. 6.22. Bridge stage for PWM inverter.

(which reduces the "switch" forward drop), reduced cost, and fewer components. However, the basic circuit does not provide input-to-output isolation. A transformer could be connected across E_0 for voltage scaling and isolation, but two major disadvantages result. The transformer must support the rms voltage at the fundamental frequency, thus significantly increasing size and weight. The core losses in the transformer must be considered at the switching frequency, thus requiring thin "laminations." The frequency of the output pulse train is the same as the carrier frequency, as opposed to twice the carrier frequency for the previously discussed circuits.

6.4.3 Control Functions and Features

Referring to Fig. 6.21, the master oscillator may be some multiple of the desired fundamental frequency. By choosing the appropriate carrier frequency, 50- or 60-Hz operation is achieved by simply changing the counter output and the sine wave reference filter characteristics. The same approach applies for 50-, 60-, or 400-Hz operation. For the latter, a 3600-Hz switching frequency produces 72 pulses per half cycle at 50 Hz, 60 pulses per half cycle at 60 Hz, and 9 pulses per

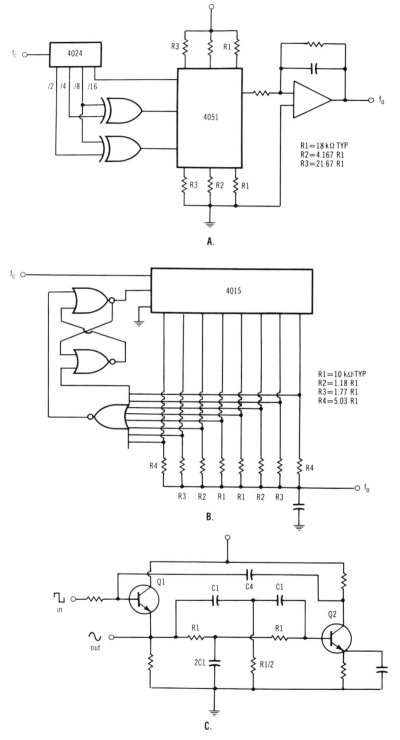

Fig. 6.23. Circuits to provide a sine wave reference.

half cycle at 400 Hz using double-edged modulation with the circuits of Figs. 6.19 and 6.20. As the number of pulses per half cycle decreases, say below 12, it is advisable that the number be an odd integer.

It is also noted the fundamental frequency need not be locked to the carrier frequency. In fact, an external sine wave reference may be fed to the control circuit, producing an excellent low frequency and highly efficient power amplifier. Operation from 0.01 Hz and higher is achieved with an inverter stage whose size and weight are typically that of one operating at the carrier frequency.

To match the high performance of the inverter, a precision sine wave reference is needed. Three circuit methods to provide this reference are shown in Fig. 6.23. The "multiplexer" circuit (Fig. 6.23A) operates at a clock frequency of 16 times the fundamental. The resultant stepped waveform without filtering is attenuated 25 dB at the 15th and 17th harmonics. With the operational amplifier pole at $0.6f_0$, all harmonics are attenuated by more than 45 dB. The 8-bit shift register and gates (Fig. 6.23B) produce essentially the same precision wave. The twin-T filter (Fig. 6.23C) operates from a fundamental square-wave input. The square wave is fed to the high input impedance of Q_1. The twin-T filter is notched at the fundamental and therefore passes all harmonics which are then amplified by Q_2. The inverted and amplified harmonics are fed back to the input via C_4 to cancel the input harmonics, leaving a low distortion sine wave at the emitter–follower output.

It is also noted that the harmonic feedback technique of reducing harmonics in the low-level circuit of Fig. 6.23C may also be applied to high-power circuits to reduce input current distortion and improve input power factor on units operating from ac sources, as discussed in Section 19.2.

6.4.4 Modulation Analysis

The following analysis provides a basic understanding of the high-frequency pulse-demodulated technique, which was developed by Bennett [6] and explained by Black [7]. The sampling of the waveform may be uniform or natural, the difference being that natural sampling produces a spectrum which has no lower harmonics of the modulating frequency f_v. The method of natural sampling may be single-edge modulation, as discussed in Section 6.4.1, or double-edge modulation, as discussed in Section 6.4.2. For further analysis, double-edge modulation is assumed due to the inherent advantages. The double Fourier series for the pulse train of Fig. 6.19H is summarized by

$$F_f = M \cos(\omega_v t) + 2 \sum_{m=1}^{\infty} \frac{J_n(mM\pi)}{m} \cos(m\omega_c t + n\omega_v t)\sin\left(\frac{n\pi}{2}\right) \quad (6.9)$$

where

M = modulation index, $0 < M < 1$
ω_v = angular modulation frequency
ω_c = angular carrier frequency
n = number of harmonics of f_v
m = number of harmonics of f_c
J_n = Bessel function of the first kind

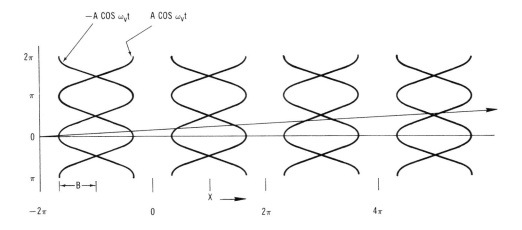

Fig. 6.24. Fourier analysis of HFPD waveform.

TABLE 6.3. Calculated Amplitude Distribution of a Double-edge Modulated Waveform Versus Modulation Index

Frequency	Relative Amplitude Modulation Index (M)				
	1.0	0.9	0.8	0.7	0.6
f_v	100.0	90.0	80.0	70.0	60.0
$f_c \pm f_v$	19.2	26.0	31.7	35.4	37.0
$f_c \pm 3f_v$	20.8	17.4	13.8	10.3	7.3
$f_c \pm 5f_v$	2.7	2.4	1.3	1.0	0.5
$f_c \pm 7f_v$	0.2	0.2	0.2	0.0	0.0
$2f_c \pm f_v$	6.6	10.2	10.4	6.5	0.4
$2f_c \pm 3f_v$	1.2	6.4	11.6	13.7	13.7
$2f_c \pm 5f_v$	11.9	10.7	8.3	5.1	4.2
$2f_c \pm 7f_v$	4.7	3.2	1.7	0.6	0.5
$3f_c \pm f_v$	3.9	5.8	2.9	2.7	6.9
$3f_c \pm 3f_v$	2.1	5.0	4.9	3.1	4.0
$3f_c \pm 5f_v$	2.7	1.6	5.7	7.4	7.0
$3f_c \pm 7f_v$	6.0	6.9	5.9	4.1	2.3

The equation is plotted in Fig. 6.24. The relative amplitude of the modulating and carrier frequencies versus modulation index is shown in Table 6.3 for the waveform of Fig. 6.22 [8]. Further data for the waveform of Fig. 6.19 is shown in Table 6.4. It should be noted that the repetition rate is $2f_c$.

6.4.5 Sine Wave Output Filter

Discussion and data developed in the modulation analysis are of great value in designing the output filter to yield a sine wave with minimum distortion and losses. From Table 6.4, the maximum amplitude of the carrier sidebands occurs at $2f_c \pm f_v$. Therefore a harmonic trap tuned to $2f_c$ is desirable to attenuate this frequency. The overall harmonic distribution may be further attenuated with a low-pass filter.

A typical output filter attenuation curve, and f_v range are shown in Fig. 6.25. The parallel combination of L_1 and C_1 is tuned to resonate at $2f_c$, thus offering a high impedance at the output, with respect to $2f_c$. The pole of L_2 and C_2 should be greater than an octave above f_v. Design of the filter is accomplished by the methods discussed in Section 6.1.5. Inductors L_1 and L_2 should have a low impedance at f_v to minimize voltage drop and a high impedance at $2f_c$ to minimize high-frequency current in C_2. The L_1 core material should be MPP, Supermalloy, or ferrite for low core loss at the carrier frequency. Capacitor C_1 should have a low dissipation factor to conduct high-frequency currents in the trap. Since the trap attenuates the high frequency, the core material of L_2 is not so critical. Typical impedance ratios at f_v compared to Z_L (load impedance) are $X_{L1}/Z_L = 1.5\%$, $X_{L2}/Z_L = 4\%$, and $X_{C2}/Z_L = 20$. The impedance of L_1 and C_1 at $2f_c$ is typically $2Z_L$.

Example

An inverter is required with an output of $E_0 = 120$ V at 60 Hz, $P_0 = 1440$ VA, and $R_L = 10$ Ω. The switcing frequency f_c is chosen as 2400 Hz and $f_c/f_v = 0.025$. Using double-edge modulation the trap frequency (L_1 and C_1 of Fig. 6.25) will be 4800 Hz. Then $X_{L1} = 0.015 \times 0.15$, $L_1 = 0.15/2\pi f_v = 0.4$ mH and $C_1 = 1/(4\pi f_c)^2 L_1 = 2.75$ μF. Choose $C_1 = 2$ μF and $L_1 = 0.55$ mH. Now $X_{L2} = 0.04 \times 10 = 0.4$, $L_2 = 0.4/2\pi f_v = 1.1$ mH, and $X_{C2} = 1/2\pi f_v X_{C2} = 13$ μF; use

**TABLE 6.4. Typical Amplitude Distribution
of a Double-Edge Modulated Waveform**

Frequency	Relative Amplitude, Modulation Index 0.85
f_v	100.0
$2f_v$	0.6
$3f_v$	1.2
f_c	3.3
$2f_c \pm f_v$	32.0
$2f_c \pm 2f_v$	8.0
$2f_c \pm 3f_v$	17.0

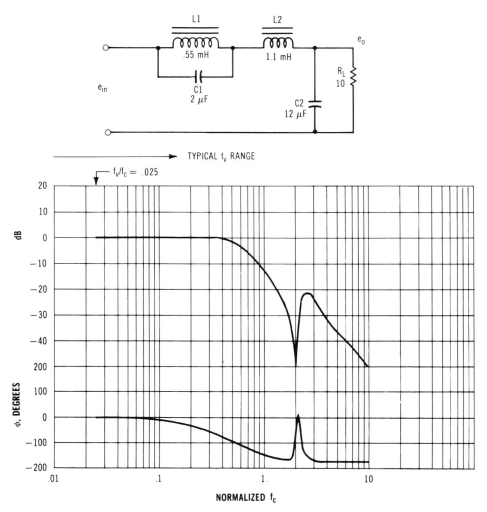

Fig. 6.25. HFPD inverter output filter and typical response.

12 μF. The filter and resulting filter response are shown in Fig. 6.25. Note that at $2f_c$ or the 4800-Hz pulse train, the attenuation is at least 50 dB, resulting in 0.4 V_{rms} at 4800 Hz for a 60-Hz fundamental of $120V_{rms}$.

6.4.6   HFPD Inverter Performance Characteristics

The double-edge pulse-demodulated inverter provides several advantages as compared to other inverter approaches. Typical electrical performance parameters include 1% voltage regulation, 2% worst case THD, 1-ms transient response, and 5% amplitude deviation to step changes of 20% line and 100% load. Typical mechanical parameters for ratings from 2 to 5 kVA are 40 VA output per pound weight and 1 VA output per cubic inch volume. Three inverter stages, with appropriate control circuitry, may be used for three-phase output, where each phase is independently regulated for unbalanced loads, or may readily be paralleled for full-power, single-phase output. Reference Section 6.6.4.

6.5 INVERTERS WITH FERRORESONANT TRANSFORMERS

A ferroresonant or constant-voltage transformer is frequently used for inverter designs up to a few kilowatts output and may be a very cost-effective approach for sine wave output. The ferroresonant transformer contains a resonant circuit and leakage inductance to perform the output filtering and inherently provide voltage regulation and current limiting. A typical power stage is shown in Fig. 6.26A. The control and logic are quite simple since Q_1 and Q_2 are driven by square waves from an oscillator, flip-flop, and drive stage. Output frequency is a function of the oscillator setting and stability. Output voltage regulation is typically 5% for line and load changes. Total harmonic distortion is typically 5–8% for line and load changes. Output voltage is selected by tapping the a coil and is therefore fixed. The output capacitor may be calculated from $C = 4.5 \ VA_0/\omega V_c^2$ F. Transformer size, ratings, and flux densities are discussed in Chapter 3.

Depending on the energized state of the transformer when the inverter was previously turned off, presaturation and high transistor currents may be developed when the inverter is again turned on. If Q_2 is conducting and the inverter is turned off near the end of the half cycle, the volt-seconds sets the core at a certain flux density. When the inverter is turned on again and if Q_2 starts conduction near the beginning of the half cycle, the total volt-seconds (from the previous energized state plus the present state) exceeds the saturation flux density before the half cycle ends and forward biased second breakdown occurs in the transistor. The circuit of Fig. 6.26B minimizes presaturation by always providing a $\frac{1}{4}$ cycle start when the inverter is turned on. Signals C and $\overline{C}$ may be from the input undervoltage and overvoltage sensing or from another delayed source. The $\frac{1}{4}$ cycle start feature may also be used in other types of inverters to minimize starting problems. Signals C and $\overline{C}$ and added logic may also be used to provide a $\frac{1}{4}$ cycle stop when the inverter is turned off.

6.6 THREE-PHASE INVERTERS

Three-phase outputs from inverters are frequently required. The following discussion relates to transistors even though thyristors (SCRs) are normally employed for high-power inverters and for uninterruptible power systems. However, the basic concepts apply to SCR stages with appropriate commutation. Typical outputs are 220/380 V at 50 Hz, 120/208 V at 60 Hz, and 115/200 V at 400 Hz. High-power outputs may also be 277/480 V at 60 Hz. To minimize load losses, the desired waveform is a sine wave with 120° phase displacement.

Inverter design concepts include square wave, quasi–square wave, harmonic cancellation, HFPD, and step wave. The latter was not previously discussed since this approach is inherently applicable to three-phase requirements. Since the inverter must be clocked to maintain good frequency and phase regulation,

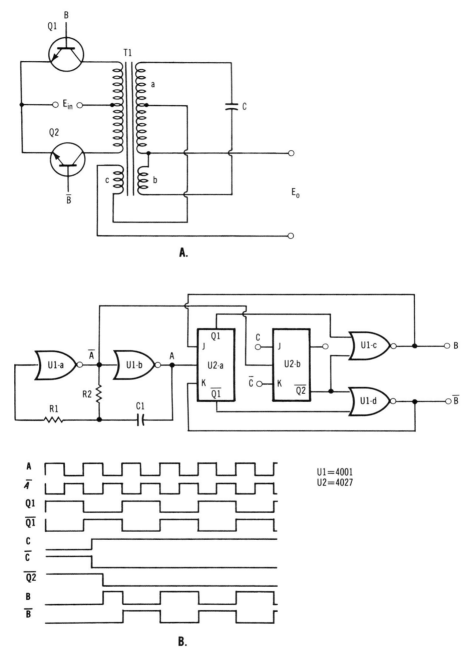

Fig. 6.26. Ferroresonant transformer inverter stage and logic waveforms for starting.

saturating transformer inverters are not practical. Ferroresonant transformer inverters are seldom used due to the phase shift versus load change effect.

6.6.1 Square-Wave and Quasi-Square-Wave Inverters

The clocked square-wave inverter concept of Fig. 6.27 is suitable for some applications. Referring to Fig. 6.13, a 30° displacement in the square wave produces a quasi–square wave with no third harmonic. When the power stage (Fig. 6.26A) is switched with the square waves (Fig. 6.25B), three-phase quasi–square waves are produced. The oscillator or clock frequency is six times the output frequency. The output phase relation (ABC or CBA) must be maintained for many loads (especially for rotation-sensitive motors). The square waves may be obtained from three gated J-K master–slave flip-flops (such as the CMOS 4095) or other suitable divide-by-6, phase-locked circuits.

This approach is especially suited for motor loads up to a few horsepower. Even though the voltage waveform is quasi–square wave, the current waveform is nearly sinusoidal due to the reactance of the motor. The rectifiers, inherent in certain bipolar Darlington and power MOSFET transistors, conduct the reactive motor currents. Voltage and speed regulation may be achieved by employing tachometer feedback to a converter to control the dc input. Variable voltage–variable frequency techniques may be used to develop variable speed and high torque at low speeds.

6.6.2 Inverters with Harmonic Cancellation

As discussed in Section 6.3, additional switching of the basic square wave within each half cycle is another method to reduce harmonics. Again, the output voltage is expressed by Equation (6.8). For a three-phase output, the third harmonic and multiples thereof are inherently eliminated in the line-to-line voltage.

6.6.2.1 *Push–Pull Primary with Wye Output*   Figure 6.28 shows push–pull stages and one line-to-neutral output waveform of the three-phase wye output. In order for this waveform to contain no 5th and 7th harmonic, the following equality applies: $1 - 2\cos 5\theta_1 + 2\cos 5\theta_2 = 1 - 2\cos 7\theta_1 + 2\cos 7\theta_2$, from which $\theta_1 = 16.25°$ and $\theta_2 = 22.07°$. For a clock frequency of 45 times the fundamental, appropriate counters and logic will provide $\theta_1 = 16°$ and $\theta_2 = 24°$. This results in 7% 5th harmonic and 3% 7th harmonic, and the first dominant harmonic is the 11th. Total harmonic distortion is therefore 3.2 times less than the true square wave. The switching sequence ensures that one transistor is always conducting, which provides a low primary impedance for reactive current flow in the secondary.

6.6.2.2 *Bridge Inverter with Delta Output*   The bridge-stage inverter of Fig. 6.27A is an ideal topology for three-phase outputs and is especially suitable for cancellation of the 5th and 7th harmonics. The dominant harmonics are then the 11th, 13th, 17th, and 19th since the 3rd, 9th, and 15th are inherently eliminated. The switching cycles require that $\theta_1 = 16.25°$ and $\theta_2 = 22.07°$, as in the previous section. These exact values are obtained by control circuitry that produce the waveforms of Fig. 6.29. A soft-start feature is also discussed. Referring to Fig.

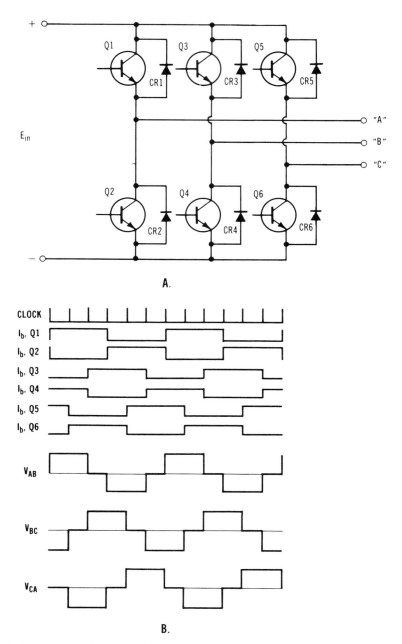

Fig. 6.27. Three-phase inverter with square-wave switching and quasi-square-wave output.

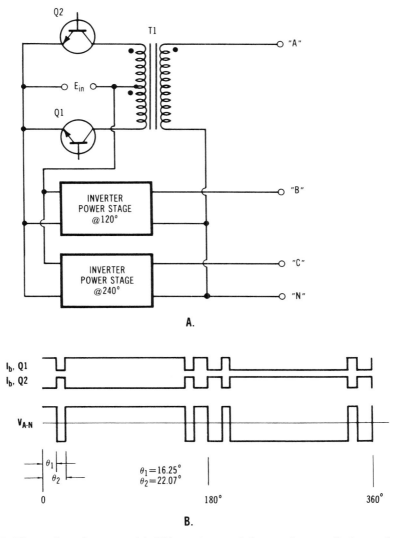

Fig. 6.28. Three-phase inverter with fifth- and seventh-harmonic cancellation and no third harmonic in the line voltage.

6.29A, the three-phase square waves are represented as A, B, and C for the respective phases. Triangle waveforms T_A, T_B, and T_C are generated for the respective phases and are inputs to comparators. Two precise voltages, V_1 and V_2, are other inputs to the comparators. The resulting comparator outputs, labeled C_{1A}, C_{1B}, and C_{1C}, produce a pulse train of $\theta_1 = 16.25°$ and $\theta_2 = 22.07°$, as established by V_1 and V_2. A ramp voltage V_3 and the triangle waveforms result in other comparator outputs, labeled C_{2A}, C_{3A}, C_{2B}, C_{3B}, C_{2C}, and C_{3C}. Associated logic then generates the transistor drive waveforms labeled Q_{1-6}. These waveforms are repeated in Fig. 6.29B, along with the resulting line-to-line output voltage

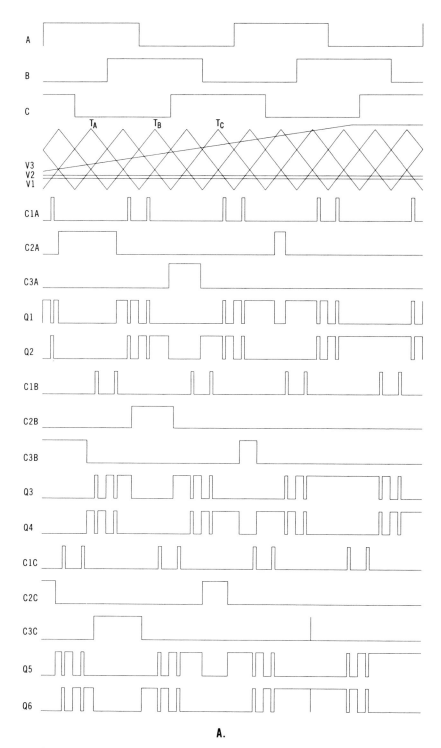

A.

Fig. 6.29. (A) Three-phase bridge inverter waveforms with third-, fifth-, seventh-, and ninth-harmonic cancellation plus soft-start feature. (B) Line–line waveforms, three-phase bridge inverter with third-, fifth-, seventh-, and ninth-harmonic cancellation with soft start.

316

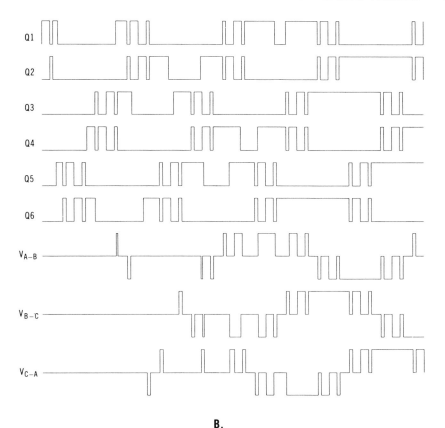

B.

Fig. 6.29. (*Continued*)

waveforms. When V_3 exceeds the triangle peak, the output voltage reaches steady state. The output waveforms do not contain opposite-polarity pulses within a half cycle, as in the topology of Fig. 6.28.

Computer analysis of the output voltage waveform of Fig. 6.29B shows the following harmonic content: fundamental 100%, 11th 14%, 13th 26%, 17th 23%, and 19th 12%. The duty cycle is 120/180 = 0.6667. Thus the rms output is

$$E_{\text{rms}} = \sqrt{D}\,E_{pk} = 0.816 E_{pk} \tag{6.10}$$

A low-pass *LC* filter can be designed to limit the output harmonics to 5% maximum. Since the 13th harmonic is 26%, an attenuation of 5.2, or 14 dB is required. The *LC* filter attenuates at 12 dB per octave, requiring 1.2 octaves. Thus a filter cut-off frequency of $f_{\text{co}} = f_{13\text{th}}/2.4 = 5.4\,f_0$ is required, where f_0 is the fundamental operating frequency.

6.6.3   Step-Wave Inverters

A significant harmonic reduction in the output waveform is achieved in the step-wave inverter design while maintaining basic square-wave switching. Multiple switching stages are employed whereby waveforms in the voltage scaling and isolation transformers are vectorially summed to produce the step wave. The waveform may be 6, 12, or 24 steps, and the two lowest odd harmonics are the number of steps ± 1.

A 12-step, three-phase inverter is shown in Fig. 6.30. Two power stages operate from a common dc bus, which may be a battery source as in a UPS (uninterrupted power system). The switching of power stage I is the same as in Fig. 6.27A, while the switching of power stage II is delayed by 30°. A phasor diagram of the transformer voltages is shown in Fig. 6.31A. The secondary voltages are summed by the appropriate secondary connections, and the number of turns on T_1 is $\sqrt{3}$ times the number of turns on T_2. When vectorially summed. $V_{A-N} = V_{A1-2} +$

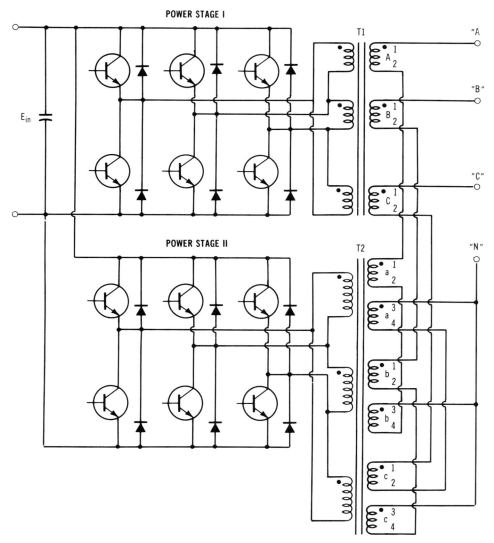

Fig. 6.30. Twelve-step, three-phase inverter.

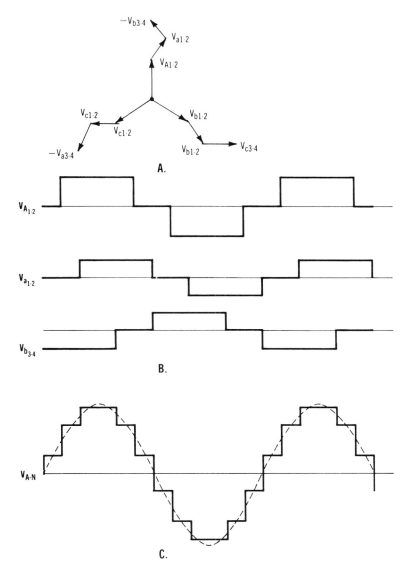

Fig. 6.31. Twelve-step, three-phase phasor diagram and inverter waveforms.

$(V_{a1\text{-}2}\cos 30° - V_{b3\text{-}4}\cos 30°)/3 = 2V_{A1\text{-}2}$. Waveforms for output phase A are shown in Fig. 6.31B. The individual secondary voltages produce the resultant output (Fig. 6.31C) with the superimposed sine wave. Phases B and C are displaced by 120° and 240° from A. From the secondary voltages, $V_{A1\text{-}2} = 1$, $V_{a1\text{-}2} = 0.577$, and $V_{b3\text{-}4} = 0.577$. In normalized form, the output waveform is

$- V_{b3\text{-}4}$ (from 0 to 30°),
$- V_{b3\text{-}4} + 1.73V_{A1\text{-}2}$ (from 30 to 60°), and
$- V_{b3\text{-}4} + 1.73V_{A1\text{-}2} + V_{a1\text{-}2}$ (from 60 to 90°).

The expression for the fundamental at $90°$ is

$$E\omega_1 = 4E_{pk}[1(\cos 0° - \cos 30°) + 2.73(\cos 30° - \cos 60°)$$

$$+ 3.73(\cos 60° - \cos 90°)]/(3.73)$$

$$= (4E_{pk} \times 3)/(3.73\pi) = 1.024E_{pk} \tag{6.11}$$

The 11th-harmonic content is 9.1% and the 13th is 7.7% of the fundamental.

For sine wave filtering, the low-pass filter of Fig. 6.8B may be used. To minimize peaking effects at no load, an additional inductor may be added in shunt with C_1 for damping. However, the output impedance should remain capacitive to supply lagging power factor loads. Typical component impedances for the circuit of Fig. 6.8B are $X_{L1} = 0.1R_L$, $X_{L2} = 4R_L$, and $X_C = 2R_L$. Reference Section 7.7.2 for a high-power step-wave thyristor inverter filter.

Voltage control and regulation may be achieved by two methods. First, the dc input may be controlled in response to ac output sensing to maintain a constant output voltage. Second, the delay angle between power stage I and power stage II may be varied to change the shape of the inverter output waveform to maintain a constant-output voltage. In the latter case, additional harmonics are produced.

Since each output phase is developed from a common transformer, independent phase regulation cannot be achieved. Thus, the step-wave inverter cannot tolerate a large unbalanced load, normally greater than 20%, without voltage regulation and phase balance being adversely affected. For high-power inverters, where paralleled transistors might be required, the same total number of transistors may be used in the step-wave inverter to develop additional steps and reduce output filter requirements.

6.6.4 HFPD Inverters

The double-edge high-frequency pulse-demodulated inverter discussed in Section 6.4.2 and shown in Figs. 6.19 and 6.20 provides one output phase. Two additional power stages, voltage sensing and control circuitry, and a three-phase sine wave reference provide a three-phase output. The sine wave reference, discussed in Section 6.4.3, may be applied to the circuit of Fig. 6.32. The phase shift circuit utilizes an inverting operational amplifier with unity gain and $60°$ phase shift. For unity gain, $e_0 = e_{in}$ and

$$A_f = R_2 \Big/ \Big[\Big(R_1\sqrt{1 + \omega^2 R_2^2 C^2}\Big)\Big] = 1$$

$$\angle - \tan^{-1} \omega R_2 C \tag{6.12}$$

For $60°$ phase shift at unity gain, $\tan 60° = 2\pi f R_2 C = \sqrt{3} = 1.732$. Resistors R_1 and R_2 and C should be precision (1% accuracy or better) components. Due to the $180°$ phase shift in the amplifier, the resultant output is $240°$ with respect to the input. These two signals are fed to Q_1 and Q_2 with emitter–follower outputs. The difference in the two signals produces a third waveform at the collectors; this third signal is displaced $120°$ from the other two.

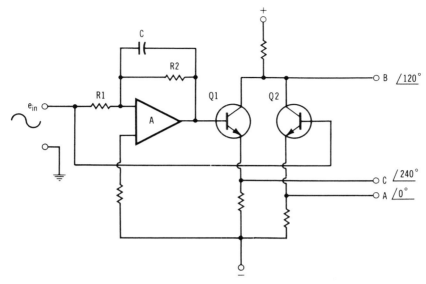

Fig. 6.32. Generating a three-phase sine wave reference from a single-phase sine wave input.

Since individual inverters are used for each phase and phase balance is maintained under all condition, the inverter can tolerate 100% unbalanced loads. The control and power stage circuitry becomes complex and component count increases, as compared to previously discussed inverter topologies. However, the inverter achieves the high-performance characteristics discussed in Section 6.4.6.

REFERENCES

1. G. H. Royer, "A Switching Transistor, AC to DC Converter," *AIEE Transactions* (July 1955).
2. J. Jensen, "An Improved Square-Wave Oscillator Circuit," *IRE Transactions on Circuit Theory* (September 1957).
3. J. R. Nowicki, "New High Power D-C Converter Circuits," *Mullard Technical Communications* (April 1960).
4. P. W. Koetsch, "Reduce Static Inverter Weight and Cost by Harmonic Neutralization," *Electronic Design News* (January 15, 1971).
5. F. G. Turnbull, "Selected Harmonic Reduction in Static DC–AC Inverters," *AIEE Transactions* (July 1964).
6. W. R. Bennett, "New Results in the Calculation of Modulation Products," *The Bell System Technical Journal*, N.Y., Vol. 12 (April 1933).
7. H. S. Black, *Modulation Theory* (D. Van Nostrand, New York, 1953).
8. R. E. Tarter, "Development and Fabrication of an Advanced Static Inverter," Varo, Inc./USAMERDC, Contract No. DA-44-009-AMC-992(T), May 1968.

7

THYRISTOR INVERTERS

7.0 Introduction

7.1 Classification of Inverter Circuits

7.2 Parallel Inverter
    7.2.1 Square Wave
    7.2.2 Triggering Circuit
    7.2.3 Sine Wave

7.3 Half-Bridge Inverter

7.4 Full-Bridge Inverter

7.5 Impulse-Commutated Full-Bridge Inverter, Quasi–Square Wave
    7.5.1 Power Stage
    7.5.2 Power Components
    7.5.3 Inverter Control Circuitry

7.6 Impulse-Commutated Full-Bridge Inverter, Pulse Width Modulated

7.7 Three-Phase Inverters
    7.7.1 Quasi-Square-Wave Inverter
    7.7.2 Step-Wave Inverter
    7.7.3 PWM Inverter
    References

7.0 INTRODUCTION

The function of an inverter is to change a dc input voltage to a symmetrical ac output voltage of desired magnitude and frequency, and to supply a load with the desired power. As in Chapter 6, the inverters discussed in this chapter also provide ac outputs in the 50 to 500-Hz frequency range. Higher frequency switching inverter stages used in dc–dc converters are discussed in Chapter 9.

The ac output waveform may be a square wave, suitable for some low-power applications, or may be a low distortion sine wave, more commonly desired in medium-power applications and required in high-power applications. For a sine wave output, the odd harmonics inherent in the step function switching to convert dc to ac are filtered and minimized, thus reducing load heating effects due to these harmonics. Since thyristors (SCRs) have the capability to conduct much higher

TABLE 7.1. Inverter Characteristics[a]

Inverter Type[b]	Advantages	Disadvantages
Parallel, push–pull (7.2)	Simple, most economical	No voltage regulation (unless preceded by a preregulator); large output filter required for low distortion; dc input must be interrupted to turn off the inverter
Half bridge (7.3)	Better transformer utilization than parallel inverter	Same as above
Full bridge (7.4)	Pulse width modulation provides voltage regulation	Higher values of commutating components than types below
Impulse-commutated full bridge, quasi–square wave (7.5)	Voltage regulation achieved by independent half-cycle control; high efficiency at all load conditions; can be turned on and off without interrupting the dc input	Requires auxiliary SCRs for commutation; complex control circuit required
Impulse-commutated full bridge, PWM (7.6)	Same as above except output filter size is reduced	Higher switching losses than above type; gating sequence and minimum pulse width requires careful attention
Step-wave inverter, three phase, impulse commutated (7.7.2)	Low distortion output; applicable for high-power outputs	Multiple power switching stages required; regulation achieved by a second set of power stages or by a preregulator
PWM inverter, three phase, impulse commutated (7.7.3)	Same as for impulse-commutated full bridge, PWM except regulation is unaffected by unbalanced loads	Same as for impulse-commutated full bridge, PWM

[a]Sine wave output, single and three phase.
[b]Number in parentheses are section numbers.

currents than transistors, the majority of inverters discussed in this chapter will apply to high-power applications with single- or three-phase sine wave outputs.

As discussed in Section 2.1.4, once the thyristor is gated on, the device will continue to conduct current after the removal of the gate signal and will conduct until the current drops to a very low value, or until the voltage across the device is reversed. Limiting initial turn-on current and providing repeatable recovery to the forward-blocking state is the essence of reliable operation for all power conversion equipment. An understanding of the turn-on and turn-off mechanism as well as the dynamic gating at turn-on in the thyristor has greatly improved the reliability of SCR inverters from the early design days when the definition of an SCR inverter was "a unit which is always ten microseconds from failure" or "if it can be started, it will run forever." Considerable attention will therefore be addressed to turn-off commutation means.

Notations are the same as those of Table 6.1. A comparison of the most useful types of inverters is presented in Table 7.1. Advantages, features, and disadvantages provide an initial selection or choice for a particular application.

7.1   CLASSIFICATION OF INVERTER CIRCUITS

The basic classification of inverter circuits suitable for symmetrical ac output is by methods of turn-off as described below [1].

Class A: Self-commutated by Resonating the Load. A typical circuit is shown in Fig. 7.1. Capacitors C_1 and C_2 form a half-bridge divider across E_b. When Q_1 is triggered, current flows from C_1 through Q_1, L_1, and the parallel combination of C_3, L_3, and the load. A positive half-cycle voltage is developed across the load. Capacitor C_3 charges to E_b, the current then reverses and flows through L_1, CR_1, and back to C_1 and Q_1 turns off. Then Q_2 is triggered and a negative half-cycle voltage is developed across the load. The resonant tank circuit produces a sine

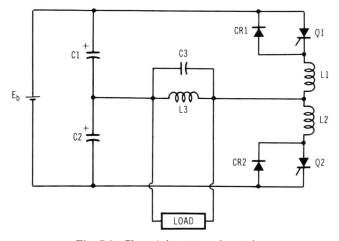

Fig. 7.1. Class A inverter schematic.

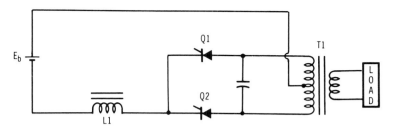

Fig. 7.2. Class B inverter schematic.

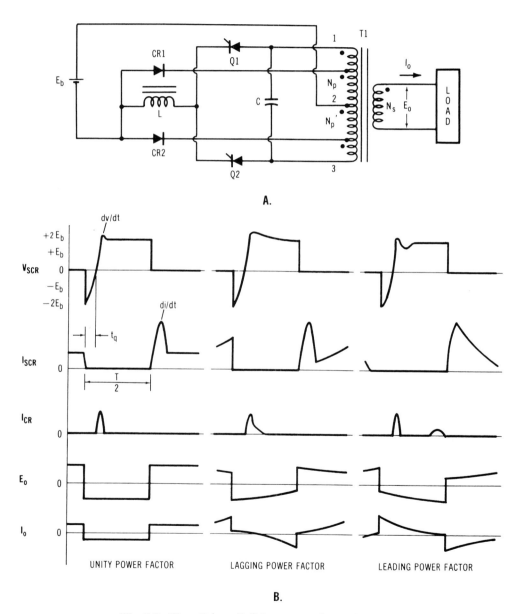

A.

B.

Fig. 7.3. Class C (parallel) inverter and waveforms.

wave output voltage. The inverter has good load regulation with a resistive load. Output voltage regulation for changes in input voltage can be achieved by varying the frequency, but this is not appropriate for fixed output frequency requirements. Thus the class A inverter is not applicable to the types of inverters to be discussed.

Class B: Self-commutated by an LC Circuit. A basic circuit is shown in Fig. 7.2. Inductor L_1 and C_1 form a resonant circuit to automatically commutate Q_1 and Q_2 when the devices are triggered to develop alternate half cycles. The inverter has poor reactive load regulation and no means to recover stored energy in the reactive elements unless additional components are added. The complexity then becomes a disadvantage and the class B inverter is not applicable to the types of inverters discussed.

Class C: C or LC Switched by a Load-carrying SCR. The parallel inverter is shown in Fig. 7.3 and the bridge inverter is shown in Figs. 7.5 and 7.6. Operation is discussed in Section 7.2.

Class D: C or LC Switched by an Auxiliary SCR. The reduction in SCR cost and in turn-off time has made this approach quite effective in high-power applications and will be discussed in more detail than the other classes.

7.2 PARALLEL INVERTER

7.2.1 Square Wave

A class C parallel inverter of the well-known *McMurray–Bedford* type is shown in Fig. 7.3A. Operation of this push–pull circuit is as follows. Assume Q_2 to be on and Q_1 to be off. Voltage is transformed from T_1 to R_L and the top of capacitor C is charged to twice the supply voltage, or $2E_b$, since $N_p = N_{p'}$ and remains at $2E_b$ until Q_1 is turned on. When Q_1 is triggered on, current flows through N_p, Q_1, and L. The voltage induced in $N_{p'}$ causes current to momentarily flow through $N_{p'}$, C, Q_1, and L; C thus charges to $2E_b$. At the end of the half cycle, Q_2 is triggered on. Since the voltage across C cannot change instantaneously, $V_L = 2E_b$ and $V_{Q1} = -2E_b$. Thus Q_1 turns off and the current through C is diverted to Q_2 causing C to discharge and then reverse charge. The reverse voltage duration across Q_1 determines the turn-off time or recovery time of Q_1. At the end of the half cycle, Q_1 is again triggered and the operation repeats for the opposite half cycle. Typical one-cycle waveforms are shown in Fig. 7.3B for unity power factor and reactive power factors.

Rectifiers CR_1 and CR_2 assist the inverter in handling a wide range of loads and power factors, and the value of C may be reduced since the capacitor does not have to carry reactive currents. When supplying inductive loads, energy transferred to the load at the end of one half cycle is returned to the source at the beginning of the next half cycle. When supplying capacitive loads, energy is transferred to the load at the beginning of one half cycle and is returned to the source later in the same half cycle. Energy is transferred to the source by CR_1 and CR_2 rather than being stored in C.

The rectifiers are shown connected to taps on each primary. If the rectifiers are connected to the outer ends of the primary, the no-load to full-load regulation may be improved, but excessive circulating current flows through the rectifiers and efficiency is reduced. If the rectifiers are connected to taps at 15% from the end, the no-load to full-load regulation is approximately 20%, and the no-load losses are reduced. Satisfactory tapping points lie between 5 and 20% from the outer ends of the winding [2]. Overall regulation is further compounded by the fact that no line regulation is provided. For a fixed load, the output voltage is directly proportional to the input voltage.

The rms current rating required for the SCRs may be calculated by including the capacitor charging current and the transformer primary current for a half cycle. The minimum voltage rating required for the SCRs is two times the maximum input voltage, plus the voltage produced by the leakage reactance between the primary halves. The transformer primary should therefore be wound bifilar to minimize leakage reactance. The rating of T_1 is governed by the output power, input voltage, and operating frequency. Reference Sections 3.10 and 3.11 for low-frequency transformer design.

For a desired output voltage and neglecting losses for simplicity,

$$E_0 = nE_b \tag{7.1}$$

where

$$n = N_s/N_p \qquad N_p = N_{p'}$$

For a resistive load, R_L reflected to the primary and considering efficiency, the load current flowing through the SCR at time of turn-off or commutation is

$$I_{pk} = E_b/(\eta n^2 R_L) \tag{7.2}$$

where η is percent efficiency divided by 100.

The values of commutating inductor L and commutating capacitor C are a function of the load current, input voltage, and turn-off time of the SCRs, plus some safety factor to ensure SCR commutation. Since the input voltage is known and the peak current is calculated from Equation (7.2), the optimum values of L and C for a *resistive* load are

$$C = t_q I_{pk}/(1.7E_{b,min}) \qquad \mu F \tag{7.3}$$

$$L = t_q E_{b,min}/(1.7I_{pk}) \qquad \mu H \tag{7.4a}$$

where t_q is the rated turn-off time in microseconds.

The value of t_q (time from the SCR current decaying to zero to time of reapplied anode voltage, as shown in Fig. 7.3) is usually specified by the manufacturers' data sheets. Silicon-controlled rectifier costs are normally inversely proportional to turn-off time, but minimum total component cost is normally achieved by choosing the shortest turn-off time available for a particular SCR family, since the savings in capacitor or inductor cost more than offsets the increased cost of the SCR.

If the load is *inductive*, the commutating inductor value increases to

$$L = t_q E_{b,min}/0.425 I_{pk} \qquad \mu H \tag{7.4b}$$

Referring again to Fig. 7.3, the dv/dt (rate of anode voltage rise after commutation) and di/dt (rate of anode current flow at turn-on) should not exceed the SCR rating.

7.2.2 Triggering Circuit

Since the parallel inverter must be clocked at symmetrical half-cycle rates, the simplest form of triggering is by an oscillator output divided by a flip-flop, which is applied to a drive stage to trigger the SCRs. The oscillator may be a UJT, a stable multivibrator, timer, or CMOS oscillator. The flip-flop may be any convenient divide-by-2 circuit or device. The drive stage may be in the form of a one-shot multivibrator driving a pulse transformer. The transformer secondary is center tapped and connected to the common cathodes in Fig. 7.3, with the outside terminals connected through resistors to each gate. The drive stage may also be a square wave, with the outer terminals of the pulse transformer connected through a capacitor and resistor to each gate to apply a differentiated pulse on which to gate the SCRs.

7.2.3 Sine Wave

The square-wave output of the inverter of Fig. 7.3 may be filtered to provide a sine wave, as shown in Fig. 7.4A, by the filter design from Ott [3]. This filter provides good load regulation while maintaining a capacitive load to the inverter over a wide power factor. The capacitive load reflected to the inverter side aids SCR commutation. The input impedance increases as the load impedance increases.

The solid lines shown in the Smith chart of Fig. 7.4B are those of the normalized filter load impedance, $Z_D \angle \phi$. The dotted lines are normalized $Z_{in} \angle \theta$. The magnitude and phase values are circled for identification. The following equations solve filter components where ω is the operating radian frequency:

$$C_1 = 1/6Z_D\omega \qquad C_2 = 1/3Z_D\omega$$
$$L_1 = 9Z_D/2\omega \qquad L_2 = Z_D/\omega \tag{7.5}$$

Solving for Z_{in} and R_{in}, the required peak square-wave voltage into the filter is

$$E_{pk} = (\sqrt{2}\,\pi/4)|Z_{in}|\sqrt{P_0/R_{in}} \qquad V \tag{7.6}$$

For the example of an inverter to supply 240 VA at 120 V, 400 Hz, and 0.8 lagging power factor,

$$Z_L = E_0^2/240 = 60 \ \Omega \qquad \angle Z_L = \cos^{-1} 0.8 = 37°$$
$$P_0 = 240 \times 0.8 = 192 \ W \qquad \omega = 2\pi \times 400 = 2513$$
$$R_L = 60\cos 37° = 48 \ \Omega \qquad X_L = 60\sin 37° = 36 \ \Omega$$

For the filter design impedance $Z_D \leq Z_L/2 \leq 30 \ \Omega$.

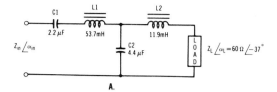

A.

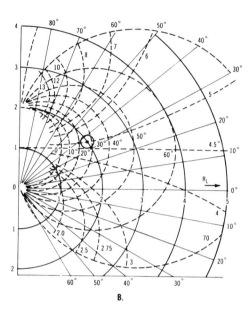

B.

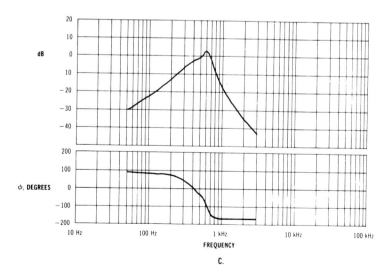

C.

Fig. 7.4. Parallel inverter filter, impedance chart, and frequency response.

Substituting in Equation (7.5),

$$C_1 = 1/(6 \times 30 \times 2513) = 2.2 \ \mu F$$

$$C_2 = 1/(3 \times 30 \times 2513) = 4.4 \ \mu F$$

$$L_1 = 9 \times 30/(2 \times 2513) = 53.7 \ mH$$

$$L_2 = 30/(2513) = 11.9 \ mH$$

Using a normalized Z_D of $2\angle 37°$, the normalized Z_{in} from Fig. 7.4B (circled point) is $5\angle -27°$. Thus $Z_{in} = 30Z_D = 150 \ \Omega \angle -27°$, $R_{in} = 134 \ \Omega$, and $X_{in} = 69 \ \Omega$ capacitive reactance. The filter frequency response is shown in Fig. 7.4C. From Equation (7.6), $E_{pk} = 199$ V.

Referring to Fig. 7.3A for the inverter design with a 28-V dc input and 85% efficiency,

$$n_{T1} = 199/28 = 7:1$$

$$VA_{in} = VA_0/\eta = 240/0.85 = 283 \ W$$

The calculations for the commutating components differ from those of Equation (7.3) due to the reflected capacitive load and are related by

$$I_{pk} = 4E_b\sqrt{C/L} \qquad t_q = (2\pi/3)\sqrt{LC}$$

Rearranging terms, $\sqrt{L} = 4E_b\sqrt{C}/I_{pk} = 3t_q/(2\pi\sqrt{C})$.

Also, $I_{pk} = VA_{in}/E_b = 283/28 = 10.1$ A. Choosing $t_q = 10 \ \mu s$,

$$C = 3_{tq}I_{pk}/(8\pi E_b) = 3 \times 10 \times 10.1/(8 \times \pi \times 28) = 0.44 \ \mu F$$

$$L = 16E_b^2 C/I_{pk}^2 = 16 \times 28^2 \times 0.44/10.1^2 = 54 \ \mu H$$

7.3   HALF-BRIDGE INVERTER

For higher input voltages and better transformer utilization, the class C half-bridge inverter (McMurray–Shattuck) of Fig. 7.5 may be used. Capacitors C_1 and C_2 may be eliminated if a center-tapped dc source is available. When Q_1 is triggered on, current flows from C_1 through Q_1, L_1, T_1 and also through C_4, which charges to E_{in}. The voltage across Q_2 is E_{in} and the voltage across T_1 is $\frac{1}{2}E_{in}$. At the end of the half cycle, Q_2 is triggered on. Since the current through L_1 and the voltage across C_4 cannot change instantaneously, C_4 induces a voltage, E_{in}, across L_2 and in turn across L_1, producing $-E_{in}$ across Q_1, and Q_1 turns off. Current now flows through T_1, L_2, Q_2, and also through C_3, which charges to E_{in}. At the end of the half cycle, the process repeats. Also, note that since each SCR is fired on alternate half cycles, the output is a square wave, so substantial filtering is required to produce a sine wave.

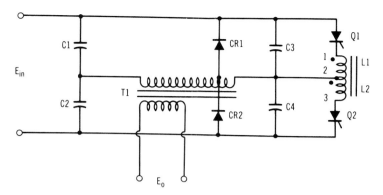

Fig. 7.5. Half-bridge inverter.

Values of $C_3 = C_4$ and $L_1 = L_2$ are given by

$$C = t_q I_{pk}/0.425 E_{b,\,min} \qquad \mu F \qquad (7.7a)$$

$$L = t_q E_{b,\,min}/0.425 I_{pk} \qquad \mu H \qquad (7.7b)$$

where t_q is the rated turn-off time in microseconds.

The values of C and L for the half-bridge inverter are four times the values of C and L for the parallel inverter of Fig. 7.3A. This results from C of Fig. 7.3A charging to $2E_b$ while C_3 and C_4 of Fig. 7.5 charge to $\frac{1}{2}E_{in}$, assuming the same input voltage and power level. For this reason, the half-bridge inverter is used where higher input voltages are available. Rectifiers CR_1 and CR_2 at the tap on the primary of T_1 provide a voltage discharge loop to absorb the unused energy from L_1 and L_2, as discussed in Section 7.2.1. The energy absorbed is passed to the load or back to the dc source. Again, CR_1 and CR_2 are normally located between 10 and 20% taps on the transformer, tending toward 10% if the input voltage is high and if the output frequency is low. Above 400 Hz, the trapped energy problem adversely affects the operating efficiency such that more complex circuits are justified.

7.4 FULL-BRIDGE INVERTER

The class C full-bridge inverter shown in Fig. 7.6 provides several advantages over the half-bridge inverter. The SCR currents are halved, the center-tapped requirement is eliminated (although a filter capacitor across E_{in} is normally required), the SCR switching may be sequenced to provide a quasi-square-wave output, which inherently requires less filtering to produce a sine wave. The required values of C_1–C_4 and L_1–L_4 are the same as those of Equation (7.7) but since I_{pk} is halved, the resulting L and C values are halved for the same input voltage and required power as that of Fig. 7.5.

The SCR switching is similar to the sequence of that for the transistor inverter discussed in Section 6.2 (reference Fig. 6.11) and is shown in Fig. 7.6B. When Q_1 is

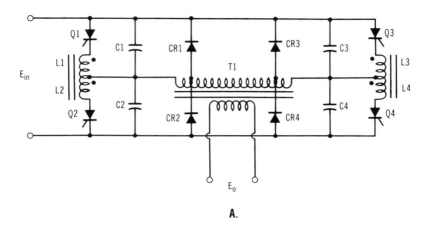

A.

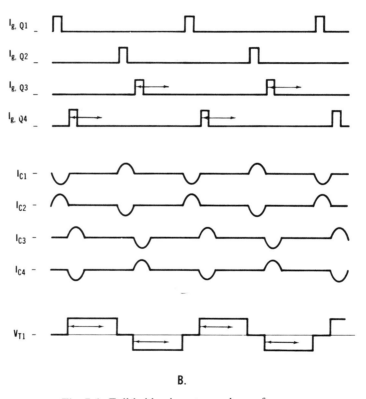

B.

Fig. 7.6. Full-bridge inverter and waveforms.

triggered on (or fired), current flows through Q_1, L_1, and C_2. When Q_4 is fired, current flows through C_3, L_4, Q_4, and also through T_1 to produce an output voltage at E_0. When Q_2 is fired, Q_1 is commutated off, and a virtual short exists across the transformer whereby circulating or reactive currents can flow through CR_2 and Q_4 or through Q_2 and CR_4. When Q_3 is fired, Q_4 is commutated off and a negative voltage is produced across T_1. The switching sequence proceeds, with Q_1 and Q_2 being turned on and off each half cycle. Also, Q_3 and Q_4 are turned on and off each half cycle but are delayed with respect to the triggering of Q_1 and Q_2. Figure 7.6B also shows the normal limits of the duty cycle, $D = t/0.5T$, to prevent commutation overlap. The maximum and minimum on-time of Q_1 and Q_4 or of Q_2 and Q_3 should be limited to

$$t_{on,max} = 0.5T - 4t_q \qquad (7.8a)$$

$$t_{on,min} = 4t_q \qquad (7.8b)$$

From the desired operating frequency and the turn-off time of the selected SCR, the minimum and maximum duty cycle may be calculated. The control circuitry can then be designed to limit the maximum duty cycle, as well as to ensure a minimum duty cycle. For a 60-Hz inverter using 20-μs devices, $t_{min} = 80$ μs $= 1\%$, and $t_{max} = 8.25$ μs $= 99\%$. In this case, the 80-μs on-time or 1% duty cycle should limit the SCR current to a level below the maximum current which can be commutated during a sustained output short circuit, since the rate of current flow through the commutating inductor is limited.

7.5 IMPULSE-COMMUTATED FULL-BRIDGE INVERTER, QUASI–SQUARE WAVE

The class D auxiliary impulse-commutated or McMurray inverter has the following distinct advantages over other means of commutation: (1) independent half-cycle control allows PWM (pulse width modulation) operation; (2) with a proper gating sequence, the inverter stage may be started and stopped without opening the dc line and without commutating the dc line; (3) a portion of the commutating energy is stored in an inductor and returned to the commutating capacitor; (4) higher efficiency at light loads is thus achieved; (5) efficiency at full load typically exceeds 90% for low-frequency (50–400 Hz) operation. The advantages more than offset the inherent requirement for auxiliary thyristors, but the current rating of the auxiliary thyristors is normally much less than the current rating of the power thyristors.

7.5.1 Power Stage

Rather than assume an initial running condition, the inverter will be started with appropriate gating and at a low duty cycle via a soft-start circuit. As the number of periods increase, the duty cycle and the output voltage increase to the operating point.

Referring to the diagram of Fig. 7.7A and the waveforms of Fig. 7.7B, at $t_{start}Q_1$ is triggered on but no current flow results. Shortly after Q_1 is triggered and with a

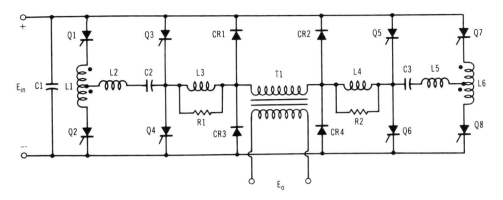

(A) Schematic.

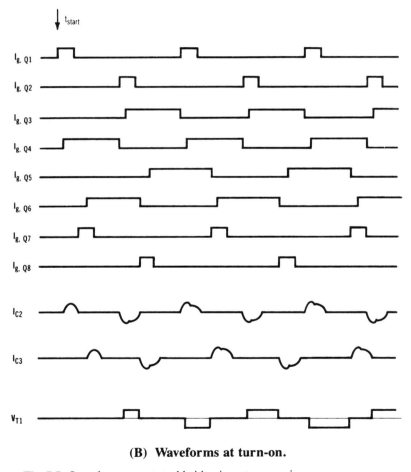

(B) Waveforms at turn-on.

Fig. 7.7. Impulse-commutated bridge inverter, quasi–square wave.

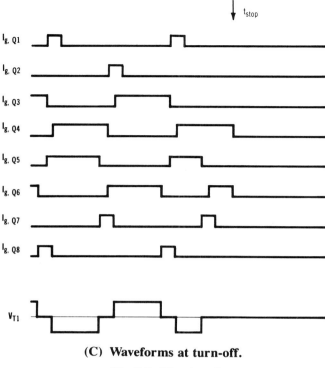

(C) Waveforms at turn-off.

Fig. 7.7. (*Continued*)

gate signal still applied to Q_1, Q_4 is triggered on. Current flows through Q_1, L_1, L_2, C_2, and Q_4 to precharge C_2. Due to the underdamped circuit, C_2 charges to $2E_{in}$ and Q_1 turns off. Next, Q_7 is triggered on but no current flow results. Shortly after Q_7 is triggered and with a gate signal still applied to Q_7, Q_6 is triggered on. Current flows through Q_7, L_6, L_5, C_3, and Q_6 to precharge C_3. As with C_2, C_3 charges to $2E_{in}$ and Q_7 turns off. Now Q_4 and Q_6 are on, but no current flow through transformer T_1 results. Next, Q_2 is fired and C_2 discharges through L_2, L_1, Q_2, CR_2, and L_3. But shortly after Q_2 was fired. Q_3 is fired and charging current flows through Q_3, C_2, L_2, L_1, and Q_2 to again precharge C_2. The gate signal is still applied to Q_6 and load current flows through Q_3, L_3, T_1, L_4, and Q_6 to produce the output voltage shown. Positive voltage is indicated by Q_3 and Q_6 being on, while negative voltage is indicated by Q_5 and Q_4 being on. The gate signals to the commutating thyristors Q_1, Q_2, Q_7, and Q_8 are from pulse transformers, with 25 μs typical duration. The gate signals to the power thyristors Q_3–Q_6 are from transformers which apply the gate signal continuously for the half-cycle conduction period. This ensures continued thyristor conduction if a momentary discontinuous anode current flow results from reflected load conditions. The firing sequence to Q_1–Q_4 is fixed. The firing sequence to Q_5–Q_8 is also fixed but is delayed with respect to Q_1–Q_4, producing a variable pulse width output.

Next, Q_8 is fired. Since the right side of C_3 is positive, current flows through C_3, L_5, L_6, Q_8, CR_3, and L_4. When the commutating current through C_3 reaches the value of the load current, the load current is diverted from Q_6 through C_3, L_5, L_6, and Q_8. Then Q_6 becomes reverse biased and turns off. To precharge C_3, Q_5 is then fired and when the sinusoidal current through C_3 decreases to zero, Q_8 turns off. Now C_3 is charged positive on the left. The current through T_1 ceases and a half cycle of operation is complete. Next, Q_1 is fired to turn off Q_3 and C_2 discharges through L_3, CR_1, Q_1, L_1, and L_2 until Q_4 is fired to again precharge C_2. Since a gate signal has been applied to Q_5, when Q_4 is fired, load current flows through Q_5, L_4, T_1, L_3, and Q_4 to produce the negative half cycle of output voltage. Then Q_7 is fired to turn off Q_5. Current flows through C_3, L_4, CR_2, Q_7, L_6, and L_5. During the time CR_2 is forward biased, I_{C3} exceeds I_{T1} and Q_5 is recovering its blocking capability. At the point where I_{C3} decreases below I_{T1}, the turn-off of Q_5 is completed and Q_6 is then fired. Since Q_6 was fired before I_{C3} reached zero, Q_7 is still on and the commutating current appears as an additional partial sinusoid. The gate firing sequence proceeds as shown in Fig. 7.7B and the duty cycle increases until the desired output voltage is reached at which point the control circuitry controls the amount of delay to Q_5–Q_8 to maintain a regulated output.

Also, note that during periods following a positive half cycle, Q_3 and Q_5 are on and a virtual short circuit exists across T_1. Reactive currents can flow through Q_3, L_3, T_1, and CR_2 or through Q_5, L_4, T_1, and CR_1. In like manner following a negative half cycle, Q_4 and Q_6 are on and a virtual short again exists across T_1.

At some point in time, it is desired to turn the inverter off. If the turn-off is a manual operation, turning the inverter off by a control switch may be preferred as opposed to turning off the input circuit breaker or contractor in series with the dc input. If automatic turn-off (such as when an input overvoltage or undervoltage occurs) is desired to protect the inverter, a means to stop thyristor gating and leave the inverter in a nonconducting state would be desirable. This operation may be accomplished as shown by the gating signals of Fig. 7.7C. The inverter has been operating at a high duty cycle. During the second negative half cycle when Q_4 and Q_5 are conducting, the gating signal to Q_7 is advanced to turn off Q_5 and shortly after this, Q_6 is fired. In this state, Q_4 and Q_6 are on and a virtual short exists across T_1. At t_{stop}, all gate signals cease and the inverter is left in a stable nonconducting state. Turn-on may again be accomplished by applying the gate signals of Fig. 7.7B at t_{start} for a soft-start condition. Now that the desired gate signals are known, the technique of developing the firing signals is discussed in Section 7.5.3.

7.5.2 Power Components

Various commutating circuits may omit L_1 and L_6 or may omit L_2 and L_5. In the former case, high dv/dt thyristors must be used for Q_1, Q_2, Q_7, and Q_8 or highly dissipative snubbers must be added. When Q_1 (or Q_7) is fired, the rate of voltage rise across Q_2 (or Q_8) could be very high. In the latter case, high forward voltage ratings must be used for Q_1, Q_2, Q_7, and Q_8. When Q_1 (or Q_7) is fired, the voltage across L_1 (or L_6) begins to reverse after the current through C_2 (or C_3) reaches a peak, which applies a minimum of $3E_{\text{in}}$ to Q_2 (or Q_8) when the current flow ceases. If C_2 (or C_3) were overcharged, as discussed in the above section, the thyristor forward-blocking voltage rating would increase accordingly. It is normally cost effective to split the commutating inductors as shown in Fig. 7.7, with L_1 (and L_6) equal to $\frac{1}{3}$ and L_2 (and L_5) equal to $\frac{2}{3}$ the total required inductance.

Inductors L_3 and L_4 serve two functions. They limit the rate of rise of current, di/dt, when the power thyristors are fired. They provide a reverse voltage across the power thyristors during turn-off. Referring to Fig. 7.7, when Q_7 is fired to turn off Q_5, the current flowing through CR_3 and L_4 produces a voltage across L_4 to reverse the voltage across Q_5, which reduces the rated thyristor turn-off time, as compared to a thyristor with reverse voltage clamped at the forward conduction value of CR_3. This increase in manufacturers' turn-off time rating is also due to the reduction in dv/dt as compared to the abrupt or step change in forward voltage across Q_5 when the reverse clamping rectifier ceases conduction. Resistors R_1 and R_2 may be used to damp the high-frequency transients across L_3 and L_4.

The commutating current must exceed the load current for an interval longer than the thyristor turn-off time, and the optimum values of commutating inductance and capacitance plus the natural frequency and pulse width are [4]

$$C = 0.893 I_{pk} t_q / E_c \qquad \mu\text{F} \tag{7.9a}$$

$$L = 0.397 E_c t_q / I_{pk} \qquad \mu\text{H} \tag{7.9b}$$

where

$$I_{pk} = \text{peak load current}$$

$$t_q = \text{rated thyristor turn-off time, } \mu\text{s}$$

$$E_c = E_{in,\,min} / \varepsilon^{-\pi/4Q}$$

$$Q = \omega L / R \text{ (typically 10–20)}$$

$$L = L_2 \text{ plus half the inductance of } L_1$$

$$\quad = L_5 \text{ plus half the inductance of } L_6$$

$$f_c = 1/(2\pi\sqrt{LC}) = 0.267/t_q \tag{7.10}$$

$$t_p = \pi\sqrt{LC} = 1.87 t_q \tag{7.11}$$

The above values are based on a peak commutating current of 1.5 times the peak load current. Using 20-μs turn-off devices, $f_c = 13.4$ kHz and $t_q = 37$ μs. Due to the high peak commutating currents, the commutating capacitors must have a low impedance. Various manufacturers supply capacitors specifically for SCR commutating circuits, and the capacitors are usually rated in terms of peak voltage, rms current, and current pulse width (reference Section 2.3.4). The relations of rms to peak current for the capacitor and the commutating SCR are shown in Chapter 1, Table 1.2, items 7 and 8, respectively. Since the output voltage waveform is pulse width modulated, Table 1.2 (items 6 and 12) shows the relation of rms to peak current typical in the power SCRs. The energy per pulse versus operating frequency must also be considered when choosing both the commutating and power thyristors, as discussed in Section 2.1.4.3 and shown in Fig. 2.24. The inductors must have sufficient volt-second capability to withstand the operating voltage at maximum input, plus conduct the required current without saturating. Since the voltage capability is directly proportional to turns and the inductance is directly proportional to turns squared, a "large" number of turns on a core results in a large air gap in the core to obtain a relatively low inductance value. The high natural frequency also produces high core losses. For these reasons, all the inductors may be air core types (reference Section 3.16). The rms current rating of the rectifiers is typically 33% of the power thyristor rms current rating.

Power transformer T_1 of Fig. 7.7 is designed for the input voltage, the output frequency, and the output power desired. Output filtering may be achieved by the methods discussed in Section 6.2.3. In fact, harmonic neutralization by the methods discussed in Section 6.3 may be used to minimize output filter component size. The penalty in this case is that the added commutations in each half cycle will reduce the operating efficiency somewhat.

7.5.3   Inverter Control Circuitry

Output voltage regulation, current limiting, and thyristor timing are discussed by referring to Fig. 7.8 [5]. For the analog operation of Fig. 7.8A, a sample of the output voltage is applied to the inverting input of U_9 and at reference voltage (REF. 1), adjustable to adjust the inverter output voltage, is applied to the noninverting input of the op-amp. The resulting error signal is applied to a soft-start circuit. The soft-start circuit holds the error signal low until it is desired to turn the inverter on, at which time the error signal ramps positive, as shown in Fig. 7.8B. The error signal is applied to U_{1B} and U_{4B}. For current sensing, a current transformer is normally inserted in series with power transformer T_1 of Fig. 7.7A. The secondary is full-wave rectified and applied to U_{4A}. If the power thyristor current exceeds a preset value (determined by Ref. 2), which must always be less than the maximum current which can be commutated by the power thyristors, the appropriate commutating thyristor is fired to turn off the power thyristor.

For the digital operation of Fig. 7.8A, an oscillator operates at twice the desired output frequency. The oscillator output is applied to U_{1A} whose pulse width is adjusted to 90% to set the maximum duty cycle and prevent commutation overlap. The error signal from U_9 controls the pulse width of U_{1B} which triggers U_{2A}. When the error signal exceeds REF. 3 of U_{4B}, as shown in Fig. 7.8B, the set signal is removed from U_{3B}, which is reset by $U_{1A,\,Q}$. Then U_{3B} resets U_2. The output signals from U_2 are applied to U_5 which forms a quad monostable. The U_5 outputs are applied to a trigger stage whose pulse transformers gate the commutating thyristors. Also U_5 resets U_7 whose delayed outputs are inverted by U_8 and applied to a trigger stage whose drive transformers gate the power thyristors. Before the inverter was turned on, U_{3B} disabled U_8 and U_7 via U_6, due to the tristate mode of U_7. As the error signal increases in amplitude, the pulse width of U_{1B} increases, which increases the duty cycle of the power stage until regulation of the output voltage is achieved. If an overcurrent condition occurs, U_{4A} sets U_{3A} which triggers U_{2A}, which in turn triggers the appropriate commutating thyristor to turn off the power thyristor for the remainder of the half cycle. Then U_{3A} is reset by U_{1A}. To turn off the inverter, the error signal is rapidly ramped low and when the error

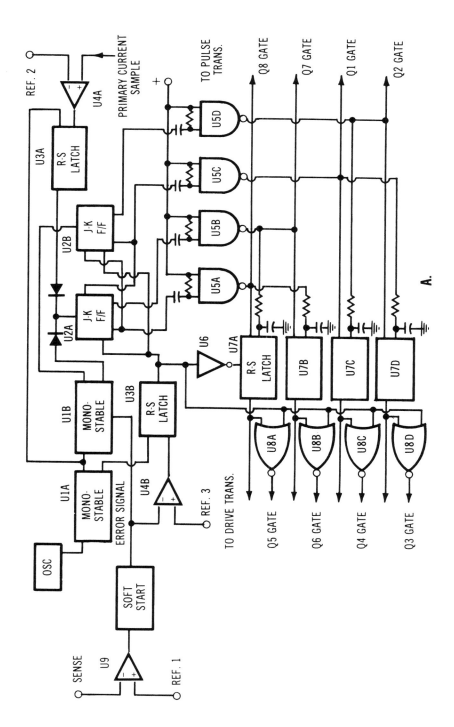

Fig. 7.8. Impulse-commutated bridge inverter control circuit and waveforms.

339

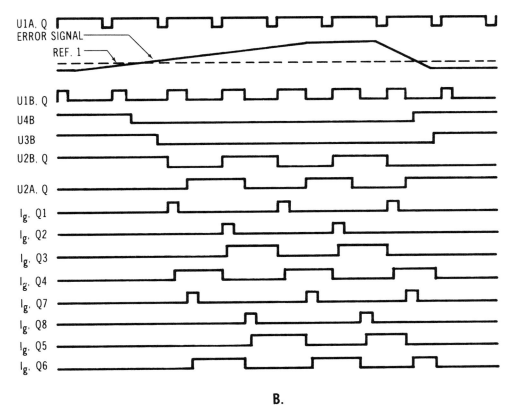

B.

Fig. 7.8. (*Continued*)

signal drops below REF. 3, U_{4B} sets U_{3B} to terminate all gate signals, as shown in Fig. 7.8B.

7.6   IMPULSE-COMMUTATED FULL-BRIDGE INVERTER, PULSE WIDTH MODULATED

Pulse width modulation techniques may also be used for thyristor inverters. Although power stage switching occurs at some multiple of the fundamental outpout frequency, PWM operation differs from the HFPD (high-frequency pulse-demodulated) operation discussed in Section 6.4. The true HFPD inverter employs primary and secondary switching semiconductors to demodulate the high-frequency transformer secondary voltage into a pulse train which can be filtered to a sine wave output. The power transformer in the HFPD inverter is designed to operate at the switching frequency. For high-power PWM inverters, the secondary switching devices are eliminated while still maintaining a high-frequency secondary pulse train, which can again be filtered to a sine wave output. The penalty for this change is the power transformer must now be designed to operate at the funda-

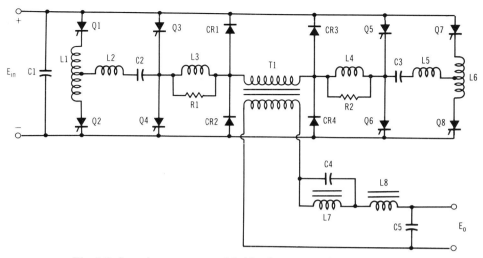

Fig. 7.9. Impulse-commutated bridge inverter, pulse demodulated.

mental or desired output frequency. However, the output filter components are much smaller than the quasi-square-wave inverter described in Section 7.5.

The impulse-commutated inverter stage of Fig. 7.7 is repeated in Fig. 7.9 with an output filter. However, operation of the PWM inverter is more complicated than the basic quasi-square-wave inverter, but the commutation techniques described in Section 7.5.2 still apply and are not repeated here. The basic control circuit and waveforms are shown in Fig. 7.10. In Fig. 7.10A, the sine wave output voltage is sampled and applied to an op-amp whose other input is a stable sine wave reference at the fundamental output frequency. The resulting output error signal (E) is applied to the input of comparators. A triangle waveform (T) and its complement ($\overline{T}$) are generated at the desired switching frequency and also applied to the comparators with waveforms as shown in Figs. 7.10B and C. The outputs of the comparators, shown in Fig. 7.10D, are applied to the proper drive stages to fire the power thyristors corresponding to Fig. 7.9. Again, the signals to the commutation thyristors are developed by means discussed in Section 7.5.2. The resulting power transformer voltage, Fig. 7.10E, is then filtered to a sine wave. The transformer volt-second capability must be sufficient to support the fundamental frequency. The output pulse train frequency, however, is at twice the switching frequency, which reduces output filter component size. The output filter consists of a harmonic trap, L_7 and C_4, to attenuate the high frequency and a low pass filter, L_8 and C_5, to attenuate lower harmonics. The switching methods of Fig. 7.10 are one of several techniques which may be employed to achieve the desired result.

7.7   THREE-PHASE INVERTERS

In high-power systems, three-phase power is normally desired. Regardless of the switching technique, the auxiliary impulse-commutated inverter is again an optimum choice for reasons previously discussed. The quasi-square-wave bridge in-

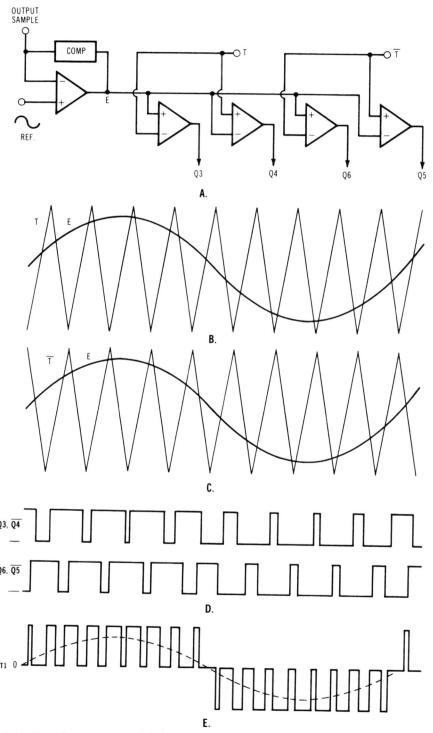

Fig. 7.10. Impulse-commutated bridge inverter, basic control circuit, and waveforms for pulse-demodulated inverter.

verter, the step-wave inverter, and the PWM inverter are presented with diagrams and typical operating waveforms.

7.7.1 Quasi-Square-Wave Inverter

Three inverter stages of Fig. 7.7A are shown in Fig. 7.11A connected to a transformer for three-phase output. In response to the square-wave switching of the power thyristors (even-numbered devices) of Fig. 7.11B, the resulting three-output waveforms are shown in Fig. 7.11C, and since $\theta = 30°$, the output waveform contains no third harmonic. Reference Section 6.2.3 for the waveform analysis. From Chapter 6, Fig. 6.13, the primary rms voltage is 81.6% of the dc input voltage. The output line-to-neutral voltage is:

$$E_{0, L-N} = 0.816 E_{dc} n_s \big/ \big(n_p \sqrt{3} \big) \tag{7.12}$$

The power thyristors are turned off by firing the appropriate commutating thyristor (odd-numbered devices). Commutating component values are obtained from Equation (7.7). A three-phase current transformer may be installed in the primary of the power transformer for current sensing, and in the event of an overload, the overall inverter switching may be terminated at the next clock pulse. The inherent square-wave switching operation does not allow output voltage regulation directly. In this case, the dc input must be controlled as a function of output voltage sensing to maintain regulation. High-power dc–dc thyristor converters are discussed in Section 9.9.

7.7.2 Step-Wave Inverter

The step-wave inverter approach discussed in Section 6.6.3 is a very effective means of providing a low-distortion sine wave output while achieving small filter size and weight, as compared to other switching and filtering techniques. A typical power stage and output filter are shown in Fig. 7.12. The power stage driving transformer T_2 lags the power stage driving transformer T_1 by 30°. The phasor diagram, phase voltage, and resulting output waveform of phase A are shown in Figs. 7.13A, B, and C, respectively. The expression for the fundamental voltage is given in Equation (6.9).

The 12-step output waveform contains no harmonics below the 11th. The 11th harmonic content is 9.1% and the 13th harmonic content is 7.7% (from Section 6.6.3). The low pass filter with added shunt inductor is an ideal filter for the step-wave inverter. Typical component impedances for the circuit of Fig. 7.12 and Fig. 7.14 are

$$X_{L1} = 0.1 R_L \qquad X_{L2} = 4 R_L \qquad X_C = 2 R_L \tag{7.13}$$

Example

A filter is to be designed for a 40-kVA, 3-phase, 120/208-V, 60-Hz inverter. The load current is $I_0 = 40000/(3 \times 120) = 111$ A, and $R_L = 120/11 = 1.08$ Ω. From Equation (7.13), $X_{L1} = 0.1 \times 1.08 = 0.11$ Ω, $X_{L2} = 4 \times 1.08 = 4.3$ Ω, and

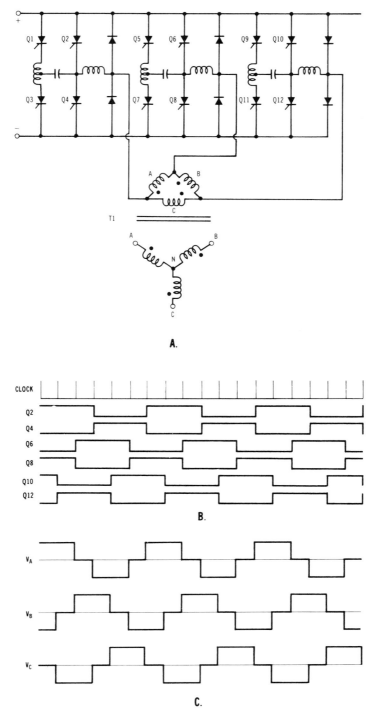

Fig. 7.11. Three-phase impulse-commutated square-wave inverter with quasi-square-wave output.

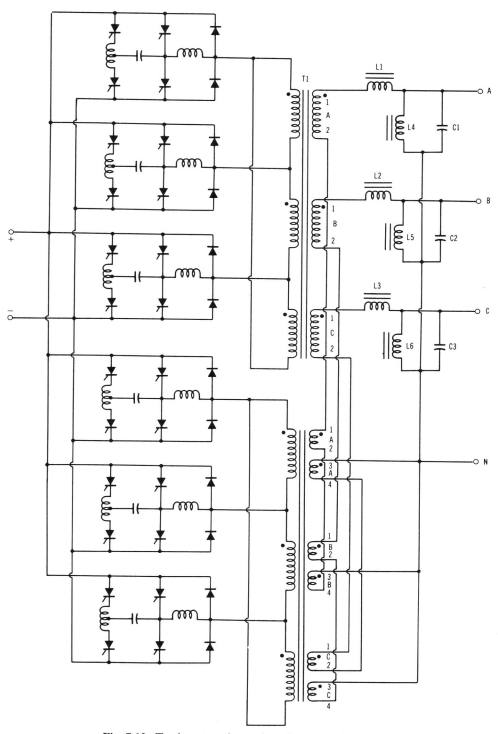

Fig. 7.12. Twelve-step, three-phase inverter schematic.

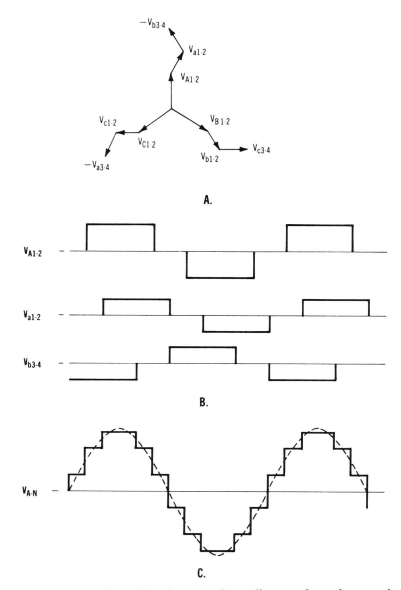

Fig. 7.13. Twelve-step, three-phase inverter, phasor diagram of transformer voltage.

$X_{C1} = 2 \times 1.08 = 2.2 \ \Omega$. Then $L_1 = 0.29$ mH, $L_2 = 11$ mH, and $C_1 = 1200 \ \mu$F. These filter components and the filter response are shown in Figs. 7.14A and B, respectively. At the 11th harmonic, or 660 Hz, the attenuation is -14 dB, or 20% of the fundamental. Since the 11th harmonic of the step wave is 9.1%, the 11th harmonic content at the load equals $0.2 \times 9.1\% = 1.8\%$. The 13th harmonic content in the output is even less.

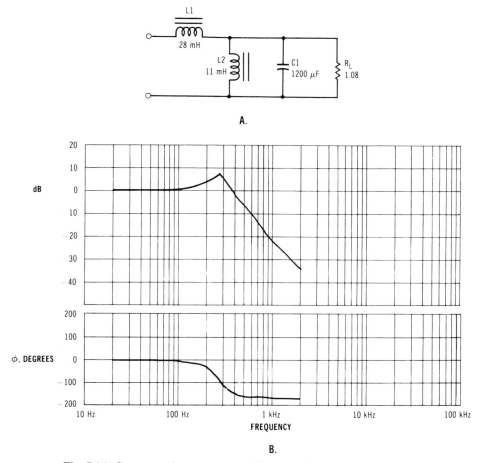

Fig. 7.14. Step-wave inverter output filter and frequency response.

Since the gate signals to the SCRs are fixed and the delayed switching between T_1 and T_2 is fixed, alternate means of voltage regulation must be used. A high-power dc–dc SCR converter, again employing impulse commutation, may be used as an input stage to the inverter. The converter output varies to maintain a regulated ac output, with the feedback loop controlling the converter.

7.7.3 PWM Inverter

Three separate stages of the PWM inverter shown in Fig. 7.9 may be connected for three-phase output. Operation is described in Section 7.6. A three-phase sine wave reference is required, and the sampling technique utilizes a triangle waveform as shown in Fig. 7.10. Each side of the resulting pulse train is then modulated and the effective crossover of the voltage waveform is fixed with respect to a point in time to maintain the 120° phase balance.

REFERENCES

1. *SCR Manual*, 6th ed. (General Electric Co., Auburn, NY, 1979).

2. *Silicon Controlled Rectifier Designers Handbook* (Westinghouse Electric Corp., Young-wood, PA, 1964).

3. R. R. Ott, "A Filter for Silicon Controlled Rectifier Commutation and Harmonic Attenuation in High Power Inverters," *AIEE Communications and Electronics* (May 1963).

4. B. D. Bedford and R. G. Hoft, *Principles of Inverter Circuits* (Wiley, New York, 1964).

5. R. E. Tarter, "A Switching Power Supply for Klystron Power Amplifiers," POWER-CONVERSION '79, Munich, West Germany, September 1979.

8

SWITCHING REGULATORS

8.0 Introduction

8.1 Topologies
    8.1.1 Buck Regulator
    8.1.2 Boost Regulator
    8.1.3 Buck–Boost Regulator
    8.1.4 Ćuk Regulator

8.2 Pulse Width–Modulated Control Circuits
    8.2.1 1524B
    8.2.2 594
    8.2.3 5560
    8.2.4 1846
    8.2.5 1825
    8.2.6 Other Devices

8.3 Buck Regulator, Fixed Frequency
    8.3.1 Switching Parameters
    8.3.2 Output Filter Considerations
    8.3.3 Autotransformer Buck Regulator

8.4 Boost Regulator
    8.4.1 Continuous Current Mode
    8.4.2 Discontinuous Current Mode

8.5 Buck–Boost Regulator
    8.5.1 Continuous Current Mode
    8.5.2 Discontinuous Current Mode

8.6 Protection for Switching Regulators

8.7 Input Filter for Switching Regulators
    References

8.0 INTRODUCTION

This chapter discusses switching regulators which in effect convert an available dc input voltage, normally unregulated, to a regulated dc output voltage. The switching regulators have a common input and output terminal, as compared to the isolated input to output converters discussed in Chapter 9. Regulation is normally achieved by pulse width modulation, which controls the conduction period or duty

349

cycle of an active device. Regulation may also be achieved by fixed pulse width, variable frequency means. A dc source is normally utilized as the input voltage. However, a transformer designed to operate at the utility frequency may be used where the secondary voltage is scaled, rectified, and filtered to provide the dc input, and the transformer provides isolation for safety purposes. This latter approach is not volumetrically efficient due to the size of the transformer, but it is highly power efficient as compared to the linear regulators as discussed in Chapter 17. Topologies of the more popular switching regulators are presented, control circuitry employing PWM (pulse width modulation) are discussed, and the regulators are then analyzed in detail. Input filter considerations to the regulators are also discussed.

8.1   TOPOLOGIES

The four topologies discussed are (1) buck regulator, (2) boost regulator, (3) buck–boost regulator, and (4) Ćuk regulator/converter. Referring to Fig. 8.1, the four topologies of the power handling components are shown with the corresponding basic waveforms. Ideal components are assumed and the current in the inductor is assumed to be *continuous*. The switch is understood to be some form of active semiconductor, usually a bipolar transistor or MOSFET. Principles of operation are discussed and the advantages and disadvantages of the various configurations are listed in Table 8.1.

8.1.1   Buck Regulator

A more detailed discussion of the buck regulator as opposed to the other topologies is presented due to popularity of the buck regulator and the fact that the operation of the buck regulator is analogous to the half-bridge and full-bridge converters discussed in Chapter 9, and the utilization of these latter topologies in switch mode power supplies discussed in Chapter 17. For the buck regulator of Fig. 8.1A, the output voltage E_0 is less than the input voltage E_{in}, hence the name *buck*. When switch S is closed, input current flows through the filter inductor (L), the filter capacitor (C), and the load. When S is opened, the voltage across L reverses since e_L becomes a voltage source ($e = -L \, di/dt$), and the energy stored in L is delivered to the load. Since the current in L cannot change instantaneously, the current flowing through S at the time S is opened now flows through L, C, the load, and flyback diode CR. When S is again closed, the current which was flowing through CR now flows through S and the cycle repeats. The average output voltage is $E_0 = E_{in}D$, where D is the duty cycle. The duty cycle is the ratio of the switch on-time to the period, or t_{on}/T. Since $E_{in}I_{in} = E_0I_0$, the average input current is $I_{in, avg} = I_0D$. It is obvious that when $D = 100\%$, $E_0 = E_{in}$ and $I_0 = I_{in}$. Conversely, when the duty cycle approaches 0%, the output voltage becomes very small and the peak input current becomes very large.

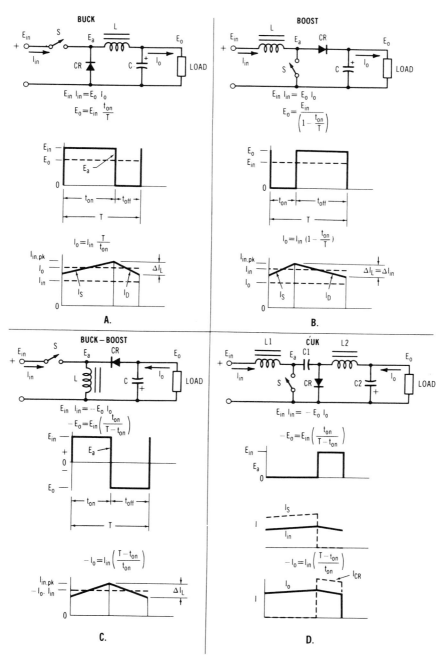

Fig. 8.1. Switching regulator topologies.

TABLE 8.1. Switching Regulator Characteristics

Regulator Type	Advantages	Disadvantages
Buck	High efficiency; simplicity; easy to stabilize; current limiting and output short-circuit protection easy to implement	Input current discontinuous; smoothing input filter normally required; crowbar required to protect critical loads from pass transistor shorts
Boost	Provides an output voltage greater than the input voltage without a transformer; high efficiency; input current continuous	High peak currents in power components; poor transient response; difficult to stabilize; output short-circuit protection requires an additional active device in series input
Buck–Boost	Provides an output voltage polarity reversal without a transformer; high efficiency; current limiting and output short-circuit protection easy to implement	Input current discontinuous; high peak currents in power components; poor transient response
Ćuk	Highest efficiency; input and output current continuous; low switching losses.	High peak currents in power components; high ripple currents in capacitors

8.1.2   Boost Regulator

For the boost regulator of Fig. 8.1B, the output voltage is greater than the input voltage, hence the name *boost*. With input voltage applied, current flows through L, CR, C, and the load. When S is closed, current flows through L and S and in effect, the voltage across L is the input voltage. When S is opened, the induced reverse voltage in L is then in series-aiding with the input voltage to increase the output voltage, and the current which was flowing through S now flows through L, CR, C, and the load. The energy stored in L is transferred to the load. When S is again closed, CR becomes reverse biased, the energy in C supplies the load voltage, and the cycle repeats. The average output voltage is $E_0 = E_{in}/(1 - D)$, where D is the duty cycle. The duty cycle is the ratio of the switch on-time to the period, or t_{on}/T. Since $E_{in}I_{in} = E_0I_0$, the average input current is $I_{in, avg} = I_0/(1 - D)$. It is obvious that when $D = 0$, $E_0 = E_{in}$. Conversely, when D approaches 100%, the output voltage does not necessarily approach *infinity* because the conducting operation of the semiconductor switch S produces a peak current which will quickly exceed the safe operating area (SOA) limit.

8.1.3   Buck–Boost Regulator

The buck–boost regulator shown in Fig. 8.1C is referred to by many names. The buck–boost terminology will be used since the output voltage may be less than, or greater than, the input voltage. The converter is sometimes referred to as a flyback converter, but this terminology will be used to distinguish the transformer isolated topology, as discussed in Chapter 9. The flyback designation is appropriate due to the inherent action of the inductor. This action in itself is sometimes referred to as a *ringing-choke* regulator. Also, the topology is sometimes referred to as an inverting regulator, since the output voltage polarity is opposite the input voltage polarity.

When switch S is closed, current flows through L since CR is reverse biased. When S is opened, the current which was flowing in S now flows through L, CR, C, and the load, and the energy stored in L is transferred to the load. When S is again closed, the current which was flowing through CR now flows through S, and CR becomes reverse biased. The average output voltage is $-E_0 = E_{in}D/(1 - D)$, where D is the duty cycle. The duty cycle is the ratio of the switch on-time to the period time, or t_{on}/T. Since $E_{in}I_{in} = E_0I_0$, the average input current is $I_{in, avg} = I_0D/(1 - D)$. It is obvious that when D equals 50%, $-E_0 = E_{in}$ and $-I_0 = I_{in}$.

8.1.4 Ćuk Regulator

The Ćuk regulator, its principles of operation, and extensions of the basic topology have been well documented [1]. The basic circuit is shown in Fig. 8.1D. The output voltage may be greater than, or less than, the input voltage, and the output voltage polarity is opposite the input voltage polarity. These features then imply a boost–buck inverting operation. With input voltage applied, current flows through L, C_1, and CR, and C_1 charges to the input voltage. When S is closed, input current flows through L_1 to charge L_1 and the discharge current from C_1 flows through C_2, the load, and L_2. When S is opened, the energy stored in L_1 is transferred to C_1 since the current which was flowing in S now flows through C_1 and CR. Also, the current which was flowing in L_2 now flows through CR. The average output voltage is $-E_0 = E_{in}D/(1 - D)$, where D is the duty cycle. The duty cycle is the ratio of the switch on-time to the period, or t_{on}/T. Since $E_{in}I_{in} = E_0I_0$, the average input current is $I_{in, avg} = I_0D/(1 - D)$. The inductors may also be coupled by a common core, as discussed in the cited reference.

8.2 PULSE WIDTH–MODULATED CONTROL CIRCUITS

To achieve the high operating efficiency in switching regulators, the active power device is driven to saturation during conduction and is driven to cutoff for nonconduction. To achieve output voltage regulation, three methods of time ratio control may be used, and each control method employs some form of voltage feedback for closed-loop operation.

First, the switching frequency is fixed by a master oscillator and the conduction time is varied to maintain a regulated output. This PWM technique has in itself two modes of operation, depending on the control circuit design: (1) the point at which the transistor turns on is synchronized with the oscillator waveform, while the point at which the transistor turns off varies as a function of input voltage and load current; (2) the point at which the transistor turns off is synchronized with the oscillator waveform, while the point at which the transistor turns on varies as a function of input voltage and load current.

Second, the conduction time of the transistor is fixed, such as in a one-shot multivibrator, and the operating frequency varies as a function of input voltage and load current. This mode of operation requires a voltage-controlled oscillator (VCO) and has the characteristics of a voltage-to-frequency converter. (Reference Chapter 10 for resonant mode converters.)

Third, the off-time of the transistor is fixed and the operating frequency varies (implying a variable transistor conduction period) as a function of input voltage and load current. Again, a VCO is used. However, this method has a major disadvantage and will not be discussed in further sections. Due to the variable on-time of the transistor and since the switching implies an inductive element is required for energy storage, the peak currents in the inductor and the transistor may vary over a wide range and the peak current requirement of the transistor and the size of the inductor will normally be greater than that of the variable frequency, fixed on-time method.

Various integrated circuits are available for *single-chip* control of switching regulators and converters. The performance of the single PWM IC has been improved to so-called fourth-generation devices and further improvements will undoubtedly occur. Table 8.2 lists characteristics and performance parameters of some popular devices, which are further discussed below. The generic part numbers reference the wide operating temperature devices. These devices are also offered as commercial grade with a narrower operating temperature range with a change in prefix number or suffix letter. Reference manufacturers' published information for other pertinent data to calculate switching frequency, set maximum duty cycle, and utilize feedforward and current mode control. These devices can also be used as the control element in converters, inverters, and power supplies.

8.2.1   1524B

A block diagram of the SG1524B/SG2524B/SG3524B PWM is shown in Fig. 8.2. This device features improved performance over the earlier 1524 version and is a pin-for-pin replacement. As compared to the 1524, the 1524B has a higher accuracy reference voltage, double-pulse suppression logic, improved current sense amplifier, and an undervoltage lockout. The uncommitted outputs provide either a common emitter or an emitter–follower configuration. A timing diagram is shown in Fig. 8.6A, and the output waveforms are for a common emitter circuit with pull-up resistors at each of the collectors.

The internal error amplifier is a transconductance type with high output impedance, as shown in Fig. 8.2B. The transconductance g_m is typically 0.002 A/V and the RC sets the corner pole at 200 Hz. Thus, the open-loop gain is $A = g_m R_L = 0.002 \times 5 \times 10^6 = 10,000 = 80$ dB and the crossover frequency is 2 MHz. The gain plot is shown in Fig. 8.2C, e/a. The amplifier can only source or sink 100 μA and this limits the slew rate. For instance, assume the RC values shown in Fig. 8.2D, which produce a pole at $f = 1/(2\pi RC) = 1$ kHz. If the error amplifier output (e/a) required a step change of 1.2 V, the response time would be $dt = C\, dv/i = 10^{-8} \times 1.2/10^{-4} = 120$ μs. Also, the low value of R severely limits

TABLE 8.2. Characteristics of Pulse Width–Modulated Integrated Circuits by Device Type

	1524B	594	5560	1846	1825
Features					
Single ended			×		
Double ended	×	×		×	×
Parallelable	×	×			
Totem pole output				×	×
Feedforward			×	×	external compensation
Low-voltage lockout	×	×	×	×	×
Internal soft start		×	×		×
Dead-time control		×	×	×	×
Shutdown input	TTL		TTL	350 mV	1.4 V
Recommended operating conditions					
Maximum input voltage, V	40	40	18	40	30
Collector voltage, V	60	40	18	40	30
Output current, mA	100	200	40	100	500
Rise and fall time, μs	0.2	0.2	0.3	0.3	0.06
Reference					
Voltage range, V	4.95–5.05	4.95–5.05	3.65–3.85	5.05–5.15	5.05–5.15
Line regulation, mV	20	25	—	20	20
Load regulation, mV	30	15	—	15	20
Temperature stability	50 mV	1%	—	0.4 mV/$^\circ$C	0.4 mV/$^\circ$C
Oscillator					
Maximum frequency, kHz	400	300	100	500	1000
Initial accuracy, %	10	10	5	9	4
Voltage stability, %	1	0.1	—	2	2
Temperature stability, %	2	4	—	−1	5
Error amplifier					
Input offset, mV	5	10	—	5	10
Input offset, μA	1	0.25	—	0.25	1
Open-loop gain, dB	72	70	54	80	60
Common-mode reject, dB	70	65	—	75	75
Common-mode range, V	2.3/5.2	−0.3/Ein-2	—	0/Ein-2	
Gain bandwidth, MHz	1	0.8	—	0.8	3
Current limit amplifier					
Sense voltage, V	0.18–0.22	Amp	0.4–0.56	0.25–0.40	1
Pulse by pulse	—	—	×	×	×
Package (DIL)	16 pins	16 pins	16 pins	18 pins	16 pins

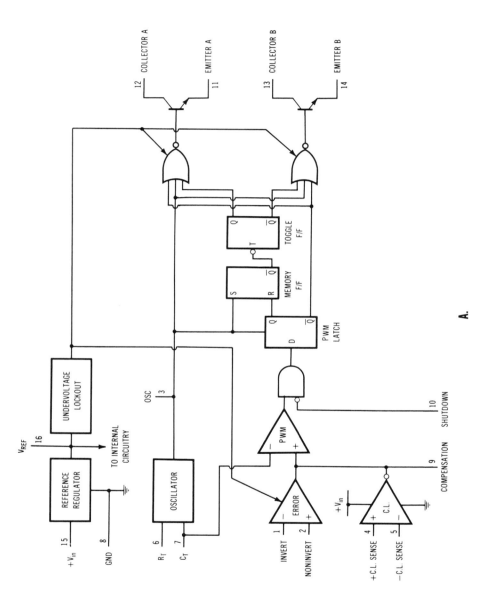

Fig. 8.2. Block diagram of SG1524B. (Courtesy Silicon General)

A.

356

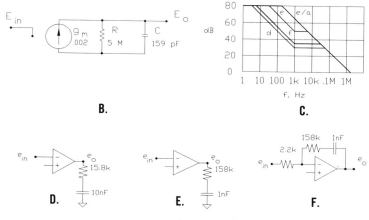

Fig. 8.2. (*Continued*)

the gain above 1 kHz to $A_v = g_m R_L = 0.002 \times 15,800 = 31.6 = 30$ dB, as shown in Fig. 8.2C, *d*. An order of magnitude improvement in response time and a 20-dB increase in gain below 1 kHz is achieved if the *RC* values are changed to those shown in Fig. 8.2E and as plotted in 8.2C, *e*. A popular series *RC* feedback connection is shown in Fig. 8.2F and the gain curve is shown in Fig. 8.2C, *f*.

8.2.2   594

A block diagram of the TL593, TL594, and TL595 PWM is shown in Fig. 8.3. The TL593 has a current limit amplifier with 80 mV offset voltage. The TL594 has an additional error amplifier. The TL595 is identical to the TL594 except an on-chip 39-V zener diode in the TL595 may be used where V_{CC} is higher than 40 V. The uncommitted outputs provide either a common emitter or an emitter–follower configuration. A timing diagram is shown in Fig. 8.6B, and the output waveforms are for a common emitter circuit with pull-up resistors at each of the collectors.

Notice that in this device, the output pulse terminates with the oscillator frequency, while the start of the output pulse varies as a function of the error signal. This is the reverse of the 1524. A further analysis of this mode, for a regulator or converter system, reveals that with a step change decrease in output current, the TL594 will produce a higher output voltage overshoot than the SG1524B. Conversely, with a step change increase in output current, the TL594 will produce a lower output voltage undershoot than the SG1524B. Either of these conditions may be immaterial since the response time of the feedback loop is not instantaneous, but the conditions may be important at very high switching frequencies.

8.2.3   5560

A block diagram of the NE5560/SE5560 PWM is shown in Fig. 8.4. This device is popular for controlling switching regulators, flyback converters (reference Section

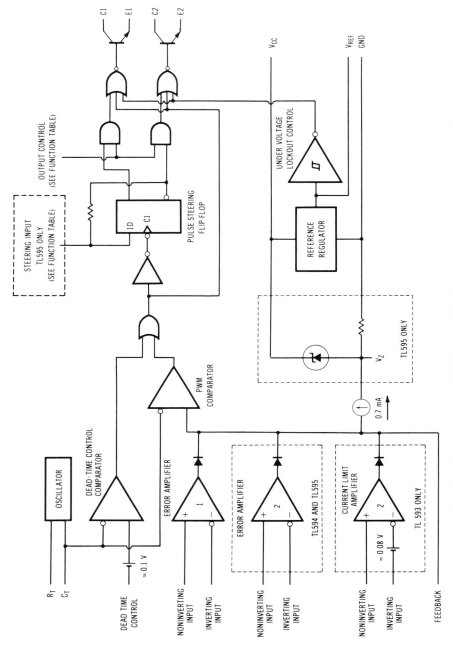

Fig. 8.3. Block diagram of TL593, 594, 595. (Courtesy Texas Instruments, Inc.)

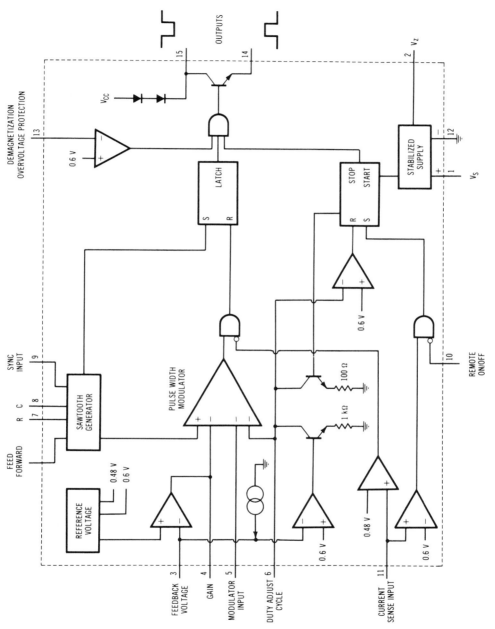

Fig. 8.4. Block diagram of NE/SE5560. (Courtesy Signetics)

359

9.4), and forward converters (reference Section 9.5). In this latter case, a demagnetization sense input can be used to prevent saturation of the forward converter output transformer. With external components, this same input can be used to output overvoltage protection. A timing diagram is shown in Fig. 8.6C, and the output waveform is for a common emitter circuit with a pull-up resistor at the collector. In the emitter–follower mode, the maximum emitter voltage is limited to +5 V. The feedforward input controls the slope of the sawtooth to make the duty cycle inversely proportional to an input voltage or current signal. For operation into push–pull and bridge power stages, external components are required to develop a double-ended output.

8.2.4   1846

A block diagram of the UC1846 PWM is shown in Fig. 8.5. This device provides all of the necessary features to implement control schemes with a minimum of external parts. Various housekeeping functions are beneficial in many applications. The output stage is a dual totem pole for sinking and sourcing external drive stage currents. The primary features of this device are considered to be the double-ended output plus a feedforward, current mode control. The former allows operation into push–pull or bridge power stages, while the latter eases the constraints of compensating the feedback loop for high stability. Pertinent waveforms are shown in Fig. 8.6D.

8.2.5   1825

Characteristics of the UC1825 are listed in Table 8.2 and a block diagram of the IC is shown in Fig. 8.7. This device is optimized for high-frequency operation to 1 MHz. Propagation delay for current sensing is 80 ns maximum and the error amplifier slew rate is 6 $V/\mu s$ minimum. The totem pole output can source and sink 1.5 A peak for driving MOSFETs and IGBTs. However, the maximum switching frequency is normally limited by the power dissipation rating of the IC.

The versatile 1825 can be used in voltage mode, feed-forward mode, or current mode PWM applications or in fixed pulse width, frequency-modulated resonant mode applications with the addition of external components. Figure 8.8A shows a voltage mode circuit where the ramp input is connected to the oscillator and timing capacitor. A feed-forward circuit is shown in Fig. 8.8B where the charging rate of C_{ff} is governed by the input bus voltage. Thus the slope of the ramp makes the duty cycle inversely proportional to the input. This is similar to the waveforms of Fig. 8.6C except the output pulses alternate each cycle. Fig. 8.8C shows a current mode circuit where a portion of the sensed current in the switch or transformer is applied to the ramp input to control the comparator switching.

A fixed pulse width output circuit is shown in Fig. 8.8D. The capacitor and resistor are chosen such that the charging voltage on C_R reaches the 1-V level of the shutdown input at the desired pulse width. Then the output switches and the NOR gate causes Q_1 to discharge C_R each half cycle. This application is beneficial

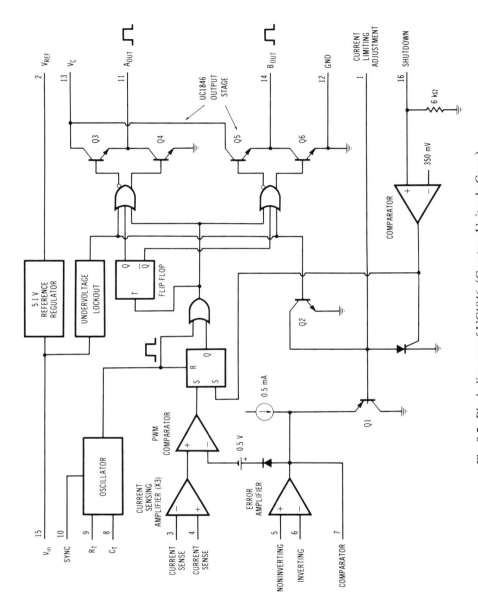

Fig. 8.5. Block diagram of UC1846. (Courtesy Unitrode Corp.)

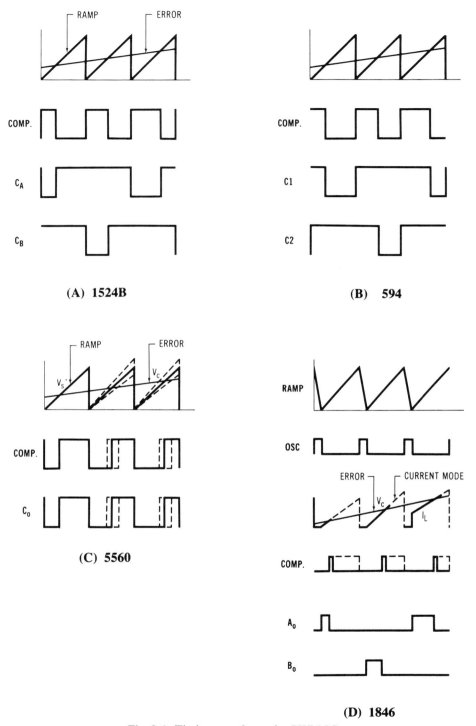

Fig. 8.6. Timing waveforms for PWM ICs.

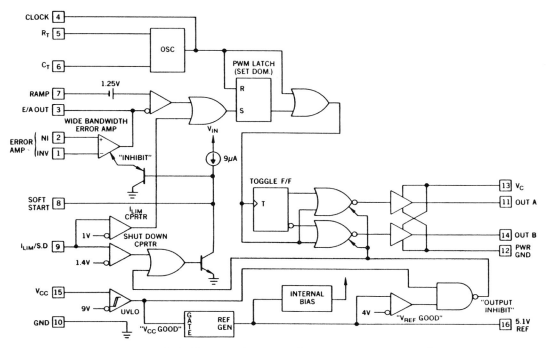

Fig. 8.7. Block diagram of UC1825. (Courtesy Unitrode Corp.)

in quasi-resonant topologies (reference Chapter 10). In this case, regulation is achieved by frequency modulation where the ramp voltage of the oscillator, and thus the period, is voltage controlled, as in the conventional VCO (voltage-controlled oscillator) or VFO (variable frequency oscillator).

It is imperative that certain rules be observed when using the 1825. As with any high-speed device, the 1825 is sensitive to noise. Pin connections and layout with a ground plane are mandatory. The GND and PWR GND pins (reference Fig. 8.7) must be connected to their respective return lines. Schottky diodes should be connected to each output pin and to the ramp input pin to prevent negative voltage spikes at these terminals.

8.2.6   Other Devices

Integrated circuit PWM control devices are not limited to the five families just discussed. Other multipurpose devices find wide usage in regulators, converters, inverters, and power supplies. Refer to manufacturers' data for devices such as ZN1066, SG1525, SG1526, SG1527, UC1840, MC3420, 78S40, NE/SE5561, NE/SE5563, TDA4700, and Am6301. The latter two equivalent devices provide double-ended output, feedforward, symmetry control, overvoltage, and undervoltage sensing, plus other housekeeping functions. They are packaged in a 24-pin DIL and may therefore consume considerable space in compact equipment.

Other integrated circuits, such as the UC1860, MC34066, and CS-3805/LD405 find direct applications in various types of resonant mode topologies. These

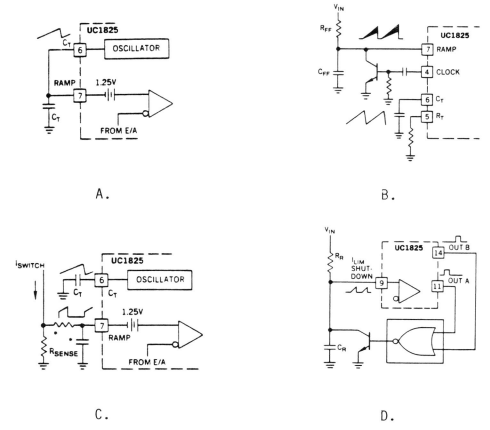

Fig. 8.8. Circuit topologies using the 1825 IC: (A) voltage mode, (B) feedforward, (C) current mode, (D) fixed pulse width (resonant applications).

devices employ a VCO or VFO for frequency modulation in excess of 1 MHz switching frequency. Specific details are discussed in Section 10.2.

8.3 BUCK REGULATOR, FIXED FREQUENCY

The buck regulator of Fig. 8.9A provides an output voltage which is less than the input voltage. Here Q_1 is the switching transistor, L_1 is the filter inductor with a dc resistance R_1 (R_1 may also be the combined dc resistance of L_1 plus a current-sensing resistor for current-limiting purposes), C_1 is the filter capacitor with an ESR R_2, CR_1 is the catch or flyback diode, and the load resistance is R_3. Operating waveforms for a continuous current mode in L_1 are shown in Fig. 8.9B. The waveforms assume the circuit has already established a steady-state output voltage, after a soft-start turn-on. Switching of Q_1 is controlled by a PWM IC operating, in this case, at a fixed frequency. The PWM senses a portion of the output voltage and controls the duty cycle or conduction time of the transistor.

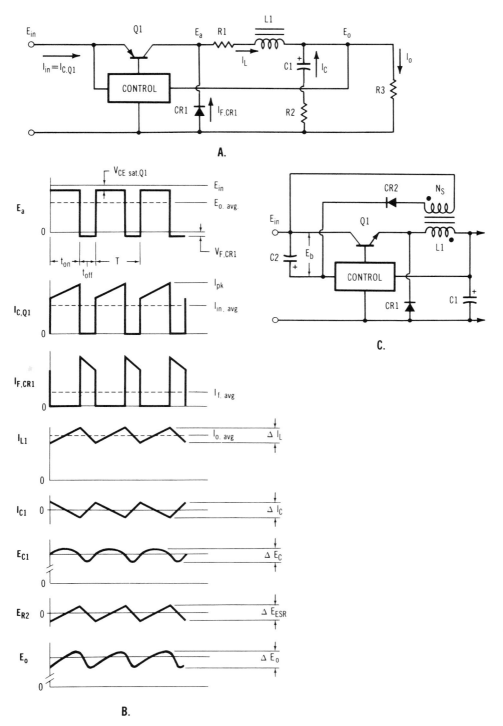

A.

B.

C.

Fig. 8.9. Buck regulator circuit and waveforms.

Current sensing, current limiting, and soft-start features are also assumed since any worthwhile design should employ these features to prevent component over-stress.

8.3.1 Switching Parameters

When Q_1 turns on, current flows through Q_1, R_1, L_1, C_1, R_2, and R_3. When Q_1 turns off, the voltage across L_1 reverses but the current flowing in L_1 cannot change instantaneously. The fall time t_f of Q_1 should be low to minimize switching losses. Then, CR_1 becomes forward biased and conducts the current which was flowing in Q_1 and is still flowing in L_1. Diode CR_1 should have a low forward recovery time to prevent a voltage increase across Q_1 when Q_1 turns off. For example, if CR_1 does not conduct until a few microseconds after Q_1 has turned off (a rather poor quality diode), the induced voltage across L_1 is $e_1 = -L\,di/dt$ and the transistor voltage is E_{in} pulse e_L. This occurrence becomes more prevalent at high input and output voltages; the voltage across the transistor should be observed in test and the voltage rating then selected accordingly. When Q_1 is again turned on, Q_1 suddenly conducts the current which was flowing in CR_1 and is still flowing in L_1. Also, CR_1 should have a low reverse recovery time and low peak reverse current to prevent a current increase in Q_1 at turn-on. For example, if CR_1 does not recover to the blocking state until a few microseconds after Q_1 has turned on (again a rather poor quality diode), the source voltage looks into a low dynamic impedance and Q_1 must conduct a high current in the forward-biased state. If a MOSFET is used for Q_1, the inherent fast rise time dictates a high-speed diode for CR_1. For either type of transistor, the reverse recovery time t_{rr} of CR_1 should be less than one-half the rise time t_r of the transistor. For switching frequencies above 40 kHz, the switching losses in Q_1 usually dominate over the conduction losses. For switching frequencies above 80 kHz, the switching losses in CR_1 usually dominate over the conduction losses. The use of high-speed switching devices can then produce significant increases in efficiency, as compared to low-speed devices. If the appropriate match of switching times cannot be accomplished, the operating efficiency may be increased by adding a ferrite bead to the lead from CR_1. The added low inductance will then support the input voltage for a few nanoseconds and reduce the reverse current in CR_1.

Additional parameters become important at switching frequencies of several hundred kilohertz where a power MOSFET and a very fast rectifier would be employed. When switching 10 A in 20 ns through a wiring inductance of 100 nH, a transient voltage of $e = -L\,di/dt = -100 \times 10/20 = -50$ V is produced. If physical component size is large or the lead length of a snubber circuit across a component is long, the snubber may be ineffective due to the inductance in the snubber leads. Skin effects resulting from high-frequency switching currents may also produce effects not normally considered as part of a particular design. For instance, high-frequency currents through a stud-mount rectifier may raise the effective resistance of the anode post to the point where the power dissipation causes the temperature of the post to rise to a level at which the solder junction melts, or the device is destroyed by excessive junction temperature. Special mechanical configurations, such as those employed in RF device packaging, should then be considered. Transmission line and strip line technology may become appropriate considerations.

8.3.2 Output Filter Considerations

The output voltage of the buck regulator is determined by input voltage and duty cycle, and the load regulation is inherently good, as compared to other regulator topologies. Since the inductor is essentially a volt-second component, changing the duty cycle or the input voltage affects this parameter, as does the rate of change in current. However, the average inductor current is equal to the output current, which results in a simplified control characteristic as compared to the boost and buck–boost regulators (reference Sections 8.4.1 and 8.5.1). For low output voltage ripple, nearly all of the ΔI_{L1} or peak-to-peak inductor current shown in Fig. 8.9B flows in C_1 and R_2.

The output voltage and the component values of L_1 and C_1 are calculated in the following manner:

$$e_I = L\, di/dt \qquad \Delta I_L = \Delta I_L^+ + \Delta I_L^- = e_L T/L_1$$

where

$$\Delta I_L^+ = (E_{\text{in}} - E_0)t_{\text{on}}/L_1 \qquad \Delta I_L^- = E_0 t_{\text{off}}/L_1$$

for $\Delta I_L^+ = \Delta I_L^-$ and

$$(E_{\text{in}} - E_0)t_{\text{on}}/L_1 = E_0 t_{\text{off}}/L_1$$

Then

$$E_0 = E_{\text{in}} t_{\text{on}}/(t_{\text{off}} + t_{\text{on}}) = E_{\text{in}} t_{\text{on}}/T$$
$$= E_{\text{in}} D \tag{8.1}$$

where D is the duty cycle $(= t_{\text{on}}/T = t_{\text{on}} f)$ and f is the switching frequency in hertz. Also,

$$T = t_{\text{on}} + t_{\text{off}} = \left[\Delta I_L^+ L/(E_{\text{in}} - E_0)\right] + (\Delta I_L^- L/E_0)$$
$$= \Delta I_L E_{\text{in}} L/\left[E_0(E_{\text{in}} - E_0)\right]$$

Then

$$L_1 = E_0(E_{\text{in}} - E_0)/(E_{\text{in}} f \Delta I_L) \tag{8.2}$$

From the diagram and waveforms of Fig. 8.9, $I_0 = I_L + I_C$. Assuming ΔI_0 to be negligible, $\Delta I_L = \Delta I_C$. The average capacitor current, I_C, flows for $\frac{1}{2}t_{\text{on}} + \frac{1}{2}t_{\text{off}}$ and $I_{C,\text{avg}} = \frac{1}{4}\Delta I_L$. Also, $e_C = (1/C)\int i\, dt$ from which

$$\Delta e_C = \Delta I_L(\tfrac{1}{2}t_{\text{on}} + \tfrac{1}{2}t_{\text{off}})/4C = \Delta I_L T/8C$$

Then

$$C_1 = \Delta I_L / 8 f \Delta e_C \qquad (8.3)$$

However, $\Delta I_L = \Delta I_C$ also flows through R_2, and the rms value of this current produces a ripple voltage across R_2 as well as power dissipation which produces a temperature rise in C_1. The rms value of ΔI_C (from Chapter 1, Table 1.2, part A, item 14) is

$$\Delta I_{C,\text{rms}} = 0.577 \Delta I_L / 2 = 0.29 \Delta I_L \qquad (8.4)$$

The rms current should not exceed the capacitor rating and the rms value is slightly higher than the average value by the factor $0.29/0.25$.

The output ripple voltage is of most interest. The capacitor voltage is reactive, while the ESR voltage is resistive. Since the waveform of Fig. 8.9B is approximately sinusoidal, the output ripple is

$$\Delta e_0 = \sqrt{(\Delta e_C)^2 + (\Delta e_{R2})^2} \qquad (8.5a)$$

However, Δe_{R2} is normally much greater than Δe_C and a close approximation of peak-to-peak output ripple is

$$\Delta e_0 \cong \Delta I_L R_2 \qquad (8.5b)$$

From Equations (8.9)–(8.12), the following observations are apparent and so-called trade-offs must be analyzed for their effect on the desired performance parameters. For a decrease in ΔI_L:

1. The peak transistor current is decreased.
2. The capacitor current is decreased.
3. The output voltage ripple is decreased.
4. The inductance value and thus the size of the inductor increases.
5. The transient response to step load changes is degraded if the inductor value is increased, since $e = -L \, di/dt$.
6. The minimum output current (while maintaining voltage regulation) is decreased.

The relation of ΔI_L to I_0 is analyzed in the buck regulator example at the end of this section.

Referring to Fig. 8.9B, the output current is equal to the collector current of Q_1 and

$$I_{c,\text{pk}} = I_0 + \tfrac{1}{2} \Delta I_L \qquad (8.6)$$

Also, $I_0 = I_L + I_C$. Neglecting losses, $E_{\text{in}} I_{\text{in}} = E_0 I_0$, and

$$I_{\text{in,avg}} = I_0 D \qquad (8.7)$$

Up to this point, losses have been neglected for simplicity. In actuality, the duty cycle will automatically increase to overcome the losses and voltage drops to maintain a regulated output. As compared to Equation (8.1):

$$e_L = E_{in} - V_{CE, sat} - I_0 R_1 - E_0 = L \, \Delta I_L^+ / t_{on} \quad (\text{during } t_{on})$$

$$e_L = E_0 + V_F + I_0 R_1 = L \, \Delta I_L^- / t_{off} \quad (\text{during } t_{off})$$

Since $\Delta I_L^+ = \Delta I_L^-$,

$$(E_{in} - V_{CE, sat} - I_0 R_1 - E_0) t_{on} = (E_0 + V_F + I_0 R_1) t_{off}$$

Since $t_{off} = T - t_{on}$, the duty cycle becomes

$$D = t_{on}/T = (E_0 + V_F + I_0 R_1)/(E_{in} - V_{CE, sat} + V_F) \tag{8.8}$$

and

$$E_0 = D(E_{in} - V_{CE, sat} + V_F) - V_F - I_0 R_1 \tag{8.9}$$

The minimum input voltage must then be sufficient to allow for the voltage drops without running out of duty cycle while maintaining a regulated output.

The operating efficiency is

$$\eta = P_0/P_{in} = P_0/(P_0 + P_{loss}) \tag{8.10}$$

where

$$P_{loss} = P_{Q1} + I_0^2 R_1 + P_{CR1} + I_C^2 R_2 + P_{L1, core}$$

Power losses for Q_1 and CR_1 may be calculated from the equations of Section 2.1. The losses for L_1 may be calculated from Sections 3.5.2 and 3.5.3.

For the semiconductors, appropriate safety margins should be added to the following minimum ratings

$$V_{CE_0} = E_{in, pk} \qquad I_{c, pk} = I_0 + \tfrac{1}{2} \Delta I_L \quad \text{for } Q_1$$

$$V_{RRM} = E_{in, pk} \qquad I_{F, avg} = I_0 (1 - D) \quad \text{for } CR_1$$

Due to a varying duty cycle, the rms current rating required for CR_1 is more applicable than the average current rating. Manufacturers' data sheets provide this rating, which is equivalent to the dc current rating. The rms rating is found from Chapter 1, Table 1.2, part B, waveform 12, where

$$t = t_{off}$$

$$t/T = 1 - D$$

$$A = I_0 - \tfrac{1}{2} \Delta I_L$$

$$Pk = I_0 + \tfrac{1}{2} \Delta I_L.$$

As previously stated, the output current, and thus the inductor current, was assumed continuous. If the inductor current becomes zero, output voltage regulation will not be maintained and the peak output voltage will rise to the value of the input voltage, since the filter then becomes a capacitive input filter. For a no-load condition on the output and to maintain a continuous current in the inductor, a bleeder resistor may be added in parallel with C_1. For high-voltage (a few hundred volts) low-current outputs, the voltage-sensing divider may provide sufficient minimum bleeder current. For low-voltage high-current outputs, the bleeder resistor is not practical due to the excessive power dissipation. Refer to Section 4.3.1 for a discussion of critical inductance. If the output voltage rises above the regulation limit and if overvoltage sensing is used, the control circuit will shut off the regulator for an overvoltage condition which could occur at no load or very light loads. The regulator will then restart in a soft-start mode until the overvoltage condition is again reached and the cycling process continues until corrected. Increasing the inductance value of L_1 reduces ΔI_L which allows a lower minimum output current; however, this degrades the transient response of the regulator. Since I_0 minimum equals $\frac{1}{2}\Delta I_L$, and I_0 minimum equals $(E_{in} - E_0)t_{on}/2L_1$, using Equation (8.2) and substituting

$$I_{0,\,min} = E_0(E_{in} - E_0)/(2E_{in}Lf) \tag{8.11}$$

Transient overshoot and undershoot and recovery time to step load changes and step line (input) changes are important performance parameters in buck regulators. Since the current in the filter inductor cannot change instantaneously, the transient response is inherently inferior to the linear regulator. The recovery time to step changes in line and load is governed by the characteristics of the feedback loop and by the feedforward loop, if used. Transient overshoot and undershoot resulting from step load changes may be analyzed and calculated in the following manner. The ac output impedance is $Z_0 = (e_{in} - E_0)/\Delta I_{load}$. Since $e_L = -L\,di/dt$ and $I_0 = C\,dv/dt$, $Z_0 = L\,\Delta I_0/[(E_{in} - E_0)C]$.
For an increasing current,

$$\Delta E_0^- = \Delta I_0 Z = L\,\Delta I_0^2/[(E_{in} - E_0)C] \tag{8.12a}$$

For a decreasing current,

$$\Delta E_0^+ = L\,\Delta I_0^2/E_0C \tag{8.12b}$$

The circuit of Fig. 8.7A shows a *PNP* transistor (Q_1) switched by the control circuit. Low saturation voltage is achieved for low-power dissipation in the transistor, but the control circuit must conduct the base current which does not flow to the load. If an *NPN* transistor were substituted, the added drop across a drive resistor, plus the base-to-emitter voltage, would be the minimum saturation voltage which could be achieved. The saturation voltage of an *NPN* transistor may be decreased by the circuit of Fig. 8.9C. An additional winding (N_s) is added to L_1, and CR_2 and C_2 are added. When Q_1 turns off, the voltage across L_1 is essentially the output voltage and this voltage is induced in N_s, which forward biases CR_1 and charges C_2. The voltage across C_2 is connected "piggy-back" to the input to provide a boost voltage to allow the control circuit to drive Q_1 to a low saturation voltage. When the regulator is turned on, a high saturation voltage will be experienced for a few cycles until C_2 is charged. Also, at high duty cycles the voltage across C_2 tends to decrease since the conduction time of CR_2 decreases.

Example

Consider the circuit of Fig. 8.9 for the following conditions: $E_{in} = 28$ V (24–32 V range), $E_0 = 15$ V, $I_0 = 3$ A (0.3–3 A range), $P_0 = 45$ W, $f = 25$ kHz, $\Delta E_0 = 100$ mV. Then $\Delta I_L = 0.6$ A for $\Delta I_L/I_0 = 0.2$. Also, let $\Delta E_c = 0.2 \Delta E_0 = 0.02$ V. The nominal duty cycle $D = 15/28 = 0.54$. From Equations (8.2)–(8.4), $L_1 = 15 \times (28 - 15)/(28 \times 25{,}000 \times 0.6) = 0.46$ mH at 3 A and a dc resistance of 0.1 Ω. Then $C_1 = 0.6/(8 \times 25{,}000 \times 0.02) = 150$ μF at a 30-V rating.

Select a capacitor with an ESR and a ripple current rating of $R_2 = \Delta E_0/\Delta I_L = 0.1/0.6 = 0.17$ Ω; $I_{C,\mathrm{rms}} = 0.29 \times 0.6^2 = 0.1$ A. The filter corner frequency (double-pole) is $f_p = 1/(2\pi\sqrt{L_1 C_1}) = 604$ Hz. The filter zero (due to ESR) is $f_z = 1/(2\pi R_2 C_1) = 6.2$ kHz. The minimum output current is $I_{0,\mathrm{min}} = 0.2 \times \frac{3}{2} = 0.3$ A. The transient undershoot for a load change from 0.3 to 3 A is $\Delta E_{0,\mathrm{min}} = 0.00046 \times (3 - 0.3)^2/(0.00015 \times 13) = 1.7$ V. The transient overshoot for a load change from 3 to 0.3 A is $\Delta E_{0,\mathrm{max}} = 0.00046(3 - 0.3)^2/(0.00015 \times 15) = 1.5$ V. The peak transistor current and voltage is $I_{c,\mathrm{pk}} = 3 + (0.6/2) = 3.3$ A; $V_{ce} = 32$ V; use 50 V.

For current-limiting purposes, let $I_{c1} = 4$ A and let the PWM threshold be 0.2 V. Then $R_1 = 0.2/4 + 0.1 = 0.15$ Ω where the 0.1 Ω is the dc resistance of L_1.

The rms rectifier current and peak voltage is

$$I_{\mathrm{rms}} = \sqrt{0.54(3.3^2 + 3.3 \times 2.7 + 2.7^2)/3} = 2.2 \text{ A};$$

$V_{\mathrm{RRM}} = 32$ V; use 50 V.

For the transistor and rectifier selection, choose a 2N3879 transistor and a 1N3880 diode. Use the circuit of Fig. 8.7C for driving Q_1. Then, $V_{CE,\mathrm{sat}} = 0.8$ V at $I_c = 3.3$ A at $I_b = 0.4$ A, and $V_F = 1.2$ V at $I_F = 2.2$ A. Also, $I_0 R_1 = 3 \times 0.15 = 0.45$ V. From Equation (8.8), $D_{\mathrm{nom}} = (15 + 1.2 + 0.45)/(28 - 0.6 + 1.2) = 0.59$, $D_{\mathrm{max}} = 0.68$, and $D_{\mathrm{min}} = 0.51$. Without losses, D_{nom} was $15/28 = 0.54$.

Transistor power loss = $3.3 \times 0.8 \times 1.5 = 4$ W (50% is for switching losses); rectifier loss = $2.2 \times 1.2 \times 1.1 = 3$ W (10% is for switching losses); resistive loss = $3^2 \times 0.15 = 1.35$ W; L_1 core loss = 0.65 W; and the capacitor loss is negligible (< 0.1 W). The efficiency, allowing 1.9 W for control and drive power, is then $\eta = 45 \times 100/[45 + (4 + 3 + 1.35 + 0.65 + 0.1 + 1.9)] = 80\%$.

8.3.3 Autotransformer Buck Regulator

As previously stated, $E_0 = DE_{in}$, $I_{in} = DI_0$, and $P_{in} = P_0$ for average values, assuming no losses. However, the peak input current is equal to the peak output current and the switching transistor must conduct this peak current. For large ratios of E_{in}/E_0, the duty cycle becomes small and the peak currents are high. An effective method of obtaining output voltages much smaller than input voltages, where the input and output share a common return, is shown in Fig. 8.10A. The

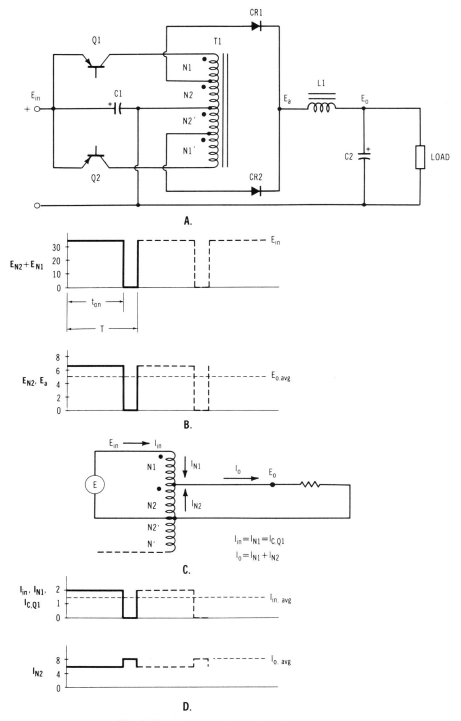

Fig. 8.10. Autotransformer buck regulator.

topology is similar to the push–pull converter except no isolation is provided. The transistors alternately switch at a high duty cycle and the voltage developed across windings N_2 and N_2' of T_1 is full-wave rectified and filtered by L_1 and C_2, and the output voltage is applied to the load. Transformer turns are symmetrical, $N_1 = N_1'$ and $N_2 = N_2'$, with N_1 and N_1' effectively being the primary and N_2 and N_2' being the secondary. In any transformer primary volts per turn equals secondary volts per turn, primary ampere turns equals secondary ampere turns and thus, primary volt-amperes equals secondary volt-amperes. These equalities should not be overlooked and, in fact, make the autotransformer attractive for this application. The following equations also apply: $E_{in} = E_{N1} + E_{N2}$; $E_0 = DE_a = DE_{N2}$; and $I_0 = I_{N1} + I_{N2} = I_{in} + I_{N2}$, where $I_{in} = I_{c,Q1}$. Also, the turns ratio $n = E_{N1}/E_{N2} = I_{N2}/I_{N1}$. Since E_{in}, E_0 and I_0 are usually known values, substituting for these relations gives

$$E_{N2} = E_0/D \tag{8.13a}$$

$$E_{N1} = E_{in} - E_0/D \tag{8.13b}$$

$$I_{N2} = nI_0/(1 + n) \tag{8.13c}$$

$$E_{N1}I_{N1} = E_{N2}I_{N2} \tag{8.13d}$$

Example

A 5 V output at 8 A is desired from a 25-V source. In the conventional buck regulator, the duty cycle would be $\frac{5}{25} = 0.2$, and the peak current in the switching transistor would be 8 A, even though the average input current would be 1.6 A. With the push–pull stage and autotransformer shown in Fig. 8.10A, the resulting waveforms are shown in Figs. 8.10B and D, and the current paths are shown in Fig. 8.10C. The nominal duty cycle is chosen to be 0.8 to allow for voltage drops plus variations in line and load. From the equations of (8.13), $E_{N2} = 5/0.8 = 6.25$ V, $E_{N1} = 25 - (5/0.8) = 18.75$, and $n = 18.75/6.25 = 3:1$. Here $I_{N2} = 3 \times 8/(1 + 3) = 6$ A, and $I_{N1} = 8 - 6 = 2$ A. These currents are peak values and the average values at 0.8 duty cycle are 4.8 and 1.6 A, respectively, which results in a transformer rating of $1.6 \times 18.75 = 4.8 \times 6.25 = 30$ VA for a 40-W input and output. As compared to the standard buck regulator: (1) the peak transistor current is reduced by a factor of 4; (2) the circuit is much less susceptible to an output overvoltage condition since T_1 will saturate if a switching transistor shorts; (3) since the transistors conduct alternately, the same basic switching frequency doubles the frequency of the pulse to the output filter, which reduces the output filter size. The penalty, of course, is an additional transistor, diode, and autotransformer.

A high-power full-bridge version of a similar technique is discussed in Section 18.7. In that case, an SCR bridge develops a boost voltage which is connected in series with the input (battery) voltage, and the sum output voltage is regulated.

8.4   BOOST REGULATOR

The boost regulator of Fig. 8.11 provides an output voltage which is greater than the input voltage. The switching transistor is Q_1, L_1 is the charging inductor with a dc resistance R_1, C_1 is the filter capacitor with an ESR R_2, CR_1 is the catch or

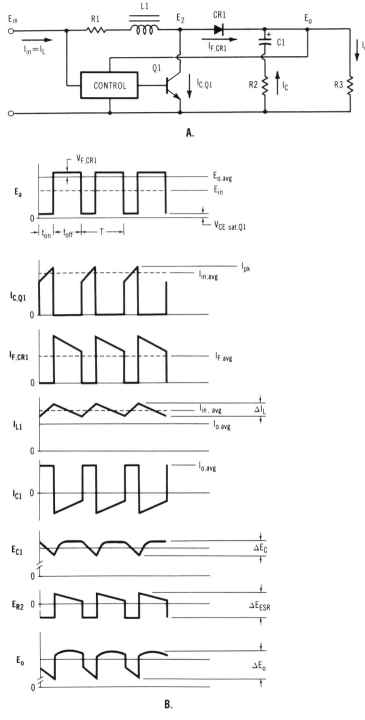

Fig. 8.11. Boost regulator, continuous current mode.

flyback diode, and the load resistance is R_3. Two modes of operation will be discussed: the continuous current mode (where current always flows in L_1 and thus in either Q_1 or CR_1, as shown in Fig. 8.11) and the discontinuous current mode (where current does not flow in either L_1 or Q_1 or CR_1 for a period of time, as shown in Fig. 8.12). The waveforms assume the circuit has already established a steady-state output voltage. Switching of Q_1 is controlled by a PWM IC or discrete components. The PWM control circuit senses a portion of the output voltage and controls the duty cycle or conduction time of the transistor to maintain regulation.

Current sensing and limiting of the transistor current and a soft-start mode are also assumed since these features are desired for device protection purposes. However, caution must be exercised in turning on the regulator. If input voltage is applied with Q_1 off, the output voltage will equal the input voltage as a minimum. In fact, if input voltage is applied when a light load exists on the output (minimum damping), the oscillatory nature of L_1 and C_1 will produce an output voltage approaching $2E_{in}$ (Section 1.6.1.2), which may be higher than the desired output voltage. Another major disadvantage of the boost regulator is lack of output short-circuit protection, since the power transistor is in "shunt" with the load. For this requirement, a fuse could be inserted in series with the input or better yet, an additional transistor could be inserted in series with the input and the transistor turned on and off by command.

8.4.1 Continuous Current Mode

Operating waveforms for a continuous current mode are shown in Fig. 8.11B, and the transfer function is quite different from the buck regulator. When Q_1 turns on, the current flowing in L_1 then flows through Q_1, and CR_1 is reverse biased. The reverse recovery, as well as the forward recovery time of CR_1, is important for the same reasons as discussed in the buck regulator. The current in L_1 increases until Q_1 is turned off, at which time the reversal of voltage across L_1 causes the current which was flowing in Q_1 to flow through CR_1, C_1, and R_3. When Q_1 again turns on, the current through CR_1 ceases and the capacitor must supply the load current. This characteristic causes three major problems. First, a high peak-to-average ratio of current flow in CR_1, and a much higher rms current flows in C_1 resulting in the use of a larger filter capacitor and a larger filter inductor than in the buck regulator. Second, the average output current is less than the average inductor current by the factor $1 - D$. For a dynamic condition where the duty cycle increases in order to increase the output voltage, I_{CR1} will initially decrease and E_{C1} will also initially decrease. Reference Equations (8.14) and (8.15). Third, output stability problems exist due to a right half plane zero in the transfer function, which leads to an additional 90° of phase lag that is additive to the 180° lag from the double pole of the output filter [1]. The inductor current and the rectifier current will finally increase, restoring the filter capacitor voltage and equilibrium. It is also evident that if the duty cycle is allowed to approach 1, say in response to a step increase in load current, I_{CR1} may actually become zero and accelerate the decline in output voltage.

Operation of the boost regulator, operating in the continuous current mode, is discussed by referring to the waveforms of Fig. 8.11B:

$$e_L = L\,di/dt \qquad \Delta I_L = (\Delta I_L^+) + (\Delta I_L^-) = e_L T/L_1$$

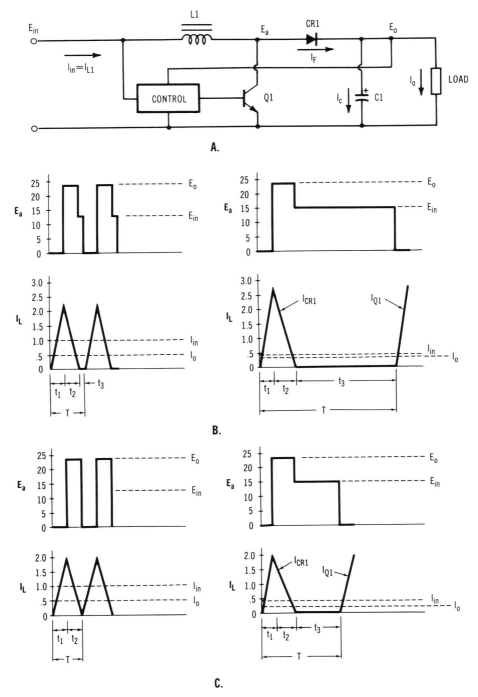

Fig. 8.12. Boost regulator, discontinuous current mode.

where

$$\Delta I_L^+ = E_{in} t_{on}/L_1 \qquad \Delta I_L^- = (E_0 - E_{in})t_{off}/L_1$$

For $\Delta I_L^+ = \Delta I_L^-$ and $E_{in} t_{on} = (E_0 - E_{in})t_{off}$,

$$E_0 = E_{in}(1 + t_{on}/t_{off}) = E_{in}T/t_{off}$$
$$= E_{in}/(1 - D) \qquad (8.14)$$

where duty cycle $D = t_{on}/T = t_{on}f$.

The input current I_{in} is equal to the inductor current I_L, and the output current I_0 is equal to the diode current I_{CR1} plus the capacitor current I_{C1}. Assuming no losses, $E_{in}I_{in} = E_0 I_0$. The ratio of average input to output current is then

$$I_{in} = I_0/(1 - D) \qquad (8.15)$$

The above equations then show the effect of a step change in duty cycle. First, a step increase in E_{in} causes a step increase in duty cycle. But the increase in duty cycle causes I_0 to decrease. Also, a step increase in I_0 causes a step increase in duty cycle. But the increase in duty cycle causes E_0 to increase. This phenomenon produces instability, also indicated by the right half plane zero, as discussed at the beginning of this section. Methods to achieve stability are discussed in Section 11.3.3.

The component values of L_1 and C_1 are calculated in the following manner:

$$t_{on} = \Delta I_L^+/E_{in} \quad \text{and} \quad t_{off} = \Delta I_L^-/(E_0 - E_{in})$$

Then

$$T = t_{on} + t_{off} = L\,\Delta I_L\,E_0/[E_{in}(E_0 - E_{in})] = 1/f$$

and

$$L_1 = E_{in}(E_0 - E_{in})/(E_0 f \Delta I_L) \qquad (8.16a)$$

But ΔI_L has a different relation to I_0 than in the buck regulator since

$$I_{L,pk} = I_{in} + \tfrac{1}{2}\Delta I_L \quad \text{and} \quad I_0 = I_{in}(1 - D)$$

and the factor k will be introduced where

$$k = \Delta I_L/I_{in}$$

Then, since $I_{in} = I_0/(1 - D)$ and $(1 - D) = E_{in}/E_0$,

$$L_1 = E_{in}^2(E_0 - E_{in})/(E_0^2 f I_0 k) \qquad (8.16b)$$

where k typically ranges from 0.05 to 0.4.

The filter capacitor supplies the load when the transistor is on, and $\Delta E_c = I_c t_{on}/C$ and $I_c = I_0$, and $t_{on} = t_{off}(E_0 - E_{in})/E_{in}$, from which

$$C = I_0(E_0 - E_{in})/(f E_0 \Delta E_c) \qquad (8.17)$$

As in the case of the buck regulator, the ESR of C_1 must be considered, where $\Delta E_0 = \sqrt{(\Delta E_c)^2 + (\Delta E_{R2})^2}$ (assuming a sinusoid) and ΔE_c is typically 10% of ΔE_0.

Up to this point, losses have been neglected for simplicity. In actual conditions, the duty cycle will automatically decrease to overcome the losses and to overcome the voltage drops in order to maintain a regulated output. During t_{on}, the inductor voltage is

$$e_L = E_{\text{in}} - V_{CE,\text{sat}} - I_{\text{in}} R_1 = L \, \Delta I_L^+ / t_{\text{on}}$$

During t_{off}, the inductor voltage is

$$e_L = -E_{\text{in}} + I_{\text{in}} R_1 + V_F + E_0 = L \, \Delta I_L^- / t_{\text{off}}$$

Since $\Delta I_L^+ = \Delta I_L^-$,

$$(E_{\text{in}} - V_{CE,\text{sat}} - I_{\text{in}} R_1) t_{\text{on}} = -(E_{\text{in}} - I_{\text{in}} R_1 - V_F - E_0) t_{\text{off}}$$

Since $t_{\text{off}} = T - t_{\text{on}}$, the duty cycle becomes

$$D = (E_0 + V_F + I_{\text{in}} R_1 - E_{\text{in}}) / (E_0 - V_{CE,\text{sat}} + V_F) \tag{8.18}$$

and

$$E_0 = [E_{\text{in}} - I_0 R_1 - V_F + D(-V_{CE,\text{sat}} + V_F)] / (1 - D) \tag{8.19}$$

The minimum input voltage must then be sufficient to allow for the voltage drops without running out of duty cycle while maintaining a regulated output.

The operating efficiency is

$$\eta = P_0 / P_{\text{in}} = P_0 / (P_0 + P_{\text{loss}}) \tag{8.20}$$

The power losses increase significantly as the duty cycle increases. This is primarily due to the increasingly higher ratio of peak current to average current in the power components.

8.4.2   Discontinuous Current Mode

The boost regulator may be operated in the discontinuous current mode where a "dead time" exists when neither the transistor nor the diode conduct and, therefore, the inductor current is also zero during this time. The output filter capacitor then produces a single pole, the right half plane zero (associated with the continuous current mode) is eliminated, and the closed loop is easier to stabilize. The diagram of the boost regulator is repeated in Fig. 8.12, with waveforms for the discontinuous mode.

Two methods of control may be used. First, the on-time of the transistor may be fixed and the frequency varied to maintain a regulated output. The control circuit is then essentially a VCO (voltage-controlled oscillator) with a fixed conduction period of Q_1. Second, the peak current may be fixed and the frequency varied to maintain a regulated output. The control circuit then operates by turning off the transistor when the inductor current reaches a preset value. Both modes of operation are discussed by the following analysis and examples and by referring to the waveforms of Fig. 8.12. Also, ideal components will be assumed for simplicity.

During period t_1, the transistor conducts and current flows through L_1 and Q_1. When the transistor turns off, the current which was flowing in Q_1 now flows through L_1, CR_1, C_1, and the load for period t_2. When the inductor current decays to zero, C_1 supplies current to the load for period t_3. Rather than use the duty cycle ratio (where $D = t_{on}/T$), the individual periods will be analyzed. The total period and operating frequency is

$$T = t_1 + t_2 + t_3 = 1/f \tag{8.21}$$

During t_1, $e_{L1} = L\,di/dt = LI_{pk}/t_1 = E_{in}$.
During t_2, $e_{L1} = L\,di/dt = LI_{pk}/t_2 = (E_0 - E_{in})$.
During t_3, $e_{L1} = $ zero and $I_{C_1} = I_0$.

Since ideal components have been assumed, $E_{in}I_{in} = E_0I_0$. Equating these relations,

$$E_0 = E_{in}(t_1 + t_2)/t_2 \tag{8.22a}$$

$$I_{pk} = E_{in}t_1/L_1 = (E_0 - E_{in})t_2/L_1 \tag{8.22b}$$

$$I_{in} = I_0(t_1 + t_2)/t_2 \tag{8.22c}$$

However, the effect of t_3 must be considered. Also, several unknown quantities exist in the relations, specifically t_1, t_2, L_1, and I_{pk}. The normally known or given quantities are the desired regulated output voltage, the input voltage range, and the output current range.

8.4.2.1 *Fixed On-Time*

For the fixed on-time waveforms shown in Fig. 8.12B, the minimum period (maximum frequency) occurs at minimum input voltage and maximum output current. To ensure discontinuous current in L_1, let t_3 equal 20% of t_1 at $E_{in, min}$ and at $I_{0, max}$, which will allow for device voltage drops. Then $T = 1.2t_1 + t_2$. From the current waveforms (and from Chapter 1, Table 1.2, part B, waveform 10), the average currents are

$$I_{in} = I_{pk}(t_1 + t_2)/2T \tag{8.23a}$$

$$I_{0, max} = I_{pk}t_2/2T \tag{8.23b}$$

Example

It is desired to provide 24 V output at a load current range of 0.25–0.5 A from an input voltage range of 12–15 V. At minimum input voltage and maximum load current, $I_{in} = E_0I_{0, max}/E_{in, min} = 24 \times 0.5/12 = 1$ A. Due to the unknown periods, select a value of 20 μs for t_1. Then $t_3 = 4$ μs and $t_2 = E_{in, min}t_1/(E_0 - E_{in}) = 12 \times 20/(24 - 12) = 20$ μs, $T_{min} = 20 + 20 + 4 = 44$ μs, $f_{max} = 10^6/44 = 22.7$ kHz. The peak current is $I_{pk} = 2I_0T/t_2 = 2 \times 0.5 \times 44/20 = 2.2$ A. The filter inductance is $L_1 = E_{in}t_1/I_{pk} = 12 \times 20/2.2 = 109$ μH. The voltage and current waveforms are shown at the left side in Fig. 8.10B. When the input voltage is a maximum (15 V) and the load current is a minimum (0.25 A), $I_{pk} = E_{in,max}\, t_1/L_1 = 15 \times 20/109 = 2.75$ A. The average input current $I_{in} = E_0 I_0/E_{in} = 24 \times 0.25/15 = 0.4$ A. Also, $t_2 = E_{in}t_1/(E_0 E_{in}) = 15 \times 20/(24 - 15) = 33.3$ μs. The total period $T = I_{pk}t_2/2I_0 = 2.75 \times 33.3/(2 \times 0.25) = 183.3$ μs, and $t_3 = 183.3 - (20 + 33.3) = 88$ μs. The minimum operating frequency is $10^6/183.3 = 5.46$ kHz. The voltage and current waveforms are shown at the right side in Fig. 8.12B.

8.4.2.2 *Fixed Peak Current*

For the fixed peak current waveforms shown in Fig. 8.12C, the minimum period (maximum frequency) again occurs at minimum input voltage and maximum output current. However, since the peak current is fixed, t_3 may be allowed to go to zero.

Example

Equations (8.22) and (8.23) will be used to analyze the preceding example where $E_0 = 24$ V, $E_{in} = 12$–15 V, and $I_0 = 0.25$–0.5 A. At minimum input voltage and maximum output current, let $t_1 = 20$ μs. Then $t_2 = E_{in}t_1(E_0 - E_{in}) = 12 \times 20/(24 - 12) = 20$ μs, $T = 40$ μs, and $f = 10^6/40 = 25$ kHz maximum. Also, $I_{in} = E_0 I_0/E_{in} = 24 \times 0.5/12 = 1$ A. Then, $I_{pk} = 2I_0 T/t_2 = 2 \times 0.5 \times 40/20 = 2$ A. Inductor $L_1 = E_{in}t_1/I_{pk} = 12 \times 20/2 = 120$ μH. The waveforms are shown at the left in Fig. 8.12C. When the input voltage is at maximum (15 V) and the output current is at minimum (0.25 A), $t_1 = I_{pk}L_1/E_{in} = 2 \times 120/15 = 16$ μs, and $t_2 = E_{in}t_1/(E_0 - E_{in}) = 15 \times 16/(24 - 15) = 26.7$ μs. But, $T = I_{pk}t_2/(2I_0) = 2 \times 26.7/(2 \times 0.25) = 106.7$ μs, from which $t_3 = 106.7 - (16 + 26.7) = 64$ μs. The minimum operating frequency is $10^6/106.7 = 9.37$ kHz. Again, the average input current is $I_{in} = E_0 I_0/E_{in} = 24 \times 0.25/15 = 0.4$ A. The waveforms are shown at the right in Fig. 8.12C.

In the discontinuous current mode (either the fixed on-time or the fixed peak current), the output filter capacitance required to limit the output ripple is best analyzed analytically by the relation $i = C\,dv/dt$, and by referring to Fig. 8.12. During t_1 and during t_3, $I_{C1} = I_0$. During t_2, $I_{C1} = I_{L1} - I_0$, and the average inductor current is $\frac{1}{2}I_{pk}$. The filter capacitor voltage change during t_2 is $\Delta E_c = (\frac{1}{2}I_{pk} - I_0)t_2/C_1$. During t_1 and t_3, the capacitor must supply the total output current and $\Delta E_c = I_0(t_1 + t_3)/C_1$. Equating these relations

$$C_1 = \left[t_2\left(\tfrac{1}{2}I_{pk} - I_0\right) + (t_1 + t_3)I_0 \right]/2\,\Delta E_c \qquad (8.24)$$

8.5   BUCK–BOOST REGULATOR

The buck–boost regulator shown in Fig. 8.13 provides an output which may be greater than, or less than, the input voltage, and the output polarity is opposite the input voltage. The switching transistor is Q_1, L_1 is the charging inductor with a dc resistance R_1, C_1 is the filter capacitor with an ESR R_2, CR_1 is the catch or flyback diode, and the load resistance is R_3. Two modes of operation will be

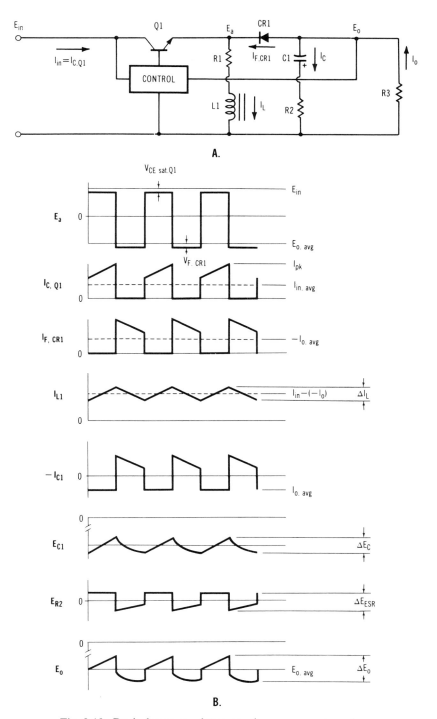

Fig. 8.13. Buck–boost regulator, continuous current mode.

discussed: the continuous current mode (where current always flows in L_1 and thus in either Q_1 or CR_1, as shown in Fig. 8.13) and the discontinuous current mode (where current does not flow in either L_1 or Q_1 or CR_1 for a period of time, as shown in Fig. 8.14). The waveforms assume the circuit has already established a steady-state output voltage. Switching of Q_1 is controlled by a PWM IC or discrete components. The control circuit senses a portion of the output voltage and controls the duty cycle or conduction time of the transistor to maintain regulation.

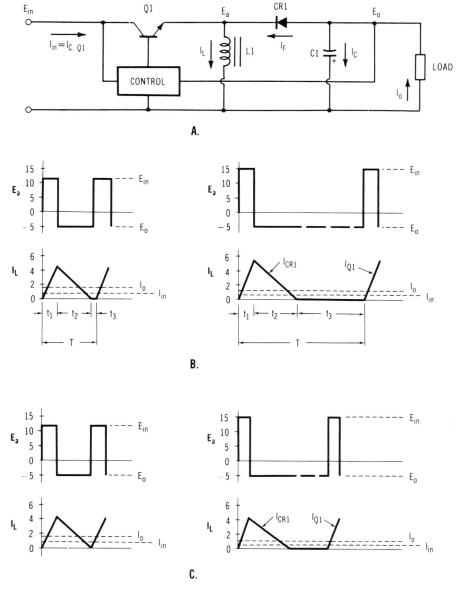

Fig. 8.14. Buck–boost regulator, discontinuous current mode.

Current sensing and limiting of the transistor current and soft-start mode are also assumed, since these features are desired for device protection purposes. When Q_1 turns on, the inductor is charged from the input voltage and the diode current is zero. When Q_1 turns off, the flyback action of L_1 causes the current which was flowing in Q_1 to now flow through C_1, the load, and CR_1. When Q_1 again turns on, the current which was flowing in CR_1 now flows through Q_1, and CR_1 is reverse biased. When operating in the continuous current mode with continuous current in the inductor, this action causes the same three major problems similar to the boost converter of Section 8.4; that is, the high rectifier current, the high capacitor current, and the right half plane zero. The continuous current mode and discontinuous current mode are discussed.

8.5.1 Continuous Current Mode

Referring to Fig. 8.13, operation is analyzed in the following manner:

$$e_L = L\, di/dt \qquad \Delta I_L = \Delta I^+ + \Delta I_L^-$$

where

$$\Delta I_L^+ = E_{in} t_{on}/L_1 \qquad \Delta I_L^- = E_0 t_{off}/L_1$$

For $\Delta I_L^+ = \Delta I_L^-$,

$$E_0 = E_{in} t_{on}/t_{off}$$
$$= E_{in} D/(1 - D) \qquad (8.25)$$

where $t_{off} = T - t_{on}$ and duty cycle $D = t_{on}/T = t_{on} f$. Then

$$T = (L_1 \Delta I_L/E_{in}) + (L_1 \Delta I_L/E_0)$$
$$= \Delta I_L (E_{in} + E_0)/(E_{in} E_0) = 1/f$$

and

$$L_1 = E_{in} E_0/[f \Delta I_L (E_{in} + E_0)] \qquad (8.26)$$

The filter capacitor supplies the load when the transistor is on (as in the case of the boost regulator), and $\Delta E_c = I_c t_{on}$, $I_c = I_0$, and $t_{on} = E_0 t_{off}/E_{in}$, from which

$$C = E_0 I_0/[\Delta E_c f(E_{in} + E_0)] \qquad (8.27)$$

As in the case of the buck regulator, the ESR of C_1 must be considered, where $\Delta E_0 = \sqrt{(\Delta E_c)^2 + (\Delta E_{R2})^2}$ and ΔE_c is typically 10% of ΔE_0.

Up to this point, losses have been neglected for simplicity. In actuality, the duty cycle will automatically increase to overcome the losses, and the voltage drops to maintain a regulated output. During t_{on}, the inductor voltage is

$$e_L = E_{in} - V_{CE,\,sat} - I_0 R_1 = L \Delta I_L^+/t_{on}$$

During t_{off},

$$e_L = E_0 + I_0 R_1 + V_F = L \, \Delta I_L^- / t_{\text{off}}$$

Since $\Delta I_L^+ = \Delta I_L^-$,

$$(E_{\text{in}} - V_{CE,\text{sat}} + E_0 + V_F)t_{\text{on}} = (E_0 + V_F + I_0 R_1)T$$

Then

$$D = (E_0 + V_F + I_0 R_1)/(E_{\text{in}} + E_0 + V_F - V_{CE,\text{sat}}) \qquad (8.28)$$

and

$$E_0 = [V_F + I_0 R_1 - D(E_{\text{in}} + V_F - V_{CE,\text{sat}})]/(1 - D) \qquad (8.29)$$

The minimum input voltage must then be sufficient to allow for the voltage drops without running out of duty cycle while maintaining a regulated output.

The operating efficiency is

$$\eta = P_0/P_{\text{in}} = P_0/(P_0 + P_{\text{loss}}) \qquad (8.30)$$

8.5.2 Discontinuous Current Mode

The buck–boost regulator may be operated in the discontinuous current mode (in a manner similar to the boost regulator) where a "dead time" exists when neither the transistor nor the diode conducts, and, therefore, the inductor current is also zero during this time. The output filter capacitor then produces a single pole, the right half plane zero (associated with the continuous current mode) is eliminated, and the closed loop is easier to stabilize. The diagram of the buck–boost regulator is repeated in Fig. 8.14, with waveforms for the discontinuous mode.

As with the boost regulator, two methods of control may be used. First, the on-time of the transistor may be fixed and the frequency varied to maintain a regulated output. The control circuit is then essentially a VCO with fixed conduction period of Q_1. Second, the peak current may be fixed and the frequency varied to maintain a regulated output. The control circuit then operates by turning off the transistor when the inductor current reaches a preset value. Both modes of operation are discussed by the following analysis and examples, and by referring to the waveforms of Fig. 8.14. Also, ideal components will be assumed for simplicity.

During period t_1, the transistor conducts and current flows through Q_1 and L_1. When the transistor turns off, the current which was flowing in Q_1 now flows through L_1, C_1, the load, and CR_1 for period t_2. When the inductor current decays to zero, C_1 supplies current to the load for period t_3. Rather than use the duty cycle ratio (where $D = t_{\text{on}}/T$), the individual periods will be analyzed. The total period and operating frequency is

$$T = t_1 + t_2 + t_3 = 1/f$$

During t_1, $e_{L1} = L \, di/dt = LI_{\text{pk}}/t_1 = E_{\text{in}}$.
During t_2, $e_{L1} = L \, di/dt = LI_{\text{pk}}/t_2 = E_0$.
During t_3, $E_{L1} = $ zero, and $I_{C1} = I_0$.

Since ideal components have been assumed, $E_{in}I_{in} = E_0 I_0$. Equating these relations,

$$E_0 = E_{in}t_1/t_2 \qquad (8.31a)$$

$$I_{pk} = E_{in}t_1/L_1 = E_0t_2/L_1 \qquad (8.31b)$$

$$I_{in} = I_0t_1/t_2 \qquad (8.31c)$$

However, the effect of t_3 must be considered. Also, several unknown quantities exist in the relations, specifically t_1, t_2, L_1, and I_{pk}. The normally known quantities are the desired regulated output voltage, and input voltage range, and the output current range.

8.5.2.1 *Fixed On-Time*   For the fixed on-time waveforms shown in Fig. 8.14B, the minimum period (maximum frequency) occurs at minimum input voltage and maximum output current. To ensure discontinuous current in L_1, let t_3 equal 20% of t_1 at $E_{in,min}$ and at $I_{0,max}$, which will allow for device voltage drops. Then $T = 1.2t_1 + t_2$. From the current waveforms (and from Chapter 1, Table 1.2, part B, waveform 10), the average currents are

$$I_{in} = I_{pk}t_1/2T \qquad (8.32a)$$

$$I_{0,max} = I_{pk}t_2/2T \qquad (8.32b)$$

Example

It is desired to provide -5 V output at a load current range of 1–1.5 A from an input voltage range of 12–15 V. At minimum input voltage and maximum load, $I_{in} = E_0 I_{0,max}/E_{in,min} = 5 \times 1.5/12 = 0.625$ A. Due to the unknown periods, select a value of 20 μs for t_1. Then $t_3 = 4$ μs, and $t_2 = E_{in,min}t_1/E_0 = 12 \times 20/5 = 48$ μs, $T_{min} = 20 + 48 + 4 = 72$ μs, $f_{max} = 10^6/72 = 13.9$ kHz. The peak current is $I_{pk} = 2I_{in}T/t_1 = 2 \times 0.625/72/20 = 4.5$ A. The filter inductance is $L_1 = E_{in,min}t_1/I_{pk} = 12 \times 20 \times 10^{-6}/4.5 = 53.3$ μH. The voltage and current waveforms are shown at the left side in Fig. 8.14B. When the input voltage is at maximum (15 V) and the load current is at minimum (1 A), $I_{pk} = E_{in,max}t_1/L_1 = 15 \times 20/53.3 = 5.63$ A. The average input current $I_{in} = E_0 I_0/E_{in} = 5 \times 1/15 = 0.333$ A. Also, $t_2 = E_{in}t_1/E_0 = 15 \times 20 \times 10^{-6}/5 = 60$ μs. The total period $T = I_{pk}t_1/(2I_{in}) = 5.63 \times 20 \times 10^{-6}/(2 \times 0.333) = 169$ μs, and $t_3 = 169 - (20 + 60) = 89$ μs. The minimum operating frequency is $10^6/169 = 5.91$ kHz. The voltage and current waveforms are shown at the right side in Fig. 8.14B.

8.5.2.2 *Fixed Peak Current*   For the fixed peak current waveforms shown in Fig. 8.14C, the minimum period (maximum frequency) again occurs at minimum input voltage and maximum output current. However, since the peak current is fixed, t_3 may be allowed to go to zero.

Example

Equations (8.31 and 8.32) will be used to analyze the preceding example where $E_0 = -5$ V, $E_{in} = 12$–15 V, and $I_0 = 1$–1.5 A. At minimum input voltage and maximum output current, let $t_1 = 20$ μs. Then $t_2 = E_{in}t_1/E_0 = 48$ μs, $T = 68$ μs, and $f = 10^6/68 = 14.7$ kHz maximum. Also, $I_{in} = E_0I_0/E_{in} = 5 \times 1.5/12 = 0.625$ A. Then $I_{pk} = 2I_0T/t_2 = 4.25$ A. Inductor $L_1 = E_{in}t_1/I_{pk} = 12 \times 20 \times 10^{-6}/4.25 = 56.6$ μH. The waveforms are shown at the left in Fig. 8.14C. When the input voltage is at maximum (15 V) and the output current is at minimum (1 A), $t_1 = L_1I_{pk}/E_{in} = 56.5 \times 10^{-6} \times 4.25/15 = 16$ μs, and $t_2 = E_{in}t_1/E_0 = 15 \times 16 \times 10^{-6}/5 = 48$ μs. But $T = I_{pk}t_2/2I_0 = 4.25 \times 48 \times 10^{-6}/2 \times 1 = 102$ μs, from which $t_3 = 102 - (16 + 48) = 38$ μs. The minimum operating frequency is $10^6/102 = 9.8$ kHz. Again, the average input current is $I_{in} = E_0I_0/E_{in} = 5 \times 1/15 = 0.333$ A. The waveforms are shown at the right in Fig. 8.14C.

In the discontinuous current mode (either the fixed on-time or the fixed peak current), the output filter capacitance required to limit the output ripple voltage is best analyzed analytically by the relation $i = C\,dv/dt$, and by referring to Fig. 8.14. During t_1 and t_3, $I_c = I_0$. During t_2, $I_c = I_L - I_0$ and the average inductor current is $\frac{1}{2}I_{pk}$. The filter capacitor voltage change during t_2 is $\Delta E_c = (\frac{1}{2}I_{pk} - I_0)t_2/C_1$. During t_1 and t_3, the capacitor must supply the total output current, and $\Delta E_c = I_0(t_1 + t_3)/C_1$. Equating these relations,

$$C_1 = \left[t_2\left(\tfrac{1}{2}I_{pk} - I_0\right) + (t_1 + t_3)I_0 \right]/2\,\Delta E_c \tag{8.33}$$

8.6   PROTECTION FOR SWITCHING REGULATORS

Switching regulators should not fail under normal operating conditions. This may be considered an obvious statement but some designs seem to overlook this requirement and the cause of failure is usually component overstress, either voltage or current, or excessive temperature rise. Switching regulators should also be immune to any type of failure and should remain in a safe mode for the following abnormal conditions: (1) input overvoltage, (2) input undervoltage, (3) input overcurrent, (4) input reverse polarity. The first two conditions are normally met by sensing the input voltage, either at the input terminal or following an input filter, and turning the regulator off if the input exceeds or drops below a safe or preset level. Certain PWM (pulse width modulated) ICs have this feature built in (reference Section 8.2). The third condition may be met by sensing the input current, switching transistor current, or the output current and turning the regulator off if the current exceeds a preset level. The fourth condition may be met by various means described below.

Figure 8.15A shows a typical method of reverse-polarity protection, as well as protection for the on/off switch and the regulator. When S_1 is closed, CR_1 conducts, CR_2 is reverse biased, and voltage is applied to the regulator and to the output load when the regulator turns on. When reverse polarity is applied, CR_2 is forward biased, but CR_1 is reverse biased, and no current flows from the input if the leakage current of CR_1 is considered insignificant. Even if a leakage current exists, CR_2 protects the input filter capacitor and the regulator from reverse voltage. This circuit has one major disadvantage: the forward voltage drop of CR_1 produces a power dissipation in CR_1 which results in decreased operating efficiency. To eliminate the voltage drop, CR_1 may be replaced by a fuse in those applications where fuse replacement is easy. When reverse polarity is applied, CR_2

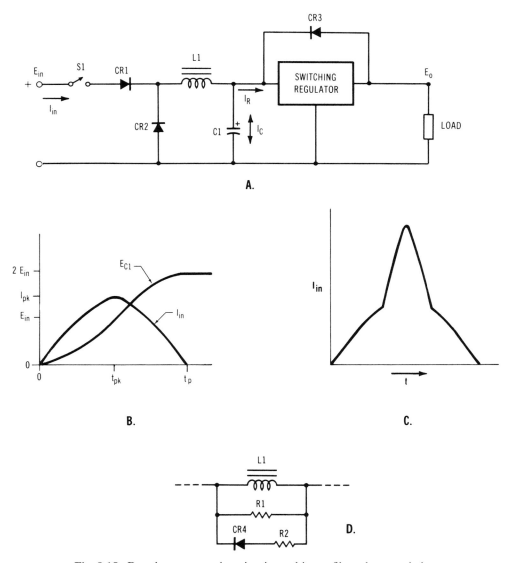

Fig. 8.15. Regulator protection circuits and input filter characteristics.

conducts a high transient current, the fuse opens, and the circuit is protected assuming the $I^2 t$ (ampere-squared second) rating of the diode is sufficiently higher than that of the fuse. For buck and buck–boost regulators, CR_3 may also be required to protect the switching transistor from reverse voltage when S_1 is opened. This is especially true if the input filter energy storage is less than the output filter energy storage, and the regulator is turned off at light load.

In the case of the LC input filter of Fig. 8.15A, CR_2 also serves an additional purpose. With CR_2 omitted, suppose a fault occurs in the regulator under normal operating conditions. The current increase in L_1 produces an energy storage in L_1

such that when S_1 is opened (or an input fuse opens), the resulting induced voltage in L_1, from $e = -L\,di/dt$, is aiding the input voltage, and an abnormally high voltage exists across the switch or fuse. In the case of the switch, severe arcing can occur and, in fact, a sustained arc may develop, depending on the contact clearance of the switch. In the case of the fuse, the induced high voltage may sustain the arc across the "open" fuse. This phenomenon is most prevalent in high-power systems where S_1 may be a circuit breaker, and where the induced voltage may sustain an arc across the breaker if the breaker does not have sufficient voltage rating margin. The insertion of CR_2 serves the same function as the flyback or clamping diode in the buck regulator when the switch is opened; that is, CR_2 serves as a path for the current and clamps the voltage across L_1. The maximum voltage across the switch is then equal to the input voltage plus the forward voltage drop of CR_2. It is also noted that CR_2 should be located at the input to the filter (as shown), not at the input to the regulator. If CR_2 were connected across C_1, the voltage across L_1 would not be clamped when S_1 opens.

8.7 INPUT FILTER FOR SWITCHING REGULATORS

An input filter is normally used to decouple source voltage transients from the switching regulator and to reduce the ripple current drawn from the source. This reduces conducted EMI and improves input susceptibility. The filter usually consists of a single-stage LC inductor input filter, as shown in Fig. 8.15A, although cascade stages may be used in certain applications. A capacitor input filter is not recommended due to the high inrush current when source voltage is applied, either by closure of a switch or relay contact or by the turn-on of an active device.

When the input voltage is applied, a voltage-peaking effect may occur on the input filter capacitor, due to an underdamped condition in the filter, which raises the voltage on the filter capacitor above the capacitor voltage rating. Depending on the dynamic impedance of the filter, a large inrush current may also occur when input voltage is applied. During switching regulator operation, the transfer function of the input filter may react with the transfer function of the switching regulator output filter to produce instability. This latter condition has the tendency to become most prevalent during abrupt changes in line (input) voltage or load (output) current. All of these effects are in conflict with the purposes for which the filter is intended, and compromise the design and performance of the input filter.

Referring to Fig. 8.15A, L_1 and C_1 comprise the input filter. When S_1 is closed, current flows through CR_1, L_1, and C_1. A finite time may elapse before the regulator switching action begins and C_1 may charge to $2E_{in}$ at $t_p = \pi\sqrt{L_1 C_1}$, as discussed in Section 1.6.1.2, and as shown by Fig. 8.15B. The capacitor voltage remains at this level until the regulator starts, since CR_1 is reverse biased. Also, the input current may reach $I_{in,pk} = E_{in}\sqrt{C_1 L_1}$, also shown in Fig. 8.15B. The inductance value of L_1 may be increased to limit the input current surge, but the physical size of L_1 increases. A swinging choke (reference Section 3.11.3) design will reduce the size of L_1, but current peaking will occur when L_1 saturates, as shown in Fig. 8.15C. Critical damping with series or shunt resistors will produce excessive power dissipation in the resistors. An alternate damping method is shown in Fig. 8.15D. If the resistance value of R_1 is small enough, R_1 will carry a portion of the inrush current while preventing L_1 from saturating, and while limiting the inrush current to charge C_1. Then CR_4 and R_2 may be used to clamp the induced voltage, dissipate the excess stored energy in L_1, and prevent an overshoot of the voltage on C_1. During normal operation, the ripple current flowing through R_1 is small, which results in negligible power dissipation.

Voltage and current relations for the regulator during normal operation were previously discussed. For the input filter components, the following analysis applies:

$$I_{\text{in, avg}} = P_0/\eta E_{\text{in}} = I_{R,\text{avg}} \tag{8.34a}$$

$$I_{R,\text{pk}} = P_0/\eta E_{\text{in}} D \tag{8.34b}$$

$$I_C = I_{\text{in}} - I_R = P_0(1/D - 1)(\eta E_{\text{in}}) \tag{8.34c}$$

However, the rms current rating of the capacitor must be observed and

$$
\begin{aligned}
I_{C,\text{rms}} &= \left[\frac{1}{T} \int_0^{t_{\text{on}}} \left(\frac{P_0}{\eta E_{\text{in}}} \right)^2 \left(\frac{t_{\text{off}}}{t_{\text{on}}} \right)^2 dt + \frac{1}{T} \int_{t_{\text{on}}}^T \left(\frac{P_0}{\eta E_{\text{in}}} \right)^2 dt \right]^{1/2} \\
&= \frac{P_0}{\eta E_{\text{in}}} \sqrt{\frac{t_{\text{off}}}{t_{\text{on}}}} \\
&= \frac{P_0}{\eta E_{\text{in}}} \sqrt{\frac{1}{D} - 1} \tag{8.34d}
\end{aligned}
$$

where

η = efficiency

t_{on} = time regulator conducts current

t_{off} = time regulator is nonconducting

$D = t_{\text{on}}/T$ (duty cycle of regulator)

The negative input resistance of switching regulators (as the input voltage increases, the input current decreases) may produce instability when an input filter is placed ahead of the regulator [1, 2]. From the cited references, the suggested design criteria are (1) the input filter cutoff frequency should be lower than the cutoff frequency of the regulator output filter; (2) the input filter output impedance should be smaller than the regulator input impedance. Referring to Fig. 8.15A and the resulting waveforms of Fig. 8.15B, the transient input current pulse width at turn-on is $t_p = \pi\sqrt{L_1 C_1}$ and a double pole exists in the transfer function at the cutoff frequency, $f_0 = 1/2t_p$. It is desired to add a resistor R_1, as shown in Fig. 8.15D, across L_1 to damp the turn-on oscillation and limit the peak current. However, the addition of R_1 will introduce a zero at $f_z = R_1/2\pi L_1$. A resistance value of R_1 low enough to provide critical damping may not significantly reduce the peak current at turn-on and may not provide sufficient attenuation of the input current ripple at the switching frequency of the regulator. A compromise is to allow a slight turn-on oscillation while maintaining filtering from L_1. The frequency domain transfer function with R_1 is

$$\frac{e_0}{e_{in}} = \frac{X_C}{X_L R/(X_L + R) + X_C} \tag{8.35a}$$

where

$$X_L = 2\pi f L_1 \qquad X_C = 1/2\pi f C_1$$

Using the Laplace transform and operator $s = j\omega$ (reference Section 1.8.1),

$$\frac{e_0}{e_{in}} = \frac{sL/R + 1}{s^2 LC + sL/R + 1} \tag{8.35b}$$

$$= \frac{1 + jf/f_1}{1 - (f/f_0)^2 + jf/f_1} \tag{8.35c}$$

where

e_0 = filter output voltage
e_{in} = filter input voltage
f = frequency response, Hz
$f_0 = 1/2\pi\sqrt{L_1 C_1}$
$f_1 = R_1/2\pi L_1$

For the time domain, Equation (8.35b) is of the form

$$(s + b)/\left[(s + a)^2 + \beta^2\right]$$

and the inverse transform is

$$\left[\sqrt{\beta^2 + (a - b)^2}\right]\varepsilon^{-at}\sin(\beta t + \phi)/\beta$$

and

$$E_0 = E_{in}\left[1 - \frac{\sqrt{\beta^2 + (a - b)^2}}{\beta}\varepsilon^{-at}\sin(\beta t + \phi)\right] \tag{8.36}$$

where

$a = 1/2R_1 C_1$
$b = L_1/R_1$
$\beta = \sqrt{(1/L_1 C_1) - (a - b)^2}$
$\phi = \tan^{-1}[\beta/(a - b)]$

Example

It is desired to design an input filter for a switching regulator (reference Fig. 8.15A except CR_1 is deleted) operating from an input of 25 V and providing 15 V output

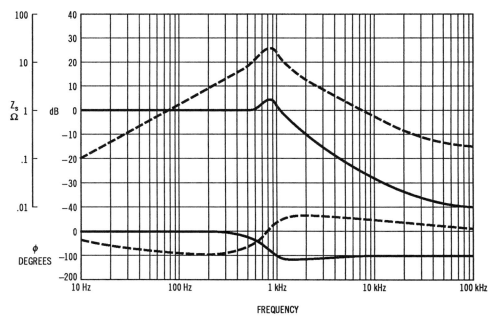

Fig. 8.16. Frequency response of damped input filter.

at 3 A. Neglecting losses (including the dc resistance of L_1 and the ESR of C_1), $P_{in} = P_0 = 45$ W, and $I_{in, avg} = 45/25 = 1.8$ A. The input resistance of the regulator is $R_{in} = 25/1.8 = 13.9$ Ω. The regulator's output filter cutoff frequency is known to be 1200 Hz, and the switching frequency is known to be 50 kHz. Let the input filter cutoff frequency be $f_{ci} = f_{co}/1.5 = 800$ Hz. Let the output impedance of the input filter be $R_{in}/1.5 = 13.9/1.5 = 9.3$ Ω. Then the filter impedance $Z = 9.3$ $\Omega = \sqrt{L_1/C_1}$, from which $L_1 = 9.3^2 C_1$. Also, $\sqrt{L_1 C_1} = 1/2\pi 800$. Equating these relations, $C_1 = 1/(2\pi \times 9.3 \times 800) = 22$ μF, and $L_1 = 22 \times 10^{-6} \times 9.3^2 = 1.9$ mH. A damping resistor will be connected across L_1, and let $R_1 = R_{in} = 13.9$ Ω; use a value of 15 Ω.

Assigning 0.08 Ω for the series resistance of L_1 and 0.16 Ω for the ESR of C_1 (typical value for a CSR-21-style tantalum) produces the frequency response shown in Fig. 8.16. A slight overshoot in the filter response results from the underdamped circuit, even with R_1, but attenuation is provided at 1200 Hz, which is the regulator output filter cutoff frequency. Since the regulator switching frequency is 50 kHz, 35 dB attenuation is provided at this frequency. If R_1 were deleted, the attenuation at this frequency would be approximately 70 dB. To meet the conducted emissions requirements at higher frequencies, an EMI filter should be installed at the input.

The output impedance Z_s of the filter is shown by the dashed line in Fig. 8.16. Again, the peaking of the impedance is minimal due to the damping provided by R_1. The maximum output impedance of the filter is given by [1]

$$Z_{s(max)} = \frac{\dfrac{Z_0^2}{R_s}\sqrt{1 + (R_s/Z_0)^2}}{1 + Z_0^2/R_s R_p} \tag{8.37}$$

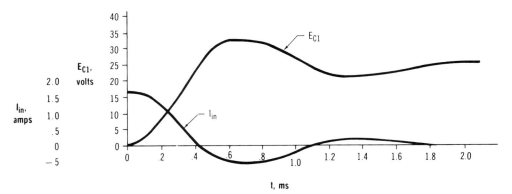

Fig. 8.17. Transient response of damped input filter.

where $Z_0 = \sqrt{L_1/C_1}$, R_s is the series resistance of L_1, and R_p is the parallel damping resistance. This equation clearly shows the effect of R_1 (or R_p) on the maximum filter output impedance. As the resistance increases above Z^2, the denominator decreases significantly and the impedance increases dramatically.

The 1.5 factor used in the above analysis for calculating R_1 and f_{co} is an arbitrary number. This number, however, does provide results that meet the stability and transient criteria.

When a step input voltage is applied to the filter, the capacitor voltage increases as shown in Fig. 8.17. The transient condition produces an overshoot of 8 V at 0.65 ms and the waveform dampens to 25 V in about 2 ms. Again, the waveforms assume that CR_1 of Fig. 8.15A is deleted. With CR_1, the capacitor voltage would remain at 33 V. Also, the instantaneous input current at turn-on is limited to $E_{in}/R_1 = 25/15 = 1.67$ A, which is less than the normal 1.8-A operating current. The input current waveform is also shown in Fig. 8.17. Without R_1, the peak input current would be $I_{in} = E_{in}/Z = 25/9.3 = 2.7$ A, and the peak capacitor voltage would be $V_C = 2E_{in} = 2 \times 25 = 50$ V.

REFERENCES

1. S. Ćuk and R. D. Middlebrook, *Advances in Switched-Mode Power Conversion*, Vols. I and II (Teslaco, Pasadena, CA, 1981).

2. N. O. Sokal, "System Oscillations Caused by Negative Input Resistance at the Power Input Port of a Switching Mode Regulator, Amplifier, DC/DC Converter," 1973 IEEE Power Electronics Specialists Conference Record, pp. 138–140 (IEEE Publication 73 CHO 787–2 AES).

9

DC–DC CONVERTERS

9.0 Introduction
9.1 Topologies
    9.1.1 Free-Running Converters
    9.1.2 Clocked Converters
9.2 Blocking Oscillator
9.3 Square-Wave Converters
    9.3.1 Royer
    9.3.2 Jensen
    9.3.3 Multivibrator
9.4 Flyback Converter
    9.4.1 Low-Voltage Outputs
    9.4.2 High-Voltage Outputs
9.5 Forward Converters
    9.5.1 Single-Ended Converter
    9.5.2 Bridge Converter
9.6 Push–Pull Center-Tapped Converter
9.7 Half-Bridge Converter
9.8 Full-Bridge Transistor Converter
9.9 SCR Converters
9.10 Current-Fed Converters
    9.10.1 Discontinuous Current Mode
    9.10.2 Continuous Current Mode
9.11 Multiple-Output Converters
    9.11.1 Cross-Coupled Filter Inductors
    9.11.2 Current Limiting
    9.11.3 Voltage Sensing
9.12 Input Filters for dc–dc Converters
    9.12.1 Square-Wave Converters
    9.12.2 PWM Converters
References

9.0 INTRODUCTION

This chapter discusses dc–dc converters which, in effect, convert an available dc input voltage to a dc output voltage or multiple output voltages, with isolation between the input and the output. The converters normally utilize high-frequency switching techniques for small transformer size, which in turn provides isolation and voltage scaling. Square-wave inverter stages, as discussed in Chapter 6, may be used, where the secondary voltage of a transformer is rectified and filtered to provide the desired dc output. With the square-wave inverter, the output voltage is not regulated and is a function of input voltage and load current. However, for fixed input voltages and fixed loads the output voltage may well be maintained to a desired voltage tolerance. Where output voltage regulation is required due to input voltage and output load variations, regulation is conveniently achieved by pulse width modulation which controls the conduction period or duty cycle of an active

TABLE 9.1. dc–dc Converter Characteristics

Converter Type	Advantages	Disadvantages
Free running, square wave	Simple; high efficiency; no control circuit required; ideal for fixed-input voltage, fixed-output load	No output voltage regulation
Flyback	Simple; low parts count; inherent output short-circuit protection; wide operating range	Poor transformer utilization; high-output ripple; transistor voltage is fn of transformer turns ratio; above 100 W, other topologies more favorable
Forward	Simple; low parts count; low output ripple; single-ended version may be driven directly from control circuitry	Poor transformer utilization; critical transformer design; transistor voltage is fn of transformer turns ratio; above 500 W, other topologies more favorable
Resonant	Lowest switching losses; reduced switching EMI	Poor transient response; high peak transistor current; high input current at no load
Push–pull	Simple; may be driven from control circuitry; transistor current rated at I_{in}/D	Fair transformer utilization; transistor voltage is $2E_{in}$
Half bridge	Good transformer utilization; transistor voltage is E_{in}; inherent capacitor filter, divider for off-line SMPS	High parts count; transistor current is $2I_{in}/D$
Full bridge	Best transformer utilization; optimum for high-power output; transistor voltage is E_{in}; transistor current is I_{in}/D; SCRs used for very high power	High parts count; requires 4 active devices; dc blocking capacitor has high ripple requirement

device. A dc source may be utilized, but most frequently, a rectified and filtered ac mains input serves as the dc source, as in the case of off-line switch mode power supplies. In the latter case, the converter forms the basic switching and control stage to provide a regulated output. Low-power converters typically employ bipolar transistors or MOSFETs while very high power converters normally employ SCRs as the switching element. The high-power converters imply a desired input voltage that may be several hundred volts in order to restrict the input current to a reasonable value. In this latter case, SCRs or IGBTs are a natural choice, since these devices are available with voltage ratings well over 1000 V.

First, basic topologies of the power-handling elements are discussed with the advantages and disadvantages of each approach. Second, each topology is discussed in detail with design equations and examples. Advantages and disadvantages of each type are presented in Table 9.1.

9.1 TOPOLOGIES

Basic principles of operation of the more popular topologies are presented. Viable converter designs are by no means limited to these configurations. Input–output isolation is provided by a transformer, and multiple outputs can be provided with proper secondary windings. The discussion concentrates on the voltage-fed converter. Current-fed converters are discussed in Section 9.10.

9.1.1 Free-Running Converters

The term free running implies the operation of the converter is controlled by volt-second parameters of magnetic components. The switching frequency and the output voltage vary as a function of input voltage and load current. These converters are ideal for low-power requirements and for operation from a fixed voltage source and into a fixed load.

9.1.1.1 Blocking Oscillator   The basic circuit for a blocking oscillator (frequently referred to as a ringing-choke converter) is shown in Fig. 9.1A, with waveforms in Fig. 9.1B. It is a simple circuit with a minimum of components. When Q_1 turns on, current increases linearly in the primary of T_1 by $i = (1/L)\int e\,dt$. A voltage is induced in N_b to increase base current. When the base current can no longer support collector current, Q_1 turns off and a voltage is induced in the secondary to forward bias CR_1, and the energy stored in N_p while Q_1 was on is transferred to C_1 and the load. The average output voltage $E_0 = E_{in}t_{on}N_s/(t_{off}N_p)$, and the average input current is $I_{in} = I_0 t_{on}N_s/(t_{off}N_p)$. The switching frequency $f_s = 1/(t_{on} + t_{off})$ and varies as a function of E_{in} and E_0.

9.1.1.2 Square-Wave Converter   A square-wave converter with saturating base drive transformer is shown in Fig. 9.2A, with waveforms in Fig. 9.2B. Each transistor conducts for 50% of the period, and the frequency is controlled by the

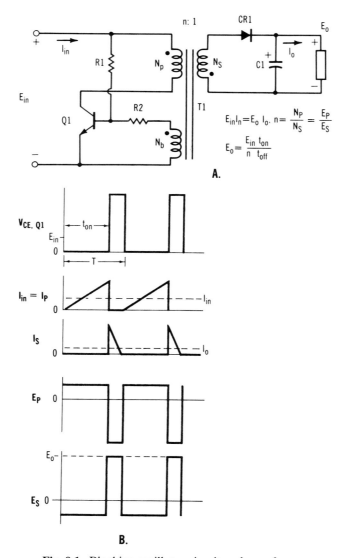

Fig. 9.1. Blocking oscillator circuit and waveforms.

volt-second capacity of T_2. The output voltage is determined by the turns ratio of T_1; $E_0 = E_{in} N_s/N_p$ and $I_{in} = I_0 N_s/N_p$.

9.1.2 Clocked Converters

The term clocked converter implies the switching frequency is controlled. Square-wave operation is seldom used unless the input voltage is fixed, since the output voltage would be proportional to the input voltage. Instead of square waves, output voltage (or current) regulation is achieved by a feedback loop which senses the

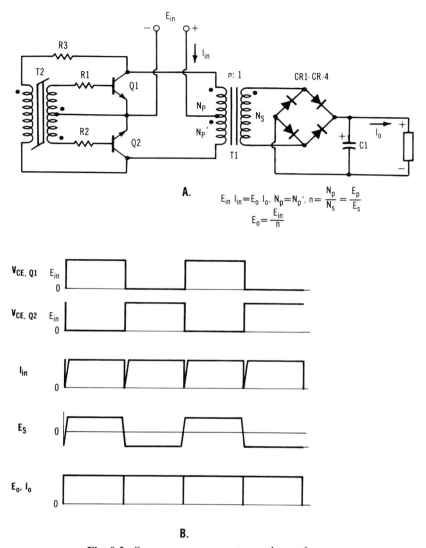

$$E_{in} \ I_{in}=E_0 \ I_0. \quad N_p=N_p'. \quad n=\frac{N_p}{N_s}=\frac{E_p}{E_s}$$

$$E_0=\frac{E_{in}}{n}$$

Fig. 9.2. Square-wave converter and waveforms.

output voltage and controls the duty cycle of the switching stage to maintain regulation.

9.1.2.1   Flyback Converter
The flyback converter of Fig. 9.3 is a voltage isolated version of the buck–boost regulator with T_1 providing isolation as well as voltage scaling. Also, the primary inductance (L_1) of T_1 is determined from the circuit parameters. When Q_1 is on, current flows through the primary of T_1 at a rate determined by $i = (1/L)\int e \, dt$. When Q_1 turns off, the resulting voltage reversal across the primary forward biases CR_1 and the energy stored in the primary inductance is transferred to C_1 and the load. The ratios of input to output voltage

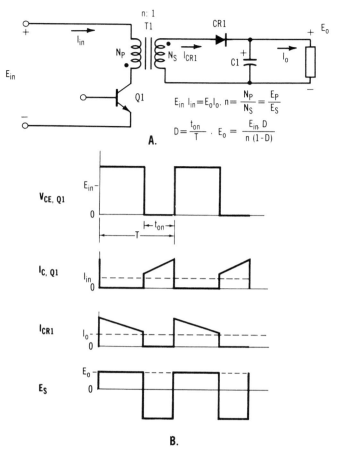

Fig. 9.3. Flyback converter and waveforms.

and current are the same as those of the buck–boost converter of Section 8.5, except for the transformer turns ratio. The average output voltage $E_0 = E_{in} N_s D / [N_p (1 - D)]$, and the average input current $I_{in} = I_0 N_s D / [N_p (1 - D)]$, where $D = t_{on} / T$.

9.1.2.2 *Forward Converter* The forward converter of Fig. 9.4 is a voltage isolated version of the buck converter, with T_1 providing isolation as well as voltage scaling. When Q_1 is on, the voltage across the primary of T_1 is coupled to the secondary and CR_2 is forward biased. Current then flows from the secondary through CR_2, L_1, C_1 and the load. When Q_1 turns off, the induced voltage reverse biases CR_2, and the current which was flowing through CR_2 now flows through CR_3. Also, CR_1 and winding N_c serve important functions. When Q_1 turned off, the primary current was interrupted, and the induced voltage in the primary would result in a high voltage across the transistor. However, the induced primary voltage is coupled

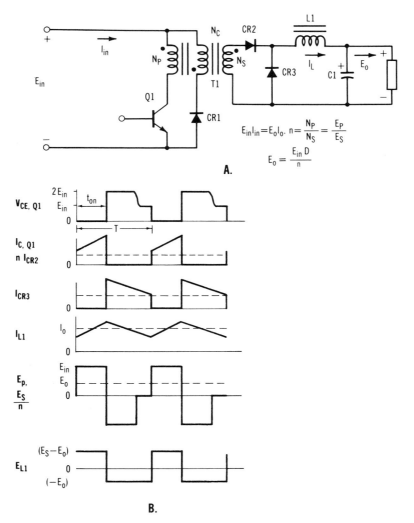

$$E_{in}I_{in} = E_0 I_0. \quad n = \frac{N_P}{N_S} = \frac{E_P}{E_S}$$

$$E_0 = \frac{E_{in} D}{n}$$

A.

B.

Fig. 9.4. Forward converter and waveforms.

to the clamp winding and CR_1 becomes forward biased. The trapped energy in the primary is fed back to the source, and the transistor voltage is limited to $E_{in}[1 + (N_p/N_c)]$, neglecting leakage inductance between N_p and N_c. The ratios of input to output voltage and current are the same as the buck converter except for the turns ratio of the transformer. The average output voltage is $E_0 = E_{in} D N_s / N_p$, and the average input current $I_{in} = I_0 D N_s / N_p$.

An additional transistor and rectifier may be added on the primary side to form a forward bridge converter, as discussed in Section 9.5.2. This approach is very effective for higher voltage inputs, such as rectified and filtered 220 V or 380 V ac input mains. The clamp winding is eliminated, as compared to Fig. 9.4. The ratio of input to output voltage and current are the same as those of Fig. 9.4.

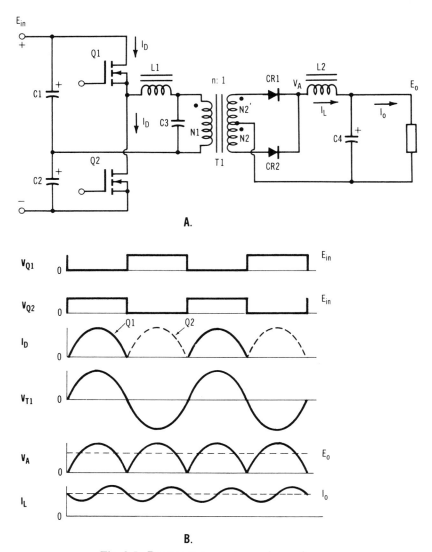

Fig. 9.5. Resonant converter and waveforms.

9.1.2.3 Resonant Converter A resonant converter is shown in Fig. 9.5. Inductor L_1 and C_3 are designed to resonate at a chosen frequency with the power transformer connected as a load across C_3. Transistors Q_1 and Q_2 are clocked to switch at a much higher frequency. When Q_1 turns on, current flows through C_1, Q_1, L_1, C_3, and the primary of T_1. The current is sinusoidal, as shown by the waveforms. When Q_1 is turned off, Q_2 is turned on and sinusoidal current flows through Q_2 to produce the negative half-cycle voltage across T_1, as shown by the waveforms. Since the switching frequency is higher than the resonant frequency, L_1 appears as a high impedance and C_3 appears as a low impedance. This divider ratio produces a voltage across T_1 which is lower than the voltage across C_1. If the switching frequency is decreased, the voltage across T_1 increases. The primary voltage is coupled to the secondary and is full-wave rectified and applied to an LC filter which averages the sinusoidal waveform to a dc output. Thus, regulation can be achieved by frequency modulation as opposed to the pulse width modulation techniques previously discussed.

The major advantage of the resonant converter is in the sinusoidal current flow in the transistors. At turn-on and turn-off, the load current is essentially zero. Thus switching losses are minimal and EMI is significantly reduced. The use of the fast switching MOSFETs eliminates overlap conduction when Q_1 and Q_2 switch. Operating frequencies exceeding 1 MHz are achievable. At these high frequencies, the power transformer size is significantly reduced. The major disadvantages are the high peak sinusoidal current and a relatively high input current at no load. At no load, the output voltage is maintained at the peak of the secondary voltage, and the primary voltage is the secondary voltage times the turns ratio. In effect, the frequency increases by the $\sqrt{2}$, which is the inverse ratio of peak to rms. To provide this primary voltage, current must flow through the reactive elements, L_1 and C_3. The impedance of this network will still allow an appreciable current flow, as compared to other topologies. A short circuit on the output may also present a problem. The short is reflected to the primary and then L_1 must support the input voltage. The current is no longer sinusoidal and the short circuit output current is a function of the current through L_1 times the turns ratio of the transformer. An additional topology of a high-power resonant converter is discussed in Section 9.9.

9.1.2.4 Push–Pull Converters

The term *push–pull* is implied for circuits where two or more power switches conduct on alternate half cycles to apply a symmetrical ac voltage to the primary of a power transformer. These include:

1. Center-tapped primary with two switches.
2. Half-bridge circuit with single primary winding and two switches.
3. Full-bridge circuit with single primary winding and four switches.

However, common terminology for the center-tapped circuit is push–pull, while the other two topologies are referred to as *half bridge* and *full bridge*. These circuits are derived from the conventional buck regulator of Section 8.3.

In the push–pull converter of Fig. 9.6A, Q_1 and Q_2 alternately conduct each half cycle at a duty cycle determined by the input voltage, transformer turns ratio, and desired output voltage. The push–pull approach is effective for lower voltage inputs since one forward switch drop results, and the resulting switch voltage of $2E_{\text{in}}$ at turn-off, or when the opposite switch is conducting, is readily met without the need for high-voltage devices. The average output voltage is $E_0 = 2E_{\text{in}}DN_s/N_p$, where D is the duty cycle. The duty cycle is the ratio of the switch on-time t_{on} to the *total* period T. Thus D is a maximum of 50%. The center-tapped primary provides an additional benefit. The resulting secondary voltage pulse train input to L_1 is twice the basic switching frequency, and the filter requirements are reduced, as compared to the buck regulator of Section 8.3.

The half-bridge converter of Fig. 9.6B is perhaps the most popular topology used in medium-power switch mode power supplies. This reason is threefold. First,

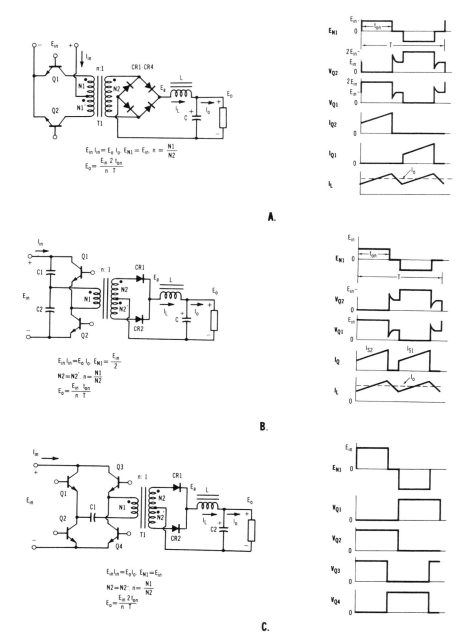

Fig. 9.6. Push–pull converters and waveforms.

capacitors C_1 and C_2 provide the filtering of the rectified input. Second, the series capacitor connection allows full-wave rectification from a 230-V input, and voltage-doubler rectification from a 115-V input without reconnection, as discussed in Section 4.2.5. Third, the maximum switch voltage is limited to the peak input voltage, neglecting leakage inductance in the transformer. The transformer primary voltage is one-half the input voltage, thus the primary current and the switch current are twice the peak input current. The average output voltage is $E_0 = E_{in}DN_s/N_p$, where D is again a maximum of 50%.

For high-power outputs, the full-bridge converter of Fig. 9.6C is an ideal choice. Four switches are required, and the switch voltage is limited to the input voltage, neglecting leakage inductance in the transformer. The switch current is equal to the peak input current but is one-half the current of the half bridge since the primary voltage is equal to the input voltage. Capacitor C_1 serves to provide an ac path and eliminate any dc component in the transformer, which would tend to saturate the transformer or develop unsymmetrical half-cycle conduction periods. The average output voltage is $E_0 = 2E_{in}DN_s/N_p$, where D is again a maximum of 50%.

The waveforms of Fig. 9.6C show Q_1 and Q_2 conduct alternately for a full half cycle. Q_3 and Q_4 also conduct alternately for a full half cycle but are phase displaced from Q_1 and Q_2. This PSM (phase shift modulation) mode is not a specific requirement for nonreactive loads. In most cases, the outputs of a PWM integrated circuit are applied to a single drive transformer stage whereby Q_1 and Q_4 conduct simultaneously for a portion of a half cycle, with Q_2 and Q_3 conducting simultaneously for a portion of the opposite half cycle. For this latter case, a single-drive transformer with four secondaries may be adequate.

9.2   BLOCKING OSCILLATOR

The blocking oscillator is a simple, single-ended flux-switching converter which can be effectively used for low-power requirements. Referring to Fig. 9.1, when Q_1 is turned on by the voltage divider of R_1 and R_2, current I_p increases in N_p of T_1, and a voltage is coupled to N_b to saturate Q_1. Collector current increases until the base current can no longer maintain Q_1 in saturation. The collector current decreases and a voltage is induced in N_b to turn off Q_1. The energy stored in the inductance of N_p is released in the collapsing magnetic field, and the voltage reversal in N_s forward biases CR_1 and current I_s flows through CR_1, C_1, and the load. Waveforms are shown in Fig. 9.1B.

The equation for output voltage given in Fig. 9.1A is deceptively simple. Neglecting losses, the following relations apply.

When Q_1 is conducting, let the peak primary and transistor current be

$$I_{pk} = 2I_{in}T/t_{on} = 2E_0I_0T/(E_{in}t_{on}) \qquad (9.1)$$

The primary inductance is

$$L_p = E_{in}t_{on}/I_{pk} \qquad (9.2)$$

Let the base winding and base resistance be

$$N_b = 2N_p V_{BE}/E_{in} \tag{9.3}$$

$$R_2 = V_{BE}/I_B = V_{BE} h_{FE}/I_{pk} \tag{9.4}$$

When Q_1 turns off, the primary and secondary voltages are

$$E_{p^0} = V_{CE} - E_{in} \quad \text{and} \quad E_{s^0} = E_0 \tag{9.5}$$

where V_{CE} is a selected parameter, depending on desired transistor voltage rating, and

$$n = N_p/N_s = E_{p^0}/E_{s^0} = (V_{CE} - E_{in})/E_0 \tag{9.6}$$

$$t_{off} = L_p I_{pk} N_s/(E_0 N_p) \tag{9.7}$$

Since $T = t_{on} + t_{off}$, summing equations gives

$$t_{on}/T = nE_0/(E_{in} + nE_0)$$

Rearranging the terms,

$$
\begin{aligned}
L_p &= \frac{E_{in}^2 t_{on}^2}{2E_0 I_0 T} \\[2mm]
&= \frac{E_{in}^2 t_{on}^2}{2E_0 I_0 T^2 f} \\[2mm]
&= \frac{E_{in}^2}{2E_0 f}\left[\frac{n^2 E_0^2}{E_{in}^2 + 2nE_0 E_{in} + n^2 E_0^2}\right]
\end{aligned}
\tag{9.8}
$$

Thus, the turns ratios and the primary inductance can be determined from the known voltages, currents, and desired operating frequency. For starting, R_1 and R_2 form a voltage divider and sufficient base voltage must be available for turn-on of Q_1. Since R_2 has been determined,

$$R_1 = R_2(E_{in} - V_{BE})/V_{BE} \tag{9.9}$$

However, the reverse base voltage on Q_1 at turn-off must be considered. Neglecting the effect of R_1,

$$V_{BE} = -(V_{CE} - E_{in})N_b/N_p \tag{9.10}$$

The output capacitor must be capable of storing sufficient energy to supply the load and minimize output ripple when Q_1 is on. The basic equation is

$$C = I_0 t_{on}/\Delta e_0 \tag{9.11}$$

Example

It is desired to provide a -5-V output at 0.3 A from a fixed 12-V input. Using a blocking oscillator, let $f = 33$ kHz, $T = 30$ μs. Let $V_{CE} = 52$ V when Q_1 turns off; use a 100-V rating for the transistor. Then $n = (52 - 12)/5 = 8$; $L_p = [12^2/(2 \times 1.5 \times 33,000)] \times 5^2 \times 8^2/[12^2 + (2 \times 12 \times 5 \times 8) + (5^2 \times 8^2)] = 0.86$ mH; $t_{on}/T = 5 \times 8/52 = 0.77$. So, $t_{on} = 23$ μs and $t_{off} = 7$ μs. Now, $I_{pk} = 12 \times 23 \times 10^6/(0.86 \times 10^{-3}) = .32$ A. Then $I_{in, avg} = 1.5/12 = 0.125$ A. For $I_{pk} = I_C = 0.32$ A, let $I_B = 0.02$ A. For $V_{BE} = 2$ V, $N_b = 2 \times 2N_p/12$, and $R_2 = 2/0.02 = 100$ Ω. For $V_{BE} = 1.5$ V at turn-on, $R_1 = 100(12 - 1.5)/1.5 = 700$ Ω; use 680 Ω. At turn-off, $V_{BE} = -40N_b/N_p$. A core (preferably a square loop, tape wound type) can then be selected and the transformer designed to provide the necessary turns to support the voltage, and the necessary inductance for energy storage in the converter. For an output ripple voltage of 0.1 V, $C_1 = 0.3 \times 23 \times 10^{-6}/0.1 = 69$ μF; use a 10-V rating for the capacitor. For the -5-V output, the positive side of C_1 is connected to the circuit common side of the load. For simplicity, the example assumes no losses. Increasing N_s to 130% of the calculated value and decreasing L to 80% of the calculated value (I_{pk} will increase to 120%) should provide the required output.

9.3   SQUARE-WAVE CONVERTERS

9.3.1   Royer

The Royer or double-ended flux-switching square-wave converter is discussed in Section 6.1.1 and is shown in Fig. 6.2. High-frequency operation and single or multiple dc outputs may be obtained with the corresponding transformer design and with secondary rectifiers and filters. For low-power converters, where the input voltage varies over an appreciable range, the circuit of Fig. 6.3, with secondary rectifiers and a filter capacitor, may be applicable for providing semiregulated outputs. High-frequency switching requires a thin tape wound core with very square loop characteristics to maximize efficiency. When the transformer saturates and the transistor starts to turn off, the peak current may be several times greater than the average input current. The high current spike will cause high conducted EMI emissions unless an appropriate input filter is used.

9.3.2   Jensen

The Jensen or saturating base drive transformer (discussed in Section 6.1.2 and shown in Fig. 6.4) is not limited in basic design for dc–dc converters. In Fig. 9.7A, additional windings are added to nonsaturating transformer T_1, and the voltage developed across these windings opposes the voltage developed across the secondary of T_2. Resistors R_1 and R_2 allow converter starting. Resistors R_3 and R_4 limit the base currents of Q_1 and Q_2. For high-frequency switching, the storage time of the bipolar transistors may be a problem since reverse bias is not applied to the conducting transistor until the collector current begins to decrease. For instance, assume Q_1 is conducting. Base current is $I_B = (E_d - E_f - V_{R3} - V_{R2} - V_{BE})/(R_2 + R_3)$. When T_2 saturates, E_d and I_B decrease to zero but the collector

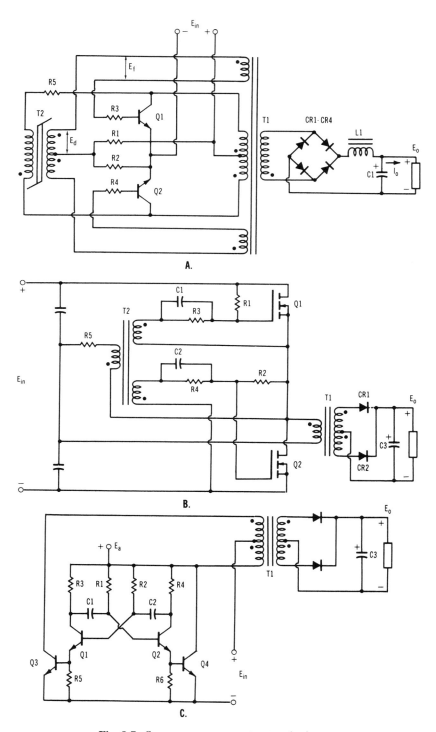

Fig. 9.7. Square-wave converter topologies.

current may continue to flow. But now E_f applies a reverse bias to Q_1 and $V_{BE} = V_{R3} + V_{R2} - E_f$. As discussed in Section 6.1.2, T_2 is designed such that the primary voltage and the voltage across R_5 are each equal to the input voltage. The secondary voltage of T_2 must be sufficient to offset the opposing voltage of T_1 auxiliary winding. The output voltage is a function of the input voltage and the primary to secondary turns ratio of T_1.

Saturating base drive circuitry is not limited to the push–pull topology. Fig. 9.7B is a half-bridge version of Fig. 9.7A, but uses MOSFETs for high-frequency switching. Since the MOSFETS switch very fast and storage time delay does not occur, switching losses may be significantly decreased. Resistors R_1 and R_2 provide converter starting. Initially, say Q_1 turns on and $V_{GS} = E_{in}R_3/[2(R_1 + R_3)]$. Since I_G is negligible, except for charging and discharging C_{GS}, R_3 and R_4 may be large in resistance value as compared to a bipolar circuit. Thus R_1 and R_2 may also be large in resistance value to minimize power dissipation. When switching commences, C_1 and C_2 provide a low impedance to charge and discharge the gate capacitance. Again, the output voltage is proportional to the input voltage and the turns ratio of T_1.

9.3.3 Multivibrator

The RC timing is used in the a stable multivibrator circuit shown in Fig. 9.7C. Operating efficiency is increased and overall size is decreased, compared to the saturating circuits previously discussed. The operating frequency is independent of input voltage and this feature reduces the design requirements of T_1. Transistors Q_1 and Q_2 conduct on alternate half cycles for $t = 0.693R_1C_1$. Since $f = 1/(t_1 + t_2)$, $f_s = 0.72/(R_1C_1)$. Transistors Q_1 and Q_3 (as well as Q_2 and Q_4) form a Darlington stage to decrease the collector current in Q_1 and Q_2. The major disadvantage of the circuit is a possible imbalance in half-cycle conduction times due to differences in transistor characteristics. MOSFETs could be used which would reduce any imbalance and would increase the resistance values to reduce power dissipation.

9.4 FLYBACK CONVERTER

The flyback converter is an ideal topology for power ratings under 100 W. The single-ended circuit is simple, can provide multiple outputs, and has an additional feature for high-voltage outputs, discussed later in Section 9.4.2.

9.4.1 Low-Voltage Outputs

Figure 9.8A shows a triple output flyback converter. Operation is similar to the buck–boost regulator of Section 8.5, except the transformer turns ratio now allows wide variations between input and output voltage. The circuit is controlled by U_1, a PWM IC. Associated components control the oscillator frequency and provide soft-start plus dead time capability. The transistor current is sensed across R_6, and U_1 turns off Q_1 if the current exceeds a preset value. When power is applied to the circuit, current flows through R_1, charging C_1 and providing a voltage to U_1, which

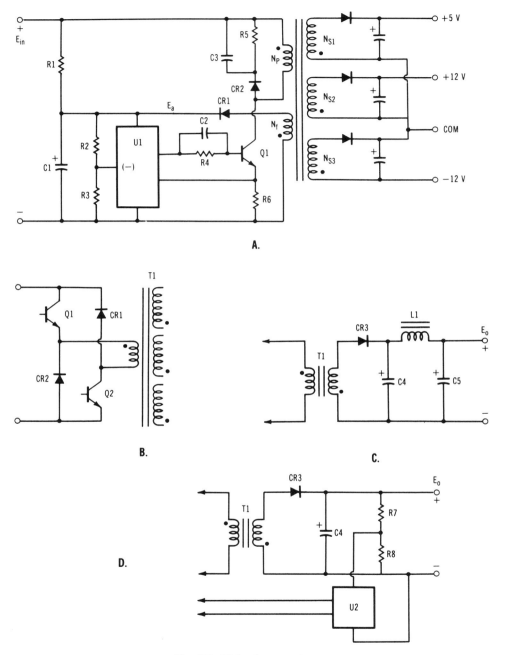

Fig. 9.8. Flyback converters.

then turns on Q_1 and current flows in N_p. When Q_1 turns off, a voltage is induced in the remaining windings and as switching continues, the voltage across N_f then supplies a bootstrap voltage to U_1 and maintains a charge on C_1. This voltage is also the feedback voltage which is applied to sensing divider R_2 and R_3. If the output load currents remain fairly constant, good output voltage regulation is achieved. If the output current changes, the load regulation is primarily a function of IR changes in the transformer secondary windings. However, R_1 must be large enough to allow an adequate load on the feedback winding. If the current through R_1 is I_x, the current from N_f is I_f, and the control circuit current is I_a; then $I_a = I_x + I_f$. Letting $I_x = I_f$:

$$R_1 = 2(E_{in} - E_a)/I_a \qquad (9.12a)$$

$$P_{R1} = (E_{in} - E_a)^2/R_1 \qquad (9.12b)$$

Variations in I_x also have an effect on line regulation. If the input voltage increases, I_x will increase and I_f will decrease, since E_a is constant. The voltage drop across CR_1 decreases, the voltage drop across N_f decreases, and the output voltages increase. Line regulation is improved by feedforward circuitry to U_1.

In the switching mode, the following voltage relations apply. When Q_1 is on,

$$E_p = E_{in} - V_{CE,sat} = E_f N_p/N_f = E_s N_p/N_s$$

$$V_{RRM} \text{ (diode reverse voltages)} = E_a + E_f = E_0 + E_s$$

When Q_1 turns off,

$$E_{p^0} = V_{CE,max} - E_{in} = (E_a + V_F)N_p/N_f = (E_0 + V_F)N_p/N_s$$

For the continuous current mode, the transformer core is not reset while Q_1 is off, and when Q_1 turns on, the current which was flowing in N_s is reflected to N_p and then flows in Q_1. A step increase in current occurs until the core is reset and then the current increases linearly. This produces the trapezoidal current waveform in Fig. 9.3B. This mode produces a higher power capability without increasing I_{pk}. Equations governing operation are the same as those developed in the buck–boost regulator analysis of Section 8.5, except for the turns ratios of T_1. The following relations apply for $P_0 = P_{in} = E_0 I_0 = E_{in} I_{in}$:

$$E_0 = E_{in}(N_s/N_p)D/(1 - D) \qquad (9.13)$$

$$I_{in} = I_0(N_s/N_p)D/(1 - D) \qquad (9.14)$$

$$\varepsilon = \tfrac{1}{2}LI^2 = \tfrac{1}{2}L(I_{pk}^2 - I_{min}^2) = P_{in}T \qquad (9.15)$$

Equating, for $\Delta I^2 = \tfrac{1}{2}(I_{pk}^2 - I_{min}^2)$

$$L = E_{in}t_{on}/\Delta I = P_{in}T/\Delta I^2$$

Then

$$\Delta I = P_{in}T/(E_{in}t_{on}) = I_{in}/D \tag{9.16a}$$

$$I_{pk} = I_{in}/D + \tfrac{1}{2}\Delta I = 1.5I_{in}/D \tag{9.16b}$$

$$I_{min} = I_{in}/D - \tfrac{1}{2}\Delta I = 0.5I_{in}/D \tag{9.16c}$$

Substituting for the above relations,

$$L = P_{in}T/\Delta I^2 = D^2 P_{in}T/I_{in}^2 = D^2 E_{in}T/I_{in}$$

Multiplying by E_{in}, the required inductance can be calculated independent of the current:

$$L = (DE_{in})^2 T/P_{in} \tag{9.17}$$

For a particular requirement, E_{in}, E_0, P_0, and P_{in} are usually known quantities. Choosing a desired switching frequency $(1/T)$ and a duty cycle, the inductance value, the peak current, and the turns ratio of the transformer may be calculated, as well as other desired parameters. The transformer primary normally requires a low inductance value and a high current capability. This implies a low permeability core material should be used, which is also more cost effective than high permeability materials. The high permeability materials require a larger air gap to achieve the same low inductance.

As opposed to the discontinuous current mode, the continuous current mode can provide higher power without increasing the peak current, but the primary inductance of T_1 must increase to increase the stored energy for the same operating frequency.

The output filter capacitor must supply the load current during the time Q_1 is on and

$$C = I_0 t_{on}/\Delta e_0 \tag{9.18}$$

This capacitance does not include the effect of ripple voltage caused by the capacitor ripple current flowing in the capacitor ESR. For typical circuits, the required capacitance may be five times the value calculated above.

In Fig. 9.8A, the transistor voltage is equal to the input voltage plus the induced voltage across the primary of T_1. For higher values of input voltage, where the voltage required for Q_1 would be excessive, the circuit of Fig. 9.8B could be used. Both transistors are switched in-phase. When Q_1 and Q_2 are on, current flows in the primary to store energy. When Q_1 and Q_2 are turned off, the induced voltage is clamped by CR_1 and CR_2 which provide a current path to the source, and the transistor voltage is limited to the input voltage.

Figure 9.8C shows a filter inductor and an additional filter capacitor added which significantly reduces the total capacitance for the same output voltage ripple. Now

$$\Delta e_0 = \Delta e_{C4} X_{C5}/(X_{L1} + X_{C5}) \tag{9.19}$$

Example

A converter output of 12 V at 3 A has a ripple requirement of 50 mV maximum. The switching frequency is 25 kHz and the duty cycle is 50%. Then, $t_{on} = 20 \mu s$. From Equation (9.11), $C_4 = 3 \times 20 \times 20^{-6}/0.05 = 1200 \mu F$. With L_1 and C_5 added, let $\Delta e_{C4} = 0.2$ V. For X_{C5} equal to 10% of the load resistance, $X_{C5} = 0.1 \times 12/3 = 0.4 \Omega$. Capacitor $C_5 = 1/(2\pi 25,000 \times 0.4) = 16 \mu F$. Now $C_4 = 1200 \times 0.05/0.2 = 300 \mu F$. From Equation (9.19), $X_{L1} = X_{C5}(\Delta e_{C4}/\Delta e_{C5} - 1) = 0.4(0.2/0.05 - 1) = 1.2 \Omega$. Then, $L_1 = 1.2/(2\pi 25,000) = 8 \mu H$. Even though C_4 is reduced by a factor of 4, the capacitor ripple current will increase (as is the case with pi filters) and must not exceed the capacitor rating.

9.4.2   High-Voltage Outputs

The flyback converter has one major advantage over other converter topologies when high-voltage outputs are required. This advantage is the inherent overcurrent protection of the primary switching transistor, independent of what happens in the load or secondary, and is described as follows. The high-voltage output may be subject to momentary arcing, especially in vacuum tubes such as CRTs and TWTs (traveling-wave tubes), even though the tube has undergone a substantial period of processing during manufacture. Figure 9.8D shows a typical output stage with voltage divider R_7 and R_8 supplying a sense voltage to U_2, which is a sense regulator and optical isolator. The isolated feedback voltage is then applied to U_1.

During the on period of Q_1, CR_3 is reverse biased. If an arc or momentary short occurs in the load (vacuum tube), C_4 discharges but CR_3 does not conduct and the current in Q_1 continues at the normal linear increase. Thus Q_1 is oblivious to momentary output shorts or arcs if they are of short duration. Compare this feature to other topologies such as the forward, push–pull, or bridge converter where an output short during the conduction period of the power transistor is reflected directly as a short across the primary of the power transformer. (An output filter inductor will provide a momentary impedance in these topologies.) The collector current rapidly increases, requiring fast current sensing and shut off of the power transistor to prevent failure due to forward-biased second breakdown.

For the flyback converter, an output short and resulting output voltage decrease actually decreases the transistor voltage at turn-off. The feedback loop then increases the duty cycle (after the short disappears) until output regulation is again achieved. A typical output power requirement in the case of the TWT is 80 W (8 kV at 10 mA), which is within the ideal power range of the flyback converter.

Flyback converters are ideal for energy storage capacitor charging. This applies to high-energy pulse applications of lasers, lamp flashers, and capacitor discharge circuits. The discontinuous current mode is desired to ensure reset of the core after each pulse. For charging a load capacitor, the following applies.

For a 50% duty cycle triangle waveform in the primary,

$$I_{pk} = 4I_{in} = 4P_{in}/E_{in, min} \tag{9.20}$$

$$P_{in} = P_0/\eta \tag{9.21}$$

The output power delivered to the load is

$$P_0 = \tfrac{1}{2}C_0(E_0)^2/\Delta t_c \qquad (9.22)$$

The conduction period of the transistor is

$$t_{\text{on}} = 2TP_0/\eta E_{\text{in}} I_{\text{pk}} \qquad (9.23)$$

The required primary inductance is

$$L = E_{\text{in}} t_{\text{on}}/I_{\text{pk}} \qquad (9.24)$$

The maximum turns ratio and transistor voltage is

$$N_s/N_p = (E_0/E_{\text{in}})(1 - D)/D \qquad (9.25)$$

$$V_{CE,\,\text{max}} = E_{\text{in}} + E_0(N_p/N_s) \qquad (9.26)$$

where I_{in}, E_{in}, and E_0 are average values and

$$I_{\text{pk}} = \text{peak current in the primary}$$
$$\Delta t_c = \text{output capacitor charge time}$$
$$T = \text{switching period,} = 1/f$$
$$\eta = \text{efficiency}$$
$$D = t_{\text{on}}/T$$

Example

It is desired to use a flyback converter to charge an external 100-μF capacitor to 800 V in 1 s. The capacitor will then be discharged into a load after which the capacitor will again be charged. The cycle continues at a 1-Hz rate. The input is a rectified ac input and the dc input to the converter is 150 V nominal. Here $P_0 = \tfrac{1}{2} \times 10^{-4} \times 800^2/1 = 32$ W. For $\eta = 71\%$, $P_{\text{in}} = 32/0.71 = 45$ W. Then $I_{\text{in}} = 45/150 = 0.3$ A, $I_{\text{pk}} = 4 \times 0.3 = 1.2$ A. Let $T = 120$ μs (8.33 kHz), then $t_{\text{on}} = 2 \times 1 \times 120 \times 10^{-6} \times 45/(150 \times 1.2) = 60$ μs. Inductance $L = 150 \times 60 \times 10^{-6}/1.2 = 7.5$ mH. The turns ratio $N_s/N_p = (800/150)(0.5/0.5) = 5.33$; use 5. Then, $V_{CE,\,\text{max}} = 150 + (800/5) = 310$ V; use a 500-V transistor with a current capability of 2 A minimum. For the transformer, use a low permeability core material and a size capable of 50 W at 8 kHz. The secondary rectifier peak current is $I_{\text{pk}}(N_p/N_s) = 1.2/5 = 0.24$ A. The peak rectifier reverse voltage is $E_0 + E_{\text{in}}(N_s/N_p) = 800 + 750 = 1550$ V. Select a rectifier with a reverse voltage rating of 2.5 kV and an average current rating of 100 mA. Also of note is the stored energy in the primary. $\varepsilon = \tfrac{1}{2}L(I_{\text{pk}})^2 = 0.5 \times 0.0075 \times 1.2^2 = 5.4$ mJ. This energy is stored during each pulse and the number of pulses in a 1-s period is 8333. For a check, the total input power is $P = \varepsilon f = 0.0054 \times 8333 = 45$ W which is the same value as previously calculated.

9.5   FORWARD CONVERTERS

Two forward converter topologies are shown in Fig. 9.9. Although the power waveforms of Fig. 9.4 still apply, the converters in Fig. 9.9 are more detailed for circuit operation.

9.5.1   Single-Ended Converter

The single-ended, dual-output version of Fig. 9.9A uses a power MOSFET (Q_1) which is controlled by PWM IC (U_1). Resistor R_1 provides power to U_1 and the

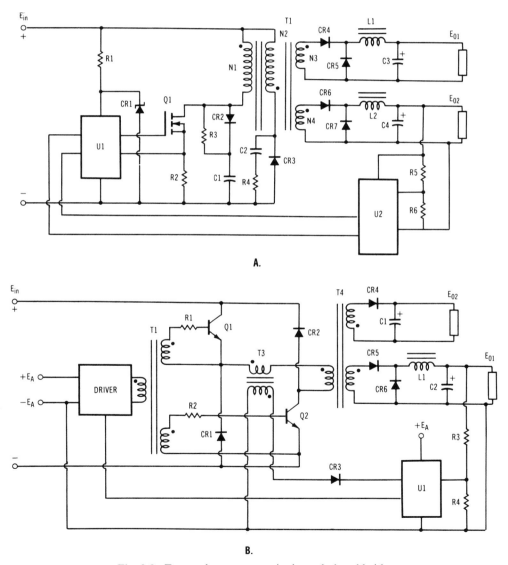

Fig. 9.9. Forward converters, single-ended and bridge.

control voltage is regulated by CR_1. Snubbers (reference Section 13.4) are used across Q_1 and CR_3 and consist of R_3, CR_2, C_1, C_2, and R_4. When Q_1 is turned on, the input voltage is applied to N_1 and a voltage is coupled to N_4 (N_3) which forward biases CR_6 (CR_4). Current flows through L_2 (L_1), C_4 (C_3) and the load, and output voltage E_{02} (E_{01}) is developed across the load(s). Here E_{02} is sensed by divider R_5 and R_6 and applied to U_2, which is an amplifier and optical isolator. The feedback voltage to U_1 controls the on-time or duty cycle of Q_1 to provide a regulated output, after a few cycles of operation following a soft-start mode. The current through Q_1 is sensed by R_2 for current limiting.

When Q_1 turns off, the stored energy in the primary inductance of T_1 produces a voltage across N_2 which forward biases CR_3, and the trapped energy is transferred back to the source. Also, the current which was flowing in CR_6 (CR_4) and L_2 (L_1) now flows through CR_7 (CR_5) in the same manner as a buck regulator. The transistor voltage at turn-off is

$$V_{DS} = E_{in}[1 + (N_1/N_2)] \tag{9.27}$$

Since Q_1 switches extremely fast, a voltage spike may be produced by the leakage inductance in the primary and the high di/dt. Close coupling between N_1 and N_2 is desired and normally N_1 and N_2 are wound bifilar to maximize the mutual coupling. This results in a transistor voltage of $2E_{in}$. However for low-voltage inputs and high-voltage MOSFET, a quadfilar winding could be wound. One winding is the primary and the other three windings, of equal but smaller diameter wire, are connected in series aiding. This produces a transistor voltage of $4E_{in}$. Also, the snubber aids in load line shaping and absorbs some of the power which would otherwise be dissipated in Q_1.

Voltage and current relations are the same as the buck regulator (reference Section 8.3) except for the turns ratio of T_1 and the maximum duty cycle D, which is limited to 50% in order to reset the core.

The following relations apply:

$$E_0 = DE_{in}N_4/N_1 \tag{9.28}$$

$$I_{in} = DI_0N_4/N_1 \tag{9.29}$$

The output filters (L_2 and C_4) are selected by the methods of Section 8.3. The power transformer core material is typically ferrite or MPP (molybdenum Permalloy powder). Since T_1 acts as a coupled transformer (as opposed to the inductive action in the flyback converter), higher permeability core materials are used and minimum air gap is desired. The high permeability results in very low magnetizing current in the transformer.

Example

A power transformer is required for a forward converter with the following characteristics: $E_{in} = 150$ V; $E_0 = 12$ V; $I_0 = 15$ A; desired efficiency = 80%. Choose $f_s = 100$ kHz; $D = 40\%$ nominal; use a ferrite core. From Fig. 3.32, an effective core volume of 6.5 cm^3 would be adequate for a push–pull converter delivering 180 W at 100 kHz. For the forward converter, V_e must be increased due to the decreased flux swing, by approximately 40%. Then $V_e = 6.5 \times 1.4 = 9.1$ cm³, which is close to a PQ32/20 or an EC41. Choosing the EC41, which has a core area of 1.21 cm², and operating at a flux density of 1.2 kG, the primary turns are

$$N_1 = E_{in} \times 10^8/(4B_{ac}A_e f)$$
$$= 150 \times 10^8/(4 \times 1200 \times 1.21 \times 10^5) = 25.8$$

Use 26 turns for N_1. The average input primary current is 180 W/(150 × 0.8) = 1.5 A. The approximate rms current during t_{on} is $I = 1.5/0.4 = 3.75$ A, and the rms current for one cycle is $I = 3.75\sqrt{0.4} = 2.4$ A. At 450 cmil/A, use AWG 20. For simplicity, use 26 turns AWG 20 for N_2, and wind N_1 and N_2 bifilar. From Equation (9.28). $N_s = E_0 N_p/(DE_{in}) = 12 \times 26/(0.4 \times 150) = 5.2$; use 6 turns which will allow a 1.85-V drop in the rectifier and filter choke. The approximate rms current in the secondary is $15/\sqrt{0.4} = 24$ A. This is 10 times the primary rms current. To minimize skin effects, wind the 6 turns with 10 strands of AWG 20. (With regard to skin effects, a smaller wire diameter than AWG 20 might be required in the actual application to reduce R_{eff}, reference Sections 3.6.5 and 3.7.5.) The total number of turns of AWG 20 is then 26 + 26 + (10 × 6) = 112. From Table 3.8, AWG 20 will wind with 850 turns/in.². The required bobbin area is 112/850 = 0.132 in², and the EC41 bobbin area is 0.225 in.², which should allow sufficient area for insulation and margin. The turns per layer and number of layers could also be calculated for winding ease.

The EC41 bobbin average mean turn length = 0.2 ft, and the resistance of AWG 20 = 10 Ω/1000 ft. The resistance of the primary and clamp winding each = $26T \times 0.2$ ft/$T \times 10$ Ω/1000 ft = 0.052 Ω each. The resistance of the secondary = (6 × 0.2/10) × 10/1000 = 0.0012 Ω. Actually, if the primary is wound first, the primary resistance will be less than, and the secondary resistance will be greater than, the values calculated. The copper loss in the primary = 0.052 × 2.4² = 0.3 W, and the copper loss in the clamp will be approximately one-third this value. The copper loss in the secondary = 0.0012 × 24² = 0.69 W. Then P_{cu} = 0.3 + 0.1 + 0.69 = 1.1 W. From typical core loss curves, P_c = 0.2 W/cm³ (at 100 kHz and 1.2 kG) = 0.2 × 10.8 = 2.2 W. The decreased flux swing in the forward converter will decrease the eddy current and hysteresis losses, and the actual core losses will probably be reduced by one-third. Thus the total loss in the transformer is 1.1 + 1.5 = 2.6 W. The total surface area of the transformer is 9 in.² approximately. From Fig. 3.24, at 2.6/9 = 0.29 W/in.² power density, the temperature rise = 37° C in a 25° C ambient.

Further calculations may be made to determine the primary inductance and the magnetizing current, by methods discussed in Chapter 3.

9.5.2   Bridge Converter

The bridge converter of Fig. 9.9B uses two bipolar power transistors and two clamp rectifiers. The circuit is also referred to as a double-ended topology, since only two active devices are used. This topology is effective for voltage inputs higher than that of the single-ended version of Fig. 9.9A, especially where 380-V three-phase mains are rectified and filtered. In this case, the transistor voltage is limited to the rectified dc voltage, which is 515 V nominal, excluding transients. If the single-ended converter were used with a 1:1 turns ratio between the primary and clamp winding, the transistor voltage rating would be 1030 V minimum. For high-power outputs, additional transistors may be paralleled to conduct the desired current.

Assume Q_1 and Q_2 are on and current is flowing through the primary of T_4. The voltage across the primary is $E_{in} - 2V_{CE, sat}$. This voltage is coupled to the secondary which forward biases CR_5 and current flows through the output filter and the load. When Q_1 and Q_2 are turned off, the current which was flowing through T_4 must have a path to prevent excessive voltage spikes. This current now flows through CR_2, back to the source, and through CR_1. Thus the clamp or demagnetizing winding required in the single-ended converter is eliminated in the bridge circuit.

An isolated drive for the transistors is supplied by T_1 which is switched by an appropriate driver stage. This transformer isolation allows the PWM IC, U_1, to be connected to the output return. Here E_A is an auxiliary voltage for control and drive power. Transformer T_3 provides transistor current sensing, and U_1 will turn off the transistors if the current exceeds a preset value.

Calculations of required performance for output E_{01} are the same as given in Section 9.5.1. Also shown in Fig. 9.9B is an additional winding on T_4 which is connected in the flyback mode to supply E_{02} from output filter C_1. Since the converter duty cycle is 50% maximum, E_{02} will be less than $E_{in}N_s/N_p$, and the turns ratio may be selected accordingly. The load regulation of this output is poor, since the regulating loop is closed around E_{01}. However, for low output power, a series regulator may be added to regulate E_{02}. In fact, a fourth winding, additional rectifier, and filter capacitor and a second series regulator could be added to provide a triple output of say +5 V at E_{01} and ±12 V at outputs E_{02} and E_{03}.

9.6 PUSH–PULL CENTER-TAPPED CONVERTER

The push–pull center-tapped converter of Fig. 9.6A is ideal for low voltage, as well as moderate voltage, inputs where the transistor voltage ratings are greater than twice the maximum input voltage. Only one forward drop results on the primary side, which is important at a 12-V input, and especially for converting a 5-V input to a desired output voltage. Figure 9.6A shows a full-bridge rectifier on the output, as would be used if the desired output is several hundred volts. Each half of the primary winding consists of equal turns, and the primary is usually wound bifilar, for convenience as well as coupling. It is understood that reverse rectifiers are connected across each transistor in the figure shown. When Q_2 turns off, the induced voltage in N_1 will forward bias the rectifier across Q_1, and a current path is provided for the energy stored in the leakage inductance of the transformer primary.

Parameters governing operation are essentially the same as those of the buck converter of Section 8.3, except for the transformer and the output filter design. The output voltage may be greater than or less than the input, since the transformer provides scaling as well as isolation. The output voltage is

$$E_0 = 2E_{in}t_{on}/nT \qquad (9.30a)$$

where t_{on} is the conduction time of each transistor, $T = $ *total* period time, and $n = N_1/N_2 = N_p/N_s$.

For convenience, the duty cycle is normally referred to as the period of conduction each half cycle (D is a maximum of 50%) and then

$$E_0 = D'E_{in}/n = D'E_s \qquad (9.30b)$$

where

$$D' = 2t_{on}/T = t_{on}/\tfrac{1}{2}T \qquad E_s = E_{in}/n.$$

Considering voltage drops in the semiconductors, the following relations apply:

$$E_0 D'(E_s - V_F) = D'(E_p/n - V_F)$$
$$= D'[(E_{in} - V_{sat})/n - V_F]$$
$$E_0 = (E_{in} - V_{sat} - nV_F)D'/n \qquad (9.30c)$$

Since the input voltage range and the desired output voltage are usually known, a maximum duty cycle at minimum input is selected, and the transformer turns ratio is calculated by

$$n = D'(E_{in} - V_{sat})/(E_0 + D'V_F) \qquad (9.31)$$

where D' is the duty cycle (per half cycle), V_{sat} is the transistor saturation voltage, and V_F is the rectifier forward voltage drop.

The average input current and the peak transistor currents are

$$I_{in, avg} = DI_0/n \qquad (9.32a)$$
$$I_{in, pk} = (2I_0 + \Delta I_L)/2n \qquad (9.32b)$$

The flux swing in the transformer varies from B^+ to B^-, which affords maximum utilization of the transformer. The transformer may be designed to support a constant volt-second product since the duty cycle will decrease when the input voltage increases. The output filter components will be smaller than those of the buck regulator, output power being equal, since the switching frequency applied to the filter is twice that of the buck regulator. The following relations for the filter components apply:

$$L_1 = E_0(E_s - E_0)/2f_s \Delta I_L E_s$$
$$= E_0(E_{in} - nE_0)/2f_s E_{in} \Delta I_L \qquad (9.33)$$
$$C_1 = \Delta I_L/16f_s \Delta e_C \qquad (9.34)$$

where f_s is switching frequency in hertz, ΔI_L is the peak-to-peak current in L_1, and Δe_C is the peak-to-peak voltage across C_1.

The control circuitry follows that of other converters, and the transistors may be driven from a PWM IC or driver stage without transformer isolation. The control circuit is then common to the input return, which allows ease of starting, since the control circuit can be powered directly from the input. This also allows ease of current sensing and current limiting with a resistor in the return line, and the positive side of the resistor connected to the PWM current sense input. An overload or short-circuit condition on the output will be reflected to the primary which will increase the input current; the PWM will turn off the conducting transistor if the current exceeds a preset level.

For high input currents which exceed the rating of a single transistor, paralleled transistors or Darlington transistors may be used. However, current sharing becomes a problem in paralleled transistors, and emitter balancing resistors would dissipate excessive power. For high-power circuits, the coupled inductor (T_1) shown in Fig. 9.10A could be used to balance the currents through Q_1 and Q_2. An additional coupled inductor (T_2) is also connected in the emitters of the opposite pair of transistors. Inductors T_1 and T_2 are in effect interphase transformers which force current sharing. If the current in each half of the inductor is equal, the ampere-turns cancel and no voltage is developed across the winding, except for the IR drop in the dc resistance. If the current in Q_1 is greater than the current in Q_2, the voltage developed across winding A induces a voltage in winding B, which is of opposite polarity and in series aiding with the input voltage. The voltage across Q_2 must then decrease, which allows the current in Q_2 to increase and again achieve current sharing.

To further increase the efficiency, the current-sensing resistor may be replaced by a current transformer (T_3), as shown in Fig. 9.10A. The center-tapped secondary is first wound on a toroid core. In final assembly, each of the power leads passes through the core, in opposite directions, to form a one-turn primary. Since the ampere-turns are equal, the voltage developed in the secondary is rectified and a burden resistor is used to scale the voltage, as discussed in Chapter 3. This voltage is then applied to the PWM current sense input for current limiting.

The disadvantage of using a Darlington transistor in high current requirements is the high saturation voltage, as opposed to a single transistor. However, the transistor voltage drop can be decreased by a two-transistor "Darlington stage," as shown in fig. 9.10B. The primary winding has added turns on the outer extremities, which are of opposite polarity to the connected primary. Since each half of the circuit is symmetrical, only Q_1 and Q_3 will be discussed. First, consider N_d does not exist and the collectors of Q_1 and Q_3 are common. Then the transistor voltage drop for typical junction voltages is

$$V_{CE,\,\text{sat},\,Q_3} = V_{CE,\,\text{sat},\,Q_1} + V_{BE,\,\text{sat},\,Q_3} = 1 + 2 = 3 \text{ V}$$

With Q_1 connected to N_d, the voltage across N_d boosts the collector voltage of Q_1 by $E_d = E_p N_d / N_p$. Then

$$V_{CE,\,\text{sat},\,Q_3} = V_{CE,\,\text{sat},\,Q_1} + V_{BE,\,\text{sat},\,Q_3} - E_d$$

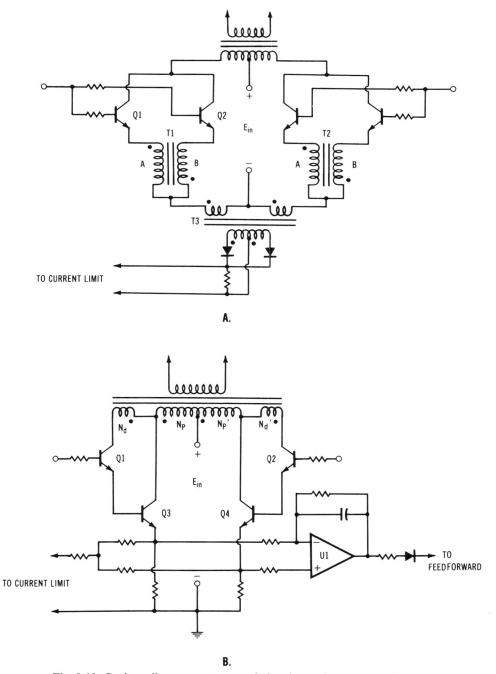

Fig. 9.10. Push–pull converter current balancing and current sensing.

If E_d is 1.5 V, using the above junction voltages, $V_{CE,\,sat,\,Q_3} = 3 - 1.5 = 1.5$ V, and the power dissipation in Q_3 is halved. Care must be exercised in choosing the proper turns for N_d. For instance, if E_d is large, the base-to-collector junction of Q_3 becomes forward biased and a high current will flow through Q_1, probably causing forward-biased second breakdown failure. On this basis, the transformer design may prevent use of this circuit. Using a 1.5-V value for E_d and a minimum one turn for N_d, the volts-per-turn is 1.5. Transformers used in converters switching at high frequency normally have a much higher volts-per-turn since less turns are required to support the volt-second product.

One of the problems associated with the push–pull center-tapped topology is the impending saturation of the power transformer due to volt-second imbalances. Unsymmetrical half-cycle conduction periods caused by control circuit timing or by switching time of the transistors will change the time that voltage is applied across the transformer. Imbalances in the saturation voltage or the "on" resistance of the transistors will change the effective voltage applied across the transformer. Under these conditions, the flux imbalance will cause the core to "walk-up" the *BH* curve toward saturation. The increase in current may cause forward-biased second breakdown in the transistors. In Fig. 9.10B, the voltage developed across the current sense resistors is applied to the current sense inputs in the control circuit for fast response to an overcurrent condition, which then turns off the corresponding transistor for the remainder of the half cycle. If no further dynamic adjustments are made, the transformer saturation and the overcurrent condition are likely to repeat each half cycle. In Fig. 9.10B, the current sense signals are also applied to U_1, which is an amplifier/integrator. The output of the amplifier is applied to the feedforward function in the control circuitry (reference Section 8.6) to change the slope of the triangle waveform which will tend to increase the conduction time of the transistor having the lowest current and to decrease the conduction time of the transistor having the highest current and will restore a balanced condition over a period of time. Other methods which may be used to minimize this effect are discussed in the cited Refs. [1–3].

For low-power converters, the transistors of Fig. 9.10 may be driven directly from the control circuit if common returns are used. Drive circuitry for the bipolar transistors may also be provided by an isolation transformer with proportional base drive, as discussed in Section 2.1.2.4. With MOSFETs, the proportional gate drive is not particularly advantageous. With either bipolars of MOSFETs, it is important to maintain an effective short across the primary of the drive transformer during the dead time to eliminate false conduction, as discussed in the referenced section. Individual transformers to drive each transistor may also be used, as discussed in the referenced section.

9.7 HALF-BRIDGE CONVERTER

The half-bridge converter of Fig. 9.6B is perhaps the most popular topology used in medium power switch mode power supplies. The normally required input filter capacitors comprise half the bridge, while only two transistors are needed for the other half of the bridge. When either transistor is on, the primary voltage is one-half the input voltage, and the primary current is twice the peak input current.

The output voltage is

$$E_0 = \tfrac{1}{2}E_{in}D'/n \tag{9.35a}$$

Considering voltage drops in the semiconductors, the following relations apply: $E_0 = D'E_a$; $E_a = E_s - V_F$; $E_s = E_p/n$; $E_p = \tfrac{1}{2}E_{in} - V_{sat}$. By combining these terms,

$$E_0 = D'[(E_{in} - 2V_{sat})/2n - V_F] \tag{9.35b}$$

where

$D' = 2t_{on}/T = t_{on}/\tfrac{1}{2}T$
$n = N_1/N_2 = N_p/N_s$
V_{sat} = collector-to-emitter saturation
V_F = rectifier voltage drop

Normally, E_0 and E_{in} are known. Rearranging terms the turns ratio of the transformer is

$$n = D'[E_{in} - 2V_{sat}]/[2(E_0 + D'V_F)] \tag{9.36}$$

Equations (9.35) and (9.36) neglect the IR voltage drops of the transformer and the output filter inductor, which should be minimal.

Considering voltage drops and losses, the average input current and the peak transistor current can be determined in the same manner as the calculation of output voltage. However, a short-cut approach may be used by assuming an overall efficiency. Then

$$I_{in, avg} = I_0E_0/\eta E_{in} \tag{9.37a}$$

$$I_{pri, pk} = (2I_0 + \Delta I_L)/n \tag{9.37b}$$

where efficiency $\eta = P_0/P_{in}$.

Example

For $E_{in} = 150$ V, $E_0 = 5$ V, $I_0 = 60$ A, $D' = 70\%$, $\eta = 75\%$, $V_{sat} = 1$ V, $V_F = 1$ V; find n, I_{in}, and the peak current in the transistors. From Equation (9.36), $n = 0.7(150 - 2)/[2(5 + 0.7)] = 9.09$; use a turns ratio of 9. For 300 W output, $I_{in, avg} = 300/0.75 \times 150 = 2.67$ A. For $\Delta I_L = 0.4I_0$, $I_{pri, pk} = 2.4 \times 60/9 = 16$ A. The voltage and current rating of the transistor must allow for variations in input voltage and some form of current limiting, which will be higher than the calculated peak current. Choose a transistor with a voltage rating of 250 V and a current rating of 25 A.

The inductance and capacitance values of the output filter are the same as those for the push–pull converter, except $E_{pri} = \tfrac{1}{2}E_{in}$, and are given by Equations (9.33) and (9.34).

As in the case of the push–pull converter, a flux imbalance in the transformer may be caused by unequal conduction periods of Q_1 and Q_2, or by unequal switching times of Q_1 and Q_2. To eliminate an imbalance in the transformer, a dc blocking capacitor may be inserted in series with the transformer primary. If C_1 and C_2 are part of a single-phase input rectifier filter, the capacitance values may be large since the capacitance is normally determined by the desired ripple filtering. If the input is dc or is a three-phase rectified voltage, the values of C_1 and C_2 may be reduced, since the input ripple voltage is lower as compared to a single-phase input. In this case, the capacitor in series with the primary is not required, if C_1 and C_2 can accommodate an unbalanced voltage. For a voltage imbalance of Δe_{dc} and for $Z = \sqrt{L/2C}$, the current is

$$I_1 = \Delta e_{dc}\sqrt{2C/L} \qquad (9.38a)$$

Also, the primary current increases as

$$I_2 = E_{dc}t_{on}/4L \qquad (9.38b)$$

where

$\quad I_1$ = current flow at circuit impedance
$\quad I_2$ = primary magnetizing current
$\quad E_{dc}$ = dc voltage across each capacitor
Δe_{dc} = change in capacitor voltage
$\quad C$ = capacitance, $C_1 = C_2$, F
$\quad L$ = primary inductance, H

Combining the currents for $I = I_1 + I_2$,

$$\Delta e_{dc}\sqrt{2C/L} = (4IL - E_{dc}t_{on})/4L$$

Solving for the capacitance value,

$$C = (4IL - E_{dc}t_{on})^2/32L(\Delta e_{dc})^2 \qquad (9.39)$$

Example

In the circuit of Fig. 9.6B, find the values of C_1 and C_2 for $E_{in} = 300$ V ($E_{dc} = 150$ V), $L = 5$ mH, $t_{on} = 20$ μs, and $I_{in} = 3$ A. Let $\Delta e_{dc} = (5\%$ of $E_{dc}) = 7.5$ V and let $I = (10\%$ of $I_{in}) = 0.3$ A. Then, for C_1 and C_2, $C = (4 \times 0.3 \times 5 \times 10^{-3} - 150 \times 20 \times 10^{-6})^2/(32 \times 5 \times 10^{-3} \times 7.5^2) = 1$ μF.

9.8   FULL-BRIDGE TRANSISTOR CONVERTER

The full-bridge transistor converter of Fig. 9.6C is more effective for higher power outputs than the half-bridge topology, especially where two transistors must be paralleled in the half-bridge converter. Since the transformer primary voltage of the full-bridge circuit is essentially the same as the input voltage, the primary current is one-half that of the half-bridge circuit. If four transistors are required, the full-bridge circuit will eliminate the problems of transistor paralleling inherent in the half-bridge circuit, for the same output power. However, for very high output power, transistors could be paralleled in the bridge circuit to conduct the required current.

The dc blocking capacitor, C_1 of Fig. 9.6C, prevents saturation of the power transformer as discussed in the preceding section. The value of C_1 may be determined from Equation (9.39), with the corresponding voltage and current parameters. Capacitor C_1 must have a low impedance at the switching frequency and must have an rms current rating greater than the rms current in the primary of the power transformer.

The waveforms of Fig. 9.6C show that Q_1 and Q_2 conduct alternately for a full half cycle. Also, Q_3 and Q_4 conduct alternately for a full half cycle but are phase displaced from Q_1 and Q_2, as in the PSM (phase shift modulation) mode. This means that during the dead time (no current drawn from the input), either Q_1 and Q_3, or Q_2 and Q_4, are on. Since reverse-polarity rectifiers are assumed to be connected across the transistors (inherent in MOSFETs), C_1 is, in effect, connected in parallel with N_1 during the dead time. Then C_1 discharges through N_1. When input voltage is again applied to N_1, the dv/dt of the capacitor is halved, since dv is halved, as opposed to the full swing of voltage across C_1 if all transistors are off during the dead time. However, this may not be an important consideration if the voltage imbalance is small. But this timing technique may result in an undesirable effect, namely the common, or overlap, conduction of either Q_1 and Q_3 or of Q_2 and Q_4 when these transistors switch. Storage time in bipolar transistors may cause one of the series transistors to remain conducting while the other transistor is turning on. This places a short across the input and high peak currents may cause transistor failure due to forward-biased second breakdown, or due to reverse-biased second breakdown. Snubbers and load line shaping can absorb some of the energy which would otherwise be dissipated in the transistors, but these additions will not improve operating efficiency.

In most cases, the outputs of a dual-ended PWM IC, such as the 1524, 1846, and 494, are applied to a single drive stage with one transformer which has four secondaries. The driver stage then causes Q_1 and Q_4 to conduct for a portion of one half cycle with Q_2 and Q_3 conducting for a portion of the opposite half cycle. During the dead time, none of the transistors conduct. This timing method eliminates the common conduction problem associated with the square-wave drive discussed in the preceding paragraph. As was the case of the half-bridge circuit, the primary of the drive stage transformer should be effectively shorted during the actual dead time.

Following the same analysis as given for the half-bridge circuit, the voltage, current, and transformer turns relations for the full-bridge circuit are

$$n = D(E_{in} - 2V_{sat})/(E_0 + DV_F) \qquad (9.40a)$$

This is exactly one-half the turns ratio required in the half-bridge circuit, given the same saturation voltages and the same rectifier forward drops.

If the output rectifier is a full-wave bridge for higher voltage output, two rectifier drops occur and

$$n = D(E_{\text{in}} - 2V_{\text{sat}})/(E_0 + 2DV_F) \tag{9.40b}$$

The average input current is given by Equation (9.37a), and the peak transistor current is one-half the value given by Equation (9.37b).

The inductance and capacitance values of the output filter are the same as those for the push–pull converter, and are given by Equations (9.33) and (9.34).

Output voltage regulation is achieved by sensing a portion of the output voltage and feeding this signal to the control circuit to control the duty cycle. The control circuit normally consists of a PWM integrated circuit, or components providing the same functions as the PWM IC. Since isolation is normally desired from input to output, a method of indirectly sensing output voltage is shown in Fig. 9.11, which allows the control circuitry to be common with the input return. Flyback rectifier CR_1 is similar to the push–pull, half-bridge, full-bridge, or forward converter; N_p is the "normal" winding on the filter inductor. When the power switching transistors are not conducting, the voltage across this primary is $E_p = E_0 + V_{F1}$. The additional winding on L_1 forms a secondary and the coupled voltage in this winding is $E_s = E_{\text{fb}} + V_{F2}$, where CR_2 is forward biased when the power switching transistors are not conducting; E_{fb} is the feedback signal to the control circuit. If the IR drops in the windings are small and good coupling exists between primary and secondary, the turns ratio gives

$$n = E_p/E_s = N_p/N_s = (E_0 + V_{F1})/(E_{\text{fb}} + V_{F2})$$

Solving for the feedback voltage,

$$E_{\text{fb}} = (N_s E_0 + N_s V_{F1} - N_p V_{F2})/N_p \tag{9.41a}$$

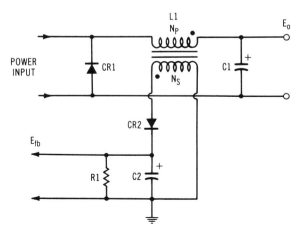

Fig. 9.11. Sensing converter output voltage indirectly.

If the turns ratio and the forward drops are such that $N_s V_{F1} = N_p V_{F2}$, then

$$E_{fb} = E_0 N_s / N_p \qquad (9.41b)$$

Thus the output voltage can be indirectly sampled in the same manner as the flyback converter of Fig. 9.8A.

9.9   SCR CONVERTERS

A full-bridge converter using SCRs is shown in Fig. 9.12. The output power is limited primarily by the SCR ratings and can be 100 kW or higher. This converter is essentially the same as the impulse-commutated inverter of Fig. 7.10 except the sine wave output filter is replaced by a bridge rectifier and LC smoothing filter for dc output. Output regulation is achieved by pulse width modulation. Output voltage is scaled by the turns ratio of the transformer and is a function of input voltage and duty cycle. The previous equations of current, voltage, and filter components also apply. Operation of the commutating mechanism and equations governing selection of components is discussed in Section 7.5, and a typical control circuit is shown in Fig. 7.9. Due to the turn-off time required for the SCR and to maximize efficiency, the maximum operating frequency is typically 20 kHz when using devices with turn-off times of 10 μs. An additional bridge circuit, shown in Fig. 18.8, is effective for boosting an input voltage where input to output isolation is not required.

The SCR converter may also be configured in a half-bridge topology where two series input filter capacitors are used. The half-bridge circuit is very cost effective since it eliminates one set of SCRs and the corresponding commutating components. In fact, the circuit of Fig. 18.8A could be changed to a half-bridge topology

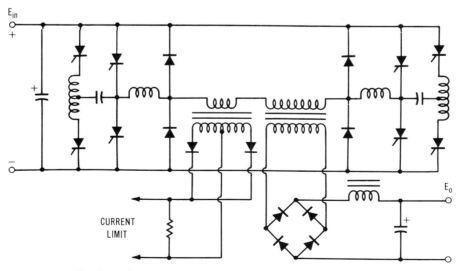

Fig. 9.12. Impulse-commutated SCR full-bridge converter.

by center tapping the battery voltage, which is typically 250 V, for higher power requirements. However, the current in the SCRs, the commutating components, and transformer primary will be twice that of the full bridge, as is the case for the transistor converter.

Another thyristor converter which resembles the previously discussed resonant converter is the topology developed by Mapham [4]. In the original configuration, the output is a sine wave. Referring to Fig. 9.13A, L_1 and C_1 resonate at a frequency which is higher than the desired switching frequency. A sine wave

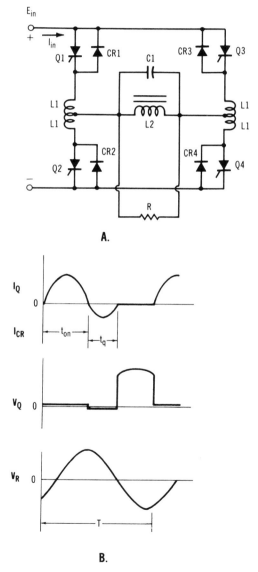

Fig. 9.13. SCR resonant converter.

voltage is developed across R, the output load. When Q_1 and Q_4 are triggered, input current flows through L_1 and C_1 to charge C_1 to approximately $2E_{in}$. As C_1 discharges, the current reverses and flows back through CR_1 and CR_4 to the source. This reverse biases Q_1 and Q_4 and provides turn-off time to the SCRs. Then Q_2 and Q_4 are triggered for the negative half cycle. Waveforms are shown in Fig. 9.13B. Parameters governing the basic operation are the ratio of load resistance R, to the circuit impedance $\sqrt{L_1/C_1}$; the ratio of resonant frequency f_r to the switching frequency f_s; and the ratio of L_1 to L_2. These parameters are documented in the cited reference for various ratios. If R is replaced by a transformer primary whose secondary voltage is rectified and filtered, a dc output is provided. Regulation can be achieved by changing the switching frequency. Output voltage decreases as the switching frequency decreases. However, the switching frequency must change considerably for normal changes in input voltage as well as light load, since the output filter capacitor will tend to charge to the peak of the secondary waveform.

Carrying the topology a step further, the circuit of Fig. 9.14A uses two half-bridge stages and two power transformers. The SCRs are triggered at a fixed frequency, but the trigger signals to Q_3 and Q_4 are delayed with respect to the trigger signals to Q_1 and Q_2. Thus E_{s1} is fixed in time, while E_{s2} is phase shifted. When these two voltages are vectorially summed and applied to the bridge rectifier, the output filter smooths the full-wave rectified sine wave to dc, as shown in the waveforms of Fig. 9.14B. This may seem a laborious approach but the circuit has distinct advantages for high-voltage outputs. First, the decreased dv/dt of the secondary voltage allows use of slower recovery rectifiers in the bridge. Second, the decreased dv/dt minimizes high transient currents flowing in the transformer winding capacitance, which could be appreciable at high voltages. Typical parameter ratios for the inverter stage are given in the cited reference as

$$f_r/f_s = 1.5 \qquad R/\sqrt{L_1/C_1} = 5 \qquad L_2/L_1 = 100$$

$$I_{in,\,avg}\sqrt{L_1/C_1}\,/E_{in} = 0.101 \qquad I_{Q,\,pk}\sqrt{L_1/C_1}\,/E_{in} = 0.7$$

$$E_{p,\,rms}/E_{in} = 0.71 \qquad t_q\big/\big(\pi\sqrt{L_1C_1}\big) = 0.6$$

where

$$f_r = 1/2\pi\sqrt{L_1C_1}$$
$$f_s = \text{switching frequency}$$
$$I_{Q,\,pk} = \text{peak SCR current}$$
$$t_q = \text{available turn-off time for the SCR}$$

Since $I_{in,\,avg} = P_0/\eta E_{in}$ and since f_r and f_s are selected values, L_1 and C_1 can be readily determined, from which other parameters can be calculated. One of the disadvantages of the circuit is apparent when comparing $I_{Q,\,pk}$ to $I_{in,\,avg}$. From the above values, this ratio is about $7:1$. However, the SCR can inherently conduct high peak currents if the rms current is within the device rating.

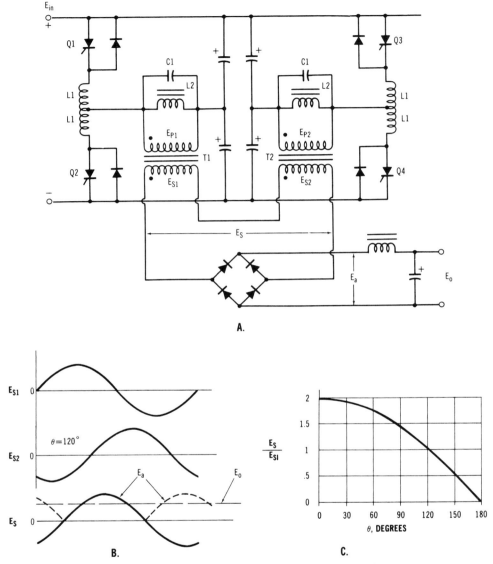

Fig. 9.14. Dual resonant converter with voltage summing.

Since the magnitude of the secondary voltages is equal, the sum of the secondary voltages is given by

$$E_{s,\text{rms}} = E_{s1}\sqrt{(1 + \cos\theta)^2 + (\sin\theta)^2} \qquad (9.42)$$

where θ is the phase angle between E_{s1} and E_{s2} and $E_{s1} = E_{p,\text{rms}}N_s/N_p = E_{s2}$.

A plot of this equation is shown in Fig. 9.14C. In an actual application, θ would be a low value at minimum input voltage and full load. Also, θ would increase as the input voltage increased and as the load current decreased to maintain a regulated dc output voltage. Thus, regulation is achieved by sensing a portion of the output voltage and varying the firing angle to Q_3 and Q_4, in the same manner as a pulse width–modulated topology.

9.10 CURRENT-FED CONVERTERS

The current-fed converter utilizes an inductor between the source and the switching inverter stage. This topology differs from the previously discussed voltage-fed converters wherein the source was directly applied to the inverter stage. Since the current cannot change instantaneously in an inductor, the current-fed topology offers inherent current limiting by $i = 1/L \int e \, dt$. Two modes of operation are possible. In the first mode, the active devices in the inverter stage alternately conduct for a portion of each half cycle and the input current is discontinuous, which is typical of the voltage-fed inverter stage. In the second mode, the active devices conduct for a period longer than a half cycle, thus periods exist where all transistors conduct. This switching action produces an overlap conduction which is normally avoided in most designs due to the catastrophic results. In the current-fed converter, this mode offers advantages, and the series inductor limits current during overlap conduction. This mode allows continuous input current. Comparisons of advantages and disadvantages of the current-fed inverter stage versus the voltage-fed inverter stage are listed in Table 9.2.

9.10.1 Discontinuous Current Mode

A typical push–pull current-fed converter is shown in Fig. 9.15A. The transistors switch in the pulse width–modulated mode and a quasi-square wave is developed across T_2. The input inductor is the primary of T_1 and is transposed from the normal output filter inductor of the voltage-fed converter. Waveforms are shown in

TABLE 9.2. Voltage-Fed Converter and Current-Fed Converter Characteristics

Converter Type	Advantages	Disadvantages
Voltage fed	Highest power-to-weight ratio; highest power-to-volume ratio.	Poor reactive load capability; high peak currents, requires fast current sensing and fast semiconductor turn-off; asymmetrical operation produces transformer flux imbalance; component overstress conditions can occur
Current fed	Operates into reactive loads; peak current limiting readily implemented; magnetics operate in linear region; cascade failures can be eliminated	Additional energy storage input inductor required; inductor specifications are demanding; low power-to-weight ratio

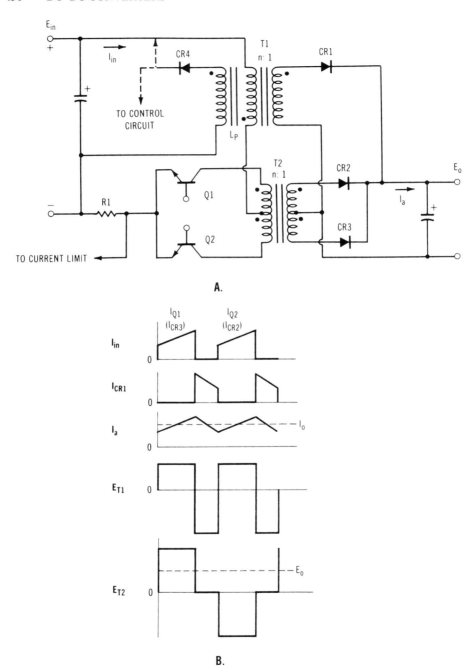

A.

B.

Fig. 9.15. Current-fed push–pull converter.

Fig. 9.15B. When Q_1 is turned on, input current flows through the primaries of T_1 and T_2. Since CR_1 (and CR_4) are reverse biased, energy is stored in L_{p1}. The voltage across N_{p2} is coupled to the secondary, rectified and applied to the filter capacitor and the load. When Q_1 turns off, the induced voltage across L_{p1} is coupled to the secondary, and the current flows through CR_1 and the load. This flyback action is synonymous with the flyback action of the flyback–converter transformer, and of the output filter inductor in the buck regulator. Diode CR_1 could be omitted and the secondary of T_1 could be connected through CR_4 to the source. This would feed the stored energy back to the source but since output power is desired at the output, connecting CR_1 to the output is more desirable. With CR_1 connected, the third winding could be coupled through CR_4 to the control circuit for indirectly sensing the output voltage for regulation purposes, as was the case of the coupled inductor of Fig. 9.11.

From Fig. 9.15, the following relations apply. When one of the transistors is conducting, $E_{p2} = E_{in} - E_{p1} = n_2 E_0$, and $E_{p1} = E_{in} - n_2 E_0$. When both transistors are off, $E'_{p1} = n_1 E_0$. The volt-seconds during the positive portion of the half cycle equals the volt-seconds during the negative portion of the half cycle or, $E_{p1} t_{on} = E'_{p1} t_{off}$. Since $t_{on} + t_{off} = T$ and the duty cycle is defined as $D = t_{on}/T$, D can be expressed as $D = E'_{p1}/(E_{p1} + E'_{p1})$. Substituting the relations for E_0,

$$D = n_1 E_0 / [(E_{in} - n_2 E_0) + n_1 E_0]$$

This is an unwieldy equation but is easily simplified if $n_1 = n_2$ to

$$E_0 = DE_{in}/n \qquad (9.43)$$

where

$$n = N_{p1}/N_{s1} = N_{p2}/N_{s2} \qquad D = t_{on}/\tfrac{1}{2}T$$

This is the same basic equation as that for the buck regulator, the forward converter, and the voltage-fed push–pull converter. By the same analysis,

$$I_{in,\,avg} = DI_0/n \qquad (9.44)$$

Example

For $E_{in} = 48$ V, $E_0 = 12$ V and $I_0 = 37.5$ A ($P_0 = P_{in} = 450$ W): calculate the turns ratio for a nominal duty cycle of 62.5%; calculate the input current; find the peak transistor voltage. From Equation (9.43), $n = DE_{in}/E_0 = 0.625 \times 48/12 = 2.5$. Also $I_{in} = 0.625 \times 37.5/2.5 = 9.375$ A. During the time when one of the transistors is on, the opposite transistor voltage is $V_{CE} = 2E_{in} - E_{p1} = (2 \times 48) - 18 = 78$ V. When both transistors are off, $V_{CE} = E_{in} + E_{p1} = 48 + 30 = 78$ V. However, as the duty cycle increases, V_{CE} increases and approaches $2E_{in}$ and the transistor voltage rating should be greater than $2E_{in}$ for safety margin and protection from voltage spikes.

Current-fed inverter stages tend to lose appeal in SCR converters, which are usually very high power, and the inductor becomes quite large. Also, the slower switching rate in the SCR converters means the inductance values of the magnetics are larger than for fast switching transistors. This means the di/dt values are lower, and the increase in time to a certain current allows sufficient time for current sensing and SCR turn-off.

9.10.2 Continuous Current Mode

Referring to the circuit and waveforms of Fig. 9.16, each transistor conducts for a period greater than a half cycle. The period of overlap conduction is defined as t_{oc} which is the equivalent of t_{on} for storing energy in the inductor. During t_{oc}, $E_L = E_{in}$, and the output filter capacitor supplies current to the load. When only one transistor is conducting, the induced voltage across the inductor plus the input voltage is applied across the transformer primary. Then, $E_p = E_{in} + E_L = nE_0$. Considering the duty cycle relation,

$$E_0 = E_{in}/n(1 - D) \tag{9.45}$$

where

$$n = N_p/N_s \qquad D = 2t_{oc}/T$$

Thus, the current-fed converter operating in the continuous current mode performs in the same manner as the boost regulator (reference Section 8.4) and also provides input-to-output isolation. In fact, this topology is frequently referred to as push–pull boost. Since the input current is continuous, EMI suppression is more readily achieved than in the discontinuous current mode. The penalty of this mode is the requirement for higher voltage transistors. Neglecting losses and voltage drops, the transistor voltage rating of the boost regulator is equal to or greater than the output voltage. In the push–pull circuit of Fig. 9.16, the transistor voltage is equal to or greater than

$$V_{CE} = 2nE_0 \tag{9.46}$$

Examples

Case 1. In Fig. 9.16B, $D = \frac{2}{6} = 0.333$. For $E_{in} = 48$ V and $E_0 = 12$ V, $n = E_{in}/E_0(1 - D) = 48/(12 \times 0.666) = 6$. When only one transistor is conducting, the voltage across the opposite transistor is $2nE_0 = 2 \times 6 \times 12 = 144$ V.

Case 2. In Fig. 9.16C, $D = \frac{4}{6} = 0.667$. For $E_{in} = 48$ V and $n = 6$, $E_0 = 48/(6 \times 0.333) = 24$ V. The output current is half that of Fig. 9.16B. Since the output power of the two circuits is the same, the input current is the same. The transistor voltage is now $2 \times 6 \times 24 = 288$ V, even for a 48-V input and a 24-V output! In an actual design, both D and n should be reduced.

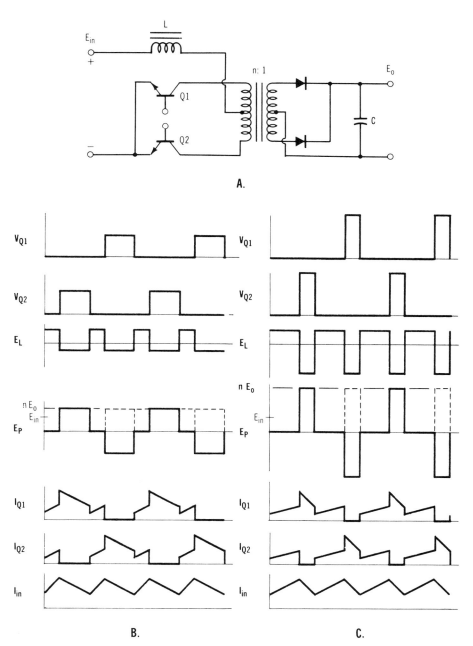

Fig. 9.16. Current-fed converter with continuous current.

9.11 MULTIPLE-OUTPUT CONVERTERS

Multiple-output converters find wide applications in powering multiple loads of varying voltages. The desired outputs may range from TTL logic voltages to CRT (cathode ray tube) anode voltages. Triple outputs of $+5$ V and ± 12 V (or ± 15 V) are popular for powering digital and linear devices. Pulsed power outputs may be required to handle the surge current required for printer drivers or the initial starting of disk drives. In this latter case, a semiregulated output may power the disk drive, while a well-regulated output may provide power to a CRT converter. Figure 9.17 shows the more popular multiple output topologies. The power transformer is connected to the proper primary switching circuit. Output voltage regulation is achieved by sensing one of the outputs, normally the high-current output, and feeding a portion of this voltage to the control circuit. Also, batch regulation can be used where all outputs are sampled from individual voltage dividers. This analysis will be discussed later. The flyback circuit of Fig. 9.17A has two secondary windings, with the second winding center tapped; this winding could be wound bifilar which allows ease of winding. The same transformer design applies to the forward converter circuit of Fig. 9.17B, and the push–pull circuit of Fig. 9.17C.

9.11.1 Cross-Coupled Filter Inductors

The circuit of Fig. 9.17B uses a separate filter inductor for each output. The circuit of Fig. 9.17C uses a cross-coupled inductor with a common core, which could also be used in Fig. 9.17B. The advantage of the cross-coupled inductor is to balance the ampere-turn product of the inductor windings to enhance load regulation. For instance, if the load resistance of E_{02} in Fig. 9.17B becomes large (light load), the current in L_2 could become zero during a portion of the on time of CR_3. This would cause E_{02} to rise to the value of E_2. The important point again is that the volt-second product of an inductor storing energy is the same as the volt-second product of the inductor delivering energy. If no current flows, the inductor is not storing or delivering energy. As previously discussed, $E_L t_{on} = -E_L t_{off}$ for continuous current, and $E_L = 0$ for $I = 0$. Now consider the coupled inductor of Fig. 9.17C. If the load resistance of E_{02} becomes large, the current in L_{1b} could become zero during a portion of the on time. But current is still flowing in L_{1a} and the voltage across L_{1a} is $E_1 - E_{01}$. The voltage coupled to L_{1b} is the ratio of the "turns ratio" to E_{L1a}, and this voltage is of a polarity to oppose E_2. Thus E_{02} will not rise to E_2. The coupled transformer action of the inductor also applies to all windings. For instance, if I_{01} becomes small, the voltage across L_{1b} will couple to L_{1a} to oppose E_1. Also, observe the polarity of L_{1c} winding, which is connected to a negative output. Current must flow in the same flux direction in all windings. Size and weight are also reduced, since one inductor rated for total energy storage will be smaller than the sum of the individual inductors with a sum total of the same energy storage.

   Just as the turns ratios of the power transformer determine the output voltages, the turns ratios of the inductor and this relation to the turns of the transformer are important. For the forward and the push–pull converter, $E_0 = DE_{in}$, where E_{in} is the transformer secondary voltage. During t_{on}, $E_L = E_{in} - E_0$ for each winding of

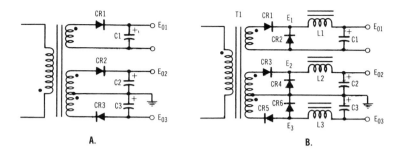

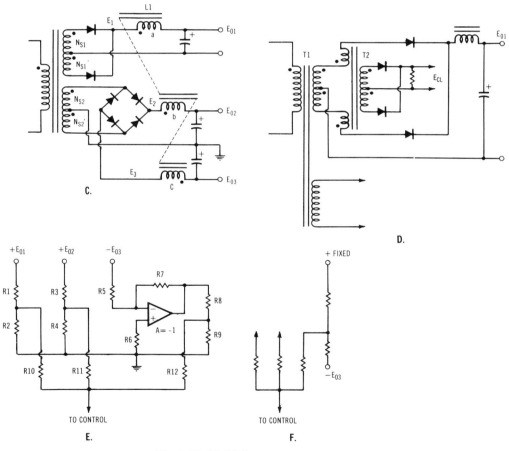

Fig. 9.17. Multiple-output converters.

L_1. Neglecting rectifier drops, $E_1 = E_{s1}$ and $E_2 = E_{s2}$ during t_{on}. Also, $E_{s2} = E_{s1}N_{s2}/N_{s1}$. Equating these relations, $E_{s1} = E_{01}/D = E_{02}N_{s1}/(DN_{s2})$, and $E_{02} = N_{s1}/N_{s2}$. Also, $E_{La} = E_{s1} - E_{01}$ and $E_{Lb} = E_{s2} - E_{02}$. During t_{off}, $E_{La} = -E_{01}$ and $E_{Lb} = -E_{02}$. Equating these relations and including E_{03} by similar analysis:

$$N_{s1}/N_{s2} = N_{La}/N_{Lb} \qquad (9.47a)$$

$$N_{s1}/N_{s3} = N_{La}/N_{Lc} \qquad (9.47b)$$

Thus the turns ratios of the inductor should be the same as the turns ratios of the transformer. In an exact analysis, this requirement will change slightly since the variations in rectifier voltage drops should also be considered, especially if one of the outputs is a low voltage and the nominal voltage between outputs is a high ratio. For the coupled filter inductor, these relations can impose unusual restrictions on the practical use. For instance, the individual inductance values are initially calculated based on voltage, current, and frequency. But if the turns ratios are considered and since the inductance value is proportional to the turns squared, the actual inductance value in one of the windings may be an order of magnitude higher than the conventional calculation. This may cause an additional operational problem. The cutoff frequency of the LC filter in this output will probably be lower and the transient response will be degraded, due to the higher inductance value. In these cases, and especially if the load current in one of the outputs remains fairly constant, a separate inductor should be used for that output.

9.11.2   Current Limiting

All high-performance converters should have a means to limit current in an overload condition and to protect the converter from output short circuits. For single outputs, the primary current or the current in the switching transistors is sensed, and the transistors are turned off if the current exceeds a preset value. For multiple-output converters, simply limiting the primary transistor current is not sufficient. Consider a triple-output converter rated at 200 W: typically 5 V at 20 A, $+12$ V at 4 A, -12 V at 4 A. If each output is supplying half the load and if primary current limiting only is used, a heavy overload on the $+12$-V output will protect the primary transistors, but the rectifiers in the $+12$-V output will probably fail from excessive junction temperature caused by excessive current. The heavy overload is normally more severe than a short circuit on the output since the rectifiers can safely conduct high peak currents if the duty cycle is sufficiently low. This would normally be the case since a high overcurrent condition reflected on the primary side would allow the control circuit to turn off the transistors, and the converter would then restart in a soft-start mode after a period of delay. If the short circuit were still present, the converter would cycle in the described mode until the short disappeared.

Current sensing with a current transformer on the secondary side is shown in Fig. 9.17D. The power leads from T_1 pass through T_2 and the equal ampere turns produce a voltage which is rectified and loaded for voltage scaling. This voltage is applied to the control circuit which turns off the transistors if the current exceeds a preset level. The current transformer and rectifier are employed in each of the secondary windings, thus providing protection for all outputs.

9.11.3 Voltage Sensing

Regulation is achieved by sensing a portion of the output voltage and controlling the duty cycle to maintain the desired output. In multiple-output converters, sensing a single output will normally provide sufficient line regulation, and will provide load regulation with respect to changes in voltage drops from the input through the primary of the power transformer. However, if the load current in one of the remaining outputs has a wide variation, the *IR* drop in the transformer secondary supplying this output, the *IR* drop in the filter inductor, and the change in rectifier forward drop will be outside the feedback loop and load regulation for this output will be degraded. This effect can be overcome by sampling all outputs with a divider network, and is normally referred to as *batch regulation*. Thus, instead of one sensed output having a $\pm 0.5\%$ load regulation and the remaining outputs having a $\pm 5\%$ load regulation, all outputs can be made to provide approximately $\pm 2\%$ regulation. In fact, the amount of feedback on each output may be scaled for individual sensitivity.

Direct sensing is accomplished if all outputs share a common return. Fig. 9.17E shows a typical network. The circuit is designed such that all divider voltages are equal without R_{10}, R_{11}, and R_{12}. The negative output is inverted by a unity gain amplifier. Then R_{10}, R_{11}, and R_{12} may be equal or may be scaled for sensitivity, and the summing junction applied to the control circuit. If the amplifier is undesirable, the circuit of Fig. 9.17F could be substituted if a positive fixed regulated voltage is available, and if the return line is common with the output return. However, care must be exercised with this circuit. When the converter is first turned on and before the output voltage appears, the voltage level of the fixed source could produce a positive voltage higher than the normal feedback voltage, which would then result in no output voltage since the duty cycle would remain at a minimum or the converter would cycle on and off.

9.12 INPUT FILTERS FOR dc–dc CONVERTERS

An input filter is normally used to decouple input source voltage transients from the converter to improve input transient susceptibility and to reduce the ripple current drawn from the source. These are power-type filters as opposed to the EMI filters (reference Section 14.1.2), which may also be required. The filter may consist of a single capacitor across the input terminals, or an *LC* stage inductor input filter, depending on the converter topology. In the case of the capacitor input filter, the effect of the inrush current at turn-on must be considered. Even with an *LC* filter, inrush current is limited only by the filter impedance ($Z = \sqrt{L/C}$), and voltage peaking on the filter capacitor can result due to undamped oscillatory nature caused by minimal dc resistance (reference Section 8.7).

9.12.1 Square-Wave Converters

For square-wave converters, the steady-state input current may be considered continuous except for the time when the transistors switch. In cases where saturating transformers control the switching, the current will increase when the transformer saturates. The input current "ripple" will be small at full load, but since the peak current does not change, the peak-to-peak input ripple may be high at light loads. In clocked converters, the storage delay time of bipolar transistors may cause a conduction overlap resulting in a high peak input current. A capacitor input filter is normally adequate for these types of converters, and the capacitor should be located close to the power stage terminals to minimize effects of internally induced voltage transients.

9.12.2 PWM Converters

The peak-to-peak input current variation to pulse width modulated, variable duty cycle converters previously discussed (except for the current-fed converter in the continuous current mode) is from zero to a peak current which is determined by the load requirements. Normally, an *LC* input filter is required to provide energy storage and minimize input current ripple. When the input voltage is applied, a voltage-peaking effect may occur on the capacitor. If the converter output filter is also an *LC* stage, the input filter and the output filter transfer functions may react to produce instability during normal operation. Reference the discussion of Section 8.7 for a detailed analysis.

REFERENCES

1. R. Patel, "Detecting Impending Core Saturation in Switch-Mode Power Converters," Proceedings of Powercon 7, March 1980.
2. K. O'Meara, "Passive Balancing of Transformer Flux in Power Converters," Proceedings of Powercon 10, March 1983.
3. I. M. Gottlieb, *Regulated Power Supplies* (Howard W. Sams, Indianapolis, IN, 1981), p. 193.
4. N. Mapham, "An SCR Inverter with Good Regulations and Sine-Wave Output," *IEEE Transactions on Industry and General Applications* (March/April 1967).

10

RESONANT MODE CONVERTERS

10.0 Introduction
10.1 Classification
10.2 Resonant Mode Control Circuits
    10.2.1 1860
    10.2.2 33066
    10.2.3 Other Devices
10.3 Zero-Current Switching
    10.3.1 Quasi-Resonant Regulators
    10.3.2 Quasi-Resonant Converters
10.4 Series Resonance
    10.4.1 Full Wave
    10.4.2 Full-Wave Bridge, Pulsed Load
10.5 Parallel Resonance
10.6 Series–Parallel Resonance
10.7 Zero-Voltage Switching
    10.7.1 ZVS, Phase Shift–Modulated Converter
    10.7.2 Optional Phase Shift Topology
    References

10.0 INTRODUCTION

Resonant mode converters and inverters use or produce sinusoidal switching waveforms, as opposed to the square and trapezoidal waveforms inherent in PWM systems. Semiconductor devices can switch at zero current or zero voltage for minimal switching losses. Conventional PWM converters are power–density limited because they trade off efficiency against operating frequency, since switching losses are a major problem as the switching frequency is increased. Resonant mode topologies can therefore switch at higher frequencies than PWM types while achieving the same overall efficiency. The higher switching frequency reduces the overall size and weight of magnetic devices. The parasitic inductances and capacitances inherent in any system (and undesired in most PWM topologies) can be used to an advantage in resonant mode topologies. However, components and switching devices must conduct higher peak currents than the devices in PWM

TABLE 10.1. Resonant Mode Converter Characteristics

Converter Type	Advantages	Disadvantages
SRC	Acts as a current source; applicable to high-voltage outputs; no output filter magnetics required; low device stress during overload or short circuit; suitable for high power	CCM causes high component stress as f_s approaches f_r; poor regulation at or near no-load conditions transfer function is very nonlinear
PRC	Acts as a voltage source; operates at no load; applicable to low-voltage outputs; full cycle, DCM is similar to PWM buck converter operation	Requires overload and short-circuit protection; snubbers required at $f_s <$ f_r; constant resonant circulating current independent of load
QRC, ZCS Fixed on time	Minimizes turn-off losses; recycles leakage inductance energy	Wide FM range required; high peak currents in components
QRC, ZVS Fixed off time	Eliminates switching losses; recycles leakage inductance energy	Operates above resonant frequency; small FM range required; high peak voltage across devices
SPRC	Low device stress during overload or short circuit; operates at no load	Additional passive components required unless parasitic elements can be used
MRC, ZVS (Modified Class E)	Eliminates switching losses; narrow range of f_s	Moderate peak voltage across devices; high peak currents; boost topology has a right-half-plane zero
PWM, clamped	Fixed-frequency switching; simple control loop; lower voltage stress; can use current mode control	Natural commutation achieved only at heavy load; turn-on transition losses can be high
PWM, ZVS	Fixed-frequency switching; minimizes switching losses in MOSFETs and secondary rectifiers; achieves high efficiency at highest switching frequency	ZVS difficult to achieve at light load; this creates switching noise

Abbreviations: SRC, series resonant converter; PRC, parallel resonant converter; QRC, quasi-resonant converter; ZCS, zero-current switching; ZVS, zero-voltage switching; SPRC, series–parallel resonant converter; MRC, multiresonant converter; PWM, pulse width modulation; DCM, discontinuous current mode; CCM, continuous current mode.

topologies while conducting the same rms current. Also, the added resonant components increase the complexity and often increase the cost.

First, basic topologies or classifications are discussed. Advantages and disadvantages of the most popular types are presented in Table 10.1. Various resonant mode control circuits are discussed, most of which use frequency modulation to achieve regulation and stability. Specific topologies are then presented.

10.1   CLASSIFICATION

The following classifications are considered the most relevant in resonant mode converters: (1) series or parallel loading of the resonant circuit, (2) fixed- or variable-frequency operation, (3) continuous or discontinuous resonant current flow, and (4) zero-current switching (ZCS) or zero-voltage switching (ZVS). Various topologies exist within each of these modes, such as (1) class A, (2) class D, (2) quasi-resonant, (3) clamped mode, and (4) class E or ZVS.

The class A topology is shown in Figs. 7.1 and 9.13 for an inverter using thyristors. The parallel resonant "tank" is driven by square-wave switching. Control is achieved by varying the switching frequency. For a converter, the load could be supplied by a transformer with secondary rectifiers and a filter for dc output, as shown in Fig. 9.14. Here, operation is at a fixed frequency and control is achieved by varying the phase angle between the two stages. The class A topology is primarily used in high-power systems.

The class D topology is shown in Fig. 7.7 for thyristor commutation. The topology for bipolar or MOSFET switches is similar to the half bridge for voltage-switching, series resonance and is similar to the push–pull stage for current-switching, parallel resonance. For a converter, the load could be supplied by a transformer with secondary rectifiers and a filter for dc output. The converter operates at a variable frequency for control of the output voltage, and the switching frequency is typically above the resonant frequency.

The quasi-resonant topology is the most popular and universal. The term *quasi-resonant* applies to switching operations that can occur before or after the natural commutation or resonant period. Under these conditions, the current can be continuous mode (CCM) or discontinuous mode (DCM), respectively. The placement of the load in the quasi-resonant topology can be in series (QSR) or in parallel (QPR) with the resonant LC components. A three-element passive resonant circuit can provide quasi–series–parallel (QSPR) operation, which eliminates some disadvantages of both the QSR and the QPR. Four and five elements can further enhance operation, albeit more complexity. In a quasi-resonant topology, the circuit components and the operating frequency can produce ZCS or ZVS. In addition, a single switch will produce half-wave operation while two or more switches will produce full-wave operation. Most of this chapter is devoted to the quasi-resonant topology.

The clamped-mode topology resembles that of a PWM topology but with an L and C connected in series resonance with the load. The converter operates at a fixed frequency, and control is achieved by PWM techniques. However, in a full-bridge configuration, each of the in-line transistors switches at 50% duty cycle, and control is achieved by phase shift modulation (PSM) between the pairs of transistors. This operation is similar to the PSM techniques for reactive power-handling capability in inverters, as discussed in Section 6.2. In addition, current mode control can effectively be used by sensing the sinusoidal current (inherent high ΔI) in the inductor. The parasitic capacitance effects that produce waveform spikes in PWM current-sensing topologies do not occur in the resonant topology.

The ZVS topologies are derived from the class E inverter but without its narrow-band, tuned circuit. Zero-voltage switching eliminates the power dissipation inherent in having a MOSFET discharge its shunt capacitance, commonly called *charge dumping*. This is accomplished by circuit design that forces the voltage across the MOSFET to zero at turn-on. Power dissipation at turn-off is significantly reduced by delaying the rise of voltage until the current has fallen. In a double-ended configuration, the ZVS topology delivers energy to the load in a manner similar to a forward converter when the MOSFET is on and in a manner similar to a flyback converter when the switch is off. In single-ended circuits, ZVS requires a fixed off-time for the MOSFET. The major disadvantage of ZVS in these single-ended circuits is the inherent high-voltage requirement of the MOSFET, typically several times the input voltage. This high-voltage requirement can be reduced by multiresonant techniques but with added complexity. The high-voltage requirement also can be overcome by using bridge topologies, where device voltage is clamped to the input voltage and a complementary device acts as a synchronous switch. These latter topologies, called multiresonant or resonant transition, operate at fixed frequency and allow quasi-resonant action during the turn-on and turn-off periods but operate similar to PWM topologies during the rest of the cycle.

10.2   RESONANT MODE CONTROL CIRCUITS

To achieve the high operating efficiency in resonant mode or quasi–resonant mode converters, the active power device or devices are driven to saturation during conduction and are driven to cutoff for nonconduction. To achieve output voltage regulation, two methods of time ratio control may be used. Each control method employs some form of timing circuit and voltage feedback for closed-loop operation.

First, the conduction time of the transistor is fixed, such as in a one-shot multivibrator, and the operating frequency varies as a function of input voltage and load current. This mode of operation requires a voltage-controlled oscillator (VCO) and has the characteristics of a voltage-to-frequency converter. This method is used for topologies employing ZCS.

Second, the off-time of the transistor is fixed and the operating frequency varies as a function of input voltage and load current. Again, a VCO is required. This method is used for topologies employing ZVS.

Various integrated circuits are available for "single-chip" control of resonant converters, in either the ZCS or ZVS mode, and are discussed by generic part numbers. Reference the manufacturers' published data for pertinent calculations of switching frequency, conduction period, nonconduction period, soft start, and fault logic.

10.2.1 1860

A block diagram of the UC1860 is shown in Fig. 10.1. The IC has a variable-frequency oscillator (VFO) with emitter-coupled logic (ECL) and can operate at switching frequencies from 1 kHz to 3 MHz. A one-shot multivibrator sets a fixed pulse width, or on-time output, and dominates the maximum frequency of the VFO. The error amplifier has a 20-MHz gain bandwidth. Special housekeeping functions add versatility such as soft start, undervoltage lockout (UVLO), hiccup mode and programmable restart, and zero-current detection. Each totem-pole output provides 2-A peak-current drive pulses with typical rise and fall times of 20 ns into a 1-nF load.

10.2.2 33066

A block diagram of the MC33066 is shown in Fig. 10.2. The IC has a VFO that can operate at switching frequencies to 1 MHz. It can operate in three modes: fixed on-time, variable frequency; fixed off-time, variable frequency; and combinations that change from fixed on-time to fixed off-time as the frequency increases. This latter feature allows a QPR converter to operate in the DCM for high input

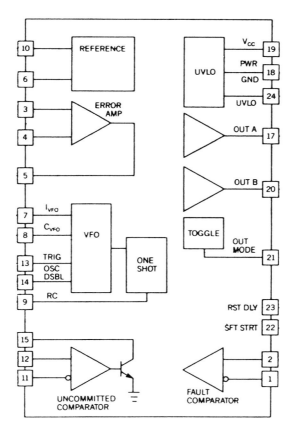

Fig. 10.1. Block diagram of UC1860. (Courtesy Unitrode Corp.)

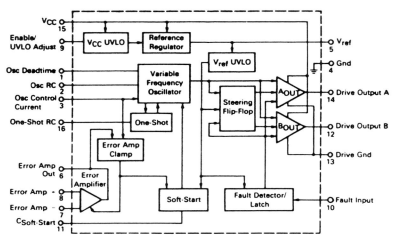

Fig. 10.2. Block diagram of MC33066. (Courtesy Motorola, Inc.)

voltages and in the CCM for low input voltages. Various housekeeping functions add versatility. Each totem-pole output can sink or source 1.5 A with typical rise and fall times of 20 ns into a 1-nF load.

10.2.3 Other Devices

Other multipurpose devices are available. The UC1861/64/65 ICs have features similar to the UC1860 except the maximum switching frequency is 1 MHz and each device is tailored to a specific topology. The 1861 has alternating, fixed off-time outputs suitable to drive dual-switch ZVS converters. The 1864 has parallelable, inverted outputs suitable to drive single-switch ZVS converters. The 1865 has alternating, fixed on-time outputs suitable to drive dual-switch ZCS converters.

The ML4815 is a single-ended ZVS controller that can operate up to 2 MHz and has a 3-MHz unity-gain bandwidth. This IC has 2 A totem-pole output, versatile housekeeping functions, and current mode control features and feedforward features in voltage mode control. The ML4816 is a dual-output ZCS controller that can operate up to 2.5 MHz and has versatile housekeeping functions.

The MC34067 is a ZVS controller that can operate at switching frequencies up to 1 MHz. The IC is optimized for double-ended push–pull or bridge-type converters. The fixed off-time, frequency-modulated IC has versatile housekeeping functions.

The UC 1875 and the ML4818 are quad output, phase-shifted controllers suitable for multiresonant ZVS bridge topologies. These fixed-frequency ICs significantly simplify the control circuit design and allow full 180° phase shift range. These devices can operate up to 1 MHz, provide up to 1.5 A drive outputs, and have versatile housekeeping functions. A programmable turn-on delay time can accommodate the specific resonant transition times of various systems. The 1875 allows the delay of the $A–B$ outputs to be separately programmed from the $C–D$ outputs. This feature accommodates the different transition times that can occur in each leg of a full-bridge topology, as discussed in Section 10.7.1.

10.3 ZERO-CURRENT SWITCHING

The term *zero-current switching* will be used to describe quasi-resonant and multiresonant topologies in half-wave and full-wave configurations [1]. Units providing low output power may appear similar to buck or boost regulators, but with an added C or LC to form a resonant or quasi-resonant circuit. Single-ended and double-ended converters use a transformer for isolation, and the leakage inductance of the transformer forms a part of the resonant inductor.

10.3.1 Quasi-Resonant Regulators

A half-wave, quasi-resonant regulator and operating waveforms are shown in Fig. 10.3. When gate voltage is applied to Q_1, a half-sinusoid current I_{L1} flows through L_1, C_1, the output filter, and the load. Here Q_1 is naturally commutated off at t_p, which produces ZCS, and the drain voltage V_{DS} rises. During and after the pulse, the capacitor voltage V_{C1} rises and decays at a rate depending on the load resistance and the output voltage is the average of this waveform. (In a conventional PWM topology, the waveform across CR_2 would be a rectangular waveform and E_0 would be the average of that waveform.) A quantity of energy is transferred from the input to V_{C1} during each pulse. Output voltage regulation is achieved by varying the switching frequency, as in frequency modulation. Neglecting the effects of initial load conditions, the approximate relations are given as

$$M = E_0/E_{\text{in}} \tag{10.1a}$$

$$Z_r = \sqrt{L_1/C_1} \tag{10.1b}$$

$$Q_r = R_1/Z_r \tag{10.1c}$$

$$f_r = 1/2\pi\sqrt{L_1 C_1} \tag{10.1d}$$

$$t_p = 1/2f_r = \pi\sqrt{L_1 C_1} \tag{10.1e}$$

$$\gamma = f_s/f_r \quad (\text{switching frequency}/\text{resonant frequency}) \tag{10.1f}$$

$$I_{\text{pk}} = E_{\text{in}}/Z_r \tag{10.1g}$$

$$V_{C1,\text{pk}} = 2E_{\text{in}} \tag{10.1h}$$

The voltage conversion ratio M versus γ is plotted in Fig. 10.3C for various values of Q_r. It can be seen that M is very sensitive to load variations. At light load (high value of R_1), the unused energy is stored in C_1 and the decay slope of V_{C1} increases, which increases the average output voltage unless the switching frequency is reduced. At no-load conditions, the switching frequency may be quite low, resulting in a wide operating frequency range.

A full-wave, quasi-resonant regulator and operating waveforms are shown in Fig. 10.4. The relations of Equation (10.1) apply. However, the excess energy stored in C_1 after each pulse is now transferred to the source via a current flow through the body diode of Q_1. (A bipolar transistor would require an antiparallel diode.) At light load, the reverse current increases and at no load a full cycle of circulating current flows in L_1. The voltage conversion ratio M is plotted in

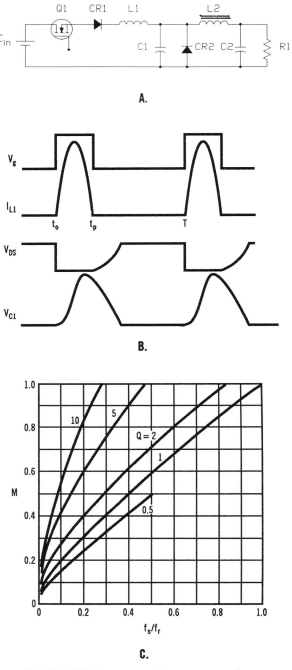

Fig. 10.3. Half-wave, quasi-resonant regulator.

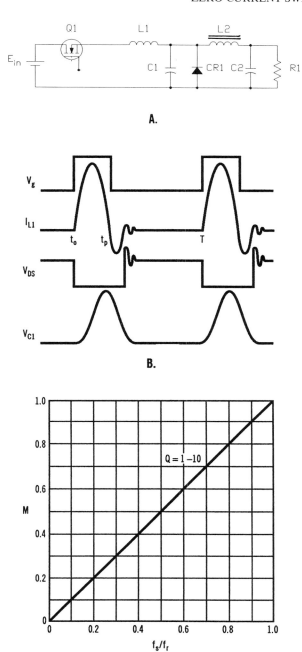

A.

B.

C.

Fig. 10.4. Full-wave, quasi-resonant regulator.

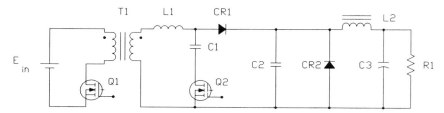

Fig. 10.5. Single-ended ZCS forward converter. (Patent to Vicor Corp.)

Fig. 10.4C. Here M is a linear function of γ and is independent of Q_r. Thus, the operating frequency range of the full-wave topology is much narrower than the half-wave topology.

10.3.2   Quasi-Resonant Converters

The converter shown in Fig. 10.5 is Vicor Corporation's patented implementation of a forward converter switching at zero current [2]. In this topology, each switch cycle delivers a "quantized" packet of energy to the converter output with Q_1 turning on and off at zero current. Thus the converter delivers a constant power output. Since the output voltage is not related to transformer voltage, regulated, multiple outputs cannot be achieved. An LC tank circuit composed of L_1 (the controlled leakage inductance of T_1) and C_2 forms an energy transfer between input and output during the conduction time of Q_1. Bidirectional energy flow cannot occur with this topology, because CR_1 will only permit "half-wave energy transfer" within the tank circuit. Thus, the performance and waveforms of this converter are similar to Fig. 10.1 except the current and voltage will decrease and increase, depending on the turns ratio of T_1. An additional feature of this topology is Vicor Corporation's patented implementation of a magnetizing-current mirror for core reset, which is accomplished by C_1 and Q_2 [3]. This technique provides better utilization of the transformer flux and allows a duty cycle higher than the 50% maximum inherent with the conventional forward converter.

The converter and waveforms shown in Fig. 10.6 are AT&T's patented implementation of a double-ended, multiresonant, push–pull converter switching at zero current [4]. Inductors L_1 and L_2 represent the leakage inductance of T_1 and C_3 represents the resonant capacitor. The voltage waveform across C_3 is similar to the waveform shown in Fig. 10.4. Also, C_1 and C_2 form an additional resonance with the magnetizing inductance of T_1. Besides the push–pull circuit shown, half-bridge and full-bridge topologies are readily implemented.

10.4   SERIES RESONANCE

The term *series resonance* will be used to describe resonant and quasi-resonant topologies in series resonant converters (SRCs). As the name implies, the SRC has its load in series with the resonant circuit elements and acts as a current source

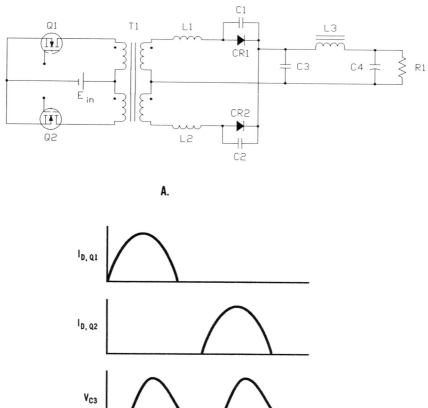

A.

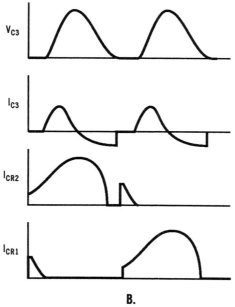

B.

Fig. 10.6. Double-ended, quasi-resonant converter. (Patent to AT&T).

with a high-impedance output. The SRC has several modes of operation, including two categories of CCM, two categories of DCM, and a switching frequency below or above the natural resonant frequency.

10.4.1 Full Wave

A SRC is shown in Fig. 10.7 as a full-bridge topology with IGBTs as active devices. The most useful operating mode for the SRC is in the range $0.5 < f_s/f_r < 1$, where f_s in the switching frequency and f_r is the natural resonant frequency. This is because the IGBT stops conducting as the current passes through zero in the DCM. Output voltage control is achieved by varying f_s. However, the ratio of the resonant circuit impedance to the reflected load impedance has a significant effect on operating parameters. Various parameters in the CCM are defined as

$$M = N_p E_0/N_s E_{in} = nE_0/E_{in} \tag{10.2a}$$

$$Z_r = \sqrt{L_r/C_r} = 2\pi f_r L_r = 1/2\pi f_r C_r \tag{10.2b}$$

$$Q_r = Z_r/n^2 R_L \tag{10.2c}$$

$$f_r = 1/2\pi\sqrt{L_r C_r} \tag{10.2d}$$

$$t_p = 1/2f_r = \pi\sqrt{L_r C_r} = \pi L_r/Z_r = \pi C_r Z_r \tag{10.2e}$$

$$\gamma = f_s/f_r \tag{10.2f}$$

$$1/M = \sqrt{Q_r^2(f_s/f_r - f_r/f_s)^2 + 1} \tag{10.2g}$$

The conversion ratio of M versus γ for various values of Q_r, with appropriate device voltage drops and computer analysis and the combination of Equation (10.2g), is plotted in Fig. 10.8. The boundary between CCM and DCM is at $Q_r = (4/\pi)f_s/f_r = 1.27\gamma$. As R_L increases, Q_r decreases, and the converter may enter the DCM and the output voltage can no longer be controlled without a decrease to $f_s < 0.5$. In this DCM and between $0.2 < f_s/f_r < 0.5$, the approximate relation is

$$M = (4/\pi Q_r)(f_s/f_r) = 1.27Q_r\gamma \tag{10.3}$$

The peak current in the IGBTs, the resonant components, and the transformer primary is a function of the resonant impedance and the reflected load impedance and is given by

$$I_{pk} = E_{in}/Z_r + nE_0/n^2 R_L = (E_{in} + nE_0 Q_r)/Z_r \tag{10.4}$$

This equation is plotted in Fig. 10.9 for various values of Q_r. This figure shows the peak current increases significantly as R_L decreases, resulting in an increase in Q_r, and as γ increases in an uncontrolled manner. However, for $Q_r = 1.27$, I_{pk} remains relatively unchanged as γ increases.

The current waveforms for various operating conditions are shown in Figs. 10.7B–E. In Fig. 10.7B, the converter is operating in the DCM at $\gamma < 0.5$.

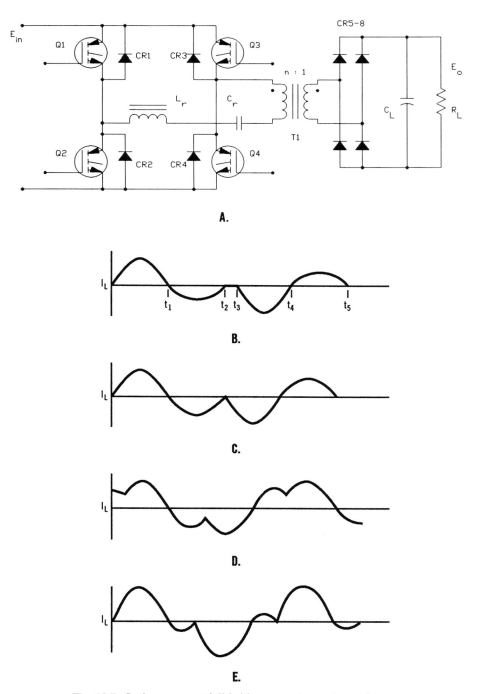

Fig. 10.7. Series resonant full-bridge converter and waveforms.

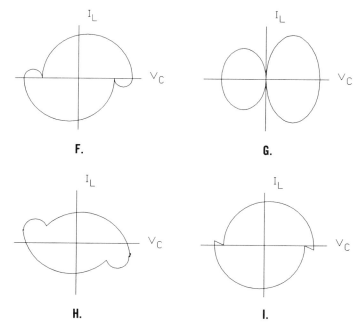

F.   G.

H.   I.

Fig. 10.7. (*Continued*)

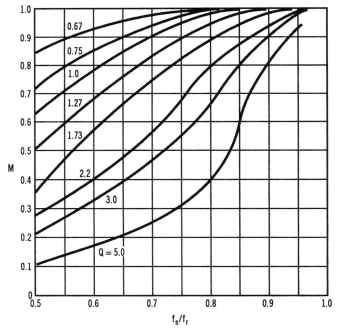

Fig. 10.8. Series resonant converter voltage conversion ratio versus frequency ratio for various Q.

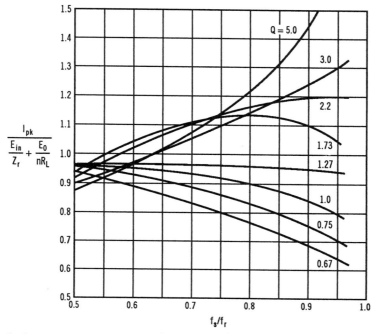

Fig. 10.9. Series resonant converter peak current ratio versus frequency ratio for various Q.

From 0 to t_1, Q_1 and Q_4 conduct. At t_1, the current reverses, as in a commutating or "ring-over" manner, and the current flows through CR_1 and CR_4 and back to the source, until t_2. From t_2 to t_3, all transistors are gated off and no current flows. From t_3 to t_4, Q_2 and Q_3 conduct, and from t_4 to t_5, current flows through CR_2 and CR_3 and back to the source. Thus, ZCS is achieved and the transistors are naturally commutated. In Fig. 10.7C, the converter is operating in the CCM at $\gamma = 0.5$ and into an output short-circuit condition. The peak current in the rectifiers is almost equal to the peak current in the transistors as no energy in the resonant tank is delivered to the load but is returned to the source. Thus, at $\gamma = 0.5$, the SRC can operate indefinitely into an output short circuit, at ZCS, and with minimal stress on components. This mode of operation is most useful when starting a converter with a capacitive load and the capacitor will charge at a faster rate than for $\gamma < 0.5$. In Fig. 10.7D, the converter is operating in the CCM at $\gamma > 0.5$. In this mode, a high value of Q_r results in a considerable duration of ring-over current in an in-line rectifier (CR_3 or CR_4), which is still flowing when the in-line transistor (Q_4 or Q_3) turns on. In this forced commutation, ZCS is achieved only at transistor turn-off, and considerable power dissipation can occur at transistor turn-on, due to the recovery time of other devices.

Zero-current switching at turn-on, and inherent at turn-off, can be achieved by reducing Z_r to reduce Q_r, as shown by the waveform of Fig. 10.7E. Reducing Z_r results in a higher peak current in the transistors but reduced current in the rectifiers. This reduction in Z_r also allows the commutating current to return to zero before the transistors are turned on. Thus, switching losses are practically eliminated and high efficiency is obtained by choosing devices with a low saturation voltage. Care must be exercised to limit γ; otherwise a loss of voltage control will occur as M approaches unity. However, this approach is ideal for charging capacitors in pulsed systems and is further discussed in Section 10.4.2.

At the beginning of the switching cycle, the capacitor voltage is given by

$$V_C = \pi Q_r M^2 E_{\text{in}} f_r / 2 f_s \tag{10.5}$$

but the peak voltage $V_{C,\text{pk}}$ is a function of the CCM parameters and the voltage increase caused by the ring-over current that flows back to the source. Computer analysis of $V_{C,\text{pk}}/E_{\text{in}}$ versus γ for various values of Q_r, with appropriate device voltage drops, is plotted in Fig. 10.10. This figure shows the capacitor voltage can approach infinity as R_L decreases or approaches a short circuit as γ increases in an uncontrolled manner. However, for $\gamma = 0.5$, the maximum capacitor voltage is twice the input voltage, regardless of Q_r and/or R_L. For $Q_r = 1.27$, the peak capacitor voltage remains relatively unchanged as γ increases, which is also the condition for the peak current.

A state-plane trajectory (V_c vs. I_L) is very useful for determining or observing operating conditions. When viewed by an oscilloscope, the trajectory can show start-up transients, subharmonic oscillations, and peak component stresses. The state-plane trajectory is shown in Figs. 10.7F–I for the conditions producing the current waveforms of Figs. 10.7B–E, respectively. Note that the waveforms of

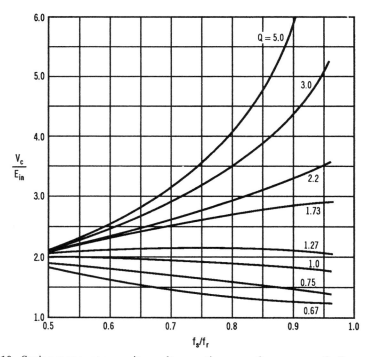

Fig. 10.10. Series resonant capacitor voltage ratio versus frequency ratio for various Q.

Figs. 10.7F, G, and I show ZCS conditions. The trajectory shown in the output short-circuit condition of Fig. 10.7G is a true series resonant, resonant-impedance-limited plot of the circulating current where the current and voltage are displaced 90°.

10.4.1.1 Component Characteristics   The semiconductor switches of a SRC conduct a higher peak current than a PWM converter of equal output power rating. In a ZCS topology, IGBTs are an ideal choice since conduction losses are less than those of MOSFETs. The reduced switching losses also allow operation at switching frequencies equivalent to PWM topologies using MOSFETs. Also, the capacitance of the IGBT is much less than that of the MOSFET, which reduces the effects of the charge dumping loss when not operating at ZVS. The resonant components L_r and C_r in the SRC must be rated to conduct an rms current that is relatively high, especially in the ZCS mode. The peak current is calculated from Equation (10.4) and the rms value is calculated from the equations of Table 1.2, part A, waveform. The capacitor is typically a polypropylene or mica formulation with inherent low ESR. The inductor is typically wound with Litz wire. Ferrite core material can be used for low-power converters, but caution must be exercised due to the high-frequency harmonic voltages across the inductor and the strong magnetic field established in the air gap (if used), which in essence makes the inductor an induction heater. For high-power converters, an air core toroid is an ideal choice. In this case, only conduction losses occur and the magnetic field is contained within the winding.

10.4.1.2 Power Transformer Characteristics   The power transformer is designed for power-handling capability and operating frequency, as in the PWM converter. Leakage inductance, considered a problematic parasitic in PWM topologies, forms a portion of the overall resonant inductance, perhaps up to 20% of the total requirement. However, a transformer with a very high percentage of leakage inductance generates high core loss, and the high magnetic field increases eddy current losses in the winding. The effects of transformer capacitance are minimized, due to the sinusoidal current flow in the resonant tank.

10.4.2   Full-Wave Bridge, Pulsed Load

The SRC of Fig. 10.7A is an ideal topology for charging capacitors in pulsed systems. In this case, R_L is replaced by a pulsed load such as a laser, TWT, klystron, or flash lamp, and C_L is the energy storage element. The converter replenishes the energy storage in the capacitor after each output load pulse. In fact, the SRC recharges the capacitor to the required voltage and then ceases switching, or switches intermittently to maintain the charge, until the next pulse occurs. Computer analysis shows that ZCS can be achieved under all operating conditions by reducing Q_r to 0.9 approximately. The value of γ (f_s/f_r) can remain relatively high as M approaches 1 since a comparator in the control circuit turns off the converter when the desired output voltage is reached.

The control ICs listed in Section 10.2 inherently terminate their output pulse when a fault or over/under voltage occurs. Terminating the pulse in the middle of the half-sinusoid pulse could produce excessive voltage across a transistor at

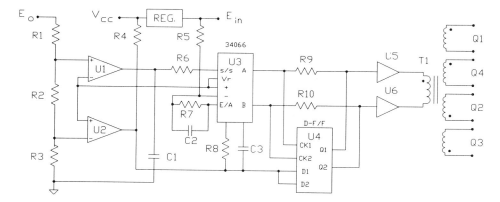

Fig. 10.11. ZCS control circuit for SRC.

turn-off. In the ZCS mode, it is most desirable to complete the half-sinusoid current pulse and stop the switching after the current has crossed through zero. The simplified control circuit in Fig. 10.11 shows a 34066 control IC (U_3) with a dual D-type flip/flop (U_4) added to ensure ZCS at turn-off. Here U_1 and U_2 are comparators and U_5 and U_6 are driver ICs for T_1, which gates the bridge transistors. The maximum operating frequency is set by R_8 and C_3. With no output voltage, the output of U_1 is low and R_6 sets $\gamma = 0.5$. The load capacitor initially acts as a short circuit, but the converter operates at ZCS and the primary current flow is as shown in Fig. 10.7C. When the output voltage reaches approximately 70% of final value, the output of U_1 goes high and the switching frequency increases to its maximum value (typically $\gamma = 0.8$). The primary current is then as shown in Fig. 10.7E and ZCS is maintained. When the desired output voltage is reached, the output of U_2 goes low, removing the D inputs to the flip-flop. The output of U_4, Q_1 or Q_2, will remain high for the duration of the half-sinusoid pulse and then go low and remain low as long as the output of U_2 is low. Hysteresis can be added to U_2 to maintain a desired incremental output voltage. In the event of a short circuit or overload, the output of U_1 will go low and the converter will immediately revert to switching at one-half the resonant frequency for component protection. Feedforward is accomplished by R_5 and the error amplifier of U_3. This changes f_s inversely proportional to changes in a sampled portion of E_{in}.

Example

For the bridge converter of Fig. 10.7A, the following parameters are given: $E_{\text{in}} = 270$–330 V, $E_0 = 2$ kV average, and $P_0 = 5$ kW. An output capacitor delivers a 25-A, 40-μs pulse to a load with an output droop voltage not to exceed 20 V. Thus, $C_0 = i\,dt/dv = 25 \times 40 \times 10^{-6}/20 = 50$ μF. The pulse repetition frequency (PRF) is 2200 Hz, or a 454-μs period. The switching may be blanked during and shortly after the pulse and the recharge time is selected as 400 μs. The capacitor is recharged at $i = C\,dv/dt = 50 \times 20/400 = 2.5$ A. The output power

$P_0 = E_0 I_c = [(2010 + 1990)/2] \times 2.5 = 5$ kW. Alternatively, the output power $P_0 = E/T = 0.5C(E_1^2 - E_2^2)T = 0.5 \times 50 \times 10^{-6}(2010^2 - 1990^2)/(400 \times 10^{-6}) = 5$ KW. The equivalent output load $R = E^2/P = 2000^2/5000 = 800$ Ω, which will be reflected to the primary.

Now, the primary-side parameters can be calculated for the given parameters: $f_r = 100$ kHz, $M_{max} = 0.93$, $\gamma_{max} = 0.74$, and $Q_r = 0.9$ for ZCS. From Equation (10.2a), $n = E_{in}M/E_0 = 270 \times 0.93/2000 = 0.125$. From Equation (10.2b), $Z_r = n^2 R_L Q_r = 0.125^2 \times 800 \times 0.9 = 11.2$ Ω. Initially, the switching frequency is 50 kHz to provide ZCS. The control circuit will increase the switching to 74 kHz, at $E_{in} = 270$ V, when C_0 exceeds approximately 1400 V and until C_0 recharges to 2010 V, while maintaining ZCS. To maintain a relatively constant recharge rate for variations of input voltage, f_s may be controlled by the feedforward circuit of Fig. 10.11. At $E_{in} = 330$ V, $M = 270 \times 0.93/330 = 0.76$. From Fig. 10.8, $\gamma = 0.58$ and $f_s = 58$ kHz.

The transistor conduction time is constant at $t_p = 1/2f_r = 5$ μs. From Equation (10.2e), $L_r = t_p Z_r/\pi = 5 \times 10^{-6} \times 11.2/\pi = 18$ μH and $C_r = t_p/\pi Z_r = 5 \times 10^{-6}/11.2\pi = 0.14$ μF. From Equation (10.4), the peak current in the primary is $I_{pk} = (330 + 0.125 \times 0.93 \times 2000)/11.6 = 48.5$ A. This is also the peak current in the IGBTs. The rms current rating for the capacitor and transformer windings can be calculated from the equations of Table 1.2, part A. The peak capacitor voltage is obtained from Fig. 10.10 as $V_c = 1.9 \times 330 = 627$ V. [Equation (10.5) gives the initial voltage and does not account for the charge produced by the ring-over current.]

10.5 PARALLEL RESONANCE

A parallel resonant converter (PRC) is shown in Fig. 10.12 as a half-bridge topology with IGBTs as active devices. As the name implies, the PRC has its load connected across the resonant-circuit capacitor and acts as a voltage source with a low-impedance output. This requires a *LC* output filter, as in the buck-derived converter. The resonant capacitor may be located across the transformer secondary; however, very high ripple-current-handling capability will be required for low-voltage outputs. Also, locating the capacitor across the primary provides a common primary-side parts configuration for a family of equal-power converters. Only the transformer secondary, rectifiers, and output filter need change for different output voltages.

The PRC has several modes of operation, including CCM, DCM, and a switching frequency below or above the natural resonant frequency. Operation above resonance is best suited for MOSFETs where ZVS can be used to eliminate "charge-dump" losses of the device's capacitance. Operation in the CCM mode, either below or above resonance, produces a very nonlinear conversion ratio at high values of Q. Operation in the DCM mode produces ZCS and a linear conversion ratio, at the expense of higher peak currents in the primary. However, IGBTs can readily conduct the higher current, as compared to MOSFETs of the same die size, and ZCS minimizes switching losses. Various parameters are

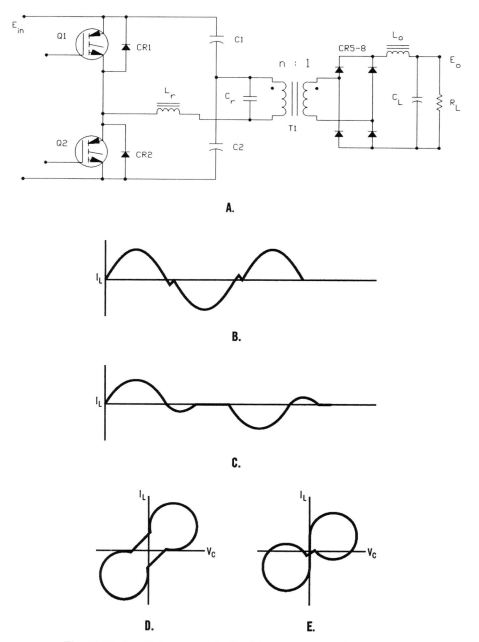

Fig. 10.12. Parallel resonant, half-bridge converter and waveforms.

defined as

$$M = N_p E_0 / N_s E_{in} = n E_0 / E_{in} \tag{10.6a}$$

$$Z_r = \sqrt{L_r / C_r} = 2\pi f_r L_r = 1/2\pi f_r C_r \tag{10.6b}$$

$$Q_p = n^2 R_L / Z_r \tag{10.6c}$$

$$f_r = 1/2\pi\sqrt{L_r C_r} \tag{10.6d}$$

$$t_p = 1/2 f_r = \pi\sqrt{L_r C_r} = \pi L_r / Z_r = \pi C_r Z_r \tag{10.6e}$$

$$\gamma = f_s / f_r \tag{10.6f}$$

$$1/M = \sqrt{(f_s/f_r)^2/Q_p^2 + \left[1 - (f_s/f_r)^2\right]^2} \quad (\gamma > 0.5) \tag{10.6g}$$

The conversion ratio of M versus γ for various values of Q_p is plotted in Fig. 10.13. For the DCM ($\gamma < 0.5$), M is a linear ratio. The inductor current waveform and the state-plane trajectory for two DCM operating conditions are shown in Figs. 10.12B–E. Operation at $\gamma = 0.45$ is represented by Figs. 10.12B and D for a high value of output current. The voltage across the capacitor is a sine wave and the current in the transformer primary and secondary is a square wave. Operation at $\gamma = 0.3$ is represented by Figs. 10.12C and E for a lower value of output current. The voltage across the capacitor is a series of alternating half sinusoids with dead time. The current in the transformer primary and secondary is a trapezoid wave with dead time. At a no-load condition, $\gamma = 0.15$ approximately and a full-cycle, circulating resonant current flows for $2t_p$. The state-plane trajec-

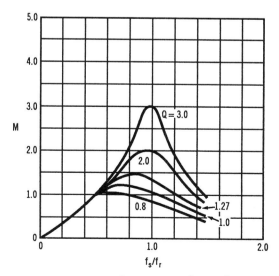

Fig. 10.13. Parallel resonant converter voltage conversion ratio versus frequency ratio for various Q.

tory at no load is the same as the SRC under a short-circuit condition, as shown in Fig. 10.7G.

Control of the PRC is accomplished by any of the FM ICs discussed in Section 10.2. A f_s range of 3 : 1 is typical for full load to no-load conditions. The output filter is designed in the same manner as the buck regulator of Section 8.3.2. Stability, compensation, and closed-loop performance are discussed in Section 11.6.

Example

For the half-bridge converter of Fig. 10.12A, the following parameters are given: E_{in} = 250–300 V, E_0 = 28 V, P_0 = 700 W, R_L = 1.12 Ω, f_r = 400 kHz, M_{max} = 0.9, γ_{max} = 0.45, and Q_p = 1. From Equation (10.6a), $n = E'_{in}M/E_0 = (250/2) \times 0.9/28 = 4$. From Equation (10.6c), $Z_r = n^2R_L/Q_p = 16 \times 1.12/1 = 18 \ \Omega$. At 400 kHz, the pulse period is 1.25 μs. From Equation (10.6e), $L_r = 1.25 \times 10^{-6} \times 18/\pi = 7.2 \ \mu$H and $C_r = 1.25 \times 10^{-6}/18\pi = 22$ nF. At full load, $f_s = 0.45 \times 400 \times 10^3 = 180$ kHz and f_s decreases to approximately 60 kHz at no load. The maximum peak current in the IGBTs is $2E'_{in}/Z_r = 2 \times 125/18 = 13.9$ A. The maximum capacitor voltage is $2E'_{in} = 2 \times 150 = 300$ V.

10.6 SERIES–PARALLEL RESONANCE

A series–parallel resonant converter (SPRC) can be obtained by adding a parallel inductor to the capacitor of a SRC [5, 6]. This three-element topology will improve the light-load operating characteristics, as compared to the PRC, albeit more complexity. In fact, it is possible to operate from no load to full load with a small range of f_s. A half-bridge SPRC (*LLC*) is shown in Fig. 10.14A. The following relations apply:

$$L_e = L_sL_p/(L_s + L_p) \tag{10.7a}$$

$$f_r = 1/2\pi\sqrt{L_eC_p} \tag{10.7b}$$

$$Z_r = \sqrt{L_e/C_p} \tag{10.7c}$$

$$Q_r = Z_r/n^2R_L \tag{10.7d}$$

$$f_p = 1/2\pi L_pC_p \tag{10.7e}$$

The voltage conversion ratio M versus f_s/f_r for various values of Q_r is plotted in Figs. 10.14B and C for $L_p = 2.5L_s$ and $L_p = L_s$, respectively. The parallel resonance of L_p and C_p produces $M = 0$ at

$$f_r/f_p = \sqrt{1 + L_p/L_s} \tag{10.8}$$

This topology significantly narrows the frequency control range, as shown in Figs. 10.14B and C, especially as L_p is decreased. However, high-frequency dynamics can cause a beat frequency between the "multiple resonance" and the

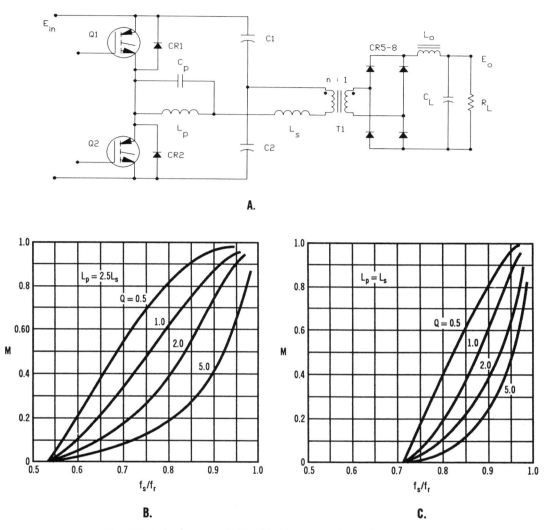

Fig. 10.14. Series–parallel half-bridge converter and waveforms.

switching frequency, which in turn has a strong impact on loop compensation and transient response.

10.7 ZERO-VOLTAGE SWITCHING

Over the past several years, a superfluity of ZVS topologies has been presented, but few have been applicable to a viable commercial product. The basic concept offers desirable features for switching frequencies in the 2–20-MHz range for quasi-resonant converters (QRCs) or multiresonant converters (MRCs). Operation in this frequency range renders the possibility of higher power densities. Zero-voltage switching eliminates the power dissipated by charge dumping of device capacitance in MOSFETs and Schottky rectifiers. Thus in an MRC, all semiconductor devices switch at zero voltage. However, some ZVS topologies offer such a poor power factor to the transformer that little volume reduction can be achieved for a reasonable efficiency. In other cases, ZVS at switching frequencies greater than the 2-MHz range is applicable only to low-power applications. This is due to the fact that the inherent device capacitance C_{oss} in high-current, high-voltage MOSFETs is a larger value than the calculated value required for the resonant circuit. Thus, the switching frequency must be reduced if these high-current devices are to be used.

Conversely, a reduction in C_{oss} is accompanied by an increase in $R_{DS,on}$ and efficiency is degraded. Device-to-device capacitance tolerance further impedes product reproducibility. The major disadvantage of ZVS in single-ended topologies is that MOSFET voltage rating must be several times greater than the input voltage value. This factor prevents use of this topology in power supplies operating from 115/230 V input. Lower device voltage ratings are achieved in half-bridge and full-bridge topologies, where the voltage of the nonconducting MOSFET is clamped to the dc bus. Low-power, high-density ZVS converters find applications in distributed power systems, especially in 48-V systems. These topologies have been analyzed in-depth in the cited references [7, 8].

10.7.1   ZVS, Phase Shift–modulated Converter

The ZVS topology presented is a fixed-frequency, PSM, full-bridge converter. This topology uses the parasitic elements of device capacitance and transformer leakage plus magnetizing inductance to accomplish a resonant-transition period [9]. Each MOSFET is turned on during this period, when its drain–source voltage is zero and when its body diode or parasitic capacitor is conducting. The converter diagram is shown in Fig. 10.15A with MOSFETs connected from the positive bus rail to the negative bus rail in a left leg and a right leg. The control circuit typically

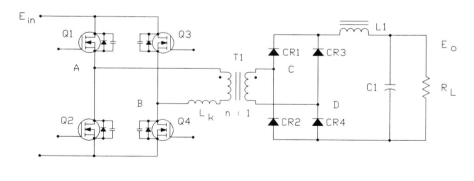

A.

Fig. 10.15. (A) Phase shift–modulated ZVS bridge converter. (B) Gate drive, drive current, drain voltage, transformer current and voltage. (C) Transformer current and voltage, rectifier current and voltage.

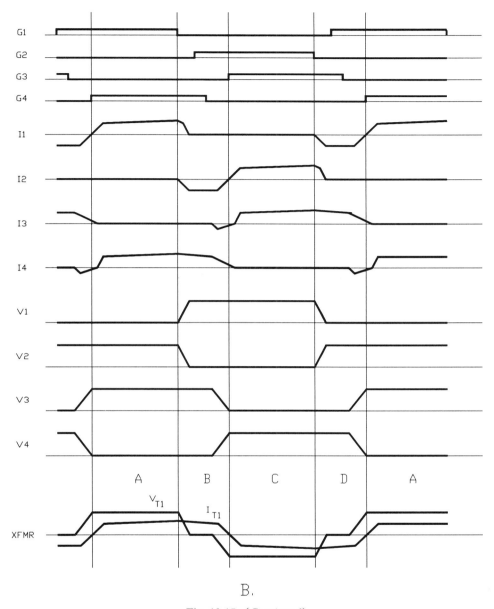

B.

Fig. 10.15. (*Continued*)

consists of an 1875 or 4818 PSM IC, which are discussed in Section 10.2.3. The control IC outputs are typically transformer coupled to the gates of Q_{1-4}. The body diode and the output capacitance of the MOSFETs are shown as parasitic elements, as is the transformer leakage inductance L_k.

Waveforms of primary-side operation are shown in Fig. 10.15B. The gate drive signals (G_{1-4}) are similar to the phase-shifted quasi-square-wave inverter base

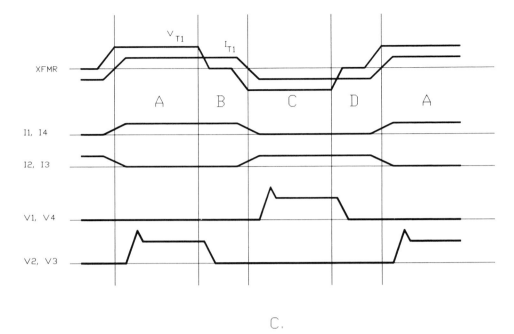

C.

Fig. 10.15. (*Continued*)

drive signals shown in Fig. 6.11A. Output voltage regulation is achieved in like manner by shifting the phase of one leg with respect to the other leg. In Fig. 10.15B, a fixed delay time between Q_1–Q_2 and between Q_3–Q_4 switching provides the resonant-transition period between C_{oss} of the MOSFET and L_k of the transformer. The inductive energy must be sufficient to drive the capacitors to the opposite bus rail during the transition time. The magnetizing inductance of the transformer and the reflected output filter inductance also affect operation, as discussed later. The MOSFET current (I_{1-4}) and voltage (V_{1-4}) plus the transformer current and voltage are shown. The negative regions of I_{1-4} indicate current is flowing in either the body diode or the device capacitance. In region A, Q_1 and Q_4 conduct to produce a positive voltage across the primary. In region C, Q_2 and Q_3 conduct to produce a negative voltage across the primary. Regions A and C are the power delivery periods. The secondary voltage is full-wave rectified and filtered to the desired dc output in the same manner as the buck-derived converter. Regions B and D are the resonant-transition periods At the start of B, $+I_4$ is flowing and as Q_1 is turned off, $+I_1$ decreases and the current commutates to $-I_2$, which indicates current flow through the body diode or C_{oss} of Q_2. Voltage V_1 rises to E_{in}. The transformer primary is shorted by Q_2 and Q_4, which is still conducting. Then Q_2 is turned on with $V_2 = 0$. When Q_4 is turned off, V_4 rises to E_{in} and $+I_4$ commutates to $-I_3$ and flows back to the source, through the body diodes and capacitance of Q_2 and Q_3 and the transformer primary. This allows the leakage energy to be returned to the input, as in a lossless snubber, rather than being dissipated in the elements. Next, Q_3 is turned on at $V_3 = 0$ to enter region C. Operation in region D is in the reverse order of operation in region B with the diagonal pairs of devices conducting. Note that the *left-leg* transition time (during $-I_1$ and $-I_2$) is longer than the *right-leg* transition time (during $-I_3$ and $-I_4$) and is a maximum at minimum output current. The difference in time is due to the leakage inductance L_k versus the reflected output filter inductance n^2L_1. The transition time t_t in the left leg is a function of L_k and C_{oss} of Q_1 and Q_2 plus the initial current in L_k. In the right leg, t_t is a function of n^2L_1, which is large enough to maintain primary current during the final portion of regions B and D, and the capacitors are charged quickly.

The relationship of the various inductances influences the operation of the converter. A large value of L_m increases the time required to reverse the transformer current during the initial portions of regions A and C. A large value of L_m adversely affects the ZVS conditions at light load. Inductance L_m can be decreased by placing an air gap in the transformer core, which in turn increases the magnetizing current and the MOSFET peak current. As a trade-off, the magnetizing current can be one-half of the maximum load current [10]. The added losses caused by this current increase are minimal. A large value of L_k increases the slope of the current waveform when voltage is applied to the primary. This increase in slope time reduces the effective duty cycle of the secondary current. Inductance L_k also affects the power stage transfer function by introducing additional damping to the output filter. The effective resistance is given by $R_d = 4L_kf_s/n^2$ [11]. A ratio of $R_d = 0.1R_L$ significantly reduces the peaking of the second-order LC output filter in the control-to-output transfer function.

Figure 10.15C shows the transformer secondary current and voltage plus the current in the secondary rectifiers and the voltage across these rectifiers. In Region B, the transformer voltage becomes negative and the current commutates from CR_1–CR_4 to CR_2–CR_3. The rectifiers should be Schottky or ultrafast recovery types. Snap-recovery rectifiers produce excessive reverse voltage upon recovery from the reverse current.

10.7.2 Optional Phase Shift Topology

An optional topology employing phase shift modulation is shown in Fig. 10.16. This converter has the advantage of operating from a standard PWM control IC, such as a 1525 or an 1846. The converter does not provide true ZVS as described in Section 10.7.1. However, it does eliminate the requirement for RC snubber networks common in PWM topologies. Elimination of the snubbers improves efficiency and reduces temperature rise. In PWM converters, the snubbers are typically placed across the MOSFETs or across the power transformer primary (C–D) to dissipate the energy stored in the transformer leakage inductance when a transistor turns off. In the subject converter, one of the MOSFETs in each leg is always conducting. This provides a current path to return the energy stored in the leakage inductance back to the source input. The current path is provided by the channel of one device and the body diode of an opposite device. The duty cycle in each leg varies as the pulse width changes in response to the control circuit. However, Q_1 and Q_3 each conduct for more than 50% of the cycle period.

Referring to the schematic and waveforms of Fig. 10.16, the primary of T_1 (A–B) can be connected directly to the A and B outputs of a 1525. The secondaries drive the small-signal MOSFETs as well as charge the low-value

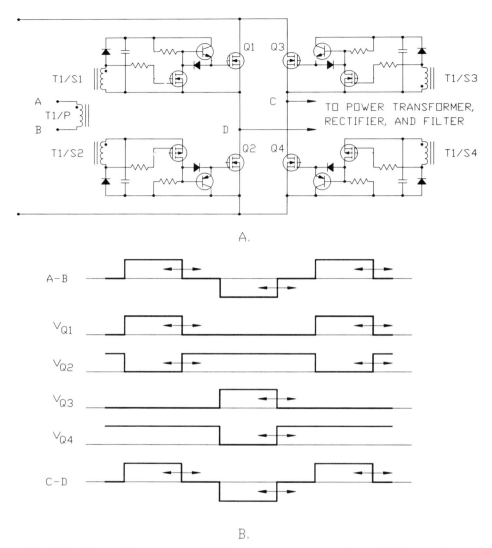

Fig. 10.16. Phase shift–modulated bridge converter operating from PWM control.

capacitors by bootstrap. These capacitors provide stored energy to charge C_{iss} of the power MOSFETS, Q_{1-4}. The low-current MOSFETs are chosen for minimum C_{iss} rather than minimum $R_{DS,\,\text{on}}$. The low C_{iss} reduces the load on T_1. In fact, the peak primary pulse current in T_1 is typically 500 mA. The threshold voltage of these devices provides good noise immunity. The emitter–follower bipolar junction transistors (BJTs) act as current amplifiers to charge or discharge the gates of Q_{1-4}.

As shown by the waveforms, Q_1 and Q_3 are held on by the *NPN* BJTs for a major portion of the cycle and are turned off by the *N*-channel MOSFETs. For example, if the half-period duty cycle at $A\!-\!B$ is 60%, Q_1 and Q_3 each conduct for

70% of the full cycle while Q_2 and Q_4 each conduct for 30% on alternate half cycles. Transistors Q_2 and Q_4 are turned on by the P-channel MOSFETs and are turned off by the PNP BJTs. Proper selection of component values can prevent simultaneous conduction of the power MOSFETs in each leg of the converter.

REFERENCES

1. F. C. Lee, "High-Frequency Quasi-Resonant and Multi-Resonant Converter Technologies," *IEEE Proceedings of the International Conference on Industrial Electronics* (October 1988).

2. Vicor Corporation, "Forward Converter Switching at Zero Current," U.S. Patent 4,415,959, 1983.

3. Vicor Corporation, "Optimal Resetting of the Transformer's Core in Single Ended Forward Converters," U.S. Patent 4,441,146, 1984.

4. AT & T Bell Laboratories, "High-Frequency Resonant Power Converter," U.S. Patent 4,823,249, 1989.

5. E. X. Yang and F. C. Lee, "Small-Signal Modeling of LLC-Type Series Resonant Converter," *HFPC Conference Proceedings* (May 1992).

6. R. Severns and I. Batarseh, "Resonant Converter Topologies with Three and Four Energy Storage Elements," *HFPC Conference Proceedings* (May 1992).

7. M. M. Jovanović and F. C. Lee, "DC Analysis of Half-Bridge Zero-Voltage-switched Multiresonant Converter," *IEEE Transactions on Power Electronics* (April 1990).

8. W. J. B. Hefferman and P. D. Evans, "Comparative Assessment of Zero Voltage Switched Off-line Power Convertor Topologies," *IEE Proceedings-B, Electric Power Applications* (March 1992).

9. General Electric Company, "Full-Bridge Lossless Switching Converter," U.S. Patent 4,864,479, 1989.

10. L. H. Mweene, C. A. Wright, and M. A. Schlecht, "A 1 kW 500 kHz Front-End Converter for a Distributed Power Supply System," *IEEE Transactions on Power Electronics* (July 1991).

11. V. Vlatković, J. A. Sabaté, R. B. Ridley, and F. C. Lee, "Small-Signal Analysis of the Phase Shifted PWM Converter," *IEEE Transactions on Power Electronics* (January 1992).

11

MODULATOR AND CONTROL ANALYSIS

11.0 Introduction

11.1 Amplifiers and Compensation
    11.1.1 Pole–Zero Pair
    11.1.2 Zero–Pole Pair
    11.1.3 Dual Zero-Pole Pair

11.2 Stability Analysis
    11.2.1 Frequency Response
    11.2.2 Bode Plots
    11.2.3 Nyquist's Criterion
    11.2.4 Conditionally Stable Systems
    11.2.5 Transient Response

11.3 PWM Voltage Mode Control
    11.3.1 Buck Converter
    11.3.2 Flyback Converter
    11.3.3 Boost Converter

11.4 PWM Feedforward Control
    11.4.1 Buck Converter
    11.4.2 Flyback Converter
    11.4.3 Boost Converter

11.5 PWM Current Mode Control
    11.5.1 Buck Converter
    11.5.2 Flyback Converter
    11.5.3 Boost Converter

11.6 Resonant FM Mode
    References

11.0 INTRODUCTION

This chapter discusses control and stability analysis of modulators typically used in regulators, converters, and inverters. The control compensation is achieved with *RC* networks, operational amplifiers, and feedback loops. The feedback loop(s) is analyzed from a stability view. The modulator controls a power stage that provides voltage (or current) scaling and commonly provides voltage isolation between input and output. The power stage feeds a filter that smooths the switching waveform, rich in harmonics, into the desired ac or dc output. A portion of the output is sensed and fed back to a difference (or error) amplifier with appropriate compensation to provide stable operation for all operating conditions. Stability criteria and performance analysis of various topologies is presented. Bandwidth (crossover) frequency is discussed relative to switching frequency and complex pole–zero considerations. The objective is to cross unity gain at a high frequency while maintaining good phase margin. Also, the performance of a converter or inverter is determined by its closed-loop output impedance (load changes) and audio susceptibility (line changes).

The discussion in this chapter applies primarily to equipment providing a dc output. However, the same principles of modulation and control apply to inverters providing a regulated ac output and operating in a quasi-square-wave mode or operating in a high-frequency pulse-demodulated mode with sine wave reference.

The pulse width modulator compares a control signal (or amplified error voltage) with a sawtooth (or triangle) waveform. The resulting output is a train of pulses of variable duty cycle and normally fixed frequency that controls a power switch or power stage in a standard PWM manner. Power stage PWM topologies are discussed in Chapters 6 through 9.

The four basic PWM topologies, as discussed in Section 8.1, may be controlled by the following three basic means. The advantages of direct duty cycle, feedforward, and current mode operation are analyzed and important design considerations are discussed. In addition, each topology may operate in the continuous current mode (CCM) or discontinuous current mode (DCM). However, the boost derived and buck–boost (flyback) derived topologies operating in the continuous current mode incur the right-half-plane zero that severely limits the bandwidth and makes compensation virtually impossible above the corner frequency of the inductor and load resistance.

For recognized practically, discussion of control in PWM topologies will be limited to the following:

1. Buck derived: CCM; voltage, feedforward, and current modes.
2. Flyback derived: DCM; voltage, feedforward, and current modes.
3. Boost derived: CCM; voltage, feedforward, and current modes.

The frequency modulator converts a control signal (or amplified error voltage) into a train of pulses of variable frequency and normally fixed pulse width. This is achieved via a voltage-controlled oscillator (VCO) that controls a power switch or power stage in a standard FM manner. The frequency modulator is used primarily in resonant mode systems, which are discussed in Chapter 10.

11.1 AMPLIFIERS AND COMPENSATION

The common operational amplifier (op-amp) is a direct-coupled high-gain integrated circuit amplifier that uses feedback to control its performance characteristics. The voltage gain is a function of the differential signals between the two inputs of the amplifier. The versatile op-amp can be used to convert voltage to

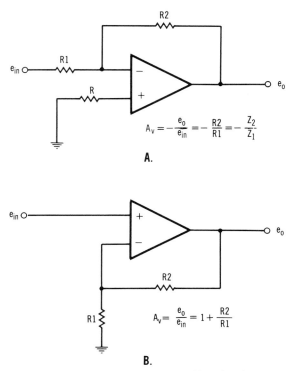

Fig. 11.1. Operational amplifier circuits.

voltage, with either an inverted output or noninverted output; voltage to current, as in certain PWM ICs; current to voltage; or current to current. A review of op-amp circuits is presented. The overall subject of op-amps has been aptly covered in other texts.

Referring to Fig. 11.1, the inverting op-amp of Fig. 11.1A has a gain of $A = -R_2/R_1$, and the noninverting op-amp of Fig. 11.1B has a gain of $A = (R_1 + R_2)/R_1 = 1 + R_2/R_1$. In closed-loop systems, compensation is normally required to achieve stable operations. Resistor–capacitor networks are used to provide a desired lag (pole), lead (zero), or combinations of lag–lead and vice versa, and the resistors of Fig. 11.1 are replaced with desired impedances. Several methods of circuit analysis are available: (1) the transfer function can be defined, the equations solved by calculator, and the gain versus frequency data plotted on semi-log or on log-log paper; (2) the circuit performance can be analyzed by computer software circuit analysis programs, and the transfer function (gain and phase) automatically plotted; (3) impedance graphs can be used to locate frequency break points, as in Section 1.8.

11.1.1 Pole–Zero Pair

An op-amp with a pole–zero pair for feedback is shown in Fig. 11.2A. The gain is $A = e_0/e_{in}$ in numeric value, and $A = 20 \log(e_0/e_{in})$ in decibels. Also, $A =$

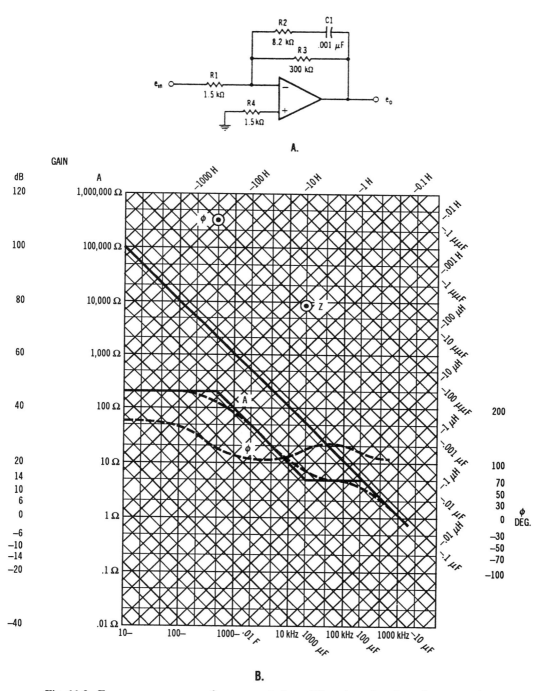

A.

B.

Fig. 11.2. Frequency response of compensated amplifier plotted on impedance graph.

$-Z_{fb}/Z_{in}$, where Z_{fb} is the feedback impedance and Z_{in} is the input impedance. Using the operator s, where $s = j\omega$, gain is given by

$$A = \frac{R_3}{R_1}\left(\frac{sR_2C + 1}{s(R_2 + R_3)C + 1}\right) \tag{11.1}$$

This equation has a numeric dc gain, a pole in the denominator, and a zero in the numerator. The following relations apply:

$$A_{dc} = R_3/R_1 \tag{11.2a}$$

$$f_1 = 1/2\pi(R_2 + R_3)C \tag{11.2b}$$

$$f_2 = 1/2\pi(R_2C) \tag{11.2c}$$

$$f_2/f_1 = (R_2 + R_3)/R_2 \tag{11.2d}$$

The frequency response for gain and phase is shown in Fig. 11.2B. The dc gain is easily calculated. The frequencies for the pole and zero can be calculated or can readily be located on the impedance graph and the gain rolls off at a -20-dB/decade slope between these points.

Example

Consider the circuit of Fig. 11.2A. It is desired to have (1) a dc gain of 200 (46 dB); (2) a pole at 500 Hz; (3) a zero at 20 kHz; (4) a commonly used op-amp. The impedance chart of Fig. 11.2B will be used to select component values. Choosing $C_1 = 1$ nF, the intersection of 1 nF and 20 kHz produces $R_2 = 8.2$ kΩ. Circle this point Z. For the desired pole at 500 Hz, the intersection of 1 nF and 500 Hz produces $R_2 + R_3 = 318$ kΩ. Circle this point P and choose $R_3 = 300$ kΩ. Since the desired dc gain is 200, $R_1 = 300$ k$\Omega/200 = 1.5$ kΩ. Next, the open-loop gain of the op-amp is drawn. The typical gain is 100,000 with a pole at 10 Hz and unity gain crossover at 1 MHz. A pole due to the op-amp occurs at 200 kHz, the gain decays at -20 dB/decade above 200 kHz. The phase curve also may be closely approximated on the impedance chart. The phase at 10 Hz is 180° and will be 135° at 500 Hz, 20 kHz, and 200 kHz and will approach 90° at 3 kHz. These points can be connected by the dashed line in Fig. 11.2B.

11.1.2 Zero–Pole Pair

The response of a higher bandwidth op-amp with a pole at the origin and a zero–pole pair for feedback is shown in Fig. 11.3. The circuit has a constant gain and a phase boost approaching 90° between f_1 and f_2. The gain is unity at f_3. The following relations apply:

$$A_1 = R_2/R_1 \tag{11.3a}$$

$$f_1 = 1/2\pi R_2C_1 \tag{11.3b}$$

$$f_2 = (C_1 + C_2)/2\pi R_2C_1C_2 \approx 1/2\pi R_2C_2 \tag{11.3c}$$

$$f_3 = (C_1 + C_2)/2\pi R_1C_1C_2 \approx 1/2\pi R_1C_2 \tag{11.3d}$$

$$f_2/f_1 \approx C_1/C_2 \tag{11.3e}$$

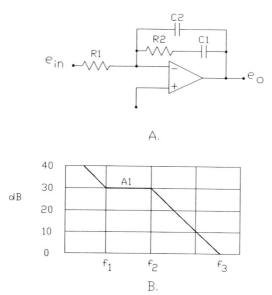

Fig. 11.3. Amplifier response with zero–pole pair.

This circuit is ideally suited to PWM topologies where the output filter is a single pole and the circuit is designed such that the open-loop crossover frequency of the system occurs between f_1 and f_2. Here, the op-amp gain A_1 is made equal to the reciprocal of the modulator gain at the desired crossover frequency and the phase boost compensates for the phase lag of the output filter at system crossover. This method inherently provides a high overall phase margin.

11.1.3 Dual Zero–Pole Pair

Two circuits and the response for a dual zero–pole pair is shown in Fig. 11.4. The following relations apply for the circuit of Fig. 11.4A:

$$A_1 = R_3/(R_1 + R_2) \tag{11.4a}$$

$$A_2 = R_3/R_2 \tag{11.4b}$$

$$f_1 = 1/2\pi R_1 C_1 \tag{11.4c}$$

$$f_2 = 1/2\pi R_3 C_2 \tag{11.4d}$$

$$f_3 = 1/2\pi R_2 C_1 \tag{11.4e}$$

$$f_4 = (C_2 + C_3)/2\pi R_3 C_2 C_1 \approx 1/2\pi R_3 C_3 \tag{11.4f}$$

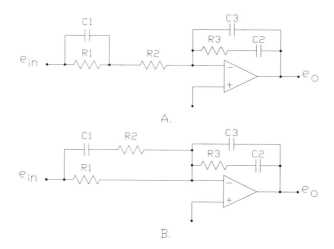

A.

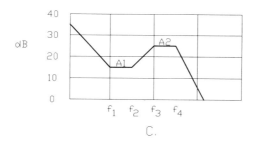

B.

C.

Fig. 11.4. Amplifier response with dual zero–pole pair.

The following relations apply for the circuit of Fig. 11.4B:

$$A_1 = R_3/R_1 \tag{11.5a}$$

$$A_2 = R_3(R_1 + R_2)/R_1 R_2 \approx R_3/R_2 \tag{11.5b}$$

$$f_1 = 1/2\pi R_3 C_2 \tag{11.5c}$$

$$f_2 = 1/2\pi(R_1 + R_2)C_1 \tag{11.5d}$$

$$f_3 = 1/2\pi R_2 C_1 \tag{11.5e}$$

$$f_4 = (C_2 + C_3)/2\pi R_3 C_2 C_3 \approx 1/2\pi R_3 C_3 \tag{11.5f}$$

This circuit is ideally suited to PWM topologies where the output filter has a double pole (LC) and operation is in the CCM. Typically, f_1 and f_2 are set near the resonant frequency of the LC output filter. Then f_3 and f_4 are set logarithmically equidistant above the desired crossover frequency that f_1 and f_2 are below the crossover frequency. This provides good phase margin over a wide operation range, as discussed in Section 11.2.

11.2   STABILITY ANALYSIS

The stability analysis ideas presented are basic to automatic control systems. There are two types of automatic control systems, open loop and closed loop. Also, there are several methods to analyze loop stability including (1) Laplace transforms (solving for the time response of the closed-loop transfer function), (2) frequency response (plotting the magnitude and phase of the open-loop function), (3) Nyquist criterion (analysis of polar plots for closed-loop plus inverse-polar plots for open loop and closed loop), (4) Bode representation (frequency response method applied to closed loop), (5) root-locus method (frequency response and transient response), (6) Nichols chart (best used when feedback is unity), and (7) state-space methods (matrix and vector representation of sets of differential equations). Methods 2, 3, and 4 are discussed herein.

Most power conversion equipment change an input of varying magnitude and frequency to a specific and controlled output. Thus, to control the output, a closed loop is required. The actual output is measured, and a signal corresponding to this measurement is fed back to be compared with the input in some manner. An error signal is produced that is usually the amplified difference between the output and the input. This error signal then drives a power stage to correct automatically any discrepancy between the desired and the actual output. This implies negative feedback is required but does not ensure stability.

Negative feedback exists in an open-loop system if the error output changes in a negative direction in response to a positive change in the input. However, phase shifts in the open-loop system can cause a negative feedback system to exhibit instability (positive feedback) at some frequencies. If the phase shift reaches $-180°$ at a particular frequency, the feedback will reinforce or amplify the applied input at this frequency and the system will be unstable if the magnitude of the open-loop gain is equal to unity at this frequency.

A single control loop is shown in Fig. 11.5A where G represents the overall system gain and H represents the feedback parameter. The following relations apply:

For open-loop response,

$$v_0/v_i = GH \tag{11.6}$$

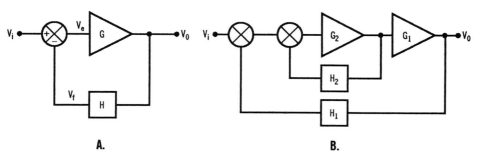

Fig. 11.5. Control loops with gain and feedback: (A) single loop and (B) double loop.

For closed-loop response,

$$v_f = v_0 H = v_e GH \tag{11.7a}$$

$$v_0 = v_e G = v_i G - v_0 GH \tag{11.7b}$$

$$v_0/v_i = G(1 + GH) \tag{11.7c}$$

However, Equation (11.7c) can be multiplied by $1/H$ to obtain the following result:

$$v_0/v_i = GH/(1 + GH)(1/H) \tag{11.7d}$$

This implies that closed-loop performance can be represented in normalized form from the analysis of open-loop response (GH) and the inverse of the feedback parameter $(1/H)$. This technique is used in Section 11.2.2 and Fig. 11.7. The overall gain G is the arithmetic product of the power stage and the output filter and applies to all converters, inverters, phase controllers, and regulators.

A double control loop is shown in Fig. 11.5B. This figure could be a representation of a current mode control where the outer loop is the voltage control and the inner loop is the current control, as discussed in Section 11.5. The overall gain is given by

$$\frac{v_0}{v_i} = \frac{G_1 G_2/(1 + G_2 H_2)}{1 + G_1 G_2 H_1/(1 + G_2 H_2)} \tag{11.8}$$

11.2.1 Frequency Response

Second-order lag networks (reference the LC output filter of Fig. 1.10D and Fig. 11.6A) have a 180° phase shift and a frequency roll-off of -40 dB per decade (-2 slope) above the resonant frequency. A 180° phase shift at unity gain in an open-loop system will cause oscillations in a closed-loop system. As discussed in Section 11.1, operational amplifiers with zero ($+1$ slope) and pole (-1 slope) compensation are used to tailor the overall response to provide the desired phase margin at unity gain. Impedance charts are helpful although computer analysis using Spice (or versions thereof) provide important information on complex parameters. Computer analysis can provide inverse-polar plots where the closed-loop response is the same as the open-loop response with a unity shift in the origin.

11.2.1.1 Impedance Chart Analysis
The circuit of Fig. 11.6A can be analyzed by impedance graphs in the same manner as that used in Section 1.8.4. The impedance graph is used to determine component values for compensation, by the intersection of these parameters with frequency, and to plot the frequency response. The frequency response plot also shows the crossover frequency f_c. At dc, the phase margin is 180° (90° for a pole at the origin). When a pole is encountered, the phase margin changes by $-\tan^{-1}(f_c/f_p)$ or by $-2[\tan^{-1}(f_c/f_{p2})]$ for a double pole. When a zero is encountered, the phase margin changes by $+\tan^{-1}(f_c/f_z)$. The

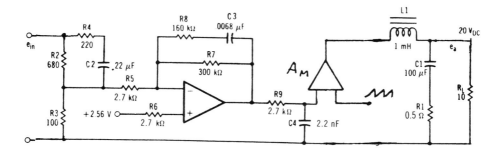

A.

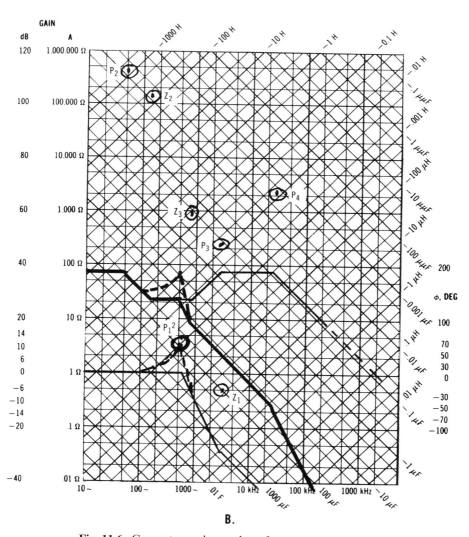

B.

Fig. 11.6. Converter and open-loop frequency response.

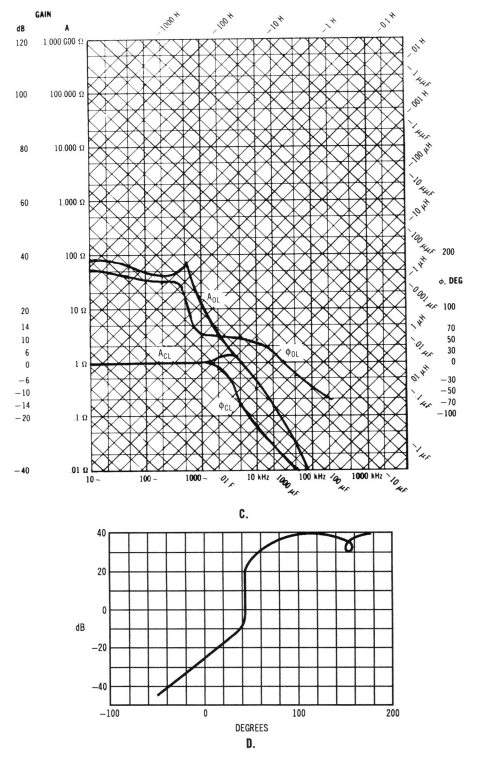

C.

D.

Fig. 11.6. (*Continued*)

general expression for phase margin is

$$\phi_M = 180° + \sum \tan^{-1}\frac{f_c}{f_z} - \sum \tan^{-1}\frac{f_c}{f_p} - \sum 2\left(\tan^{-1}\frac{f_c}{f_{p2}}\right) \qquad (11.9)$$

Example

In the circuit of Fig. 11.6A, the selected or calculated component values are as shown, the modulator has a gain of 5 (14 dB), and the modulator switching frequency is 50 kHz. The desired crossover frequency is midway between the output filter resonant frequency and the switching frequency or $f_c = \sqrt{f_r f_s}$. Locating the 1-mH line and the 100-μF line, the intersection is at 500 Hz, which is the resonant frequency. Circle this point as P_1^2, indicating a pole with a -2 slope. Follow the 100-μF line down to the intersection with 0.5 Ω, at 3.2 kHz and circle this point Z_1, indicating a zero with a $+1$ slope. Starting at unity gain, draw a line horizontal to 500 Hz, then at -2 slope to 3200 Hz, then at -1 slope on down. This curve is then the response of the output filter, except for the underdamping of $R_L = 3.16\sqrt{L/C}$, resulting in about a 10-dB increase due to peaking, as shown by the dashed line.

The input attenuator has a gain of $100/(100 + 680) = 0.128 = -17.8$ dB and the error amplifier has a gain of $300/2.7 = 111 = 40.9$ dB, resulting in a stage gain of $0.128 \times 111 = 14.24 = 23.1$ dB, along with a modulator gain of $5 = 14$ dB. Thus the overall dc gain of the system is $14.2 \times 5 = 71 = (23 + 14)$ dB $= 37$ dB. However, it is desired to reduce the gain peaking of the filter by 9 dB at 500 Hz, while maintaining 37 dB gain in the amplifier at dc. Moving down to 28 dB gain at an octave or two either side of 500 Hz, a zero is desired at 150 Hz, and a pole is then required at approximately 53 Hz. (Note that $53 = 150/(10^{9/20})$.) A pole and zero may be introduced in the amplifier by adding R_8 and C_3 across R_7 (reference Section 11.1.1), and the pole will exist at the intersection of $R_7 + R_8$ and C_3 at 53 Hz. Choosing $C_3 = 6.8$ nF, $R_7 + R_8 = 468$ kΩ, from Equation (11.2b). Then, $R_8 = 468 - 300 = 168$ kΩ; use 160 kΩ. The pole is then at the intersection of 460 kΩ and 6.8 nF, or 51 Hz. Circle this point P_2. The zero will occur at the intersection of 160 kΩ and 6.8 nF, or 146 Hz, from Equation (11.2c). Circle this point Z_2. Starting at 37 dB, draw a line horizontal to 51 Hz, then at -1 slope to 146 Hz, then horizontal to 500 Hz.

To offset the double pole and the -2 slope of the output filter, a zero–pole will be added across R_2, with R_4 and C_2. Choosing 800 Hz for the zero and $C_2 = 0.22$ μF, the intersection results in an R of 900 Ω. Since $R_2 = 680$ Ω, $R_4 = 900 - 680 = 220$ Ω, which is the calculated value from Equation (11.4c). Circle this intersection Z_3. Solving for f_p from Equation (11.4e), a pole occurs at 3.3 kHz, also the intersection of 0.22 μF and 220 Ω. Circle this point P_3. Now, continue the horizontal line from 500 to 800 Hz, then at a $+1$ slope to 3.3 kHz, then horizontal. Thus, this $+1$ slope will offset the -2 slope of the output filter and with the zero in the output, the crossover slope should be -1 for stability. Now an additional pole will be added near one-half the switching frequency by R_9 and C_4. For $C_4 = 2.2$ nF and $R_9 = 2.7$ kΩ, the intersection is 26.8 kHz. Circle this point P_4. This additional pole may not be required since the -1 slope coincides with an error amplifier bandwidth of 2 MHz.

The overall frequency response is drawn as the heavier line from the combination of the filter and amplifier/modulator response. The curve shows definite break points, except for the simulated peaking of the output filter, and the unity gain crossover is approximately 6 kHz.

11.2.1.2 Computer Analysis A Spice computer analysis of the open-loop gain, A_{OL} and phase ϕ_{OL} of the circuit of Fig. 11.6A is shown in Fig. 11.6C. The crossover frequency is 5000 Hz, the phase margin is 45°, and the gain margin is 25 dB. The phase margin is more graphically represented by plotting gain versus phase (a substitute for the chart technique developed by N. B. Nichols), as shown in Fig. 11.6D. Observe that an adequate phase margin of 45° minimum is maintained for gain variations of ± 12 dB or an arithmetic variation of 15.8 : 1. In the exact Nichols chart, an overlay plot is used to transform the open-loop response into a closed-loop response.

The circuit of Fig. 11.6A also complies with the parameters for switching frequency suppression. A suggested rule-of-thumb is that the open-loop gain should be at least -40 dB at $2f_s$.

11.2.2   Bode Plots

The important analysis by H. W. Bode applies to linear systems whose poles and zeros are all in the left half of the s plane. The analysis showed that the phase angle part of the frequency response was uniquely related to the magnitude part. Thus *Bode plots* of the normalized closed-loop response can be plotted from the open-loop frequency response. The circuit of Fig. 11.6A is redrawn in Fig. 11.7 as a closed-loop equivalent for computer analysis. Here, a signal, e_{in}, from a sweep generator is injected into the loop after the feedback network. An identical feedback network with inverting amplifier is added to the output at e_a. Installation of R_X and R_Y complete the loop. The added circuitry represents the function $1/H$ of Equation (11.7). The closed-loop response, taken from the inverting amplifier output at e_0, is shown in Fig. 11.6C as A_{CL} and ϕ_{CL}. Slight peaking occurs near the crossover frequency due to the effective damping factor, but the system is stable. Note that the initial statement in this section referred to all poles and zeros being in the left-half plane. Thus using Bode analysis for boost and

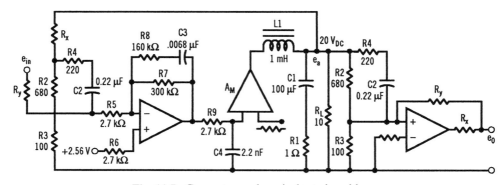

Fig. 11.7. Converter and equivalent closed loop.

buck–boost converter topologies is quite complicated due to the right-half-plane zero with its 90° phase lag.

Bode plots and stability tests can be readily performed with proper instrumentation. Referring to Fig. 11.6A, e_0 is connected to e_{in} in an actual closed-loop system. A sinusoidal signal from a network analyzer is magnetically coupled into the error amplifier output (in series with R_9). This node provides a low impedance source and a high impedance load, as seen by the analyzer. The analyzer measures the magnitude ratio and the phase difference between its input and the output signals, and this information can be automatically plotted. For high-performance analysis, a dynamic signal analyzer can provide information, based on high-speed calculations, in the time domain, frequency domain, and modal domain by using its periodic random noise source rather than a sinusoidal source. In fact, signals in different domains can be viewed simultaneously. The time domain provides a record of signals versus time. The frequency domain provides a spectrum of a signal versus frequency and can easily resolve small signals in the presence of large ones. The modal domain can provide a three-dimensional view of a signal versus distance and can also derive time domain data.

11.2.3 Nyquist's Criterion

A graphical technique, developed by H. Nyquist, uses a polar coordinate graph to plot the locus of the transfer function GH in the complex plane. Then the amplitude is the magnitude of the vector and the phase is the angle of the vector. When the analysis of circuits such as those of Fig. 11.6A is performed by computer, the real part of the vector can be plotted on the abscissa (x axis) and the imaginary parts of the vector can be plotted on the ordinate (y axis).

The stability criteria is that the loci must not enclose the $-1 + j0$ point. The loci of a simple and stable system is shown in Fig. 11.8A. The phase margin is measured as the angle of loci trajectory as it intercepts the unity gain circle. The gain margin is measured at the loci trajectory as it intercepts the abscissa. A system represented by the loci of Fig. 11.8B is unstable. A system represented by the loci of Fig. 11.8C is conditionally stable. In this latter case, a decrease or an increase in overall gain will cause the system to oscillate.

In an open-loop system, the loci begins along the abscissa at the real magnitude of open-loop voltage gain A_v with $\omega = 0$ (dc and thus zero phase) and progresses toward the origin as ω approaches infinity. In a closed-loop system, the real magnitude of the vector is unity for $\omega = 0$, and the loci proceeds from $+1 + j0$ to the origin. Applying computer analysis of this technique to the circuit of Fig. 11.7A results in the graph of Fig. 11.8D for closed-loop performance and shows a stable system. Again, the real part of the output voltage is plotted on the x axis and the imaginary part of the voltage is plotted on the y axis. This graph also shows an overshoot factor of 1.4 or 2.9 dB, which agrees with that shown by the frequency response curve A_{CL} of Fig. 11.6C.

11.2.4 Conditionally Stable Systems

Conditionally stable systems, as shown in Fig. 11.8C, will oscillate when the overall gain is within a range that produces a phase shift of 180° or more from input to

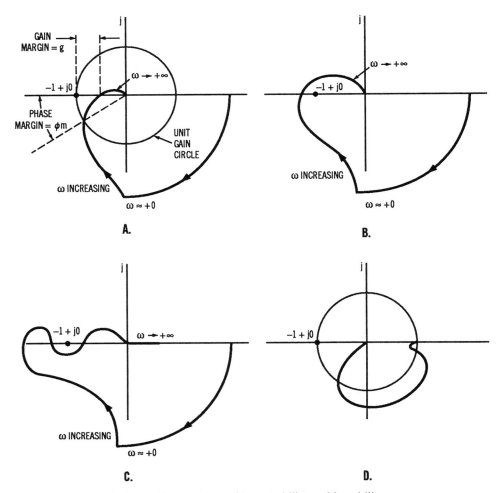

Fig. 11.8. Nyquist plots of loop stability and instability.

output. A hypothetical open-loop circuit is shown in Fig. 11.9A with poles or zeros occurring each half decade. The output L_1–C_1 filter produces a double pole at 100 Hz. The output R_2–C_2 results in a pole at 316 Hz. A zero from R_1–C_1 occurs at 1 kHz and a zero from R_3–C_2 occurs at 3.16 kHz. Thus at 100 Hz, the system roll-off is at a -2 slope (-40 dB per decade) to 316 Hz, then a -3 slope to 1 kHz, then a -2 slope to 3.16 kHz, then a -1 slope above 3.16 kHz.

Rather than plotting the frequency response, the phase versus magnitude plot of the transfer function for various values of gain indicates more dramatic results, as shown in Fig. 11.9B. By interpretation of the curves, any value of dc gain A_v between 6 dB (2) and 52 dB (398) will cause a phase shift exceeding 180° at unity gain output. Thus the system will oscillate in a closed-loop mode for these specific conditions. Due to the higher values of gain and the steep gain slope, the transient recovery time is normally shorter and the slew rate is higher than for unconditionally stable systems. However, purposely designing a conditionally stable system for

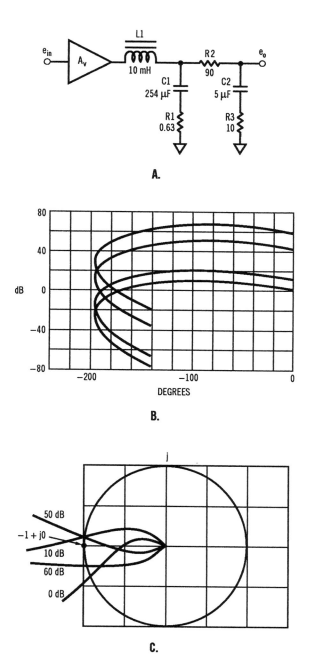

Fig. 11.9. Amplifier and output filter with phase versus gain plot and Nyquist plot.

normal operating conditions is poor practice and usually produces instability during start-up or during periods of abnormally low line input.

The curves of Fig. 11.9C result from the real and imaginary vectors of the Nyquist plot. Again, any value of dc gain between 6 and 52 dB will enclose the $-1 + j0$ point, producing instability.

11.2.5 Transient Response

Transient response is the ability of a system to recover from a step change in input voltage or output load. The maximum transient overshoot and undershoot amplitudes for the buck converter are given by Equation (8.12) and can be derived for other topologies. The transient recovery time is a function of the phase margin, the bandwidth, and the damping factor of the output filter. Insufficient phase margin (typically below 15°) will cause the output voltage to ring or oscillate due to underdamping. A high phase margin (typically above 75°) will result in an overdamped condition, possibly producing a long recovery time. The damping factor and also the phase margin can be calculated from closed-loop data obtained from test or from the simulated closed-loop analysis of Figs. 11.6C and 11.7. A filter response and phase shift versus normalized frequency is shown in Fig. 6.9 for various damping ratios. The increase in gain due to peaking (for $\delta < 0.707$) is given by

$$A_{pk} = 20 \log\left(1/2\delta\sqrt{1 - \delta^2}\right) \qquad \text{dB} \qquad (11.10)$$

The phase margin is given by

$$\phi_M = \tan^{-1}\sqrt{4\delta^2/\left[(4\delta^4 + 1)^{0.5} - 2\delta^2\right]} \qquad (11.11)$$

A rule-of-thumb for transient recovery time as a function of bandwidth f_c and for a phase margin above 60° (critical damping) is given by

$$t_{rec} = 0.707/f_c \qquad (11.12)$$

Example

The circuit of Fig. 11.6A was analyzed in the example of Section 11.2.1.1 for open-loop operation and in Section 11.2.2 for closed-loop operation. The closed loop of Fig. 11.6C shows $A_{pk} = 1.4 = 2.9$ dB. From Equation (11.10), $\delta = 0.387$ and from Equation (11.11), $\phi_M = 41°$ as compared to a 45° observed from the open-loop phase plot. From Equation (11.12), the approximate recovery time for a step change in load is $t_{rec} = 0.707/5000 = 141$ μs. The peak amplitude of overshoot and undershoot, as calculated by Equation (8.12), is below 5% for a 50% load change.

11.3 PWM VOLTAGE MODE CONTROL

Voltage mode (duty ratio) control has been the most common method of modulation. This is a single-loop topology in which the output voltage is sensed and compared to a reference voltage. Pulse width modulation integrated circuits abound for this function. Duty cycle and dc relations as well as design of compensation networks are normally straightforward. The open-loop dc load regulation is good, although dynamic line regulation (susceptibility) is poor. Duty ratio control requires more error amplifier gain bandwidth than with current mode control. The second-order output filter pole with its 180° phase shift limits dynamic response since control changes must propagate through the filter. Also, load capacitance added by the end user will lower the output filter pole frequency and may cause a conditionally stable system.

11.3.1 Buck Converter

Power stage operation of the buck regulator and buck-derived converters is discussed in Chapters 8 and 9, respectively. A forward converter and frequency response waveforms are shown in Fig. 11.10. The ramp (sawtooth) voltage is defined as v_s and the control (error) voltage is defined as v_c. The drive voltage is v_D, which controls the duty cycle of the power switch. The basic duty cycle control and dc control is given by

$$D = v_c/v_s = nE_0/E_{in} \qquad (11.13)$$

and the control to output dc gain is given by

$$v_0/v_c = E_{in}/nv_s \qquad (11.14)$$

The output filter has attenuation slope changes at the following frequencies:

$$f_p = 1/2\pi\sqrt{L_0C_0} \qquad (11.15)$$

$$f_z = 1/2\pi R_c C_0 \qquad (11.16)$$

Various heuristic parameters apply also. A dual zero–pole pair around the error amplifier is used for compensation and the two zeros typically coincide with the output filter double pole. Alternately, the first zero is placed slightly lower than the pole to avoid a conditionally stable system. The second zero is placed higher than the pole to obtain adequate phase boost. One of the amplifier poles may coincide with the ESR of the filter capacitor and the other pole may also be coincident or may be approximately $\frac{1}{3}f_s$ for high-frequency rolloff. The error amplifier should have gain at the output filter double pole, otherwise input line modulations at this frequency will pass through without attenuation, and the system will not meet conducted susceptibility requirements. It is desirable to have the bandwidth (crossover frequency) at least midway between the filter double pole and the switching frequency, or $f_c \geq \sqrt{f_p f_s}$. Circuit analysis and compensation is best discussed by the following example.

Example

For the circuit of Fig. 11.10, the following parameters are known: $E_0 = 12$ V, $I_0 = 0.5$–20 A, $R_0 = 0.6$–24 Ω, $E_{in} = 120$–160 V, $f_s = 80$ kHz, $D = 0.5$ max,

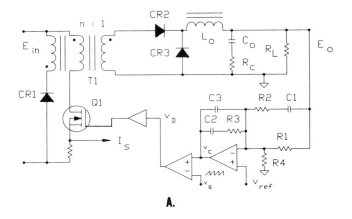

A.

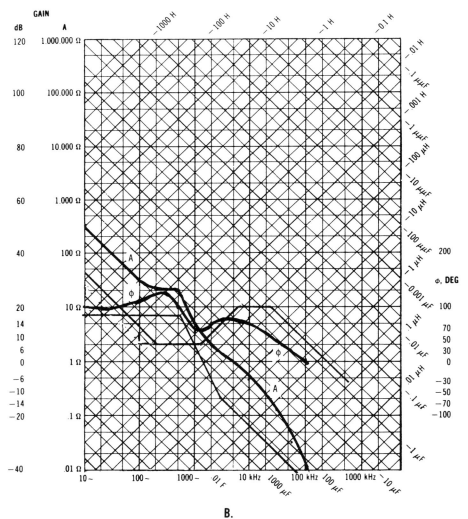

B.

Fig. 11.10. Voltage mode control, buck converter.

$L_0 = 90\ \mu\text{H}$, $C_0 = 1000\ \mu\text{F}$, $R_c = 0.05\ \Omega$. One output of a dual output PWM IC with a 2.5 V ramp is used, which relates to 5 V full scale. To allow for margin and losses, let $D = 0.4$. From Equation (11.13), $n = 0.4 \times 120/12 = 4$. The modulator gain is $v_0/v_c = E_{\text{in}}/4 \times 5 = 6$ (16 dB) to 8 (18 dB). From Equations (11.15) and (11.16), the filter break points are $f_p = 531$ Hz and $f_z = 3.2$ kHz. The desired bandwidth is $f_c = \sqrt{f_p f_s} = 6.5$ kHz. The modulator and filter response is plotted in Fig. 11.10B and shows -21 dB amplitude at f_c. Therefore, the error amplifier must have a $+21$-dB gain at f_c, or $A_2 = 21$ dB $= 11.2$. Using Equation (11.4), the following parameters are calculated. Let $R_3 = 30$ kΩ, then $R_2 = 3.3$ kΩ, and $C_1 = 6.2$ nF for $f_3 = 6500$ Hz. Drawing a line at a $+1$ slope with decreasing frequency, the zeros of the error amplifier cannot be placed at the filter double pole, otherwise the error amplifier would have no gain at this frequency. A 6-dB (2) gain is selected, which places the second zero near 1400 Hz. Then, $R_1 = 15$ kΩ. Now, the first zero is placed, say, an octave below the filter double pole, or 250 Hz. Then, $C_2 = 22$ nF. Adding a pole at 25 kHz, $C_3 = 270$ pF. The response of the error amplifier is shown in Fig. 11.10 with zeros at 250 and 1400 Hz and poles at 6500 and 25 kHz. The overall system response (magnitude and phase) is shown by the heavy lines of Fig. 11.10 and indicates a phase margin of 70° at f_c, with a phase margin above 50° up to 20 kHz or up to a 15-dB increase in overall gain.

11.3.2 Flyback Converter

Power stage operation of the flyback converter is discussed in Section 9.4. A flyback converter and frequency response waveforms are shown in Fig. 11.11 for operation in the DCM. The DCM mode is typically chosen over the CCM mode, which has a right-half-plane zero. (This right-half-plane zero typically reduces the bandwidth of the converter and is more difficult to compensate than a converter operating in DCM.) The ramp (sawtooth) voltage is defined as v_s and the control (error) voltage is defined as v_c. The drive voltage is v_D, which controls the duty cycle of the power switch. The basic duty cycle control and dc control is given by

$$D = v_c/v_s = nE_0/(E_{\text{in}} + nE_0) \tag{11.17}$$

and the control to output dc gain of the modulator is given by

$$\frac{v_0}{v_c} = \frac{E_{\text{in}}}{nv_s}\sqrt{\frac{n^2 R_L}{2L_p f_s}} = \frac{E_{\text{in}}}{v_s}\sqrt{\frac{R_L}{(2L_p f_s)}} \tag{11.18}$$

The output filter capacitor and load has attenuation slope changes at the following frequencies:

$$f_p = 1/2\pi R_L C_0 \tag{11.19}$$

$$f_z = 1/2\pi R_c C_0 \tag{11.20}$$

A single pole around the error amplifier is used for compensation and may

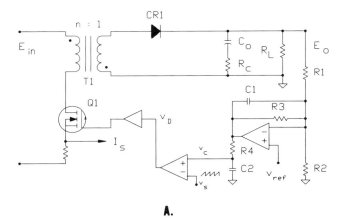

A.

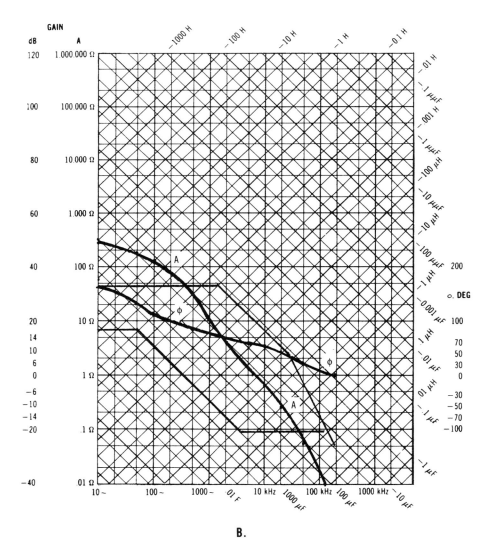

B.

Fig. 11.11. Voltage mode control, flyback converter.

coincide with the filter zero at a slightly lower frequency for increased dc gain. Circuit analysis and compensation is best discussed by the following example.

Example

For the circuit of Fig. 11.11, the following parameters are known: $E_0 = 12$ V, $I_0 = 3$ A, $R_0 = 4$ Ω, $E_{in} = 120$–160 V, $f_s = 80$ kHz, $D = 0.5$ max, $L_p = 1.1$ mH, $C_0 = 800$ μF, $R_c = 0.06$ Ω, $v_s = 2.5$ V, $f_c = 6.5$ kHz, dc gain = 50 dB. (Note that E_{in}, E_0, f_s, and f_c are the same as in the example of Section 11.3.1.) For a maximum switch on time of 6 μs, $D = 6/12.5 = 0.48$. From Equation (11.17), $n = DE_{in}/E_0(1 - D) = 9.2$. The modulator gain is $v_0/v_c = (E_{in}/2.5)\sqrt{4/2 \times 0.0011 \times 80{,}000} = 7.24(17.2$ dB) to 9.65 (19.7 dB). From Equations (11.19) and (11.20), the filter capacitor and load break points are $f_p = 40$ Hz and $f_z = 3.2$ kHz. The modulator and filter response is plotted in Fig. 11.11B and shows a -21-dB gain at $f_c = 6.5$ kHz. Therefore, the error amplifier must have a $+21$-dB (11.2) gain at f_c. However, the desired dc gain is 50 dB, requiring an amplifier gain of $(50 - 17.2) = 32.8$ dB $= 43.6$, which is also the ratio R_2/R_1. Let $R_2 = 130$ kΩ; then $R_1 = 3$ kΩ. Drawing a horizontal line from 32.8 dB (43.6) and a -1 slope line through 6.5 kHz, the intersection is at 1500 Hz for the amplifier pole. Then, $C_1 = /1(2\pi \times 130{,}000 \times 1500) = 820$ pF. Also, a pole may be added at the output of the amplifier, say at 28 kHz, to attenuate high frequencies. Choosing $R_4 = 10$ kΩ, $C_2 = 560$ pF. The overall system response (magnitude and phase) is shown by the heavy lines of Fig. 11.11 and indicates a phase margin of 55° at f_c.

11.3.3   Boost Converter

Power stage operation of the boost converter (actually, a boost regulator) is discussed in Section 8.4.1 for operation in the continuous current mode (CCM). The boost–CCM mode has the advantage of continuous input current from the source and is popular for power factor correction topologies, even with the inherent right-half-plane zero. The basic duty cycle control and dc control is given by

$$D = v_c/v_s = 1 - E_{in}/E_0 \tag{11.21}$$

and the control to output dc gain of the modulator is given by

$$v_0/v_c = (E_{in}/v_s)(E_0/E_{in})^2 \tag{11.22}$$

A boost regulator diagram is shown in Fig. 8.11. The components are redesignated for present purposes. The output filter and load has attenuation slope changes at the following frequencies:

$$f_p = (1 - D)/2\pi\sqrt{L_{in}C_0} \tag{11.23}$$

$$f_z = 1/2\pi R_c C_0 \tag{11.24}$$

$$f_{rhp} = (1 - D)^2 R_L/2\pi L_0 = R_L E_{in}^2/2\pi L_{in} E_0^2 \tag{11.25}$$

High values of input and output voltage, with moderate voltage ratios, have a significant effect on the modulator operation. Also, the bandwidth is severely restricted, due to the right-half plane zero. In fact, single-loop voltage mode operation may not be a viable topology and current mode operation may be required, as discussed in Section 11.5.3

Example

For $E_0 = 380$ V, $P_0 = 760$ W, $R_L = 190$ Ω, $E_{in} = 150$–340 V, $L_{in} = 2$ mH, $C_0 = 500$ μF, $R_c = 0.2$ Ω, the following tabulation is calculated from Equations (11.21)–(11.25):

E_{in}	D	f_p	f_{rhp}	v_0/v_c	f_z
150	0.6	64 Hz	2.4 kHz	103 dB	1.6 kHz
380	0.1	144 Hz	12.2 kHz	45 dB	1.6 kHz

A few observations can be made. The desire to place f_z below f_{rhp} to offset its $-90°$ shift results in a large value for C_0, or either a large value of R_c; possibly requiring the addition of a discrete resistor. Plotting these parameters on an impedance chart shows a gain variation from 8 to 52 dB at 1 kHz, probably a maximum value for f_c. Amplifier compensation thus becomes difficult, even with a dual zero–pole (lead–lag) pair. Lag compensation may be used but this inherently produces a narrow bandwidth.

11.4 PWM FEEDFORWARD CONTROL

Feedforward may be used to compensate for changes in input voltage, thereby improving input line regulation. The dynamic feedforward also minimizes the effects of ripple and random disturbances on the input and therefore has good audio susceptibility performance. Referring to Fig. 8.6C, the ramp voltage is dv_s/dt and the error signal is v_c, which is constant for a fixed load. Feedforward changes the slope of the ramp for input changes (as shown by the dashed lines) by a factor K. Thus, $E_{in} = KT\,dv_s/dt$. At $dt = t_{on}$, $v_s = v_c$, and for the buck converter, $T = t_{on}/D$. Rearranging,

$$K = DE_{in}/v_c = E_0/v_c \tag{11.26}$$

Thus if E_{in} increases (decreases), the duty cycle will automatically decrease (increase) to maintain a constant volt-second product and to maintain output regulation. Inherent fast response is achieved since this compensation is an inner loop with the closed-loop feedback. For regulators and converters, the line change may be due to changes in other load currents and may be due to high source impedance. For power supplies, the rectified and filtered ac input produces a ripple voltage on the input filter capacitor. In power supplies designed to operate from 50/60 Hz mains, the loop gain of the modulator may be sufficiently high (perhaps 40 dB) at 100/120 Hz to regulate out the input ripple effects. For power supplies designed to operate from 400-Hz sources, the gain may not be sufficient (perhaps only 20 dB) to regulate out the 800-Hz ripple, and the dynamic feedforward compensation of v_s will substantially decrease the low-frequency component in the output ripple.

11.4.1   Buck Converter

Feedforward could be implemented in the buck converter of Fig. 11.10, from input to v_s, and the PWM could be an 1846 or a 5560. (Reference Fig. 8.6.)

Example

Operation is described by referring to the example of Section 11.3.1. A typical value of v_c is 3.5 V since the feedforward increases the ramp by approximately 40% ($2.5 \times 1.4 = 3.5$). The turns ratio of T_1 must be considered for a voltage divider and the modulator gain at dc is $K = DE_{\text{in}}/nv_c = 0.4 \times 120/4 \times 3.5 = 3.4 = 10.7$ dB. This is 5.3 dB lower than the 16 dB of the subject example. Thus the gain of the amplifier in this application must be increased by 5.3 dB to achieve the same overall performance. The recalculation of component values for the desired gain and break frequencies is straightforward.

11.4.2   Flyback Converter

Feedforward could be implemented in the flyback converter of Fig. 11.11, from input to v_s, and the PWM could be an 1846 or a 5560. (Reference Fig. 8.6.)

Example

Operation is described by referring to the example of Section 11.3.2. Using $v_s = 3.5$ V and $n = 9.2$, $K = DE_{\text{in}}/nv_c = 0.48 \times 120/9.2 \times 3.5 = 1.8 = 5$ dB. This is 12.2 dB lower than the 17.2 dB of the subject example. Thus the gain of the amplifier in this application must be increased by 12.2 dB to achieve the same overall performance. The recalculation of component values for the desired gain and break frequencies is straightforward.

11.4.3   Boost Converter

Feedforward in a boost converter does not overcome the inherent problems of voltage mode control in CCM, as discussed in Section 11.3.3. For this reason, current mode control is used for optimum performance. (Reference Section 11.5.3.)

11.5   PWM CURRENT MODE CONTROL

Current mode control (CMC) provides an inner loop control by sensing a current that is proportional to the current in an inductor and using this signal to control duty cycle. (The outer voltage loop controls the output voltage for regulation.) The current in the inductor is proportional to both the output current ($I_{in} = DI_0$ for a buck converter) and the input voltage ($E_{in} = -L\,di/dt$). If this current can be used to control duty cycle, an instantaneous open-loop response can be achieved for input voltage changes (inherent in feedforward) *and* output current changes. This inner feedback loop changes the inductor into a current source that reduces the normal two-pole, second-order output LC filter to the single pole of the output filter capacitor and load. The filter loop response is then a phase lag of 90° (instead of 180°), which simplifies closed-loop compensation. However, slope compensation of the peak current ramp is required to maintain stability, especially at duty cycles greater than 0.5. Also, the current loop has a pole that is related to the switching frequency, the slope compensation, and the duty cycle. The phase lag of this pole adds to the phase lag of the right-half-plane zero in the boost and flyback topologies and can cause a loss of phase margin or can cause instability. A number of advantages result from peak CMC. Transient response is improved. Line regulation improved (comparable to feedforward). Flux balancing is automatic in push–pull topologies. Peak current is instantaneously limited on a pulse-by-pulse basis. Several CMC units can be paralleled with equal current sharing between units. Average CMC is commonly used in programmable converters and power supplies with adjustable constant-voltage and constant-current features.

A few disadvantages result with CMC, which can be overcome with good design practices. A current sensor (resistor or current transformer) is required; although this is usually required in voltage mode control for overload sensing. Loop subharmonic instability can occur at duty cycles greater than 50%; although this can be overcome by slope compensation to the ramp. Peak CMC suffers from poorer noise immunity than average current control or voltage mode control. This is primarily due to noise and parasitic effects on the peak current sensed waveform and can be overcome with detail to circuit layout and lead lengths.

11.5.1   Buck Converter

Current mode control can be used in forward and push–pull derived topologies. Current mode control is implemented in the forward converter of Fig. 11.12 and can be compared to the voltage mode control of Fig. 11.10. A current transformer senses the primary current for inner-loop control of the compensated ramp. The basic duty cycle control function is given by Equation (11.13). Slope compensation analysis is presented, though the duty cycle is limited to 0.5 in Fig. 11.12. The system transfer functions with CMC has been analyzed by Middlebrook and Ridley [1, 2]. These cited references consider the exact effects of duty cycle, slope, and time constant as well as the basic dc ratio of load resistance to sensed-current resistance. As compared to prior analysis, the cited studies show CMC can cause a decrease in the dc gain and an increase in the corner frequency of the output R_L and C_0. The approximate control to output dc gain of the modulator, with the influence of the current loop gain A is given by

$$v_0/v_c = (nR_L/n'R_s)/A \tag{11.27}$$

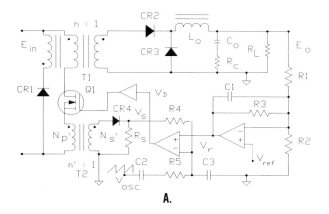

A.

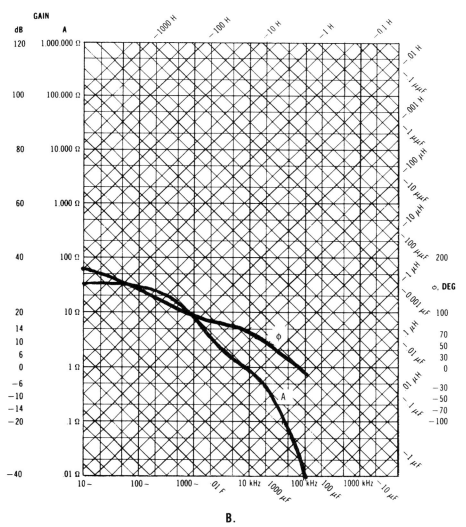

B.

Fig. 11.12. Current mode control, forward converter.

where

$$n = N_p/N_s \text{ for the power transformer}$$

$$n' = N_p/N_s \text{ for the current transformer}$$

$$A = 1 + R_L T_s[m(1 - D) - 0.5]/L_0 \qquad (11.28a)$$

$$m = 1 + s_{osc}/s_{on} \qquad (11.28b)$$

The numerical parameter m relates the slope of the oscillator ramp to the slope of the current ramp during the *on time* as dv_s/dt. These slopes plus the slope of the current ramp during the *off time* are

$$s_{osc} = v_{osc}/T_s \qquad (V/\mu s) \qquad (11.29a)$$

$$s_{on} = n'R_s(E_{in}/n - E_0)/nL_0 \qquad (V/\mu s) \qquad (11.29b)$$

$$s_{off} = n'R_s E_0/nL_0 \qquad (V/\mu s) \qquad (11.29c)$$

The slope of the injected ramp from v_{osc} to v_r is typically selected as *one-half* the slope of the inductor current ramp during the off time, as reflected to the primary at v_s. This parameter optimizes the transient response and improves audio susceptibility and is set by the ratio of R_5 to R_4. Then, $s_{osc}/s_{off} = 0.5$ and $s_{osc}/s_{on} = 0.5E_0/(E_{in}/n - E_0)$. With these known quantities, m can be determined. The ratio of R_5 to R_4 is then

$$R_5/R_4 = s_{osc}/0.5s_{off} \qquad (11.30)$$

The sawtooth is ac coupled to the ramp by C_2 and C_3 filters initial voltage spikes that are due to parasitic capacitances in T_1. The output voltage is divided by R_1 and R_2. The gain of the error amplifier and the compensation by a single pole plus the modulator response determine overall system performance.

The output filter inductor acts as a current source, which eliminates the double pole of the output LC filter. The exact effects of CMC on the frequency response are analyzed in the cited references [1, 2]. In general, the output pole frequency will increase slightly from the nominal value given as

$$f_p = 1/2\pi R_L C_0 \qquad (11.31a)$$

and the output zero occurs at

$$f_z = 1/2\pi R_c C_0 \qquad (11.31b)$$

However, an additional pole occurs at the crossover of the current loop and is given by

$$f_c = f_s/2\pi m(1 - D) \qquad (11.31c)$$

Example

The parameters and values listed in the example of Section 11.3.1 are used for CMC analysis. In summary, $E_0 = 12$ V, $R_0 = 0.6$ Ω, $E_{in} = 120$ V, $f_s = 80$ kHz,

$T = 12.5\ \mu s$, $T_s = 6.25\ \mu s$, $L_0 = 90\ \mu H$, $C_0 = 1000\ \mu F$, $R_c = 0.05\ \Omega$, and $D = 0.4$. First, n' and R_s must be calculated. Since $I_0 = 20$ A and $n = 4$, $I_p = 5$ A. The desired voltage at V_s is 1.5 V. Here $N_{p'}$ is almost always 1 turn, and for $N_{s'} = 100$ turns, $n' = 0.01$ and $I_{s'} = 0.05$ A. Then, $R_s = 1.5/0.05 = 30\ \Omega$. From Equation (11.28), $m = 1.5$ and $A = 1 + 0.6 \times 6.25 \times 10^{-6}[1.5(1 - 0.4) - 0.5]/90 \times 10^{-6} = 1.017$. From Equation (11.27), $v_0/v_c = 4 \times 0.6/(0.01 \times 30)/1.017 = 8/1.017 = 7.87 = 18$ dB. From Equation (11.31), $f_p = 1/2\pi \times 0.6 \times 0.001 = 265$ Hz, $f_z = 1/2\pi 0.05 \times 0.001 = 3.18$ kHz, and $f_c = 80,000/2\pi \times 1.5(1 - 0.4) = 14$ kHz. From Equation (11.29), $s_{osc} = 2.5/6.25 = 0.4$ V$/\mu s$, $s_{on} = 0.01 \times 30(120/4 - 12)/4 \times 90 = 0.015$ V$/\mu s$, and $s_{off} = 0.01 \times 30 \times 12/4 \times 90 = 0.01$ V$/\mu s$. Then, $R_5/R_4 = 0.4/0.5 \times 0.01 = 80$. For $R_4 = 1.5$ kΩ, $R_5 = 120$ kΩ. When using a control IC with an internal current amplifier, the gain of the amplifier must be considered.

A gain of 10 is chosen for the error amplifier, with a pole at 3.18 kHz to offset the zero in the output capacitor. Thus $R_1 = 2$ kΩ, $R_3 = 20$ kΩ, and $C_3 = 2.4$ nF. The overall system response (magnitude and phase) is shown in Fig. 11.12B and indicates a bandwidth of 10 kHz, a phase margin of 60°, and 20 dB attenuation at half f_s. The current loop f_c, calculated as 14 kHz, is higher than the voltage loop f_{co}. For the voltage divider, $R_2 = V_{ref}/[E_0 - V_{ref})/R_1] = 2.5/[12 - 2.5)/2000] = 526$, use a 523-$\Omega$ resistor.

11.5.2   Flyback Converter

Current mode control does not offer a significant advantage in flyback converters operating in DCM. The modulator gain is typically lower, requiring a higher error amplifier gain, as in the case of feedforward in DCM. Continuous current mode operation incurs the right-half plane zero and bandwidth is usually reduced, either in current mode or voltage mode control.

11.5.3   Boost Converter

All boost topologies are inherently "current fed" because of the inductor in series with the input. Operation in CCM can significantly reduce the input current ripple, as compared to DCM. Current mode control does not eliminate the right-half zero inherent in CCM operation, although it does eliminate the LC output double pole. The system transfer functions with CMC has been analyzed by Middlebrook and Ridley [1, 2]. These cited references consider the exact effects of duty cycle, slope, and time constant as well as the basic dc ratio of load resistance to sensed-current resistance. A boost regulator with CMC is shown in Fig. 11.13. The basic duty cycle control is given by

$$D = 1 - E_{in}/E_0 = 1 - I_0/I_{in} \tag{11.32}$$

The approximate control to output dc gain of the modulator is given as

$$v_0/v_c = (1 - D)[2\pi f_c L_c \| R_L/2]/R_s \tag{11.33}$$

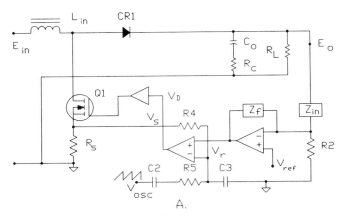

Fig. 11.13. Current mode control, boost regulator.

where

$$f_c = f_s/\pi n(1 - D) \qquad L_e = L_{in}/(1 - D)^2 \qquad n = 1 + 2s_{osc}/s_{on}$$

The pole at f_c is the crossover frequency of the current loop. The numerical parameter n relates the slope of the oscillator ramp to the slope of the current ramp during the on time as dv_s/dt. These slopes plus the slope of the current ramp during the off time are

$$s_{osc} = v_{osc}/T_s \tag{11.34a}$$

$$s_{on} = E_{in}/L_{in} \tag{11.34b}$$

$$s_{off} = (E_0 - E_{in})L_{in} = DE_0/L_{in} \tag{11.34c}$$

The slope of the injected ramp from v_{osc} to v_r is typically selected as *equal* the slope of the inductor current during the off time, or $s_{osc} = s_{off}$. This parameter optimizes the transient response and improves audio susceptibility and is set by the ratio of R_5 to R_4. Equating these parameters,

$$n = 1 + 2DE_0/E_{in} = (1 + D)/(1 - D) \tag{11.35}$$

$$R_5/R_4 = s_{osc}/s_{off} \tag{11.36}$$

The value of n in Equation (11.35) can be substituted into Equation (11.33). The sawtooth is ac coupled to the ramp by C_2 and C_3 filters initial voltage spikes that are due to parasitic capacitances. The output voltage is divided by R_1 and R_2.

The exact effects of CMC on the frequency response are analyzed in the cited reference [1]. The initial pole is given as

$$f_p = 1/2\pi(2\pi f_c L_c \| R_L/2)C_0 \tag{11.37a}$$

and can be approximated by

$$f_p = 1/\pi R_L C_0 \qquad (11.37b)$$

and an output zero occurs at

$$f_z = 1/2\pi R_c C_0 \qquad (11.38)$$

The current loop pole f_c is given following Equation (11.33). The right-half-plane zero is

$$f_{rhp} = R_L(1 - D)^2/2\pi L_{in} \qquad (11.39)$$

An additional zero is related to the switching frequency and the right-half-plane zero [1]. This frequency can be approximated by

$$f_{zs} = (f_s + f_{rhp})/\pi D \qquad (11.40)$$

Example

The parameters and values listed in the example of Section 11.3.3 are used for CMC analysis in the boost regulator of Fig. 11.13. In summary, $E_0 = 380$ V, $P_0 = 760$ W, $R_L = 190$ Ω, $E_{in} = 150$–340 V, $L_{in} = 2$ mH, $C_0 = 500$ μF, and $R_c = 0.2$ Ω. The switching frequency is chosen at $f_s = 50$ kHz and $T_s = 20$ μs. At minimum input, $D_{max} = 1 - 150/380 = 0.6$ and $I_{in} = 2/(1 - 0.6) = 5$ A. Allowing 10% for ΔI and for $V_s = 1$ V, $R_s = 1/5.5 = 0.18$ Ω.

From Equation (11.35), $n = 1.6/0.4 = 4$ and from Equation (11.33), $v_0/v_c = 0.4(2\pi 10{,}000 \times 0.0125 \| 190/2)/0.18 = 0.4 \times 85/0.18 = 189 = 46$ dB. Note that $f_c = 10$ kHz. From Equations (11.37b) and (11.38), $f_p = 1.7$ Hz and $f_z = 1.6$ kHz. The right-half plane zero occurs at $f_{rhp} = 2.4$ kHz and the additional zero occurs at $f_{zs} = 28$ kHz. The slopes of the current and oscillator waveforms and the values of R_4 and R_5 can be determined in the same manner as the example of Section 11.3.1. To meet the desired design performance, the error amplifier compensation can be determined after analyzing other parameters. The crossover frequency is typically selected at one-third the frequency of the right-half-plane zero, or 800 Hz in this case. However, f_z is in close proximity to f_{rph} and this condition causes a $+2$ change in slope with no phase change. The overall system can be modeled by CAE programs or the actual circuit tested with a dynamic signal analyzer, as discussed in Section 11.2.2.

11.6 RESONANT FM MODE

Frequency modulation (FM) is typically used in resonant mode or quasi-resonant converters to control the output in response to input line and output load changes. These topologies operate in a fixed on-time, variable off-time and switch at zero current or zero voltage. Various ICs use a voltage-controlled oscillator to perform these functions and the ICs are discussed in Section 10.2. The discussion in this section is devoted to the parallel and the series–parallel resonant mode converter.

The series resonant converters must operate over a very wide frequency range if the load varies over a wide range. For this reason, the series resonant converter is most applicable to pulsed-power systems, as described in Section 10.4. Also, the zero voltage switching, phase-shifted topology, and the clamped-mode topology switch at fixed frequency and use PWM for control.

A parallel resonant, half-bridge converter is shown in Fig. 10.12. The voltage transfer functions of the parallel and series–parallel resonant converters are shown in Figs. 10.13 and 10.14, respectively. The transfer function can be quite linear by selecting a Q between 0.8 and 1.2 for $f_s < f_r$. The gain of the modulator of Fig. 10.12 is

$$A_M = E_{in'}/nf_r \quad \text{V/MHz} \quad (11.41)$$

where

$E_{in'}$ = voltage applied to resonant circuit

$n = N_p/N_s$

$f_r = 1/2\pi\sqrt{L_r C_r} \quad \text{MHz}$

The gain of the control circuit IC is the slope of the frequency versus error amplifier voltage and is expressed as a function of the RC timing components:

$$A_c = 1/R_T C_T = f_s/V_{c/a} \quad \text{MHz/V} \quad (11.42)$$

The overall gain of the converter is the product $A_c A_M$.

The output filter has an LC double pole, which is compensated by the error amplifier to provide the overall desired response. The compensation network is typically a dual zero–pole pair, as discussed in Section 11.1.3.

Example

The parameters and values listed in the example of Section 10.5 for a parallel resonant converter are used for this analysis. In summary, $E_{in} = 250$ V ($E_{in'} = 125$ V for the half-bridge), $n = 4$, $f_r = 0.4$ MHz, and $f_s = 0.18$ MHz. From Equation (11.41), $A_M = 125/(4 \times 0.4) = 78$ V/MHz. The control IC operates at 180 kHz and the error amplifier voltage swing is 3 V. From Equation (11.42), $A_c = 0.18/3 = 0.06$ MHz/V. The overall gain is $78 \times 0.06 = 4.7 = 13$ dB.

REFERENCES

1. R. D. Middlebrook, "Modeling Current Programmed Buck and Boost Regulators," *IEEE Transactions on Power Electronics* (January 1989).
2. R. B. Ridley, "A New, Continuous-Time Model For Current-Mode Control," *IEEE Transactions on Power Electronics* (April 1991).

12

PULSE-FORMING NETWORKS AND MODULATORS

12.0   Introduction
12.1   Pulse-Forming Networks
     12.1.1   Waveform Analysis
     12.1.2   Component Value Analysis
12.2   Line-Type Modulators
     12.2.1   Circuit Timing and Analysis
     12.2.2   Energy Regulator
12.3   Magnetic Modulators
     12.3.1   Operational Characteristics
     12.3.2   Design Parameters
     Reference

12.0   INTRODUCTION

This chapter discusses networks and modulators used in a wide variety of high-power communications, transmitters, radars, and electronic warfare (EW) equipment. Various magnetic devices are used as passive elements but applications differ from those discussed in Chapter 3. Capacitors are used as energy storage elements and tend to be of reconstituted mica dielectric to minimize dissipation factor and meet high-voltage isolation requirements (reference Section 2.3.6). Power semiconductors are usually limited to thyristors and fast recovery rectifiers, both of which can conduct high values of current for short periods of time.

The mere fact that pulsed magnetic devices and capacitors are used in the transfer of energy indicates an oscillatory or even quasi-resonant topology exists. The principles involved here are the same as those discussed in Section 10.1 except in these cases, saturable cores perform more exotic functions. Also, impedance matching between stages becomes important for optimum or desired waveform preservation.

12.1   PULSE-FORMING NETWORKS

There are three general types of waveforms associated with pulse-forming networks (PFN): (1) rectangular, (2) trapezoidal; and (3) flat top with parabolic rise and fall. Transmission line theory and Fourier series analysis are helpful in analyzing PFN performance. This is the better method to analyze PFNs whereas the transient response method is most applicable to pulse transformers, discussed in Section 3.18. An open-circuit transmission line having a one-way propagation time $(\frac{1}{2})\tau$, a characteristic impedance Z_n, and an open-circuit voltage E_n may be discharged into its own characteristic impedance. The rectangular output voltage pulse amplitude is then $(\frac{1}{2})E_n$ and the pulse duration is τ. However, meeting bulk transmission cable requirements are often impractical. This ideal rectangular pulse has an infinite rate of rise and fall and can only be achieved with an infinite number of odd harmonics of a sine wave. Any departure from a true step function appears as oscillations or ringing near the step.

All networks have passive elements of inductance and capacitance that inherently produce sinusoidal waveforms as shown in Section 1.6. Thus, a network of lumped inductances and capacitances with finite rise and fall times and pulse duration can be selected. If discontinuities can be eliminated, the Fourier series will have a property of uniform convergence throughout the pulse. This property ensures that overshoots and oscillations in the pulse can be reduced by using a sufficient number N of L and C sections and that the pulse of trapezoidal or similar shape will have finite and equal rise and fall times.

12.1.1   Waveform Analysis

A trapezoidal waveform is shown in Fig. 12.1A. The equation for current during the linear rise and fall time is

$$i_t = I_{pk}t/at_p \tag{12.1a}$$

The overall pulse width is

$$t_p = 2at_p + t_1 \tag{12.1b}$$

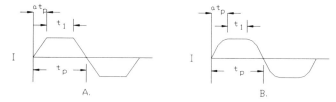

Fig. 12.1. (A) Trapezoidal pulse and (B) parabolic pulse with rise time, fall time, and pulse width.

Since the Fourier series contains only sine terms, i_t is an odd function of the form

$$i_t = I_{pk}(4/v\pi)[\sin(v\pi a)/v\pi a][\sin(v\pi t)/\tau]$$

where $v = 1, 3, 5, \ldots$ and a is the slope.

A flat-top waveform with parabolic rise and fall time is shown in Fig. 12.1B. The equation for current during the parabolic rise and fall time is

$$i_t = 2I_{pk}t/at_p - (t/at_p)^2 \tag{12.2a}$$

The overall pulse width is

$$t_p = 2at_p + t_1 \tag{12.2b}$$

In this case, i_t is an odd function of the form

$$i_t = I_{pk}(4/v\pi)\left[\sin(v\pi a/v\pi)^2[\sin(v\pi t/\tau)]\right]$$

Each term of the Fourier series consists of a sine wave amplitude $4/v\pi$ and frequency $v/2t_p$. This is the familiar undamped current waveform for a series LC circuit, as given in Table 1.3, and is

$$i_t = (E/\omega L)\sin(\omega t) \tag{12.3a}$$

$$= (E/Z)\sin(t\sqrt{LC}) \tag{12.3b}$$

where $Z = \sqrt{L/C} = \omega L = 1/\omega C$ and

$$t_p = \pi\sqrt{LC} \tag{12.4}$$

Note that, by definition, t_p is the pulse width from the initial rise in current until the fall of current back to zero and is consistent with the notations used in Chapter 1. In Section 12.1.2, a pulse width t_w is defined as the pulse width at 70.7%, or $(\frac{1}{2})^{1/2}$, of the waveform peak with a resistive load. This typically conforms to the pulse width at one-half the peak current or peak power in a high-power pulsed microwave tube, such as a magnetron.

The resultant network required to produce the desired wave shape consists of a number of resonant LC sections in parallel. These networks, sometimes referred to as parallel admittance, are often inconvenient for practical use. The inductors have distributed capacitance, which disrupts the pulse shape, and the capacitors have a wide range of values.

Various types of voltage-fed networks to achieve the desired PFN performance have been analyzed by Guillemin [1]. A two-section, parabolic pulse type A is shown in Fig. 12.2A along with the corresponding Fourier series coefficients and the output pulse shape. In all cases, the actual component values are obtained by multiplying the values of inductances by t_pZ_N/π and the values of capacitance by $t_p/Z_N\pi$. The parabolic rise $a = 0.33$.

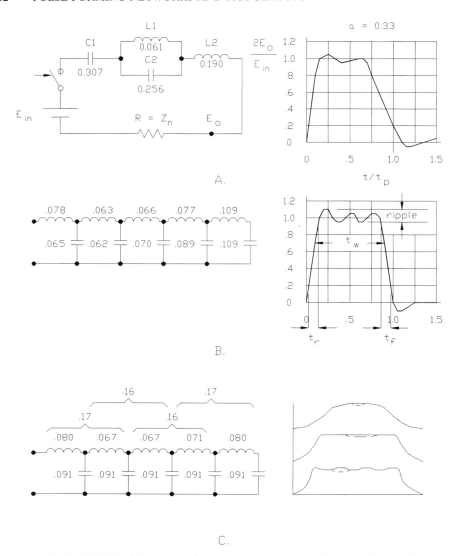

Fig. 12.2. Typical PFNs: (A) two-section A type, parabolic rise; (B) five-section B type, trapezoidal pulse; (C) five-section E type, trapezoidal pulse. Normalized values are Fourier series coefficients.

Faster rise time and/or longer pulse width can be achieved with the B and E types shown in Figs. 12.2B and C, respectively, along with the coefficients. The waveforms of Fig. 12.2B define the pulse characteristics. Rise time applies to the linear portion of the wave front and is measured from 10 to 90% of the average peak amplitude. When powering microwave tubes, a long rise time results in poor spectrum characteristics, and too short a rise time can result in sparking over-shoots. Ripple is defined as the excursions above and below the average peak amplitude and is typically less than 5%. Again, the pulse width t_w is normally measured at 70% of the average amplitude. The fall time is approximately three times the rise time for a E-type PFN.

These five-section PFNs produce a trapezoidal pulse and have an inherent rise time of approximately 8–10%; reference Equation (12.12). The B type has individual inductors and multiple capacitance values. From a production standpoint, it is highly desirable to have all capacitances of equal value. Therefore, the PFN of Fig. 12.2C is more desirable, but the mutual inductances must be considered since the capacitors are connected to appropriate taps on a continuous winding. An explanation of the waveforms of Fig. 12.2C is in order. Rise time increases as turns before the first tap increase. The dotted curve shows the result of greater separation between second and third taps. As a practical guide, the taps are located to obtain equal inductance for all sections except the ends, which have approximately 25% more self-inductance, as indicated by the Fourier coefficients of Fig. 12.2C.

The inductor is essentially constructed as an air core solenoid. Proper wire size is chosen as follows. The charging current has the same average value as the discharge current, but the discharge time is much shorter than the charging time. The conductor wire size must be rated for rms current and the ratio of average to rms for a rectangular pulse is $(t/T)^{1/2}$, as given in Table 1.2 part A, waveform 6. Thus for a duty cycle of 0.5%, the rms value is 14 times the average value. For such low duty cycles, the effects of charging current can normally be neglected.

Turns of tinned-cooper or silver-plated wire can be wound on a threaded-rod mandrel made from nonconducting material. The mandrel serves as a permanent, physical mounting structure. The depth and pitch of the threads are chosen to comply with the diameter of the wire conductor and the separation between turns. The ratio of winding diameter to width (d/w) is chosen by a method involving Nagoaka's function to give a mutual inductance that is 15% of the self-inductance of each center for d/w equal to 0.9. However, Equation (3.63) and Figs. 3.34 and 3.35 can be used to design the single-layer solenoid. For other values of d/w equal to 0.5, 1.0, and 1.5, the mutual inductance is approximately 9, 16, and 21%, respectively. In general, if the percentage of mutual inductance is multiplied by a certain factor, the percentage ripple will be multiplied by the square of that factor and simultaneously, the rise time will be divided by the square root of the same factor.

The optimum overall results are usually obtained with the following number of sections for the corresponding pulse width ranges: 2 for less than 0.1 μs; 2–5 for 0.1–2 μs; 3–8 for 2–6 μs. For PFNs of less than five sections, it is usually difficult to obtain satisfactory pulse shapes by using the same inductance per section. Of course, the rise time increases and the ripple amplitude increases as the number of sections decrease. Final positioning of the tap connections to the capacitors is normally done experimentally for optimum results.

12.1.2 Component Value Analysis

In the design of a PFN with three or more sections, the actual values of capacitance and inductance to be used for a specific pulse duration and a matching impedance are based on the following discussion. Note that in Equations (12.1)–(12.3) the pulse width t_p is measured from the instant the voltage starts to rise to the instant it finally reaches zero, such as sinusoidal current waveforms charging the PFN. For this case, the following relations apply:

$$Z_N = \sqrt{L_N/C_N} \tag{12.5}$$

$$t_p = \pi\sqrt{L_N C_N} = \pi C_N Z_N = \pi L_N/Z_N \tag{12.6}$$

$$L_N = t_p Z_N/\pi \tag{12.7}$$

$$C_N = t_p/\pi Z_N \tag{12.8}$$

As discussed in Section 12.1.1, the pulse width t_w in a trapezoidal pulse is measured at the 70.7% value of peak current in a matched resistive load ($ZN = PL$) and the following relations apply:

$$t_w = 2\sqrt{L_N C_N} = 2C_N Z_N = 2L_N/Z_N \tag{12.9}$$

$$L_N = \tfrac{1}{2}t_w Z_N \tag{12.10}$$

$$C_N = t_w/2Z_N \tag{12.11}$$

In the design of the PFN, the desired pulse width and rise time are given. The PFN impedance can be calculated from the equivalent reflected resistive load, as discussed in Section 12.2. Thus, the total inductance and capacitance can be determined. For the type E PFN, the calculated value of L_N may require an additional 15% to offset mutual coupling. The number of sections required is given by

$$N = t_w/2t_r \tag{12.12}$$

where t_r is the rise time from 10 to 90%.

The total inductance L_N is divided into the individual coefficients for proper tap placement. All capacitance values are of equal value as C_N divided by N. Imperfections in the PFN normally result in a fall time t_f roughly three times the rise time.

Examples

1. A PFN is required to provide a 0.5-μs output pulse into a 30-Ω load. Impedance of the PFN is matched to the load impedance and the required load pulse voltage is 200 V. Thus, the PFN must initially be charged to 400 V. A two-section PFN, as shown in Fig. 12.2C is selected where $R = Z_N = 30 \ \Omega$. The Fourier coefficients must be multiplied by $t_p Z_n$ for the inductors and by t_p/Z_n for the capacitors. Thus, $C_1 = 0.307 \times 0.5 \times 10^{-6}/30 = 5.1$ nF, $C_2 = 0.256 \times 0.5 \times 10^{-6}/30 = 4.3$ nF, $L_1 = 0.061 \times 0.5 \times 10^{-6} \times 30 = 0.91 \ \mu$H, $L_2 = 0.19 \times 0.5 \times 10^{-6} \times 30 = 2.85 \ \mu$H.

Using these values as a starting point, a computer analysis of the circuit produced the waveform shown in Fig. 12.2A but with the values of C_1 and C_2 changed to 6.2 and 3.3 nF, respectively. For this design, $E_{in} = 400$ V, $E_0 = 204$ V, $t_p = 0.5 \ \mu$s, $t_w = 0.35 \ \mu$s, $t_r = 65$ ns, $t_f = 135$ ns. The result of increasing the capacitors to experimentally obtain a flat pulse produced a 2% increase in desired output voltage. This nominal increase may actually be desirable in the final application.

2. A PFN is required to provide a 5-μs output pulse into a 2.5-Ω load. The pulse width applies to a 70% amplitude. Impedance of the PFN is matched to the load impedance and the required load pulse voltage is 500 V. Thus, the PFN must initially be charged to 1000 V. A five-section PFN, as shown in Fig. 12.2C is selected where $Z_n = R = 2.5$ Ω. From Equations (12.10) and (12.11), $L_N = 5 \times 2.5/2 = 6.25$ μH and $C_N = 5/2 \times 2.5 = 1$ μF. The individual inductances are proportioned by the coefficients of Fig. 12.2C as 1.4, 1.1, 1.1, 1.25, and 1.4 μH for 6.25 μH total. Each capacitor value is $\frac{1}{5} = 0.2$ μF.

Computer analysis of this PFN produced the waveform shown in Fig. 12.2B where $E_0 = 515$ V, $t_p = 6$ μs, $t_w = 4.9$ μs, and $t_r = 350$ ns.

If the Fourier coefficients shown in Fig. 12.2C are multiplied by $t_p Z_n$ for the inductors and by t_p/Z_n for the capacitors than $L_N = 4.55$ μH and $C_N = 0.736$ μF. Analysis showed $t_p = 5$ μs, which is the result predicted by using the coefficients. Also, $E_0 = 530$ V and $t_w = 3.95$ μs.

12.2   LINE-TYPE MODULATORS

The purpose of a modulator is to provide a pulse voltage or current waveform that will allow a load to operate properly. The load may be a high-power microwave tube in various types of radars, radio telescopes, and linear accelerators. The pulse duration is normally short compared to the period or pulse repetition frequency (PRF). The modulator may be solid-state controlled, controlled by switching magnetics, or controlled by vacuum tubes. The modulators discussed in this section are solid state and are frequently called line-type modulators.

All modulators have a common characteristic. They contain a means of storing energy, normally in a capacitor, and a means to switch this stored energy into a desired load during the desired pulse. The energy in the storage element is then replenished before the next pulse occurs. A typical line-type modulator with solid-state switching is shown in Fig. 12.3A and consists of an energy storage element and regulator, pulse forming network, output pulse transformer, and thyristors for controlling system timing. Voltage E_d is normally a mains rectified and filtered dc source and is subject to voltage variations from the utility. Thyristor Q_1 is triggered to charge the capacitors in the PFN in a resonant manner via the inductance of T_1 primary and then Q_1 becomes reverse biased. Thyristor Q_2 is triggered to initiate the output pulse by discharging the PFN across the pulse transformer.

12.2.1   Circuit Timing and Analysis

Operation is aided by referring to the waveforms of Fig. 12.3. Assuming the PFN has previously been charged to $2E_d$, a master trigger source gates Q_2 on at T_g. The inductance of the first section of the PFN may limit the di/dt to a safe rate; otherwise, a saturable reactor is required in series with Q_2 to delay the rate of current rise in Q_2 until the gate current has spread across the die. Peak primary

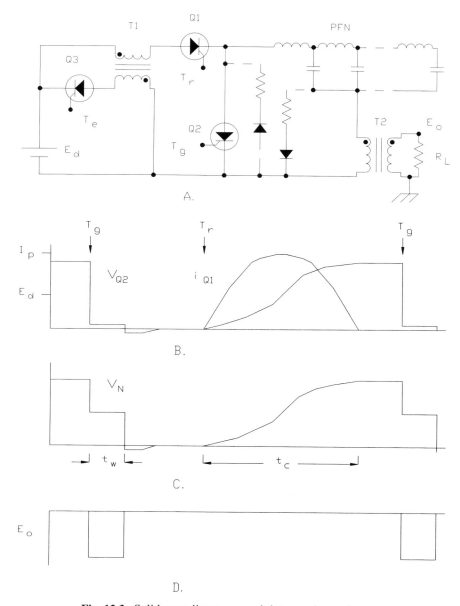

Fig. 12.3. Solid-state line-type modulator and waveforms.

currents of several hundred amperes are typical in magnetic modulators, but the duty cycle is quite low. Since the characteristic impedance of the PFN has been matched to the reflected load impedance by initial design, one-half the PFN voltage is applied to the T_2 primary. This voltage is stepped up in the secondary to produce $-E_0$ with the transformer polarity shown. The pulse duration t_w is governed by the PFN characteristics, as given by Equation (12.9). When the output pulse decays or tries to reverse, Q_2 turns off due to backswing voltage of the pulse transformer. Frequently, a normally reverse-biased rectifier and series resistor are placed across the pulse transformer primary to dissipate unused energy trapped in the transformer by any reflected load mismatch. Also, a normally reverse-biased rectifier and series resistor may be placed across Q_2 to provide a current path during transformer voltage reversal. These components are connected by the dashed line shown in Fig. 12.3A.

In the system control circuitry, a timer is triggered simultaneously with Q_2 at T_g. After a preset delay, the timer output drives a pulse transformer that triggers Q_1 to recharge the PFN capacitors. The recharge time and the peak current are governed by the undamped conditions of Table 1.3 where

$$t_c = \pi\sqrt{L_{T1}C_{\mathrm{PFN}}} \tag{12.13}$$

$$I_p = E_d\sqrt{C_{\mathrm{PFN}}/L_{T1}} \tag{12.14}$$

The charging current has practically no effect on T_2 since this peak current is typically one or two orders of magnitude less than the peak primary current when Q_2 is triggered. Normally, t_c is approximately one-half the period T where $T = 1/\mathrm{PRF}$. This allows all thyristors to recover to steady-state conditions while awaiting the next master trigger. All the laws of physics apply and the input power from E_d equals the output power of T_2 and the recharge energy in the primary of T_1 equals the stored energy in the PFN capacitors, assuming no losses.

Example

The PFN discussed in Section 12.1.2, Example 2, has an impedance of 2.5 Ω and uses a total capacitance of 1 μF charged to 1000 V. From the analysis of this example, the parabolic pulse current into T_2 is 200 A for 5 μs at 500 V. Also, the pulse transformer discussed in Section 3.18 is used as a load. It is desired to design a charging inductor, T_1 of Fig. 12.3A, to match the system requirements. The PRF is 500 Hz and the recharge time is selected at one-half the period, or 1 ms. To verify results, calculate the power and energy relations of this design. From Equation (12.13), the required primary inductance of T_1 is $L_p = (t_c/\pi)^2/C_N = (10^{-3}/\pi)^2/10^{-6} = 101$ mH. This is quite straightforward since the analysis started with the known load conditions and worked back to the input. Other observations may also be made. The impedance of the charging circuit is $Z = \pi L/t_c = t_c/(\pi C) = 317$ Ω. The peak value of the charging current pulse is $I_p = E_d/Z = 500/317 = 1.57$ A. Since t_c was chosen at 1 ms, the duty cycle is 50%. The average dc input current is $I_{\mathrm{avg}} = 1.57 \times 0.5 \times 2/\pi = 0.5$ A. The average input power is $E_d I_{\mathrm{avg}} = 500 \times 0.5 = 250$ W. The peak power delivered to T_2 is $P_p = 500 \times 200 = 100$ kW. The average output power is $P_{\mathrm{avg}} = P_p t_w \mathrm{PRF} = 10^5 \times 5 \times 10^{-6} \times 500 = 250$ W also. The energy stored in the PFN capacitance is $\mathcal{E} = \frac{1}{2}CE^2 = 0.5 \times 10^{-6} \times 500^2 = 0.125$ J. The energy in the charging inductor (T_1 primary) is $\mathcal{E} = \frac{1}{2}LI^2 = 0.5 \times 0.101 \times 1.57^2 = 0.125$ J also. The number of turns on the primary of T_1 must be sufficient to support the 500-V input for 1 ms. The wire size is selected based on the rms input current where $I_{\mathrm{rms}} = 1.57 \times 0.707 \times 0.707 = 0.78$ A. If laminations or a C core is used, the gap width is then calculated to obtain the required inductance.

12.2.2   Energy Regulator

Variations in E_d produce corresponding variations in E_o unless some form of regulation is provided. Voltage regulators, either series or shunt, could be used to regulate the PFN voltage after recharge, but these devices are inherently dissipative. An open-loop, nondissipative regulator is implemented with Q_3 and a secondary winding on T_1, as shown in Fig. 12.3A. The detailed waveforms of the charging cycle are shown in Fig. 12.4 and illustrate the energy regulation for

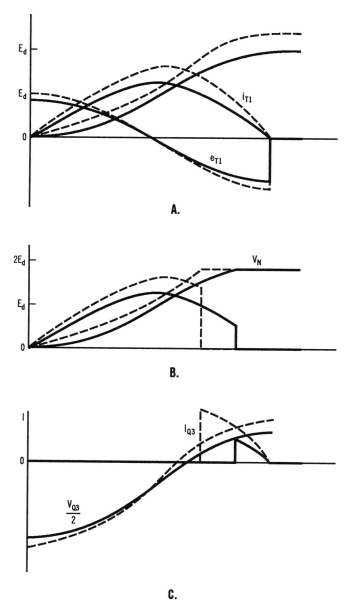

Fig. 12.4. Energy regulator waveforms.

variations in input voltage. The solid lines represent a minimum input (low-line) condition while the dashed lines represent a maximum (high-line) input. This topology is called an energy regulator since the charge on the capacitors is neither depleted nor replenished prior to T_g.

When Q_1 is triggered, a half-sinusoid current flows through T_1 primary and attempts to charge the PFN to $2E_d$. The polarity of T_1, as shown, induces a reverse-bias voltage across Q_3, in addition to the reverse-bias voltage from E_d. A resistor divider across Q_2 senses the PFN capacitor voltage rise after Q_1 is triggered. A portion of this voltage, with respect to the input common, is applied to the positive input of a comparator. A preset reference voltage is applied to the negative input of the comparator. When the desired PFN voltage is attained, the sensed voltage exceeds the reference level. The comparator output switches high and applies a voltage to a pulse driver that triggers Q_3. By this time, Q_3 has become forward biased since the voltage on the primary of T_1 has reversed. The current now switches from Q_1 to Q_3. By the induced voltage across T_1, Q_1 now becomes reverse biased and turns off. Thus the unused or trapped energy in T_1 is returned to the source and the PFN capacitor remains at the desired voltage.

For the polarity shown in Fig. 12.3A, the associated transformer and anode-to-cathode thyristor voltages excursions are

$$V_p = E_d - V_N \tag{12.15}$$

$$V_s = -V_p N_s / N_p = -(E_d - V_N)/n = (V_N - E_d)n \tag{12.16}$$

$$V_{Q1} = E_d \quad \text{to} \quad V_{Q1} = E_d - V_N \quad \text{(range)} \tag{12.17}$$

$$V_{Q2} = V_N \quad \text{to} \quad V_{Q2} = -0.1V_N \quad \text{(typical range)} \tag{12.18}$$

$$V_{Q3} = -E_d + V_s = (V_N - E_d)n - E_d \tag{12.19}$$

At the start of the recharge cycle, the peak reverse voltage stress on Q_3 may be substantial, as given by Equation (12.19), and Q_3 must be selected accordingly. To ensure turn-off of Q_1 when Q_3 is triggered, the turns ratio of T_1 is a function of the input voltage variation and is typically $n = 1/2.5$. After the PFN is charged, the anode of Q_1 is then approximately $-E_d$ with respect to the cathode. When Q_2 is triggered at T_g, the anode of Q_1 abruptly switches to $+E_d$ since its cathode is now at the input common. A snubber may be required across Q_1 to limit the dv/dt to a rate less than that specified for the device.

12.3   MAGNETIC MODULATORS

Magnetic modulators refer to switching reactors that are employed as a saturating power stage between an intermediate thyristor and the output transformer. The switching reactors are employed in many very high power, short-pulse systems, and provide a significant improvement in reliability, as compared to hydrogen thyratrons. They are also used where thyristors become impractical as a output switch due to the peak powers involved plus current capability, practical voltage stress, fast turn-on time, or combinations thereof. The reactors utilize toroids with square-loop core materials, have a high initial inductance, a low saturated inductance and can rapidly switch between these two parameters.

In a particular application, the peak output power may be enormous while the average power may be relatively low. This is evident from the waveforms of Fig. 12.3 where a long energy storage period t_c is condensed into a very short output pulse period t_w. This characteristic is numerically illustrated in the example of Section 12.2.1 where the peak output power is 100 kW while the average power (input and output) is 250 W. In these cases, the high peak power is achieved with topologies using pulse compression techniques.

12.3.1 Operational Characteristics

A typical magnetic modulator is shown in Fig. 12.5A and similarities to Fig. 12.3A are evident. In this case, the energy regulator has been omitted and the PFN is from Fig. 12.2A where L_2 of that figure is now the saturated inductance of T_1 secondary. Referring to the waveforms of Fig. 12.5, Q_1 is triggered at t_0 to charge C_1 to $2E_d$. The half-sinusoid input current ceases at t_1 and Q_1 turns off. The energy stored in C_1 is $\mathcal{E} = \frac{1}{2}CE_d^2$. In response to an external signal at t_2, Q_2 turns on and essentially becomes a forward-biased rectifier at t_3. The inductance of L_2 supports the voltage across C_1 and no appreciable current flows initially, which limits the di/dt in Q_2 until the gate carriers combine and until L_2 saturates at t_4. Now, the full voltage on C_1 is applied across T_1. This voltage level is stepped up in the secondary and produces a current i_c to charge C_2 from t_4 to t_5. The energy stored in C_1 is transferred to C_2, which is the main PFN capacitor. The voltage produced across L_3 and T_2 during this period is quite low. At t_5, the voltage on C_2 appears across the secondary of T_1, which supports this voltage until t_6. Then T_1

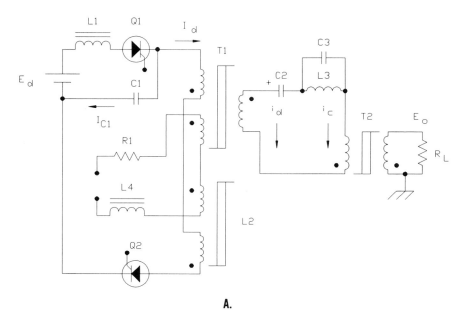

A.

Fig. 12.5. Solid-state magnetic modulator and waveforms.

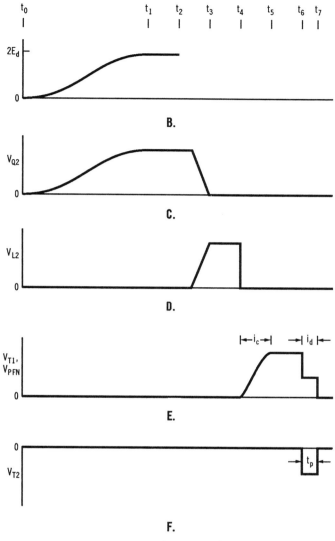

Fig. 12.5. (*Continued*)

saturates and the saturated inductance becomes a critical parameter of the PFN (L_2 of Fig. 12.2A, in fact). Now, the PFN produces a discharge current i_d and the full voltage appears across T_2, which is stepped up in the secondary to provide the desired output pulse t_p. At t_7, the pulse transformer saturates and the pulse terminates.

The toroid tape-core reactors and transformers use square-loop materials, as discussed in Section 3.13. A typical pulse transformer winding and the corresponding schematic for powering a high-voltage vacuum tube are shown in Fig. 3.21. These magnetics have very fast "core switching" times, typically 20 ns. Design parameters are based primarily on the volt-microsecond product and the saturated inductance value. These parameters are listed in Table 3.16 for various sizes of square-loop materials.

Example

Inductor L_2 of Fig. 12.5A may be required to support 400 V (the voltage across C_1) for 2.85 μs (the time from t_3 to t_4) to delay the rapid current rise in Q_2. This product is 1140 V-μs. From Table 3.16, a 5772 core has a 190-V-μs$/T$ rating, thus requiring 6 turns. However, the saturated inductance must be low to maximize the voltage across the primary of T_1. Again, Table 3.16 lists the K_0 factor for the cores and the saturated inductance is given by Equation (3.57) as $L_0 = 6^2 \times 14.7 \times 10^{-10} = 53$ nH. These two parameters are even more critical in the design of T_1. In this case, the required volt-microseconds is a function of the PFN voltage, the pulse width, and the compression ratio (essentially the ratio of delay time to pulse width). From these parameters, the required number of turns can be calculated. However, the selected core must meet the required K_0 factor that determines the required saturated inductance for the desired PFN performance. Some trial and error methods may be required before an optimum parameter balance is achieved.

In the discussion so far, the magnetics operated from zero flux to a positive saturation. However, the reset windings shown in Fig. 12.5A are used for flux reversal between pulses. Typically 1 to 5 ampere-turns are sufficient to drive the cores into negative saturation. The bias supply is normally a 5-V source and R_1 limits the current. A high-impedance linear inductor L_4 is placed in series with R_1 to support induced voltages that appear on the reset windings of T_1 and L_2 and to prevent current from being driven back into the source during power pulses.

12.3.2   Design Parameters

In the design of a magnetic modulator, the known requirements for a particular application are peak output voltage and current, pulse width, pulse repetition frequency, and available input voltage. The design process normally begins with the known output parameters and proceeds to work back toward the source, while calculating the passive component values. The total energy and power relations are

$$\varepsilon = P_0 t_p = E_0 I_0 t_p \tag{12.20}$$

$$P_{\text{avg}} = E_0 I_0 t_p (\text{PRF}) \tag{12.21}$$

A turns ratio for the pulse transformer is selected (typically in the range of $1:3$ to $1:6$). For matched impedances, the PFN must be charged to twice the primary voltage of T_2 (reference Section 12.1.2). The primary voltage, PFN voltage, and PFN capacitance are calculated from

$$E_p = nE_0 = \tfrac{1}{2}(N_p/N_s)V_N \tag{12.22}$$

$$C_N = 2E_0/V_N^2 \tag{12.23}$$

Typically, the PFN capacitance is increased 10% from the calculated value to offset losses in the system. The reflected load resistance and PFN inductance relations are

$$R_p = (E_0/I_0)/n^2 = \sqrt{L_N/C_N} \tag{12.24}$$

$$L_N = \tfrac{1}{2}R_p t_p \tag{12.25}$$

The parameters for the switching reactor T_1 are a function of the rated capacity of the selected core, expressed in volt-microseconds per turn, and the saturated inductance factor K_0. These values are listed in Table 3.16 or are obtained from core manufacturers data. The pulse compression is a ratio of T_1 delay time versus output pulse time and is typically about $1:5$. Then, for T_1, the compression ratio (CR) is

$$\text{CR} = t_d/t_p \tag{12.26}$$

$$\text{V-}\mu\text{s} = \tfrac{1}{2}V_N t_p(\text{CR}) \qquad (t_p \text{ in } \mu\text{s}) \tag{12.27}$$

$$N_s = \text{V-}\mu\text{s}/(\text{V-}\mu\text{s}/T) = \sqrt{L_0/K_0} \tag{12.28}$$

Now, the process reverts to the known input voltage and the charging capacitor voltage will be $V_{C1} = 2E_d$. Since V_N has been calculated and V_{C1} is now known, the turns ratio of T_1 and the required value of C_1 are

$$n_{T1} = N_p/N_s = E_p/E_s = V_{C1}/V_N = 2E_d/V_N \tag{12.29}$$

$$C_1 = C_N/n^2 = C_N/(V_N/V_{C1})^2 \tag{12.30}$$

The inductance of the di/dt (delay) reactor, L_2, is determined from the calculated time delay and is

$$L_2 = (t_d/\pi)^2/C_1 \tag{12.31}$$

In some designs, the calculated value of L_2 at this point may be to low to be practical. If this is the case, the compression ratio is increased, which requires a large core to provide the increased volt-microsecond capacity.

The peak discharge current through Q_2 is

$$I_d = \pi E_d C_1/t_d \tag{12.32}$$

The remaining item is the charging inductor L_1. The inductance for a selected charge time is

$$L_1 = (t_c/\pi)^2/C_1 \tag{12.33}$$

Since the charging time is much longer than any delay or pulse periods, the charging current through Q_1 is quite low:

$$I_c = E_d\sqrt{C_1/L_1} \tag{12.34}$$

REFERENCE

1. E. A. Guillemin, "A Historical Account of the Development of a Design Procedure for Pulse Forming Networks," RL Report No. 43, October 1944.

13

PROTECTION AND SAFETY

13.0 Introduction
13.1 Fuses
13.2 Circuit Breakers
13.3 Transient Protection
    13.3.1 *RC* Networks
    13.3.2 Silicon Transient Suppressors
    13.3.3 Metal–Oxide Varistors
    13.3.4 Spark Gaps
13.4 Snubbers
    13.4.1 Dissipative Snubbers
    13.4.2 Nondissipative Snubbers
13.5 Safety Requirements
    13.5.1 Underwriters' Laboratories (UL)
    13.5.2 Canadian Standards Association (CSA)
    13.5.3 International Electrotechnical Commission (IEC)
    13.5.4 British Standards Institute (BSI)
    13.5.5 Japanese Industrial Standards Committee (JISC-MHW)
    13.5.6 Other

13.0 INTRODUCTION

This chapter discusses protection of components, equipment, and personnel from various anomalies. Conditions ranging from internal faults to lightning strikes can have a devastating effect on electronic equipment and can cause a safety hazard if not properly addressed. The protective devices discussed are classified into two groups: current sensitive and suppression. Fuses and circuit breakers are in the first class, while transient suppressors, snubbers, and line conditioners are in the latter class. National and international safety requirements, which primarily establish isolation requirements for personnel protection from unintentional contact with unsafe voltages, are presented.

13.1   FUSES

Fuses protect equipment from overload or fault conditions. After an excessively
high current, the fuse opens or clears. Referring to Fig. 13.1, the fused ac input
line is supplying normal load current and voltage. (It is understood dc operation is
also applicable.) A fault then occurs in the load and the current suddenly
increases. Without a fuse, the current would increase to a peak current I_p. But the
internal heating in the fuse causes the fuse to melt during time t_m, and the current
is limited to I_{plt}, the peak let-through value. When the fuse melts, an arc develops
across the fuse. The sudden increase in fuse impedance causes the current to
decrease but also results in an arc voltage being formed across the fuse. The heat
generated vaporizes the fuse element, which increases the arc length to further
decrease the current during t_a. However, the faster the fuse clears, the higher the
arc voltage becomes. Melting time is a function of rms current, ambient temperature, and service factor (short overloads). The sum of the melt time and the arc
time is equal to the clearing time t_c, which is a function of power factor during the
fault condition.

Fuse categories are divided into (1) slow (frequently referred to as slo-blow), (2)
standard, (3) fast (for protection of semiconductor devices), (4) current limiting,

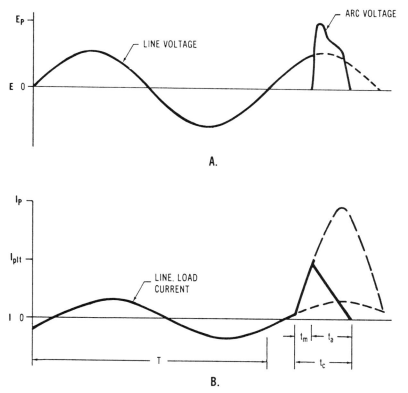

Fig. 13.1. Fuse clearing characteristics from fault current.

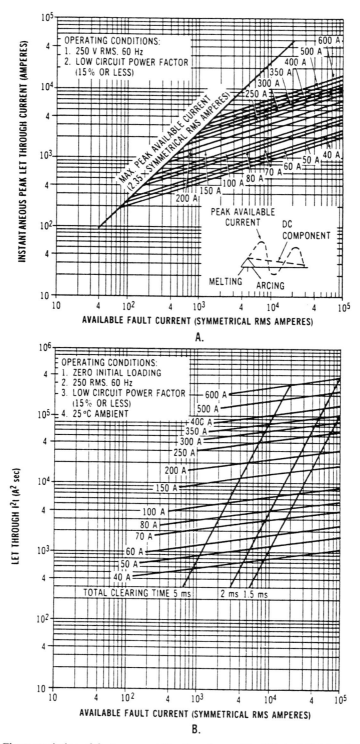

Fig. 13.2. Characteristics of fast, current-limiting fuses, 40–600 A, 250-V ratings. (Courtesy International Rectifier.)

and (5) high voltage. This section concentrates on fast-acting fuses for protection and safety of power conversion equipment and on current-limiting fuses for protection of semiconductors. The term *current limiting* applies to a fuse which will limit the peak current to a value lower than that which would otherwise flow in a fault condition. In power conversion equipment, various internal and external source impedances are usually very low to maximize efficiency. The low impedances allow fault currents to flow which are well in excess of current ratings of the devices to be protected. Proper coordination of fuse characteristics, such as shown in Fig. 13.2, with those of semiconductors will prevent damage to these devices from various types of malfunctions. However, a fuse with an interruptive rating less

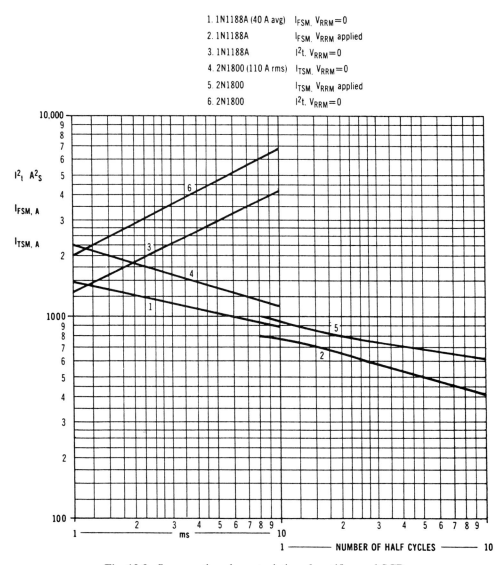

Fig. 13.3. Surge-rating characteristics of rectifier and SCR.

than the fault capability of the system or source may not be able to interrupt a short-circuit current within its clearing I^2t (A^2-s) and I_{pk} ratings, resulting in damage to the semiconductor. Therefore, the time-versus-current let-through characteristics of the fuse must be less than the time–current rating of the semiconductor. Also, the maximum arc voltage rating must not be exceeded. This rating is typically 1.7 times the rms voltage rating of the fuse.

Rectifiers and thyristors are capable of sustaining high overload currents for short periods of time. Surge characteristics of a 1N1188A rectifier rated at 40 A average and a 2N1800 SCR rated at 70 A average are shown in Fig. 13.3. The JEDEC SCR is shown for reference only since various manufacturers offer devices with an equivalent average current rating but higher surge current ratings with their own manufacturer part numbers. Bipolar transistors are gain dependent and are limited in peak-current capability. A condition causing excessive transistor current will usually result in device failure, due to forward-biased second break-down, before the fuse clears. The MOSFETs are transconductance devices and do not suffer from forward-biased second breakdown. The peak current is typically three times the continuous rating. A fast acting fuse *may* protect the MOSFET if the rated peak current is not exceeded and if the fuse clears fast enough. As the operating frequency of switch mode power conversion equipment increases, an additional consideration must be given to fuses. For ac voltage sources, the first half cycle following a fault should contain enough I^2t to melt the fuse.

An additional application of fused protection is in large-capacitor banks. A group of capacitors may be paralleled for filtering rectified ac voltages in high-power equipment or for filtering harmonics in high-power sine wave output inverters. Variations in capacitor impedance, ESR, and leakage current may produce excessive ripple currents in one of the capacitors. This capacitor then overheats and fails, usually in the shorted condition, possibly in an explosive mode if the short-circuit energy is high. The resulting short-circuit condition will probably clear a master input fuse or circuit breaker, but equipment operation will cease. If individual fuses are placed in series with each capacitor, the resulting short-circuit failure in a particular capacitor clears that particular fuse, and the equipment continues to operate, providing desired performance in most cases. The open-fuse condition can be electronically indicated, and the proper maintenance or component replacement can be performed at a more convenient time.

The following procedures are suggested for proper fuse selection to protect rectifiers and thyristors:

1. Calculate the maximum available fault current.
2. From fuse tables, use the maximum fault current to find I_{plt}.
3. From another fuse table, find I^2t let-through and t_c for a fuse with a current rating greater than I_{rms} of the load.
4. Compare these factors with the I^2t and I_{FSM} ratings of the semiconductor.
5. These ratings for the semiconductor must be higher than those of the fuse for protection.

Factors affecting t_c are available short-circuit current, input voltage, and power factor at the fault condition. The worst case total clearing I^2t will occur at the highest short circuit current, maximum voltage, minimum power factor, and t_a, which commences before or near E_p. Other data pertaining to power are frequently given as X/R ratios, where data for dc operation are given as L/R ratios. In a fault condition, the resistive component is normally the low resistance of the wiring, and the power factor is typically less than 0.15.

The maximum fault or short-circuit current is a function of the source impedance and the operating voltage, $I = E/Z$. If the source is a transformer, as is the case of any utility or mains line feeding the facility or equipment, the short-circuit current is given by

$$I_{sc} = VA/(Z_{pu}E) \tag{13.1}$$

where VA is the power rating of the transformer, Z_{pu} is the per-unit impedance (percent/100), and E is the transformer secondary voltage.

Typical parameters for fast-acting, current-limiting fuses are shown in Fig 13.2 for fuses rated at 250 V and current ratings from 40 to 600 A. Once the available fault current is established, the peak let-through current, the let-through I^2t, and the clearing time can be determined for various fuse current ratings. Clearing time can also be calculated (if data are not available), since the fault current approaches a triangle wave. From Equation (1.9b), $I_{rms}^2 = \frac{1}{3}I_{pk}^2$. Then the clearing time for a given I^2t rating is

$$t_c = 3I^2t(\text{rated})/I_{plt}^2 \tag{13.2}$$

Example

Two SCRs are connected to the secondary of a power transformer in a phase control circuit. A fuse is to be connected in series with each SCR for protection . The transformer rating is 18 kVA, with a per-unit impedance (Z_{pu}) of 3%. The secondary voltage is 240 V and the load current is 75 A. A 2N1800 is chosen for the SCR. Voltage and current ratings are $I_{avg} = 70$ A, $I_{rms} = 110$ A, and $V_{DRM} = 600$ V. Surge ratings are $I^2t = 5900$ A^{2}s at 8.33 rms; $I_{TSM} = 1200$ A at 8.33 ms. Note that these parameters are for V_{RRM} reapplied $= 0$, which will be the case if the fuse opens within a half cycle. Initially, an 80-A, 250-V fuse is chosen. From Equation (13.1), $I_{sc} = 18,000/(0.03 \times 240) = 2500$ A. From Fig. 13.2A, for an 80-A fuse at an available fault current of 2500 A, $I_{plt} = 1000$ A. From Fig. 13.2B, $I^2t = 3200$ A^{2}s and $t_c = 4.8$ ms. Since this time is less than 8.33 ms, the $I^2\sqrt{t}$ factor must be used for the SCR. The $I^2\sqrt{t}$ rating of 64,500 for the 2N1800 assumes no reapplied voltage, which will be the case. Then at 4.8 ms, the I^2t rating is $64,500\sqrt{0.0048} = 4469$ A^{2}s, which is 40% greater than the I^2t rating of the fuse. Also note that the half-cycle peak current rating of the SCR is 1400 A at 4.8 ms, which is greater than the 1000 A peak let-through current rating of the fuse. Since these two conditions are satisfied, the SCR will be protected by the fuse.

As a *general* rule of thumb, selecting a fast current-limiting fuse with a current rating (the rated current is an rms value) equal to or less than the average current rating of the rectifier or thyristor will normally provide adequate protection for the semiconductor.

The preceding analysis applied to ac circuits. For dc circuits, additional considerations apply with inductive components. In a fault condition, the L/R time constant may be long. Thus the current increase may be slow, which increases the melting time of the fuse. Also, the arcing time and the arc voltage increase. This is due to the inductance trying to sustain the current flow plus the induced voltage developed by the inductance when the current does decrease. In addition, the voltage never goes to zero, as opposed to ac circuits, and extinguishing the arc is more difficult. For circuits operating from dc voltages, the voltage rating of the fuse should typically be 1.5 times the equivalent ac rms voltage.

13.2   CIRCUIT BREAKERS

The tripping action of a circuit breaker provides overload and safety protection for the entire electrical system—power source, equipment, switching devices, and wiring. Circuit breakers provide a means of manually switching equipment on and off. In addition, standard breakers are trip free, which means they cannot be held on during an overload condition. There are essentially two types of circuit breakers, thermal and magnetic. The thermal breakers operate on the principle of temperature change. Current flow heats a bimetallic strip which bends to open a set of contacts for circuit interruption. The time required to heat the strip and consequently the tripping time are relatively long. Thermal breakers are used where overloads can be tolerated for an extended time, such as in commercial wiring and in the protection of motors, transformers, and tungsten lamp loads. The long temperature-dependent delay means the thermal breakers are not applicable to electronic systems and power conversion equipment protection.

By contrast, magnetic breakers operate on the principles of flux change and of hydraulics and are electrically analogous to current-sensitive relays. A solenoid coil is wound around a tube containing a spring-loaded armature and the tube is filled with a silicone fluid. Increased current flow through the solenoid coil moves the armature against the spring and the fluid and toward a pole piece. This reduction in reluctance further attracts the armature, causing the mechanism to trip and open a set of contacts for circuit interruption. The current interruption deenergizes the solenoid and the spring moves the armature to its original position. This describes the action of the "series-trip" breaker, where the coil and the contacts are in series. The trip current is dependent on ampere-turns, the characteristics of the damping fluid, and the spring design. The tripping response time to overload conditions is a function of design and is relatively fast. In fact, this time delay can be made to vary from instantaneous to a variety of inverse time delays. The inverse time delay means the trip time is inversely proportional to the overload current value. In addition, circuit breakers are available for specific applications in dc operation as well as in 50- and 400-Hz operation. For these reasons, magnetic circuit breakers are used extensively in electronic equipment.

The desired time delay is primarily a function of the type of equipment to be protected. Typical delay times versus percent rated current are shown in Fig. 13.4 for 250-V, 60-Hz circuit breakers, with current ratings up to 100 A and an interrupting capacity of 1500 A. The interrupting capacity rating is a very important parameter. If the breaker cannot clear the fault or short circuit let-through

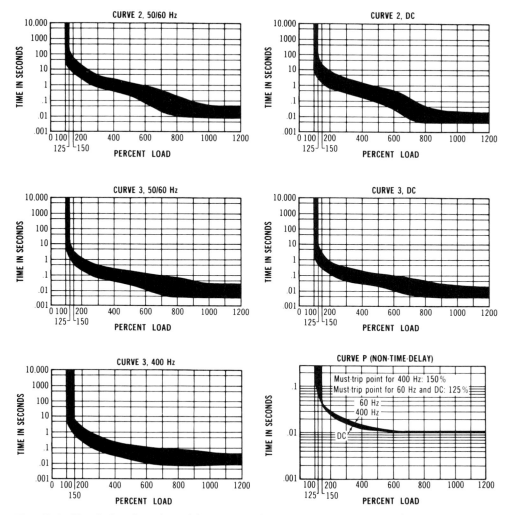

Fig. 13.4. Circuit breaker time delay curves, 0.1–100-A, 250-V ratings. (Courtesy Heinemann Electric Company.)

current available from the source, severe damage to the breaker and the equipment can result. For instance, the example following Equation (13.2) showed an available short-circuit current of 2500 A from the transformer. If a circuit breaker were used in this application, the interrupting capacity rating of the breaker chosen must be well in excess of 2500 A.

Instantaneous or non-time-delay breakers, Fig. 13.4, curve *P*, provide tripping on overloads of any magnitude and are therefore sensitive to inrush current at turn-on as well as vibration and shock. Transformers or rectifier filter capacitors (which inherently have a short-duration, high-inrush current) are located at the input of most power conversion equipment. Therefore, use of this breaker should be carefully evaluated. Nuisance tripping can cause as many problems as the breaker solves. Curves 2 and 3 of Fig. 13.4 are typical of breakers with time delay. Curve 3 is normally applicable to semiconductors, and a breaker with this characteristic will trip on an overload current of five times the current rating within 200 ms maximum.

Example

An input rectifier bridge uses 1N1188A devices rated at 40 A average. The 60-Hz ac input current is 26 A, and a 30-A circuit breaker with curve 3, 50/60 Hz, Fig. 13.4, characteristics is initially chosen. Will the breaker protect the rectifier in an overload condition? From Fig. 13.3 and manufacturers' data, the 1N1188A has the following surge characteristics for V_{RRM} reapplied: $I_{FSM} = 800$ A for one half cycle; for 12 half cycles (100 ms), the peak current rating is 400 A; for 48 half cycles (400 ms), the peak current rating is 275 A. From Fig. 13.4, the breaker will trip at 240 A in 100 ms or 90 A in 400 ms. However, these values are rms versus the peak rating of the rectifier. Thus the peak interrupting current through the breaker will be $240\sqrt{2} = 340$ A at 100 ms, or $90\sqrt{2} = 127$ A at 300 ms. Thus the surge ratings of the rectifier are higher than the interrupting ratings of the breaker, and the breaker will protect the rectifier.

The interrupting capacity of circuit breakers is an important but often overlooked parameter in breaker selection. If the breaker is not rated to interrupt the available fault current, sustained arcing across the breaker contacts will result in breaker damage, wiring damage, load damage, fire, explosion, or combinations thereof. As previously mentioned, the curves of Fig. 13.4 represent circuit breakers with an interrupting capacity of 1500 A. This requires a source impedance of 0.16 Ω or greater at an input of 240 V. Circuit breakers are available with interrupting capacities of 100,000 A for high-power applications.

13.3 TRANSIENT PROTECTION

Since any line has a finite inductance, transient voltages occur whenever a line carrying current is interrupted. The magnitude of the voltage transient is proportional to the inductance and the rate at which the current is interrupted, from $e = -L\,di/dt$. These transients are an everpresent hazard to semiconductor devices as well as a source of EMI. Transients external to power conversion equipment may originate in the input source of a utility line. Lightning strikes can induce several kilovolts into utility lines. These higher voltage transients can damage equipment and cause a breakdown in wiring. Other sources of transients are electrostatic discharge (ESD), electromagnetic pulse (EMP), and nuclear magnetic resonance (NMR). Susceptibility requirements for equipment, as pertaining to EMI, are given in Section 14.1.4.

Domestic specifications for high-voltage transients and surges, which equipment may be required to meet, are tabulated below. The $A \times B$ factor denotes A as the rise time in (t_r, microseconds) of the pulse to a peak value and B as the pulse time (t_d in microseconds) with an exponential decay from the peak value to 50% of the peak value.

ANSI C62.41: 0.5-μs, 100-kHz oscillatory voltage waveform, 6 kV peak, each peak is approximately 60% of the preceding peak.

ANSI C62.41: 1.2 × 50 impulse wave for high-impedance devices, 6 kV peak.

ANSI C62.41: 8 × 20 current impulse wave for low-impedance devices, 3 kA peak.

REA (Rural Electrification Administration) PE-60: 10 × 1000, 1 kV, peak.

FCC Docket 19528, Part 68: 10 × 160, 1.5 kV peak and 10 × 560, 800 V peak.

DOD-STD-1399, Shipboard Systems: 1.5 × 40, 2.5 kV peak.

The selection of transient-suppression components is primarily a function of the energy (or voltage) in the transient, the transient duration, and the source impedance of the circuit. The transient suppressors highlighted are (1) resistor–capacitor networks, used for suppression of arcs occurring across the contacts of electromechanical devices; (2) silicon suppressors, used in control and logic circuitry and for semiconductor protection; (3) metal–oxide varistors, used across utility lines and inductive units; and (4) spark gaps, used in high-voltage lines and high-voltage equipment. Characteristics of various networks, components, and buffer isolation methods to protect equipment are discussed.

13.3.1 *RC* Networks

This section applies to electromechanical devices as opposed to the snubber networks for semiconductors discussed in Section 13.4. The simple combination of a series resistor and capacitor connected across the contacts of a relay or switch is an effective means to reduce arcing when switching inductive loads. The capacitor absorbs the energy in the load when the contacts open and reduces arcing across the contacts. The resistor limits the discharge current of the capacitor when the contacts again close, thereby preventing possible welding of the contacts. The selection of resistor and capacitor values is given by the following empirical interrelation:

$$R = \frac{E}{10\left[3.16\sqrt{C}\,\right]^{(1+50/E)}} \quad \Omega \tag{13.3}$$

where E is the rms supply voltage and C is the capacitance in microfarads.

Example

A switch makes and breaks a load current of 5 A in a 120-V, 60-Hz inductive load. It is desired to connect an *RC* network across the contacts to reduce arcing. Since the source is ac, the maximum current through the capacitor (to the load) when the switch is open is to be limited to 1 mA. Then $I = E/X_C = E(2\pi f C)$; $C = I/[E(2\pi f)] = 0.001/120 \times 377 = 0.022$ μF. Now, $R = 120/10(3.16\sqrt{0.022}\,)^{(1+50/120)} = 35$ Ω; use 33 Ω. If the switch will not be actuated in a rapid sequence, a $\frac{1}{2}$-W resistor should be sufficient since its voltage rating is 350 V. If the switch is again closed when the capacitor is charged to the peak input voltage, or 170 V, the peak current through the switch will be $170/33 = 5.2$ A plus the instantaneous load current. It is obvious that other combinations of C and R may be selected or calculated for other conditions.

13.3.2 Silicon Transient Suppressors

Silicon PN junction devices, frequently referred to as transient-absorbing zeners or Transzorbs,* are characterized by their high-surge-current capability, fast response time, small size, and low on-resistance above the breakdown voltage. These transient voltage suppressors are ideal for low-voltage analog and digital circuit protection. Voltage ratings from 5 to 200 V are available in peak power ratings to 1.5 kW in both unidirectional (voltage clamping in one polarity direction) and bidirectional (clamping in either polarity direction). Bidirectional devices have ratings up to 700 V in peak power ratings to 15 kW for suppressing transients on utility mains. Proper voltage selection of the latter devices will provide transient voltage clamping at approximately 2.5 times the rms line voltage.

Silicon transient suppressors clamp voltage transients by providing a shunt path for current produced by excessive voltage. As a shunt device, an impedance is required between the source of the transient and the equipment to be protected. In this manner, the suppressor is analogous to the zener diode. However, the zener diode is a voltage regulator and seldom has surge ratings specified (except for transient thermal impedance). Also, the zener diode typically has a higher on resistance at the peak transient current which increases the clamping voltage at high reverse currents.

Figure 13.5A shows pulse power ratings versus pulse duration for a 1.5-kW rated device. The 1.5-kW dissipation point occurs at 1 ms. Figure 3.5B shows a typical capacitor discharge test waveform parameter for percent of I_{pp} versus time. The $t_r \times t_d$ factor is typically 10×1000. This means the rise time to I_{pp} is 10 μs and the pulse duration, from I_{pp} to $\frac{1}{2}I_{pp}$, is 1000 μs.

Example

A control assembly operates from a $+12$-V power supply at a remote distance from the power supply and draws 3 A from this source. In the electrical environment, transients may be generated by the switching of magnetic components or by lightning strikes on the power line. The control assembly is to be protected from these transients, such as a 2-kV, 200-μs transient which has a source impedance of 50 Ω. The control assembly voltage is to be clamped at 20 V maximum. From manufacturers' data, a 1N6043A has the following characteristics: $BV = 14.3$–15.8 V (typically 15 V) at 1 mA; $V_c = 21.2$ V at $I_{pp} = 71$ A. Since V_c is higher than the desired clamp voltage, I_{pp} must be reduced. The on resistance $R_{on} = \Delta E / \Delta I = (21.2 - 15)/71 = 0.087$ Ω. The current available from the transient is $2000/50 = 40$ A, which will be the suppressor current. Then the control voltage is limited to $V_c = 15 + (0.087 \times 40) = 18.5$ V. The power dissipation in the suppressor is $P = 18.5 \times 40 = 740$ W, well within the rating for a 200-μs pulse. The energy absorbed by the suppressor is $\varepsilon = Pt = 740 \times 200 \times 10^{-6} = 148$ mJ, or about 10% of the device rating.

---

*Transzorbs is a trademark of General Semiconductor Industries.

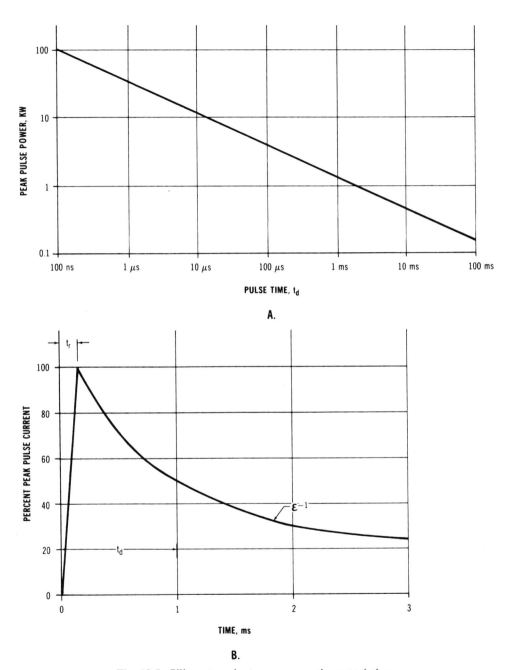

Fig. 13.5. Silicon transient suppressor characteristics.

13.3.3 Metal–Oxide Varistors

Varistors are nonlinear variable-impedance devices consisting of bulk metal–oxide particles separated by an oxide film for insulation. As the applied voltage increases, the film becomes conductive and current flows through the device as

$$I = KV^\alpha \tag{13.4}$$

where K is a constant, V is the voltage rating, and α is a figure of merit and is the slope of the VI characteristic:

$$\alpha = \frac{\log(I_2/I_1)}{\log(V_2/V_1)}$$

Depending on material, processing, size, and design parameters, α is typically between 30 and 40. Obviously for high values of V, the constant K is a very small number.

Varistors are characterized by high-peak-current capability plus the ability to absorb the high-energy levels in utility line transients. Voltage ratings from 6 V to hundreds of volts are available in the GE-MOV* series.

Manufacturers' data typically list energy and current capability, VI graphs, and pulse ratings, which in turn determine the number of pulses a device can withstand and thus its life.

Example

An SCR phase-controlled power supply is shown in Fig. 13.6. The unit operators from a 480-V, 60-Hz input, the transformer is rated at 13 kVA, and the secondary voltage is 48 V. When the circuit breaker is opened, observed transients are induced across the secondary which cause semiconductor failure, even though the devices are conservatively rated. The high-voltage transient is due to the energy storage in the transformer. A varistor connected on the secondary side of the transformer will absorb the transient and protect the semiconductors. The energy $\varepsilon = \frac{1}{2}L_m I_m^2$, where L_m is the inductance of the primary, and I_m is the magnetizing current. A typical value of I_m is 4% of the full-load current, or $I_m = 0.04 VA/E_{in} = 0.04 \times 12,000/480 = 1$ A, and $I_{m,pk} = 1 \times \sqrt{2}$. Then $X_L = E_{in}/I_{m,pk} = 480/\sqrt{2} = 339$ Ω, and $L_m = X_L/\omega = 339/337 = 1$ H. The energy in the transformer when the breaker is turned off is $\varepsilon = 0.5 \times 1 \times 2 = 1$ J. The peak secondary voltage is $48 \times \sqrt{2} = 68$ V. Choose a varistor with a minimum rating of 1.3 times this value for line variations, or 88 V. The clamping voltage will typically be twice this value, or 176 V. Since 1 J of energy is to be absorbed for a secondary voltage of 176 V and a secondary current of 10 A, the time duration is $t = \varepsilon/EI = 1/176 \times 10 = 0.57$ ms. Therefore a varistor with a clamp voltage of 176 V and capable of conducting 10 A for 0.57 ms and absorbing 1 J can be selected from manufacturers' data.

*GE-MOV is a trademark of General Electric Company.

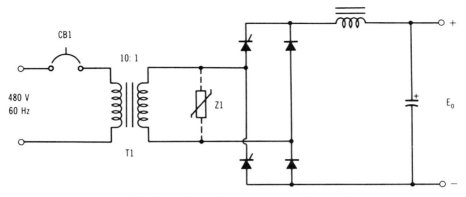

Fig. 13.6. Typical transient protection with MOV.

13.3.4 Spark Gaps

The term *spark gap* is used to define gas-filled discharge devices which have an extremely high impedance at normal working voltages but which can conduct very high current when the breakdown (or striking) voltage is exceeded. The breakdown voltage is the value at which the gap will ignite and is physically limited to 75 V minimum. Depending on construction and gap length, the maximum breakdown voltage can be many kilovolts.

The simplest form of a gas discharge device is a neon bulb. The more exotic forms of gas discharge devices include triggered spark gaps and surge arrestors. Even the nonexotic automotive spark plug has been used for safety protection from arcs and voltage breakdowns in high-voltage power supplies (a rather nonscientific but very cost effective approach).

The voltage and current characteristics of a gas discharge spark gap during a transient are shown in Fig. 13.7. The breakdown voltage V_B is the device rating, but in most cases, this voltage is subject to dv/dt of the applied transient. For dv/dt up to 10 V/μs, V_B is not affected and increases exponentially to approximately $4V_B$ at 10 kV/μs. Actual manufacturers' data should be consulted for specific parameters. Here V_G is the glow voltage when the current increases. When an arc is sustained, the voltage drops to V_a, and the current increases significantly to I_p. As the transient decays, the spark gap continues to draw current until the arc extinguishes at V_e, and the current can no longer be sustained at I_s. Note that the supply voltage must be lower than V_e; otherwise the device will continue to draw a follow-on current. In this case, the circuit must be interrupted by an auxiliary device to allow recovery. As in the case of all shunt mode devices, an input impedance must be present, and in the case of a spark gap, a resistance of a few tenths of an ohm may be required in series with the spark gap to limit current.

Spark gaps with V_B ratings of several hundred volts and I_p ratings of several thousand amperes can readily suppress high-energy, short-duration transients. An additional important parameter is the coulomb capacity of the spark gap which determines the number of transients which can be suppressed and, thus, the life of the device. These ratings typically refer to an 8×20 transient (8 μs rise time, 20 μs pulse time). For instance, if the peak current is 5000 A and the half-width pulse

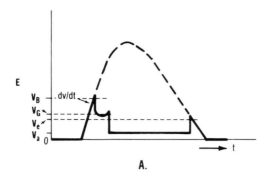

A.

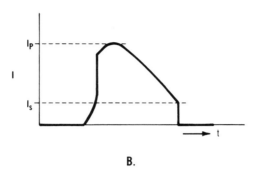

B.

Fig. 13.7. Spark-gap characteristics.

time is 10 μs, the charge contained in the pulse is $5000 \times 10^{-5} = 0.05$ Coulomb. If the capacity of the device is 100 Coulombs, the device is capable of 2000 discharges.

13.4 SNUBBERS

Snubbers are used to suppress undesirable transients and to eliminate ringing in switching circuits which contain parasitic inductive and capacitive elements. If not suppressed, the transient voltages can exceed the rating of semiconductors, resulting in device failure. The ringing or oscillations can produce EMI (electromagnetic interference) which is conducted or radiated to other sensitive equipment. Snubbers are also used to control the load line of switching transistors to prevent reverse-biased second breakdown and to decrease applied dv/dt to thyristors. In performing these functions, energy is transferred from the switching device to the snubber, and considerable power may be dissipated in the resistor of a simple RC snubber at high frequencies. Nondissipative snubbers are more complex and normally require inductive and capacitive components which store the energy produced by the switching transient and then transfer this energy back to the input source for maximum operating efficiency.

13.4.1 Dissipative Snubbers

An effective snubber for lamp loads with triac control is shown in Fig. 13.8A. Although C_1 and L_1 dissipate negligible power, the reactances reduce conducted EMI on the input lines, since the switching of Q_1 "looks into" the LC filter. The RC network of Fig. 13.8C is also connected across Q_1. Typical emissions with and without the snubber are compared to a common EMI limit. Without the snubber the emissions at 15 kHz may be 120 dBμV. From Equation (14.3a), this is equivalent to 1 V. Addition of the snubber will achieve 40 dB attenuation to an equivalent 0.01 V.

Figure 13.8B shows an inductive load controlled by a triac, although back-to-back SCRs could be used for high-power loads. The waveforms show a high commutating dv/dt when Q_1 turns off. At turn-on, di/dt is limited by the inductance of the load. In heater loads, the inductance may only be that of the wiring. When the triac turns on, a high di/dt results. The rate of current rise may be calculated and the inductance increased to limit di/dt. To reduce dv/dt at turn-off, the RC circuit of Fig. 13.8C is connected across the triac. When the triac turns off, a current path is provided by C_1. At turn-on, C_1 may be charged to the peak input voltage, and a high current limited only by R_2 causes a high di/dt, as shown in the waveforms. Thus, the snubber will contribute a di/dt problem. To reduce di/dt, the circuit of Fig. 13.8D includes CR_1 to rapidly charge C_1 to further reduce dv/dt at turn-off, while L_2 is inserted in series with R_2 to reduce di/dt at turn-on. In this case, the gate signal must be maintained long enough to allow sufficient current increase in L_1 and to overcome any oscillatory current produced by C_1 and L_2 at turn-on. Analysis of series RLC circuits is discussed in Section 1.6.2. The following analysis assumes a worst case power factor of 0.5 (60° current lag). The following relations apply: $R_1 = Z \cos 60°$, $X_{L1} = Z \sin 60°$, and $L_1 = X_{L1}/\omega_{in}$. The circuit resonance will occur at f_r and maximum dv/dt will occur at $\omega_r E_{pk}$. Then $f_r = (dv/dt)/2\pi E_{pk}$, where dv/dt is less than the rated commutating dv/dt of Q_1. Also, the circuit will be critically damped for $R^2 = L/C$ but underdamped for $R^2 < L/C$. For this analysis $R^2 = 2L/C$ for a nominal 30% overshoot. Generally, an oscillatory condition is undesired since this contributes to EMI. In the case of devices with low dv/dt ratings, the resonant frequency of the LC circuit will be below the minimum frequency of EMI specifications. Also, $f_r = 1/2\pi\sqrt{LC}$. Then the component values for Fig. 13.8C are

$$C = \frac{1}{(2\pi f_r)^2 L} = \left[\left(2\pi \frac{dv/dt}{2\pi E_{pk}} \right)^2 L \right]^{-1}$$

$$= \frac{E_{pk}^2}{L(dv/dt)^2} \tag{13.5}$$

$$R = \sqrt{2L/C} \tag{13.6}$$

Example

A 4-kVA inductive load with a power factor of 0.5 operates from a 208-V, 60-Hz input and is controlled by a triac with a maximum dv/dt rating of 10 V/μs. Select

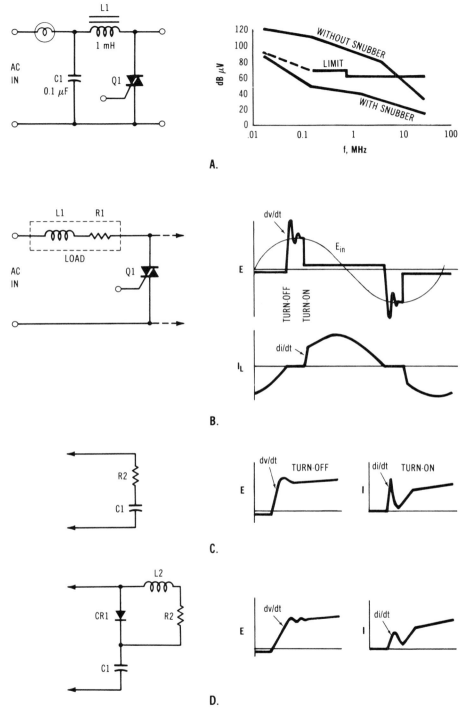

Fig. 13.8. Thyristor snubbers.

an *RC* snubber (Fig. 13.8C) for the triac. The input current $I = VA/E = 4000/208 = 19.2$ A. The load impedance $Z = 208/19.2 = 10.8$ Ω. Thus, $R_1 = 10.8 \times \cos 60° = 5.4$ Ω, and $X_{L1} = 10.8 \sin 60° = 9.35$ Ω. Then, $L_1 = X_{L1}/\omega = 9.35/377 = 24.8$ mH. Also, the peak voltage $E_{pk} = 208 \times \sqrt{2} = 294$ V. From Equation (13.5), $C_1 = 294^2/0.0248(10 \times 10^6)^2 = 0.035$ μF. From Equation (13.6), $R_1 + R_2 = \sqrt{2 \times 0.0248/(0.035 \times 10^{-6})} = 1.2$ kΩ ($R_2 = 1.2$ kΩ). For safety, C_1 could be increased to 0.047 μF. For critical damping, R_2 would be $\sqrt{4 \times 10^6 \times 0.0248/0.047} = 1.5$ kΩ. If the triac is turned on at E_{pk}, the maximum current from the snubber is then $294/1500 = 0.02$ A.

The inverter stage of Fig. 13.9A is an impulse-commutated SCR circuit. Assume Q_1 is conducting and the left side of C_1 is charged positive. When Q_3 is gated on, current flows through L_2, C_1, L_1, and CR_1. When this current exceeds the load current, Q_1 turns off, as shown in the waveforms. Detailed operation is discussed in Section 7.5. The anode voltage then increases at dv/dt, which must not exceed the device rating. To meet this requirement, a snubber is normally required, especially in high-frequency circuits with fast-switching SCRs. (Frequently, component values, especially L_1, can be calculated to allow the anode voltage to become slightly positive at a dv_{rec}/dt rate to enhance SCR turn-off.) To reduce dv/dt, the snubber of Fig. 13.9B may be used. Component parasitic elements are shown and must be considered in high-frequency operation. Inductor L_3 is the wiring inductance to R_1 and CR_3, L_4 is the wiring inductance to R_2, and C_2 has internal ESL and ESR plus wiring inductance L_5. These inductive paraistics must be minimized; else the snubber will become a high impedance at high transient frequencies and will become ineffective in reducing dv/dt.

Generally, the maximum dv/dt of inverter-rated SCRs is sufficiently high to require a critically damped or an overdamped snubber circuit to minimize EMI in the specified frequency range. For critical damping and from previous analysis in this section, the following relations for Fig. 13.9B apply:

$$C_2 = \frac{E_{pk}^2}{L(dv/dt)^2} \tag{13.7}$$

$$R_1 = 2\sqrt{L_1/C_2} \tag{13.8}$$

It is important to note that L_1 (or the total inductance of L_1 plus the parasitics) may be quite low in value, which results in a low value for R_1. This in turn would result in a high di/dt at turn-on, except CR_3 is installed to prevent this current flow. Since C_2 must be discharged, R_2 is installed to provide a current path. In this case, an inductor (L_4) in series with R_2 may even be desirable to reduce di/dt if R_2 is a low-resistance value. The power rating of the resistors is a function of operating frequency and the energy stored in C_2. Since $\varepsilon = \frac{1}{2}CE^2 = Pt$ (t is the total period since current flows through one resistor at turn-off and through the other resistor at turn-on), the power dissipation in each resistor can be found from

$$P_R = \tfrac{1}{2}CE^2 f_s \quad \text{W} \tag{13.9}$$

where C is the capacitance of C_2, E is the maximum capacitor voltage, and f_s is the switching frequency.

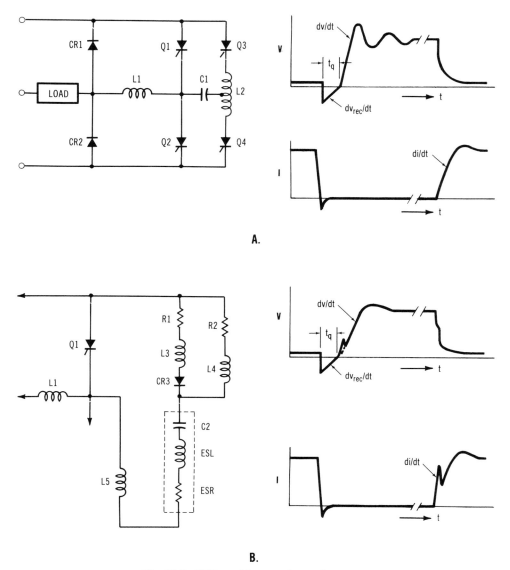

Fig. 13.9. SCR snubbers and waveforms.

Power dissipation given by the above equation is independent of the value of resistance. This can be compared to the following analysis. The current waveform through the resistors closely approximates an exponential decay. From Equation (1.15b), $I_{rms} = 0.316 I_{pk}(5\tau/T)^{1/2}$, where $I_{pk} = E/R$ and $T = 1/f$. Since $P = I^2 R = (0.316 E/R)^2(5RC/T)R$, the power dissipation $P = (0.1 E^2 \times 5RCf/R^2)R = \frac{1}{2}CE^2f$, which is the same as Equation (13.9). This is obviously a more laborious method, but the analysis allows a calculation of I_{rms} if desired.

A high-frequency switching circuit using MOSFETs is shown in Fig. 13.10A. The inherent fast turn-off time of Q_1 can produce an excessive voltage across the

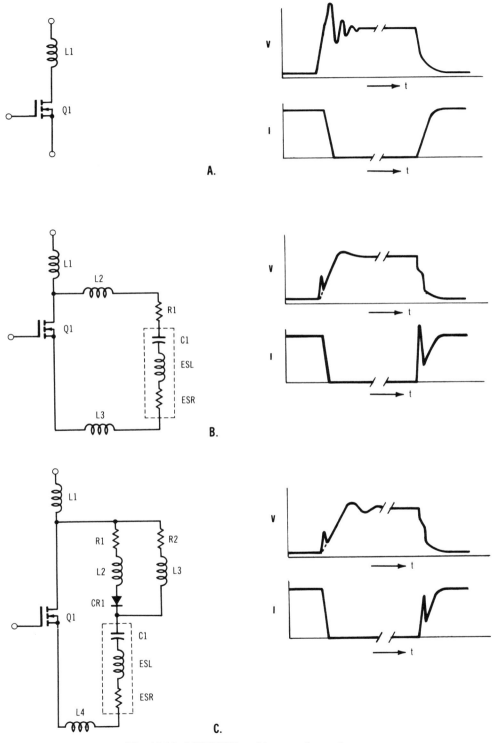

Fig. 13.10. MOSFET snubbers and waveforms.

transistor when switching inductive loads or when the load is a transformer with its associated leakage inductance. This voltage may be clamped by auxiliary devices, but a snubber is normally used, as shown in Fig. 13.10B. Snubber analysis is similar to the SCR circuit of Fig. 13.9, except the fast switching time of the MOSFET places added restrictions on snubber components and connections. Parasitic inductive elements (L_2 and L_3) in the snubber circuit must be minimized. This requires the shortest possible lead length in R_1 and C_1, plus C_1 must have a low ESL. Equations (13.7) and (13.8) may not be applicable due to unknown or interrelated values of L_1 and the restrictions on R_1 and C_1. The value of R_1 must be sufficiently large to limit the current from C_1 through the transistor when Q_1 turns on. Also, Q_1 must conduct the current flow through the winding capacitance of the load transformer at turn-on. A transistor switching time of 50 ns is equivalent to 40 MHz, which is a typical resonant frequency of 0.003-μF film or ceramic capacitors (reference Fig. 14.7). Thus any appreciably larger capacitance will be ineffective. An empirical relation for R_1 and C_1 is

$$\tau f_s = 0.01 \tag{13.10}$$

where $\tau = R_1 \times C_1$ ($= 1\%$ of the total period), $R_1 = E_{C1}/I_{max}$, and f_s is the switching frequency.

Since current flows through R_1 at turn-off and at turn-on, the power dissipation is twice that of Equation (13.9):

$$P_{R1} = CE^2 f_s \quad \text{W} \tag{13.11}$$

Example

A snubber is desired for a MOSFET in a bridge inverter stage operating from 150 V dc and switching at 100 kHz. The snubber current at transistor turn-on will be limited to 0.5 A. Then $R_1 = 150/0.5 = 300 \ \Omega$. From Equation (13.10), $C_1 = 0.01/300 \times 100,000 = 333$ pF; use a 330-pF ceramic capacitor. From Equation (13.11), $P_{R1} = 330 \times 10^{-12} \times 150^2 \times 100,000 = 0.74$ W; use a 300-Ω, 2-W, noninductive resistor.

The snubber circuit of Fig. 13.10C may be used if the snubber capacitance value is restricted by the preceding analysis of Fig. 13.10B. The addition of CR_1 (and possibly R_1) will allow an increase in the value of C_1 and also an increase in the value of R_2. However, the addition of CR_1 and R_1 increases the lead inductance of the snubber, and the total series inductance should be low enough to allow a high resonant frequency with C_1. Analysis is comparable to that of Fig. 13.9B and the component values are given by previous equations as

$$C_1 = \frac{E_{pk}^2}{L(dv/dt)^2} \tag{13.12}$$

$$R_1 = 2\sqrt{L_1/C_1} \tag{13.13}$$

$$P_{R1} = \tfrac{1}{2}CE^2 f_s = P_{R2} \tag{13.14}$$

The value of R_2 is determined by the conduction time of the transistor and the value of C_1. In order to discharge C_1 within $t_{on} = 3\tau$,

$$R_2 = 0.33t_{on}/C_1 \qquad (13.15)$$

where t_{on} is the minimum on time of Q_1.

As previously discussed, L_3 may even be desirable to limit di/dt when Q_1 turns on. The lead length of R_2 is immaterial, and in fact, a ferrite bead on the lead may be used. The recovery time of CR_1 is not particularly significant since CR_1 will probably be reverse biased by the voltage across C_1 before Q_1 turns on.

Snubbers connected across rectifiers can reduce voltage transients and reduce EMI. As shown in Fig. 13.11, the center-tapped secondary voltage of an inverter stage is rectified and filtered for a dc output. An *RC* network is connected across the rectifiers in Fig. 13.11A. In Fig. 13.11B, and *RC* network is connected across the transformer secondary. The optimum connection depends on the function of the circuit and the type of rectifiers used; this will be discussed after the purpose of the snubber is presented. Referring to Fig. 13.11A, when E_{s1} goes positive, CR_1 conducts and C_2 charges to $2E_s$ through R_2 and N_s. The voltage rating of the rectifiers must obviously be greater than $2E_s$. When the secondary voltage decreases, C_2 discharges through the load, N_{s2}, and R_2. However, the secondary leakage inductance induces a voltage reversal across N_{s1}, and this voltage transient could exceed the voltage rating of CR_1 unless a current path is provided to absorb

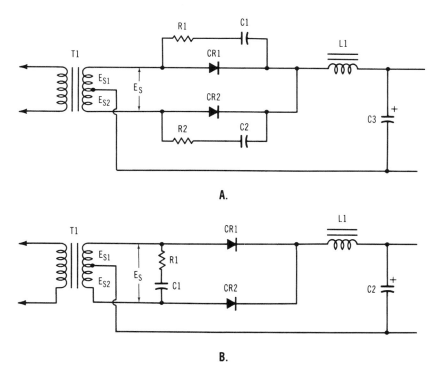

Fig. 13.11. Rectifier snubbers.

the energy in the transient. When the voltage reverses, current can flow through CR_2 (also through R_2 and C_2), C_1, R_1, and N_s. The reverse occurs on the opposite half cycle, where C_2 and R_2 limit the transient voltage across CR_2. In Fig. 13.11B, R_1 and C_1 absorb the energy stored in the secondary leakage inductance.

The circuit of Fig. 13.11A is recommended for a high output current; CR_1 and CR_2 are Schottky rectifiers, regardless of the current value. The higher depletion capacitance in the Schottky, as compared to the standard rectifier, may produce resonances with the leakage inductance; this could cause excessive voltage across the device if these transients are not damped by a snubber (reference Section 2.1.1.6). The snubber values for R and C can be determined using previously discussed analysis. Where Schottky rectifiers are used and the switching frequency is greater than 50 kHz, an empirical relation analogous to Equation (13.10) is

$$\tau f_s = 0.002 \qquad (13.16)$$

The power dissipation in each of the resistors is given by Equation (13.14).

The circuit of Fig. 13.11B is recommended for low output currents and/or high-voltage outputs. It is also ideal for suppression across the secondary of current transformers used to sense semiconductor current for limiting and protection purposes. In this latter case, very fast recovery signal diodes are used, and the output filter is replaced by a burden resistor. In either case, the values of R_1 and C_1 can be found from Equation (13.10), and the power dissipation in the resistor can be found from Equation (13.11).

13.4.2 Nondissipative Snubbers

Power dissipation can be essentially eliminated if the resistor in the snubber network is replaced with reactive components and steering rectifiers. These snubber networks are more complex, but they increase operating efficiency and can be cost effective in high-power equipment. In Fig. 13.12, Q_1 conducts current through a load which has a leakage inductance (L_1), and a voltage transient will be produced by this inductance when Q_1 turns off. The transformer load may be that of a bridge inverter stage, in which case additional semiconductors are used. The single stage is shown for simplicity. The transformer may also be that of a forward or flyback converter. In these latter cases, the transistor voltage is $V_{CE} = E_{in} + V_x + V_{L1}$, where V_x is the clamp voltage across the primary and V_{L1} is the voltage produced by the leakage inductance when Q_1 turns off.

For the bridge topology, assume Q_1 is initially conducting and the left side of C_2 has charged to $+E_{in}$. When Q_1 turns off, the current flows through L_1, C_2, CR_1, and the load. The energy stored in L_1 is transferred to C_2, which reverses in voltage. Also, if V_{C2} exceeds E_{in} when Q_1 turns off, current will flow through C_2, L_1, the load, C_1, CR_2, and L_2 until $V_{C2} = E_{in}$. When Q_1 again turns on, current flows through C_2, R_1, CR_2, and L_2. The energy in C_2 is transferred to L_2 as C_2 again reverses polarity. When the current decays, the trapped energy in L_2 causes a current flow through L_2, CR_1, C_1, and CR_2. Thus the excess energy is fed back to the input source, and the circuit operates at a high efficiency. However, this simplified analysis is fraught with danger. The first time power is applied or Q_1 is turned on, C_2 is not charged. When Q_1 turns off, the induced voltage in L_1 will

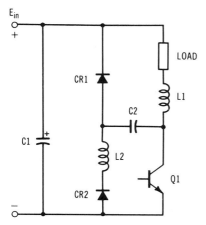

E_{in}

LOAD

CR1

C2

L1

C1

L2

CR2

Q1

Fig. 13.12. Nondissipative snubber.

allow current flow through C_2, but V_{CE} of Q_1 will increase significantly (and may exceed the transistor voltage rating) as compared to a precharged condition of C_2, where V_{L1} and V_{C2} are series aiding. Since $e = -L\,di/dt$, this voltage can be reduced by turning off Q_1 before the current reaches the full-load value. This can readily be accomplished in the control circuit by soft-start means. Also, additional devices may be paralleled with CR_1 to allow current flow from the source through C_2 and Q_1 to precharge C_2 during the on time. (This is analogous to the impulse-commutated SCR inverter stages discussed in Section 7.5 which provide a precharge on the commutating capacitor before or during load current flow.) Also, the transistor must be rated to conduct the instantaneous sum of the discharge current of C_2 and the load current. During normal operation, the minimum on time of Q_1 is determined by the time to recharge C_2 plus the time for the current decay through L_2 and CR_1. The former is given by $t_p = \pi\sqrt{L_2 \times C_2}$, and the latter is given by $di/dt = E_{L2}/L_2$.

Since the energy stored in L_1 $(\frac{1}{2}LI^2)$ during the on time must be transferred to C_2 $(\frac{1}{2}CE^2)$ during the off time, C_2 can be found by

$$C_2 = L_1(I_{L1})^2/(E_{in})^2 \tag{13.17}$$

For the forward converter topology, the leakage inductance of the primary may be low if close coupling between the primary and clamp winding is achieved. This will reduce the value of C_2 or may not warrant the complexity of this type of snubber.

For the flyback converter, snubber parameters are governed by the discontinuous current mode versus the continuous current mode in the inductance of the primary. The continuous mode will be addressed due to the advantages of this mode as discussed in Section 9.4. When Q_1 turns off, the primary voltage is $E_0 N_p/N_s$ and C_2 charges to this value during the off time. During the on time, C_2 is charged to $-E_{in}$. Again equating the energy relations,

$$C_2 = \frac{L_1(I_{L1})^2}{E_{in}E_0 N_p/N_s + (E_0 N_p/N_s)^2} \tag{13.18}$$

The complexity of the snubber circuit may not warrant justification of the normally low power flyback converter at low output voltages. At high-voltage outputs, the leakage inductance of the primary may be significant, since close coupling may not be feasible due to the voltage isolation required. In this case, the nondissipative snubber may significantly improve efficiency as compared to the power dissipated in an *RC* snubber.

13.5   SAFETY REQUIREMENTS

Safety has become nationally and internationally important for all types of electric and electronic equipment. This section discusses various safety requirements of UL, CSA, and IEC organizations as related to power conversion equipment. This equipment may operate alone or may be a subsystem within a larger system, specifically office machines and data processing equipment. UL 1950 has replaced UL 478 and CSA 950 will replace CSA 22.2. The IEC specifications are common to the European Economic Community, having replaced the Verband Deutscher Elektrotechniker (VDE) and other individual-country specifications. The primary objective of these specifications is to establish isolation requirements to provide personnel protection from unintended contact with input voltage mains and/or from voltages within the equipment itself. A summary of these requirements is given in Table 13.1. A multitude of other specifications exist, such as MIL Standards, ANSI and IEEE documents, which govern operational safety, hazards, safety from transients and surges, and protection for rectifier circuits and transformers. These various documents also detail the requirements for mandatory interlocks on high-voltage equipment to prevent exposure to lethal potentials. For low-voltage equipment, protection covers and warning labels are normally sufficient.

TABLE 13.1. Safety Requirements

Specification	Line-to-Ground Leakage Current (mA)	Creepage (mm)	High Potential Test Voltage (typically 1 min)	Equipment Type[a]
UL 1950	5.0	2.4[b,c]	1000[b,d]	DP
CSA 22.2, No. 154, CSA 950 (9-93)	5.0	2.4[b,c]	1000[b,d]	DP
EN60950				
Class I	3.5	2.5[b]	1500[b]	OM
Class II	0.25	5.0[a]	3000[a]	OM
UL 114	0.25	2.4[b,c]	1500[b,d]	OM
UL 1012	5.0	2.4[b]	1500[b,d]	PS
≤ 250 V rms		6.4[c]		

[a]Abbreviations: DP, data processing equipment; OM, office machines; PS, power supplies.
[b]Between line and ground.
[c]Between primary and secondary.
[d]Input to output.

13.5.1   Underwirters' Laboratories (UL)

UL 1950   Electronic Data-Processing Units and Systems

UL 114   Electric Office and Business Machines

UL 1012   Standard for Power Supplies: Requirements cover portable, stationary, and fixed power supplies that are rated at 600 V or less, dc and ac, and are divided into two classes—those rated 10 kVA or less and those rated more than 10 kVA

UL 544   Electric Medial and Dental Equipment

UL 94   Test of Flammability of Plastic Materials for Parts in Devices and Appliances: Has no European counterpart although European manufacturers offer components to meet this requirement

13.5.2   Canadian Standards Association (CSA)

CSA 22.2 #234   Safety of Component Power Supply

CSA 22.2 #950   Similar to UL 1950 and EN60950

13.5.3   International Electrotechnical Commission (IEC)

EN60950   Replaced IEC 380/435 and VDE 0804/0806, in general

13.5.4   British Standards Institute (BSI)

BS5850   Similar to EN60950

BS6301   Connection to telecommunications networks

13.5.5   Japanese Industrial Standards Committee (JISC-MHW)

T1001–1004   Electrical and Medical Equipment

13.5.6   Other

ANSI/IEEE C62.41   Surge Voltages in Low-Voltage ac Power Circuits (Reference Section 10.3)

MIL-STD-882   System Safety Program Requirements: Primarily a plan as related to different categories of "failures"

MIL-STD-454   Standard, General Requirements for Electronic Equipment: Requirement 1—Safety (Personnel Hazard)

14

ELECTROMAGNETIC COMPATIBILITY AND GROUNDING

14.0 Introduction
14.1 Electromagnetic Compatibility
    14.1.1 The Nature of EMI
    14.1.2 Conducted Emissions
    14.1.3 Radiated Emissions
    14.1.4 Susceptibility Requirements
14.2 Grounding Techniques
    14.2.1 Printed Circuit Boards
    14.2.2 Circuit Components
    14.2.3 Subassemblies and Modules
    14.2.4 System Grounding
    References

14.0 INTRODUCTION

This chapter discusses various national and international electromagnetic compatibility (EMC) limits and methods of analysis which must be considered to meet these requirements. Grounding techniques are also presented for minimizing the effects of assembly and system generated noise.

Designers of electronic equipment, specifically power conversion equipment, should be aware of the applicable EMC specification which in turn affects other design parameters. The extra filtering required to reduce emissions may utilize a significant portion of the volume and weight budget of a unit. The waveforms of switching circuits must be free of ringing and need to be damped to minimize harmonics. Both of these tend to either decrease efficiency or transfer energy from existing components to added components. The wire routing and shielding techniques necessary for EMC and ground noise reduction will influence the layout and packaging of the equipment. Both structural and electrical ground connections must be considered in the initial design phase to assure safety and proper operation of the equipment or system.

14.1 ELECTROMAGNETIC COMPATIBILITY

Electromagnetic compatibility is defined as the ability of all types of equipment which emit intentional or unintentional signals at radio frequencies (RF) to operate in a manner that is mutually compatible. Electromagnetic interference (EMI) is the undesirable signals which must be suppressed in order that the equipment be compatible. This section discusses EMC primarily in terms of minimizing EMI. Electromagnetic compatibility has become nationally and internationally important in minimizing conducted and radiated emissions from all types of electrical equipment, specifically power conversion equipment and data processing equipment. There are four reasons for minimizing EMI. First, the equipment must be designed to minimize noise or pollution being *conducted* to the input utility or the mains power to prevent interference and noise from being injected into other equipment. Second, the equipment must be designed to minimize interference *radiated* into the atmosphere by electric and magnetic fields to prevent interference with various communications equipment. Third, both conducted and radiated interference must be minimized for *security* reasons to prevent undesired operational monitoring. Fourth, both types of equipment should also be *susceptibly immune* to external interference, such as input voltage transients and/or external radiation from magnetic fields and transmitters. International cooperation and the exchange of information on EMC requirements is leading to the standardization of EMI limits by the governing organizations. Information on some of these organizations is tabulated at the end of this chapter. It is important that power conversion designers know the end use of the equipment or system because the various specifications have different limits, depending on final usage. This applies to both commercial and military equipment.

The FCC (Federal Communications Commission) regulations for conducted emissions apply to electronic equipment marketed in the United States and were primarily adopted in response to the increasing number of documented cases of interference from various sources. To name just a few cases [1]:

The state police of a western state complained that coin-operated electronic games were causing harmful interference to highway police communications at 42 MHz.

Purchasers of a popular computer marketed to the general public complained that this computer completely disrupted TV reception in their own and neighbors' homes.

Interference to aeronautical safety communications at an east coast airport was traced to an electronic cash register in a drug store about one mile from the airport.

14.1.1 The Nature of EMI

A pure sine wave (no distortion) of a certain frequency and a certain amplitude does not contain harmonics above the fundamental frequency. If the sine wave is rectified, even harmonics are produced. A square wave, on the other hand, consists of a fundamental frequency and contains all odd harmonics of this frequency. In between these extremes are rectangular, trapezoidal, and oscillatory waveforms of a pulse or transient nature which have combinations of odd and even harmonics. When a current or voltage abruptly changes amplitude with respect to time, the dv/dt and di/dt changes produce unwanted harmonics of an electrical or magnetic nature. These waveforms are generally produced by the input rectifiers, output rectifiers, switching stages, or switch driver stages of power conversion equipment. If the frequency spectrum of the exciting waveforms is known, the frequency spectrum of the response may then be predicted and the response itself determined. This requires a conversion from the conventional time domain display of an oscilloscope to the frequency domain of a spectrum analyzer.

An arbitrary waveform $f(t)$, which is a function of time, can be converted to a function of frequency by Fourier transform as

$$F(\omega) = \frac{1}{2\pi} \int_{-\infty}^{+\infty} f(t)\varepsilon^{-j\omega t} \, dt \tag{14.1}$$

The absolute value of $|F(\omega)|^2$ is proportional to the energy of the signal $f(t)$ that is associated with the angular frequency ω. Relations between the function $f(t)$ and its frequency distribution are as follows:

1. The longer the duration of a signal, the greater is the energy which is concentrated in its low-frequency components.
2. The faster the rise and fall time of the signal, the greater is the energy which is concentrated in its high-frequency components.

This latter effect leads to the conclusion that only those signals, pulses, or transients which have sharp wavefronts will cause appreciable EMI at high frequencies. Waveforms in the time domain and the corresponding waveforms in the frequency domain are shown in Fig. 14.1.

Equation (14.1) can be expanded by referring to the trapezoid pulse in Fig. 14.1. The area under the pulse is proportional to the total energy in the pulse. The points at which $f(t) = \frac{1}{2}$ for $t = 0$ and $t = T$ are fixed. The slope of the rise time is k_1 and the slope of the fall time is k_2. Then

$$F(\omega) = \frac{1}{2\pi} \left[\int \left(\frac{1}{2} + k_1 t \right) \varepsilon^{-j\omega t} \, dt + \int_a^b \varepsilon^{-j\omega t} \, dt + \int_b^c \left(\frac{1}{2} - k_2(t - D) \right) dt \right]$$

Since $a = \frac{1}{2}k_1$, $b = T - \frac{1}{2}k_2$, $c = T + \frac{1}{2}k_2$, integrating results in

$$F(\omega) = \frac{1}{2\pi\omega^2} \left[k_2 \sin D \sin \frac{\omega}{2k_2} + j \left(k_2 \cos \omega D \sin \frac{\omega}{2k_2} - k_1 \sin \frac{\omega}{2k_1} \right) \right]$$

For $\omega = 2\pi f$, $T =$ pulse period. If $k_1 = k_2 = 1/t$, the absolute value of the frequency function is

$$|F(\omega)| = \frac{1}{\pi\omega^2 t} \sin \frac{\omega t}{2} \sin \frac{\omega T}{2} \tag{14.2}$$

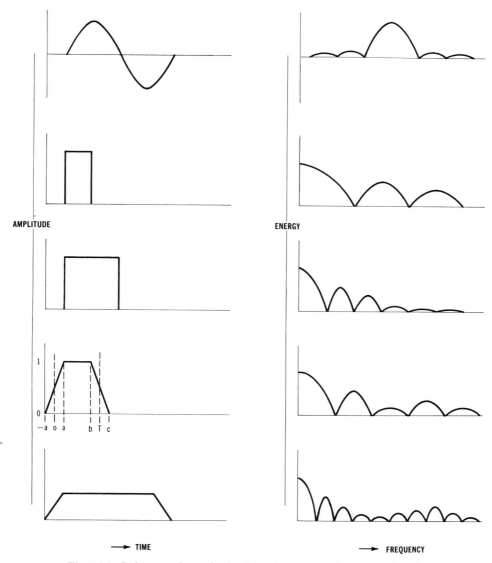

Fig. 14.1. Pulse waveforms in the time domain and frequency domain.

Inspection of this equation reveals that the first term decays as f increases; the second term is a function of the rise and fall times t, but also varies in magnitude as f increases; and the last term is a function of the period time T. Also, the last term is unity at odd harmonics of f and is zero at even harmonics of f, relative to T.

Magnitudes shown in Fig. 14.1 are relative. Since the very high frequency harmonics may be low in amplitude, these values are normally converted to decibels of appropriate units for display on a spectrum analyzer.

14.1.2 Conducted Emissions

Conducted emission limits apply to ac and/or dc power leads connected to a particular unit. These limits may also apply to signal leads and to output leads from the unit. When these emissions flow on wires outside the unit, these wires may also act as antennas which radiate this interference to other equipment. The conducted emissions limits of four major organizations are presented in the next sections. Filtering emissions, minimized the source effects of EMI and testing techniques are also discussed.

14.1.2.1 Specifications   Electromagnetic interference limits for the specifications of FCC, Verband Deutscher Electrotechniker (VDE), Comite International Special des Pertubations Radioelectriques (CISPR), and MIL-STD-461 are shown in Fig. 14.2. The emission limits are expressed either in decibels above one microvolt (dBμV) or decibels above one microampere (dBμA). These parameters can also be converted to absolute units by the following factors:

$$dB\mu V = 20 \log(\mu V) \tag{14.3a}$$

$$dB\mu A = 20 \log(\mu A) \tag{14.3b}$$

$$\mu V = 10^{(dB\mu V/20)} \tag{14.4a}$$

$$\mu A = 10^{(dB\mu A/20)} \tag{14.4b}$$

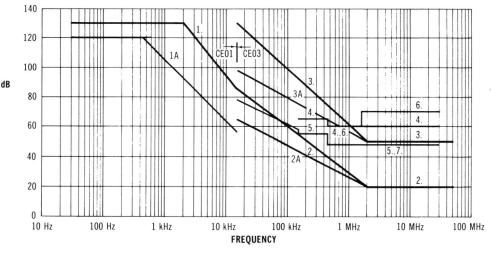

1.	MIL-STD-461C, CE01, parts 2, 3, and 4.	Narrowband, dBμA (30 Hz–15 kHz)
1A.	MIL-STD-461C, CE01, parts 5 and 6.	Narrowband, dBμA (30 Hz–15 kHz)
2.	MIL-STD-461C, CE03, parts 2, 3, and 4.	Narrowband, dBμA (15 Hz–50 kHz)
2A.	MIL-STD-461C, CE03, parts 5 and 6.	Narrowband, dBμA (15 Hz–50 kHz)
3.	MIL-STD-461C, CE03, parts 2, 3, and 4.	Broadband, dBμA/MHz
3A.	MIL-STD-461C, CE03, parts 5 and 6.	Broadband, dBμA/MHz
4.	VDE 0871A, C and CISPR.	dBμV (150 kHz–30 Mhz)
5.	VDE 0871B.	dBμV (15 kHz–30 MHz)
6.	FCC Docket 20780, part 15 J, Class A.	dBμV (450 kHz–30 MHz)
7.	FCC Docket 20780, part 15 J, Class B.	dBμV (450 kHz–30 Mhz)

Fig. 14.2. Conducted emissions limits for FCC, VDE, and military equipment.

Examples

The 48-dBμV limit of Fig. 14.2 is equivalent to an emission level of $10^{(48/20)} = 251$ μV, and the 80-dBμA limit is equivalent to an emission level of $10^{(80/20)} = 10,000$ μA = 10 mA. Conversely, 1 mV or 1000 μV is equal to 20 log 1000 = 60 dBμV.

The nonmilitary parameters are measured for a matching line or system impedance with an LISN (line impedance stabilization network) which can be either 50 or 150 Ω. The parameters can then be converted to units of power, either microwatts or decibels above one milliwatt (dBm) where 0 dBm = 1 mW. The following relations apply:

$$P_{(watts)} = E^2/Z = \left[10^{-6} \times 10^{dB\mu V/20)}\right]^2 \Big/ Z \qquad (14.5a)$$

$$P_{(dBm)} = 10 \log(1000 P_{(watts)}) \qquad (14.5b)$$

Examples

It is desired to convert 66.8 dBμV to units of power for a 150-Ω system. Here $P = [10^{-6} \times 10^{(66.8/20)}]^2/150 = 0.0319$ μW = 10 log(1000 × 0.0319 × 10^{-6}) = -45 dBm. Also note that 0.0319 μW = 31,900 pW = 10 log 31,900 = 45 dBpW.

Actually, the conversion factors just presented are in reverse order. The unit of decibel is applied to power measurements but may also be expressed as *ratios* of voltage or of current. Direct conversion of dBμV to dBμA must be made at a constant impedance, just as the scaling factors of Section 1.10 used impedance as a base. However, the following conversion factor is applicable to convert a measured value of dBμA to a value of dBμV across a 50-Ω LISN: add +18 dB at 10 kHz; decrease linearly at -10 dB/decade to -2 dB at 1 MHz; add -2 dB from 1 to 30 MHz.

A. FCC Requirements. The FCC regulates radio frequency usage in the United States. FCC Docket 20780, Part 15, Subpart J regulations are divided into two categories, class A and class B. Class A (for industrial devices) encompasses units intended for use in commercial, industrial, and business environments. Class B (for consumer products) applies to units intended for use in home or residential environments. The interference limits of class B are more stringent than those of class A. In the case of class B, the main concern of the FCC is the potential interference of computers and electronic games designed to connect to a television receiver, and their effect on other communications. Class A devices are less likely to interfere with communications due to the normal physical separation between commercial and residential districts.

Both class A and class B cover the frequency range of 450 kHz–30 MHz, which encompasses standard and international AM broadcasting and the citizens band (CB). Since the switching frequency of most power conversion equipment is below 450 kHz, an EMI filter need only suppress the higher harmonic components, and the class A requirement is easy to meet. The limits of class A and class B are shown in Fig. 14.2 for a comparison to other conducted emissions standards.

Docket 20718, Part 18 applies to ISM (industrial, scientific, and medical) equipment using RF energy directly for purposes other than communications, and is also divided into an industrial and a consumer class, depending on the environment. Ultrasonic cleaners and induction heaters are in the industrial class, while microwave ovens are in the consumer class. Each class covers the frequency range from 10 kHz to 30 MHz and when implemented, these limits will be the most strict of all limits. Switch mode power supplies operating in the hundreds of kilohertz region may then be classified as RF energy devices, and the filters required to meet the specified EMI limits may have a serious impact on size and weight.

B. VDE Requirements. The VDE-conducted emissions limits of Germany are typical of Western European requirements. These limits are based on the recommendations of the IEC (International Electrotechnical Commission). The VDE specifications, 0875 and 0871, apply to the type of interference generated by noncommunication equipment, as opposed to the FCC specifications which apply to equipment operating in different environments.

The VDE-0875 applies at frequencies from 150 kHz to 30 MHz and covers accidental interference produced by devices operating at frequencies between dc and 10 kHz, such as consumer appliances and hand-operated tools. Because no RF sources are present in this frequency range, the interference limits may be defined as broadband. For power conversion equipment operating below 5 kHz, a filter need only suppress the higher harmonics, and VDE-0875 may be met with relative ease. Also, the allowable interference increases from 5 to 30 MHz.

The VDE-0871 requirements are more encompassing than 0875 and cover the frequency range from 10 kHz to 30 MHz. Power conversion equipment operating at frequencies above 10 kHz therefore contains an RF source, and the interference limits may be defined as narrow band. Substantial filtering is normally required in switch mode power supplies to meet this requirement, and the volume required by the filter normally consumes 10% of the total volume. However, in large data processing systems using multiple power supplies and other electronic assemblies, one filter on the utility or mains input is usually the most volumetrically efficient and the most cost effective. VDE-0871 is further divided into three categories which pertain to system approval. For 0871-A or 0871-C, each system must be tested to verify compliance. For 0871-B, which is the more stringent requirement, system compliance is accorded "general approval" after once successfully meeting the requirements, and individual system tests are waived. Interference limits for VDE-0871-A, -B, and -C are shown in Fig. 14.2. Limits are in dBμV, as is the case for the FCC requirements.

C. EEC Requirements. European Economic Community (EEC) requirements for EMC compliance are directed by CENELEC (European Committee for Electrotechnical Standardization) and are distinguished by EN5xxx standards. These standards are similar or equivalent to VDE, IEC, and CISPR specifications. However, products require the CE mark and methods of compliance are self-certification, third-party certification, and technical construction file. Third-party certification is obtained from certified testing laboratories.

D. Military Requirements. Military requirements are primarily governed by MIL-STD-461C (although some applications specify the more stringent requirements of 461A) with test methods per MIL-STD-462 and the definitions of MIL-STD-463. The requirements are divided by class and applicable part, depending on equipment type. The requirements are further divided into categories of frequency range, narrow-band emissions, and broadband emissions. These limits are shown in Fig. 14.2 with narrow-band amplitudes in dBμA, and the broadband amplitudes in dBμA/MHz. The spectral energy of narrow-band interference is confined to specific frequencies, such as the switching frequency of an inverter stage, and all odd harmonics of that frequency. The spectral energy of broadband interference occurs over a wide frequency range, and amplitude measurement is a function of the selected bandwidth of the test equipment.

Recommended as a reference is MIL-HDBK 241B, titled "Design Guide for Electromagnetic Interference (EMI) Reduction in Power Supplies." This document tabulates important considerations that can be analyzed in the design phase to reduce EMI.

The parameter dBμA, as opposed to the previously discussed dBμV, is due to the use of a current probe to measure EMI levels for the military requirement, plus the use of feedthrough capacitors for decoupling purposes and the chassis grounding of the unit. In addition to input EMI filters, internal bypass capacitors from the power leads to the chassis are effective means to suppress conducted interference on these lines. This latter means is not normally available to manufacturers of commercial equipment, especially power supplies, due to the likelihood of an unpolarized ac input or the lack of a "chassis," as in the case of "open frame" units. Also, the bypass capacitor approach is not applicable to medical equipment, due to the low specified limit of line-to-ground leakage current. Reference Section 13.5 for limits.

14.1.2.2 Filtering Conducted Emissions   Filter manufacturers present data of insertion loss (dB) versus frequency instead of attenuation characteristics. The insertion loss applies to frequency-independent sources and loads and is usually measured with 50-Ω terminations. This leaves the designer with the responsibility of ensuring that equipment operating frequencies and intermodulation harmonics will not excite or produce filter resonances, which could result in instability. These possible resonances can also produce excessive current in capacitors and excessive voltage across active devices. Characteristics of common input line filters for lower power applications are shown in Fig. 14.3. Various manufacturers offer these and other combinations of EMI filters. Curves 1, 2, and 3 are 0.01, 0.1, and 1 μF feedthrough capacitors, respectively. These capacitors normally have a film dielectric and are bulkhead or panel mounted. Increasing the capacitance increases the attenuation up to a certain frequency, at which point the self-resonance of the capacitor occurs. The equation for insertion loss for other capacitance values at 50-Ω impedance is given by

$$dB = 20 \log \sqrt{1 + (50 \pi f C)^2} \tag{14.6}$$

When operating from ac sources, the steady-state capacitor current will be inversely proportional to the capacitance. MIL-STD-461A does not establish a limit on capacitance. MIL-STD-461C states that the use of line-to-ground filters shall be minimized and limits this capacitance to 0.1 μF for 60-Hz inputs and to 0.02 μF for 400-Hz inputs. Such filters establish low-impedance paths for structure

1.   .01 μF FEEDTHROUGH CAPACITOR

2.   0.1 μF FEEDTHROUGH CAPACITOR

3.   1 μF FEEDTHROUGH CAPACITOR

4.    5 A RATING (LC)

5.   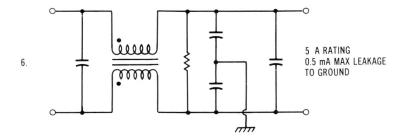 OR   5A RATING (π OR T)

6.   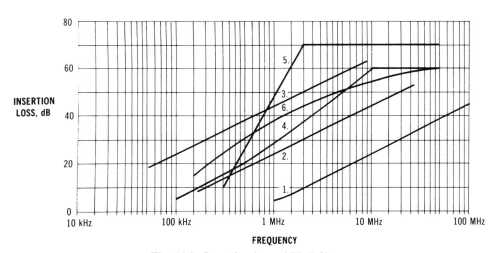 5 A RATING 0.5 mA MAX LEAKAGE TO GROUND

Fig. 14.3. Insertion loss of EMI filters.

currents which can cause system interference problems in other equipment using the same ground plane.

When input power is applied, a high inrush current occurs to charge the capacitor. The peak current is essentially independent of capacitance value, but the energy content is proportional to capacitance. The short duration of the charging current will not normally affect the input source, but caution should be exercised if the input source is present at a connector plug when the plug is mated to a receptacle on the equipment. In this latter case, arcing and pitting of the connector pins is likely to occur.

Curve 4 is an L-type filter rated at 5 A. Curve 5 is a pi filter or T filter rated at 5 A providing 70 dB attenuation from 2 to 50 MHz. Lossy ferrite material is used for a series impedance, while a high K ceramic or film dielectric is used for the shunt capacitance. These filters may also have a substantial capacitance to ground.

Curve 6 is a typical multistage filter rated at 5 A, but the capacitance to ground is low to limit the leakage current to 0.5 mA maximum at 120/240 V input. This filter provides approximately 30 dB attenuation at 450 kHz, which is the low-frequency limit for FCC requirements. The balun inductor (note the polarity) provides common mode filtering.

A discrete component power line filter is shown in Fig. 14.4A. This filter is designed to suppress both common mode and differential mode conducted emissions. Inductor L_1 is a common core balun inductor, with each winding of equal turns. With the polarity of the inductor connected as shown, equal currents in the line and neutral (differential mode) cancel the flux in the core, which prevents saturation due to ac line current, and the full inductance is available for common mode suppression. Inductors L_2 and L_3 are individual inductors for differential mode suppression, as are capacitors C_1 and C_4. Capacitors C_2 and C_3 suppress common mode emissions. International standards (specifically IEC and VDE 0565) divide these capacitors into two groups. The X capacitors (C_1 and C_4) are used only in positions where a failure of the capacitor would not expose personnel to electric shock, and apply to capacitors connected across the supply lines. The Y capacitors (C_2 and C_3) are connected from one side of the supply line to "earth ground" and could present a potential hazard if the ground line should be open circuit. For this reason, Y capacitors are limited in capacitance value and the manufacturing process minimizes the risk of electrical breakdown.

The equivalent circuits for the common mode and the differential mode are shown in Figs 14.4B and 14.4C, respectively, with the component parasitics. These parasitics, specifically the winding capacitance of the inductors and the series inductance of the capacitors, cannot be ignored in the RF region of EMI. For instance, at some very high frequency, the capacitors become inductors and the inductors become capacitors and no attenuation is realized. Two parameters not shown in these figures and which may be ignored are the parallel resistance of the inductors, which would be equivalent to the core loss, and the series dc resistance of the inductors. In the common mode equivalent, the line and neutral are, in effect, shorted and the windings of L_1 are in series, L_2 and L_3 are in parallel, C_2 and C_3 are in parallel, and C_1 and C_4 do not contribute. The maximum capacitance value of C_2 and C_3 is limited by the safety requirements of line current flowing to ground at the power line frequency, since $I = E(2\pi fC)$. In the differential mode, all components contribute, except that the inductance of L_1 is now only the leakage inductance between the windings.

A further analysis of L_1 is of interest. The windings must be rated to conduct the required ac input current but since the flux is canceled by this current, a high permeability core material, with little or no air gap, may be used. This allows a high inductance value in a compact volume. Since each winding is of equal turns, bifilar turns offer ease of winding. However, the bifilar turns increase the inter-

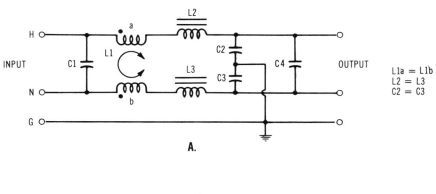

A.

$$L1a = L1b$$
$$L2 = L3$$
$$C2 = C3$$

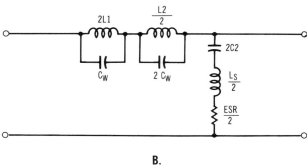

B.

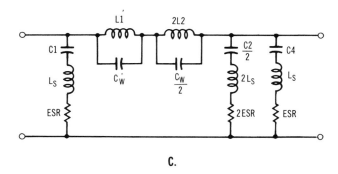

C.

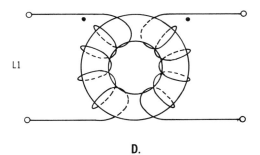

D.

Fig. 14.4. Typical characteristics of EMI filters.

winding capacitance, and the inductor could resonate at a relatively low frequency. The suggested winding method for a toroid core is shown in Fig. 14.4D as a 180° traverse, where the windings are physically separated to maximize the resonant frequency. Higher power units could utilize a C core with a core-type winding as shown in Fig. 3.28B. For the C core, the winding polarity must be correctly observed.

Example

An input line filter is desired to be placed between the 120-V, 60-Hz mains and the equipment. The circuit of Fig. 14.4 will be used. After the component value calculations which follow, the circuit is shown in Fig. 14.5. First, C_2 is calculated to limit the current to ground to 0.5 mA, and $C_2 = C_3 = I/(2\pi f E_{in}) = 5 \times 10^{-4}/(377 \times 120) = 0.01$ μF. The cutoff frequency of the filter for common mode is selected as 18 kHz. From Fig. 14.4B, $L = 1/[C(2\pi f)^2] = 1/[2 \times 10^{-8}(2\pi \times 18,000)^2] = 4$ mH. Then each winding of L_1 will be 2 mH since the contribution of L_2 and L_3 will probably be insignificant. Next, C_4 is selected as 1 μF. The cutoff frequency for differential mode is selected as 16 kHz. From Fig. 14.4C, $L =$

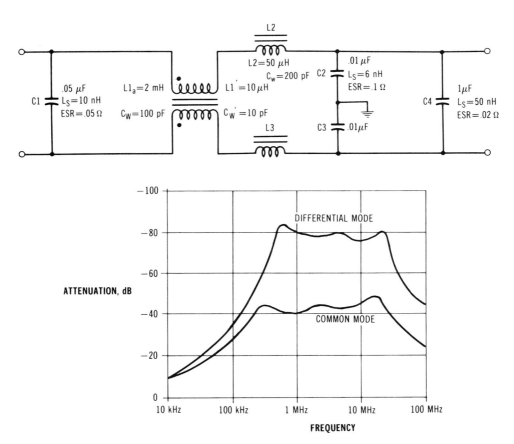

Fig. 14.5. Discrete filter attenuation characteristics.

$1/[C(2\pi f)^2] = 1/[10^{-6}(2\pi \times 16,000)^2] = 100 \ \mu$H. Then $L_2 = L_3 = 50 \ \mu$H. At the input, C_1 is selected as 0.05 μF. Values of the parasitics are also shown in Fig. 14.5A. The series inductance of the capacitors is calculated from Fig. 14.7. The attenuation characteristic of the circuit of Fig. 14.5A is shown in Fig. 14.5B for a source and load impedance of 50 and 100 Ω for the common mode and differential node, respectively.

Typical filters for three-phase inputs with three-wire and four-wire lines are shown in Figs. 14.6A and 14.6B, respectively. Usually the power line impedance is low, while the load may be low or high impedance. In the case of three-phase inputs for power conversion equipment followed by an input rectifier, the load impedance is low. For this reason both filters use a T configuration, since both the source and the load are typically low impedance. An additional inductor is frequently placed in the ground line to provide optional ground current isolation. If the filter of Fig. 14.6A is followed by a three-phase full-wave rectifier, auxiliary power for control circuits may be obtained from a single-phase transformer

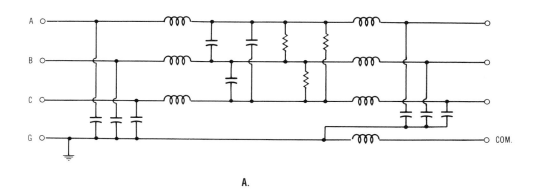

A.

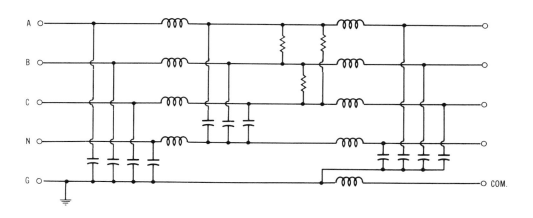

B.

Fig. 14.6. EMI filters for three-phase inputs.

connected line to line, or from a converter operating from the dc bus. This eliminates the usage of the neutral. When the neutral is used and filtering is required on the neutral, the circuit of Fig. 14.6B may be used. The design and analysis considerations are the same as those of Fig. 14.5.

14.1.2.3 *Attacking the Source of Emissions*

Some form of filtering is obviously required for the input and output(s) lines of equipment. However, the optimum design approach is to minimize EMI at the source of the emissions rather than use a brute force filter approach. That is, prevent EMI generation rather than suppress it. This reduces the size and volume of the filter and reduces the chance EMI being radiated internally to other sensitive components in the equipment. Thus, the designer must have a plan of attack. Generally, EMI increases as dv/dt increases or abruptly changes. But in power-switching circuits, a high dv/dt plus minimum dc resistance results in high efficiency. The underdamped nature of the power circuitry results in EMI, due to oscillations and ringing which occur during device switching. The active device switching produces odd harmonics while the rectifiers and filters produce even harmonics. Thus a compromise must be made. During the breadboard or prototype test stage, all waveforms can be observed with an oscilloscope, and any ringing or abrupt changes in voltage (or current) should be eliminated by modifying the circuitry or by critically damping. Snubbers can tailor the switching waveforms of semiconductors. Ferrite beads are effective in low-impedance switching circuits. When placed around conductors connected to power transistors, SCRs, and rectifiers, the beads will reduce EMI above 1 MHz by eliminating parasitic oscillations. The beads are especially effective on rectifiers which produce emissions in the 20- to 40-MHz region during recovery. Faraday shields in transformers will reduce capacitive coupling between primary and secondary (input and output). The shield termination must also have a low-imped-ance path to ground. Double and triple shielding in ultra isolation transformers is effective in reducing low-frequency EMI, and in reducing transients from the power line to the equipment and vice versa.

Minimizing capacitance from circuits to ground will reduce common mode EMI. Consider, for instance, a bridge-inverter stage with TO-3 devices isolated from a chassis-mounted heat sink. The collector (or drain or anode) of the "top" device is connected to the positive dc bus which is a fixed potential relative to ground (except at the low frequency of the power line). The collector (or drain or anode) of the "bottom" device is connected to the load, and the collector potential switches from the positive bus (when the top device is on) to the negative bus (when the bottom device is on). Current at the switching frequency then flows from the bottom device case through the interface mounting capacitance to the heat sink and to ground. This analysis also applies to single device switching stages with a collector load. The TO-3 case to heat sink capacitance is typically 220 pF with a mica or film insulator, but the capacitance (and the current) will be reduced to approximately 25 pF with a BeO (beryllium oxide) insulator. If the capacitance is still too high, the emitter of the bottom device, which is connected to the negative bus, may be connected to the heat sink, and the heat sink then isolated from ground. Also, a TO-3 shielding screen may be placed between two insulators, and the screen connected to the negative bus in the shortest possible length. The bottom transistor interfaces with the opposite side of one insulator while the heat sink, which may now be grounded, interfaces with the opposite side of the other insulator. In high-power units, high current SCRs and rectifiers are mounted directly to a heat sink for maximum heat transfer. The heat sink is then well isolated from ground by insulating bushings or brackets, and the heat sink to ground capacitance is negligible.

Example

It is desired to limit common mode emissions to 50 dBμV (316 μV) above 150 kHz (from Fig. 14.2) in a switch mode unit. If the voltage of the switching stage is 316 V, this required an attenuation factor of 10^6. The common mode insertion loss of the EMI line filter may provide 40 dB (100 : 1) attenuation leaving a required factor of 10^4. The first relevant harmonic may be approximately 10% of the switching voltage, leaving a factor of 1000. If the common mode capacitance in the filter is 0.02 μF (as in the example of Section 14.1.2.2), the capacitance to ground in the circuit cannot exceed $C = 0.02 \times 10^{-6}/1000 = 20$ pF. If the capacitance to ground is 200 pF, the required common mode insertion loss of the EMI filter is 60 dB (1000 : 1).

Eliminating emissions at the source applies to transformer cores (including ferrite) clamped to ground or grounded. In transformers where the primary potential is higher than the secondary potential, the capacitance from primary to core to ground provides a current path. If the primary is connected to the positive bus, as in the case of single-device switching stages, the core also should be connected to the positive bus. In bridge-switching stages, both sides of the primary change in voltage potential, and the core should be well isolated from ground. In transformers where the secondary potential is higher than the primary, the secondary capacitance to ground should be minimized in like manner. (Reference Section 3.5.6 for further discussion on termination of transformer windings. Reference Section 14.1.3.2 for suppressing radiation from transformers with a copper shading ring.)

Wiring harnesses should be physically located away from the case of the unit. Control and power leads should be physically separated. Dual lines to input filters, transformers, and subassemblies should be twisted pair or parallel to minimize inductance. Component leads of snubbers, which are connected across devices, should be as short as possible. As opposed to single-point grounding to reduce noise, multipoint grounding is recommended to reduce EMI by reducing inductance. The inductance of a straight, round conductor is given by [2]

$$L = (\mu/2\pi)[\ln(4h/d)] \qquad \text{H/m} \tag{14.7a}$$

where $h > 1.5d$ is the distance above the ground plane, d is the diameter of the round conductor, and $\mu = 4\pi \times 10^{-7}$ H/m is the permeability of free space.

Changing units to nanohenrys per inch,

$$L = 5[\ln(4h/d] \qquad \text{nH/in.} \tag{14.7b}$$

The line inductance of two parallel conductors, separated by distance D, and carrying uniform current in opposite directions is given by

$$L = 10[\ln(2D/d)] \qquad \text{nH/in.} \qquad (14.8)$$

Applying this equation to typical insulated hook-up wires, an inductance rule-of-thumb factor is $L = 1 \ \mu H/m$.

Internal bypass capacitors are an effective means to suppress differential mode EMI. The capacitors reduce external high-frequency current flow by shunting the current path. Bypassing should not be confused with decoupling which acts as a low pass filter and does not isolate equally in both directions. Capacitors for waveform and ripple filtering generally have a low resonant frequency, and above this frequency the capacitor acts as an inductor (reference Fig. 2.34D). Electromagnetic interference can be reduced by using low ESR capacitors and by shunting large value capacitors with lower value capacitors which have a higher resonant frequency. Typical resonant frequencies for various capacitor types and values are shown in Fig. 14.7. For instance, a typical power supply output filter capacitor may be 100 μF. To reduce high-frequency ripple, a designer parallels this capacitor, probably near the output terminals, with a 1-μF ceramic capacitor. This may be effective in reducing high-frequency ripple but since the resonant frequency of this ceramic capacitor is approximately 3 MHz, it will have little effect on EMI reduction above 10 MHz. The addition of the third capacitor of 2.2 nF may mean

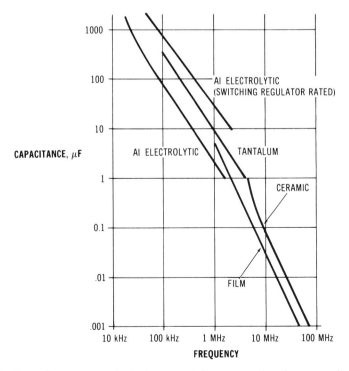

Fig. 14.7. Capacitance versus typical resonant frequency of various capacitor types.

the difference in meeting EMI requirements. The very low dissipation factor of ultrastable NPO-type capacitors, as discussed in Section 2.8.5, are most applicable. This analysis also applies to other internal capacitors used for filtering rectified ac, both the main filter capacitor and filter capacitors for auxiliary or control voltages.

While on the subject of output ripple of power supplies, the following notes may be of interest. As compared to EMI, which is measured in the frequency domain, ripple is measured in the time domain. A unit which meets a 50-mV ripple requirement will not necessarily meet the EMI requirement. Also, erroneous ripple measurements will be indicated by a low bandwidth scope and by improper scope intensity control.

To further reduce differential mode EMI, wiring to transformers should be short, twisted-pair leads to minimize electric and magnetic coupling or radiation. If the transformer is mounted on a printed circuit board (PCB), the traces from the transformer terminals should run back to back on a double-sided board. The twisted-pair technique also applies to wires on the ac or dc input line and on lines from the rectifier to the filter capacitor. For transformers with center-tapped secondaries, these wires should be twisted, triple lines. Control wiring should be physically separated from power wiring. Where this is impractical, the control and power leads should cross each other at right angles.

14.1.2.4 Testing   Testing procedures have been well documented in various specifications. Computer-automated test equipment rapidly performs measurements and plots results. Various manufacturers offer instrumentation and facilities for testing. For FCC and VDE testing of three-wire, single-phase power lines, a line impedance stabilization network (LISN) is connected between the power source and the equipment under test (EUT) to isolate the line from the EUT. Three standardized LISN circuits and their corresponding impedance characteristics are shown in Figs. 14.8A and B for FCC tests and for CISPR tests with currents exceeding 25 A. The older 50-Ω, 5-μH network may be used if the proper correction factors are applied. The newer 50-Ω, 50-μH network is normally recommended. The characteristic impedance closely matches the impedance of the utility or mains power line in the RF region [3]. For CISPR tests below 25 A, a 150-Ω network and its characteristic impedance of Fig. 14.8C applies. For most of the military requirements, standardized 10-μF feedthrough capacitors are connected as shown in Fig. 14.8D to isolate the line from the EUT. Notice the case of the capacitors and the case of the EUT is securely fastened to a ground plane. A wide-band current probe is used to measure the emission current in each input lead in turn.

If test facilities are not readily available, a spectrum analyzer and a current probe may be used for initial evaluation of a breadboard or prototype, as shown in Fig. 14.8D. The spectrum analyzer measures power (dBm) in its 50-Ω termination. Since 0 dBm = 1 mW = $E^2/R = E^2/50$, then $E = \sqrt{50/1000} = 0.224$ V for 0 dBm. Given a probe factor of 1 (0 dBm), the current in the probe would be $I = 0.224$ A $= 0.224 \times 10^6$ μA = 107 dBμA. With the proper attenuation factors and bandwidth settings on the analyzer and from the displayed amplitude at corresponding frequencies, the narrow-band emission level in dBμV or dBμA is found when the 107-dB factor plus the probe correction factor is applied to the measurement indicated. For instance, if the displayed reading at 1 MHz is -70 dB

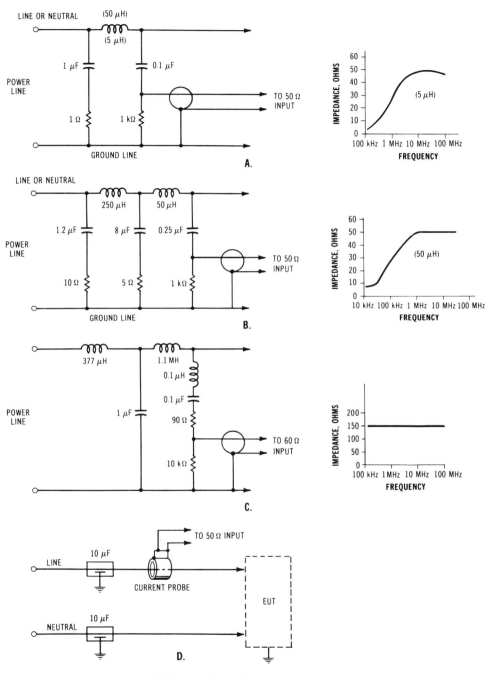

Fig. 14.8. LISNs and feedthrough capacitor isolation.

and the probe correction factor is $+10$ dB, the actual emission level is $-70 - 10 + 107 = 27$ dB. For wide-band measurements, correction factors must be applied to obtain dBμA/MHz. For a spectrum analyzer bandwidth setting of 10 or 100 kHz, add $\cong 37$ or $\cong 17$ dB, respectively, to the displayed measurement.

14.1.3 Radiated Emissions

An electromagnetic field impinging on electrical conductors will induce currents in the capacitors. When a break, opening, or high impedance occurs between conductor bonds or shields, the current decreases and a voltage develops on the inside surface, which is then conducted to the outside surface. The break or opening then acts as an antenna to radiate energy. The radiated EMI consists of electric fields E, which are proportional to voltage, and magnetic fields H, which are proportional to current, and their behavior is governed by Maxwell's equations. The wave impedance is defined as E/H. As the distance from a radiating component approaches $\lambda/2\pi$ wavelengths, the impedance approaches 377 Ω, which is the intrinsic impedance of free space. If the radiating device has an impedance much greater than 377 Ω, such as from high-frequency fast switching circuits, electric field radiation will dominate. If the radiating device has an impedance much lower than 377 Ω, such as from a transformer, magnetic field radiation will dominate.

14.1.3.1 Specifications Radiated emissions limits for FCC 20780, VDE 0871 and MIL-STD-461C are shown in Fig. 14.9. Parameters are decibels above one microvolt per meter, dBμV/m. For the FCC requirement, class A equipment (in-

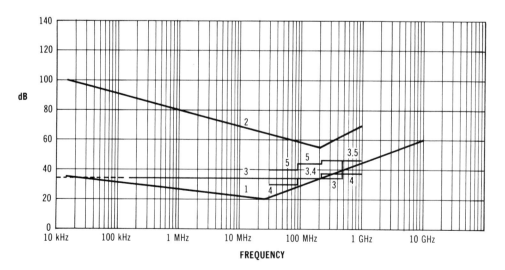

1. MIL-STD-461C, RE02. Narrowband, dBμV/m (d = 1 m). U.S. Army shown, USAF and USN limit $\approx$ 10 dB greater.
2. MIL-STD-461C, RE02. Broadband, dBμV/m/MHz (d = 1 m). U.S. Army shown, USAF and USN limit $\approx$ 10 dB greater.
3. VDE 0871B. dBμV/m (10 kHz–30 MHz, d = 30 m; 30 MHz–1 GHz, d = 10 m)
4. FCC Docket 20780, part 15, subpart J, class A; dBμV/m (d = 30 m)
5. FCC Docket 20780, part 15, subpart J, class B; dBμV/m (d = 30 m)

Fig. 14.9. Radiated emissions limits for FCC, VDE, and military equipment.

dustrial) is measured at a distance of 30 m, while class B equipment (consumer) is measured at a distance of 3 m. Even though the limits for class B are 10 dB higher than those of class A, class B is actually more difficult to meet due to the shorter distance. At a distance of 30 m, the limit is 20 dB less than curve 5. Also, note the distance for VDE-0871B changes above 30 MHz. In addition to the variations in limits and distance, comparisons between FCC and VDE are further complicated by the slightly different receiver bandwidth specifications and quasi-peak detection.

14.1.3.2 Shielding Radiated Emissions   Radiated electromagnetic energy can be suppressed by metallic shields, good electrical bonding, and minimizing openings in enclosures or cabinets. The absorption and reflection characteristics of shielded cables, panel riveting, honeycomb structures, and EMI gaskets for shielding have been aptly covered in other documents and general considerations only are discussed herein. The above techniques reduce the radiation emitting from a unit and decrease its susceptibility to externally radiated energy by absorbing and reflecting the energy contained in the field. The effective attenuation depends on the impedance of the shield being low as compared to the impedance of the electromagnetic wave, and is also a function of absorption and reflection factors. An electric field may be reflected by a magnetic or nonmagnetic material. Steel enclosures are effective in reflecting low-frequency electrical waves and provide some absorption to magnetic fields. Aluminum enclosures are more effective than steel in reflecting high-frequency waves due to the higher electrical conductivity of aluminum.

Shielded cables are available in spiral wrap, foil, braid, and combinations of each. The shielding effectiveness is compared in Fig. 14.10. Shielding reduces noise along input lines and reduces capacitive coupling in the cable. In addition, shields also absorb magnetic fields and reflect electric fields. Care must be exercised when using a foil shield with a pigtail drain or ground wire. The inductance of this lead, even for a short distance, will negate any effect of the shielded cable. The ground lead or tab should have a width of five times the thickness of the tab and a width of at least 20% of the length, and the length should be as short as possible. The shield should be grounded at both ends of the cable for EMI reduction. (Single-point grounding at the source is the optimum approach for low-level, audio-type signals where noise reduction is of importance.)

However, for decreased capacitance to ground to meet conducted emissions, twisted-pair cables may be a better choice than a two-conductor shielded cable. The twisted-pair cables equally distribute capacitance along the length of the conductors. This reduces capacitive coupling and maintains a high common mode rejection ratio. This is especially true of applying gate-to-cathode signals to SCRs in high-power inverter stages. The cathode of the "top" SCR in a bridge stage alternately switches from the positive bus to the negative bus, which produces currents in the cable capacitance. The twisted pair provides both electromagnetic and electrostatic isolation.

Magnetic energy can radiate from transformers and inductors through an aluminum enclosure. For high-frequency operation, the solution is to add a copper shading ring around the outside of the magnetic component to reflect the energy. A low-frequency magnetic field may be difficult to suppress because the shield must absorb the wave. High permeability materials such as mu-metal or Permalloy

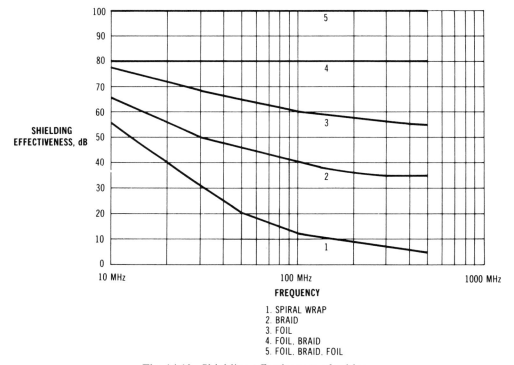

Fig. 14.10. Shielding effectiveness of cables.

may be used around the outside of the magnetic component, and the required thickness is a function of the magnetic field strength. The radiated field is also a function of component orientation for devices with an air gap.

Low-power equipment, such as open-frame switching power supplies, may be enclosed in perforated aluminum for ventilation and cooling. The shielding effectiveness of square panels with round openings for electric and magnetic fields is given by [4]

$$S_e = 20 \log(c^2 l/d^3) + 41.8t/d + 2.68 \qquad \text{dB} \qquad (14.9a)$$

$$S_m = 20 \log(c^2 l/d^3) + 32.0t/d + 3.83 \qquad \text{dB} \qquad (14.9b)$$

where

S_e = electrical shielding effectiveness
S_m = magnetic shielding effectiveness
c = hole spacing
l = length of panel hole pattern
d = hole diameter
t = panel thickness

For rectangular panels,

$$l = \sqrt{\text{panel hole pattern area}}$$

(all dimensions are in inches).

These equations show that shielding effectiveness is independent of frequency and is applicable if d is less than $\lambda/2\pi$. The thickness factor treats each hole as a waveguide below cutoff. The perforated panels can provide 50 dB attenuation from 50 kHz to 1 GHz.

High-power equipment is normally housed in a metal enclosure, such as a "rack mount" cabinet or National Electrical Manufacturers Association (NEMA) cabinet with doors, covers, and cooling openings. Honeycomb structures are essentially hexagonal waveguides which can be used to fill larger openings and reduce radiation. These multiwaveguides have a relatively constant attenuation below the cutoff frequency, which is a function of cell width. A length (or depth) to cell width ratio of 3 : 1 provides approximately 50 dB of shielding effectiveness from 0.1 MHz to 2 GHz. Both perpendicular and parallel fields may be suppressed by a double stacked honeycomb with parallel and perpendicular directions of foil.

Electromagnetic interference gaskets, consisting of a knitted wire mesh, provide many closely spaced points of contact to "short circuit" the voltage developed between mating surfaces. In high-frequency shielding, gasket resiliency is most important in order to maximize the number of contact points, and to keep the slot length less than $1/100$ wavelength. In low-frequency magnetic shielding, gasket width, conductivity, and permeability are most important. The surface material in contact with the gasket must be free of paint, oxides, and other insulators. A conductive finish, such as alodine, irridite, or chromate is mandatory for high electrical conductivity, and these finishes also provide environmental protection. Frequent removal and reinstallation of covers or panels may deform the gasket and actually increase emissions above the level which would occur without the gaskets. Gaskets have also been painted by uninformed maintenance personnel, resulting in increased emissions.

The high cost and weight of metal housings have prompted the use of plastic housings for low-power equipment. But the plastic enclosures (used in systems equipment housing, among other assemblies, power conversion equipment) are insulators and are transparent to EMI and can neither shield nor absorb electromagnetic energy. Being insulators, the plastic also presents problems in grounding and static charge dissipation. These problems have been successfully resolved by the following methods:

1. Conductive coatings such as silver, copper, or nickel can be painted on the interior surface to achieve a shielding effectiveness of 20–50 dB [5]. However, silver is usually cost prohibitive. Copper may oxidize and lose effectiveness with age. Nickel is a slightly inferior electrical conductor compared to silver and copper, but nickel can absorb more emissions due to its magnetic permeability.

2. Metal coatings can be applied by electroless plating, vacuum metallization, or sputtering to achieve a shielding effectiveness exceeding conductive paints. A shielding effectiveness of 50–70 dB is typical for a coating thickness of 0.2 mil.

However, the process may be expensive and damage in handling and service could reduce the effective shielding.

3. Zinc arc spray offers the highest shielding effectiveness, typically 70 dB minimum, and essentially creates a metal shell, approximately 5 mils thick, inside the plastic housing. Achieving a good adhesion may require special surface preparation. Flaking or cracking due to thermal expansion or impact may significantly reduce the effectiveness of the shield.

4. Metal-fiber-filled composites incorporated into a molding compound overcome many of the problems associated with painting and plating. Uniform distributions of the metallic conductive particles can be achieved with conventional transfer and injection molding techniques. A shielding effectiveness of 30–50 dB can be achieved. Researchers have estimated that an attenuation of 30 dB would be adequate for 80–90% of all shielding applications [6].

14.1.4 Susceptibility Requirements

Equipment may be susceptible to external sources of interference, both radiated and conducted which, in turn, causes undesirable responses or degradation of performance. However, the equipment is normally required to provide specified performance when subjected to certain levels of interference. Radiated susceptibility is a function of the equipment type and a multitude of other considerations. For this reason, only conducted susceptibility, applied to the input terminals, will be discussed.

Susceptibility requirements are almost nonexistent for industrial and commercial equipment. The typical utility or mains voltage range is 105–130 V, 210–260 V, 420–520 V, or a variation of $\pm 13\%$ from nominal. Switch mode power supplies can operate from an input voltage range of 95–132 V. However, these steady-state limits do not encompass line transients and brown-out conditions. Since no standard specification exists for maximum utility voltage excursions, designers frequently include a transient suppressor at the input of the equipment for "insurance." Battery-operated converters and UPS (uninterruptible power systems) typically have a $\pm 14\%$ input variation, but the battery represents a low impedance source to transients. However, if other dc loads located at some distance from the battery switch on and off and these loads are adjacent to the equipment, transients can be generated from the inductance of the power distribution cables. High-voltage transients and surges, which high-power and low-power equipment may be required to meet, are listed in Section 13.3.

For military equipment, MIL-STD-461C (or 461A) specifies applicable requirements of the following:

CS01: an external voltage and frequency superimposed on the input terminals as shown in Fig. 14.11.

CS02: 1 V at a frequency from 50 kHz to 400 MHz from a 50-Ω source superimposed on the input terminals except the input power will not exceed 1 W.

CS06: 200 V peak for 10 μs (typical) superimposed on the input terminals.

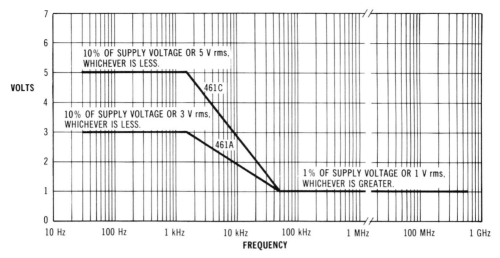

Fig. 14.11. MIL-STD-461 susceptibility requirements.

A specific design parameter must be addressed for CS01 compliance and overall stability in switch mode power supplies and converters. These power conversion units have a negative input resistance and can become unstable at low-frequency input modulation (susceptibility). However, compliance and stability is assured if the output impedance of the input filter (parallel resonance) is less than the reflected input impedance of the output filter (series resonance). This topic is discussed in Section 8.7.

Although not described as a susceptibility requirement, the general equipment specification for military equipment may require compliance with MIL-STD-704D (or the longer duration conditions of 704A). For normal operation on 115-V ac inputs, the maximum input is 180 V rms for 10 ms, linearly decaying to steady-state limits in 80 ms. For abnormal operation, the maximum input is 180 V rms for 50 ms, exponentially decaying to steady-state limits in 1 s. On 28-V dc inputs, the maximum input is 50 V dc for 45 ms, exponentially decaying to steady-state limits in 1 s. On 270-V dc inputs, the maximum input is 470 V dc for 27 ms, exponentially decaying to steady-state limits in 1 s. These parameters require sufficient component voltage rating and safety margin, especially for semiconductors and capacitors.

14.2   GROUNDING TECHNIQUES

From the National Electrical Code: "System and circuit conductors are grounded to limit voltages due to lightning, line surges, or unintentional contact with higher voltage lines, and to stabilize the voltage to ground during normal operation. Systems and circuit conductors are solidly grounded to facilitate overcurrent device operation in case of ground faults [7].

Destructive grounding system fault currents and noise generation due to ground loops have plagued electronic equipment and have had costly safety and economic consequences. Once a circuit, unit, or system is connected and installed, problems arising from poor grounding techniques are difficult to trace (and even more difficult to analyze) through the maze of wiring and architectural diagrams and electrical schematic diagrams. The best solution is, of course, to accurately consider the optimum grounding approach during design, assembly, and installation and to adequately document the current paths to earth ground, structural grounds, electrical grounds, power commons, and shielding terminations.

One approach to eliminate grounding problems is to use a single point ground where each unit is connected to a single termination point by its own independent ground path. In a complex system, this may be physically impossible. Power conversion equipment, especially power supplies, may be connected to various units via a common line and the benefit of low output impedance is lost. Also, the noise voltage between the unit ground and the master ground is coupled back through each of the ground return lines. Major units of the system may be distributed to a set of cabinets or throughout an aircraft or ship, and each of these units may contain components which are locally grounded. An appreciable differential ground noise may develop between the locally grounded unit and the system master ground.

In the following sections, approaches to optimum grounding techniques will begin with printed circuit boards and individual circuit components and progress to subassemblies and system grounding. This is actually the *reverse* of a well-planned grounding scheme but is presented in that sequence to show the importance of interfaces between low-level units, as well as interfaces between low-level units and power units operating at different voltage potentials with respect to ground. Once these parameters are identified, the grounding scheme can be approached from a system standpoint.

14.2.1 Printed Circuit Boards

Printed circuit boards (or printed wiring boards) are commonly the lowest operating level within a system hierarchy. Circuit common and supply voltage layout is important in both digital and analog circuits. A frequently used, but poor, layout is shown in Fig. 14.12A. Here the supply trace runs across the top of the circuit board and the return or common trace runs across the bottom of the board, with the vertical traces distributed to rows of integrated circuits. This results in undesired high trace self-inductance and greater circuit crosstalk. The typical impedance of circuit traces 1 in. in length is 1 Ω at 10 MHz. Due to trace inductance, the input bypass capacitor is practically useless at high frequencies. To preclude these problems, the layout of Fig. 14.12B is recommended. Here, the power supply and return traces are located close together to form a transmission line which significantly reduces the distribution impedance and minimizes the trace inductance.

For analog circuits, an example of incorrect connections is shown in Fig. 14.13A. The input power lines are connected to the lowest level stage first. Each of the traces to the higher level stages will have some impedance. The di/dt effects of the power switching transistor will produce ripple and inject noise into the amplifier and comparator circuit common lines. The bypass capacitors will not

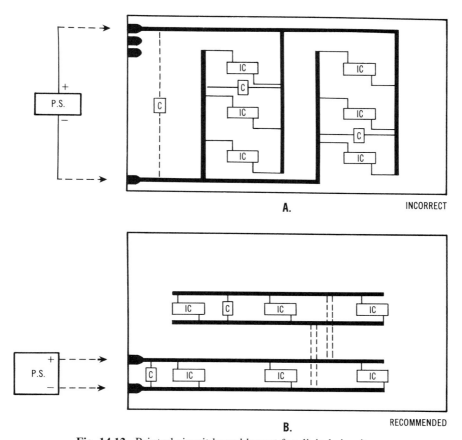

Fig. 14.12. Printed circuit board layout for digital circuits.

improve noise immunity if they are located on traces removed from the devices. In Fig. 14.13B, the input supply lines are first routed to a bypass capacitor, and then to another bypass capacitor located at the highest level stage (in this case, the output stage). Traces from this bypass capacitor are then routed to other bypass capacitors and then to the devices located near these capacitors. The ripple current from any given stage does not flow through the supply lines connecting it to any lower level stage. Also, connections to the common terminals of the individual components within a stage are made by traces close together before being connected by a trace to the ground bus or to the power supply return.

Multilayer printed circuit boards are frequently used in dense packaging. Two or more layers may be used for interconnections, while individual layers with inherent low impedance may serve for B^+, B^-, common, and ground plane distribution.

From the above discussion, it should be apparent that the analysis followed in these interconnections is not restricted to printed circuit boards. The considerations also apply to interconnections between subassemblies and modules within a system, and even in the system wiring itself.

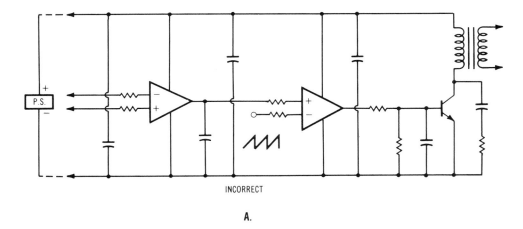

INCORRECT

A.

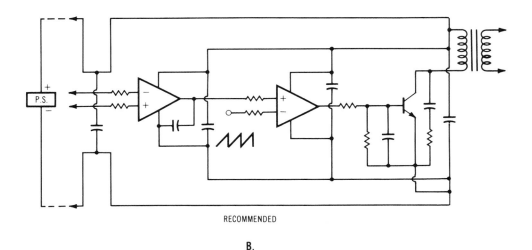

RECOMMENDED

B.

Fig. 14.13. Printed circuit connections for analog circuits.

14.2.2 Circuit Components

The circuit of Fig. 14.14A is an open-loop op-amp which amplifies remote input voltage e_{in}. The input is grounded at the source and the amplifier and load are grounded elsewhere. The ground loop voltage e_g couples directly into the amplifier. Connecting each end of a coaxial shielded cable to ground will not significantly reduce the ground loop voltage, because e_g is usually a low impedance source. An improvement is shown in Fig. 14.14B where a common mode voltage e_{cm} appears on both inputs at the source, and the common mode rejection of the amplifier then reduces the ground loop interference.

However, very few op-amps operate open loop, and some form of feedback is used to control the transfer function (gain and frequency response), which then alters the common mode rejection. Fig. 14.14C shows a noninverting op-amp similar to Fig. 14.14B, except R_3 is added. The equation for output voltage is

$$e_0 = e_{in}(1 + R_3/R_2) + e_{cm} \tag{14.10a}$$

which shows that the overall common mode voltage actually approaches unity. Even if the voltage gain is 40 dB to amplify low-level signals, the common mode voltage between the ground points may obscure the signal. The same analysis is true for an inverting op-amp where the output voltage is

$$e_0 = -e_{in}R_3/R_2 - e_{cm} \tag{14.10b}$$

In the circuit of Fig. 14.14D, resistor R_4 is added to the noninverting amplifier to produce a bridge circuit and the equation for output voltage is

$$e_0 = e_{in}\left[\frac{1 + R_3/R_2}{1 + R_1/R_4}\right] + e_{cm}\left[\frac{R_3}{R_2} - \frac{1 + R_3/R_2}{1 + R_1/R_4}\right] \tag{14.11a}$$

From this equation, it is observed that the common mode voltage will be zero if

$$\frac{R_3}{R_2} = \left(1 + \frac{R_3}{R_2}\right)\bigg/\left(1 + \frac{R_1}{R_4}\right)$$

$$= \frac{R_4(R_2 + R_3)}{R_2(R_4 + R_1)} = \frac{R_4}{R_1}$$

Then

$$e_0 = e_{in}R_3/R_2 \tag{14.11b}$$

The noninverting op-amp gain is then reduced slightly from Equation (14.10a) for the same values of R_3 and R_2, or the gain may be the same as Equation (14.10a) if R_3 is increased slightly or R_2 is reduced slightly. Also, the voltage divider effect of R_4 and R_1 loads the signal source as compared to the normal high input impedance of the amplifier, but the value of R_4 may be quite high since the ratio of R_4 to R_1 is the important consideration. Note that R_1 also includes the internal impedance of the signal source.

Example

It is desired to eliminate the common mode voltage effect from the circuit of Fig. 14.14D which has the following characteristics.

For $e_{in} = 5$ mV and $R_{in} = 100$ Ω, the desired $e_0 = 1$ V. Then, $R_3/R_2 = 1/0.005 = 200$. Let $R_2 = 100$ Ω, then $R_3 = 20$ kΩ. Also, $R_4/R_1 = 200$. Choosing $R_4 = 20$ kΩ, $R_1 = 20$ k$\Omega/200 - R_{in} = 100 - 100 = 0$, since $R_1 = R_{in}$.

Further considerations are important in the circuits of Figs. 14.14C and D. The amplifier power supply common should be common to the output terminal instead of to the input signal ground. The components in the bridge must be carefully balanced to maintain a desired gain, eliminate unwanted signals, and prevent oscillations. Component value tolerances of 1% may be adequate for some circuits, while 0.1% tolerances may be required for most applications.

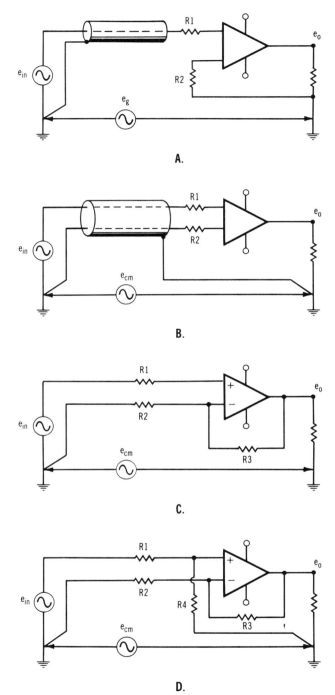

Fig. 14.14. Operational amplifier connections.

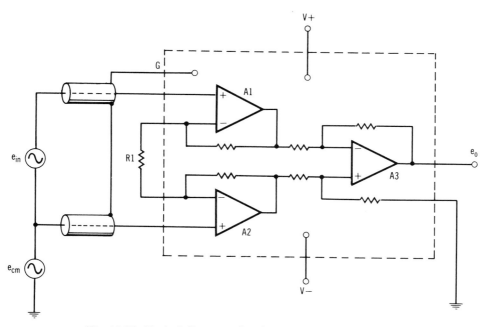

Fig. 14.15. Typical diagram of an instrumentation amplifier.

A method of eliminating common mode signals and noise, plus grounding and circuit common problems for single-ended outputs, is through the use of an instrumentation amplifier, normally an integrated circuit. The input impedance typically exceeds 100 MΩ, and both the CMRR (common mode rejection ratio) and the PSRR (power supply rejection ratio) typically exceed 100 dB. In the simplified schematic of Fig. 14.15, A_1 and A_2 are noninverting stages, while A_3 is a differential to single-ended output stage. The overall gain is adjusted by external resistor R_1. Another feature of many instrumentation amplifiers is the guard drive output, shown in Fig. 14.15, which drives or charges the capacitance of the shielded input cable. The shielded input cables reduce noise pick-up, and guard drive maintains the shield at the common mode voltage.

Single-conductor, single-shield coaxial cables are shown in Figs. 14.14 and 14.15. Other cables which may be used include triax, a single conductor with separate signal return (inner) and ground (outer) shields; twinax, a twisted-pair conductor with a single shield; quadrax, a twisted-pair conductor with separate ground and signal shields. Twinax cable is effective against crosstalk and low-frequency magnetic fields, is relatively inexpensive, and is becoming increasingly popular. Twinax can be used for frequencies up to 15 MHz and can be attached to a standard shielded connector which is a modified and polarized BNC type.

14.2.3 Subassemblies and Modules

A differential line driver and receiver, as shown in Fig. 14.16A, may be used to transmit signals between subassemblies and modules in high-noise environments.

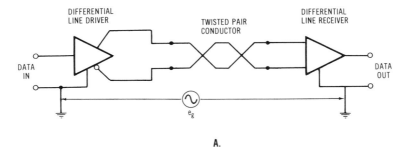

A.

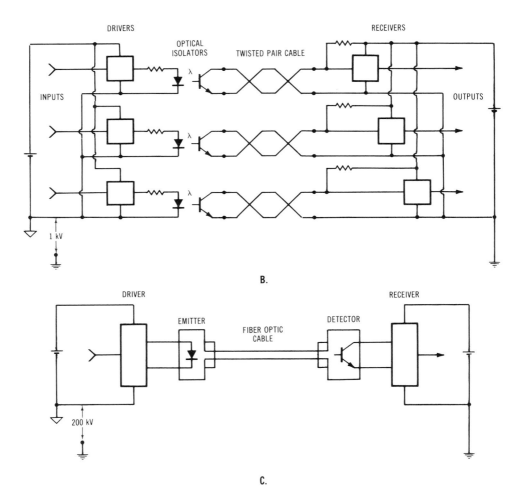

B.

C.

Fig. 14.16. Isolation and ground loop elimination techniques.

This technique is most common in data bus lines but it is also applicable to signal and command interfaces to power conversion equipment. A single-ended input, referenced to ground at the source end is applied to the line driver. The differential output is then connected to the receiver by a twisted-pair balanced transmission line. The receiver may also have an internal termination resistor for the twisted pair. A ground voltage e_g (which may exist between the source ground and the load ground) induced on the transmission line will appear equally on both inputs of the receiver. The receiver responds only to the differential signal and delivers a single-ended output referenced to the ground at the receiving end.

Optical isolators are effective in transmitting signals from one assembly to another when the voltage potential between the units is higher than the rating of line drivers and receivers. One unit may operate at ground potential, while another unit may operate at several hundred volts with respect to ground. Referring to Fig. 14.16B, status or command signals are applied to driver stages which supply current to the photodiode which controls the conduction of the phototransistor. The signals from the phototransistor are transmitted to the receiver stages at another location. The input to output capacitance of the optical isolator is typically 1 pF maximum. Twisted-pair lines are again used in the connecting cable. Notice that individual signal line pairs are used as opposed to individual signal lines with a common return. The line pairs provide noise immunity and decrease crosstalk effects between signals. Obviously, the process can be reversed. Signals at near ground potential can be transmitted to modules operating above or below ground potentials. In applications where these functions are combined, separate twisted-pair cables of sufficient insulation should be used. If both sending and receiving line pairs were combined in a single cable, dv/dt effects could couple between the lines and inject noise or transient voltages on the signals.

Fiber optics are an ideal method to transmit signals from one point to another. This is applicable where the sending and receiving ends are subject to a high-noise environment. It is especially applicable for signals transmitted between units where the potential difference is 100 kV, or higher. In Fig. 14.16C, circuitry at the driver end is powered by an internal low-voltage power supply and this circuitry monitors the status of the unit operating at 200 kV off ground. A signal from this circuitry is applied to an emitter and transmitted through the fiber optic cable to the detector of another unit operating at near ground potential. Multiple fiber optic cables for sending and receiving signals may be grouped as a bundle, since each fiber optic cable has "near infinite" isolation.

The instrumentation amplifier, discussed in the preceding section, is also a useful component in high-voltage power supplies and converters. The high-voltage output must be regulated by sensing a portion of the output voltage, and the resistance in the voltage divider of the feedback loop is usually from several megohms to hundreds of megohms in value. The high input impedance, differential voltage sensing, and high common mode rejection inherent in the instrumentation amplifier are advantageous in these applications. These advantages are further realized when the power supply or converter powers a communication vacuum tube, such as a klystron or TWT. For the TWT, a regulated voltage is applied between the collector and cathode, and a highly regulated voltage is applied between the helix and cathode. The helix is connected to the case of the TWT, which then becomes the system ground point in a cabinet or other installation.

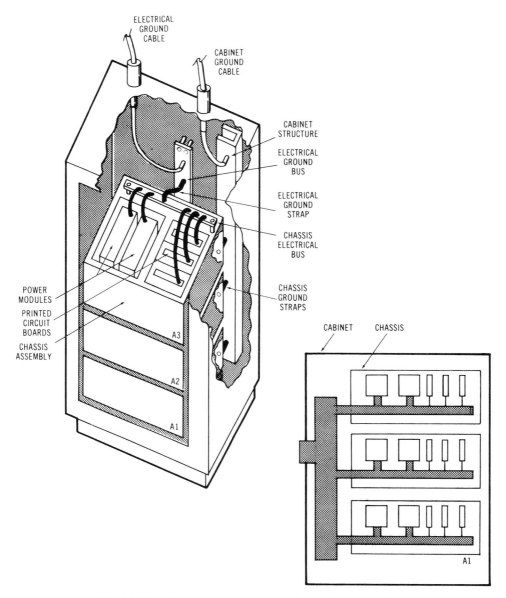

Fig. 14.17. Diagram of power conversion equipment cabinet.

Sensing both cathode voltage and helix current with respect to the helix case terminal requires careful consideration of the connections to the power supply circuit common and other quasi-grounds. The same considerations apply to the klystron, where a regulated voltage is applied between the collector and cathode; the body terminal is the electrical case of the klystron and the body current must be sensed and limited for tube protection.

14.2.4 System Grounding

A power conversion equipment cabinet with single point grounding is shown in Fig. 14.17. A pictorial view as well as an electrical diagram is shown. The electrical ground cable from the master system ground enters the cabinet and is connected to an electrical ground bus. The bus extends vertically and each chassis assembly is connected to the bus by a low impedance strap. Distribution to each assembly within the chassis is provided by separate cables from the chassis bus. A separate cabinet or earth ground cable enters the cabinet and is connected to a cabinet structure beam. The case of each chassis is then strapped to this beam.

With the system analyzed in this manner, it becomes apparent than an interface problem may exist even though single point grounding is used. Consider a circuit on a printed circuit board in chassis A1 which must sense an analog voltage in chassis A3. An appreciable noise voltage may be produced across the interconnecting power bus by high currents flowing between the power modules and the electrical ground. The noise generated may exceed the level of the voltage to be sensed. Digital circuitry may also have insufficient noise immunity to transmit and receive desired signals. This problem becomes more pronounced when the control circuitry must interface with that of another cabinet. Once these interfaces are defined, the designer then has a choice of employing methods discussed in the three preceding sections to eliminate any potential problems.

REFERENCES

1. FCC Bulletin OST 54, March 1982.
2. Henry W. Ott, *Noise Reduction Techniques in Electronic Systems* (J. Wiley, New York, 1976).
3. J. A. Malack and J. R. Engrstrom, "RF Impedance of United States and European Power Lines," *IEEE Transactions on Electromagnetic Compatibility* Vol. EMC-18, No. 1 (February 1976).
4. D. S. Bunk and T. J. Donovan, "Electromagnetic Shielding," *Machine Design* (July 6, 1967).
5. D. R. Dreger, "Plastics That Stop EMI," *Machine Design* (July 26, 1979), p. 114.
6. D. M. Bigg and D. E. Stutz, "Molded Composites as EMI Shields," *Industrial Research/Development* (July 1979), p. 103.
7. National Fire Protection Association, *National Electrical Code* (1990).

15

SEMICONDUCTOR AND EQUIPMENT COOLING

15.0   Introduction
15.1   Thermal Conduction and Resistance
    15.1.1   Semiconductors
    15.1.2   Mounting Interface
    15.1.3   Heat Sinks
    15.1.4   Thermal Circuit Design
15.2   Natural Convection and Radiation
    15.2.1   Natural Convection
    15.2.2   Radiation
    15.2.3   Finned Heat Sinks
15.3   Forced Air Cooling
    15.3.1   Volumetric Air Flow
    15.3.2   Pressure Drop
    15.3.3   Heat Sinks
    15.3.4   Fans and Blowers
15.4   Forced Liquid Cooling
15.5   Heat Pipes
15.6   Thermoelectric Coolers
    References

15.0   INTRODUCTION

This chapter presents semiconductor and equipment cooling information in terms relevant to the electronics industry. The basic quantities of heat transfer, as related to electronic equipment, are analogous to the electrical quantities. Unfortunately, designers frequently neglect complete analysis of equipment cooling either by unfamiliarity of the terms involved or by erroneous condition assumptions. Heat transfer notations used in this chapter are given in Table 15.1 and conversion factors to other units are given in Table 15.2.

The basic methods of transferring heat are (1) conduction, (2) convection, and (3) radiation. These methods are discussed in the following sections in terms of

TABLE 15.1. Notation

Notation	Interpretation	Notation	Interpretation
A_c	Cross-sectional area, in.^{2} or cm^2	p	Pressure, psi or in. Hg, in. H$_2$O
A_s	Surface area, in.^{2} or cm^2	P	Power, watts
b	Convection factor	Q	Heat flow, watts
c	Specific heat, W-s/lb-°C	t	Time, seconds
CFM	Volumetric air flow, ft^3/min	T	Temperature, °C (°F, °R, K)
d, h, s, w, t	Linear dimensions	v	Linear velocity, ft/min
f	Volumetric air flow, CFM, ft^3/min	θ	Thermal resistance, °C/W
F	Volumetric liquid flow, GPM, gal/min	ρ	Thermal resistivity, °C-in./W
g	Specific gravity, gram	ε	Radiation emissivity
k	Thermal conductivity, BTU/h-ft^2°F/ft, W/°C-in.	μ	Viscosity of liquid, lb/h-ft
l	Length (direction of heat flow, in. or cm)	α	Absorptivity
L	Life	σ	Stefan–Boltzmann constant
N	Number of ducts or fins		

TABLE 15.2. Conversion Factors

To Convert	To^a	Multiply by	Conversely, Multiply by
Power			
Watts	BTU/h	3.413	0.293
Watts	cal/s	0.2388	4.187
Power density			
Watts/in.^{2}	BTU/h-ft^2	491.5	2.035×10^{-3}
Watts/in.^{2}	cal/s-cm^2	0.037	27.01
Thermal conductivity			
Watts/°C-in.	BTU/h-ft^2°F/ft	22.753	0.04395
Watts/°C-in.	cal/s-cm^2°C/cm	0.09403	10.635
Specific heat			
Watt-min/lb-°C	BTU/lb-°F	0.03175	31.5
Watt-min/lb-°C	cal/g-°C	0.03175	31.5
Flow rate (Air at sea level)			
CFM	gal/min	7.5	0.1333
CFM	liters/s	476	2.1×10^{-3}
Pressure			
psi	in. Hg	2.04	0.491
psi	kg/cm^2	0.0703	14.23

^{a}Definitions: BTU, quantity of heat required to raise the temperature of one pound mass of water through 1°F at standard pressure; calorie, quantity of heat required to raise the temperature of one gram mass of water through 1°C at standard pressure.

heat flow in watts (instead of BTU per hour or gram-calories per second), temperature in degrees Celsius, and dimensions in inches or centimeters.

Cooling of power conversion equipment primarily involves the removal of heat from semiconductors, although heat produced by passive devices should not be overlooked. This thermal problem is common to everything from LSI devices on a densely packaged printed circuit board to individual thyristors dissipating hundreds of watts. Excessive junction temperature is the most common cause of semiconductor failure. Good thermal design increases reliability and performance by keeping the junction temperature to a safe level.

The life of materials, including semiconductors, varies logarithmically with the reciprocal of absolute temperature, and is expressed by the Arrhenius equation:

$$L = A(\varepsilon^{b/T} - 1) \tag{15.1}$$

where

L = expected life
A = constant
b = empirical constant involving the Boltzmann constant
T = temperature, kelvin

which states (after conversion to appropriate units) life is halved for every 20°C rise in temperature. Expressed in exact terms, the equation becomes a ratio of life at temperature differentials:

$$\text{Life}_{\text{hot}}/\text{Life}_{\text{cold}} = 2^{-\Delta T/20} \tag{15.2a}$$

where

$$\Delta T = T_{\text{hot}} - T_{\text{cold}} \qquad °C$$

and conversely,

$$\text{Life}_{\text{cold}}/\text{Life}_{\text{hot}} = 2^{\Delta T/20} \tag{15.2b}$$

Examination of the above equation reveals that in addition to component life doubling for a 20°C decrease in temperature, component life increases 10-fold for a 66.44°C decrease in temperature. However, practicality must prevail. If a power transistor operating at 25°C ambient is placed in a liquid nitrogen bath, a life increase of several orders of magnitude may not necessarily be achieved.

The packaging orientation of printed circuit boards and semiconductor heat sinks within a cabinet or enclosure should minimize restrictions in the direction of *air flow*. Since heat rises, vertical panels, boards, and fins are desirable in natural convection. Magnetic components and resistors can frequently operate at higher temperatures than semiconductors and can survive momentary power overloads. These components can radiate heat which raises the temperature of adjacent semiconductors beyond a safe limit. The necessity of power resistors notwithstanding, the sole result of their use is the production of heat, and their usage should be avoided wherever possible. Overheating in capacitors is caused by excessive current flow through the equivalent series resistance (ESR). The manufacturer's maximum rms current rating, which is frequency dependent, should always be observed.

Physical environmental conditions are important to users and manufacturers of power conversion equipment. For instance, manufacturers of voltage-regulated power supplies frequently rate output current as a function of ambient temperature. The manufacturers may test the supplies in a temperature chamber which invariably has an internal fan to circulate air for temperature regulation. If the power supply has been designed for natural convection and radiation, the test is not representative since a small amount of air flow will increase convection and reduce component temperature. Power conversion equipment is frequently the last item to be specified in an overall system, and mounting location may be in a subchassis, equipment rack, or any space remaining. During system tests the user finds power supplies, correctly tested and specified by the manufacturer, fail due to excessive component temperature. A prolonged analysis typically reveals (1) the power supply heat sink faces adjacent equipment of the same or higher temperature, (2) the power supply heat sink is covered by an overhead panel, or (3) the enclosure is not ventilated. In the first case, radiation cooling is restricted and in the latter cases, natural convection is restricted. Good thermal design is therefore imperative from a system's standpoint as well as a hardware standpoint.

15.1   THERMAL CONDUCTION AND RESISTANCE

Thermal conductivity is the conduction of heat through materials and is given by Fourier's equation:

$$Q_d = kA_c \, \Delta T / l \qquad (15.3)$$

where

Q_d = conducted heat flow, W
  k = thermal conductivity, W/°C-in.
A_c = cross-sectional area, in.^{2}
  l = thermal path length, in.
ΔT = temperature differential, °C

For a given heat load, the minimum temperature differential is obtained by using high conductivity materials, a large cross-sectional area, and a short conductive path. Thermal conductivity and resistivity for various materials commonly used in power conversion equipment are listed in Table 15.3.

Thermal resistance is used to describe the temperature gradient across a material or interface while dissipating power, and is given by

$$\theta = \rho l / A_c \qquad °C/W \qquad (15.4)$$

where ρ is the thermal resistivity $(1/k)$ in °C-in./W and is analogous to the electrical resistance, Equation (3.30). A model electrical and thermal equivalent

TABLE 15.3. Thermal Conductivity and Resistivity of Selected Materials

Electrical Conductors	Thermal Conductivity k (BTU/ft^2 h °F/ft)		Thermal Resistivity ρ (°C-in/W)	
	20°C	100°C	20°C	100°C
Silver	243	241	0.094	0.095
Copper	227.5	226.5	0.10	0.10
Gold	171	168.5	0.133	0.135
Aluminum (pure)	126	125	0.181	0.182
1100	128	120	0.178	0.189
2024-T6	84.6	80	0.268	0.284
6061-T6	96.8	91	0.235	0.250
6063-T6	121	114	0.188	0.2
Brass	63	61.4	0.361	0.371
BeCu	68	61.4	0.335	0.371
Zinc	64	59.2	0.356	0.384
Nickel	52.3	52	0.435	0.438
Steel, mild	38	34	0.6	0.67
Stainless steel	10	11.4	2.28	2.0
Insulators	75°C		75°C	
Air	0.016–0.019		1428–1200	
Rock wool	0.025		910	
Asbestos paper	0.040		569	
Freon	0.04–0.06		569–379	
Kapton, Mylar, Teflon	0.09–0.11		253–207	
Silicone grease	0.10–0.12		227–190	
Transformer oil	0.10–0.12		227–190	
RTV, epoxy	0.12–0.18		190–126	
Mica	0.35–0.43		65–53	
Water	0.36–0.38		63–60	
Thermal paste	0.40–0.43		57–53	
Silicon rubber	0.52–0.54		44–42	
Fiberglass	0.52–0.54		44–42	
Thermal epoxy	0.73–0.80		31–28	
Alumina, 99.5%	16		1.42	
Beryllia, 99.5%	114		0.20	
BeO, 99.5%	145 at 20°C, 109 at 10°C		0.16 at 20°C, 0.21 at 100°C	
Diamond	320–364		0.0711–0.0625	

circuit is shown in Figs. 15.1A and 15.1B, respectively. Thermal gradients for a semiconductor mounted on a heat sink in ambient conditions are shown in Fig. 15.1C.

15.1.1 Semiconductors

The intrinsic temperature limit of semiconductor materials (typically 300°C for Si) is a function of the impurity concentration and is determined by the most lightly

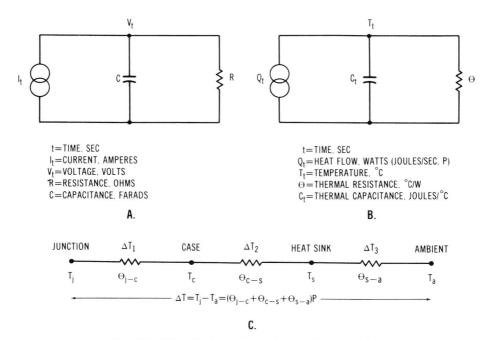

$t=$ TIME, SEC
$I_t=$ CURRENT, AMPERES
$V_t=$ VOLTAGE, VOLTS
$R=$ RESISTANCE, OHMS
$C=$ CAPACITANCE, FARADS

A.

$t=$ TIME, SEC
$Q_t=$ HEAT FLOW, WATTS (JOULES/SEC, P)
$T_t=$ TEMPERATURE, °C
$\Theta=$ THERMAL RESISTANCE, °C/W
$C_t=$ THERMAL CAPACITANCE, JOULES/°C

B.

JUNCTION ΔT_1 CASE ΔT_2 HEAT SINK ΔT_3 AMBIENT

T_j Θ_{j-c} T_c Θ_{c-s} T_s Θ_{s-a} T_a

$$\Delta T = T_j - T_a = (\Theta_{j-c} + \Theta_{c-s} + \Theta_{s-a})P$$

C.

Fig. 15.1. Electrical and thermal equivalent models.

doped region. However, rate of heat removal from the chip, solder melting temperature, and surface contamination reduce the maximum junction temperature. Typical values are 150°C for Si, and 110°C for Ge devices.

Semiconductor blocking characteristics are highly temperature dependent and the off-state or leakage current in Si devices doubles for every 10°C rise in the junction temperature. This increase in leakage current, especially in high-voltage devices, produces a power dissipation which, in addition to the forward-biased or saturated conduction losses, may induce thermal runaway conditions in bipolar transistors. The thermal runaway, or "snowball" effect, describes the positive feedback chain reaction where the temperature rise produces a current rise which, in turn, heats the junction, causing a further increase in current.

Semiconductor manufacturers normally state average current for power rectifiers at 100–150°C case temperature, rms currents for thyristors at 65–125°C case temperature, and power dissipation for transistors at 25°C case temperature. However, few transistors are ever operated with an "infinite" heat sink. The devices are therefore derated linearly to the maximum temperature limit. The inverse of the derating factor is the thermal resistance, junction-to-case, θ_{j-c}, which is normally specified by the manufacturer. Typical thermal resistances for various Joint Electron Device Engineering Council (JEDEC) case styles are tabulated in Table 15.4. Also tabulated for various styles are θ_{c-s}, θ_{j-a} and recommended mounting torque or force. Correct mounting torque is also important from a mechanical stability standpoint and for environmental conditions of vibration and shock.

TABLE 15.4. Semiconductor Thermal and Physical Parameters

JEDEC Case Type	Press-Pak Case Type	$\theta_{j\text{-}c}$ (°C/W)	$\theta_{c\text{-}s}{}^{a}$ (°C/W)	$\theta_{j\text{-}a}$ (°C/W)	Mounting Torque (in.-lb)	Mounting Force (lb)	Stud Size
			Rectifiers				
DO-41				75^{b}			Axial
DO-15, 27				70^{b}			Axial
DO-4		2–3	0.5		12–15		10–32
DO-5, 203		0.8–1.2	0.3		25–30		$\frac{1}{4}$–28
DO-8		0.35–0.4	0.15		75–125		$\frac{3}{8}$–24
DO-9		0.15–0.2	0.075		300–350		$\frac{3}{4}$–16
DO-200AA	$\frac{1}{2}$ in. thick	0.095	0.06			1000	
DO-200AB	1 in. thick	0.055	0.03			2200	
DO-200AC	1 in. thick 2.9 in. diameter	0.023	0.0075			5000	
			Thyristors				
TO-92		75		200			
TO-5, 39		5		150			
TO-64		3–3.5	0.5		12–15		10–32
TO-48		1.3–1.7	0.3		25–30		$\frac{1}{4}$–28
TO-83, 94		0.3–0.4	0.12		120–130		$\frac{1}{2}$–20
TO-93		0.13–0.15	0.075		300–350		$\frac{3}{4}$–16
TO-200AB	$\frac{1}{2}$ in. thick	0.08–0.12	0.06			1000	
TO-200AC	1 in. thick	0.04–0.06	0.03			2200	
	1 in. thick, 2.9 in. diameter	0.023	0.0075			5000	
			Transistors				
TO-92		75–125		200–350			
TO-18, 46		100		300			
TO-5, 39		18–35		150–175			
TO-202		10–20		70–80			
TO-8		6–7		75–100			
TO-59^{c}, 111		3.3–5	1	85–90	12–15		10–32
TO-66		2.3–7	0.5–0.8	60–70	4–6		
TO-220		1.6–3.1	0.8–1	65–70	4–6		
TO-3, 204		0.8–1.5	0.05–0.1	30	6–8		
TO-36(Ge)		0.8	0.4	25	6–8		
TO-63, 82		0.5–0.75	0.4		30–50		$\frac{5}{16}$–24
TO-61^{c}		1–1.5	0.3		25–30		$\frac{1}{4}$–28
TO-114			0.12				$\frac{1}{2}$–20

^{a}With thermal compound, no insulator.
^{b}Typical printed circuit board mount.
^{c}Isolated collector.

Allowing for a junction temperature safety factor to reduce failure rates, the maximum operating case temperature is given by

$$T_{c,\,\max} = T_j - \theta_{j\text{-}c} P \quad °C \qquad (15.5)$$

The "rated" power dissipation may be exceeded on a transient or low duty cycle basis for rectifiers and thyristors. Transient thermal impedance for rectifiers and thyristors at one millisecond (1 ms) conduction is typically one-tenth of the continuous duty thermal resistance. However, the peak one-cycle surge, or I^2t, ratings are usually the predominant characteristic which must not be exceeded.

Transistors, unlike rectifiers and thyristors, are gain-dependent devices and are subject to failure from nonsaturated conduction or secondary breakdown condi-

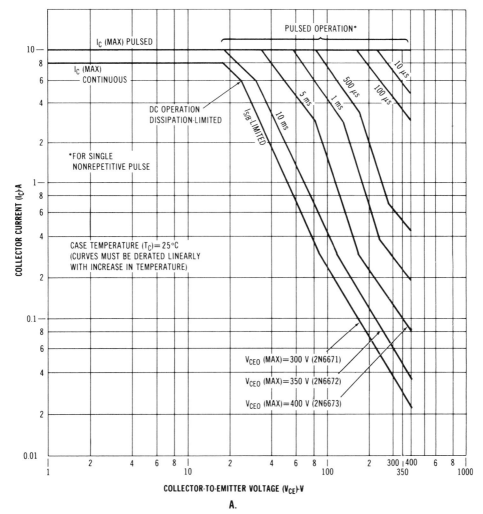

Fig. 15.2. SOA (safe operating rate) for 2N6671, 2N6672, and 2N6673. (Courtesy Harris Semiconductor.)

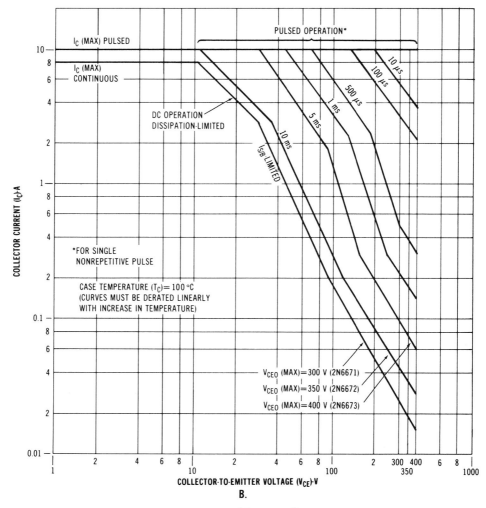

Fig. 15.2. (*Continued*)

tions. Again, transistor manufacturers' data sheets normally include a safe operating area (SOA) showing dissipation limits and pulse operation parameters. However, these figures are usually for 25°C case temperature and for a single nonrepetitive pulse! The transistor must therefore be derated for higher case temperatures. Figure 15.2 shows the SOA of 2N6671-6673 epitaxial switching transistors at $T_c = 100$°C, $T_c = 25$°C, and pulsed operation.

15.1.2 Mounting Interface

The most frequently neglected or misapplied parameter of semiconductor cooling is the thermal impedance between the case of a power device and some form of heat sink. The device may be mounted directly to the heat sink for electrical

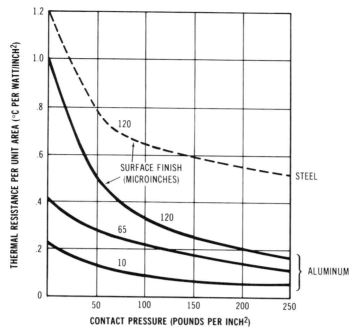

Fig. 15.3. Thermal resistance of interfaces as a function of contact pressure. (From A. W. Scott, *Cooling of Electronic Equipment*. Copyright 1964 John Wiley & Sons.)

conduction, or may be electrically isolated from the heat sink for safety or "nongrounding" purposes. In either case, surface finish, surface flatness, and mounting force of the mating parts are important. Quantitative values of thermal resistance per unit area as a function of surface finish and contact pressure are shown in Fig. 15.3 [1]. Surface flatness usually applies to larger heat sinks whose flat surface is mounted against a chassis or wall plate.

Typical values of thermal resistance from case to heat sink, $\theta_{c\text{-}s}$, for various types of JEDEC devices and various interface conditions and isolation are listed in Table 15.5. Mica requires handling care to prevent cracking and peeling. Kapton, Mylar film, and silicone rubber insulators are highly resistant to cracking and puncture. However, some silicone rubber insulators are hygroscopic (they acquire and retain moisture), and their use in high-humidity environments is not advised. Hard anodized surfaces eliminate insulator handling but are subject to crazing and chipping due to inadvertent slippage of a mounting tool. Beryllium oxide (BeO) insulators have a high dielectric strength and a thermal conductivity higher than aluminum. As compared to other insulators, the increased insulation thickness of BeO wafers decreases the interface capacitance which may be important in minimizing EMI in high-frequency switching units. Since BeO is very brittle, the insulators must be handled with care and proper surface finish and mounting force must be observed to prevent cracking. Thermal bonding of the BeO wafer to the heat sink or device also minimizes effects of cracking and enhances the ability to withstand high vibration, shock levels, and thermal cycling.

TABLE 15.5. Semiconductor Thermal Interface Parameters

| JEDEC Case Type | Thermal Resistance, $\theta_{c\text{-}s}$ (°C/W) | | | | | | | |
| | Metal to Metal | | MICA Insulator | | Film Insulator | | BeO Insulator | |
	Dry[a]	Compound	Dry[a]	Compound	Dry[a]	Compound	Dry[a]	Compound
DO-4, TO-64	0.7	0.5	3	2.5			1.5	0.5
DO-5, TO-48	0.5	0.2	2.2	1.8			0.6	0.3
TO-220	1.2	0.9	3.5	1.6	4.5	2.5		
TO-66	1	0.7	2	0.34	2	1.5	0.9	0.3
TO-3[b]	0.3	0.1	0.45	0.35	0.8	0.54	0.6	0.15
TO-36	0.4	0.2	—	0.9			0.6	0.3
TO-63	0.5	0.3	—	0.9			0.6	0.4
TO-83, -94	0.2	0.1	1.5	1.2				

Insulator	°C/W per Square inch of Area

Thermal Resistance of Insulators with Recommended Mounting Torque

0.0025 in. mica	0.34
0.002 in. film	0.47
0.007 in. silicone rubber	0.35
0.06 in. BeO	0.009
0.02 in. anodized aluminum	0.004

[a]Shown for effect. Thermal compound is recommended.
[b]With silicone rubber, $\theta_{c\text{-}s}$ = 0.65 dry = 0.45 with compound. With anodized or coated aluminum, $\theta_{c\text{-}s}$ = 1.1 dry = 0.35 with compound.

The data listed in Table 15.5 assumes a surface finish of 50 μ in., or smoother. It should be noted that the "DRY" column is listed for reference only. A very *thin* coat of thermal paste, which has approximately three times the thermal conductivity of thermal grease, should always be applied to the mating surfaces, whether they be metal or insulator. The paste fills the microscopic voids between the mating surfaces and maximizes the heat transfer. An excessive amount of thermal paste will collect contaminants over a period of time, depending on the environment, which may cause a decrease in insulation resistance or intermittent shorts between the heat sink and the device. Common commercial thermal pastes are:

Manufacturer	Type or Part Number
Aham	980 compound
Dow Corning	340 compound
General Electric	G640 compound
Thermalloy	Thermalcoate
Wakefield	120 compound

Thermal resistance $\theta_{c\text{-}s}$ for a TO-3 transistor as a function of mounting force with various insulators is shown in Fig. 15.4 [2]. However, these values are for ideal

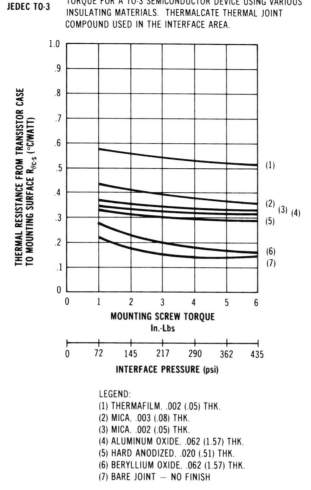

JEDEC TO-3

INTERFACE THERMAL RESISTANCE VERSUS MOUNTING SCREW TORQUE FOR A TO-3 SEMICONDUCTOR DEVICE USING VARIOUS INSULATING MATERIALS. THERMALCATE THERMAL JOINT COMPOUND USED IN THE INTERFACE AREA.

LEGEND:
(1) THERMAFILM, .002 (.05) THK.
(2) MICA, .003 (.08) THK.
(3) MICA, .002 (.05) THK.
(4) ALUMINUM OXIDE, .062 (1.57) THK.
(5) HARD ANODIZED, .020 (.51) THK.
(6) BERYLLIUM OXIDE, .062 (1.57) THK.
(7) BARE JOINT — NO FINISH

Fig. 15.4. Interface thermal resistance versus mounting force and pressure. (Courtesy Thermalloy, Inc.)

conditions and may be difficult to achieve repeatedly. As an example of interface losses, consider the TO-3 transistor mounted on a heat sink with a thermally coated film insulator. Using the thermal resistance values of Table 15.5, $\theta_{c-s} = 0.54°C/W$ typical. From Table 15.3, the thermal resistivity of aluminum is $0.18°C$-in./W typical. Thus, to conduct the same power dissipation at the same temperature differential, the thermal resistance of the 0.002-in. insulator is the equivalent of 3 in. $(0.54/0.18 = 3)$ of aluminum! For this reason, insulators are seldom used on large stud-mount devices and press-pak devices. In those cases, the heat sink is electrically isolated. The same condition may apply to a TO-3 transistor, or to paralleled transistors, as analyzed in the example of Section 15.1.4.

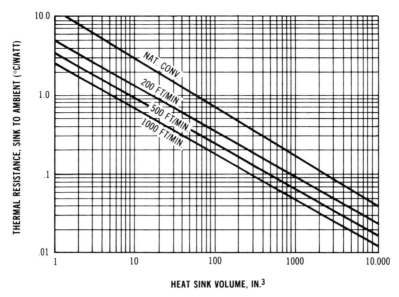

Fig. 15.5. Thermal resistance versus heat sink volume. (Courtesy EG & G Wakefield.)

15.1.3   Heat Sinks

Aluminum is primarily used as a heat sink material because of low cost and relative light weight. It can be machined, extruded, or die cast. Table 15.3 lists various aluminum alloys with 1100 and 6063 being the most common heat sink alloys, due to the higher thermal conductivity. Copper, which has 1.8 times the thermal conductivity of aluminum, may be used for dense or critical space applications. Diamond has the highest known thermal conductivity. Finned heat sinks, designed primarily for convection, are discussed in detail in Section 15.2.

Knowledge of component or assembly volume is important in packaging of power conversion equipment. Frequently, the three major volumetric items (especially in linear power supplies) are magnetic components, filter capacitors, and heat sinks. Values for the first two are dictated by electrical design. The volume of required heat sinks is primarily dictated by power dissipation and by operating ambient temperature. As a guide, thermal resistance from heat sink to ambient, θ_{s-a}, versus typical heat sink volume for a 50°C heat sink rise above ambient, is plotted in Fig. 15.5 for finned heat sink selection. The associated equations for natural convection at sea level, and accounting for radiation effects, are:

$$\theta_{s-a} = 11 \, V^{-0.644} \qquad °C/W \tag{15.6a}$$

$$V = (0.091\theta_{s-a})^{-1.55} \tag{15.6b}$$

where V is volume in cubic inches.

The thermal resistance with forced air cooling is discussed in Section 15.3.

15.1.4 Thermal Circuit Design

From the discussion of previous sections and with reference to Fig. 15.1C, the following design example is presented. The given conditions are:

(a) TO-3 transistor, $\theta_{j\text{-}c} = 1.1°\text{C/W}$.
(b) Power dissipation = 30 W.
(c) Maximum junction temperature = 200°C.
(d) Allowable junction temperature = 120°C derated.
(e) Ambient temperature = 50°C maximum.
(f) Cooling by conduction, natural convection, and radiation.

The analysis considers the relation of an insulated versus a noninsulated transistor case, required heat sink thermal resistance and volume, effects on junction temperature, and power dissipation.

Case 1. Using a film insulator:

$$T_c = T_j - \theta_{j\text{-}c}P = 120 - 1.1 \times 30 = 87°\text{C maximum}$$

$$T_s = T_c - \theta_{c\text{-}s}P = 87 - 0.6 \times 30 = 69°\text{C maximum}$$

$$\theta_{s\text{-}a} = (T_s - T_a)/P = (69 - 50)/30 = 0.63°\text{C/W}$$

$$V = (0.091 \times 0.63)^{-1.55} = 84 \text{ in.}^3$$

Case 2. Without an insulator:

$$T_c = 87°\text{C} \quad (\text{as in Case 1})$$

$$T_s = T_c - \theta_{c\text{-}s}P = 87 - 0.12 \times 30 = 83.4°\text{C}$$

$$\theta_{s\text{-}a} = (T_s - T_a)/P = (83.4 - 50)/30 = 1.1°\text{C/W}$$

$$V = (0.091 \times 1.1)^{-1.55} = 35 \text{ in.}^3$$

Thus, this heat sink occupies approximately 40% of the volume of the heat sink in Case 1.

Case 3. Without an insulator, same heat sink as Case 1, find the reduced junction temperature.

$$T_s = T_a + \theta_{s\text{-}a}P = 50 + (0.63 \times 30) = 69°\text{C}$$

$$T_c = T_s + \theta_{c\text{-}s}P = 69 + (0.12 \times 30) = 73°\text{C}$$

$$T_j = T_c + \theta_{j\text{-}c}P = 73 + (1.1 \times 30) = 106°\text{C}$$

Thus, for the same power dissipation, the junction will be 14°C cooler than Case 1. From Equation (15.2b), the life of the transistor will be 1.6 times that of Case 1.

Case 4. Without an insulator, same heat sink as Case 1, same junction temperature as Case 1, find the increase in power dissipation:

$$P = (T_j - T_a)/(\theta_{j-c} + \theta_{c-s} + \theta_{s-a})$$
$$= (120 - 50)/(1.1 + 0.12 + 0.63) = 37.8 \text{ W}$$

Thus, the power dissipation can be 7.8 W more than Case 1.

15.2   NATURAL CONVECTION AND RADIATION

Heat produced by "heat sink"–mounted power dissipating components is transferred to the surroundings by natural convection and radiation. In the case of vertically mounted heat sinks, the air near the sink is heated, then rises, and is replaced by cooler air to provide additional convection. The hot surfaces of the sink simultaneously radiate heat to the cooler surroundings.

15.2.1   Natural Convection

The amount of heat transferred by natural convection depends on the temperature differential, dimensions, and orientation of the hot surfaces, and altitude or pressure. The basic equation of natural convection is given by [3]

$$Q_c = \frac{bA_s}{h^{0.25}} (\Delta T)^{1.25} \qquad \text{BTU/h} \qquad (15.7a)$$

where

b = convection factor (see Fig. 15.6)
A_s = area of hot surface, ft^2
h = height (or width) of surface plane, ft
ΔT = temperature differential $(T_s - T_a)$, °F

Converting to units of watts, inches, °C, and considering altitude, the above equation is

$$Q_c = \frac{3.787 \times 10^{-3} bA_s}{h^{0.25}} \left[[1.8(T_s - T_a)]^{1.25} \sqrt{\frac{p}{29.9}} \right] \qquad \text{W} \qquad (15.7b)$$

where

b = convection factor (see Fig. 15.6)
A_s = area of hot surface, in.^{2}
h = height (or width) of surface plane, in.
T_s = hot surface temeprature, °C
T_a = ambient temperature, °C
p = barometric pressure, in. Hg

The convection factor b in the above equations is a function of the orientation of the heat sink or plate. The optimum orientation is shown in Fig. 15.6A, where

CONVECTION FACTOR, b, FOR

LAMINAR FLOW	TUBULENT FLOW

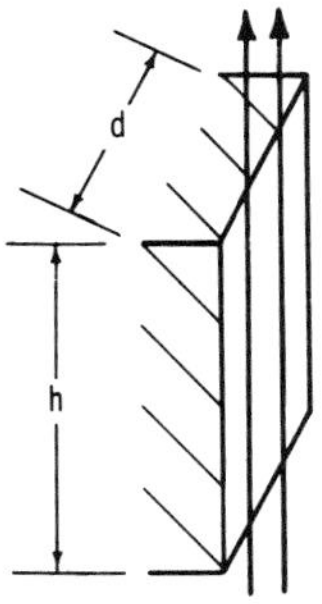

$A_S = h \times d$

.28   .30

VERTICAL SURFACES AND LARGE CYLINDERS

A.

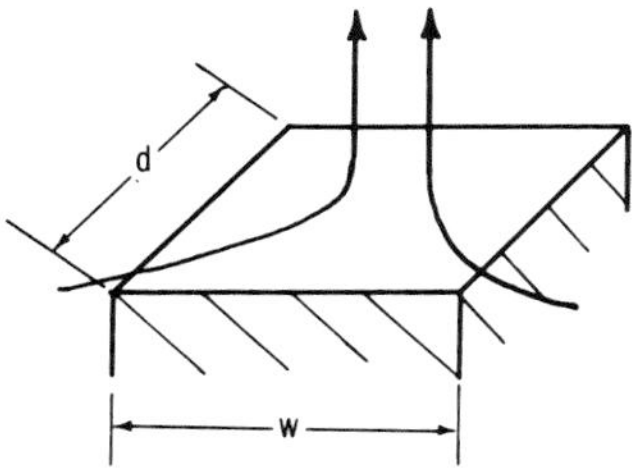

$A_S = w \times d$

.26   .28

HORIZONTAL SURFACES FACING UP AND
SMALL VERTICAL CYLINDERS

B.

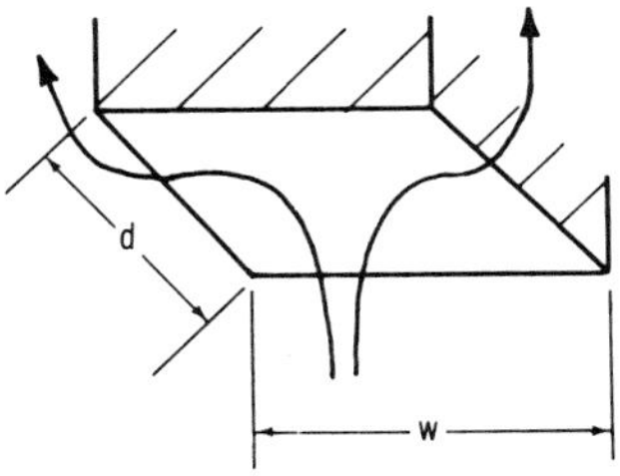

$A_S = w \times d$

.12   .12

HORIZONTAL SURFACES FACING DOWN

C.

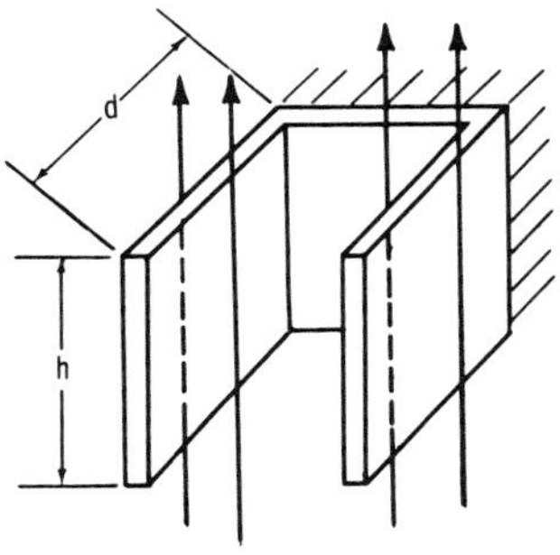

$A_S = 4\ h \times d$

.28   .30

VERTICAL MOUNTED U-SHAPED HEAT DISSIPATOR

D.

Fig. 15.6. Convection factors for heat sink orientation.

the air rises along the vertical surface and b is a maximum. In Fig. 15.6B, the air flows along the top surface and then rises. In Fig. 15.6C, the air must flow along the surface to the edge before it can rise and be replaced by cooler air. In this case, the convection factor b is a minimum. In each case, height (or width) times the depth is the surface area A_s. For enclosed equipment, Figs. 15.6A–C would represent sides (front and rear), top, and bottom, respectively. A U-shaped heat sink is shown in Fig. 15.6D. Since all four sides provide convection cooling, the total area is four times the height times the depth. The area of finned heat sinks, as shown in Fig. 15.8, is given by Equation (15.9). The air movement in natural convection is considered to range from 0.05 to 0.5 linear feet per second [4].

Natural convection also depends on barometric pressure, as given in Equation (15.7b). Convection cooling requirements, especially for forced convection as discussed in Section 15.3.1, are frequently specified in terms of air density. Barometric pressure and air density at equivalent altitude versus temperature are plotted in Fig. 15.7. The data refers to ARDC, NACA, and ICAO standard atmospheres. The hot and cold temperature profiles represent U.S. Military extreme criteria as defined by MIL-STD-210. Thus for equipment operating at 40,000 ft, the effective heat transfer is $\sqrt{(5.56/29.9)} = 43\%$ of the convection at sea level. Conversely, the maximum thermal resistance of a heat sink for the same temperature differential and power dissipation would be 233% times that of a heat sink at sea level. Furthermore, convection does not exist in a vacuum. For this reason, careful thermal analysis is imperative when power conversion or other equipment is installed aboard spacecraft. In this case, the only means to remove heat is by conduction and radiation.

15.2.2   Radiation

The amount of heat flow transferred by radiation depends on the temperature differential, dimensions, and finish of the hot surfaces, and shielding effects of adjacent surfaces. The basic equation for radiation is given by the Stefan–Boltzmann formula:

$$Q_r = \varepsilon \alpha \sigma F A_s \left(R_s^4 - R_a^4 \right) \qquad \text{BTU/h} \qquad (15.8\text{a})$$

where

ε = emissivity of hot surface

α = absorptivity of cooler surface

F = view factor, numeric

A_s = area of hot surface, ft^2

$^{\circ}R$ = degrees Rankine

σ = Stefan–Boltzmann constant, $= 1.74 \times 10^{-9}$ BTU/h ft^2 $^{\circ}$R^4

Converting to units of watts, inches, and $^{\circ}$C where $^{\circ}$R = ($^{\circ}$F + 460) = $(1.8 \times ^{\circ}$C + 492), the above equation is

$$Q_r = 3.54 \times 10^{-12} \varepsilon \alpha F A_s \left[(1.8 T_s + 492)^4 - (1.8 T_a + 492)^4 \right] \qquad \text{W} \quad (15.8\text{b})$$

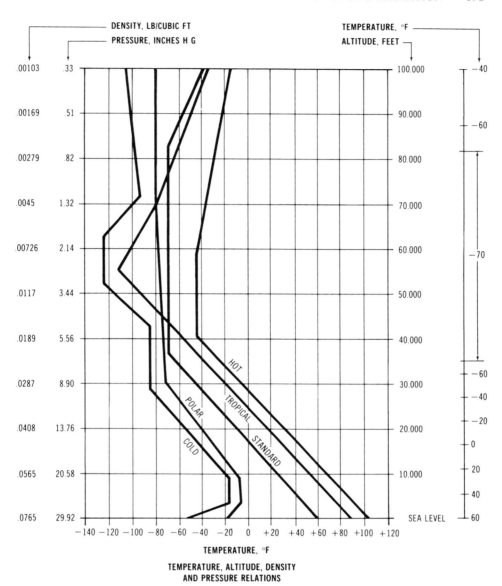

Fig. 15.7. Temperature, altitude, density, and pressure relations.

The emissivity and absorptivity factor in the above equation is tabulated in Table 15.6 for various finishes. Note that emissivity is essentially color blind, and that paint of any color improves emissivity. However, dark colors in the visible spectrum absorb more radiant energy than light colors, as shown in Table 15.6. The shielding effect determines the view factor and effective surface area. For flat surfaces, small compared to the surroundings, $F = 1$, and $A_s = dw$. For finned heat sinks, reference Equation (15.9) and Fig. 15.8.

TABLE 15.6. Absorptivity and Emissivity of Various Surface Finishes

Surface	Absorptivity	Emissivity
Polished aluminum	0.10	0.04–0.06
Irridite aluminum		0.1–0.2
Aluminum paint		0.27–0.67
Anodized aluminum		0.6–0.8
Black-anodized aluminum		0.92–0.96
Polished copper		0.03–0.07
Oxidized copper		0.6–0.8
Polished stainless steel		0.2–0.3
Enamel paints (any color)		0.85–0.87
Oil paints (any color)		0.89–0.92
White paint, 1 mil thick	0.15	
Black paint, 1 mil thick	0.97	
Clear varnish on aluminum	0.20	

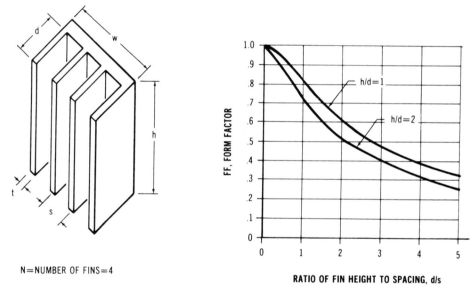

N=NUMBER OF FINS=4

Fig. 15.8. Heat sink dimensional parameters and form factor.

15.2.3 Finned Heat Sinks

Manufacturers of finned aluminum heat sink extrusions normally publish surface area, weight, and thermal resistance data per unit length. The most ambiguous term is the thermal resistance which depends on ambient temperature and power dissipation. Thus a specified thermal resistance can be deceptive unless test conditions are stated. Frequently the temperature rise above ambient is plotted versus power dissipation and is more informative. Discrepancies in specifications can be due to test methods. The designer may be misled by in-house performance testing if thermocouples are used which act as a heat sink, if the mounting surface is highly polished and finished, or if the devices are not mounted with a recommended force. The EIA (Electronic Industries Association) Bulletin No. 5 and Amendment 5-1 defines "Recommended Test Procedures for Semiconductor Thermal Dissipating Devices."

In natural convection, the total surface area contributes to heat transfer. However, this is not the case with radiation. Adjacent fins shield portions of the fin area from the surroundings. This shielding effect determines the value of F in Equation (15.8). Referring to Fig. 15.8, the total surface area (neglecting the flat mounting surface which conducts heat to the fins) is given by

$$A_s = (2\, dh N) + h(w - tN) + tN(h + 2d) \qquad \text{in.}^2 \qquad (15.9)$$

where N is the number of fins and d, h, w, and t are dimensions in inches as shown in Fig. 15.8.

A plot of F versus fin configuration is also shown in Fig. 15.8 [1]. As the number of fins increases, s decreases and F decreases. The values of F and A_s in Equation (15.8) are obtained from Fig. 15.8 and Equation (15.9), respectively. Other cited references utilize heat transfer coefficients, fin efficiency calculations, and nomographs to analyze heat transfer [3, 4].

15.3 FORCED AIR COOLING

In many cases, heat sinks cannot dissipate sufficient power by natural convection and radiation, especially in dense packages. Forced air cooling can provide an order of magnitude increase in heat transfer and an order of magnitude decrease in heat sink volume, as indicated in Fig. 15.5. The important considerations in forced convection cooling are power dissipation, temperature differential, air flow, and pressure drop. These factors also influence the type of blower, as well as acoustic noise, life, and cost of the blower.

15.3.1 Volumetric Air Flow

The forced convection equation for heat transfer at sea level and 69°F is

$$Q_{\text{fc}} = C_p W \Delta T \qquad \text{BTU/h} \qquad (15.10)$$

where C_p is the specific heat of air, 0.242 BTU/lb °F; W is the air mass flow rate at 0.075 lb/hr; and ΔT is the temperature rise in degrees Fahrenheit.

Applying the proper conversion factors, considering altitude, and rearranging terms:

$$f = (1.74 \times 29.9Q)/(p \Delta T) \qquad \text{CFM} \qquad (15.11)$$

where Q is heat flow in watts; ΔT is the air temperature rise in degrees Celsius, and p is the barometric pressure in inches of mercury.

Example

For a ducted cabinet where the power dissipation is 600 W and operating at a 30°C inlet air temperature and the outlet temperature must be limited to 50°C, a blower capable of delivering $(1.74 \times 600)/20 = 52$ CFM is required. At an altitude of 10,000 ft (reference Fig. 15.7), a blower capable of delivering $(52 \times 29.9)/20.6 = 76$ CFM would be required.

15.3.2 Pressure Drop

The amount of air flow which can be forced through an enclosure or equipment rack by a given fan or blower depends on the pressure drop of the entry and exit ports, EMI filters (mechanical as in "honeycomb" structures), ducting, finned heat sinks, and overall packaging density. Figure 15.9 shows typical empirical data for various enclosure densities. The following "rules" may be applied to scale up or down from that shown in Fig. 15.9:

1. For a given CFM, the pressure drop varies inversely as the square of the cross-sectional area perpendicular to the direction of air flow.
2. For a given CFM, the pressure drop varies directly as the distance of air flow through the enclosure.
3. For a given structure, the pressure drop varies directly as the square of CFM.

For minimum pressure drop at inlet or exit ports or at ducting ports within the enclosure, venturilike guides should be used to provide a gradual air entry, as opposed to an abrupt transition of air flow. Manufacturers of grilles, filters, guards, and heat sinks normally provide pressure drop versus air flow characteristics.

The configuration of finned heat sinks, as discussed in the following section, affects the pressure drop as well as the temperature rise. Referring to the heat sink diagram of Fig. 15.8, an approximate pressure drop is given by [1]

$$\Delta p = (f/N)^2 (1 + 0.01h/s) \times 10^{-3}/(sd)^2 \qquad (15.12)$$

where

$$\Delta p = \text{pressure, in. } H_2O$$
$$f = \text{total air flow, CFM}$$
$$N = \text{number of ducts}$$
$$s, d, h = \text{dimensions shown in Fig. 15.8, in.}$$

15.3.3 Heat Sinks

As previously discussed, limiting semiconductor junction temperature to improve reliability and performance is frequently the major concern in power conversion equipment design. Forced air cooling of semiconductor heat sinks provides an effective means of heat transfer (as does liquid cooling, discussed in Section 15.4). The amount of heat transfer depends on the dimensions of the heat sink and

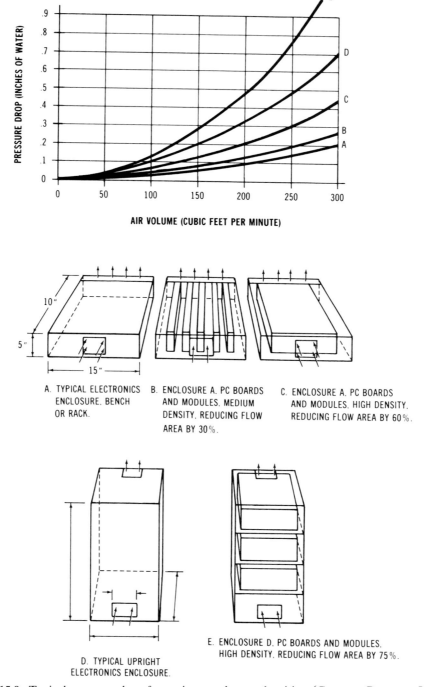

Fig. 15.9. Typical pressure drop for various enclosure densities. (Courtesy Pamotor, Inc.)

velocity of the cooling air. The velocity is given by

$$v = f/A_c \qquad \text{FPM (ft/min)} \qquad (15.13)$$

where f is the volumetric air flow in CFM and A_c is the total cross-sectional area perpendicular to air flow in square feet.

Thermal resistance characteristics for a common 4-$\times$ 4-in. finned heat sink are shown in Fig. 15.10. Without forced convection, a 5-in. length dissipating 80 W

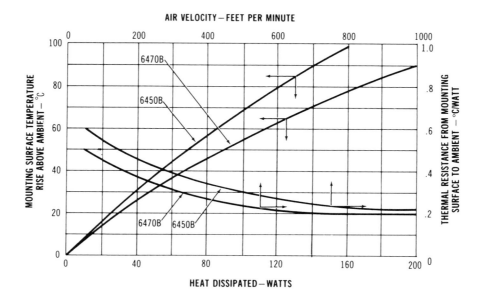

P/N	L, inches
6450 B	5.0
6470 B	7.50

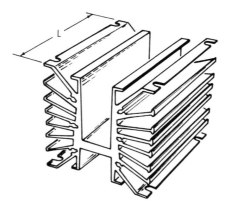

Fig. 15.10. Thermal characteristics of a 4 × 4-in. heat sink. (Courtesy Thermalloy, Inc.)

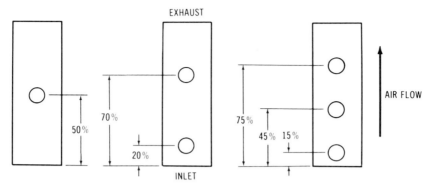

EXHAUST

70%

50%

20%

INLET

75%

45% 15%

AIR FLOW

Fig. 15.11. Optimum device mounting locations as a function of air flow direction. (From *Electronic Products Magazine*, October 1966.)

produces a temperature rise of 55°C. The thermal resistance is therefore 0.71°C/W. If a 4-× 4-in. fan delivering 75 CFM is placed in front of the heat sink, the air velocity is 675 FPM, the thermal resistance is reduced to 0.26°C/W, and the temperature rise is reduced to 21°C.

The temperature rise of a finned heat sink is also affected by the temperature rise of the air as it absorbs heat, and the temperature rise of the fins above the air. The former characteristic is important in determining the location of semiconductors on a heat sink. Figure 15.11 shows the optimum mounting location for single and multiple semiconductor devices which dissipate equal amounts of power [5]. As the air passing over the fins near the inlet is heated, the *downstream* air becomes warmer and the induced temperature rise or incremental thermal resistance of each *upstream* device must be added to the thermal resistance of the downstream devices.

The approximate temperature rise of the fins above the air is given by [1]

$$\Delta T = 140sP/(n^{0.2}d^{0.2}f^{0.8}h) \qquad °C \qquad (15.14)$$

where ΔT is the temperature rise of fins above air, P is power dissipation in watts, and s, n, d, f, and h are given by Equation (15.12) and shown in Fig. 15.8.

A detailed analysis of the discussion in this section, considering Colburn factors and Reynolds number for convection coefficients and Fanning factors, plus the Hagen–Poiseulle equation for friction factors, is presented by Steinberg [2].

15.3.4 Fans and Blowers

Though fans and blowers serve the same purpose, they differ, depending on the particular application. Fans are generally axial, low-pressure devices capable of moving from 20 CFM (typically a 3-in. blade) to 600 CFM (typically a 10-in. blade), at static pressure up to 0.5 in. of water. Air movement is parallel to the axis of the motor shaft. Blowers are generally high-pressure centrifugal or squirrel cage devices capable of moving from 50 to 800 CFM at static pressures in excess of 5 in. of water. Blowers may be single stage or dual stage.

The steps in selecting a fan or blower are:

1. Calculate the pressure drop from Equation (15.12) or from vendor data on items through which the air must travel.
2. Determine the required air flow in feet per minute or in CFM.
3. Select size, shape, and style.
4. Verify that the noise level is not objectionable.
5. Verify that the magnetic field radiation will not disturb adjacent devices, such as CRTs.

The air flow required is given by [6]

$$f = (178.4 \times T_{in}P)/(\Delta T p) \qquad \text{CFM} \tag{15.15}$$

where

T_{in} = inlet temperature, °R
P = power dissipated, kW
ΔT = temperature differential, °F
p = barometric pressure, in. Hg

For inlet air at 25°C and 29.9 in. Hg pressure, the above equation may be simplified to

$$f = (1740P)/\Delta T \qquad \text{CFM} \tag{15.16}$$

where P is power dissipated in kilowatts and ΔT is the temperature differential in degrees Celsius.

Fans are available in several types including tubeaxial, ventraxial, vaneaxial, and propeller. Perhaps the most popular fan is the tubeaxial 100 CFM unit known by such trade names as Boxer, Muffin, Cyclohm, and Pentaflow. Due to compact size ($4\frac{3}{4}$ in.^{2} by $1\frac{1}{2}$ in. thick), low cost, low noise, reliable operation, and low input power, the tubeaxial is widely used in semiconductor heat sink coolers, small heat exchangers, and high-current, low-voltage switch mode power supplies. A typical performance curve for this type fan is shown in Fig. 15.12. The fan operates from 115 V ac at 50 or 60 Hz.

Example

A heat sink to be cooled within a cabinet having a pressure drop of 0.2 in. H_2O, dissipates 200 W; the desired temperature rise is 10°C maximum. From Equation (15.16), $f = 1760 \times 0.2/10 = 35.2$ CFM. From Fig. 15.12, the tubeaxial fan would provide adequate air flow if operated from 60 Hz but would not meet the requirement if operated from 50 Hz since the fan speed would be reduced.

Note that the fan input power, Fig. 15.12, is 18 W. In general, the electrical input power to an ac fan or blower is typically 10% of the power dissipation (heat) to be removed from the equipment. A tubeaxial fan which has gained in popularity is the brushless dc motor type. Of course, dc motors are seldom used due to poor brush life, but the brushless type uses solid-state devices for commutation. The typical 120 CFM fan delivers 6.7 CFM/W, while a similar brushless dc unit, at equal air flow and static pressure, yields 30 CFM/W, quadrupling the efficiency as

MODEL NUMBER	SINTERED IRON SLEEVE	BALL BEARING	VOLTAGE (AC)	FREQUENCY (Hz)	INPUT POWER	AIR FLOW (CFM)	SPEED (rpm)	WEIGHT (OUNCES)	NOISE LEVEL dB(A)	NOISE LEVEL dB(SIL)
4600X	X		115	50/60	18 W	120	3100	20	50	40
4606X*		X	115	50/60	17.5 W	120	3170	20	51	41
4650X	X		230	50/60	17 W	120	3100	20	50	40
4656X*		X	230	50/60	17 W	120	3150	20	51	41

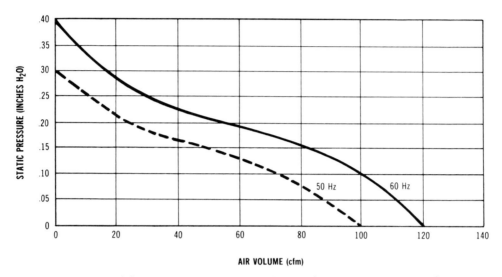

Fig. 15.12. Airflow versus pressure, tubeaxial fan. (Courtesy Pamotor, Inc.)

compared to the ac fan. In addition to better efficiency, EMI is reduced, operating noise is normally reduced, and the lower supply voltage (15–28 V) means the fan can readily meet safety codes such as UL, CSA, and VDE. In fact, the dc fan in a power supply can be powered by one of the outputs. The disadvantage is increased cost as compared to the ac fan.

For blowers, a virtual plethora of selections exists. A high-performance blower curve is shown in Fig. 15.13. The blower operates from 200 V ac, three-phase, 400 Hz input at 5400 RPM, full load power is 600 W, and weight is under 10 lb.

Example

An electronic cabinet to be cooled has the following characteristics:

1. 6 kW power dissipation.
2. Inlet air is 40°C.
3. Cabinet must operate at 10,000 ft altitude.
4. The required air flow at 10,000 ft is 12 lb/min (160 CFM) at a pressure drop of 1.2 in. H_2O.

In order to minimize heated air through the blower, the blower will push air through the cabinet as opposed to pulling air. The air density at 10,000 ft and 40°C

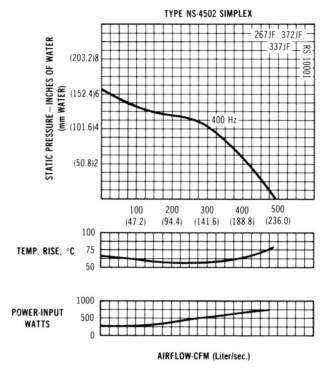

Fig. 15.13. Performance characteristics of a centrifugal blower. (Courtesy EG&G Rotron.)

is 0.048 lb/ft^3 and the required air flow is $f = $ (12 lb/min)/(0.048 lb/ft^3) = 250 CFM. The resulting static pressure at sea level is $p = (250/144)^2 \times 1.2 = 3.6$ in. H$_2$O. Thus the blower performance curve shown in Fig. 15.13 is above the desired operating point. The nominal temperature rise, from Equation (15.16), is $\Delta T = 1740 \times 6/250 = 42°C$, and the exit air temperature is $40 + 42 = 82°C$.

15.4   FORCED LIQUID COOLING

For high-power electronic equipment, forced liquid cooling is frequently employed to remove heat, especially in power semiconductors carrying very high current. In this case, the semiconductors are mounted to a cold plate or heat sink, aluminum or copper, through which the liquid is circulated to carry away the heat. Forced liquid cooling can offer an order of magnitude increase in heat transfer over forced air cooling, just as forced air cooling can provide an order of magnitude increase in heat transfer over natural convection and radiation. The advantages of forced liquid cooling are:

1. Savings in weight and volume.
2. Generated heat can be transferred outside of an equipment cabinet.
3. Acoustic noise and vibration problems are eliminated in the cabinet.
4. High-density packaging of heat-dissipating components can be achieved.

However, the heat generated in power devices or assemblies can only be transferred to another location.

Various configurations are available for liquid cooling, two of which are shown in Fig. 15.14. In Fig. 15.14A, the liquid is stored in a reservoir tank and the pump circulates the liquid through the electronic equipment. Heat is removed from the tank by convection and radiation which may necessitate using a large size tank. In

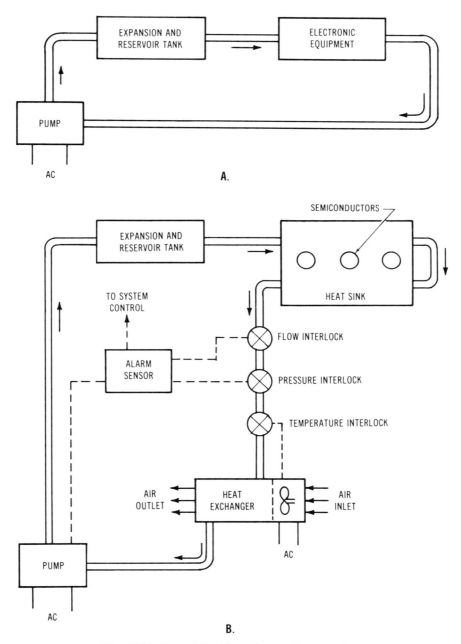

Fig. 15.14. Forced liquid cooling configurations.

Fig. 15.14B, the liquid is again stored in a reservoir tank and the pump circulates the liquid through a heat sink (with mounted semiconductors), through interlock sensors, and a heat exchanger. The liquid is cooled in the heat exchanger by forced convection from a fan or blower. The pressure and flow interlocks are safety devices which alarm the system of malfunctions, and power to the electronics may be turned off to prevent damage to either the pump or the semiconductors. The temperature interlock may also employ proportional control to maintain a desired coolant or semiconductor temperature by controlling the air flow rate through the heat exchanger, and to conserve energy. The heat exchanger may also employ an electrical heater to initially heat the coolant for start-up purposes in cold ambient temperatures.

The equations governing forced liquid cooling are similar to those of forced air cooling. The parameter most often to be evaluated is the liquid flow rate (gallons per minute in this case) required to remove a quantity of heat (power dissipation), while maintaining an acceptable temperature rise. The parameter is given by

$$F = 3.788P/(Cg\,\Delta T) \qquad \text{GPM} \tag{15.17}$$

where

P = power dissipated, kW
C = specific heat of coolant, BTU/lb-°F
g = specific gravity of coolant
ΔT = temperature rise, °C

If water is used as a liquid coolant, the above equation, with the same units, becomes

$$F = 3.79P/\Delta T \qquad \text{GPM} \tag{15.18}$$

Referring to Fig. 15.14B, the liquid path through the heat sink cools both sides of the heat sink such that the average of inlet and outlet temperatures to the "left-hand" device is the same as the average temperature to the "right hand" device. A similar technique is frequently employed for double-sided cooling of high-power press-pak devices, where the inlet liquid flow is in series through the cooling blocks on one side and returns in series through the cooling blocks on the other side of the devices. Manufacturers such as Westinghouse Electric, International Rectifier, and General Electric offer complete assemblies containing a rectifier or thyristor (SCR) in a press-pak case, cooling blocks and clamps, plus data sheets tabulating required water flow rates for average output current as a function of inlet temperature. In this case, the required or known output current relieves the designer from calculating the power dissipation in the devices and the thermal resistances of the heat sink.

Properties of various coolants are listed in Table 15.7. Characteristics of the coolants are:

1. Water is the best general coolant due to its high thermal conductivity and low viscosity. However, water has poor insulating properties and freezes at 0°C.

TABLE 15.7. Properties of Liquid Coolants

	Water		FC 75		Coolanol 45		Diala Oil AX
	25°C	60°C	25°C	60°C	25°C	60°C	25°C
Thermal conductivity, BTU/h ft °F	0.35	0.38	0.037	0.035	0.08	0.08	0.076
Specific heat, BTU/lb °F	1.0	1.0	0.25	0.27	0.45	0.49	0.44
Specific gravity	1.0	1.0	1.8	1.5	0.89	0.93	0.87
Density, lb/ft³	62.2	61.4	110	104	55.9	54.3	
Viscosity, lb/ft h	2.2	1.1	3.5	2.1	42	17	75

For long-term operation, distilled water which has been deionized should be used. In fact, an on-line deionizer is frequently installed which will raise the resulting resistivity of water to as high as 5 MΩ-cm. This allows water to be used as a coolant in systems requiring 100-kV isolation where, of course, the coolant pipe or hose material is an electrical insulator. Ethylene gycol (antifreeze) may be added to extend low-temperature operation to −40°C or lower, but the thermal conductivity will decrease. The ethylene gycol also prevents corrosion in aluminum and copper pipes.

2. FC-75 is a fluorochemical manufactured by the 3M Company. FC-75 has a viscosity slightly higher than water, but its thermal conductivity can be an order of magnitude less than water. However, the freezing point is −113°C and the usable temperature range is −65°C to 100°C.

3. Coolanol 45 is a silicone ester (hydraulic fluid) manufactured by Monsanto Chemical Company. Coolanol 45 has a usable temperature range of −65 to 175°C, but its thermal conductivity is approximately 20% that of water.

4. Diala Oil AX is manufactured by Shell Oil Company. The oil has wide usage in transformers, especially high-voltage transformers, due to its excellent insulating properties. However, the oil has a high viscosity, especially at low temperature, which increases the pressure on pumping mechanisms.

A presentation of effects due to pressure drop, laminar flow, and turbulent flow is beyond the scope of this book. These subjects are aptly treated in appropriate texts on fluids and by the cited references [1, 2].

15.5 HEAT PIPES

Heat pipes offer another method of transferring heat from one location to another location. Heat at one end of a sealed pipe causes liquid inside that end of the pipe to vaporize and travel to the cooler end where it condenses. The liquid returns to the hot end by being drawn by capillary action along a wick lining the pipe. The heat pipe functions by isothermal operation and with a very small thermal resistivity and thus small temperature differentials. In fact, the thermal conduction of heat pipes is several orders of magnitude greater than solid conductors, and the energy-to-weight ratio is also higher with the heat pipe.

The most effective usage of heat pipes is perhaps the transfer of heat from semiconductors located in a restricted space to external fins, which may be cooled by forced convection. The semiconductors are mounted on a heat sink which is an integral part of the heat pipe. The fins are also an integral part of the heat pipe.

15.6   THERMOELECTRIC COOLERS

Thermoelectric coolers are active semiconductor devices which act as small heat pumps. They can provide safe operating temperatures for discrete or integrated semiconductors, especially inside a restricted or sealed enclosure. Thermoelectric coolers require an external source of dc power with a voltage range of 0.3–8 V, and a current range of 1–15 A. If the voltage source is provided by a transformer/rectifier, filtering must be added to reduce the ripple voltage to less than 10%. The cooling process may be reversed to a heating process by reversing the input voltage polarity. Thus the thermoelectric cooler is useful in controlling or maintaining a particular device at a fixed temperature even if the ambient temperature is above or below the desired operating temperature.

However, thermoelectric coolers are very inefficient from an energy or power standpoint. Imagine using a device which requires an order of magnitude greater input power than the power or heat which it removes.

REFERENCES

1. A. W. Scott, *Cooling of Electronic Equipment* (Wiley, New York, 1974).
2. D. S. Steinbeg, *Cooling Techniques for Electronic Equipment* (Wiley, New York, 1980).
3. Joel Newberger, "Thermal System Approach to Heat Sink Selection," AHAM, Inc.
4. F. W. Gutzwiller, "Heat Sinks for Stud-mounted Semiconductors," General Electric Company.
5. W. E. Goldman, "Nine Ways to Improve Heat Sink Performance," *Electronic Products Magazine* (October 1966).
6. *Rotron Custom Airmovers*, 4th ed. (EG & G) Rotron.

16

RELIABILITY AND QUALITY

16.0   Introduction
16.1   Terminology
    16.1.1   Quality Assurance
    16.1.2   Reliability
16.2   Specifications and Documents
16.3   Reliability Equations
16.4   Mean Time between Failure
    16.4.1   Part Failure Rate Tabulation
    16.4.2   Failure Rate Analysis
16.5   Confidence Levels
16.6   Reliability Modeling of Redundant Units
    16.6.1   Two Redundant Units
    16.6.2   Three Redundant Units
16.7   MTBF Example
    References

16.0   INTRODUCTION

Reliability is defined by dictionaries as "worthy of confidence; trustworthy." *Quality* is defined as "excellent grade." High reliability means long-life, failure-free operation. Reliability must be an inherent characteristic. It is a function of the manufacturing process, the design (electrical and mechanical), the quality of materials used, and the application of an item in end use. In more realistic terms, reliability may be defined as the probability of performing a specific function under specific conditions for a specific time. Thus, reliability is a measure of performance over a period of time, as opposed to quality, which is a measure of conformance to specific standards at a given point in time.

The complexity of equipment and the development of new components have forced government and industry to invest considerable effort in establishing quality levels and in predicting reliability. Thus, quality and reliability are important considerations at all levels of electronics (plus other disciplines), from materials and workmanship to operating systems.

A typical failure-rate-versus-time plot for a component or a product has been the well-known "bathtub" curve consisting of three regions: infant mortality, useful-life (the flat portion of the curve), and end-of-life or wear-out. During the useful life, even high-quality materials and devices can fail from various environmental and electrical stresses, especially high temperature. However, there is no advantage in simply knowing that a failure might occur. The advantage comes in having the opportunity to do something to prevent the failure, by predicting results and correcting the problem. The limitation on reliability predictions is the ability to accumulate data of known validity and apply this information properly.

Reliability is not improved by testing or by stressing. Instead, the reliability testing is intended to provide data on the operating life of a part by stressing or aging the part and predicting the result. Statistical confidence levels may then be established to predict a true value of reliability. The government has an "established reliability" classification for various components. This classification is achieved from actual data of actual use.

The following guidelines are recommended for designing high-reliability power conversion equipment and other electronic equipment:

1. Use verified circuit designs.
2. Use high-quality parts and materials.
3. Use conservative stress derating practices.
4. Use low junction and hot-spot temperatures.
5. Use redundant circuitry and modules where required.

Quality, on the other hand, is a frame of mind. A quality improvement barrier is that this country has been conditioned to think that errors or defects are inevitable. At academic levels, a score of 95% is an "A" grade. However, 95% right is also 5% wrong. In industry, this figure results in 50,000 parts per million defects! Fortunately, improved quality in industry has significantly reduced this number. Quality is an inside job and cannot be delegated. Quality improvement requires a change in the culture of the company, from the chief executive officer to the soldering iron. Effective approaches to quality programs have been explained by various experts, notably Crosby and Deming [1, 2].

The International Standardization Organization (ISO) has published five standards (ISO 9000-9004) on quality systems. The ISO-9000 document has been adopted by the American National Standards Institute (ANSI) and the American Society for Quality Control (ASQC) as Q-9000. Where traditional quality control systems were focused on inspection of the final product, ISO-9000 shifts attention to the manufacturing process and building in quality. Quality system approval for a supplier or manufacturer is achieved through a "certificate of registration."

16.1   TERMINOLOGY

A great deal of reliability and quality is expressed in statistical terms and acronyms. Common plans, definitions, and requirements are listed below.

16.1.1 Quality Assurance

AQL (Acceptable Quality Level). A sampling plan used to determine statistically the maximum percentage of defective devices in a given lot at a specified assurance level. Sample size is proportional to lot size. The AQL is used primarily as an in-process control. This approach essentially specifies the manufacturers risk.

LTPD (Lot Tolerance Percent Defective). Similar to the AQL except sample size and the number of allowable rejects are independent of lot size. This approach specifies the user's risk.

ZD (Zero Defect). The ZD planning program establishes a performance standard and is highly motivational. The idea is to shock people into a new way of thinking to support a new standard of excellence. The type of quality performance projected by ZD makes the AQL and the level of LTPD passé. The semiconductor and automotive industries (domestic) have been instrumental in lowering defects from 5,000 ppm to 50 ppm in the 1980s. The electronic industry's new "Six Sigma" quality standard permits only 3.4 defects in 1 million parts. Other well-known quality programs include "First Time Right" and "Quality Is Job 1."

16.1.2 Reliability

Failure Rate, Lambda (λ). The failure rate is the number of failures experienced (or predicted to be experienced) in 10^6 h combined operation. A failure rate of 2.3 means (to reliability engineers) that, on the average, 2.3 failures will occur when 10^6 h operation have been attained. Unfortunately, vendors also use a failure rate based on percent failures per 1000 h. A value of 0.23% per 1000 h is equal to a lambda of $2.3/10^6$ h.

Mean Time between Failure, MTBF (θ). Another word for *mean* is *average*. The MTBF is the reciprocal of λ. A unit with a predicted failure rate of 50 has an MTBF of 10^6 per 50, or 20,000, h.

PDA (Percentage of Defective Devices). The percentage of defective devices that may not be exceeded during the performance of a specified test.

JAN (Joint Army–Navy). Government prefix assigned to JEDEC (Joint Electron Device Engineering Council) designated semiconductors to denote devices that have successfully passed specific qualification tests. The codes JANTX (class B) plus JANTXV and JANS (class S) denote even higher reliability devices, in that order.

Class B denotes parts that are screened to specified levels to accomplish a maximum failure rate of 0.08% per 1000 h, or 0.8 per 10^6 h.

Class S denotes parts that are subjected to additional screening to accomplish a maximum failure rate of 0.004% per 1000 h. This screening is intended to remove potential failures from space-borne equipment.

QPL (Qualified Parts List). Parts manufactured by suppliers that meet specified test and screening requirements.

16.2 SPECIFICATIONS AND DOCUMENTS

Government specifications governing discrete semiconductors, integrated circuits, and passive devices are respectively:

MIL-S-19500: Semiconductor Devices, General Specification For. Establishes the procedures that a manufacturer must follow for screening and testing. Detail requirements and characteristics are specified in detail specifications (slash numbers).

MIL-M-38510: Microcircuits, General Specification For. Establishes the procedures that a manufacturer must follow to have devices listed on the QPL. Testing to these standardized requirements, per appropriate screening level, assures the user of long-term reliability and product homogeneity.

MIL-STD-750. The detailed "how to" test methods for MIL-S-19500.

MIL-STD-883. The detailed "how to" test methods for MIL-M-38510. The standardized environmental, visual, mechanical, and electrical test methods that enable manufacturers and users to screen for a specific reliability.

MIL-X-390xx. The 39000 series of specifications govern passive devices (R for resistors and C for capacitors) that have established failure rate levels. Failure rates range from S level (0.01% per 1000 h) to M level (1% per 1000 h).

NAVMAT P4855-1A. Navy Power Supply Reliability. A "Design and Manufacturing Guidelines" document that defines (1) guidelines for program managers and (2) design and manufacturing fundamentals for power supply engineers that will result in power supplies that meet or exceed reliability specifications.

16.3 RELIABILITY EQUATIONS

The probability that a device will survive failure free for a specified number of hours is

$$R(t) = \varepsilon^{-\lambda t} = \varepsilon^{-t/(\text{MTBF})} = \varepsilon^{-t/\theta} \tag{16.1}$$

which is the exponential law of distribution. This equation shows that if a device is operated for a time equal to its MTBF, the probability of survival will be 0.37, or 37%. When operated for a time equal to 10% of its MTBF, the probability of survival is 0.90, or 90%. This equation applies only when the failure rate is constant.

The Arrhenius equation, reference Equation (15.1), shows failure rate can be expressed by

$$\lambda = \varepsilon^{A} + B/T \qquad \%/1000 \text{ h} \tag{16.2}$$

where T is temperature in degrees Kelvin (junction or hot spot) and A and B are empirical constants.

16.4   MEAN TIME BETWEEN FAILURE

Mean time between failure (MTBF) is a statistical prediction based on the total number of failures experienced by a particular group of identical products based on the total time of service. This product group may be a single component or a complete system. The MTBF is not a calculation of how long an item will operate before it fails. The MTBF prediction does not consider infant mortality failures, which are minimized by proper testing and burn-in by the manufacturer before the item is shipped. Also, the MTBF prediction does not consider end-of-life failures since product quality and stated operating life are manufacturer quality considerations.

The MTBF is given by

$$\text{MTBF} = 1/\lambda_p \qquad (16.3)$$

where λ_p is expressed in failures per million hours.

The MTBF prediction is based on failure rate data from MIL-HDBK-217E, Notice 1, which has two methods of calculation. The prediction may be based on a parts count (quick method) or may be based on a part stress analysis. The first method is frequently used to predict MTBF before a particular design is complete, as in a proposal effort. The latter method is more accurate but requires a great amount of detailed circuit analysis and calculations from the tabulated data of 217E. In this chapter, the stress analysis method is presented with tabulations of calculated failure rates of typical components used in power conversion equipment. The part failure rates for various operating temperatures are based on the established component reliability and quality and with stress ratios considered "good design practice" for a particular market. The product itself varies from the competitive commercial power supply market to ultrareliable converters for space applications.

The basic equation for part failure rate is

$$\lambda_p = \lambda_b(\pi_x) \qquad \text{failures per million hours} \qquad (16.4)$$

where λ_b is the base failure rate, selected from the appropriate table and finding the row (temperature) and the column (stress), and π_x is the product or sum of other factors.

Definitions of these factors plus values and formulas are given in Table 16.1. The environmental stress factor depends on one of the particular 20 environments listed by 217E and on the particular component type.

16.4.1   Part Failure Rate Tabulation

Typical part failure rates for common electronic components used in power conversion equipment are listed in Tables 16.2–16.4. The expression for λ_p is shown for each component group. The environment is G_B (ground, benign). Three separate ambient temperatures (junction temperature for integrated circuits and hot-spot temperature for magnetic devices) are listed.

TABLE 16.1. Summary of MIL-HDBK-217E, Notice 1 Parameters

Base Factors

λ_p Failure rate factor
λ_b Base failure rate
π_E Environmental factor
G_B Ground, benign; nonmobile, laboratory environment; includes instruments and equipment, computer complexes
G_F Ground, fixed; installation in racks, adequate cooling, ground support, and control equipment
G_M Ground, mobile; equipment installed in wheeled or tracked vehicles
S_F Space, flight; approaches ground, benign conditions; vehicle not under powered flight nor in reentry
N_S Naval, sheltered; below deck and protected equipment
A_{IC} Airborne, inhabited cargo; typical conditions as occupied by aircrew, without environment extremes
A_{RW} Airborne, rotary winged; equipment installed on helicopters
π_Q Quality factor
π_A Application
π_R Resistance factor
π_{CV} Capacitance factor
π_{SR} Series resistance (applies to CSR-style solid tantalum electrolytic)
π_C Construction factor (applies to CL- and CLR-style tantalum electrolytics)
π_T Temperature factor
π_u Utilization factor (applies to circuit breakers and some magnetics)

Semiconductor Factors

π_A Analog signal factor
π_R Rating or resistance
π_r Power or current rating factor (transistor or thyristor)
π_C Complexity
π_s Voltage stress factor
π_V Voltage stress derating factor
π_L Device learning factor
C_1 Circuit complexity factor (monolithic transistor count)
C_2 Package complexity failure rate

Component	π_E Factors				
	G_B	S_F	G_F	N_S	A_{IC}
IC, monolithic	0.38	0.45	2.5	4.0	3.0
Diodes, transistors	1.0	0.50	5.5	9.5	13.0
Opto-isolators	1.0	0.50	2.4	5.7	3.8
Resistors					
RCR	1.0	0.50	2.9	5.2	3.5
RLR, RNC	1.0	0.20	2.4	4.7	3.0
RWR	1.0	0.30	1.5	4.9	3.5
Capacitors					
Film	1.0	0.50	2.4	5.7	2.5
Ceramic	1.0	0.40	1.6	5.5	3.0
Ta, CSR	1.0	0.40	2.4	4.9	2.5
Ta, CLR	1.0	0.50	1.4	6.7	4.0
Aluminum	1.0	0.50	2.4	6.7	10.0
Inductive devices	1.0	0.50	5.7	5.7	6.0

TABLE 16.2. Typical Part Failure Rates for Semiconductors, MIL-HDBK-217E, Notice 1, Environment G_B

T_j (°C)	MIL, Class B		Hermetic Seal		Plastic	
	Digital	Linear	Digital	Linear	Digital	Linear

Monolithic ICs; Bipolar, MOS, Linear: $\lambda_p = \pi_Q(C_1\pi_T\pi_V + C_2\pi_E)\pi_L$

T_j (°C)	Digital	Linear	Digital	Linear	Digital	Linear
25	0.0029	0.0029	0.029	0.029	0.058	0.058
65	0.015	0.022	0.149	0.219	0.838	1.30
105	0.094	0.212	0.939	2.12		

T_j (°C)	MIL, JTX			Hermetic Seal			Plastic		
	Switch	Power	Fast Recovery	Switch	Power	Fast Recovery	Switch	Power	Fast Recovery

Diodes, Rectifiers[a]: $\lambda_p = \lambda_b\pi_s\pi_Q\pi_E\pi_T$

T_j (°C)	Switch	Power	Fast Recovery	Switch	Power	Fast Recovery	Switch	Power	Fast Recovery
25	0.0001	0.002	0.04	0.006	0.009	0.22	0.008	0.014	0.32
55	0.0002	0.003	0.07	0.011	0.017	0.40	0.014	0.025	0.58
85	0.0003	0.005	0.12	0.018	0.029	0.66	0.024	0.042	0.96

T_j (°C)	MIL, JTX		Hermetic Seal		Plastic	
	Signal	Power	Signal	Power	Signal	Power

Transistors, Bipolar[b]: $\lambda_p = \lambda_b\pi_A\pi_r\pi_s\pi_Q\pi_E\pi_T$

T_j (°C)	Signal	Power	Signal	Power	Signal	Power
25	0.00004	0.0015	0.0002	0.0085	0.0004	0.012
55	0.00008	0.0029	0.0004	0.016	0.0007	0.023
85	0.00014	0.0051	0.0008	0.028	0.0012	0.040

Transistors, MOSFET[b]: $\lambda_p = \lambda_b\pi_A\pi_E\pi_T$

T_j (°C)		Power		Power		Power
25		0.067		0.37		0.54
55		0.128		0.70		1.02
85		0.222		1.22		1.74

[a]Voltage stress: 35% for switching; 75% for power.
[b]Voltage stress: 30% for signal; 75% for power.

The part quality levels are as follows:

1. Integrated circuits
   a. Class B per MIL-M-38510
   b. Hermetic sealed with no screening beyond the manufacturers regular quality assurance practice
   c. Commercial parts encapsulated with organic materials (plastic)
2. Discrete semiconductors
   a. JANTX per MIL-S-19500
   b. Hermetic sealed (as in 1b)
   c. Plastic (as in 1c)
3. Resistors and capacitors
   a. Established reliability (designated failure rate per 1000 h) per MIL-X-390xx
   b. Non–established reliability (non-ER) for MIL-designated parts
   c. Lower quality (typical off-the-shelf commercial components)

TABLE 16.3. Typical Part Failure Rate for Resistors and Inductive Devices, MIL-HDBK-217E, Notice 1, Environment G_B

T_a (°C)	Composition, 20% Rated Power		Film, 20% Rated Power		Wire Wound, 40% Rated Power		
	R Level	Non-ER	R Level	Non-ER	M Level	Non-ER	Lower

Resistors, Fixed: $\lambda_p = \lambda_b \pi_E \pi_R \pi_Q$

T_a (°C)	R Level	Non-ER	R Level	Non-ER	M Level	Non-ER	Lower
25	0.00002	0.001	0.00008	0.0039	0.0097	0.048	0.15
55	0.00008	0.004	0.0001	0.0049	0.013	0.065	0.20
85	0.00025	0.0125	0.00014	0.0068	0.018	0.090	0.27

T_a (°C)	Cermet, $< 200k$, 20% Rated Power			Wire Wound, $< 2k$, 20% Rated Power		
	R Level	Non-ER	Lower	R Level	Non-ER	Lower

Resistors, Variable: $\lambda_p = \lambda_b \pi_V \pi_E \pi_R \pi_Q$

T_a (°C)	R Level	Non-ER	Lower	R Level	Non-ER	Lower
25	0.0014	0.070	0.024	0.0007	0.034	0.115
55	0.0016	0.081	0.270	0.0009	0.045	0.150
85	0.0020	0.105	0.350	0.0013	0.066	0.220

T_{HS} (°C)	105°C Rated				130°C Rated				155°C Rated			
	Power		Pulse		Power		Pulse		Power		Pulse	
	MIL	Lower	MIL	Lower	MIL	Lower	MIL	Lower	MIL	Lower	MIL	Lower

Inductive Devices: $\lambda_p = \lambda_b \pi_E \pi_Q$

T_{HS} (°C)	MIL	Lower	MIL	Lower	MIL	Lower	MIL	Lower	MIL	Lower	MIL	Lower
45	0.02	0.08	0.004	0.012	0.02	0.08	0.004	0.012	0.018	0.066	0.003	0.011
95	0.10	0.39	0.019	0.064	0.043	0.16	0.008	0.027	0.025	0.093	0.005	0.016
145									0.076	0.285	0.014	0.048

4. Inductive devices (transformers and inductors)
  a. Mil-spec (typically MIL-T-27 or MIL-T-21038)
  b. Lower quality (typical commercial processes)

Stress factors vary, depending on the part. The stress ratings used are a function of power, voltage, and current. Signal transistors, low-power resistors, and small capacitors are typically used in control circuits, and the actual stress is a small percentage of the rated parameter. For instance, 2200-Ω, $\frac{1}{2}$-W resistors with 12 V applied would dissipate 66 mW, or 13% of rating.

16.4.2 Failure Rate Analysis

From the tabulations or from MIL-HDBK-217E, Notice 1, part types may be selected for a particular design based on the lowest failure rate consistent with specified reliability, product market (commercial or military), and cost. If parts for a design have been selected, then the tables provide a quick evaluation of MTBF. An example of a MTBF calculation for a power supply is given at the end of this chapter.

TABLE 16.4. Typical Part Failure Rate for Capacitors, MIL-HDBK 217E, Notice 1, Environment G_B

T_a (°C)	85°C Rated			125°C Rated		
	M Level	Non-ER	Lower	M Level	Non-ER	Lower

Ceramic: 40% Rated Voltage, C = 0.05 μF^a

T_a (°C)	M Level	Non-ER	Lower	M Level	Non-ER	Lower
25	0.0030	0.009	0.030	0.0027	0.0082	0.027
55	0.0034	0.010	0.034	0.0030	0.0090	0.030
85				0.0032	0.0098	0.032

Film (Polypropylene): 60% Rated Voltage, C = 5 μF

T_a (°C)	M Level	Non-ER	Lower	M Level	Non-ER	Lower
25	0.0122	—	0.122	0.0113	—	0.113
55	0.0195	—	0.195	0.0120	—	0.120
85	—	—	—	0.0169	—	0.169

Tantalum (Solid): 60% Rated Voltage, C = 10 μF, R_S = 0

T_a (°C)	M Level	Non-ER	Lower	M Level	Non-ER	Lower
25				0.0086	—	0.086
55				0.011	—	0.11
85				0.02	—	0.20

Tantalum (Foil, Non-Hermetic): 60% Rated Voltage, C = 20 μF

T_a (°C)	M Level	Non-ER	Lower	M Level	Non-ER	Lower
25	0.03	—	0.30	0.022	—	0.22
55	0.06	—	0.60	0.028	—	0.28
85	—	—	—	0.050	—	0.50

Aluminum Oxide Electrolytic: 60% Rated Voltage, C = 1700 μF

T_a (°C)	M Level	Non-ER	Lower	M Level	Non-ER	Lower
25	0.07	0.21	0.70	0.030	0.090	0.30
55	0.26	0.78	2.60	0.064	0.190	0.64
85	—	—	—	0.182	0.550	1.82

^{a}For $C = 1 \mu F$, multiply by 1.4.

Other observations may be made from the tables and from 217E:

1. Carbon composition resistors have a lower failure rate than metal-film resistors when operating at ambient temperatures below 70°C and when operating at the same power derating.

2. A silicon transistor operating at 80% of rated power at an ambient temperature of 30°C has the same base failure rate as the same transistor with the same thermal impedance path operating at 40% of rated power but at an ambient temperature of 90°C.

3. The lower predicted failure rate for thyristors, as compared to the high failure rates listed in earlier handbooks, is a reasonable and significant improvement. Manufacturers have made considerable progress in understanding failure mechanisms and providing operating data for proper application of these devices. In fact, high-power SCRs are used in inherently

reliable equipment such as HVDC (high-voltage direct current) conversion, uninterruptible power systems, and large motor controls and propulsion.

4. Inductive devices rated for 130°C service have a lower failure rate than devices rated for 155°C service if the actual hot-spot temperature is below 100°C.

5. A filter inductor, with a magnetic core and a dozen turns of magnet wire, has a higher failure rate than a zener diode operating at 50% of rated power for the same operating temperature.

6. The high failure rate of aluminum electrolytic capacitors may stem from the cleaning solvents problem discussed in Section 2.3.1.1.

An additional general observation can be made about 217E. The handbook only considers static conditions and does not address effects of switching, which is an inherent operation in most types of power conversion topologies. If current-limiting circuits, soft-start circuits, snubbers, over/under voltage protection, and fail-safe protection (inherent in most worthwhile designs) are added to obtain a high actual operating reliability, the MTBF prediction, based on 217E, may be lowered significantly, due to the added component count.

16.5 CONFIDENCE LEVELS

The previous discussion on MTBF related to a "point estimate" that has a 50% confidence associated with it. A confidence interval is bounded by upper and lower limits, and the broader the limits, the higher the confidence of particular occurrences. The upper and lower limits then define a two-sided confidence interval. In most electronics equipment, a one-sided confidence level is most appropriate for reliability testing. For this case, a test is desired that will demonstrate with a specific confidence that a unit MTBF is greater than a minimum limit. Frequently, it is desired that the unit operate for a specific time before failure, and the MTBF prediction must be determined for a certain confidence level. The following formulas and the data of Table 16.5 have been developed by Ireson [3]:

$$\text{UL} = K_U m \quad \text{(upper limit)} \tag{16.5}$$

$$\text{LL} = K_L m \quad \text{(lower limit)} \tag{16.6}$$

$$K_L = 2f/\chi^2_{2f:(1-P)/2} \tag{16.7}$$

where K_L is taken from Table 16.5 and

$m = t/f =$ MTBF test
$f =$ number of failures
$t =$ time of test, h
$\chi^2 =$ chi-square distribution
$P =$ confidence level

TABLE 16.5. Multiplier K for MTBF Estimate[a]

Number of Failures f	Upper Limit K_U					Lower Limit K_L				
	95%	90%	80%	70%	60%	60%	70%	80%	90%	95%
1	28.6	19.2	9.44	6.50	4.48	0.620	0.530	0.434	0.333	0.270
2	9.2	5.62	3.76	3.00	2.43	0.667	0.600	0.515	0.422	0.360
3	4.8	3.68	2.72	2.25	1.95	0.698	0.630	0.565	0.476	0.420
4	3.7	2.92	2.29	1.96	1.74	0.724	0.662	0.598	0.515	0.455
5	3.0	2.54	2.06	1.80	1.62	0.746	0.680	0.625	0.546	0.480
6	2.73	2.30	1.90	1.70	1.54	0.760	0.700	0.645	0.568	0.515
7	2.50	2.13	1.80	1.63	1.48	0.768	0.720	0.667	0.592	0.535
8	2.32	2.01	1.71	1.57	1.43	0.780	0.730	0.680	0.610	0.555
9	2.19	1.92	1.66	1.52	1.40	0.790	0.740	0.690	0.625	0.575
10	2.09	1.84	1.61	1.48	1.37	0.800	0.752	0.704	0.637	0.585
11	2.00	1.78	1.56	1.45	1.35	0.805	0.762	0.714	0.650	0.598
12	1.93	1.73	1.53	1.42	1.33	0.815	0.770	0.720	0.660	0.610
13	1.88	1.69	1.50	1.40	1.31	0.820	0.780	0.730	0.652	0.620
14	1.82	1.65	1.48	1.38	1.30	0.824	0.785	0.736	0.675	0.630
15	1.79	1.62	1.46	1.36	1.28	0.826	0.790	0.746	0.685	0.640
16	1.75	1.59	1.44	1.35	1.27	0.830	0.795	0.750	0.690	0.645
17	1.71	1.57	1.42	1.33	1.26	0.835	0.800	0.760	0.700	0.655
18	1.69	1.54	1.40	1.32	1.25	0.840	0.805	0.765	0.710	0.660
19	1.66	1.52	1.39	1.31	1.24	0.845	0.808	0.767	0.715	0.665
20	1.64	1.51	1.38	1.30	1.23	0.847	0.810	0.768	0.719	0.675
25	1.55	1.44	1.33	1.26	1.21	0.860	0.830	0.790	0.740	0.700
30	1.48	1.39	1.29	1.23	1.18	0.870	0.840	0.806	0.756	0.720
40	1.40	1.32	1.24	1.19	1.16	0.884	0.860	0.826	0.787	0.750
50	1.35	1.28	1.21	1.17	1.14	0.892	0.872	0.847	0.806	0.770
70	1.28	1.23	1.18	1.14	1.11	0.910	0.890	0.860	0.830	0.800
100	1.23	1.19	1.14	1.12	1.09	0.924	0.906	0.880	0.852	0.830
200	1.16	1.13	1.10	1.08	1.06	0.940	0.935	0.916	0.890	0.870
300	1.12	1.10	1.08	1.06	1.05	0.955	0.942	0.930	0.910	0.895
500	1.09	1.08	1.06	1.05	1.04	0.965	0.954	0.942	0.930	0.915

[a]From ref. 3.

For the one-sided or single-tailed interval, the required MTBF prediction for a unit to operate for a desired time with a given confidence level and before the first failure is

$$\text{m} = mK_L \qquad (16.8)$$

where m is the observed MTBF.

Rearranging terms, the required or observed test time before the first failure to demonstrate with a given confidence level that a unit will meet a desired MTBF is

$$m = \text{m}/K_L \qquad (16.9)$$

Examples

Case 1. If a unit is to operate for 3 years before the first failure, what is the required MTBF prediction for an 80% confidence level? First, 3 years = 26,300 h. From Table 16.5, $K_L = 0.434$ for 80% and $f = 1$. Then, from Equation (16.8), MTBF = m = mK_L = 26,300 × 0.434 = 11,400 h.

Case 2. It is desired to demonstrate with 90% confidence that the actual MTBF of a unit is equal to or greater than 2500 h. What must be the observed MTBF if the test is terminated at the first failure? From Table 16.5, $K_L = 0.333$ for 90% confidence and $f = 1$. Then, from Equation (16.9), $m = 2500/0.333 = 7500$ h = 10 months minimum.

16.6 RELIABILITY MODELING OF REDUNDANT UNITS

Power conversion equipment of identical (or different) characteristics is seldom operated in series due to the nature of the input and output voltages. However, identical units may be operated in parallel to provide redundancy, which significantly improves reliability. The basic equation for the MTBF of a unit is the reciprocal of the failure rate. System reliability, as given by Equation (16.1), is the probability that a given system (unit, component, device) will operate without failure (probability of success) until a given time and under specified environmental conditions. These parameters are now analyzed for paralleled units from the methods presented by Arsenault and Roberts [4].

16.6.1 Two Redundant Units

Two identical units may be connected in parallel, each capable of supplying the total required output. If one unit fails and can be removed from the system (does not interfere with the operation of the other unit), then the other unit supplies the required output. Using a constant failure rate, the following relations apply:

$$\text{MTBF} = 1/\lambda + 1/\lambda - 1/(\lambda + \lambda)$$

$$= 3/2\lambda \tag{16.10}$$

$$R(t) = 1 - (1 - \varepsilon^{-\lambda t})^2$$

$$= 2\varepsilon^{-\lambda t} - \varepsilon^{-2\lambda t} \tag{16.11}$$

where λ is the failure rate per million hours and t is the operating time in hours.

Example

Two inverter stages in an uninterruptible power system are to be connected in parallel. Each inverter has full-load capability. The calculated failure rate of each stage is $\lambda = 200/10^6$ h. The MTBF is $10^6/200 = 5000$ h. From Equation (16.1), the probability that each inverter will operate failure free for 1000 h is $R(t) = \varepsilon^{-1000/5000} = 0.8178 = 82\%$. When both inverters are paralleled, the system MTBF is $5000 \times \frac{3}{2} = 7500$ h, and the probability that the system will operate failure free for 1000 h is $R(t) = 2\varepsilon^{-200 \times 10^{-6} \times 1000} - \varepsilon^{-2 \times 200 \times 10^{-6} \times 1000} = 96.7\%$.

16.6.2 Three Redundant Units

Three redundant units may be paralleled with each unit capable of supplying total output for the conditions of the above section. Using a constant failure rate, the following relations apply:

$$\text{MTBF} = \frac{1}{\lambda} + \frac{1}{\lambda} + \frac{1}{\lambda} - \left[\frac{1}{\lambda + \lambda} + \frac{1}{\lambda + \lambda} + \frac{1}{\lambda + \lambda} \right] + \frac{1}{\lambda + \lambda + \lambda}$$
$$= 11/6\lambda \tag{16.12}$$

$$R(t) = 1 - \left[1 - \varepsilon^{-\lambda t} \right]^3$$
$$= 3\varepsilon^{-\lambda t} - 3\varepsilon^{-2\lambda t} + \varepsilon^{-3\lambda t} \tag{16.13}$$

Example

For a critical mission, three power supplies, each capable of supplying the total required output, are to be paralleled. The power supplies are also decoupled such that a failure of any power supply will not affect the output. The calculated failure rate of each power supply is $\lambda = 4/10^6$ h. The MTBF is 250,000 h. The probability that each power supply will operate failure free for 5 years (43,800 h) is $R(t) = \varepsilon^{-43,800/250,000} = 83.9\%$. When the three power supplies are paralleled, the system MTBF is $250,000 \times \frac{11}{6} = 458,333$ h. The probability that the system will operate failure free for 5 years is $R(t) = 3\varepsilon^{-43,800/250,000} - 3\varepsilon^{-2 \times 43,800/250,000} + \varepsilon^{-3 \times 43,800/250,000} = 99.6\%$. It also should be obvious that the equations presented in the previous section and this section can be manipulated to determine the required MTBF of individual units to meet a desired system MTBF and system probability when the units are paralleled.

16.7 MTBF EXAMPLE

Calculate the MTBF of a "typically commercial" 300-W, multioutput switch mode power supply with EMI filter and appropriate safety features operating at 50°C ambient temperature. Allow a 5°C internal rise above ambient. Use the failure rate values from Tables 16.2–16.4. The supply consists of the following components:

2 each ICs, plastic linear ($\lambda = 5.2 \times 0.7 = 3.64$)
1 each opto-isolator ($\lambda = 1.65 \times 0.8 = 1.32$)
2 each hermetic sealed power switch transistors ($\lambda = 0.033$)
2 each plastic power switch transistors ($\lambda = 0.026$)
4 each plastic signal transistors ($\lambda = 0.0052$)
2 each hermetic sealed power diodes ($\lambda = 0.064$)
8 each plastic power diodes ($\lambda = 0.019$)
6 each hermetic sealed switch diodes ($\lambda = 0.0024$)

32 each composition resistors, non-ER ($\lambda = 0.0032$)
3 each potentiometers, commercial ($\lambda = 0.3$)
8 each pulse type magnetics, 130°C rated, 115°C hot spot, non-MIL ($\lambda = 0.44$)
12 each ceramic capacitors, commercial ($\lambda = 0.042$)
3 each film capacitors, commercial ($\lambda = 0.2$)
9 each Al electrolytics, commercial ($\lambda = 0.48$)

Semiconductors:

$$\begin{aligned}
\lambda &= (2 \times 3.64) + 1.32 + (2 \times 0.033) + (2 \times 0.026) + \\
&\quad (4 \times 0.0052) + (2 \times 0.064) + (8 \times 0.019) + (6 \times 0.0024) \\
&= 9.0332
\end{aligned}$$

Resistors:

$$\begin{aligned}
\lambda &= (32 \times 0.0032) + (3 \times 0.3) \\
&= 1.0024
\end{aligned}$$

Magnetics:

$$\begin{aligned}
\lambda &= (8 \times 0.044) \\
&= 0.352
\end{aligned}$$

Capacitors:

$$\begin{aligned}
\lambda &= (12 \times 0.042) + (3 \times 0.2) + (9 \times 0.48) \\
&= 5.424
\end{aligned}$$

Adding 10% for electromechanical components,

$$\lambda_{sum} = (9.0332 + 1.0024 + 0.352 + 5.424) \times 1.1 = 17.4$$
$$\text{MTBF} = 10^6/17.4 = 57{,}500 \text{ h}$$

REFERENCES

1. P. Crosby, *Quality is Free* (McGraw-Hill, New York, 1979).
2. W. Edwards Deming, *Out of the Crisis* (M.I.T., Center for Advanced Engineering, Cambridge, MA, 1986).
3. W. Grant Ireson, *Reliability Handbook* (McGraw-Hill, New York, 1966).
4. J. E. Arsenault and J. A. Roberts, *Reliability and Maintainability of Electronic Systems* (Computer Science Press, Rockville, MD, 1980).

17

REGULATED POWER SUPPLIES

17.0 Introduction

17.1 Power Supply Topologies

17.2 Series-Regulated Power Supplies
    17.2.1 Control
    17.2.2 Current-Limiting Protection
    17.2.3 Integrated Circuit Voltage Regulators
    17.2.4 Complementary and Tracking Regulators
    17.2.5 Overvoltage Protection
    17.2.6 Other Operating Considerations

17.3 Shunt-Regulated Power Supplies
    17.3.1 High-Voltage Output
    17.3.2 Solar Array Input

17.4 Switch Mode Power Supplies
    17.4.1 Start-up and Auxiliary Power
    17.4.2 Hold-up Time
    17.4.3 Overvoltage Protection
    17.4.4 Synchronous Rectifiers

17.5 High-Power, Low-Voltage Power Supply

17.6 High-Power, High-Voltage Power Supply

17.7 Current-Regulated Power Supplies
    17.7.1 Linear Current Regulators
    17.7.2 Current-Regulated Hybrid Power Supply
    References

17.0 INTRODUCTION

The term *power supply* is frequently used as an all-encompassing description of power conversion equipment. As a definition, a power supply converts an available ac input voltage to a desired dc output voltage or to multiple dc outputs. The voltage, current, and power ratings of power supplies cover a vast spectrum. Typical characteristics of low-voltage low-power units are operation from 115/230 V utility or mains input, providing 5 V dc at 2 A (10 W) to power digital logic assemblies. Near the upper range of power and voltage, a unit may operate from

619

13.8 kV utility service and provide an output of 100 kV dc at 100 A (10 MW). In between these extremes are units providing outputs of 10 V at 1000 A (10 kW).

Power supplies discussed in this chapter are typically rated at 2 kW or less and provide a regulated output(s) of voltage or current. The term *regulated* defines a unit which maintains a *constant* output voltage (or current) independent of changes in input voltage or frequency (over a specified range) and independent of changes in output load conditions (over a specified range). Further references are listed [1–4]. Unregulated power supplies essentially contain a transformer/rectifier/filter, and these components are discussed in Chapters 3 and 4.

The regulated power supplies discussed are a combination of various functional blocks described in previous chapters. The presentation concentrates on the combination, interface, and interdependence of these blocks to provide a power supply *system*. This is not to say that power supply design is a simple task. A successful product requires the discipline, knowledge, and proper applications of information from all previous chapters. This fact is often overlooked by those who consider the power supply an unavoidable item which can be designed, assembled, and installed at the last minute to achieve a system goal.

The electrical, mechanical, and environmental requirements of power supplies are notorious for being overlooked in system operation. During development of complex equipment, electrical or electronic subassemblies are frequently powered by laboratory power supplies for design verification of these assemblies. The assemblies are then packaged in a housing, and a power supply to power these assemblies must then be packaged in the remaining space. This approach results in some most unusual power supply physical form factors for custom applications. The environmental requirements of power supplies range from the benign environment of office equipment to the hostile environment of military aircraft. In either case, the space allotted to power supplies may have insufficient heat transfer capability, and the power supply then overheats, resulting in failure or shortened life. Also, the required current draw of the subassemblies or modules is not finalized until the system designverification is complete. This results in a rapid preparation of a specification for the power supply, and a desire for immediate procurement or delivery of the power supply from the lowest bidder.

Fortunately, various power supply manufacturers offer product lines of standardized single- and multiple-output voltage and power values in standardized configurations, with proven designs and verified high reliability in field experience. Unfortunately, compliance with these latter two parameters is not always met by all manufacturers. The very nature of the competitive OEM (original equipment manufacturers) market for low-power commercial power supplies results in a commodity item which can be designed with minimal performance margins and minimal component deratings. These supplies can be manufactured in a garage shop atmosphere with the lowest cost parts available, with minimal screening and burn-in, with cursory quality assurance, and then dumped on the market where the unwary customer or buyer chooses the lowest price unit. Subsequent failures then cause animosity between vendor and customer. The foregoing discussion is presented as a desire to eliminate the proverbial "black eye" name frequently associated with power conversion. So much for philosophy; let's return to the immediate subject.

17.1   POWER SUPPLY TOPOLOGIES

Regulated power supplies are divided into three categories: (1) linear, (2) switching, (3) hybrid. Block diagrams of the basic linear and switching types are shown in Fig. 17.1. Hybrids (applies to a topology rather than a component) are a combination of 1 and 2. Phase control (class F) may be considered a form of switching and this topology is presented in Chapter 5. Advantages and disadvantages of linear power supplies, as compared to switching power supplies, are given in Table 17.1.

A linear series regulator power supply is shown in Fig. 17.1A. The input voltage is stepped up or down by T_1, rectified by CR_1 and CR_2, and filtered by C_1. Here E_a is the input voltage to Q_1 which is a series pass transistor. Transistor Q_1 is controlled to provide a regulated output voltage, and C_2 is the output filter. As shown in the waveform, the transistor voltage is

$$V_{CE} = E_a - E_o \tag{17.1}$$

The transistor power dissipation is then

$$P_{Q1} = (E_a - E_o)I_o \tag{17.2}$$

where I_o is the output current.

For low output voltage and high output current, the efficiency is typically 50%.

A linear shunt regulator power supply is shown in Fig. 17.1B. The input voltage is stepped up or down by T_1, rectified by CR_1 through CR_4, and filtered by C_1. Voltage E_a is the input to R_1 which provides a series impedance to limit current. Shunt transistor Q_1 is controlled to provide a regulated output voltage, and C_2 is the output filter. As shown in the waveform, the transistor voltage is

$$V_{CE} = E_o \tag{17.3}$$

The voltage across the resistor is

$$V_{R1} = E_a - E_o \tag{17.4}$$

The current flowing in R_1 is $I_{R1} = I_{Q1} + I_o$. For a given input voltage, the current through R_1 remains constant. Thus the input power remains constant, independent of the output power. This topology inherently has a low operating efficiency, which may drop to 10% at low values of I_o.

A hybrid power supply is shown in Fig. 17.1C. The input voltage is stepped up or down by T_1, rectified by CR_1 and CR_2, and filtered by C_1. Voltage E_a is the input to Q_1, which is switched on and off at a high frequency by the control circuit. Output current flows through Q_1, L_1, and the load when Q_1 is on, and flows through CR_3, L_1, and the load when Q_2 is off. The average output voltage is proportional to E_a and the on-time (duty cycle) of Q_1, or $E_o = DE_a$. Thus the topology is a transformer/rectifier followed by a buck regulator. Reference Chapters 3, 4, and 8. The advantages of this approach, as compared to the linear series regulator, are (1) a higher operating efficiency (typically 75% as compared to 50%)

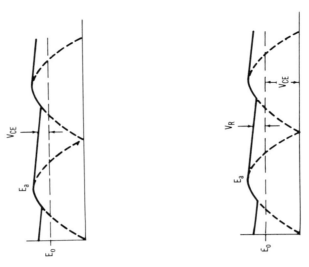

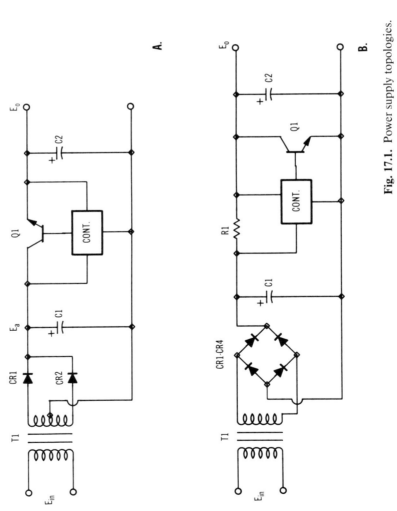

A.

B.

Fig. 17.1. Power supply topologies.

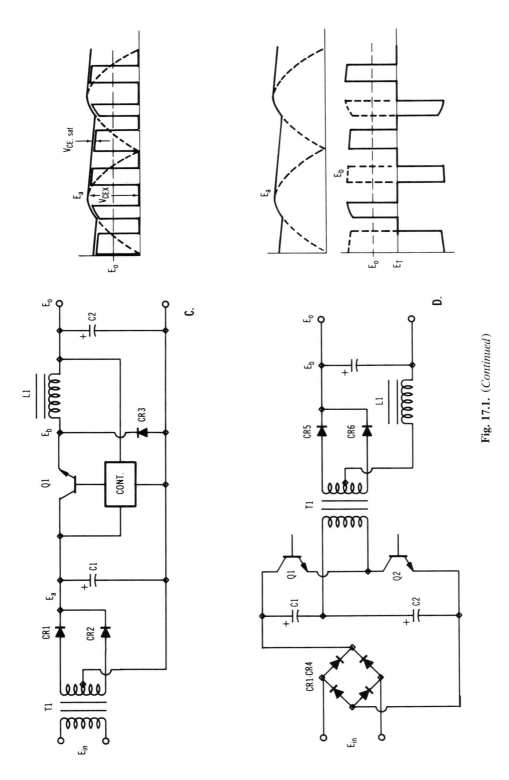

Fig. 17.1. (*Continued*)

TABLE 17.1. Comparison of Linear and Switching Power Supplies

	Linear	Switcher
Line regulation (105–130 V)	0.01% + 5 mV	0.2%
Load regulation (no load to full load)	0.01% + 5 mV	0.5%
Ripple, rms	1 mV	0.2%
Ripple, peak to peak	5 mV	1%
Transient response	50 μs, 100% load change	1 ms, 25% load change
Overshoot/undershoot	1%, 100% load change	2%, 25% load change
Hold-up time	Not applicable	17 ms
Temperature coefficient	0.01%/°C	0.02%/°C
Cross regulation (multiple outputs)	0.01%	3%, single loop control
Efficiency	50% inversely proportional to input voltage range and output voltage value	75% inversely proportional to output voltage value
Size and weight	Large, heavy inversely proportional to input frequency	Small, light

and (2) slightly less volume (heat sink requirements are reduced since the power dissipation in Q_1 is reduced).

A switching power supply is shown in Fig. 17.1D. The supply is frequently referred to as a switch mode power supply (SMPS) or simply as a *switcher*. The input voltage is rectified by CR_1 through CR_4 and filtered by C_1 and C_2. Transistors Q_1 and Q_2 are alternately switched into conduction at a high frequency, with each transistor conducting less than 50% of the cycle. The resulting voltage across T_1 is stepped up or stepped down and rectified by CR_5 and CR_6. The rectified pulse train is filtered by L_1 and C_3 to provide an average output voltage. Thus the topology is an off-line rectifier/filter followed by a half-bridge converter. The advantages of this topology, as compared to Fig. 17.1A, are (1) high operating efficiency (typically 75%) and (2) decreased size and weight (the high-frequency transformer is smaller and lighter than the transformer operating at utility supplied frequency and the required heat sink volume is reduced because of the higher efficiency). It should be understood that the converter topology is not limited to the half-bridge shown. A flyback converter (reference Section 9.4), forward converter (reference Section 9.5), or full-bridge converter (reference Section 9.8) could be used.

17.2 SERIES-REGULATED POWER SUPPLIES

As shown in Fig. 17.1A, the linear series-regulated power supply consists of a transformer, rectifier, and filter followed by a series regulator stage. Transformers are discussed in Chapter 3, while rectifiers and filters are discussed in Chapter 4. The major discussion in this section will therefore concentrate on the series

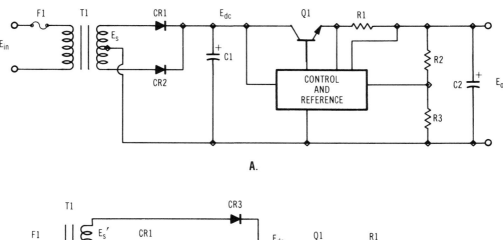

A.

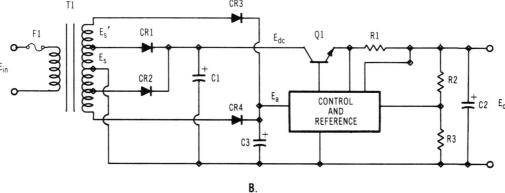

B.

Fig. 17.2. Series-regulated power supply with control power.

regulator stage. Output voltage regulation is achieved by sensing a portion of the output voltage, comparing this voltage to a stable reference voltage, and the resulting error signal then controls the conduction of the pass transistor. This control circuit may be discrete components or an integrated circuit. In fact, the complete regulator may be an integrated circuit, such as the popular "three-terminal regulators" which eliminate much of the design process. Basic circuits are first presented, followed by a discussion of these regulators.

Referring to Fig. 17.2A, the following relations apply: $E_s = E_{in}/n$, $E_{dc} = kE_s$, $E_o = (E_{dc} - V_{Q1} - V_{R1})$. Then

$$E_o = kE_{in}/n - V_{Q1} - V_{R1} \qquad (17.5)$$

where

n = turns ratio of T_1 (N_p/N_s)

k = proportional constant, depending on $\omega C E_{dc}/I_o$ and rectifier forward drop

V_{R1} = current limit sense voltage

V_{Q1} = collector–emitter voltage of Q_1

The limiting parameter in this case is V_{Q1}, which must be sufficient to maintain regulation at the minimum value of E_{dc}. Since the control circuit is also powered by E_{dc}, the voltage drop across the drive components must be considered. In Fig. 17.2B, the control circuit is powered by an auxiliary voltage (developed from additional secondary turns) which is rectified and filtered.

Some form of filtering is required to provide an energy storage during the time the rectified input voltage is below the minimum desired value. Due to the high-energy storage per unit volume, capacitive input filters are normally used for low-power units. Linear power supplies may not require an output filter capacitor for stability. Nevertheless, these capacitors reduce the ac output impedance and improve the transient response time.

Examples

Case 1. The power supply of Fig. 17.2A operates from 105 to 130 V, 60 Hz input, and provides an output of 5 V at 20 A. Calculate the power dissipation at 130 V input. Calculate the efficiency at 120 V input. Let the peak-to-peak ripple voltage across C_1 be 13%. Let the current sense voltage be 0.5 V at 20 A. Let the minimum voltage across Q_1 be 3 V. Then $E_{dc,min} = 3 + 0.5 + 5 = 8.5$ V and $E_{dc,max} = E_{dc,min}/(1 - 0.13) = 9.8$ V, from which $E_{dc,nom} = 9.1$ V. Voltage $E_{s,pk} = E_{dc,max} + V_F = 9.8 + 1.2 = 11$ V. This value of E_s is required at $E_{in} = 105$ V.

At $E_{in} = 130$ V, $E_{s,pk} = 11 \times 130/105 = 13.6$ V, $E_{dc,max} = 12.4$ V, $E_{dc,min} = 10.8$ V, and $E_{dc,avg} = 11.6$ V. Then $V_{CE,avg} = 11.6 - (0.5 + 5) = 6.1$ V. The rectifier dissipation is $P_{CR} = 1.2 \times 20 = 24$ W. The transistor dissipation is $P_Q = 6.1 \times 20 = 122$ W. A heat sink design must allow for these losses. The resistor dissipation is $P_R = 0.5 \times 20 = 10$ W, and a 20- or 25-W rating should be used.

At $E_{in} = 120$ V, $E_{s,pk} = 11 \times 120/105 = 12.6$ V. Then $E_{dc,avg} = 10.6$ V and $P_{dc,avg} = 10.6 \times 20 = 212$ W. The rectifier loss is again 24 W. Assuming a transformer efficiency of 97%, $P_{in} = (212 + 24)/0.97 = 243$ W. This allows 7 W in the transformer. The efficiency is then $\eta = 100/243 = 41\%$ at a nominal input of 120 V.

Case 2. The same input and output conditions exist except the circuit of Fig. 17.2B will be used. In this case, E_a is sufficiently high to allow a minimum saturation voltage across Q_1 of 1.2 V. Then, $E_{dc,min} = 1.2 + 0.5 + 5 = 6.7$ V, and $E_{dc,max} = 6.7/0.87 = 7.7$ V. A high-efficiency rectifier is used, and $V_F = 0.7$ V. Then, $E_{s,pk} = 7.7 + 0.7 = 8.4$ V at $E_{in} = 105$ V. At $E_{in} = 130$ V, $E_{s,pk} = 8.4 \times 130/105 = 10.4$ V and $E_{dc,avg} = 9.1$ V. Now, $V_{CE} = 9.1 - 0.5 - 5 = 3.6$ V. The rectifier power dissipation is $P_{CR} = 0.7 \times 20 = 14$ W. The transistor dissipation is $P_Q = 3.6 \times 20 = 72$ W. As compared to Case 1, the transistor power dissipation has been reduced by 50 W and the rectifier dissipation has been reduced by 10 W. The resistor power dissipation is still 10 W.

At $E_{in} = 120$ V, $E_{s,pk} = 8.4 \times 120/105 = 9.6$ V, $E_{dc,avg} = 8.3$ V, and $P_{dc} = 8.3 \times 20 = 166$ W. Again using a transformer efficiency of 97%, $P_{in} = (166 + 14)/0.97 = 186$ W. The efficiency (for this case) is $\eta = 100/186 = 54\%$ (as compared to 41% for Case 1).

17.2.1 Control

To maintain a regulated output voltage, a portion of the output is sensed to control the conduction of the pass transistor. The control circuit may consist of discrete components or may be an integrated circuit voltage regulator. The output voltage may be adjustable over a certain range if a potentiometer is included.

A simple regulator and control is shown in Fig. 17.3A. When E_a (the rectified and filtered dc input) is applied, current flows through R_1 to forward bias Q_1. When E_o increases sufficiently to forward bias Q_2, a portion of the base current to Q_1 is diverted through Q_2 and CR_1 to maintain E_o constant. The zener diode voltage is held constant by the current flow through R_2. Assuming sufficient input voltage, the output voltage is

$$E_o = (R_3 + R_4)(V_{Q2,B\text{-}E} + V_{CR1})/R_4 \qquad (17.6)$$

For output voltages greater than 10 V, improved temperature stability is achieved if CR_1 is selected such that its positive temperature coefficient is offset by the negative temperature coefficient of the base–emitter junction of Q_1. A typical voltage for CR_1 is 5.6 V. For a desired minimum zener current,

$$R_2 = (E_o - V_{CR1})/I_{CR1} \qquad (17.7)$$

The voltage across R_1 is $V_{R1} = E_a - E_o - V_{Q1,B\text{-}E}$ and the minimum current through R_1 is $I_o/h_{FE,Q1}$. Then

$$R_1 = h_{FE}(E_{a,\min} = E_{o,\min} - V_{Q1,B\text{-}E})/I_o \qquad (17.8)$$

Example

A control stage and a pass transistor stage are required for a *commercial* power supply with the following conditions: $E_a = 35\text{--}45$ V, $E_o = 28$ V, and $I_o = 0.5$ A. Also, the size of the input transformer for a single-phase center-tapped secondary and the rectifiers are to be selected. For the basic example, current limiting is not included. Referring to Fig. 17.3A, the worst-case power dissipation in Q_1 will be $P_{Q1} = 0.5(45 - 28) = 8.5$ W. A 2N6292 ($V_{CEO} = 70$ V and $I_C = 7$ A maximum) in a TO-220 package is selected. The primary consideration is to ensure that $V_{CE} = 17$ V at $I_C = 0.5$ A is within the SOA of the transistor, as is the condition for the 2N6292. For $V_{CE,\min} = 7$ V, at $I_C = 0.5$ A, $h_{FE} \cong 50$. Then from Equation (17.8), $R_1 = 50(35 - 28 - 0.7)/0.5 = 630$ Ω; use 620 Ω. For CR_1 at 5.6 V, select a 1N5524 and a minimum zener current of 10 mA. Then $R_2 = (28 - 5.6)/0.01 = 2.2$ kΩ. For $E_a = 45$ V, select Q_2 to conduct full current through R_1, and $I_C = (45 - 5.6 - 1)/620 = 62$ mA; use a 2N2222A. Transistor Q_2 must also operate within its SOA at $V_{CE} \cong E_o$ and $I_c \cong 20$ mA. Allowing 10 mA through R_3 and R_4, $R_3 + R_4 = 28/0.01 = 2800$ Ω. From Equation (17.6), $R_4 = 2800(0.6 + 5.6)/28 = 620$ Ω. Then $R_3 = 2800 - 620 = 2180$ Ω. A 2.2-kΩ resistor could be used for R_3, a 200-Ω potentiometer inserted between R_3 and R_4, and R_4 could be 560 Ω. Then, $E_{o,\max} = 6.2 \times 2960/560 = 33$ V, and $E_{o,\min} = 6.2 \times 2960/760 = 24$ V. From Chapter 4, Fig. 4.20, a 20% peak-to-peak ripple could be tolerated for a full-wave rectified input, and $\omega_s CR_L = 12$. For 60 Hz input and $R_L = E_a/I_o$,

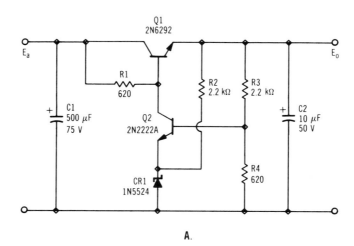

A.

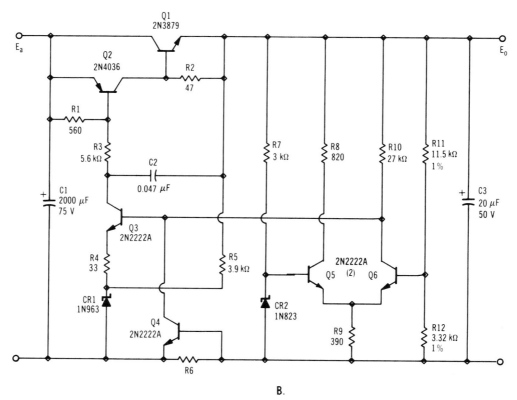

B.

Fig. 17.3. Discrete component series regulators.

$C_1 = 12 \times 0.5/(377 \times 35) = 455$ μF; use 500 μF. For C_2, use 10 μF. Fig. 17.3A shows component values. Use two each 1N4003 (200 V, 1 A) rectifiers on the input. For a nominal 40-V input at 0.5 A and allowing 1 W loss in the rectifiers, the power transformer output must be 21 VA. From Fig. 3.26, choose a square stack of EI-87 and operate at 500 cmil/A. Since this is a rectifier transformer, use M-6X material and operate at 14.4 kG.

Improved performance *and* current limiting is provided by the circuit of Fig. 17.3B. Q_1 and Q_2 form a complementary pass stage. The advantage, as compared to a Darlington *NPN* stage, is the minimum required input-to-output voltage is reduced, as discussed in the bipolar transistor drive of Section 2.1.2.4 and shown in Fig. 2.8. Conduction of Q_2 is controlled by Q_3. Transistors Q_5 and Q_6 form a "single-ended" difference amplifier, and the gain of this stage is established by R_{10}. Here CR_2 is a temperature-stable zener diode reference, biased by R_7. Resistors R_{11} and R_{12} form a voltage divider for output voltage sensing. Current limiting is provided by R_6 and Q_4. Capacitor C_2 provides compensation and allows Q_2 to turn on when input power is applied. Operation is described as follows. Transistor Q_2 turns on, turning on Q_1 and Q_3, and the output voltage increases. During this time, Q_5 is on, Q_6 is off, and Q_3 is in full conduction. When the output voltage reaches a value such that the voltage at the base of Q_6 is equal to the voltage at the base of Q_5, Q_6 conducts to decrease the base current in Q_3. If the output voltage tends to rise above this point, Q_6 conduction increases to further decrease the base current of Q_3 which, in turn, decreases the conduction of Q_2 and Q_1 to maintain output voltage regulation. The voltage of CR_1 is 12 V which causes the voltage across Q_6 to be approximately 6 V. If the output current increases due to an overload or short circuit, the voltage across R_6 causes Q_4 to conduct, which decreases the base current to Q_3 and, in turn, decreases the conduction of Q_2 and Q_1. Component values shown are typical for 28 V output at 1 A, operating from a dc voltage of 35–45 V obtained from a full-wave rectified single-phase 60-Hz input (the same input as the above example for Fig. 17.3A). Resistance tolerance for R_{11} and R_{12} is 1%. Semiconductors are hermetically sealed types. Detailed calculations of component values shown should be straightforward. A typical input transformer (42-VA rating) would be a 1.2-in. stack of EI-100 laminations, and a "current density" of 500 cmil/A (from Fig. 3.26). Two each 1N4999 rectifiers could be used for rectifying the secondary voltage. The circuit typically provides 70 dB of gain at 120 Hz, decreasing to unity gain at approximately 300 kHz. The high gain provides good static regulation (typically 0.1%) for line (35–45 V) and load (0–1 A) changes. The high bandwidth provides good dynamic response and fast transient recovery time (typically 30 μs) for the same changes in line and load.

The major disadvantage of this circuit is the high-power dissipation in the pass transistors when a prolonged output short circuit exists. In fact, the voltage across Q_1 will be the input voltage, and the current through Q_1 will be greater than the rated full-load output current. This condition will cause the 2N3879 to exceed its rated SOA unless the current is reduced. Another possible disadvantage of this circuit is the current-sensing resistor is in the negative line, which requires isolation between the input return and the output return. These problems can be overcome by the techniques discussed in the following sections.

17.2.2 Current-Limiting Protection

The most common failure in series-regulated power supplies is a collector-to-emitter short in the pass transistor. It is imperative the transistor operate within its temperature-rated SOA for both power dissipation and forward-biased second breakdown conditions in both steady-state and transient modes. Protection can be provided by either SOA current limiting or foldback current limiting. Current sensing can be accomplished in the positive line (for positive outputs) or in the negative line (for negative outputs) which allows a common input and output return. Referring to Fig. 17.4A, the pass transistor and foldback current-limiting circuitry are shown for a positive output regulator. The conduction of Q_1 is controlled by a feedback loop for voltage regulation, as previously discussed. When the voltage across R_3 increases to cause Q_2 to conduct ($E_b = E_o + V_{BE}$), base current is diverted from Q_1 through the collector of Q_2 and the output voltage decreases which, in turn, causes a further decrease in output current, as shown by the EI plot. The output short-circuit current may be 30% of rated output current. The following relations apply:

$$E_e = E_b(R_1 + R_2)/R_2 = I_k R_3 + E_o$$
$$= (E_o + V_s)(R_1 + R_2)/R_2 \qquad (17.9)$$

where I_k is the current at the knee where output current limiting begins ($I_k > I_o$) and V_s is the base–emitter voltage where Q_2 begins to conduct (typically 0.66 V). Rearranging terms,

$$R_3 = V_s(R_1 + R_2)/(I_{sc}R_2)$$
$$= \frac{E_o/I_{sc}}{1 + E_o/V_s - I_k/I_{sc}} \qquad (17.10)$$

where I_{sc} is the output current at short circuit.

This equation shows that as the foldback ratio of I_k/I_{sc} increases, the required value of R_3 increases. This requires a higher voltage at E_e and also at E_a. Thus, the operating efficiency will decrease and considerable power dissipation will occur in R_3 if I_{sc} is very low compared to I_o. However, a low value of I_{sc} is not mandatory from a power dissipation standpoint. For instance, if the pass transistor stage dissipates 30 W during normal (regulated output) operation, then 30 W could be dissipated in an overload or short-circuit condition, *provided* the forward-biased second breakdown area of the SOA is not exceeded. A possible advantage of foldback current limiting (as compared to fixed current limiting) is the low-temperature coefficient for I_k. Since V_s is normally small compared to E_o, temperature variations in V_{BE} will have minimal effect on I_k. However, in a short circuit, E_o is zero, and I_{sc} will be dependent on V_{BE} which typically has a temperature coefficient of -2.2 mV/$°$C.

Example

Let $I_o = 5$ A, $E_o = 10$ V, $I_k = 6.5$ A, $I_{sc} = 1.5$ A, and choose $R_2 = 10$ kΩ. From Equation (17.10), $R_3 = (10/1.5)/[1 + (10/0.66) - (6.5/1.5)] = 0.564$ Ω. From Equation (17.9), $E_a = 13.667 = 10.667(R_1 + R_2)/R_2$. For $R_2 = 10$ kΩ, $R_1 = 2.8$ kΩ.

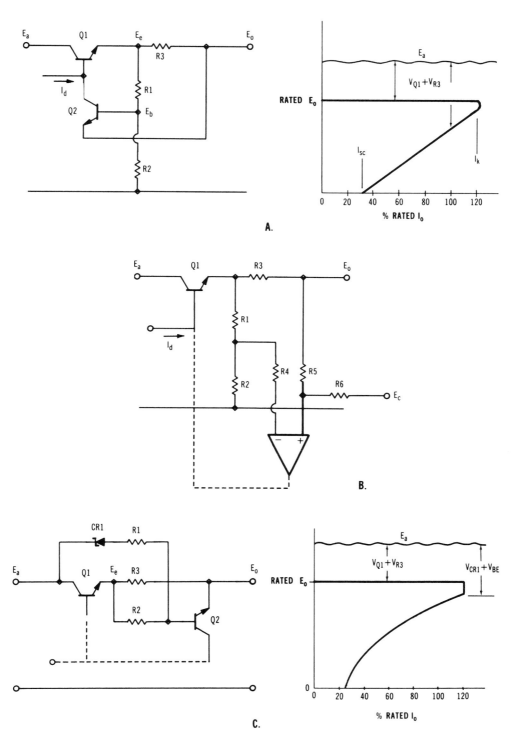

Fig. 17.4. Current-limiting and foldback for discrete component series-regulated power supplies.

To round off resistor values, if $R_3 = 0.55$ Ω, $R_1 = 2.7$ kΩ, and $R_2 = 10$ kΩ, $I_k = [(10 + 0.66)(12.7/10) - 10]/0.55 = 6.4$ A, and $I_{sc} = (0.66/0.55)(12.7/10) = 1.52$ A.

Since $R_3 = 0.55$ Ω, $E_e = 10 + (5 \times 0.55) = 12.75$ V, and allowing a 2-V drop across Q_1, the minimum input voltage at E_a is 14.75 V to maintain regulation at $I_o = 5$ A. Also, the maximum power dissipation of R_3 is $P_{R3} = I_k^2 R_3 = 6.4^2 \times 0.55 = 22.5$ W, and a 50-W rating should be used, since prolonged overloads could occur.

For high current outputs, and especially at low output voltages, considerable power is dissipated in R_3. One method of decreasing V_{R3} at I_k and at I_{sc} is to use an operational amplifier, as shown in Fig. 17.4B. The op-amp output decreases the conduction of Q_1 (through the control circuitry) for current limiting. An auxiliary regulated control voltage E_c may typically be twice the output voltage for output voltages less than 15 V. The following relations apply: $E_b = E_e R_2/(R_1 + R_2) \cong E_r = E_o + (E_c - E_o)R_5/(R_5 + R_6)$ and $E_e = E_o + V_s$. The assumption that $E_b \cong E_r$ is valid since the op-amp input impedance is high and R_3 may be a low value resistance. Solving for V_s at I_k and at I_{sc}, *respectively*, gives

$$V_s = (E_o R_6 + E_c R_5)(R_1 + R_2)/R_2(R_6 + R_5) - E_o \qquad (17.11a)$$

$$V_s = E_c R_5(R_1 + R_2)/R_2(R_6 + R_5) \qquad (17.11b)$$

With careful choice of component values (R_1 may be a very low value and R_6 may be a very high value), V_s at the knee (initial current limit) and at short circuit can be 250 mV and 50 mV, respectively.

As was shown by Equation (17.10), foldback current limiting increases the minimum input to output voltage differential, since a larger voltage drop is required across the sense resistor than for fixed current limiting. Safe operating area current limiting overcomes this problem by decreasing output current as a function of the input to output voltage differential. Referring to Fig. 17.4C, the current limit point is primarily dependent on the voltage drop across R_3, since a very small current through R_2 will cause Q_2 to conduct. Thus $I_k = V_{BE}/R_3$, provided V_{CR1} does not conduct. A descriptive analysis is presented (instead of equations) due to the dependent variables. As the load resistance decreases and the output voltage decreases, CR_1 will conduct when $E_o = E_a - V_{CR1} - V_{BE}$, since a small current through R_1 will cause Q_2 to conduct. As $E_a - E_o$ further increases, a corresponding decrease in current occurs in R_3, since the voltage across R_2 increases and V_{BE} is "fixed." Thus the current through Q_1 decreases and in fact may be zero in a short circuit. The output current in a short circuit is then the current through CR_1, R_1, R_2, and R_3. The precision of this technique is not as good as the foldback approach, but the SOA limiting does provide protection against excessive input voltage as well as excessive power dissipation during an output overload, or an output short circuit.

17.2.3 Integrated Circuit Voltage Regulators

A plethora of integrated circuits is available for *single-chip* control of linear regulators. For many years, these monolithic ICs have simplified power supply

design by reducing complexity, improving reliability, increasing ease of maintenance, and reducing overall power supply cost. Basic building blocks for these
regulators consist of (1) voltage reference, (2) biasing network, (3) error amplifier,
(4) pass transistor, and (5) current limiting. For power supply operation, some
regulators require the addition of several external components, while other regulators are a "drop-in" type with a fixed output voltage and do not require external
components (except for input and output filter capacitors). Four popular generic
families are presented below.

17.2.3.1 *The 723 Regulator*

The "ancient" μA723 voltage regulator, introduced
in the 1960s, revolutionized power supply control circuit design. Although other
devices are presently available with improved performance, the versatile 723 has
become an industry standard and is offered by almost every linear IC manufacturer. It can operate with either positive or negative output power supplies in
series, shunt, switching, or floating modes. Other features include external foldback current limiting and remote shutdown. An equivalent circuit is shown in Fig.
17.5. The biasing network for the *PNP* current sources, Q_3, Q_7, and Q_8 is
comprised of Q_1, Q_2, D_1, R_1, and R_2. The *N*-channel FET conducts a constant
current and improves line regulation, while minimizing power dissipation. The
resulting high output impedance of Q_3 provides line ripple rejection and increased
gain in the reference error amplifier. The Darlington pair of Q_4 and Q_5 achieves a
low output impedance for V_{REF} (7.15 V nominal), and C_1 provides internal
compensation for the reference loop. Transistors Q_9, Q_{10}, Q_{11}, Q_{12}, and Q_{13}
comprise the differential error amplifier. High gain is achieved by the current

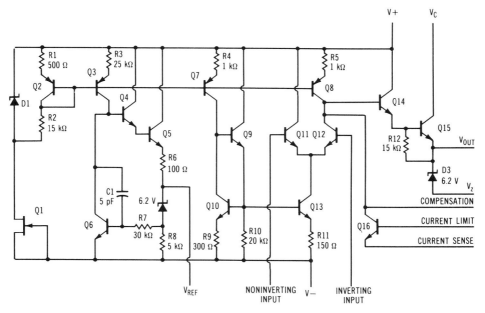

Fig. 17.5. Equivalent circuit for μA723 regulator. (Courtesy Fairchild Camera and Instru-
ment.)

source as part of the collector load for Q_{12}, and the single stage simplifies external compensation. Transistors Q_{14} and Q_{15} comprise a Darlington output stage. Transistor Q_{15} is a multiple emitter/resistor device to provide current sharing and increase the SOA and is capable of 150 mA output. The base–emitter junction of Q_{16} allows external sensing for current limiting, and Q_{16} decreases the conduction of Q_{14} in a current limit mode.

Three basic regulator circuits (perhaps 15% of the applications) are shown in Fig. 17.6. A positive voltage regulator with an external *NPN* pass transistor (Fig. 17.6A) can provide outputs from $+8$ to $+28$ V, assuming the voltage differential between input and output is 3 V minimum. Maximum differential is a function of power dissipation in the 723 and in Q_1. The circuit will typically deliver 0–1 A output, with a line and load regulation of 2 and 15 mV, respectively. A portion of the output voltage is applied to the inverting input of the 723, and the output voltage is given by

$$E_{\mathrm{o}} = V_{\mathrm{REF}}(R_1 + R_2)/R_2 \qquad (17.12)$$

Since $V_{\mathrm{REF}} = 7.15$ V, if $R_2 = 7.15$ kΩ (a standard 1% resistance value), then for a desired output voltage, R_1 may be found by

$$R_1 = E_{\mathrm{o}} - 7.15 \qquad \mathrm{k}\Omega \qquad (17.13)$$

In actual application, a potentiometer would be inserted between R_1 and R_2, with the wiper connected to the inverting input, to adjust the desired output voltage. For output voltages between 2 and 7 V, the output is connected to the inverting input, and the resistor divider delivers a portion of V_{REF} to the noninverting input. Also, the minimum input voltage for proper operations is 10 V.

Current limiting (at 25°C) occurs at

$$I_{\mathrm{limit}} = 0.66/R_3 \qquad (17.14)$$

Foldback current limiting is readily achieved by adding the external resistors shown in Fig. 17.4A with values calculated by the methods of Section 17.2.1. The external 1-nF capacitor (connects between base and collector of the internal error amplifier) provides compensation.

A negative voltage regulator with external *PNP* pass transistor (17.6B) can provide outputs from -10 to -28 V (or higher) at 0 to -1 A in a manner similar to the positive regulator. In this case, the voltage applied across the 723 is the output voltage, and the maximum input voltage is limited by the voltage rating of Q_1. A portion of the output voltage is applied to the noninverting input, while a portion of the reference is applied to the inverting input. For $R_3 = R_4$, the nominal output voltage is given by

$$|E_{\mathrm{o}}| = V_{\mathrm{REF}}(R_1 + R_2)/(2R_1) \qquad (17.15)$$

If $R_1 = 3.57$ kΩ (a standard 1% resistance value) for a desired output voltage, R_2 is given by

$$R_2 = |E_{\mathrm{o}}| - 3.57 \qquad \mathrm{k}\Omega \qquad (17.16)$$

Also, R_3 and R_4 may be 3.57 kΩ. The V_Z terminal, with a 6.2-V offset (available on the 14-pin DIP package) is connected to Q_1, and R_5 provides the

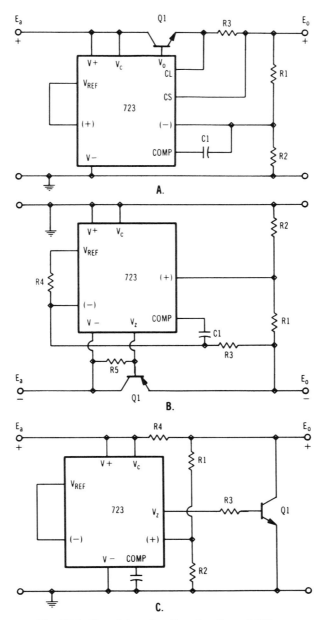

Fig. 17.6. Regulator circuits using the μA723.

bias current for Q_1. For "less negative" output voltages, V^+ and V_c must be connected to an external positive voltage to provide the required 10 V across the device.

A positive shunt voltage regulator is shown in Fig. 17.6C. Output voltage is given by Equation (17.12). The V_Z output terminal is used for a 6.2-V offset since Q_1 is a common emitter rather than the emitter follower of Fig. 17.6A. If V^+ and

V_c are connected to an auxiliary voltage source, typically 15 V, the maximum output voltage is limited only by the voltage rating of Q_1. Reference Section 17.3 for a discussion of shunt-regulated power supplies.

17.2.3.2 The 7800 and 7900 Regulators The μA78xx, μA78Mxx, and μA78Lxx series of positive regulators and the μA79xx, μA79Mxx, and μA79Lxx series of negative regulators are fixed voltage, three-terminal devices for "spot" regulation, and are offered by various manufacturers. The terminals are input, output, and common return. The xx denotes the output voltage. Available voltages are 5, 6, 8, 12, 15, 18, and 24 V (plus for 7800 and minus for 7900). The 79xx are also available in -2, -3, and -5.2 V outputs for powering MECL logic and other special devices. The maximum rated input voltage is typically 35 V. These series are designed with thermal overload protection that shuts down the circuit when subjected to an excessive power overload condition, internal short-circuit protection that limits the maximum current the circuit will pass, and output transistor SOA that reduces the output short-circuit current as the voltage across the pass transistor increases. The low-cost 78Lxx and 79Lxx series do not provide SOA protection.

The 78xx and 79xx devices are available in TO-3 or TO-220 packages with voltage tolerances of ± 2 or $\pm 4\%$, and an output current rating of 1 A. The 78Mxx and 79Mxx devices are available in TO-220 or TO-39 packages with voltage tolerances of $\pm 4\%$, and an output current rating of 0.5 A. The 78Lxx and 79Lxx devices are available in TO-39, TO-92, or 8-pin DIL packages with voltage tolerances of ± 5 or $\pm 10\%$, and an output current rating of 0.1 A. Two temperature ranges are available, 0 to $+125°C$ for the plastic packages and -55 to $+150°C$ for the hermetic seal packages. Line and load regulation for these devices is each typically 50 mV. Other than the input rectifier filter capacitor of the power supply, the recommended input and output capacitors located close to the device are typically 0.5 and 0.1 μF, respectively, for the 7800 series. For the 7900 series recommended capacitance values are typically 10 times the above values. The devices are thermally protected, but caution must be exercised when choosing input transformer power ratings and input rectifier current ratings. For instance, the 7800 maybe used in a power supply to provide a 0.6-A output and the rectifiers may be rated for 1 A average. However, the peak output current rating of the 7800 is 3.3 A in a short circuit. If this device is mounted on a low thermal resistance heat sink and is operating at a low ambient temperature, a prolonged output short circuit may cause a rectifier failure due to excessive current, even though the regulator is protected.

17.2.3.3 The 117 and 137 Adjustable Regulators The LM117, LM217, and LM317 positive regulators and the LM137, LM237, and LM337 negative regulators are also three-terminal devices but with input, output, and *adjust* terminals. These devices are essentially floating regulators, with features becoming to an industry standard. The addition of two external resistors allows the output voltage to be set over a range of 1.2–37 V. This eliminates the need for maintaining an inventory of various fixed voltage devices. Maximum input voltage is 40 V, although devices with the suffix HV are available in a 50 or 60 V rating. Package types, rated power dissipation, current ratings, and operating junction temperature range are listed in

**TABLE 17.2. Ratings for LM117 / 217 / 317 and LM137 / 237 / 337
Integrated Circuit Regulators**

Device Type	Package Type	Maximum Power Dissipation (W)	Rated Output Current (A)	Operating Junction Temperature Range (°C)
LM117, LM137	TO-3	20	1.5	−55 to +150
	TO-39	1.5	0.5	
LMM217,	TO-3	20	1.5	−25 to +150
LM237	TO-220	15	1.5	
	TO-39	2	0.5	
LM317, LM337	TO-3	20	1.5	0 to +125
	TO-220	15	1.5	
	TO-202	7.5	0.5	
	TO-39	2	0.5	
	TO-92	0.6	0.1	

Table 17.2. Line regulation is typically 0.02%/V input change (0.07%/V maximum), and load regulation is typically 20 mV or 0.3%, whichever is greater (70 mV or 1.5% maximum) from 10 mA (the minimum load current) to rated output current. Temperature stability is typically 1% over the operating temperature range. This is a significant improvement over the fixed regulators discussed in the preceding section. Protection inherent in the devices includes current limiting (which is constant with temperature), thermal overload, and SOA. This protection remains functional even if the adjustment terminal is an open circuit. Most manufacturers perform a burn-in at thermal limits on 100% of all devices to ensure reliability. The devices also have a low noise output (typically 0.003% rms) and a high ripple rejection ratio (typically 60 dB). In addition to line, load, and temperature stability, thermal regulation must be considered. Thermal regulation is the effect on output voltage of temperature gradients produced by device power dissipation and is expressed in percent change in output per watt of power change in a specified time. The specification for the 117 and 137 is 0.02%/W maximum.

Example

Find the worst-case regulation for an LM317 in a TO-220 package for the following conditions: $E_o = 15$ V; $E_{in} = 30$ V at $I_o = 10$ mA; $E_{in} = 20$ V at $I_o = 1.01$ A. The input change is 10 V and line regulation is 0.07% × 10 = 0.7%. The load regulation is 1.5%. Since the power dissipation change is 5 W, the thermal regulation is 0.04% × 5 = 0.2%. Overall output regulation (for a fixed ambient) is 2.4% or 0.36 V.

Referring to Fig. 17.7A, the LM117 develops a nominal 1.25-V reference voltage across programming resistor R_1. Since this voltage is constant, a constant current I_1 flows through R_1 and through output set resistor R_2. Also, 100 μA (maximum) current flow from the adjust terminal through R_2. The recommended value of R_1 is 240 Ω. The output voltage is given by

$$E_o = 1.25(1 + R_2/240) + I_{ADJ}R_2 \qquad (17.17)$$

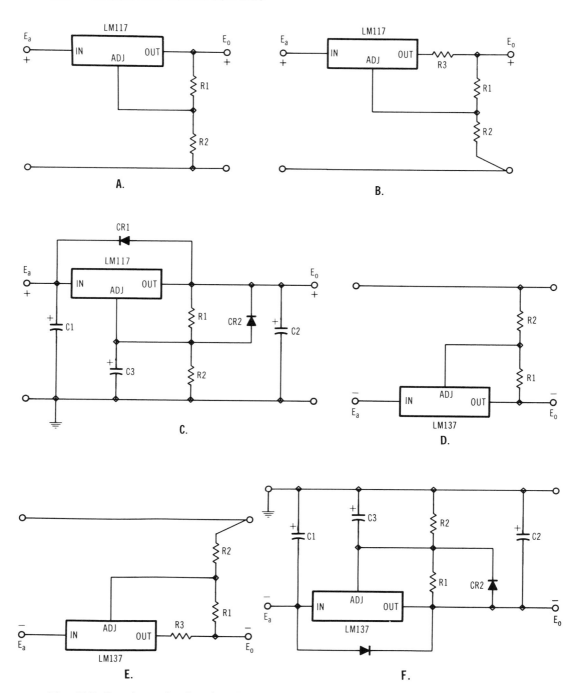

Fig. 17.7. Regulator circuit using the LM117 positive regulator and the LM137 negative regulator.

Since E_o is normally known and I_{ADJ} is very low, the resistance of R_2 for a desired output voltage is

$$R_2 = 240(E_o/1.25 - 1) \tag{17.18}$$

For optimum load regulation, the program resistor should be connected directly to the output terminal of the device instead of the load. Fig. 17.7B shows the effect of R_3 as an output line resistance to the power supply output, or to a remote sensing point. Since the reference current flows through R_3 and R_1, a voltage drop across R_3 will appear in series with the reference voltage, and degraded regulation will occur. However, load regulation can be improved and the voltage drop in the common or return line can be compensated by connecting R_2 to the power supply output or the external load.

Certain precautions should be considered when input and output filter capacitors are used. In Fig. 17.7C, C_1 should be a 1-μF tantalum if the main input filter capacitor is more than 4 in. from the regulator. When C_2 is larger than 20 μF, CR_1 protects the regulator in case the input to the regulator becomes shorted. When $C_3 > 10$ μF, CR_2 provides a discharge current path for C_3 if the output is shorted. The addition of C_3 can improve the ripple rejection by approximately 15 dB.

The mirror image diagrams of the 137 negative regulator are shown in Figs. 17.7D–F. The same considerations discussed above apply, except the recommended value of R_1 is 120 Ω. The output voltage is given by

$$-E_o = -1.25(1 + R_2/120) - I_{ADJ}R_2 \tag{17.19}$$

Since E_o is normally known and I_{ADJ} is very low, the resistance of R_2 for a desired output voltage is

$$R_2 = 120(E_o/1.25 - 1) \tag{17.20}$$

17.2.3.4 The 1834 Voltage Regulator
The UC1834, UC2834, and UC3834 high-efficiency IC regulators are packaged in a 16-pin DIL, and have all the functions typically required to control a linear power supply The block diagram is shown in Fig. 17.8. With both positive and negative precision references of 2% tolerance, either polarity of output can be implemented. A high-gain (65-dB open loop) error amplifier is used to control a pass stage that can sink or source 200 mA. A current sense amplifier with an adjustable low threshold can limit current in either the positive or negative supply lines. Special features include: fault monitoring for undervoltage and overvoltage, with an overvoltage latch and reset; a user-defined delay for transient rejection; a fault alert signal; a 100-mA source for activating an external crowbar; and compensation/shutdown control. This latter function is connected to the outputs of the error amplifier and the current sense amplifier, both of which are transconductance types. Output voltage sensing and the versatility of the uncommitted pass stage in driving external pass transistors is shown in Fig. 17.9 for positive and negative outputs.

An important advantage of the 1834, as compared to other regulators, is the low current sense voltage threshold which can be adjusted by a voltage applied from

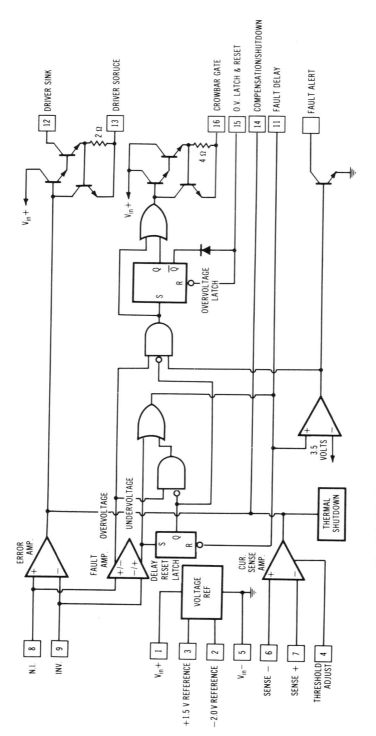

Fig. 17.8. Block diagram of the UC1834 regulator. (Courtesy Unitrode Corp.)

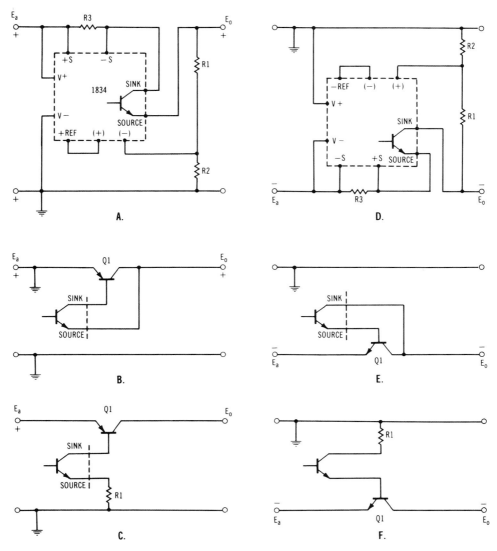

Fig. 17.9. Regulator circuits using the UC1834 regulator.

the reference pin to pin 4, and is typically 0.1 V/V to a maximum of 0.15 V. Thus, the voltage developed across an external current sense resistor can be as low as 50 mV, instead of the normal 600 mV associated with a conventional bipolar base–emitter voltage, which can significantly reduce the power dissipation of the sense resistor. As a comparison, if the sense voltage is 100 mV, the power dissipation in the external resistor (independent of maximum current) is a factor of six less than that required for a base–emitter junction. In addition, the minimum input-to-output differential voltage is reduced by 0.5 V. Also, the current limit point is independent of temperature change. Foldback current limiting is easily

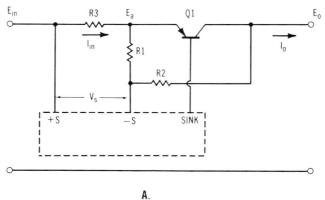

A.

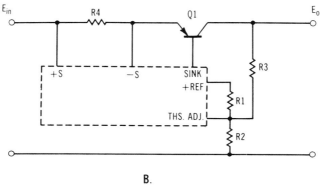

B.

Fig. 17.10. Current limit circuits with the UC1834 regulator.

implemented. Referring to Fig. 17.10A, V_s is the current sense voltage and

$$V_s = I_{in}R_3 + V_{R1} = I_{in}R_3 + (E_a - E_o)R_1/(R_1 + R_2)$$
$$= [I_{in}R_3R_2 + R_1(E_{in} - E_o)]/(R_1 + R_2)$$

Letting $V_s = 0.1$ V and solving for the maximum (knee) current and the short-circuit current:

$$I_{knee} = \frac{1}{R_3}\left[0.1\left(1 + \frac{R_1}{R_2}\right) - \frac{R_1}{R_2}(E_{in} - E_o)\right] \qquad (17.21a)$$

$$I_{sc} = \frac{1}{R_3}\left[0.1\left(1 + \frac{R_1}{R_2}\right) - \frac{R_1 E_{in}}{R_2}\right] \qquad (17.21b)$$

Thus the circuit responds to changes in $E_{in} - E_o$. To prevent latch-up, I_{sc} must be greater than zero.

In a similar manner, foldback current limiting can be implemented by the circuit of Fig. 17.10B which responds only to the change (decrease) in E_o when operating in the foldback mode. When the input current produces a voltage exceeding a preset V_s, current limiting occurs and the output voltage decreases. As the output voltage decreases, the threshold adjust input voltage decreases, which decreases V_s which, in turn, decreases the output current.

If the 1834 regulator is powered by a voltage more positive (or less negative) than the voltage applied to the external pass transistor, the minimum input-to-output differential can be as low as 1 V (0.9 V for the pass transistor, and 0.1 V for current sensing) instead of the "normal" 3 V associated with a three-terminal regulator with an *NPN* pass stage. This in turn decreases the power dissipation by a factor of 3, and is an ideal topology for precision post regulators on auxiliary outputs of multioutput switching power supplies.

17.2.4   Complementary and Tracking Regulators

Complementary regulators are used in power supplies to provide positive and negative outputs, normally from a center-tapped transformer secondary, full-wave bridge rectifier, and capacitive filter. Typical voltages are ± 5, ± 12, and ± 15 V.

Example

A schematic for a complementary power supply is shown in Fig. 17.11A for a 115/230 V ($\pm 15\%$) input, and using a LM217 and a LM237 to deliver ± 15 V outputs at 0.6 A. Resistor values are obtained from Equations (17.18 and 17.20) and are standard 1% tolerance values. Output capacitance values are nominal and input capacitance values are calculated from Fig. 4.20 of Chapter 4. Transformer size is obtained from Fig. 3.26 of Chapter 3. Maximum power dissipation in each regulator is approximately 5.3 W, and a proper heat sink can be selected, depending on the ambient temperature.

It is frequently desirable that each output track one another; that is, a change in the output voltage of one regulator should cause the other regulator to change a corresponding voltage but of opposite polarity. Thus one regulator typically becomes a *master* and the other regulator becomes a *slave*. A single adjustment of the master regulator output voltage divider will produce exactly the same change in the slave unit. Dual-tracking regulators are available in a single package with fixed outputs, or the outputs may be adjustable with external voltage-sensing resistors. External pass transistors may be added for increased current capability, and foldback current limiting can be implemented. A potential problem with certain tracking regulators occurs at turn-on of the power supply if an unbalanced load is present, or if a load exists between the positive terminal and the negative terminal without being connected to "common." One solution to this problem is to add antiparallel rectifiers across each output and across the pass regulator itself. Frequently, the positive output powers analog devices plus drive stages within the power supply, and the negative output is only required to power op-amps and the negative current is much less than the positive current.

Example

Figure 17.11B shows a dual-tracking power supply providing outputs of $+12$ V at 1 A, and -12 V at 50 mA. The positive regulator is a μA723PC driving a 2N3055,

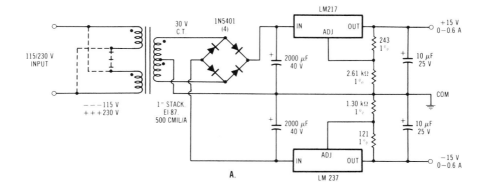

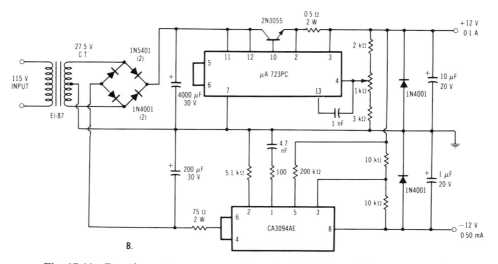

Fig. 17.11. Complementary power supply and adjustable tracking power supply.

while the negative regulator is a CA3094AE. The 3094 has a versatile feature in that the collector and the emitter of the pass transistor are available at the device pins, and this transistor can thus sink or source current. The noninverting amplifier is essentially at common, while the inverting input is connected at the junction of equal value precision resistors, one of which is connected to the positive output while the other is connected to the negative output. Thus, if the positive output increases, the voltage at the inverting terminal will increase in a positive direction, and the offset voltage will cause the amplifier in the negative regulator to increase conduction in its pass transistor to achieve the same, but opposite polarity, voltage until the inverting terminal again is essentially at zero volts. Typical resistance, capacitance, and component values are shown for the power supply. Current sensing is provided in the positive output line, while the 75-Ω resistor in the negative return line protects the 3094 from momentary output short circuits. In normal operation, this resistor will dissipate negligible power since the output current is low.

17.2.5 Overvoltage Protection

Figure 17.12 shows linear power supplies with overvoltage protection in the form of a crowbar SCR. If the pass transistor shorts, the output must be protected from excessive voltage across the input filter capacitor. If an overvoltage occurs from external sources, the output must also be clamped since the output filter capacitor may not have sufficient energy capability to absorb the transient. When an overvoltage occurs, this condition is sensed by the control circuit, which then fires Q_2 to "clamp" the output voltage to zero.

In Fig. 17.12A, if Q_1 shorts, Q_2 fires to discharge C_2 and C_1, and F_1 opens. The I_{TSM} rating of Q_2 must be sufficient to conduct the discharge current of the capacitors, and the I_{FSM} rating of CR_1 and CR_2 must be sufficient to conduct the secondary surge current until F_1 opens. Foldback current limiting is also shown. If an overvoltage occurs from external sources, Q_2 fires, but the control circuit sees this as a sustained output short circuit (the holding current of Q_2 is normally much

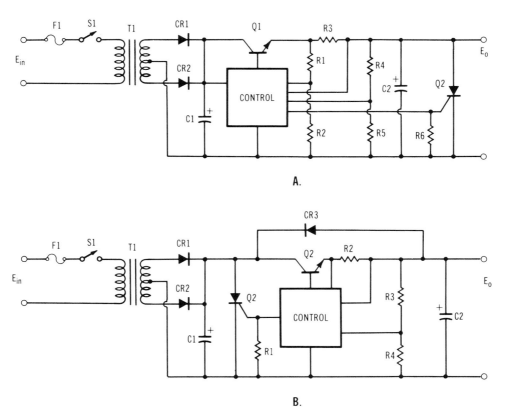

Fig. 17.12. Overvoltage protection with crowbar.

higher than the foldback short-circuit current). In this case, F_1 will not open and S_1 must be opened to allow Q_2 to cease conduction, and the unit may then be turned on again (a recycle mode). Since $i = C \, dv/dt$ and this current is low, S_1 must remain open until C_1 is discharged. Another means to achieve normal operation is by remote turn-off of the control circuit, which turns off Q_1 to allow Q_2 to recover.

Without foldback current limiting, as shown in Fig. 17.12B, Q_2 is connected across the input capacitor, and the control circuit fires Q_2 if an overvoltage occurs and F_1 then opens. If Q_2 were connected across the output and an overvoltage occurred from external sources, Q_1 could fail from forward-biased second break-down. Also, CR_3 is required to discharge C_2 and to protect Q_1 from reverse voltage.

17.2.6 Other Operating Considerations

Series-regulated power supplies do not normally require output capacitors for stability. However, an output capacitor can reduce the ac output impedance and improve transient response. A medium value capacitor can shift the dominant roll-off of the regulating loop from the regulator compensating capacitor to the output filter capacitor, plus reduce noise on the output. In addition to an input filter capacitor for filtering the rectified input voltage, a capacitor should be located physically close to the input of integrated circuit regulators. The line rejection ratio of the IC may be insignificant at 1 MHz, but the amplifier may still have gain at this frequency. If a low input impedance to the regulator is not provided, oscillations may occur and further increases in output capacitance will not achieve stability.

Oscillations can occur during current-limiting operation, even though the unit is stable for all other conditions. These oscillations may be caused by parasitic capacitances in a high-frequency pass transistor, or by an inductive resistor used for current sensing. In the latter case, a noninductive resistor should be used, or a wire-wound resistor should be paralleled with a small capacitor.

In multioutput power supplies, antiparallel rectifiers should be used across the output to provide circuit protection from accidental output-to-output shorts. Antiparallel rectifiers should also be used across the pass stage to eliminate the possibility of component damage from reverse voltage of abnormal or unexcepted operating conditions.

In high-current power supplies, pass transistors are frequently paralleled to conduct the required current. Methods for current sharing and balancing are discussed in Section 2.1.2.5.

For series-regulated power supplies with output voltages of a few hundred volts, transistor leakage current at maximum operating junction temperature must be considered, in addition to transistor voltage rating and SOA. For instance, if the load current approaches a no-load condition and the voltage across the pass transistor is high, the leakage current through the pass transistor may be sufficiently high in this forward biased condition to allow the output voltage to rise above the regulated value, even if the base current is negligible. Conversely, if the load current is at maximum rated load, the leakage current through a shunt amplifier transistor may not allow sufficient base current in the pass transistor, and the output voltage may fall below the regulated value.

17.3   SHUNT-REGULATED POWER SUPPLIES

As shown in Fig. 17.1B, the shunt-regulated power supply is derived by substituting a fixed impedance (normally a resistor if the output is dc) for the series pass transistor, and connecting the transistor in shunt (parallel) with the output. The simplest form of a shunt regulator, operating from an unregulated input, is a series resistor and a shunt zener diode, but output voltage variations due to zener tolerance, impedance changes, and temperature effects may not provide the fixed output nor the regulation desired. An integrated circuit adjustable shunt regulator is shown in Fig. 17.13A. Output voltage is given by

$$E_0 = V_{REF}(1 + R_2/R_3) + I_{ADJ}R_2 \qquad (17.22)$$

Since E_0 is normally known and I_{ADJ} is very low, the resistance ratio is

$$R_2/R_3 = E_0/V_{REF} - 1 \qquad (17.23)$$

These devices have a typical reference voltage of 2.75 V and a typical temperature coefficient of 50 ppm/°C. Figure 17.13B shows the characteristics of E_0 and I_z versus I_0. At $I_0 = 0$, $I_z = I_{in}$ and at $I_z = 0$, $I_{in} = I_0 = (E_{in} - E_0)/R_1$. With a short circuit on the output, $I_0 = I_{sc} = E_{in}/R_1$. Thus the maximum rated output current and the minimum rated input voltage determine the minimum value of R_1:

$$R_{1,min} = (E_{in,min} - E_0)/I_{in,max} \qquad (17.24)$$

The power dissipation in R_1 occurs at maximum input voltage and maximum input current and is

$$P_{R1} = (E_{in,max} - E_0)^2/R_1 \qquad (17.25)$$

Examples

Case 1. An existing power supply provides a regulated -12 V output and a shunt regulator is desired to provide a -5.2 V output at currents from -200 mA to -500 mA. An adjustable shunt regulator with increased current capability is shown in Fig. 17.13C, where Q_1 is added. If the output voltage tries to increase, U_1 increases the conduction of Q_1 to reduce the output voltage back to the regulated value. Using $V_{REF} = 2.75$ V and selecting $R_3 = 330$ Ω, from Equation (17.23), $R_2 = (5.2 \times 330)/2.75 - 330 = 294$ Ω; use 300 Ω. At -500 mA output, it is desired that the regulator conduct some current to maintain regulation, say 6 mA in U_1 and 6 mA in Q_1. Also, the current through R_2 and R_3 is $I_R = 5.2/630 = 8.25$ mA. Thus, $I_{in,max} = I_0 + I_Q + I_z + I_R = 500 + 6 + 6 + 8 = 0.52$ A. From Equation (17.24), $R_1 = (12 - 5.2)/0.52 = 13$ Ω. Also, the power dissipation in R_1 is $P_{R1} = (12 - 5.2)^2/13 = 3.6$ W; use a 7.5-W rating. Since the input current will be

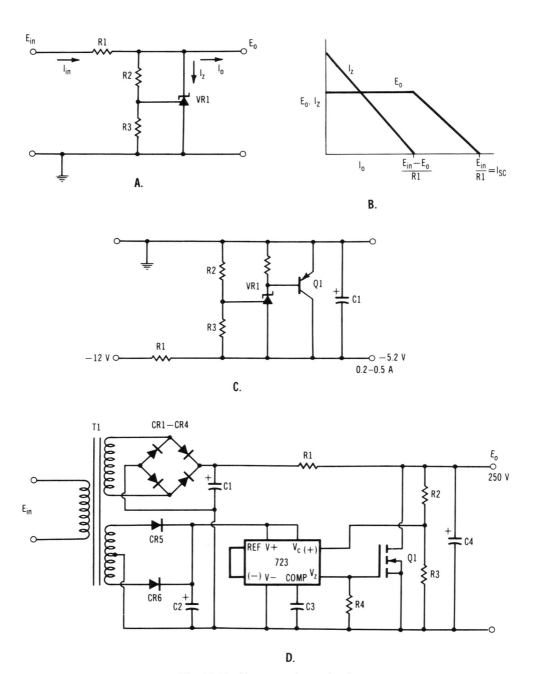

Fig. 17.13. Shunt regulator circuits.

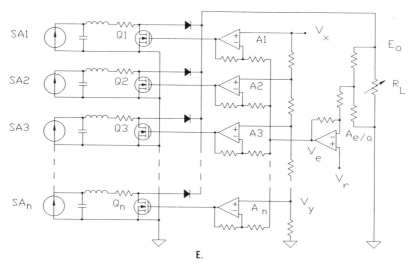

Fig. 17.13. (*Continued*)

relatively constant, the maximum current through Q_1 at minimum output current is $I_{Q1} = 0.52 - 0.20 = 0.32$ A. The maximum power dissipation in Q_1 is $P_{Q1} = 5.2 \times 0.32 = 1.66$ W. Overall efficiency at full load is $\eta = P_0/P_{in} = 5.2 \times 0.5/12 \times 0.52 = 42\%$.

Case 2. If the above shunt regulator were replaced with a series regulator, R_1 is replaced with a series pass transistor and the power dissipation in Q_1 would be $P_{Q1} = (E_{in} - E_0)I_0 = (12 - 5.2)0.5 = 3.4$ W, and the efficiency at full load would be $\eta = 5.2/12 = 43\%$.

17.3.1   High-Voltage Output

Shunt regulators are useful in providing higher output voltages. In Fig. 17.13D, a shunt-regulated power supply provides an output voltage to the screen grid of a vacuum tube, say 250 V output from 350 to 400 V input. In this case, an auxiliary voltage, say 20 V, is developed to power a 723 regulator which controls a MOSFET. The V_Z terminal is used for a 6.2-V offset. For a voltage drop of 4.8 V across the 723 pass transistor, the gate-to-source voltage available is $20 - (3.8 + 6.2) = 10$ V. Resistor R_1 is the series resistance, R_2 and R_3 sense the output voltage, and R_4 provides a current path for the 273 output, since the MOSFET has a very high input impedance. A small output filter capacitor may be used for stability. Care must be exercised with large output filter capacitors. When the power supply is turned on at the same time as the system power, the output capacitor must charge through R_1, and this time constant must be considered along with the rise in voltage of other components within the system. The MOSFET does not suffer from forward-biased second breakdown. For the same voltage and current rating, the SOA curve is broader for the MOSFET than for a bipolar transistor. With either a MOSFET or a bipolar transistor, leakage current at higher junction temperatures must be considered. For instance, at low output currents, R_1 may be quite large in value. If excessive leakage current is drawn through Q_1, the output voltage will decrease below the regulated value.

17.3.2   Solar Array Input

Shunt regulators are useful in regulating the output voltage from solar arrays. However, the system must be capable of handling the power dissipation inherent in the linear regulator. In Fig. 17.13E, the output of solar arrays SA_1–SA_n are "diode OR-ed" to E_0. Since the solar array is a constant-current source, means are required to provide a regulated output voltage to R_L. Regulation is accomplished by the conduction of Q_1–Q_n, in response to an error signal. A portion of E_0 is compared to a reference voltage V_r by $A_{e/a}$ and the resulting error voltage V_e is applied to the inverting input of A_1–A_n. A voltage divider provides inputs from V_x to V_y to the noninverting inputs. The output of A_1–A_n drives Q_1–Q_n. Again, MOSFETS are chosen as regulating elements since they do not suffer from forward-biased second breakdown. Depending on load conditions, the *upper* transistors are in the off state, and current flows to the load; the *lower* transistors will be in the on state, and current flow is shunted through the transistor. An *intermediate* transistor will be operating in the linear region, in response to the error signal. As the load current increases, additional transistors turn off to direct that solar array current to the load.

Example

In a power system similar to Fig. 17.13E, the desired output voltage is 120 V and the load current varies from 1 to 100 A for a total output power of 12 kW. A bank of 50 solar arrays with an output current of 2 A each is selected. If the load current is 50 A at some operating point, the top 24 transistors will be off, the bottom 25 transistors will be on, and the 25th transistor will be in the active region to maintain the output at 120 V. As the load current increases, additional bottom transistors turn off to divert the solar array current to the load and a lower transistor enters the active region. The power dissipation is very low in the conducting transistors, which are saturated. For $R_{DS,on} = 0.5$ Ω, $P = 2$ W. The power dissipation in a transistor operating in the linear region can be as high as $E_0 x I_{SA} = 120 \times 2 = 240$ W, but only one transistor will be in the active region at a given time. Thus if all transistors are mounted on a common heat sink, the maximum power dissipation will be $240 + (49 \times 2) = 338$ W. This represents a high efficiency for 12 kW output.

The use of shunt regulators in satellite and space station applications is intolerable due to the power dissipation of a transistor operating in the linear region. Heat transfer cannot be accomplished by convection, which results in a large and heavy heat sink to dissipate heat. In these applications, topologies such as Fig. 17.13E can be used, but duty cycle modulation at a selected switching frequency is applied to any of the transistors. Thus the average current from any solar array to the load is inversely proportional to the duty cycle of the transistor shunting that array. However, EMC problems can arise from the use of a switching regulator. The inherent capacitance associated with the solar array can resonate with the inductance of the line from the solar array, resulting in circulating EMI currents at the resonant frequency and requiring additional filtering.

17.4 SWITCH MODE POWER SUPPLIES

This introduction to SMPS is not a panacea for all requirements and applications and is not intended to be a lesson in economics, but the factors discussed are important to the success of the product. The major advantages of switchers, as compared to linears, are decreased size and weight, increased efficiency, and decreased cost. The first two factors are the result of the switching action of power transistors operating in the saturated conduction mode, as opposed to the linear mode. The first and last factors are the result of high-frequency operation which reduces passive component sizes, especially magnetic devices. The reduction in raw materials alone (specifically iron and copper) has a significant economic advantage. Hybrid devices, which combine the control functions in a single package and the switching semiconductors in a single package, offer additional size decrease and cost decrease. The combination of CMOS logic and DMOS power in a single package may result in a "drop in" regulating stage similar to the linear three-terminal regulator. The increased costs of design effort and complexity, as compared to linears, tends to be minimal once a unit is in volume production and the development costs are recovered. Cost has been a paramount consideration for OEM switchers, due to competition, followed by performance and reliability. Fortunately, end users are demanding higher quality and increased reliability through component selection and screening processes. For space applications, reliability and system effectiveness are overwhelming considerations, followed by performance and cost.

Switch mode power supplies essentially consist of an input rectifier and filter stage (discussed in Chapter 4), followed by a dc–dc converter stage (discussed in Chapters 9 and 10). The ac input is normally preceded by an EMI filter, since these power supplies generate radio frequency harmonics which range from the switching frequency up to several megahertz. Voltage regulation is achieved by pulse width modulation. Since these topics are presented in preceding chapters, the main discussion will be on the overall power supply system.

17.4.1 Start-up and Auxiliary Power

Since the ac input is rectified directly off the line and applied to the power-switching stage, means must be provided to power the control circuitry. Figure 17.14 shows a few possibilities. Referring to Fig. 17.14A, the switcher operates from either 115 or 230 V input, and the power stage is a buck-derived push–pull half-bridge converter. For 115 V input, CR_3, CR_4, C_1, and C_2 form a full-wave voltage doubler, and voltage is applied to T_{2A} for auxiliary power. For 230 V input, CR_1–CR_4, C_1, and C_2 form a full-wave bridge rectifier, and voltage is applied to the series-connected primaries of T_2. The secondary voltage of T_2 is rectified and filtered and applied to the control circuit. Output voltage is sensed and also applied to the control circuit for regulation. This circuit has the advantage

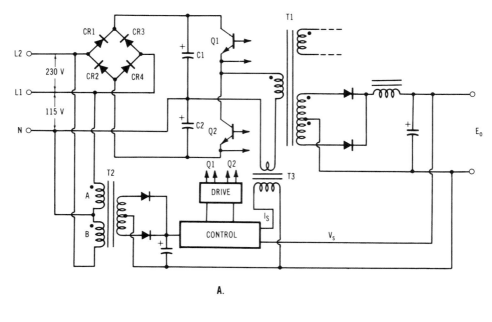

A.

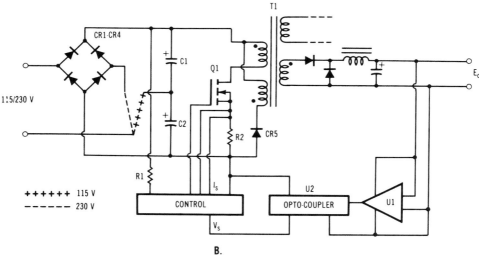

B.

Fig. 17.14. Switching power supply start-up and auxiliary power.

of a common return for the control circuit and the output. Multiple outputs are possible and in these cases, batch output sampling can improve cross regulation between outputs, if all outputs share a common return. Current sensing is accomplished by T_3. Isolation is achieved by T_1, T_2, T_3, and the transformers in the drive stage, which drive Q_1 and Q_2. The penalty incurred is the requirement for several magnetic devices. Also, T_2 must operate at the input line frequency.

In Fig. 17.14B, a dual voltage input is again accommodated but is field selectable by a jumper connection, which is also applicable to Fig. 17.14A. The power stage is a buck-derived forward converter and Q_1 is switched by the control circuit. In this case, the control circuit return is common to the power stage return, and current to the control circuit flows through R_1. Also, current is sensed across R_2 and applied to the control circuit for limiting and protection. The output voltage is sensed by U_1, which drives U_2 to apply a signal to the control circuit. The penalty incurred is the requirement for an optical isolator, plus the power dissipation in R_1 may be appreciable. However, power to the control circuit can be bootstrapped from an auxiliary winding of T_1 after the switcher starts.

17.4.2   Hold-up Time

An additional advantage of switching power supplies is the ability to provide a regulated output(s) for a period of time after the utility or mains input voltage decreases to an abnormally low value, such as in a brownout condition, or when the input is totally interrupted. Referring to Fig. 17.14, the energy storage of the input filter capacitors will provide power to the converter stage as the capacitor voltage decreases. During this time the duty cycle increases to maintain the average regulated output voltage. The energy in the capacitor is $\varepsilon = \frac{1}{2}CE_{dc}^2$. The required filter capacitance for a desired holdup time is

$$C = 2P_0 t_h / \eta \left(E_{dc1}^2 - E_{dc2}^2 \right) \qquad (17.26)$$

where

P_0 = total output power

t_h = desired holdup time

η = converter efficiency

E_{dc1} = capacitor voltage at minimum input

E_{dc2} = minimum capacitor voltage required for regulation

Example

The nominal ac input range to a power supply may be 100–130 V, 200–260 V at 50 Hz. The power supply is a half-bridge topology with two capacitors in series across the dc bus. The efficiency is 75% and the output power is 300 W. Using a full-wave rectifier for 230 V and a voltage doubler for 115 V inputs, a typical dc bus voltage corresponds to 254–332 V dc on the capacitors. (This assumes an input EMI filter is used which results in an inductive input filter.) If the converter is designed to provide a regulated output at 220 V dc minimum (the equivalent of 87/174 V ac), the energy storage in the capacitor will provide output power during a utility outage as the capacitor voltage decreases from 254 to 220 V dc. The desired holdup time is 20 ms or one cycle of the input. From Equation (17.26), the required filter capacitance is $C = 2 \times 300 \times 0.02/[0.75(254^2 - 220^2)] = 993$ μF. Since the capacitors are in series and accounting for a low tolerance of 20%, each capacitor should be $C_1 = C_2 = 993 \times 2 \times 1.2 = 2400$ μF. Using a 25% capacitor voltage derating and allowing for C_1 and C_2 to charge to the peak input voltage at light load, the required voltage rating is WVDC = $\sqrt{2} \times 260 \times 1.33/2 = 250$ V.

Carrying this example further to calculate the capacitor ripple voltage for nominal conditions of 115/230 V ac 50 Hz, and 292 V dc, and $R_L = E_{dc}^2 \eta / P_0 = 292^2 \times 0.75/300 = 213$ Ω, then $\omega C R_L = 2\pi 50 \times 1200 \times 10^{-6} \times 213 = 80$. From Chapter 4, Fig. 4.20, the peak-to-peak ripple is 3.6% or 10.5 V. Since the power supply should readily operate with a ripple voltage 3 times this value, the penalty incurred to provide the 20 ms holdup time is a capacitance value 3 times greater than that required if holdup time were a factor of 3 less, or 6.67 ms.

17.4.3 Overvoltage Protection

Switchers may not require crowbar circuitry for output overvoltage protection since the transformer isolates the input voltage and the switching transistors from the output, and the output filter capacitor can absorb external transients. However, at switching frequencies of several hundred kilohertz, the output filter components become smaller in value, and the output capacitor may not be able to absorb energy from an external voltage transient without the voltage exceeding a safe level. Also, logic devices or other critical loads may be destroyed if an accidental short occurs between that output and a higher voltage output. In these cases, inclusion of a crowbar is a minor cost penalty for protection. In fact, a single SCR and multiple rectifiers can be used for multioutput switchers which have a common return line.

Referring to Fig. 17.15, a voltage reference is applied to comparator U_1, the other input being a portion of the output voltage. Here $E_{01} < E_{02}$. If any output "exceeds" a present value, the output of U_1 goes low and Q_1 turns on to fire Q_2. Capacitor C_2 discharges through Q_2 and CR_1; C_1 discharges through CR_2, Q_2, and CR_1; C_3 discharges through C_2, Q_2, and CR_3. This configuration places demands on rectifier pulse current ratings. For instance, let $E_{01} = +5$ V, $E_{02} = +15$ V, and $E_{03} = -15$ V. The output current of E_{01} is typically greater than the output current of E_{02}, and C_1 is typically much greater than C_2. Thus CR_2 must conduct the discharge current from C_1, while CR_1 must conduct the discharge current from C_1 and C_2. Of course, Q_2 must conduct the discharge current of all capacitors. Also, a signal from Q_1 collector can be applied to the control circuit to inhibit switching, and the power supply ceases operation until some form of reset, either manual or timed automatic recovery, is again applied. The speed of this shutdown signal may not be critical if current sensing of each output is employed to provide initial protection of the switching transistors.

17.4.4 Synchronous Rectifiers

For logic families operating at a nominal 3 V, the voltage drop across rectifiers connected to the transformer secondary can become an appreciable percentage of the output voltage. For instance, if the rectifier drop is 0.51 V (a typical minimum for Schottky rectifiers), the rectifier losses are 17% of the total output. If the remaining stages of the power supply are 80% efficient, the overall efficiency will be a maximum of $\eta = (1 - 0.17)0.8 = 66\%$. To reduce the voltage drop, MOSFETs can be used as synchronous rectifiers in a third quadrant mode. Referring to Fig. 17.16A, the internal parasitic rectifier will conduct current when the source is more positive than the drain. However, this bipolar junction has a

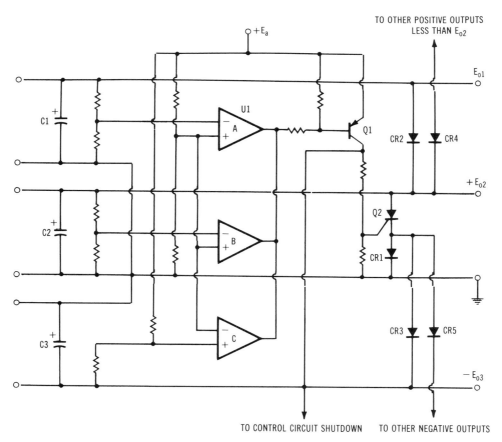

Fig. 17.15. Single crowbar overvoltage protection for multioutput switching power supplies, $E_{02} > E_{01}$.

high forward drop compared to Schottky rectifiers. But if V_{GS} is positive while V_{SD} is positive, current will flow in the channel parallel to the rectifier and $V_{SD} = V_{GD} - V_{GS}$. This eliminates the offset voltage inherent in rectifiers. Referring to Fig. 17.16B, V_{SD} decreases as V_{GS} increases for a given I_D, and the voltage drop becomes $V_{SD} = I_D R_{SD,\text{on}}$. Superimposed is the first quadrant characteristic of a Schottky rectifier, V_F versus I_F. If the MOSFET is operated at a current below the voltage crossover point, less power will be dissipated in the MOSFET than in the Schottky rectifier. A synchronous rectifier schematic is shown in Fig. 17.16C. When E_{s1} is positive, E_{s2} applies gate voltage and Q_1 conducts. When $E_{s'1}$ is positive, Q_2 conducts. If $V_{GS} = 10$ V, $R_{SD,\text{on}}$ may be 0.02 Ω and a voltage drop of 0.3 V results at $I_D = 15$ A, and the power dissipation is 4.5 W or approximately 3 W less than a Schottky rectifier. For high currents, multichip MOSFETs may be paralleled to maintain a low "on" resistance. The required voltage rating BV_{DSS} becomes small, which allows the on resistance to be low, as compared to a high-voltage device.

The same analysis can also be applied to bipolar transistors, if the voltage rating of the reverse-biased base–emitter junction is observed. This assumes a diffused

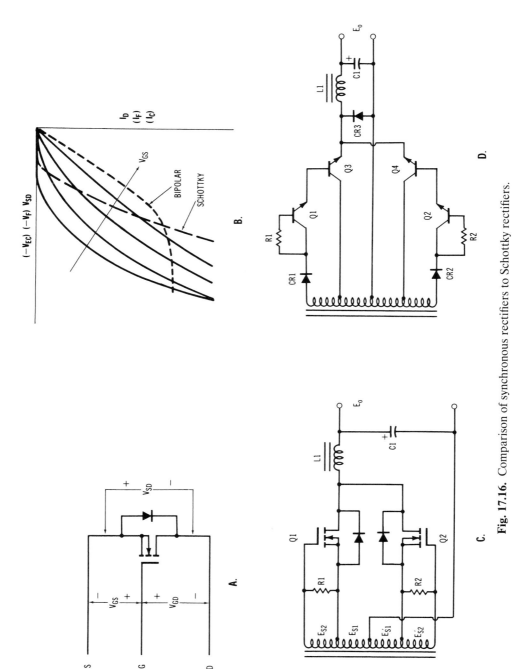

Fig. 17.16. Comparison of synchronous rectifiers to Schottky rectifiers.

656

transistor is used instead of a slow switching alloy transistor. The first quadrant output characteristic of a bipolar transistor is also superimposed in Fig. 17.16B. If the transistor is driven with sufficient base current, $V_{CE,sat}$ may be lower than the MOSFET or the Schottky rectifier voltage. A synchronous rectifier schematic is shown in Fig. 17.16D. When E_{s1} and E_{s2} are positive, Q_1 conducts to drive Q_3 into saturation. On the opposite half cycle, Q_2 and Q_4 conduct. However, the V_{EB} rating of Q_3 and Q_4 will limit the secondary voltage and thus the output voltage, as discussed below.

Example

In Fig. 17.16D, bipolar transistors are used as synchronous rectifiers in a switching power supply with 2.5 V output. For the transformer secondary voltage, $2E_{s,max} = V_{CE,sat(Q4)} + V_{EB(Q3)} + V_{BC(Q3)}$. Typical values are $2E_{s,max} = 0.2 + 7.9 + 0.9 = 9$ V, and $E_{s,max} = 4.5$ V. For a current of 15 A, the power dissipation in Q_3 is 3 W maximum. Neglecting the voltage drop across L_1, the *minimum* duty cycle $D = E_0/(E_s - V_{CE,sat}) = 2.5/(4.5 - 0.2) = 58\%$. If regulation is required at a lower duty cycle, E_s must increase, but this results in a breakdown of the emitter–base junction of the power transistor. Transistor Q_1 is driven by E_{s2} and R_1 while CR_1 prevents current flow through the reverse base junction. Since neither Q_3 nor Q_4 conduct during the absence of secondary voltage, CR_3 is required to conduct the flyback current.

17.5 HIGH-POWER, LOW-VOLTAGE POWER SUPPLY

The ac input to mainframe computers is typically three phase due to the high-power requirements, and the output current for low-voltage logic families may exceed several hundred amperes. Linear power supplies are not applicable due to low efficiency. Switching power supplies may require substantial filtering to minimize output ripple and EMI. Phase control may also require enormous filtering to minimize output ripple and reduce input waveform distortion. A hybrid approach is shown in Fig. 17.17A. A three-phase line regulator operates from the input, and provides an output voltage which is regulated within a desired range for input changes to provide line regulation. Since the output current change may be negligible, output voltage sensing for load change effects may not be required. The output from the line regulator is applied to a power transformer whose secondary voltage is 12-phase rectified for low output ripple.

The line regulator may consist of SCRs for tap switching the transformer primary. The regulator transfer function is shown in Fig. 17.17B. With three taps per phase, a $\pm 3\%$ output variation can be achieved with a $\pm 10\%$ input variation.

Since a single rectifier forward voltage drop is desired in the output line, the transformer secondary must have a neutral or center tap. The six-phase star (Chapter 4, Table 4.2 and Fig. 4.10) has excessive output ripple. However, the peak-to-peak ripple (relative to the peak voltage) in a 12-phase star will be

$$E_{r,pk\text{-}pk} = (1 - \cos 15°)100 = 3.4\% \qquad (17.27)$$

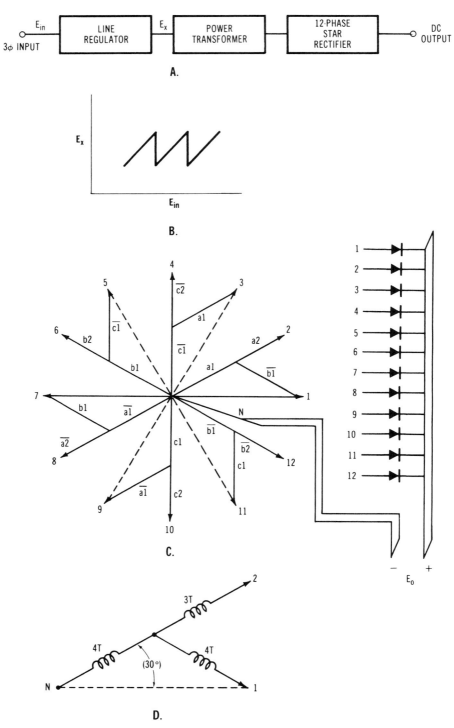

Fig. 17.17. Voltage-regulated, high-current power supply with 12-phase star rectifier.

The low ripple will require minimal filtering. An $A\phi$, $B\phi$, and $C\phi$ phasor diagram of the transformer secondary is shown in Fig. 17.17C, with each winding connected to a rectifier. Winding a_1 and winding $\bar{b}_1$ produce phase 1, winding a_1 and winding a_2 produce phase 2, and the interconnections repeat through phase 12. The phase voltages are

$$\text{Phase } 1 = \text{phase } 2 = \text{phases } 3 \cdots 12 \tag{17.28a}$$

$$\text{Phase } 1 = 2a_1 \cos 30° \tag{17.28b}$$

$$\text{Phase } 2 = a_1 + a_2 \tag{17.28c}$$

$$a_2/a_1 = (2\cos 30° - 1) = 0.732 \tag{17.29}$$

where

$$a_1 = b_1 = c_1 = \bar{a}_1 = \bar{b}_1 = \bar{c}_1$$
$$a_2 = b_2 = c_2 = \bar{a}_2 = \bar{b}_2 = \bar{c}_2$$

For low-voltage outputs, the turns ratios of the secondary windings become critical. For a normalized 1 V per turn, a $1.268 : 1.732$ ratio is required. Since the number of turns should be an integer and minimum turns are desired, the minimum turns with minimum error between phase voltages is $a_1 = 4$ turns and $a_2 = 3$ turns, as shown in Fig. 17.17D. Thus phase 1 = 6.93 and phase 2 = 7.00 for a 1% error variation in voltage amplitude. From Chapter 4, Table 4.2, $E_{s,\text{rms}}/E_{\text{do}} = 0.715$ for a 12-pulse rectifier. For a desired output voltage, the secondary turns per volt can be determined, and since the primary winding must be rated at the same volts per turn as the secondary, the core area A_e can then be determined.

Example

For an output voltage of 3.3 V dc, allowing for a 0.5-V rectifier drop, and neglecting the transformer reactance and resistance, the phase voltage is $E_{s,\text{rms}} = 0.715(3.3 + 0.5) = 2.72$ V. Since the turns are $4 : 3 : 4$, the required volts per turn is $2.72/7 = 0.388$ V/turn. If the input is a 230-V delta line, the required primary turns are $N_p = 230/0.388 = 593$. The core area can then be determined, depending on the flux density, input frequency, and power rating.

This topology does have disadvantages. Since the conduction period of each rectifier is 30°, the ratio of $I_{F,\text{rms}}/I_{\text{dc}}$ is high, and the ratio of VA_s/P_{dc} is high. However, the larger transformer size is offset by the decreased output filtering required. Decreased filtering implies that a capacitor is required to provide a low source impedance at high frequency.

This configuration could be extended to a 24-phase star transformer/rectifier which produces a voltage pulse every 15°. Neglecting line reactance, the maximum peak-to-peak ripple is less than 1% $[E_r = (1 - \cos\frac{15}{2})100 = 0.86\%]$. Output filter capacitors should not be required and, in fact, the bus bars connecting the output to the load or to distributed loads could be placed physically parallel which would provide a distributed capacitance for filtering and noise suppression.

17.6 HIGH-POWER, HIGH-VOLTAGE POWER SUPPLY

High-voltage power supplies may be designed with a flyback converter stage, as discussed in Section 9.4.2. For high-power units, the transformer, high-voltage rectifiers, filter capacitor, and resistive dividers for voltage and current sensing are normally enclosed in a container and submersed in transformer oil, as shown in Fig. 17.18A. In this "brute force" approach, the input is applied to a variable transformer whose output is connected to the primary of the high-voltage transformer. This provides adjustment of the output voltage but does not provide regulation for line and load effects. Transient protection devices are inherent, as in an "off-zero" control which prevents operation unless the variable transformer wiper is physically at zero volts.

A switching topology to provide regulation and good transient response is shown in Fig. 17.18B. The transistors and rectifier bridges form bilateral switches for ac current flow. Switching these transistors on and off, from saturation to cutoff, at a moderate or high frequency will produce a duty cycle waveform of the input voltage envelope, as shown in Fig. 17.18C. This pulse train is applied to a low pass filter to reduce high-frequency core losses in the transformer. When Q_1 is on, the input voltage is applied to the filter and transformer primary. When Q_1 turns off, Q_2 is turned on to provide a path for reactive current flow, since the transformer primary has a high leakage inductance. A portion of the output voltage is sensed and compared to an adjustable reference. An amplified error signal then controls the duty cycle of the transistors to maintain a regulated output. The waveform shows 18 pulses per cycle or a switching frequency of 1080 Hz for the bipolar transistors. A higher switching frequency, with improved response time, can be obtained with the MOSFET ac switches shown in Fig. 17.18D. In this case, the inherent reverse rectifier allows current flow when the opposite transistor is turned on, assuming the dv/dt rating is observed. In either case, a soft-start mode and current limiting provide protection on the primary side.

This topology essentially operates in a reverse cycloconverter mode with pulse width modulation. The transformer must be designed for line frequency operation, and the ripple on the dc output is at twice the line frequency. However, a topology which allows the transformer to be designed at the switching frequency may not substantially reduce overall volume, since sufficient spacing is required for high-voltage isolation between the components within the "tank."

The control circuitry consists of a master oscillator which is synchronized to the line frequency by a PLL (phase-locked loop). A triangle waveform and the error signal are applied to a comparator for pulse width modulation. At 1080 Hz, the triangle waveform provides a symmetrical conduction period at 20° intervals, as opposed to a sawtooth waveform which modulates only one edge. The average voltage across the transformer is essentially the input voltage times the duty cycle.

Many other topologies are possible. For instance, the bilateral switches of Fig. 17.19A can produce a symmetrical high-frequency pulse train, shown in Fig. 17.19B, across the primary, and the transformer can be designed for the switching frequency, to reduce transformer size. The penalty in this case is a center-tapped primary, and transistor voltage ratings are double those of Fig. 17.18B. Also, the fundamental frequency of the full-wave rectified pulse train at the capacitive filter is twice the input frequency. The switching frequency should be an even multiple

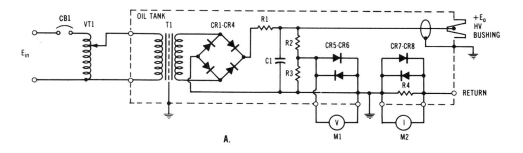

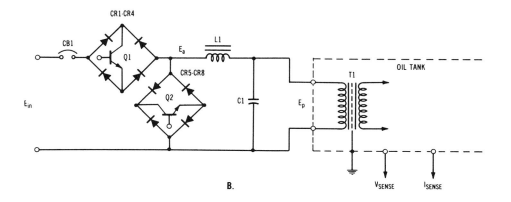

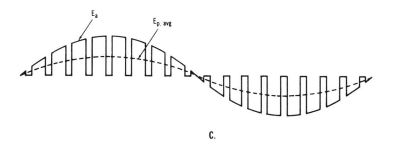

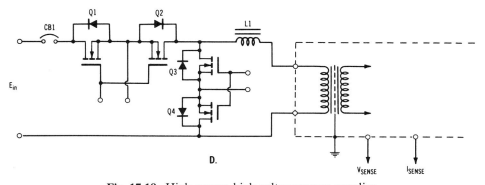

Fig. 17.18. High-power, high-voltage power supplies.

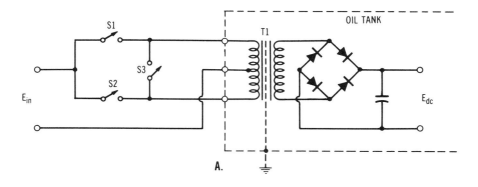

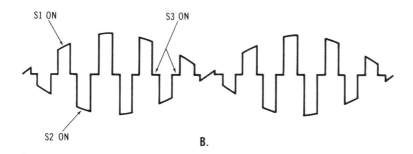

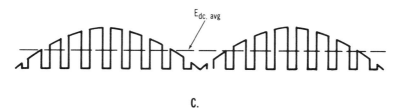

Fig. 17.19. High-frequency switching on the primary of a high-power, high-voltage transformer.

of the input frequency. In Fig. 17.19B, five positive and five negative pulses occur each half cycle, and the pulse widths are symmetrical about zero, which prevents a dc component in the transformer. The rectified output waveform and $E_{dc, avg}$ are shown in Fig. 17.19C.

17.7   CURRENT-REGULATED POWER SUPPLIES

Current-regulated power supplies are frequently required to maintain a constant current through a load, as opposed to the previously discussed voltage-regulated power supplies. The constant current is independent of load voltage or load impedance, and the load voltage may vary over a wide compliance range. Typical applications of constant current power supplies are:

1. Electromagnetic devices such as solenoids and magnet loads. In these cases the field strength is proportional to the number of turns, the magnetic permeability of the material, and the current through the coil. Since the first two parameters are normally fixed, a constant magnetic field requires a constant current. The voltage range at the load will vary as a function of coil resistance, which changes with temperature.

2. Electrochemical reactions such as battery charging and plating. In these cases the constant-current charging of a discharged battery will prevent excessive charge current, which could result from constant-voltage charging. Constant-current (or current-limited) charging over a fixed period of time restores a certain ampere-hour capacity to the battery. For controlled coulomb plating purposes, a constant current allows a predetermined plating time.

3. Plasma arc sources. The load impedance of arc sources may vary over an appreciable range, and a power supply with a high output impedance is normally required to prevent damage to the plasma source. A constant current is desired and the compliance voltage range of the power supply may vary over a wide range.

A fundamental difference between current regulation and voltage regulation is the relation

$$I = \Delta E / \Delta R \tag{17.30}$$

This simple ratio governs the voltage range of the power supply with respect to load resistance. For instance, a constant current cannot be maintained in an open circuit. A constant current of 10 A into a load resistance which varies from 2 to 10 Ω requires a compliance voltage of 20–100 V. If the load resistance increases to 500 Ω, the voltage required would be an unrealistic 5 kV.

17.7.1 Linear Current Regulators

Current regulation is obtained by sampling the load current and comparing this voltage to a stable reference voltage. The resulting error signal then controls a power stage to maintain regulation for changes in line voltage and load resistance. The power supply may be a linear type or a switching type. The 3-terminal LM117 regulator, discussed in Section 17.2.3.3, is an ideal low current linear regulator. In Fig. 17.20A, a rectified and filtered input powers the LM117 which has a 1.25 V reference. The regulated current is adjustable by R_3. Minimum output current occurs when the wiper of R_3 is at the right-hand terminal. Assuming R_2 and R_3 are much greater than R_1, the required value of R_1 is

$$R_1 = 1.25 / I_{0,\min} \tag{17.31}$$

At the maximum output current, $V_{R1} = V_{R2} + V_{R3} = 1.25 + 1.25 R_3 / R_2$. Then the maximum output current is

$$I_{0,\max} = (1.25 / R_1)(R_2 + R_3) / R_2 \tag{17.32}$$

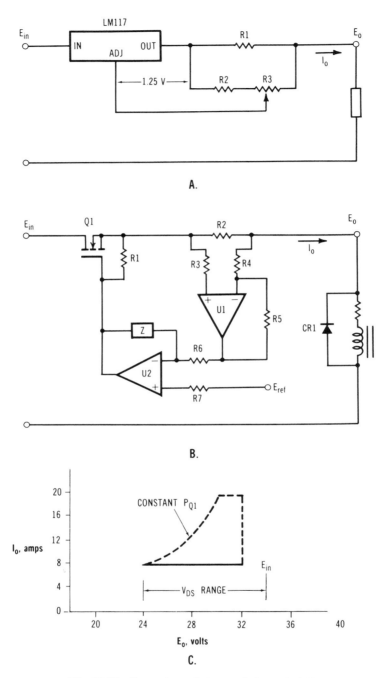

Fig. 17.20. Current regulators and characteristics.

Since the output current range is normally known, the ratio of the divider resistor is

$$R_3/R_2 = (I_{0,\max} R_1/1.25) - 1 \tag{17.33}$$

Example

An output current of 0.2–0.5 A is required. Assuming the input voltage is sufficient to overcome the voltage drop across an LM117 and R_1 and the power dissipation in the LM117 is well within the rating, find the resistance values. From Equation (17.31), $R_1 = 1.25/0.2 = 6.25$ Ω. From Equation (17.33), $R_3/R_2 = 0.5 \times 6.25/1.25 - 1 = 1.5$. Choosing $R_3 = 500$ Ω, use $R_1 = 332$ Ω, a standard 1% value. Even though a portion of I_0 and the adjustment current of the LM117 flows through R_2 and R_3, these values are sufficiently greater than R_1, to cause negligible error.

In Fig. 17.20B, a MOSFET is controlled to provide a regulated output current. The voltage is sensed across R_2, which is a current shunt. This voltage is amplified by U_1 and applied to U_2 for comparison to a reference voltage. The resulting error signal is applied to the gate of Q_1. If the output current increases, the output of U_1 will increase, causing the output of U_2 to decrease, which reduces the conduction of Q_1 to return to output current to the original value. Since Q_1 operates in the forward-biased mode, a MOSFET is chosen over a bipolar transistor. The voltage drop across Q_1 will be $E_{in} - E_0$ and the power dissipation in Q_1 is

$$P_{Q1} = (E_{in} - E_0)I_0 \tag{17.34}$$

The power dissipation must be well within the device rating. The maximum output current and the maximum output voltage are normally known, and the minimum voltage drop allowable across Q_1 can be determined. Then the minimum input voltage is

$$E_{in,\min} = (P_{Q1} + E_0 I_0)/I_0 \tag{17.35}$$

An observation of the above equation shows that for a fixed input voltage and a fixed output current, the power dissipation in Q_1 decreases as E_0 increases, which indicates the output current can also increase without exceeding the rated power dissipation. Since the load is inductive, CR_1 is connected across the load to clamp a transient voltage should the input suddenly disappear, or the load become disconnected.

Example

An output current of 8 A is required for a load resistance which varies from 3 to 4 Ω. The output voltage then varies from 24 to 32 V. Let $E_{in} = 34$ V for a 2-V drop across Q_1. From Equation (17.34), $P_{Q1} = (E_{in} - E_{0,\min})I_0 = (34 - 24)8 = 80$ W. For Q_1, choose a 2N6763 with the following ratings: $BV_{DSS} = 60$ V, $I_D = 28$ A, $P = 150$ W at $T_c = 25°C$. At 80 W, the case temperature must not exceed 83°C. (In actual design, T_c should be derated to 55°C, which results in a large size heat sink.) For R_1, use a 50-mV, 20-A shunt. If U_1 has a gain of 100, $E_{ref} \cong E_{U1} =$

$(0.05/20)I_0 \times 100 = 0.25I_0 = 2$ V at 8 A. The output characteristic is shown in Fig. 17.20C. The output current may be increased at higher output voltages without exceeding the 80-W dissipation. The area within the dashed line represents the regulator power dissipation capability without exceeding a drain current of 20 A. The reference voltage may be increased to 5 V for 20 A, but an additional circuit is required to monitor V_{DS} and limit the output of U_2 to maintain Q_1 within the SOA derating. Also, the input voltage will probably vary over some range. If E_{in} increases by 25% ($\pm 11\%$ of nominal), the maximum input voltage is 42.5 V, and the maximum transistor dissipation is $P_{Q1} = (42.5 - 24) \times 8 = 148$ W. In this case, two transistors in parallel would be required.

Magnet power supplies require good regulation and low output ripple. For the latter reason, linear power supplies are frequently used. A high current output then means several transistors must be paralleled, not only for current capability but also for SOA operation, since the voltage across the transistors will be the maximum input voltage minus the minimum compliance voltage. The absence of forward-biased second breakdown in MOSFETs makes these devices particularly attractive. The inductive nature of the load indicates that load current cannot change instantaneously, and the inherent high-speed MOSFET is not utilized. In fact, slow-speed bipolar alloy transistors may be used. The SOA comparisons of alloy, diffused, and MOSFET devices are shown in Chapter 2, Fig. 2.4. In either case, the effects of paralleling must be considered. For bipolar devices, a means for current sharing must be implemented, as discussed in Section 2.1.2.5. The high transconductance of MOSFET devices operating in the linear mode may produce parasitic oscillations due to external drive impedances.

17.7.2 Current-Regulated Hybrid Power Supply

The block diagram of a current-regulated hybrid power supply is shown in Fig. 17.21A. The input is directly rectified, filtered, and applied to a full-bridge PWM converter which is voltage regulated, but with a variable output voltage controlled by the reference of the current regulator. Current limit and soft start are inherent in the converter. The output of the converter is applied to a linear current regulator. (The converter itself could supply the regulated current by closing the loop around the current sense element. In this case, a maximum duty cycle limit or dead time control would be mandatory to protect the converter from the occurrence of a high output impedance.) The linear regulator is used to obtain a very low output current ripple, and a fast response to load changes. An auxiliary transformer, rectifier, filter, and appropriate regulators provide control and drive power. A typical load profile is shown in Fig. 17.21B.

For a desired output current, the reference voltage is adjusted to a fixed value, and the voltage across the pass stage is a function of E_{dc} and the compliance voltage of E_0. To maintain the pass stage within the rated SOA, E_{dc} is automatically decreased as I_0 is decreasingly adjusted. This also increases efficiency. From Equation (17.34), the pass stage must be designed for a power dissipation of $P = (E_{dc} - E_0)I_0$. However, the significance of Equation (17.30) is now evident. Since ΔR_L is normally fixed, ΔE_0 increases as I_0 increases. But E_{dc} must be sufficient to provide maximum E_0 at maximum I_0. If the load impedance de-

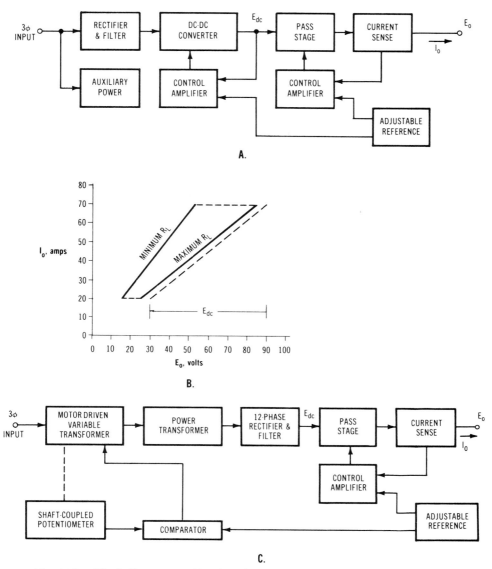

Fig. 17.21. Block diagrams and load profile of current-regulated power supplies.

creases to minimum, the voltage across the pass stage and the power dissipation of the pass stage will increase.

Example

Referring to the load profile of Fig. 17.21B, $R_{L,\min} = 0.75$ Ω and $R_{L,\max} = 1.20$ Ω. It is desired to supply a regulated current, adjustable from 20 to 70 A, to a load. Thus the maximum excursion of E_0 is 15–84 V. The converter is designed to supply 70 A at $E_{dc} = 90$ V, or 6300 W. This allows a 6-V minimum drop across the pass stage. When the reference voltage is adjusted for the desired current, the circuitry can be designed such that this 6-V difference between E_{dc} and $E_{0,\,max}$ at a desired current can be maintained. For $I_0 = 70$ A at $R_{L,\,min}$, the power dissipation in the pass stage is $P = (90 - 52.5)70 = 2625$ W. The pass stage efficiency is $\eta = 52.5 \times 70/6300 = 58\%$ minimum, and $\eta = 84 \times 70/6300 = 93\%$ maximum. However, another interesting factor is present. If E_{dc} is fixed at 90 V, the maximum power dissipation in the pass stage occurs at the "half power" point of $E_0 = 45$ V, where $I_0 = 60$ A, resulting in $P = (90 - 45)60 = 2700$ W. In fact, if E_{dc} is fixed at 90 V for an I_0 range of 40–70 A, the power dissipation in the pass stage varies from 2400 to 2700 W. Since this change is minimal compared to the 2625 W for which the pass stage must be rated when $I_0 = 70$ A, the effect of decreasing E_{dc} at lower currents results in decreased input power and increased operating efficiency.

When supplying a load, such as a high-power magnet, with precise currents up to 200 A at a load resistance of 1 Ω nominal, the output power is 40 kW. A cost-effective topology, assuming ample volume is available, is shown in Fig. 17.21C. In this case, the three-phase input could be 480 V which is applied to a motor-driven variable transformer, and this output applied to the primary of a power transformer with separate wye and delta secondaries. The secondary voltage is 12-phase rectified to produce low ripple at E_{dc}. The pass stage typically consists of a multitude of paralleled transistors mounted on a liquid cooled heat sink. To minimize dissipation in the transistors at lower output currents, the reference voltage, which controls the conduction of the pass stage, is also applied to a comparator stage whose other input is a voltage developed at the wiper of a potentiometer which is shaft coupled to the variable transformer. In response to the reference, the motor rotates the variable transformer rotor until the comparator is balanced, and then the motor stops. This minimizes the voltage difference across the pass stage by lowering (or raising) E_{dc}, thus significantly reducing the power dissipation in the pass stage at low currents, as well as decreasing the total input power.

REFERENCES

1. G. C. Chryssis, *High Frequency Switching Power Supplies* (McGraw-Hill, New York, 1984).
2. I. M. Gottlieb, *Regulated Power Supplies* (Howard W. Sams, Indianapolis, IN, 1981).
3. E. R. Hnatek, *Design of Solid-State Power Supplies* (Van Nostrand Reinhold, New York, NY, 1971).
4. A. I. Pressman, *Switching and Linear Power Supply, Power Converter Design* (Hayden Book Company, Rochelle Park, NJ, 1977).

18

UNINTERRUPTIBLE POWER SYSTEMS

18.0   Introduction
18.1   UPS Equipment and Components
18.2   On-Line System
18.3   Off-Line System
18.4   Batteries
    18.4.1   Lead–Acid
    18.4.2   Nickel–Cadmium
18.5   Battery Chargers
    18.5.1   Lead–Acid Battery Charging
    18.5.2   Nickel–Cadmium Battery Charging
18.6   Inverters
18.7   Inverter Preregulator
18.8   Transfer Switches
    18.8.1   Phase Synchronization
    18.8.2   Sensing Utility Failure
18.9   Complete System Analysis

18.0   INTRODUCTION

A consistent increase in the need for reliable and quality ac power has spurred the applications and needs for uninterruptible power systems, hereafter referred to as UPS. The Industrial and Commercial Power Committee of the IEEE defines an UPS as a system which is designed to provide power during all periods wherein the normal or prime source of power is outside acceptable limits, without causing disruption of the flow of acceptable power to the load. The historic function of an UPS was to provide backup power to critical loads to maintain operation in the event of utility mains failure, frequently defined as a blackout condition. More recently, the requirement for an UPS is a result of the declining quality of available utility-provided power. As opposed to a blackout, a brownout condition may occur during peak periods of utility usage whereby the mains voltage drops to a level well below the minimum voltage required to operate a critical load, and the duration of this condition may be seconds or hours. This is especially true of modern computer systems which are also sensitive to a variety of disturbances, such as voltage transients and noise, as well as to undervoltage or overvoltage conditions. Other data processing equipment and computer-controlled machining equipment also need reliable power for operation to protect memory contents as a minimum. The UPS then provides backup power to maintain operation until the utility power returns or until an orderly shutdown of the critical power equipment can be made. In critical or emergency conditions where power to the load must be maintained, the static UPS can supply power until a diesel-powered generator can be brought on-line.

18.1   UPS EQUIPMENT AND COMPONENTS

A solid-state (or static) UPS is comprised of the following units: (1) an energy storage device (usually a battery and thus a dc source) to provide power during utility or mains interruption; (2) a rectifier-charger as a means of restoring the energy to the storage device when the utility power is again available; (3) an inverter, which is powered by the dc source, to provide the desired and regulated ac output; and (4) an automatic transfer switch which connects the load to the inverter or to the utility line, depending on the desired configuration.

Various topologies, configurations, and installation methods are available, depending on actual user requirements. The two major topologies are on-line system, where the inverter powers the load continuously, and off-line system, where the utility is the primary power for the load.

18.2   ON-LINE SYSTEM

The on-line system (sometimes called a floating system) shown in Figure 18.1 is the most common UPS configuration used for low-, medium-, and high-power applications. In Fig. 18.1A, utility power is on and is applied to a regulated rectifier-charger which supplies dc current to the inverter and maintains a float condition or "trickle charge" on the battery. The charger normally contains a transformer for voltage scaling and also provides isolation for safety purposes. The heavy lines show the main path of current flow, and the dashed lines show a path of lesser current flow. A regulated output from the rectifier-charger is necessary to maintain the battery at full charge independent of utility voltage variations. The inverter is coupled to a transfer switch (reference Section 18.8) whose output is coupled to the load. In this mode of transfer switch operation, the system is referred to as reverse transfer.

In Fig. 18.1B, the utility power has gone off and the battery now supplies power to the inverter which continues to supply power to the load via the transfer switch. Thus the critical load has not been affected in any way by the loss of utility power. The regulated output of the inverter supplies power to the load independent of the battery voltage, which will now decrease due to the discharge current being drawn by the inverter. At some point in time, determined by the capacity or ampere-hour rating of the battery, the battery voltage will decrease to the minimum input voltage required by the inverter to maintain a regulated output. Many UPS

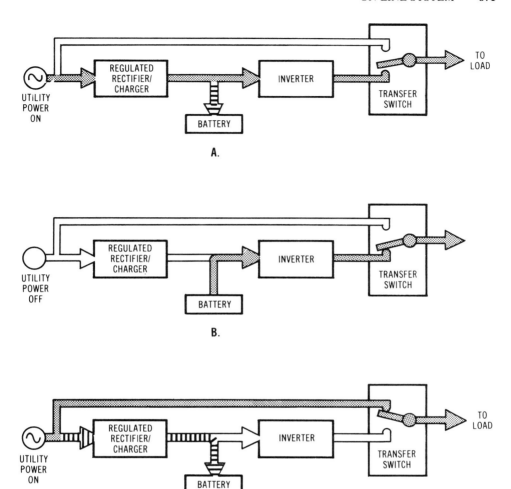

Fig. 18.1. An on-line UPS.

manufacturers offer an alarm means to indicate this condition is about to occur, which allows the operator to commence an orderly shutdown of the load if the utility power is still off. If utility power returns, as in Fig. 18.1A, before the minimum inverter input voltage is reached, the regulated rectifier-charger supplies current to the inverter and to the battery for recharge. Thus the rectifier-charger must have the output current capability to supply the required current to the inverter, plus the charging current to the battery (which is normally much less than the discharge current) at a rate recommended by the battery manufacturer. Battery charging continues at this rate until a total ampere-hours exceeding the discharge ampere-hours has been restored to the battery. The charging current then decreases to a value to maintain the float voltage on the battery ("trickle charge" in some terminology).

Static inverters have been known to malfunction even with the highest degree of quality components and design reliability. Suppose that the system is operating as in Fig. 18.1A, and the inverter ceases operation. A control circuit senses this occurrence and activates the transfer switch. Operation is now as shown in Fig. 18.1C, where the utility mains provide power to the load (hence the name reverse transfer) until the inverter operation can be restored. Of course, in this condition, no backup power source is available and the load will cease to function if the utility power is lost. The operator then has two choices: (1) ignore the loss of inverter power and hope the utility remains on until the inverter can be restored or (2) perform an orderly shutdown of the system while utility power is still on. To solve the problem of potential inverter malfunction, many manufacturers (sometimes by user insistence) offer redundant inverter stages wherein each inverter is capable of supplying full-load power, and the loss of one inverter stage does not affect the operation of the critical load.

Example

Prior to a detailed discussion, the following example is presented to clarify operation of an UPS topology of Fig. 18.1. The load requires 120 V, 60 Hz at 750 VA. For an inverter efficiency of 78%, the input to the inverter will be $750/0.78 = 962$ W. The dc voltage to the inverter is chosen as 48 V with an operating range of 42–56 V in order to use four 12-V batteries in series. The recommended float voltage on the battery is $4 \times 13.75 = 55$ V. A 20-min backup time or desired operating time when the utility power fails is desired. When the utility power fails, the battery voltage will drop abruptly (due to the inverter load) to approximately 50 V, and the voltage at near to end of discharge will be 44 V. For an average voltage of 47 V at 962 W, the battery current is $962/47 = 20.5$ A. This is equivalent to $20.5 \times 20/60 = 6.8$ Ah (ampere-hour), but the battery must be sized per manufacturer's high-rate discharge (reference Section 18.5). Since a total of 6.8 Ah was removed from the battery, approximately $6.8 \times 1.2 = 8$ Ah must be restored within say 4 h when the utility power returns. At 55 V output from the charger, the charger current is $962/55 = 17.5$ A to the inverter plus $8/4 = 2$ A to the battery, or 19.5 A from the charger. The charger peak output power is $19.5 \times 55 = 1073$ W. For a charger efficiency of 80%, the input power is $1073/0.8 = 1341$ W. The input power factor (PF) seen by the utility will be quite low due to the high peak nonsinusoidal input current, and a typical value is 0.7 PF. Thus the utility input will be $1341/0.7 = 1916$ VA. If the utility voltage range is 105–130 V, the maximum input current to the UPS is $1916/105 = 18.2$ A. The utility service to the UPS installation should therefore be at least a 20-A rating.

A cost disadvantage of the system shown in Fig. 18.1 is that the rectifier-charger must provide a regulated output of up to 1073 W, of which $17.5 \times 55 = 963$ W is required by the inverter, and only 110 W is required to recharge the battery. But the inverter is designed to operate over a wide input range. An unregulated rectifier could be used to power the inverter, with a much smaller power rectifier-charger with regulated output being used to recharge the battery. Such a configuration is shown in Fig. 18.2A. From the above example, the rectifier-charger can now be designed to supply a maximum of 2 A to the battery. A means of decoupling the battery from the inverter input is now required, due to the

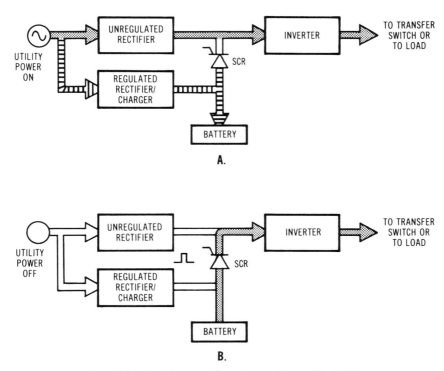

Fig. 18.2. Auxiliary rectifier-charger for on-line UPS.

condition at the low-input line where the output of the unregulated rectifier may be less than the battery voltage. This is accomplished by inserting a thyristor (SCR) between the battery and the inverter. (Note that a discrete rectifier would probably not work in this case since the rectifier could become forward biased at low values of utility voltage, and the battery would discharge into the inverter.) When the utility power fails, a sensing circuit then triggers the SCR, and current flows from the battery to the inverter, as shown in Fig. 18.2B. Power is dissipated in the SCR and the inverter input voltage will decrease by the forward voltage drop of the SCR. When the utility power is restored, the SCR becomes reverse biased and turns off to decouple the battery from the inverter.

18.3   OFF-LINE SYSTEM

For conditions where the utility power is clean and is not subject to abnormal voltage excursions or transients, an off-line system could be used to power a critical load, with an overall reduction in cost. This cost savings is primarily due to the reduced inverter performance characteristics. Referring to Fig. 18-3A, utility power is normally applied to the load. The regulated rectifier-charger need only supply recharge current flow to the battery and a small current flow to the inverter which is now operating at no load. Thus the unregulated rectifier shown in

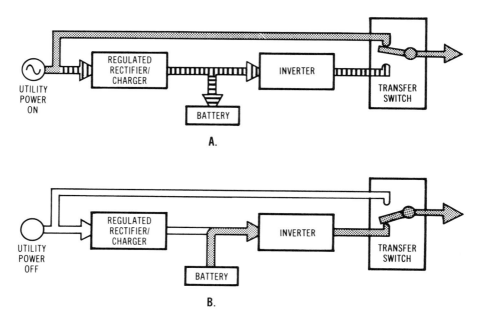

Fig. 18.3. An off-line UPS.

Fig. 18.2 is eliminated. When the utility power fails, a sensing circuit actuates the transfer switch and the load is then powered by the inverter, as shown in Fig. 18-3B. However, the sudden change in inverter load will normally cause the inverter output voltage to decrease for several cycles. In fact, the inverter voltage might not even be present at the time of transfer. Monitoring circuits should therefore indicate the occurrence of an inverter malfunction.

When the utility power returns, the transfer switch is again actuated and operation once again reverts to that of Fig. 18.3A. As mentioned at the beginning of this section, the load must be capable of tolerating the normal excursions of the utility power. This approach is therefore used primarily to protect loads from a total blackout condition. The topology is commonly known as forward transfer and is sometimes referred to as an IPS (interruptible power system).

18.4 BATTERIES

A battery electrochemically converts chemical energy into electrical energy and consists of a single cell or multiple cells connected in series. The cell consists of an anode (negative plates), cathode (positive plates), separators between the plates, and electrolyte. Anodic materials such as lead, cadmium, magnesium, and zinc are characterized by the ease with which they give up electrons, which in turn become positive charged ions (cations) in the electrolyte. Cathodic materials such as lead dioxide, nickel, mercury, and silver are characterized by the ease which they accept electrons. During discharge, the oxidation state of these positive plates is reduced, which in turn produces negative charged ions (anions) in the electrolyte. The electrolyte forms an ionic path to complete the electrical circuit.

Batteries can be grouped in two classifications, primary and secondary. The primary types, such as zinc–carbon, mercury, and lithium, have a "one-time" usage, cannot normally be recharged, and are normally restricted to powering portable devices. Secondary battery types, such as lead–acid, nickel–cadmium, and some alkaline types, can be discharged and recharged for several hundred cycles. Therefore, the secondary batteries or cells are applicable to an UPS where the battery (1) stores energy during the time when the utility or mains power is present; (2) releases this energy to an inverter or converter (which in turn powers the required load) when the utility power is absent; and (3) is recharged when the utility power returns. Lead–acid is the most common type used in an UPS, primarily due to low cost. Nickel–cadmium has a higher energy density than lead–acid but may cost three times as much. Silver–zinc has the highest energy density, typically 55 watt-hours per pound (Wh/lb) and 4.5 watt-hours per cubic inch (Wh/in.^{3}), but the extreme high cost prohibits usage except where size and weight are primary considerations.

18.4.1   Lead–Acid

Each lead–acid cell consists of a lead anode (negative plate), a lead dioxide cathode (positive plate), various alloys in the grid structure of the plates, an electrolyte concentration of sulfuric acid, and microporous rubber separators between the plates. The power or capacity rating of the cell is proportional to the surface area of the plates and the number of parallel positive plates. Generally, the number of negative plates equals the number of positive plates plus 1, since the two end plates are negative. The nominal cell voltage is 2 V. The cell is enclosed in a container and cells may be connected in series for a desired battery voltage. The chemical reaction is expressed as

$$PbO_2 + 2H_2SO_4 + Pb \underset{\text{charge}}{\overset{\text{discharge}}{\rightleftharpoons}} PbSO_4 + 2H_2O + PbSO_4 \qquad (18.1)$$

$$+\text{plate} \qquad\qquad -\text{plate} \qquad\qquad +\text{plate} \qquad\qquad -\text{plate}$$

Present lead–acid battery types are commonly referred to as lead–antimony, lead–calcium, and gelled electrolyte. Typical energy densities are 12 Wh/lb and 1 W/in^3. The high-capacity, long-life batteries specifically designated for UPS usage may be 80% of these values. Open-circuit voltage of a fully charged battery is 2.1 V per cell. Specific gravity of the electrolyte is typically 1.21 at 25°C and decreases as the battery discharges or as temperature increases. Battery voltages and inverter input voltages for various operating ranges are given in Table 18.1. As the required output power increases, the number of series cells is increased to prevent an excessive current draw. For medium output power, low backup periods, and where safety permits, the 72-V system is popular. The ac input line is directly rectified, controlled, and applied to the battery and inverter. Thus the transformer normally associated with battery chargers is eliminated. Reference Section 18.5.1 for battery chargers.

TABLE 18.1. Lead–Acid Battery and Inverter Input Voltages

	12 V	24 V	48 V	72 V[a]	125 V	250 V
Inverter input voltage range	11–15	22–30	42–56	63–85	105–140	210–280
Number of lead–acid cells	6	12	24	36	60	120
Typical float voltage/cell[b]	2.25–2.3	2.25–2.3	2.2–2.25	2.2–2.25	2.2–2.25	2.2–2.25
Typical equalize voltage/cell	2.4	2.4	2.33	2.33	2.33	2.33
Typical battery float voltage	14.4	28.8	53–54	79–81	132–135	264–270
End-of-discharge cell voltage	1.83	1.83	1.75	1.75	1.75	1.75
End-of-discharge battery voltage	11	22	42	63	105	210

[a]Applies to direct off-line controlled rectification of 120 V ac (no isolation) where the battery charger can supply power to the inverter stage and maintain the charging voltage on the battery at prolonged low line ac input conditions.
[b]Values shown are for lead–calcium. Float voltage for lead–antimony is typically 2.15 V/cell.

18.4.1.1 Battery Characteristics Pure lead grids and plates are soft and lack the necessary tensile strength for actual service. In the lead–antimony battery, the lead is alloyed with 5–10% antimony to provide the required hardness. Over a period of time, the antimony electrochemically transfers from the positive grid to the negative plate, which reduces battery life. Also, the battery gives off a fair amount of hydrogen during charging, and distilled water must be added periodically. In fact, increasing amounts of water must be added as the battery ages. The addition of 0.1% calcium (instead of antimony) provides the necessary tensile strength with less brittleness without any adverse electrochemical reaction. The amount of hydrogen released is significantly reduced. The calcium battery may therefore be sealed, as in the present automotive and gelled electrolyte batteries, which also reduces maintenance. In addition, the calcium battery has longer life, lower operating cost, and lower float current requirements (typically 10% of the antimony battery). For systems requiring high-energy capacity, the initial cost increase of the lead–calcium type must be weighted against these advantages. The gelled electrolyte batteries contain calcium plates, have a low internal resistance, are sealed, and can be shipped directly without special handling precautions.

Several electrochemical processes take place as the battery discharges. The active material changes from the original composition of lead dioxide to lead sulfate. As a result, the electrolyte becomes less concentrated with sulfuric acid and the specific gravity decreases. For long life, the discharge should be terminated before the battery reaches a deep discharge. This is readily accommodated in an UPS since the inverter has some minimum input voltage below which output regulation cannot be maintained. After providing an alarm that end of discharge is imminent, the inverter undervoltage circuitry may turn off the inverter, thus removing the load from the battery. This is in sharp contrast to the automotive battery where prolonged cranking can "run the battery down" and decrease battery life.

18.4.1.2 Battery Capacity   Battery capacity is the total amount of energy available from a fully charged battery. Since the nominal cell voltage is known, the capacity is expressed in ampere-hours. Capacity depends on the state of charge, the discharge current, battery temperature, and final cutoff voltage. The backup time (period when the battery is supplying power to the inverter) is then directly dependent on the capacity rating of the battery. To determine the required capacity of a fully charged battery for an UPS, the following parameters must be known or calculated: (1) the UPS output power to an external load; (2) the load power factor; (3) the efficiency of the inverter stage; (4) the average battery discharge current; (5) the average battery voltage during discharge; (6) the required backup time; and (7) the battery (or cell) voltage at end of discharge. The average battery discharge current can be calculated from

$$I_B = P_0(\text{PF})/(1.88\,n\eta)   \text{A} \tag{18.2}$$

where

P_0 = output power, VA

PF = load power factor

1.88 = average cell voltage during discharge

n = number of cells

η = inverter stage efficiency

Manufacturers of gelled electrolyte batteries typically rate capacity in terms of a constant discharge current over a 20-h period at 20°C and an end-of-discharge voltage of 1.72 V per cell. Battery voltages range from 2 V (single cell) to 12 V (six cells), but the higher voltage is recommended for decreased current and volumetric efficiency since the minimum operating voltage of an UPS inverter is normally 12 V. Capacity ranges from 1 to 80 Ah. Thus a 10-Ah battery will deliver 0.5 A for 20 h. However, at higher discharge rates, the voltage will decrease due to internal resistance, and the capacity will decrease. This condition occurs in an UPS where a backup time of only a few minutes may be required. Figure 18.4 shows the typical performance for 6- and 12-V batteries. The battery voltage remains fairly constant for a relatively long period before rapidly declining. The factor C denotes capacity, while the numbers are the reciprocal of the discharge rate. For instance, a battery with a 24-Ah rating discharged at $0.05C$ will deliver $0.05 \times 24 = 1.2$ A for 20 h. If the required backup time is 6 min, the battery can deliver a current for $2C$ or $2 \times 24 = 48$ A for this period. At this discharge rate, the capacity is decreased to $48 \times 6/60 = 4.8$ Ah. Some manufacturers are using the designation $C/20$ and $C/0.5$ or J_{20} and $J_{0.5}$ for discharge rates corresponding to the $0.05C$ and $2C$ factors, respectively.

Example

A gelled electrolyte battery with characteristics shown in Fig. 18.4 is to be selected for the 750 VA UPS in the example of Section 18.2. The nominal battery voltage is

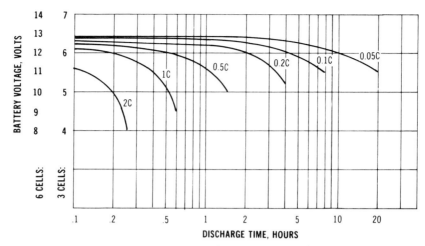

Fig. 18.4. Typical lead–acid battery (gelled electrolyte) discharge curves.

48 V, the battery current is 20.5 A, and the backup time is 20 min. The ambient temperature is 25°C. Four 12-V batteries will be connected in series. At 20 min and an end voltage of 42 V (10.5 V per battery), the discharge current is $1.2C$. The capacity is $C = 20.5$ A $\times$ 0.333 h = 6.83 Ah at a 20-min rate. However, the required capacity at a 20-h rate is $C = 20.5$ A $\times$ 20 h $\times$ 0.05/1.2 = 17.1 Ah.

Manufacturers of standard type lead–antimony and lead–calcium batteries and cells normally rate capacity at an 8-h rate. Capacities range from 25 Ah (with three cells per unit) to 4000 Ah (with a single-cell unit). As capacity requirements increase, the plate surface area and/or the number of plates are increased. A typical "fan curve" of discharge current versus ampere-hours for each positive plate is shown in Fig. 18.5. Also plotted is the cell voltage at end of discharge. Referring to Fig. 18.5, a current of 53 A for 10 min to 1.75 V/cell (the industry standard for end-of-discharge voltage) is equivalent to 53 $\times$ 10/60 = 8.83 Ah. However, at an 8-h rate the capacity is 6.25 $\times$ 8 = 50 Ah.

Example

Select a battery for an UPS rated at 30 kVA output which will provide 30 min backup time. The load power factor is 0.9 and the inverter efficiency is 87%. The battery voltage is selected as 240 V nominal (120 cells) to limit the current to a reasonable value. The input power to the inverter and the output power of the battery is 30 kVA $\times$ 0.9/0.87 = 31 kW. During discharge, the average battery voltage will be approximately 225 V and on a constant-current basis, the battery current will be 31,000/225 = 138 A. Using Equation (18.1), $I_B = 30,000 \times 0.9/(1.88 \times 120 \times 0.87) = 138$ A. From Fig. 18.5, at a discharge time of 30 min and a discharge voltage of 1.75 V per cell, the allowable current per plate is 36 A. Thus, the number of positive plates in each cell is 138/36 = 3.8. Obviously, four plates are required. This results in a current of 138/4 = 34.5 A per plate which will provide approximately 32 min backup. At this rate, the plate capacity is 34.5 $\times$ 32/60 = 18.4 Ah. However, based on an 8-h rate, the capacity of each plate is 6.25 $\times$ 8 = 50 Ah and the required battery capacity is 50 $\times$ 4 = 200 Ah.

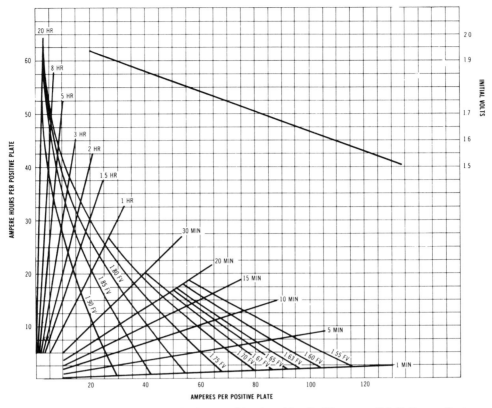

Fig. 18.5. Typical fan curve for discharge of lead–acid cell. (Courtesy C&D Power Systems an Allied company.)

Thus the battery consists of 120 individual cells with each cell rated at 200 Ah at an 8-h rate. (Note that if the inverter and battery voltage were 120 V nominal, the battery current would be $138 \times 2 = 276$ A, and 60 cells rated at 400 Ah each would be required.) Energy densities for lead–acid batteries were previously given as 12 Wh/lb and 1 Wh/in^3. Using 80% of these numbers for the high-capacity battery, total battery weight is $1.25 \times 250 \times 200/12 = 5208$ lb and total battery volume is $1.25 \times 250 \times 200/1728 = 36$ ft^3. Sufficient support and space must be allowed for the battery installation and for maintenance. The cells are to be installed on tiered racks, and the total battery and rack weight will be approximately 5400 lb and the total volume will probably be at least double that of the battery or approximately 75 ft^3. Depending on seismic zones, cabinet or two-step earthquake-braced racks may be required. This substantially increases total battery system cost as well as floor space.

18.4.2 Nickel–Cadmium

The nickel–cadmium (Ni–Cd) cell consists of a sintered cadmium hydroxide anode (negative plate), a sintered nickel hydroxide cathode (positive plate), an electrolyte concentration of approximately 33% potassium hydroxide (KOH), and woven nylon separators between the plates. The sintered plates are porous and thus have a large surface area which allows the cell to be charged and discharged at high rates. The capacity of the cell is proportional to plate surface area and the number of plates. The sintered plates are mechanically and electrochemically rugged and thus provide long life and high reliability. The electrolyte does not react with the plates, as is the case of the lead–acid battery. Thus, the Ni–Cd plates do not deteriorate, and the specific gravity of the electrolyte does not change. The nominal cell voltage is 1.2 V. The chemical reaction is expressed as

$$2NiOOH + 2H_2O + Cd \underset{\text{charge}}{\overset{\text{discharge}}{\rightleftharpoons}} 2Ni(OH)_2 + Cd(OH)_2 \qquad (18.3)$$

The cell is enclosed in a sealed or vented container. For purposes of providing backup in an UPS, the vented cell is normally used which allows high charge and discharge rates and allows gas to escape. The vent mechanism opens at a low pressure (typically 3–9 psig) but prevents entry of CO_2 from the atmosphere which could carbonate the cell. Thus the cell must periodically be replenished with distilled water.

Nickel–cadmium battery characteristics (as compared to a lead–acid battery) are (1) long operating life; (2) long storage life; (3) low internal resistance; (4) high discharge rates; and (5) wide operating temperature range ($-40°$ to $+65°$C). The latter characteristic is advantageous in military applications although the capacity at $-30°$C may be one-half the rated capacity at 25°C.

Open-circuit voltage of a Ni–Cd cell depends on several factors: (1) the state of charge; (2) the rest time following a charge; (3) the rest time following a discharge; (4) the inactive or shelf time; and (5) the cell memory phenomenon. The chemical

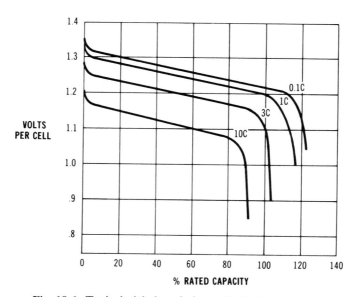

Fig. 18.6. Typical nickel–cadmium cell discharge curves.

TABLE 18.2. Nickel–Cadmium Battery and Inverter Input Voltages

	12 V	28 V	48 V
Inverter input voltage range	11–15	22–30	42–56
Number of Ni–Cd cells	10	20	38
Typical cell voltage	1.25	1.25	1.25
Typical battery voltage	12.5	25.0	47.5
Charge voltage cell	1.45	1.45	1.45
Charged battery voltage	14.5	29.0	55.1
End-of-discharge cell voltage	1.1	1.1	1.1
End-of-discharge battery voltage	11	22	42

reactions producing this latter characteristic are herein omitted. In an UPS, the condition of the cell is either fully charged, being discharged, or being charged. In the fully charged "float" mode, the cell voltage is typically 1.4–1.5 V. When discharging, the cell voltage drops from 1.3 to 1.2 V (actual values depend on the discharge current) over an extended period. If discharge continues, the cell voltage will abruptly decrease and the discharge should be terminated when the cell voltage reaches 1 V. The battery capacity (C) is expressed in ampere-hours and is normally based on a 1-h rate. A typical discharge curve for a 25-Ah battery is shown in Fig. 18.6 for various rates of discharge as a function of rated capacity. For a desired battery voltage, the recommended number of series cells is given in Table 18.2. Energy densities are typically 20 Wh/lb and 1.5 Wh/in^3. Certain precautions must be observed when charging Ni–Cd batteries, as discussed in Section 18.5.2.

18.5 BATTERY CHARGERS

Battery chargers (frequently referred to as a rectifier-charger or simply as a charger) normally operate from an ac source and provide a controlled output to charge a battery. In an UPS, the battery charger may also provide power to an additional load. The charger is in effect a power supply with special features or options and as such is subject to a wide range of designs. The most common topologies are (1) controlled ferroresonant; (2) thyristor-controlled rectification; and (3) off-line rectification followed by a series regulator or a switching regulator. However, the source is not limited to ac. Remote locations, including spacecraft and satellites, may use solar cells or fuel cells as the primary energy source to charge a battery and supply a load during sunlight, and the energy storage of the battery powers the load during darkness. In this case, the inherent high efficiency of the switching regulator is an ideal charger topology.

Block diagrams of common charger topologies are shown in Fig. 18.7. Detailed designs have been discussed in previous chapters. The chargers are voltage regulated by a feedback loop against changes in line and load. Current limiting and short-circuit protection should be standard. Additional features such as alarms and displays to indicate battery charge status, timers for two-step charging or for equalizing the battery voltage, and battery temperature monitoring to prevent overcharging may be included. In the case of Ni–Cd battery charging, the charger

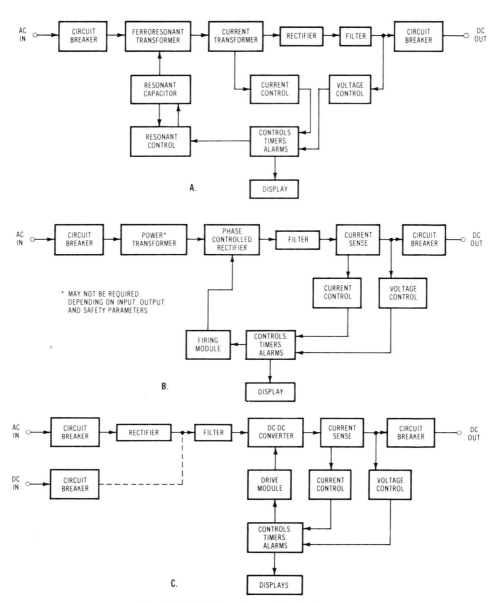

Fig. 18.7. Battery charger block diagrams.

may employ a pulsating charge current plus an alternating discharge current to prevent polarization and gassing.

The ferroresonant transformer of Fig. 18.7A inherently has good input line regulation and provides protection from output overloads and short circuits. This is accomplished by the high reactance and the tuned circuit, since a portion of the magnetic circuit is driven to saturation. In order to provide load regulation, the resonant circuit is shunted by a thyristor in series with an inductor. The firing of the thyristor controls the saturation level and thus the transformer output voltage to maintain a regulated output. The major advantages of the ferroresonant circuit are the high power factor presented to the line and the low input ac current distortion.

The phase-controlled rectifier of Fig. 18.7B is a common and simple topology for battery chargers. Design and operation are discussed in Chapter 5. In lower power UPS, the power transformer can be eliminated and the line rectified and phase controlled directly to provide a nominal 80 V dc output for changes in line and load. Reference Table 18.1. The disadvantages of phase control are the high ac input current distortion and the EMI (electromagnetic interference) conducted to the ac input (unless input filters and wave-shaping circuits are used). In high-power units operating from a three-phase source and employing 6-phase, 12-pulse controlled rectification, the input current distortion is minimized.

The topology of Fig. 18.7C is essentially a switch mode power supply operating at a high frequency. The major advantage of this approach is small size and lightweight, which may be important in charging a battery while supplying power to a switching dc–dc converter.

18.5.1 Lead–Acid Battery Charging

The recommended charging method is the limited-current, constant-voltage approach. Any of the charger topologies shown in Fig. 18.7 are applicable. The charging current (observing battery manufacturers' recommended rate) should normally be limited to a current of $0.2C$ for the lead–antimony battery and $0.05C$ for the lead–calcium battery, where C is the battery capacity in ampere-hours. The typical charging voltage per cell is given in Table 18.1. Overcharging will increase internal temperature and reduce battery life. Undercharging will allow some of the lead sulfate to remain on the plates, which will eventually reduce battery capacity. For long battery life, the peak-to-peak ripple at the charger output should not exceed 0.5% of the dc voltage. Once the battery is charged, the current to maintain the charge in the float mode is typically 1 mA/Ah at 25°C for the lead–calcium battery. The current in the float mode for the lead–antimony battery may be 5 or 10 times this value.

18.5.2 Nickel–Cadmium Battery Charging

Various charging methods, such as slow, quick, fast, and pulsed, may be used with any of the topologies of Fig. 18.7. However, the Ni–Cd cell is a complex electrochemical device. Relations between voltage, charge rate, pressure, and temperature are interdependent. The slow or "overnight" charge is a constant-current method of charging at $0.1C$ since the Ni–Cd can tolerate an overcharge at this rate, but the extended charge time is seldom applicable to an UPS. The quick-charge method at a $1C$ rate can charge a battery in a few hours. The fast-charge method at a $3C$ rate can charge a battery in less than 1 h. Both of these methods typically employ some form of battery temperature sensing to reduce the fast charge to a lower rate, or even to a slow charge, when the voltage or temperature reaches a preset value. However, an oscillating condition of slow charge to fast charge to slow charge may result as the battery cools and then heats during the charging cycle. Temperature sensing during the fast-charge method is required due to the effect of electrolysis. During charging, hydrogen and oxygen gas bubbles form on the plates. These bubbles, which increase in size with the charging process, retard the charging rate and reduce the transfer of heat between the plates and the electrolyte. The pulse method employs a duty cycle rate with average charging current decreasing with time. An additional pulse type of charger is the REFLEX,* except in this case the battery is automatically charged and discharged at a preset rate. This method depolarizes the plates and reduces the electrolysis reaction to allow a high rate charge with minimum heating and minimum gassing. The process is highly efficient and the battery can be fully charged in a few hours.

18.6 INVERTERS

The inverter is powered by the battery or battery charger and provides a symmetrical ac sine wave output to the critical load. Transistor inverters and SCR inverters are discussed in detail in Chapters 6 and 7, respectively. For low-power single-phase outputs with transistor switching, the following topologies are common: (1) pulse width–modulated, single-step waveform inverter; (2) pulse-demodulated inverter; and (3) inverter with ferroresonant transformer. For high-power three-phase outputs, SCR switching is normally used. In this case, impulse commutation is used in the inverter stage, and the power transformers are connected to produce a step-wave output which is filtered to a low-distortion sine wave.

Regardless of the topology, the inverter should have good voltage regulation, low output distortion, good frequency stability, output current limiting and short-circuit protection, plus other protection devices normally included in the design. In the case of high-power inverters, the ability to clear a load circuit breaker or fuse during a fault condition of a branch load may be important. This latter condition requires sufficient energy storage in the inverter output filter. Another important consideration is synchronizing the operating frequency to the utility frequency, as discussed in Section 18.8.1

18.7 INVERTER PREREGULATOR

In high-power systems, a preregulator may be placed ahead of the inverter to maintain output voltage regulation. This topology is applicable to a number of system designs, specifically a three-phase, step-wave inverter, where the step waves are filtered to provide a sine wave output. Since the "timing" of the inverter may be fixed, the output voltage is proportional to the input voltage plus inverter voltage drops. Two configurations are shown in Fig. 18.8. In each case, the battery voltage E_b is summed with the preregulator voltage E_a to produce output voltage E_0. Thus the preregulator need only switch a portion of the input power instead of the total power, as would be the case for an isolated preregulator stage.

---

*REFLEX is a trademark of Christie Electric Corp.

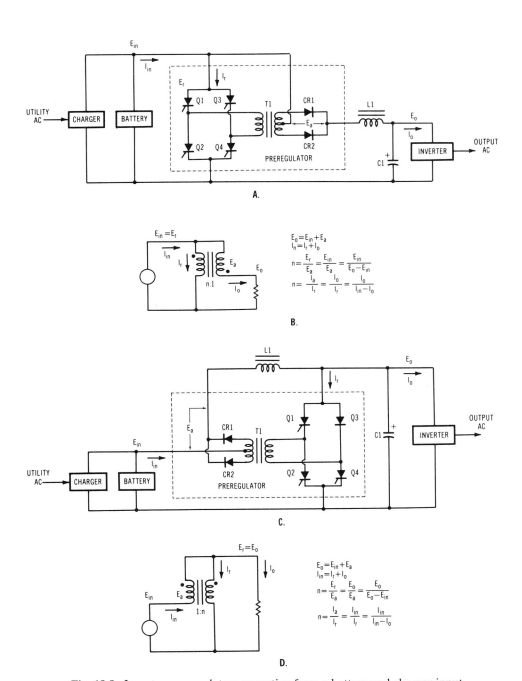

Fig. 18.8. Inverter preregulators operating from a battery and charger input.

When the utility voltage is present, the battery charger maintains the proper charge on the battery and supplies power to the preregulator, which consists of a thyristor bridge with center-tapped full-wave rectified output. The preregulator output is filtered, and a feedback loop from the inverter output controls the preregulator switching to maintain a regulated inverter output for changes in line and load. The bridge is a pulse width–modulated converter, and the commutation (not shown here) of the SCRs is discussed in Section 7.5. When utility power is interrupted, the battery voltage decreases due to the discharge current and the internal resistance of the battery. The battery voltage continues to decrease as the battery discharges, but the preregulator voltage increases to provide the necessary input voltage to the inverter to maintain the ac output voltage in regulation. In Fig. 18.8A, the bridge is powered from the battery, and the "autotransformer" equivalent circuit and ratios are shown in Fig. 18.8B. In Fig. 18.8C, the bridge is powered from the output; Fig. 18-8D is the "autotransformer" equivalent circuit and ratios. At the end of battery discharge, the duty cycle of the regulator must be less than unity to maintain regulation. Thus the minimum battery voltage determines the worst case condition for regulator design.

Referring to Figs. 18.8A and B, the following relations apply:

$$E_0 = E_{in} + E_a = E_r + E_a \tag{18.4}$$

$$I_{in} = I_r + I_0 = I_r + I_a \tag{18.5}$$

$$E_{a,avg} = DE_{a,pk} = DE_r/n \tag{18.6}$$

$$I_{r,avg} = DI_{r,pk} = DI_0/n \tag{18.7}$$

where

E_0 = output voltage

E_{in} = input (battery) voltage

E_a = regulator output voltage

E_r = regulator input voltage

I_{in} = input (battery) current

I_a = regulator output current

I_0 = output current

D = duty cycle, $= 2t_{on}f$ (full wave)

f = switching frequency of the regulator

n = turns ratio (primary/secondary) of T_1

Referring to Figs. 18.8C and D, the following relations apply:

$$E_0 = E_{in} + E_a = E_r \tag{18.8}$$

$$I_{in} = I_r + I_0 = I_a \tag{18.9}$$

$$E_{a,avg} = DE_{a,pk} = DE_r/n \tag{18.10}$$

$$I_{r,avg} = DI_{r,pk} = DI_a/n = DI_{in}/n \tag{18.11}$$

where notations are the same as in Equation (18.7).

Examples

For either topology of Fig. 18.8, an inverter requires 45 kW input to supply a critical load. A 250-V battery provides the source backup. With utility ac on, the battery float voltage is 275 V and with utility ac off, the battery end-of-discharge voltage is 215 V. For simplicity, losses are assumed negligible. To ensure regulation at minimum battery voltage, a maximum duty cycle at 85% is chosen. The inverter input voltage is 300 V nominal. The switching frequency of the regulator depends primarily on volume and efficiency requirements but may be in the range of 2–5 kHz. From the preceding formulas, the following calculations are made. Current $I_0 = P_0/E_0 = 45,000/300 = 150$ A. At $E_{in} = 215$ V, $I_{in} = 45,000/215 = 209.3$ A, $I_r = I_{in} - I_0 = 209.3 - 150 = 59.3$ A, and $E_a = E_0 - E_{in} = 300 - 215 = 85$ V. For $D = 0.85$ at $E_{in} = 215$ V, $E_{a,pk} = E_a/D = 85/0.85 = 100$ V.

For Fig. 18.8A, the fixed turns ratio of T_1 is $n = E_r/E_{a,pk} = 215/100 = 2.15:1$. Current $I_{r,pk} = I_0/n = 150/2.15 = 69.8$ A. Also, $I_{r,pk} = I_{r,avg}/D = 59.3/0.85 = 69.8$ A as a check. The power rating of the regulator is $P_r = E_r I_r = 215 \times 59.3 = 12.75$ kW. Also, $P_r = E_a I_0 = 85 \times 150 = 12.75$ kW as a check. At $E_{in} = 275$ V, $I_{in} = P_{in}/E_{in} = 45,000/275 = 163.6$ A. Current $I_r = I_{in} - I_0 = 163.6 - 150 = 13.6$ A. Voltage $E_a = E_0 - E_{in} = 300 - 275 = 25$ V. Since $n = 2.15$, $E_{a,pk} = E_r/n = 275/2.15 = 128$ V. But $E_a = 25$ V. Then $D = E_a/E_{a,pk} = 25/128 = 0.195$. Also, $I_{r,pk} = I_{r,avg}/D = 13.6/0.195 = 69.8$ A. As a check, $I_0 = I_{r,pk}n = 69.8 \times 2.15 = 150$ A. The regulator power is now $P_r = E_r I_r = 275 \times 13.6 = 3.75$ kW due to the low duty cycle. However, the peak current into the regulator is still 69.8 A and is the current which must be commutated off by the circuit.

For Fig. 18.8C, at $E_{in} = 215$ V and $D = 0.85$, I_{in} is still $45,000/215 = 209.3$ A, E_a is still $300 - 215 = 85$ V, and $E_{a,pk}$ is still 100 V. But the fixed turns ratio of T_1 is $n = E_r/E_{r,pk} = 300/100 = 3:1$. The power rating of the regulator is $P_r = E_r I_r = 300 \times (209.3 - 150) = 17.8$ kW due to the increase in E_r, as compared to Fig. 18.8A. Then, $I_{r,pk} = I_{in}/n = 209.3/3 = 69.8$ A, which is the same as Fig. 18.8A. At $E_{in} = 275$ V, I_{in} is still $45,000/275 = 163.6$ A, I_r is still $163.6 - 150 = 13.6$ A, and E_a is still $300 - 275 = 25$ V, but $E_{a,pk} = E_r/n = 300/3 = 100$ V. Then $D = E_r/E_{r,pk} = 0.25$, and $I_{r,pk} = I_{r,avg}/D = 13.6/0.25 = 54.5$ A. As compared to Fig. 18.8A, the peak regulator current ratio is $54.5/69.8 = 0.78$, which is also the ratio of the duty cycles, $0.195/0.25 = 0.78$. The peak current into the regulator is a maximum (69.8 A) at $E_{in} = 215$ V.

Comparing the two regulator configurations:

1. Both require a commutating circuit capable of 69.8 A at 215 V input, although for conditions when the utility ac is present, the circuit in Fig. 18.8C will provide additional turn-off time to the SCRs since the peak current is lower than in the circuit of Fig. 18.8A.

2. The "power rating" of the circuit in Fig. 18.8A is about 5 kW less than the circuit in Fig. 18.8C.

3. The secondary current in the transformer (I_a) and the filter inductor current of Fig. 18.8A is about 59 A less than that of the regulator circuit of Fig. 18.8C.

4. The regulator of Fig. 18.8A is therefore considered a more optimum approach, primarily due to the fact that the regulator of Fig. 18.8C must

develop a boost voltage to "feed itself." However, other considerations may be important in the choice.

18.8 TRANSFER SWITCHES

Transfer switches, shown in Figs. 18.1 and 18.3, are essentially single-pole (three-pole in three-phase systems) double-throw power devices which switch the load from the inverter output to the utility line, or vice versa. The transfer switches may be electromechanical or static (solid state). Maintenance bypass switches (manual), which bypass the UPS itself, are used to transfer the load during periodic system maintenance. Electromechanical switches, such as relays and contactors, are relatively slow and the transfer time is typically more than one cycle of the utility line. Static switches, such as triacs and back-to-back SCRs, can switch within less than one-fourth cycle, and transfer time is a function of the control circuit sensing impending loss of voltage. In the following sections, various means of synchronizing the inverter frequency and phase with the utility frequency, sensing utility, or inverter failure, and activating the transfer switch are presented. The discussion focuses on the "off-line" system of Fig. 18.3, although the "on-line" system of Fig. 18.1 is also applicable or may be preferred.

18.8.1 Phase Synchronization

Synchronization of the phase, and thus the frequency, of the utility and the inverter is an important consideration in transfer switch operation. If an imbalance in phase or frequency is present when the switch is activated, transient conditions may occur, depending on the type of load. One method of accomplishing synchronization is with a phase-locked loop (PLL). To understand the importance of phase synchronization between the utility and inverter, first consider the action of the transfer switch without a PLL. Figure 18.9A shows a relay as an open-loop transfer switch, and waveforms are shown in Fig. 18.9B. The relay coil is energized by the utility and the load is powered by the utility through the relay contacts. When the utility fails, at time t_1, the relay deenergizes during the transfer time T_{tr}, and the contacts break with the utility and make with the inverter, at time t_2. The transfer time is shown as approximately one cycle. This random switching causes a high transient current in the transformer load since approximately three-fourths of a positive half-cycle was present before the utility failed, and approximately three-fourths of a positive half-cycle is again applied by the inverter. This transient current can cause the inverter voltage to decrease or cause the inverter to current limit until a steady-state condition is achieved. Also note that this transient condition can exist even though the phase angle between the utility and the inverter is small. For these reasons, the electromechanical transfer switch is not recommended where critical loads are present, not even with PLLs.

A static switch and PLL control is shown in Fig. 18.10. In the "off-line" topology, the utility is applied to Q_1 and Q_2, which are assumed to be on, and thus the utility is applied to the load. The utility is also applied to a step-down transformer whose secondary voltage is changed to a square wave, by U_1, and applied to the signal input of U_2. Here U_2 is a 4046 PLL which has a voltage-con-

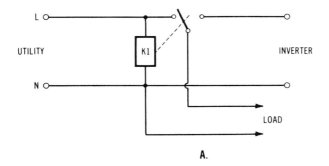

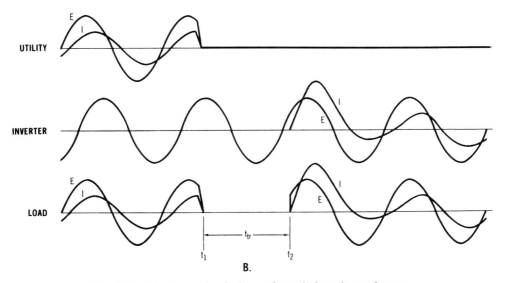

Fig. 18.9. Electromechanical transfer switch and waveforms.

trolled oscillator (VCO) and two phase comparators. Phase I is an exclusive-OR while phase II is edge triggered. For this purpose, phase I will be used. The VCO output is the clock frequency for inverter switching. The inverter output is applied to Q_3 and Q_4, which are presently off. The inverter is also applied to a step-down transformer whose secondary voltage is shifted 90°, changed to a square wave, by U_4, and applied to the comparator input of U_2. A low-pass filter is connected from phase I to the VCO input. The inherent nature of the exclusive-OR produces a symmetrical square wave of twice the input frequency when the signal and the comparator inputs are at 90°. Thus the 90° shift from the inverted output of the inverter will cause the VCO input signal to maintain the phase of the utility and the inverter within a few degrees. Also shown is the utility sense and switch control circuit which gates the SCRs.

The power waveforms are shown in Fig. 18.10B. Initially, Q_1 and Q_2 are gated on and the utility powers the load. At time t_1, the utility fails. The switch control then removes the gate signal from Q_1 and Q_2 and applies a gate signal to Q_3 and

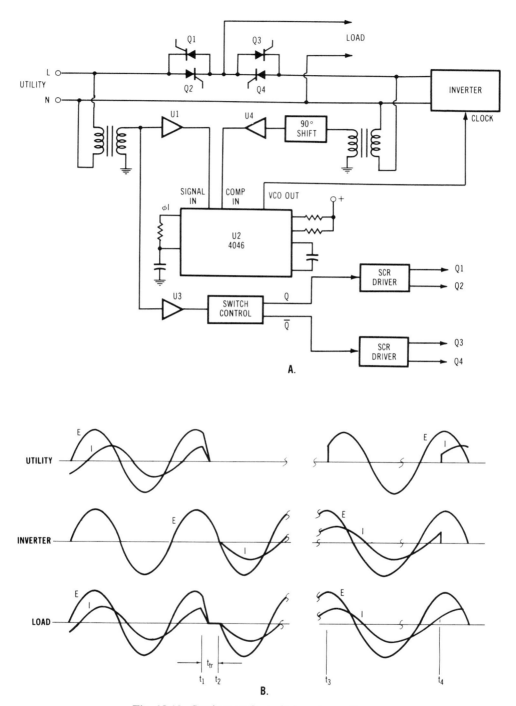

Fig. 18.10. Static transfer switch and waveforms.

Q_4 after a delay time, T_{tr}. At t_2, Q_3 and Q_4 conduct, and within less than one-fourth cycle, the voltage and current are again applied to the load. At time t_3, the utility is again present. However, the inverter and the utility may not be instantaneously in phase. The PLL then compensates the VCO input for phase synchronization at the predetermined frequency slew rate, which will not adversely affect the load. At time t_4, phase balance is achieved. Transistors Q_1 and Q_2 are then gated on, while the gate signals are removed from Q_3 and Q_4. This latter mode can be make-before-break which eliminates any disturbance in the load. In this manner, transformer and motor loads do not experience abrupt phase displacement or reversal, as might be the case without a PLL.

In a well-designed system, it can be demonstrated that the gate signals to Q_3 and Q_4 can be maintained and the inverter voltage manually increased such that the inverter reverts to a current limit mode when the inverter voltage slightly exceeds the utility, and the inverter tries to power the utility as well as the load. This is obviously an inefficient approach since the utility also supplies power to the charger and thus the battery.

Though not directly classified as an UPS, this technique is effective in supplying power to the utility if an alternate dc source is used. For instance, a solar array could be used to power the inverter (during periods of sunlight) and the inverter could directly interface with the utility to feed power (inverter power minus load power) back to the utility. Other applications include demand load switches which only activate the inverter if the load requirement is high and if an alternate power source (such as a windmill) is present. Other operational and safety switches could be used as disconnects during periods of utility outage or maintenance.

18.8.2   Sensing Utility Failure

The control circuitry must sense the condition of utility or mains power in the case of the off-line UPS of Fig. 18.3 and activate the transfer switch if the input voltage falls to a value less than the required minimum to maintain output regulation. The difference between a brownout condition or a power failure may be important in determining when to transfer the load to the inverter output, although the transfer is normally made in either case. The utility voltage may be stepped down by a transformer and the secondary voltage rectified and lightly filtered to maintain a minimum dc voltage which is applied to a comparator, the other comparator input being a reference voltage which is lower than the minimum rectified and filtered voltage. When the utility power is interrupted, the comparator switches to activate the transfer switch. However, the delay time in discharging any filter capacitor to a voltage below the comparator reference may be greater than desired for fast transfer.

A circuit which overcomes the delay from a rectified and filtered single-phase input is shown in Fig. 18.11. This is essentially the switch control block shown in Fig. 18.10A. The input voltage is stepped down by T_1 and voltage e_a is applied to a phase shift network. It is desired that e_1, e_2, and e_3 be equal in amplitude and be displaced by 60° each. When the three 60° signals are full-wave rectified by CR_1–CR_6, a six-pulse rectified waveform results, and the minimum voltage to the comparator will be 0.866 times the line-to-line voltage of e_1, e_2, and e_3 neglecting rectifier voltage drops. When E_{in} is interrupted, the rectified voltage immediately

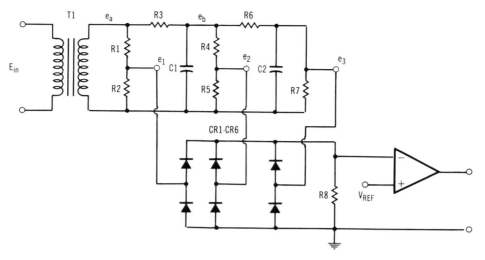

Fig. 18.11. Three-phase, full-wave rectified signal voltage with a single-phase input.

falls below V_{ref}, and the comparator switches. Referring to Fig. 18.10A, the switch control gates the SCRs in the static transfer switch within approximately 1 ms of the power interruption. The components for the 60° phase shifts are calculated by the methods of Section 1.8.1.

Example

It is desired that the input voltage to the comparator be 8 V dc nominal. From Chapter 4, Table 4.2, the line-to-neutral voltage of e_1, e_2, and e_3 is $E_{\text{dc}}/(\sqrt{3} \times 1.35) = 3.4$ V with respect to T_1 secondary return and displaced by 60° each. Choosing $R_7 = 100$ kΩ and $C_2 = 0.22$ μF, $X_{C2} = 12$ kΩ at 60 Hz. The equivalent parallel resistance and 60° phase shift is then $R_6 R_7/(R_6 + R_7) = 12{,}000(\tan 60°) = 20.8$ kΩ, from which $R_6 = 27$ kΩ. The required voltage at $e_b = e_3(R_6 + R_7)/(R_7 \cos 60°) = 8.6$ V. For $e_2 = 3.4$ V and choosing $R_5 = 8.2$ kΩ, $R_4 = (8.6R_5 - 3.4R_5)/3.4 = 13$ kΩ. In a similar manner and letting $C_1 = 1$ μF, then $R_3 = 6.8$ kΩ, $e_a = 27$ V, $R_1 = 15$ kΩ, $R_2 = 2.2$ kΩ, and $e_1 = 3.4$ V. For 116 V nominal input, the turns ratio of T_1 is then $116/27 = 4.3:1$. It is also desired that the comparator switch whenever the input voltage falls below 100 V_{rms}. Then $e_a = 100/4.3 = 23.2$, and $e_1 = 3.4 \times 23.2/27 = 2.9$ V. The minimum voltage to the comparator, allowing a 1.2-V drop for each rectifier, is then $(2.9 \times \sqrt{3} \times \sqrt{2} \times 0.866) - 2.4 = 3.75$ V, and $V_{\text{ref}} = 3.75$ V. Also, R_8 is a large value.

For three-phase systems, fast response to power interruption is obtained by using a three-phase transformer for T_1, eliminating the phase shift network and rectifying the secondary voltages directly. Providing fast response to input power interruptions in single-phase or three-phase systems may also be achieved by sensing the rate of decrease of the input voltage. This approach requires a more complicated circuit involving amplifiers, detectors, and discriminators.

18.9   COMPLETE SYSTEM ANALYSIS

The UPS, with its own power conversion stages, normally powers other equipment, which in turn have their own power conversion stages. A typical sequence of events is described. Power supplies are frequently installed in electronic equipment which manufacturers sell to an end user. The power supplies are usually switching supplies, but linear supplies may also be installed in certain equipment. For backup protection from utility failure, the end user then procures an UPS to power the complete set of equipment items. The following power path then occurs from the utility to the digital and analog loads in each of the lowest levels of equipment.

1. Utility ac is rectified to dc to charge a battery and supply power to an inverter.
2. The inverter provides an ac output to the power supplies.
3. The power supplies then power the dc loads.

To energy-conscious personnel, this ac–dc–ac–dc conversion sequence, with associated power losses in each stage, would appear wasteful. The cost of equipment to provide this power path would also appear to be high. One method of improvement would be to eliminate the second ac conversion and provide an ac–dc–dc conversion. Both topologies are analyzed in the following discussion.

The block diagram of an ac–dc–ac–dc power flow is shown in Fig. 18.12A for a hypothetical, but realistic, condition. Most of the required power is dc, the major portion provided by switching supplies and the remainder provided by linear supplies. Some ac power is also required for fans, displays, and auxiliary items. The charger, battery, and inverter is the UPS, which delivers 4.57 kW. Output power, individual stage efficiency, and losses are shown. The overall efficiency is 52.7% and the power dissipation is 2958 W. These losses (plus the 3300-W loss in the loads) are in turn converted to heat which would add an air-conditioning burden if the installation were so equipped.

The same output power requirements are shown in Fig. 18.12B, except the power supplies are replaced by converters and only a low-power inverter is required for the ac loads. The overall efficiency increases to 66.4%, and the power dissipation in the equipment is reduced to 1672 W.

Comparing the two topologies, the advantages of Fig. 18.12B are as follows:

1. From a cost standpoint, the converters should be less expensive than the power supplies (the input rectifier stage is eliminated) and a 300-W inverter is certainly less expensive than a 4.6-kW inverter.
2. The required output power of the charger is reduced by approximately 20%, resulting in a cost savings.
3. The required capacity of the battery is also reduced by 20%, resulting in further cost savings, plus decreased size and weight.
4. The input power from the utility is reduced by 1286 W. This reduces the utility cost to power the system as well as reducing the air-conditioning load. For instance, if the system operates for a period of 12 h per day, 21 days per month, the energy savings is $1286 \times 12 \times 21 = 324$ kWh/month.

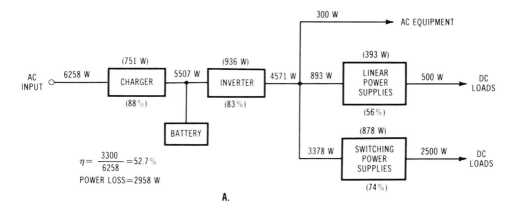

A.

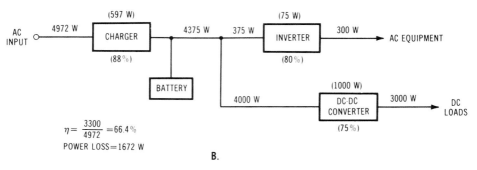

B.

Fig. 18.12. System power analysis and characteristics.

This type of analysis is not limited to that of Fig. 18.12. Block diagrams and power calculations of other system topologies can be performed in a manner similar to the above discussion. Both the equipment costs and the operating costs can be analyzed for various types of equipment and the equipment interface.

19

POWER FACTOR CORRECTION

19.0   Introduction
19.1   Power Factor Effects
19.2   Passive Correction
      19.2.1   Single-Phase Input
      19.2.2   Three-Phase Input
19.3   Active Correction
      19.3.1   Low Frequency
      19.3.2   High Frequency
      References

19.0   INTRODUCTION

Power factor (PF) correction has long been a concern by utilities supplying mains power to the user and the user's desire to maximize the available watts delivered to a desired load. The PF of a system is the ratio of real power (watts) to apparent power (volt-amperes) in linear or nonlinear loads. In linear loads the phase angle θ between the sinusoidal voltage and the sinusoidal current is related to PF as

$$PF = \text{watts}/\text{VA} = \cos \theta \qquad (19.1)$$

The PF for linear inductive loads such as motors and inductors is a lagging PF since the current lags the voltage, as discussed in Section 1.4.2. Capacitor loads produce a leading PF, as discussed in Section 1.5.3. If a capacitive reactance X_C is added across the power line supplying the motors and is equal to the inductive reactance X_L of the motors, a near-unity PF can be achieved.

The majority of power conversion equipment present nonlinear loads to the mains or power lines. The nonsinusoidal current draw caused by rectifier circuits (reference Section 4.1) and by capacitive-input filters (reference Section 4.2) can produce a discontinuous current mode (DCM), as shown in Figs. 19.2A and 4.23B. This pulse current waveform I_t contains an rms value of the fundamental I_f, plus harmonics, I_h, that do not contribute to the real power but do cause harmonic distortion. These harmonics increase the rms value of the input current and produce added heat in the conductors. The effective resistance of the conductors

695

also increases due to skin effects of the harmonic frequencies. Thus the overall I^2R losses increase without increasing the real power. The PF in the nonlinear load is then the product of both the harmonic distortion factor and the phase angle and is given by

$$PF = (I_f/I_t)\cos \theta \tag{19.2}$$

The harmonic distortion factor dominates since the phase angle of the current with respect to the voltage is usually low and slightly leading.

19.1   POWER FACTOR EFFECTS

In a typical industrial office installation, utility power is supplied by 120/208-V, three-phase, four-wire branch circuits of No. 12 American Wire Gage (AWG). Each single-phase leg may supply 12 outlet receptacles based on 150 VA per receptacle. This represents 1800 VA or 15 A available per circuit. On a 15-A rated circuit, the rms current is usually limited to 80% or 12 A to comply with UL (Underwriters' Laboratories) limits for circuit breakers, and the available apparent power is 1440 VA or 1440 W at unity PF. This office complex is installed with a multitude of personal computers, work stations, and other electronic equipment, all of which contain power supplies with input rectifiers and capacitive-input filters. The typical PF for such equipment is 0.65. Thus the actual power available is 936 W. If the power supplies within the equipment have an efficiency of 80%, the dc load power is reduced to 748 W.

These parameters are listed in Table 19.1 as an example of no power factor correction (PFC) along with the results that would be obtained with passive and active PFC. The efficiency of each type of PFC is considered. Conversely, for a desired output load of 1000 W, the rms utility current would be 16, 12, and 11 A for equipment with no PFC, passive PFC, and active PFC, respectively. Advantages and disadvantages of each type of PFC are presented in Table 19.2.

The office complex discussed above has a more insidious problem. In most cases, little or no consideration was given to voltage drop or to the effects of an overloaded condition on the neutral line. The low-power electronic equipment is typically connected from each phase to neutral. If the input currents from each phase are sinusoidal and equal, the current in the neutral will be canceled, as in

TABLE 19.1. Available Power From Standard AC Receptacle[a]

Parameter	No PFC	Passive PFC	Active PFC
Line VA	1440	1440	1440
Power factor	0.65	0.90	0.98
Efficiency, %	100	97	95
Output, W	936	1257	1340
Load, W	748	1005	1072
($\eta_{conv} = 80\%$)			

[a]120-V, 15-A circuit, $I_{rms} \leq 12$ A to comply with UL limits for circuit breaker.

TABLE 19.2. Power Factor Correction Characteristics

Type	Advantages	Disadvantages
Passive correction	Simple and rugged; high MTBF; low EMI; best suited for three-phase operation	Large size and weight; typical 0.9 PF (maximum); dual-input voltage range difficult to implement; high THD
Active, low frequency	High efficiency; simple circuitry; typical 0.96 PF moderate THD	The dc bus not regulated; moderate size and weight; holdup time changes with E_{in}
Active, high frequency	Very low THD; typical 0.98 PF; holdup time; wide E_{in} range (90–260 V ac); small size and weight	Complex circuitry; double power conversion; higher cost

balanced linear loads. If the input currents from each phase are sinusoidal but unequal, the current in the neutral will be the difference between the phase currents. Due to the capacitive-input filters of the equipment, the actual input currents from each phase to the power supplies are pulsed and discontinuous, as shown in Fig. 19.1. The current flowing in the neutral is not canceled but is additive and can approach $\sqrt{3}$ times the phase current if all phase currents are equal. Because the neutral has no circuit breaker protection, the high current will overheat the neutral conductor and can cause insulation damage and possible fire.

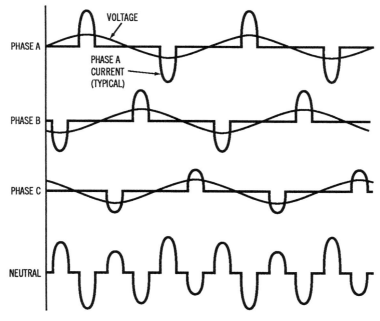

Fig. 19.1. Input current waveforms to a nonlinear three-phase load.

The high harmonic current can also overload or overheat power distribution transformers supplying the operating power from the utility. However, zig-zag windings in these transformers can cancel the neutral current that would normally flow in balanced, nonlinear loads.

The three-phase, four-wire analysis in the above paragraph is not limited to 120/208-V systems but applies to 220/380-V European mains as well. In this case, the equipment is operated from 220 V phase to neutral on a 220/380-V mains and the same problem of high neutral currents still exists. This has led the International Electrotechnical Commission to establish IEC 555-2, which limits the harmonic content of the current drawn from the mains. Class A applies to balanced three-phase equipment drawing up to 16 A and has absolute values of maximum harmonic current through the 39th harmonic. The limit of 3rd-harmonic content is 2.3 A. Class D applies to equipment having an input current with a "special waveshape" and the absolute limits are much lower than for class A equipment plus relative limits are specified that apply to power levels less than 300 W. Switching power supplies with capacitive input filters generally fall into class D unless PFC is provided or included. With PFC, the equipment will fall into class A. The emphasis of IEC 555-2 is on "public" and "household" equipment and has little impact on large industrial mains systems.

In the United States, ANSI/IEEE C57.110-1986 recommends a practice for establishing transformer capability when supplying nonsinusoidal load currents. This standard applies to distribution transformers and discusses the effects of harmonic currents and methods to calculate same. This standard has led some manufacturers to establish a "K-factor" rating for distribution transformers. These ratings are based on the transformer's ability to withstand the heating effects from harmonic (plus fundamental) currents flowing in the transformer. Studies have shown that as little as 5% total harmonic distortion can cause a 4°C rise in the transformer winding. The IEEE 519 standard is a guide for harmonic control and reactive compensation of static power converters. This standard applies to voltage harmonics at the customer service entrance.

The standard DoD-STD-1339/300, Interface Standard for Shipboard System, AC Electric Power, specifies harmonic current limits of 3% of the full-load fundamental. A high-performance PFC unit is required to meet this specification. For high-power systems, this requirement can be met with 3-phase, 12-pulse rectification to reduce line harmonics. The requirement stems from the desire to minimize harmonic currents that can be capacitively coupled to the hull, causing circulating currents in the hull of the ship and the desire to reduce sonar signatures.

19.2   PASSIVE CORRECTION

Passive PFC can be accomplished by inductive-input filter, resonant tuned series LC filter, inductive-input filter followed by a resonant tuned harmonic shunt LC trap, and ferroresonant transformer. The inductive-input filter is probably the most common passive filter used. A series-tuned LC filter requires large inductors and capacitors operating at the line frequency and ferroresonant transformers are large and heavy.

The *maximum* achievable PF for rectifier circuits with inductive-input filters is obtained from Table 4.2 as the inverse of VA_p/P_{dc}. For single-phase, full-wave rectification, $PF_{max} = 0.9$. For 3-phase, 6-pulse, rectification $PF_{max} = 0.955$, as given by Equation (19.5). For 3-phase, 12-pulse, rectification $PF_{max} = 0.99$. In single-phase inputs, the inductor can be located in series with the ac line or on the rectified dc side. In three-phase inputs, a three-phase inductor or three single-phase inductors of equal value can be located in series with the ac line.

19.2.1   Single-Phase Input

Waveforms of the ac input current to a full-wave rectified and filtered dc output, with input voltage superimposed, are shown in Fig. 19.2. Figure 19.2A represents a condition of minimal line inductance (perhaps an EMI filter) and shows the effects of the capacitive filter, as discussed in Section 4.2. Figures 19.2B–D represent conditions of increasing line inductance. Figure 19.2E represents a true inductive load with a square-wave input current waveform, as discussed in Section 4.1.1.

The inductive-input filter is conveniently analyzed by using a conduction parameter relating the inductance to the equivalent load resistance as

$$K_1 = \omega L / \pi R \tag{19.3}$$

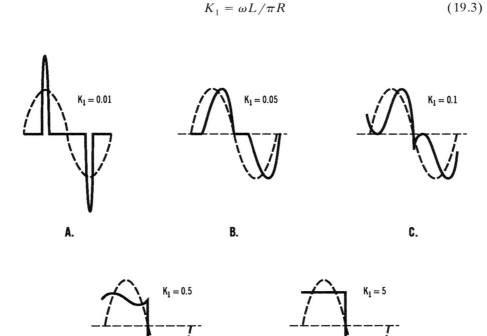

Fig. 19.2. Input current waveforms to an inductive filter as a function of K_1 for a single-phase input.

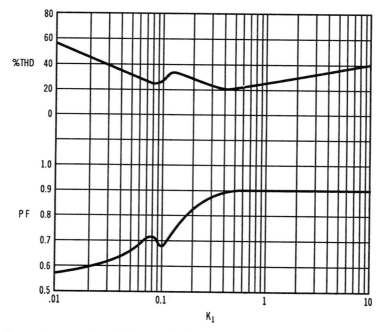

Fig. 19.3. Power factor and total harmonic distortion as a function of K_1 for a single-phase input.

For operation in the continuous current mode (CCM), the PF is defined as [1]

$$PF = 0.9/\sqrt{1 + (.075/K_1)^2} \qquad (19.4)$$

The boundary between DCM (Fig. 19.2B) and CCM (Fig. 19.2D) occurs at $K_1 = 0.1$ and at PF $= 0.72$ (Fig. 19.2C). The PF versus K_1 is plotted in Fig. 19.3, along with a computer analysis of total harmonic distortion (THD) versus K_1. As K_1 decreases below 0.1, the current waveform changes and the PF increases slightly before decreasing monotonically. The THD of the current waveform decreases from the full odd-harmonic content of the in-phase square wave of Fig. 19.2E to a more sinusoidal but phase-displaced waveform of Fig. 19.2C. As K_1 decreases further, the THD again increases due to the high third- and fifth-harmonic content of the waveform of Fig. 19-2A. The nonlinearity of THD and PF near $K_1 = 0.1$ is due to the shift between the harmonic distortion factor and the phase angle.

19.2.2 Three-Phase Input

Waveforms of the ac input current to a full-wave, six-pulse rectified and filtered dc output (with a phase-shifted input voltage superimposed for reference) are shown in Fig. 19.4. Figure 19.4A represents a condition of minimal line inductance (perhaps an EMI filter) and shows the high peak values typical of capacitive-input

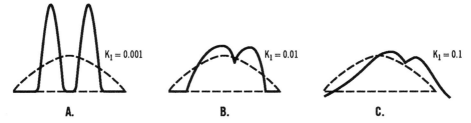

Fig. 19.4. Input current waveforms to an inductive filter as a function of K_1 for a three-phase input.

filters, as well as three DCM periods within a half cycle. Figures 19.4B and C represent conditions of increasing line reactance.

The PF and THD versus K_1 is plotted in Fig. 19.5. These results were obtained from computer analysis. Note that the PF increases as K_1 increases, reaching a maximum at $K_1 \approx 0.015$ and then decreases. This PF decrease is due to the increased phase shift produced by the inductive reactance, as shown in Fig. 19.4C. The THD decreases with increasing K_1 as the current waveform becomes more sinusoidal.

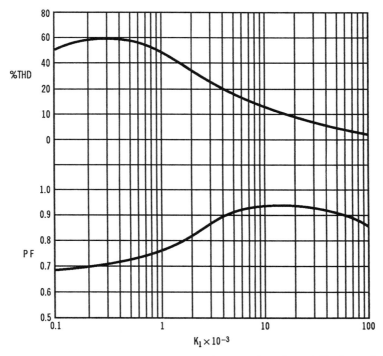

Fig. 19.5. Power factor and total harmonic distortion as a function of K_1 for a three-phase input.

The line inductors affect the dc voltage regulation at the expense of improved PF. The reactance causes a commutation problem in the rectifiers by preventing the dc load current from instantaneously transferring from one conducting rectifier to the other. This causes the output dc voltage to decrease as the load current increases. These effects are discussed in Section 4.1.3 and are shown in Fig. 4.14. the voltage decrease caused by the commutation is E_x, and the commutation angle μ is given by Equation (4.10). The PF is then given by [2]

$$PF = (3/\pi)(1 - E_x/E_{do})/\sqrt{1 - 3f(\mu)} = 0.955 \text{ maximum} \qquad (19.5)$$

In high-power systems, line reactors are rated in percent impedance X_u. As an example, the ac voltage drop across the inductor is measured at rated current and this voltage compared to the line voltage. For $V_x = 3$ V in a 120-V, single-phase system, $X_u = 3 \times 100/120 = 2.5\%$. A $\sqrt{3}$ multiplier is required in a three-phase system, and for $V_x = 6$ V in a 208-V line, $X_u = \sqrt{3} \times 6 \times 100/208 = 5\%$. Conversely, for a given percent X_u at operating current I at a line-to-neutral voltage V, the required inductance is found by

$$L = 0.01VX_u/2\pi fI \qquad (19.6)$$

19.3   ACTIVE CORRECTION

Active PFC can be implemented by techniques derived from basic PWM topologies. These include boost, flyback, and buck. The boost method has the advantage of continuous input current via the series inductor. The input current can be programmed to follow the input voltage waveform to obtain a high PF. The output dc bus is regulated at a voltage higher than the peak input voltage, and a fixed hold-up time is provided for operation after an input power outage occurs. The flyback method operates in the DCM, which causes the peak current to be several times higher than the line current. The size of the inductor is larger than either the boost or the buck topology. However, the output voltage can be lower or higher than the input. The buck method allows the input current to go to zero only when the input waveform is below the output. This topology can provide a high PF at a dc output voltage that is typically one-fourth the peak input voltage. For these reasons, only the boost topology is discussed for single-phase inputs.

Three-phase active PFC can be implemented by using three bridge rectifiers, each followed by a dc–dc converter. The converter outputs are paralleled at the load. By using resonant techniques, each converter input approximates a resistive load and the input current harmonics are substantially reduced and the triplen harmonics are canceled [3].

19.3.1   Low Frequency

Power factor correction by a quasi-boost synchronized switching topology operating at twice the utility frequency is shown in Fig. 19.6A. In this case, low-frequency operation permits the use of an IGBT, which inherently has a lower saturation voltage than a MOSFET of the same die size. The dc output voltage is applied to

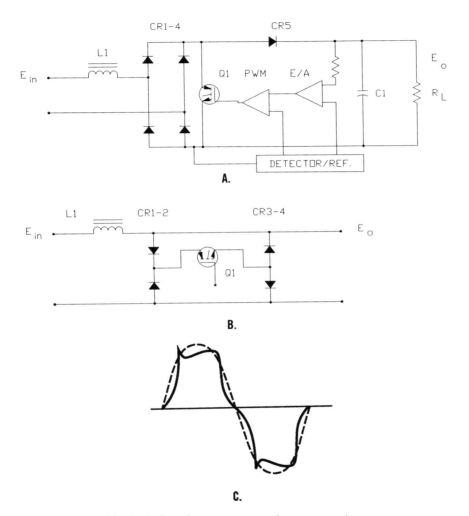

Fig. 19.6. Low-frequency power factor correction.

an error amplifier, and the error signal is applied to a PWM whose other input is a synchronized ramp voltage. The PWM drives Q_1, which is turned on at the zero-voltage crossing of E_{in} and turned off at a time determined by the load condition. When Q_1 turns off, the energy stored in L_1 is transferred to the load, as the input voltage continues to increase. This produces a pseudo-sinusoidal input current of low harmonic content and minimal phase shift, as shown in Fig. 19.6C. The output voltage is typically equivalent to 90% of the peak input voltage.

The current flow from an ac source through an *RL* circuit is given by Equation (1.27). However, in the case of Fig. 19.6, $R = 0$, $\lambda = 0$, and $\theta = 90° = 1.57$ rad. Thus, when Q_1 turns on, the current flow through L_1, the rectifier bridge, and Q_1, as a function of time, is given by

$$i = (E_{pk}/\omega L)[1 + \sin(\omega t - 1.57)] \qquad (19.7a)$$

and the peak current at the time of turn-off is related to the load current by

$$i_{pk} = 2I_o \qquad (19.7b)$$

The required inductance is given by the empirical formula

$$L = 4500/\omega P_0 \qquad (19.8)$$

The topology of Fig. 19.6B can be added in front of a power supply containing an input rectifier and capacitive filter. Here Q_1 conducts in the same manner as in Fig. 19.6A, and operation of this add-on circuit is the same except the output is an ac current of the waveform shown in Fig. 19.6C.

Computer analysis of the pseudosinusoidal input current waveform shows the PF at full load is typically 0.96 and the harmonic content is typically 4, 8, 5 and 1.5% for the 3rd, 5th, 7th, and 13th harmonics, respectively. In many cases, this simple topology is sufficient for single-phase PFC at medium- and high-power applications. However, the value of inductance is approximately 20 times the required inductance of a 50-kHz boost PFC topology.

19.3.2 High Frequency

Power factor correction with the boost regulator topology is shown in Fig. 19.7. This circuit is the most popular for PFC and has the advantages listed in Table 19.2. The following parameters govern voltage and current:

$$E_o = E_{in}/(1 - D) \qquad (19.9a)$$

$$I_o = I_{in}(1 - D) \qquad (19.9b)$$

The dc output voltage of the regulator is typically 390 V, which accommodates a peak value of a 220 V + 20% rms input at 4% duty cycle. At this output voltage, the maximum duty cycle is typically 66%, which accommodates a peak value of a 120 V − 25% rms input. (Operation of the boost regulator is discussed in detail in Section 8.4.)

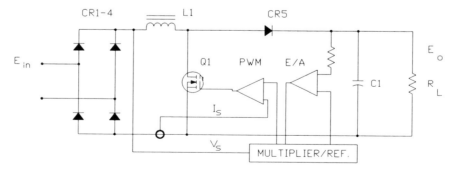

Fig. 19.7. High-frequency power factor correction, boost topology.

Two popular control circuits are the UC3854 and the ML4812. These ICs provide most of the intelligence functions and require only external programming and compensation for proper operation. The ICs include a voltage and current amplifier, an analog multiplier, a fixed-frequency PWM, and an output driver plus various housekeeping functions. The major function blocks are shown in Fig. 19.7A. The 3854 uses average current mode control while the 4812 uses peak current mode control. Average current mode control has more noise immunity and eliminates the slope compensation requirement of peak current mode control.

In the 3854, the output of the multiplier is equal to the product of the voltage amplifier output and the rectified input line voltage divided by the square of the rms line voltage. The current amplifier uses this signal to force an equivalent value following the shape of the input voltage across a line current sense element in a controlled PWM manner. The value of the inductor controls operational aspects. If the inductance is too low, the inductor current will become discontinuous (DCM) at low output currents and the PF will decrease. The typical values of L_1, C_1, and output ripple voltage, as related to switching frequency and input power, are given by [4]

$$L_1 = 25,000/f_s P_{in} \qquad H \qquad (19.10)$$

$$C_1 = 2P_{in} \qquad \mu F \qquad (19.11)$$

$$E_{r,pk} = P_{in}/2\omega C_1 E_o \qquad V \qquad (19.12)$$

In the 4812, the DCM is related to maximum duty cycle, which occurs at $E_o(1 - D_{max})$, and to minimum output power. The required inductance value is usually greater than that required for the 3854. The harmonic distortion is "noticeable" around the zero crossover of the input current. This condition can be significantly improved with an external enhancement circuit. Slope compensation of the ramp can be implemented to eliminate subharmonic oscillations. (Reference Section 11.5.3)

REFERENCES

1. California Institute of Technology, "Input-Current Shaped Ac-to-Dc Converters," Report N86-25693 for NASA LRC, May 1986.
2. Canadian General Electric Company Limited, "Power Converter Handbook," 1983.
3. Delco Research Center, General Motors Corp., "Three Phase AC to DC Voltage Converter with Power Line Harmonic Current Reduction," U.S. Patent 4,143,414, 1979.
4. Unitrode Corp., "Power Factor Correction With The UC3854," Application Note.

GLOSSARY

Brownout Protection   Protection of the supply and external loads from an excessive reduction in input voltage.

Centering   For multioutput SMPS with one regulation loop, the percentage change from nominal of the auxiliary output voltage with respect to the main (closed-loop) output voltage.

Complementary Tracking   The interconnecting of two regulated supplies whereby one supply (master) is operated to control the other (slave). The slave output voltage is made equal to the master output voltage, but of opposite polarity.

Compliance Current   Master-slave connection of two or more current-regulated supplies to increase their compliance voltage range through series connection.

Compliance Voltage   The voltage range required to sustain a given value of constant current over a range of load resistance.

Constant Current (CC)   A power supply capable of maintaining a preset current through a variable load resistance.

Constant Voltage (CV)   A power supply capable of maintaining a preset voltage across a variable load resistance.

Cooling   Refers to the method of cooling used to remove generated heat. Methods include convection (natural or forced), conduction, radiation, and combinations thereof.

Cross Regulation   In multioutput SMPS, the percentage change in auxiliary output voltage resulting from a specified change in the main (closed-loop) output current.

Crossover   Characteristic of a power supply that automatically changes method of regulation (CC to CV or vice versa) as dictated by varying loads.

Crowbar   A method of reducing (shorting) the output voltage of a power supply to protect critical external loads from an overvoltage condition.

706

Current Limiting   A current overload protection mechanism which limits the maximum output current to a preset value for conditions of load resistance or short circuit. (*See* Foldback.)

Cycling   A mode which may occur in SMPS when an overload or short circuit is applied to the output terminals. (*See* Hiccup.)

Dropout Voltage   The decreased input voltage at which the supply (or regulator) ceases to regulate the output for further decreases in input voltage.

Efficiency   The ratio of output power (watts) to input power (watts).

Error Signal   The amplified difference between the sampled output voltage and a fixed reference voltage.

Foldback   Means whereby the output current is reduced to a low percentage of rated output current for severe overloads or short circuits. Considered mandatory in linear supplies to reduce pass stage power dissipation.

Frequency Response   The measure of the power supply ability to respond to a sinusoidal effect. Normally referred to as bandwidth or the frequency at which unity gain occurs.

Gain Margin   The resulting closed-loop "gain" (negative dB) when the phase shift is 180°. A measure of stable operation.

Hiccup   An operating mode resulting from cycle-by-cycle current limiting in certain SMPSs. The output current pulsates in a short-circuit condition. (*See* Cycling.)

Holdup Time   The time during which the supply will maintain a regulated output when the input power is interrupted or removed.

Inrush Current   The maximum input current drawn by the supply when input power is applied.

Isolation Voltage   Rating which specifies the value of external voltage which can be safely applied between input to output, input to case, output to case. May be specified per safety requirements. Ref. Table 13.1.

Line Regulation (Source Effects)   Refers to the maximum change in output voltage (or current) resulting from changes in input (line) voltage, normally specified from minimum input voltage to maximum input voltage. May be expressed in percent or millivolts (or milliamperes).

Load Regulation (Load Effects)   Refers to the maximum change in output voltage (or current) resulting from changes in load resistance (load), normally specified from no load (NL) to full load (FL). May be expressed in percent or millivolts.

Margining   Built-in means to momentarily vary the output voltage around the nominal value by a specified percentage ($\pm 5\%$ typical).

Master–Slave Operation   A system of interconnecting two or more regulated supplies whereby one (master) controls the others (slave). Connections may be complementary tracking, parallel operation, series operation, compliance extension (to obtain increased output voltage for current regulation).

Minimum Load   The decreased output current at which the supply ceases to regulate for further decreases in output current.

Open Frame   A supply which is not fully enclosed.

Operating Temperature Temperature range (ambient or case) over which a supply can be safely operated at rated output. Output power may be derated to a specified value at a higher specified temperature.

Output Impedance The incremental change in output voltage divided by a forced incremental change in output current over a specified frequency range.

Output Inhibit Output remotely programmed off by external signal (or contact closure).

Overshoot (Undershoot) Peak transient output voltage resulting from a step decrease in load current. (Minimum transient output voltage resulting from a step increase in load current.)

Overvoltage Protection Protection of the supply and externally connected loads against excessive output voltage, either from internal or external causes. (*See* Crowbar.)

Pard Periodic and random deviation. Similar to peak-to-peak ripple and noise voltage appearing at the output terminals.

Phase Margin The resulting closed-loop phase angle at unity-gain crossover.

Piggy Back An auxiliary supply normally connected in series with an unregulated supply whereby the sum output voltage is regulated by feedback to the auxiliary supply.

Power Factor Input power (watts) divided by input volt-amperes (VA). Also equal to $\cos \theta$, where θ is the phase angle between voltage and current in linear systems.

Power Fail Signal Signal provided to an external monitor to indicate a drop in input voltage below the specified minimum. Time for this signal to appear may be specified, and this time should be less than the holdup time.

Programming The control of the output voltage (or current) by external means of variable resistance, variable voltage, variable conductance, or digital signals (digital-to-analog converter).

Programming Linearity Refers to the correspondence between incremental change in programming input and consequent incremental change in output voltage (or output current).

Programming Speed The time required to change the output voltage of a voltage-regulated supply from one value to another. External output filter capacitors have an effect. (Same as "recovery time" for a current-regulated supply.)

Recovery Time (Current Regulation) Time required for the output current to return to a value within the regulation limits after a step change in line or load. For load changes, recovery will be governed by the rate of change of compliance voltage across the load.

Recovery Time (Voltage Regulation) Time required for the output voltage to return to a value within the regulation limits after a step change in line or load.

Remote Sense The means by which the output voltage is directly monitored and regulated at the external load. This mode compensates for voltage drops in the output wiring leads.

Resolution   Minimum voltage (or current) increment within which the output can be set by using an internal or external control. For continuous control, the minimum increment is the voltage change resulting from one degree of shaft rotation.

Ripple   Stated either as a peak-to-peak or an rms value of an ac component (typically from dc to 1 MHz) appearing in the output voltage or current.

Self-Recovery   The ability to automatically recover to regulated operation following a short circuit, output crowbar, or overvoltage condition.

Sequencing   In multiple-output supplies, the method by which output voltages are turned on or achieve regulated limits in a specified sequence.

Short-Circuit Protection   Any automatic current limiting which enables the supply to continue operating, without damage, when a short circuit is applied across the output terminals.

Slew Rate   The maximum rate of change of output voltage (or current) in response to a step change in the programming input.

SMPS   Acronym for switch mode power supply. Also referred to as switching power supply, switcher.

Soft Start   *Input:* Limits inrush current from the line at turn-on. Normally applies to direct-line rectification and filtering. *Output:* Limits component current stress in SMPS at turn-on and prevents overshoot of output voltage at turn-on.

Stability (Long-Term Drift)   The change in output voltage (or current) as a function of time for conditions of constant line, load, and temperature.

Temperature Coefficient   The percentage change in output voltage (or current) as a result of change in ambient temperature. Normally expressed as $\%/^\circ$C.

Turn-on Time   The time required for the output voltage (or current) to reach regulation limits after input power is applied.

Warm-Up Time   The time required for the supply to meet specified operation (regulation and ripple as a minimum) after power is applied. This time may be a function of ambient temperature at the time power is applied.

INDEX

Absorptivity, 592
Admittance, 25
Air-core inductor, 200–202
Air flow, *see* Thermal
Air gap, 138, 183
AISI, 141–142
ANSI, 539
Altitude, effects, 591
Ampere's law, 132
Amplifier:
  difference, 629
  instrumentation, 569
  operational, 470, 566
Antisaturation, transistor, 67
Application guide:
  capacitors, 117
  resistors, 107
Area:
  core, 133
  effective, 133
  surface, 178, 184–185
  window, 166
Arhenius equation, 576, 608
ASTM, 139, 142
Audible noise:
  fans and blowers, 598
  magnetic devices, 174–175
Automatic tracking, 643
Autotransformer, 136, 371
Average value, waveforms:
  clipped sinusoid, 5, 7
  critically damped, 9, 11
  exponential, 9, 11–12
  rectangle, 5, 7

  rectified sine, 4, 6
  sawtooth, 7, 10
  sine, 4, 6
  square, 4, 6
  trapezoid, 8, 10
  triangle, 8, 11
AWG (American Wire Gage), 153–156

Baker clamp, 69
Banding, 185
Bandwidth, 480
Base drive, 64, 71
Batteries:
  lead–acid, 675
    capacity, 677
    characteristics, 676
    chargers, 681–683
  nickel–cadmium, 679–681
    chargers, 681–683
B–H curve, 137, 139
Bifilar windings, 149, 281
Blackout, 669
Blocking oscillator, 395, 403–404
Blowers, 597
Bobbins, 165
Bode plot, 480
Boltzmann's constant, 46
Boost regulator, 352, 373
Brownout, 669
BSI (British Standards Institute), 539
Buck–boost regulator, 352, 380
Buck regulator, 350, 364

Capacitance, transformer, 150–151

Capacitor(s):
  application guide, 117
  ceramic, 126
  charge, 13
  cleaning solvents, 119
  dielectric constant, 116
  dissipation factor, 117
  electrolytic:
    aluminum, 118
      characteristics, 118
      power dissipation, 119
      ripple current, 119
      series/parallel, 120
      stacked foil, 121
    tantalum, 121
      life, 121
      solid, 122
      wet-slug, 122
  equivalent circuit, 117
  film, 123
    metalized, 123
    temperature coefficient, 123
  input filters, 227
  mica, 127
  oil filled, 123
    dissipation factor, 124
    pulse operation, 125
  power factor, 117
  Q factor, 117
  resonant frequency, 555
  two precharged, 18
Cartesian form, 31, 39
C cores:
  banding, 185
  inductors, 187
  magnetic path length, 184
  mean turn length, 184
  surface area, 185
  transformers, 185
Chargers, battery, 681–683
Choke(s), see Inductor(s)
Circuit breakers, 520
  interrupting capacity, 522
  time delay, 520–521
Circular mils (cmil), 153
Cleaning solvents, 119
Clearance distance, 170
Closed loop, 476
Coercive force, 139
Common mode (EMI), 549
Commutation:
  rectifier, 222
  thyristor, 97, 333
Component:
  derating, 43, 45
  life, 576
  scaling, 40
Conductors, 152, 161

Confidence levels, 614
Control:
  current mode, 491–496
  feedforward, 390
  phase, 94, 211, 260
  PSM, 290, 403, 423
  PWM, 290, 364, 375, 382
  resonant, 497
  transient, 43
Convection cooling, 588, 593, 602
Converters, dc-dc:
  blocking oscillator, 395, 403
  characteristics, 394, 429
  Ćuk, 353
  current-fed, 429
  flyback, 397, 407
  forward, 398, 413
  full-bridge, 422
  half-bridge, 420
  input filter for, 437
  multiple output, 434
    current limiting, 436
    filter inductor, 434
    voltage sensing, 437
  push–pull, 401, 416
  resonant, 400, 439
  SCR, 425
  square wave, 405
Cooling, see Thermal
Core(s):
  loss, 143, 145
  material, 140–142
  saturation, 189
Corona, 170
Coulomb, 13, 528
Creepage distance, 538
Crest factor, 214
Critical inductance, 254, 273
Crossover frequency, 480
Crowbar (SCR), 102
CSA (Canadian Standards Association),
  538–539
Current density, 137, 153
Current-fed converter, 429
Current limiting:
  Converter, 436
  Power supplies, 630
  quasi-square wave inverter, 297
Current mode control, 491–496
Current sharing, 73, 84
Current surge, 233
Current transformer, 189

Damped:
  critically, 22
  overdamped, 22
  undamped, 22
  underdamped, 22

Damping ratio, 20
Darlington, 64
Delta–delta connection, 136
Deltamax, 188
Delta–wye connection, 136, 218
Derating, 43, 45
Diamond, 578
Dielectric:
 constant, 116
 field, 13
Difference amplifier, 629
Differential mode (EMI), 549
Diode, *see* Rectifiers
Dissipation factor, 117, 124
Doubler, voltage, 243, 258. *See also* Rectifier
 circuits

E core, 185
Eddy current, 143, 163
Efficiency:
 converter, 394
 inverter, 281
 power supply, 621, 649
 regulator, 369
 transformer, 146
 UPS, 693
Electromagnetic compatibility, 540
EMI:
 conducted emissions, 544
  capacitive impedance, 555
  filters, 547–552
  sources of, 541, 553
  specifications:
   FCC, 454
   military, 546
   VDE, 546
  testing, 556
  wire harness, 554
 Faraday shield, 553
 nature of, 541–543
 radiated emissions, 558
  electric field, 558
  magnetic field, 558
  limits, 558
  shading ring, 559
  shielding, 559
  specifications, 558
  susceptibility, 485, 562
Emissivity, 592
Emitter balance resistors, 73
Energy:
 second breakdown, 57
 storage, 13
 transfer, 528
Equivalent:
 capacitor, 117
 resistor, 105
 series inductance, *see* ESL

series resistance, *see* ESR
Erg, 137
ESL, 117
ESR, 116
Extended delta connection, 171

Failure rate, 609–614
Fans, 597
Faraday shield, 553
Faraday's law, 133
FCC (Federal Communications Commission),
 541, 545, 558
Feedback, 475
Feedforward, 490–491
Ferrite cores:
 core characteristics, 146
 geometries, 194
 inductors, 197
 skin effect in windings, 163
 transformers, 197
Ferroresonant, inverters, 311
Fiber optics, 571
Filters (rectifier):
 capacitor input, 227
  ripple voltage:
   singe phase, 228
   three phase, 231
  volume, 228
 inductor input, 253
  critical inductance, 254
  ripple voltage, 256
Flux:
 density, 133, 139, 185
 lines, 144, 190
Flyback converter, 397, 407
Foldback current limiting, 297,
 630
Forced air cooling, 593–594, 597
Forced convection cooling, 593
Forced liquid cooling, 600
Form factor, 214
Forward biased SOA, 47, 79, 88
Forward converter, 398, 413
Fourier:
 analysis, 294
 equation, 577
Frequency:
 converter, 171
 crossover, 480
 resonant, 21, 289, 445
 response, 476
 scaling, 41
Full-bridge:
 converter, 422
 inverter, 331
Fuses, 515
 categories, 515
 selection, 517

Gain margin, 480
Gapped inductors:
  C cores, 187
  ferrite cores, 197
  laminations, 180
Gauss, 133
Gilberts, 132
Grain-oriented, 141
Grounding, 563
  cables, 569
  circuits, 564–567
  loops, 564
  system, 573
GTO, 103

Half-bridge:
  converter, 420
  inverter, 330
Harmonic cancellation, 298–300, 313
Heat pipe, 603
Heat sinks, *see also* Thermal
  characteristics, 586
  mounting interface, 582
  orientation, 589
  thermal resistance, 586
  volume, 586
Hold-up time, 251, 653
Hysteresis:
  loops, 139
  losses, 143

IEC (International Electrotechnical
    Commission), 538, 539, 698
IGBT, 86
  gate drive, 90
  paralleling, 90
  parameters, 86
  SOA, 88
  temperature coefficient, 88
Impedance:
  graph, 34, 36, 476
  scaling, 39
Impregnate, vacuum, 170
Inductance:
  critical, 254, 273
  index, 139, 193
  leakage, 147–149
Inductor(s):
  air-core, 200–202
  air-gap, 138
  energy rating:
    C cores, 187
    ferrites, 197–199
    laminations, 180–181
  input filters, 253
    converters, 437–438
    output ripple, 256
    switching regulators, 388–391

iron powder cores, 193
  MPP cores, 190–193
  swinging choke, 183
Inrush current limiting, 233
Instrumentation amplifier, 569
Insulating materials, 168
Integrated circuits (ICs):
  linear regulator, 633, 636, 639, 549
  power factor correction, 705
  PSM, 290
  PWM, 353–363
  resonant controller, 442–444
Interface, thermal, 582
Interleaving:
  laminations, 144, 176
  windings, 149
Interphase transformer, 222
Inverter(s):
  thyristor (SCR):
    characteristics, 323
    classification, 324
    full-bridge, 331
    half-bridge, 330
    impulse commutation:
      control circuit, 338
      power components, 336
      power stage, 333
      PWM, 340
      turn-off time, 331, 337
    parallel, 326
    sine wave, 328
    three phase, 341–346
  transistor:
    characteristics, 279
    pulse demodulated, 300
      control circuits, 305–307
      modulation index, 307–308
    quasi-square wave, 290
      current limiting, 297
      waveform analysis, 294–295
    sine wave:
      ferroresonant transformer, 311
      harmonic cancellation, 298–300, 313
    square wave, 278
      clocked, 285
      output filtering, 286–290
      saturating drive transformer, 284
      saturating output transformer, 278
    step wave, 317
      transformer phasor diagram, 319
    three phase, 311
  UPS, 684
Iron powder cores, 193
Isolation:
  safety, 538
  transformers, 168–170

Jensen inverter, 284

JISC (Japanese Industrial Standards
    Committee), 539
Junction capacitance, 50, 77
Junction temperature:
    calculation, 581
    derating, 45, 579
    rectifiers, 48
    transistors, 57, 80, 88

Kapton, 168
Kraft paper, 168

Laminations, 175–182
    annealing, 176
    dimensions, 176
    magnetic path length, 175
    mean turn length, 177
    scrapless, 177
    surface area, 178
Laplace transforms, 25
Leakage current, 538, 579
Leakage inductance, 147–149
Linear:
    power supplies, 624–632
    regulators, 633–644
Line regulation, 707
Liquid cooling, 600
Litz wire, 156, 159
Load regulation, 707
Loop:
    gain, 475
    stability, 476, 481

Magnetic:
    circuit, 130
    core loss, 143
    field, 13
    materials, 134
    path length, 138
Magnetizing current, 143
Magnetomotive force, 130
Majority carriers, 52, 75
Margin:
    gain, 480
    phase, 479
Materials:
    core, 140–142
    Kapton, 168
    Nomex, 168
    thermal compounds, 584
MCT, 105
Mean time between failure, *see* MTBF
Mean turn length, 177
Metal oxide varistor, 526
Meter movement, 3, 285
Minority carriers, 52
Modulation index, 307–308
Modulators:

line-type, 505–509
    magnetic, 509–513
    PSM, 290, 403, 423
    PWM, 290, 364, 375, 382
MOSFET, 74–85
    characteristics, 75
    reverse rectifier, 84
    SOA, 79
Mounting interface, 582
MPP (molypermalloy powder) cores:
    ac voltage, 191
    core loss, 191
    dc bias, 192
    inductance index, 191
    permeability, 191
Multiplier, voltage, 244
Multivibrator, 407

Natural convection, 588
NAVMAT, 43, 608
N-channel, 75
Networks:
    lag, 31
    lag-lead, 31, 470
    lead-lag, 34, 472
    impedance, 34, 476
Nomex, 168
Nyquist, 481

Oersted, 132
Off-line:
    power supply, 651
    UPS, 673
On-line UPS, 670
Open loop, 475
Operational amplifiers, 470, 566
Orientation, heat sink, 589
Orthonol, 188
Oscillator (magnetic):
    blocking, 395, 403–404
    frequency, 407
    single transformer, 278, 405
    two transformer, 284, 405
Oscillatory, 21
Overdamped, 21
Overlap conduction, 285
Overshoot, 370, 484
Overvoltage:
    protection, 645, 654
    sensing, 102

Paralleling:
    MOSFETs, 84
    transistors, bipolar, 73
PARD, 708
P-channel, 75
Permeability:
    conductor, 161

Permeability (*Continued*)
  core material, 132, 138, 191
PFN (pulse forming network):
  five-section, 503, 505
  impedance, 501
  pulse width, 500–501
  two-section, 501–502, 504
Phase:
  control, 94, 211, 260
    critical inductance, 272
    output voltage, 260
      single phase, 261
        control circuit, 273
      three phase:
        capacitive loads, 267
        control circuit, 275
        inductive loads, 267
        resistive loads, 267
  lag-lead network, 31
  lag network, 31
  margin, 479, 487
Phase locked loop, 688
Polar form, 31, 39
Power factor, 146
  correction, 698
    active:
      high frequency, 704
      low frequency, 702
    passive:
      single phase, 699
      three phase, 700
Power rating:
  transformers, 60 Hz, 179
  transformers, high frequency, 198
Power supplies:
  comparison, 624
  regulated:
    current, 662
    series, 624
      complementary tracking, 643
      control, 627
      current limiting, 630
      IC regulators, 632–642
      overvoltage protection, 645
    shunt, 647
    switching, 651
      hold-up time, 653
      overvoltage protection, 654
      start-up, 651
      synchronous rectifier, 654
  topologies, 621
  unregulated, *see* Rectifier circuits
Preregulators, 684
Pressure drop, 594
Proportional base drive, 71
Protection:
  current limiting, 297, 436, 630

devices:
  circuit breakers, 520
  fuses, 515
  spark gaps, 527
  transient suppressors, 524
  varistors, 526
  overvoltage, 645, 654
  snubbers, 528
  switching regulators, 386
Proximity effects, 163
PSM (phase shift modulation), 290, 403, 423
Pulse transformers, 202–208
PWM (pulse width modulation), 290, 364, 375, 382
  control circuits, 354–363

Q factor, 117, 160, 202
Quality:
  definition, 605
  ISO-9000, 606
  terminology, 607

Radiated:
  EMI, 558
  heat, 590
RC circuit, 17–19
Reactance, transformer, 222
Rectangular form, 31, 39
Rectifier circuits:
  average current, 212
  commutation, 222
  crest factor, 214
  form factor, 214
  hold-up time, 653
  inrush current limiting, 233
  output voltage, 212
  peak inverse voltage, 213
  ripple voltage, 214
  rms current, 212
  single phase, 215
  synchronous, 654
  three phase:
    double wye, 218
    dual extended delta, 173
    full-wave, 218
    half-wave, 217, 218
    star, 220
  voltage doubler, 243, 258
  voltage multiplier, 244–250
Rectifiers (diodes):
  doping, 51
  fast recovery, 52
  germanium, 46
  losses, 53
  ratings, 48
    capacitance, 50
    current, 46, 54

leakage, 50
pulse, 55
surge current, 54–55
temperature coefficient, 48
voltage, 47
recovery:
  forward, 52
  reverse, 50
rms current, 213
Schottky, 46, 52, 536
synchronous, 654
Regulation:
batch, 437
cross, 706
line, 707
load, 707
Regulators:
linear:
  IC, 632–644
  series, 621, 627
  shunt, 647
  tracking, 643
switching:
  boost, 352
  buck, 350
  buck-boost, 352
  characteristics, 352
  current mode control, 491
  Ćuk, 353
  feedforward, 490
  input filters, 388
  protection, 386
  PSM control, 290
  PWM control, 484
  ringing choke, 328
tracking, 643
Reliability:
confidence levels, 614
definition, 605
equations, 608
"established", 606
failure rate:
  analysis, 612
  tabulation, 611–613
MTBF, 609
redundant units,
  616–617
specifications, 608
terminology, 606–607
Reluctance, 132
Resistance:
copper, 154
thermal, 577
wire, 156–157, 159
Resistivity:
conductors, 152–153
thermal, 577–578

Resistor(s):
application guide, 107
carbon composition:
  characteristics, 108
  power overload, 109
critical resistance, 106
equivalent circuit, 105
metal film:
  characteristics, 114
  high frequency, 110
wirewound:
  characteristics, 114
  coatings, 113
  materials, 114
Retentivity, 139
Reverse biased SOA, 57, 79, 88
RFI, *see* EMI
Ringing choke, 328
RL circuit, 14–16
RLC circuit, 20–24
RLCR circuit, 25–30
rms value, waveforms:
clipped sinusoid, 5
critically damped, 12
exponential, 12
rectangle, 5
rectified sine, 4
sawtooth, 10
sine, 4
square, 4
trapezoid, 10
triangle, 11
Royer inverter, 278

Safety, standards, 538
Scaling:
frequency, 41
impedance, 40
SCR:
crowbar, 102
gate triggering, 92
inverter rated, 97
di/dt, 92
dv/dt, 97
pulsed energy, 101
turn-off time, 97
Second breakdown, 57
Series regulator, 624
Shield:
EMI, 559–561
transformer, 151
Shunt regulator, 647
SIT, 103
Skin depth, 161
Skin effect:
conductors, 162
windings, 163

Snubbers:
  dissipative, 529
  nondissipative, 536
SOA:
  IGBTs, 88
  MOSFETs, 79
  transistors, bipolar, 56–59
Soft-start, 233, 355
Spark gaps, 527
Stability analysis, 475
  conditional, 481
Stacking factor, 133–134
Step-start, 236
Stephan–Boltzmann formula, 590
Sulfur hexaflouride, 170
Supermalloy, 189
Supermendur, 188
Suppressors, 522–526
Surge current, 233
Susceptibility, 485, 562
Swinging choke, 183
Switching regulators, *see* Regulators, switching
Synchronous rectifier, 654

Temperature rise:
  semiconductors, 578
  transformers, 173
Tesla, 133
Thermal:
  compounds, 584
  conductivity, 578
  convection:
    forced air:
      fans and blowers, 597
      flow rate, 593
      pressure drop, 594
    forced liquid, 600
      coolants, 603
      flow rate, 602
    natural, 588
  conversion factors, 575
  heat pipes, 603
  heat sinks, 586 592–593
  mounting interface, 582–584
  radiation, 590
  resistance, 577
  resistivity, 578
  runaway, 56, 579
  testing, 593
Thermistor, 236
Thermoelectric coolers, 604
Thyristors, 91. *See also* SCRs
Time constant, 14, 17
Toroid cores:
  ferrite, 194–199
  MPP, 190–192
  powder iron, 193

tape wound, 188–190
  saturated inductance, 189
Transconductance, 76
Transfer switches, 688–692
Transformer:
  audible noise, 174
  bobbins, 165
  capacitance, 150–151
  connections:
    delta–delta, 136
    delta–wye, 136, 218
    extended delta, 171
    wye–delta, 135
    wye–wye, 136, 217
  current, 134
  efficiency, 146
  impregnate, 170
  insulation, 168
  interphase, 222
  leakage inductance, 147–149
  power rating:
    ferrite, 198
    low frequency, 179
  pulse, 202–208
  reactance, 212, 222
  resistance, 212
  saturating, 510
  tape cores, 188–190
  temperature rise, 173
  winding:
    layers, 166
    mandrel, 165
    margin, 166
    shuttle, 166
    start, 165
    turns per layer, 166
Transient:
  circuit analysis, 15, 17, 21
  control, 43
  power stress, 43
  protection:
    snubbers, 528–537
    spark gaps, 527
    suppressors, 524
    varistors, 526
  response, 484
Transistors:
  bipolar (BJT):
    alloy, 56
    base drive, 64
      direct coupled, 65–68
      proportional, 71
      transformer coupled, 69–72
    Darlington, 64
    epitaxial, 56
    losses, 61
    paralleling, 73

planar, 56
SOA, 56–59
IGBT:
  gate drive, 90
  paralleling, 90
  parameters, 86
  SOA, 88
  temperature coefficient, 88
MOSFET:
  capacitance effects, 77
  characteristics, 74
  gate drive, 81
  losses, 79
  N-channel, 75
  paralleling, 84
  parameters, 75
  P-channel, 75
  reverse rectifier, 85
  SOA, 79
  temperature coefficient, 76
  threshold voltage, 79
  transconductance, 76
Triac, 103
Turn-on surge, 233

U.L. (Underwriters' Laboratories), 538–539
Underdamped, 22
Undershoot, 370, 484
Unity gain, crossover, 476
UPS:
  components, 670
    batteries, 674
      chargers, 681
      lead–acid, 675
      nickel–cadmium, 679
    inverters, 684
    preregulator, 684
    transfer switches, 688
      phase locked loop, 688

off-line system, 670
on-line system, 670
sensing utility failure, 691
system analysis, 693

Vacuum impregnate, 170
Varistors, 526
VCO (voltage controlled oscillator), 442
VDE (Verband Deutscher Elektrotechniker):
  EMI, 544, 546, 558
  safety, 538
VFO (variable frequency oscillator), 443
Viscosity, 602–603
Voltage doubler, 243, 258
Voltage multipliers, 244–251
Volt-amperes, 134, 137, 695

Waveform relations, *see* average value; rms value
Wire:
  aluminum, 160
  copper, 152–156
  current density, 153
  gages, 154–157
  Litz, 156, 159–160
  tables, 154–156
Wye–delta connection, 135
Wye–wye connection, 136, 217

X capacitors, 549

Y capacitors, 549

ZCS (zero current switching), 440–442, 453, 455
Zener diode:
  reference, 629
  shunt regulator, 647
ZVS (zero voltage switching), 440–442, 455, 461–465